Modern Information Optics with MATLAB

现代信息光学

(MATLAB)(中英双语版)

▶ 张亚萍　潘定中　著

▶ YAPING ZHANG and TING-CHUNG POON

中国教育出版传媒集团

高等教育出版社 · 北京

内容简介

本书采用中英双语逐页对照的形式排版，系统地介绍信息光学的基础理论、傅里叶光学数理知识及现代相关应用。全书内容共七章：前三章主要提供了基础光学的背景知识，包括几何光学、波动光学、信息光学的重要数学储备知识。随后，介绍了相干和非相干图像处理系统的内容（第四章），并阐述了相干全息理论（第五章）以及非相干数字全息术和计算全息中重要的现代发展（第六章）。此外，亦对诸如声光和电光调制器的光信息处理器件的原理进行了较为深入的阐述（第七章）。

本书在内容上注重基础与前沿的有机结合，满足该领域读者对基础知识和研究进展的需求，由浅入深地阐述现代信息光学的理论基础和发展应用，公式推导详尽，强调公式背后的物理意义和现象分析，每章用实例分析协助读者理解。

本书可供高等院校光学、光学工程、光信息科学技术、电子科学与技术等专业的本科和研究生进行中英文专业术语对照学习，也可为相关专业领域的英文翻译提供参考，也可供教师及科技工作者参考。

图书在版编目（CIP）数据

现代信息光学 ： MATLAB ： 汉、英 / 张亚萍，潘定中著 . -- 北京 ： 高等教育出版社，2024. 11. -- ISBN 978-7-04-063091-6

Ⅰ. O438

中国国家版本馆 CIP 数据核字第 202413TX10 号

XIANDAI XINXI GUANGXUE (MATLAB)

策划编辑　王　超　　责任编辑　柴连静　　封面设计　姜　磊　　版式设计　杨　树
责任校对　高　歌　　责任印制　赵义民

出版发行	高等教育出版社	网　址	http://www.hep.edu.cn
社　址	北京市西城区德外大街4号		http://www.hep.com.cn
邮政编码	100120	网上订购	http://www.hepmall.com.cn
印　刷	北京盛通印刷股份有限公司		http://www.hepmall.com
开　本	850mm × 1168mm　1/16		http://www.hepmall.cn
印　张	34.75		
字　数	930 千字	版　次	2024 年11月第 1 版
购书热线	010-58581118	印　次	2024 年11月第 1 次印刷
咨询电话	400-810-0598	定　价	259.00 元

物 料 号　63091-00

致谢
Acknowledgements

感谢我的父母、我的丈夫和我的女儿! 感谢生命中出现的每一位老师!

To my parents, my husband, and my daughter

——张亚萍 (Yaping Zhang)

感谢我的外孙们: Gussie, Sofia, Camden 和 Aiden!

To my grandchildren, Gussie, Sofia, Camden, and Aiden

——潘定中 (Ting-Chung Poon)

作者简介

张亚萍, 昆明理工大学教授, 主要从事全息算法及其三维目标识别和显示、光致聚合物全息膜的制备与特性、光学扫描全息术和红外全息术的相关研究。2007 年博士毕业于哈尔滨工业大学光学工程专业。2008 年至今在昆明理工大学理学院工作, 2009—2011 年在浙江大学和美国哈佛大学数学系进行博士后研究。2018—2020 年在美国弗吉尼亚理工大学进行访问学者研究。2018—2022 年教育部光电信息科学与工程专业教学指导分委员会委员, 云南省 (高校) 现代信息光学重点实验室主任, 云南省物理学会第十四届理事会理事, IEEE 高级会员。曾获 “云南省中青年学术和技术带头人” 等多项荣誉称号。

潘定中, 美国弗吉尼亚理工大学电气与计算机工程系教授, 本科、硕士和博士皆毕业于美国爱荷华大学, 长期从事信息光学、三维图像处理和光学扫描全息术 (OSH) 等研究工作。美国光学学会 (Optica)、国际光学工程学会 (SPIE)、电气电子工程师学会 (IEEE) 和英国物理学会 (IOP) 的会士 (Fellow)。因对光学扫描全息术开创性的贡献获 2016 年 SPIE Dennis Gabor 奖。独著专著一部, 合著、主编专著或教材共八部。曾担任 SCI 期刊 *Applied Optics* 的区编辑、*Chinese Optics Letters* 的副主编和 *IEEE Transactions on Industrial Informatics* 的副编辑, 并于 2021 年和 2022 年担任 Optica 年会前沿光学 + 激光科学 (FiO+LS) 的总主席。

Authors

Yaping Zhang is Professor and Director of Yunnan Provincial Key Laboratory of Modern Information Optics, Kunming University of Science and Technology. Professor Zhang is an academic leader in optics in Yunnan Province and a member of the Steering Committee on Opto-Electronic Information Science and Engineering, Ministry of Education of the People's Republic of China. She is Senior Member of the Institute of Electrical and Electronics Engineers (IEEE).

Ting-Chung Poon is Professor of Electrical and Computer Engineering at Virginia Tech, USA. Professor Poon is Fellow of the Institute of Electrical and Electronics Engineers (IEEE), the Institute of Physics (IOP), Optica, and the Society of Photo-Optical Instrumentation Engineers (SPIE). He also received the 2016 SPIE Dennis Gabor Award.

前言

本书涵盖了信息光学中使用的基础理论及一些近现代相关知识，如非相干图像处理、非相干数字全息术、计算全息的现代方法，及信息光学中用于光信息处理的器件等。这些新的内容将不断地在信息光学中发挥其重要作用。

本书可为光学、光电信息工程或应用物理专业的学生，信息领域工作的科学家和工程师等科技人员提供帮助。主要面向光学相关专业的本科三、四年级的学生以及研究生一年级的学生。对于重要的步骤，书中包括较多详细的推导，这将使学生拥有清晰的数学概念。希望本书能使本科生建立起牢固的数理基础。书中各章结尾都配有适当习题，利于研究生巩固知识并进行挑战。

本书的前三章提供了基础光学的背景知识，包括几何光学、波动光学、信息光学中重要的数学储备知识。随后，介绍了相干和非相干图像处理系统的内容，阐述了相干和非相干数字全息术以及计算全息中重要的进展。此外，亦对光学信息处理中诸如声光和电光调制器的光信息处理器件的原理进行了较为深入的阐述。

本书内容可分为两部分。第一部分即第一至五章的内容，可用于诸如“傅里叶光学”“全息术”和“现代信息光学”等课程一个学期的课程教学，亦可用于如“光学信息处理”的系列课程的上学期教学内容。第二部分内容为第六、七章 (在开始时包含对第三至五章的简单复习与回顾)，可供下学期使用。本书的重点及特点是为读者提供了常用软件工具 MATLAB® 进行理论建模和应用的经验，该软件的使用使得读者能够看到一些重要的光学现象，如衍射、光学图像处理以及全息重建等。

我们希望本书能有一个中英双语版，从而可作为单独的教材/著作，这在信息光学领域的教材/著作当中是极少见但极其重要且具有开创性的工作。信息光学是一个不断发展的领域，对此类双语版书籍将有较大的需求。此类书籍非常有助于学生和科研人员进行中英文专业术语的对照学习，亦可对信息光学领域的专业及技术翻译提供帮助。

我们要感谢刘荣平 (Jung-Ping Liu) 教授帮忙编写了书中一些 MATLAB 代码；同时感谢博士生姚勇伟、硕士生张竟原和范厚鑫帮忙绘制书中的一些图，最后但同样重要的是，感谢 Christina Poon 阅读部分手稿并提出改进意见和建议。

Preface

This book covers the basic principles used in information optics including some of its modern topics such as incoherent image processing, incoherent digital holography, modern approaches to computer-generated holography, and devices for optical information processing in information optics. These modern topics continue to find a niche in information optics.

This book will be useful for engineering or applied physics students, scientists, and engineers working in the field of information optics. The writing style of the book is geared toward juniors, seniors, and first-year graduate-level students in engineering and applied physics. We include details on most of the derivations without omitting essential steps to facilitate a clear mathematical development as we hope to build a strong mathematical foundation for undergraduate students. We also include exercises, challenging enough for graduate students, at the end of each chapter.

In the first three chapters of the book, we provide a background on basic optics including ray optics, wave optics, and important mathematical preliminaries for information optics. The book then extensively covers topics of incoherent image processing systems, digital holography, including important modern development on incoherent digital holography, and computer-generated holography. In addition, the book covers in-depth principles of optical devices such as acousto-optic and electro-optic modulators for optical information processing.

The material covered is enough for a one-semester course (Chapters $1 \sim 5$) with course titles such as Fourier optics, holography, and modern information optics or a two-course sequence with the second course covering topics from Chapters 6 to 7 (with a brief review of Chapters 3 through 5). Example of a course title would be optical information processing. An important and special feature of this book is to provide the reader with experience in modeling the theory and applications using a commonly used software tool MATLAB®. The use of MATLAB allows the reader to visualize some of the important optical effects such as diffraction, optical image processing, and holographic reconstructions.

Our vision of the book is that there is an English and Chinese version of this book that are printed together as a single textbook. It is the first of its kind in textbooks and a pioneering project. Information optics is a growing field, and there is an enormous need for pioneering books of this kind. A textbook like this will allow students and scholars to appreciate the much-needed Chinese translation of English technical terms, and vice versa. The textbook also provides them with professional and technical translation in the area of information optics.

We would like to thank Jung-Ping Liu for his help on writing some of the MATLAB codes. Also thanks are extended to Yongwei Yao, Jingyuan Zhang, and Houxin Fan for drafting some initial figures used in the book and, last but not least, Christina Poon for reading parts of the manuscript and providing suggestions for improvements.

张亚萍要感谢她的父母、丈夫以及女儿 (许馨怡)，正是他们的鼓励和支持，让她能够实现自己的梦想。她特别要感谢她的合作者潘 (Poon) 教授，他极具专业知识素养，并对英文版内容进行了多次校对及语言润色，从而使得本书更易于读者阅读和理解，并最终得以出版。与潘教授合作是一件非常愉快的事情，此次合作亦使得她的专业知识和经验得到了进一步的提升。

潘定中 (Ting-Chung Poon) 非常感谢他的父母，正是他们在精神上的鼓励和奉献，使得他能够实现自己的梦想和成就。永远缅怀和铭记他们。

本书得到了国家自然科学基金项目“面向三维目标识别的光学扫描全息技术研究 (项目编号: 62275113)”和云南省“兴滇英才支持计划”项目的支持。在此表示感谢!

张亚萍 (Yaping Zhang)，中国，云南，昆明

潘定中 (Ting-Chung Poon)，美国，弗吉尼亚，黑堡

Yaping Zhang would like to thank her parents, her husband, and her daughter (Xinyi Xu) whose encouragement and support have enabled her to fulfill her dreams. In particular, she wishes to express her appreciation to her collaborator, Professor Poon, for his professional knowledge and language polishing that made the book more readable for users, and resulted in the publication of the book. Working with Professor Poon on this project was a great pleasure and resulted in further growth in her professional experience.

Ting-Chung Poon is greatly indebted to his parents, whose moral encouragement and sacrifice have enabled him to fulfill his dreams and further his achievements. They shall be remembered forever.

Yaping Zhang, Kunming, Yunnan, PRC
Ting-Chung Poon, Blacksburg, Virginia, US

目录
Contents

第一章　高斯光学和不确定性原理

本章包含高斯光学 (*Gaussian optics*) 并采用矩阵形式来描述通过光线形成的光学图像的信息. 在光学中, 光线是光的理想模型. 然而, 在后续的章节 (第三章 3.5 节) 中, 我们会发现, 采用矩阵形式也可以描述诸如用波动光学的方法进行高斯激光束 (*Gaussian laser beam*) 的衍射分析. 采用矩阵形式进行光的传输描述的优点是, 任意光线在光学系统中传播时, 都可以通过矩阵之间的依次相乘来进行追迹, 这也方便在计算机上对其进行编程, 目前该技术已在光学元件的设计中得到了广泛的应用. 本章将基于射线方法来讨论一些诸如分辨率、焦深和景深等重要的概念.

1.1　高斯光学

高斯光学是以卡尔·弗里德里希·高斯 (Carl Friedrich Gauss) 的名字命名的, 它属于几何光学范畴, 描述了在傍轴近似下光学系统中光线的行为. 假定光轴 (*optical axis*) 沿 z 轴方向, 即 z 轴是光线传播的主方向, 我们的讨论仅限于 x-z 平面上靠近光轴的光线. 换句话说, 只有偏离光轴足够小角度范围内的光线才会被考虑. 这些光线被称为傍轴 (近轴) 光线 (*paraxial ray*). 因此, 这些角度的正切和余切值可以近似表示为其角度本身的值, 即 $\sin\theta \approx \tan\theta \approx \theta$. 的确, 由于采用线性处理, 数学运算会极大简化. 例如, 描述折射的斯涅耳定律 (*Snell's law of refraction*) $n_1 \sin\phi_i = n_2 \sin\phi_t$ 的线性形式为 $n_1\phi_i = n_2\phi_t$. 图 1.1-1 所示为满足斯涅耳定律 (*Snell's law*) 的光线折射. 其中, ϕ_i 和 ϕ_t 分别为入射角和折射角 (*angles of incidence and refraction*), 通过法线 ON 到介质 1 与介质 2 之间的界面 POQ 的角度来测得. 介质 1 和介质 2 的恒定折射率分别为 n_1 和 n_2, 图 1.1-1 也表示出了反射定律 (*law of reflection*), 即 $\phi_i = \phi_r$, 其中 ϕ_r 为反射角(*angle of reflection*). 这里要注意的是: 入射光线、折射光线和反射光线都位于由入射光线和法线所决定的入射面 (*plane of incidence*) 内.

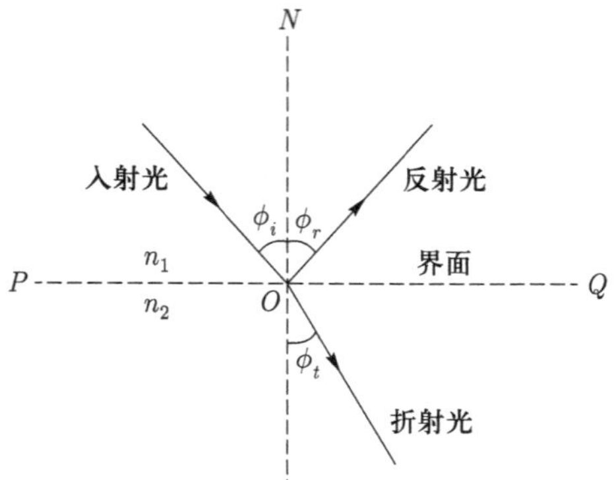

图 1.1-1　斯涅耳定律和反射定律的几何图示

1 Gaussian Optics and Uncertainty Principle

This chapter contains *Gaussian optics* and employs a matrix formalism to describe optical image formation through light rays. In optics, a ray is an idealized model of light. However, in a subsequent chapter (Chapter 3, Section 3.5), we will also see that a matrix formalism can also be used to describe, for example, a Gaussian laser beam under diffraction through the wave optics approach. The advantage of the matrix formalism is that any ray can be tracked during its propagation through the optical system by successive matrix multiplications, which can be easily programmed on a computer. This is a powerful technique and is widely used in the design of optical elements. In the chapter, some of the important concepts in resolution, depth of focus, and depth of field are also considered based on the ray approach.

1.1 Gaussian Optics

Gaussian optics, named after Carl Friedrich Gauss, is a technique in geometrical optics that describes the behavior of light rays in optical systems under the paraxial approximation. We take the *optical axis* to be along the z-axis, which is the general direction in which the rays travel, and our discussion is confined to those rays that lie in the x-z plane and that are close to the optical axis. In other words, only rays whose angular deviation from the optical axis is small are considered. These rays are called *paraxial rays.* Hence, the sine and tangent of the angles may be approximated by the angles themselves, that is, $\sin\theta \approx \tan\theta \approx \theta$. Indeed, the mathematical treatment is simplified greatly because of the linearization process involved. For example, a linearized form of *Snell's law of refraction*, $n_1 \sin\phi_i = n_2 \sin\phi_t$, is $n_1\phi_i = n_2\phi_t$. Figure 1.1-1 shows ray refraction for Snell's law. ϕ_i and ϕ_t are the angles of incidence and refraction, respectively, which are measured from the normal, *ON*, to the interface *POQ* between Media 1 and 2. Media 1 and 2 are characterized by the constant refractive indices, n_1 and n_2, respectively. In the figure, we also illustrate the *law of reflection*, that is, $\phi_i = \phi_r$, where ϕ_r is the *angle of reflection.* Note that the incident ray, the refracted ray, and the reflected ray all lie in the same *plane of incidence.*

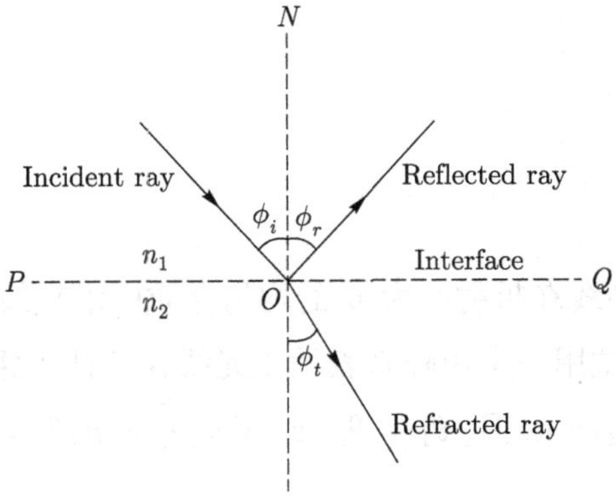

Figure 1.1-1 Geometry for Snell's law and law of reflection

考虑如图 1.1-2 所示的光学系统中一束傍轴光线的传播. 当给定 z 平面上的一束光线, 那么该光线的位置可以通过其距离光轴的高度 x 及其出射角度 θ 来确定. 按照惯例, 其角度 θ 用弧度值表示, 且从 $+z$ 轴开始沿逆时针方向为正. 数值 (x,θ) 表示给定 z 平面上该光线的坐标. 这里, 我们用另一种表示来描述该光线与 z 轴的夹角, 即用相应的关系式 $v = n\theta$ 来替换坐标中的角度 θ, 其中, n 为光线传输时介质的折射率. 以后将会发现, 这些规定可确保所有的矩阵都有正单位模 (*positive unimodular*). 单位模矩阵 (*unimodular matrix*) 是行列式为 "+1" 或 "−1" 的实数方阵, 而正单位模矩阵的行列式为 "+1".

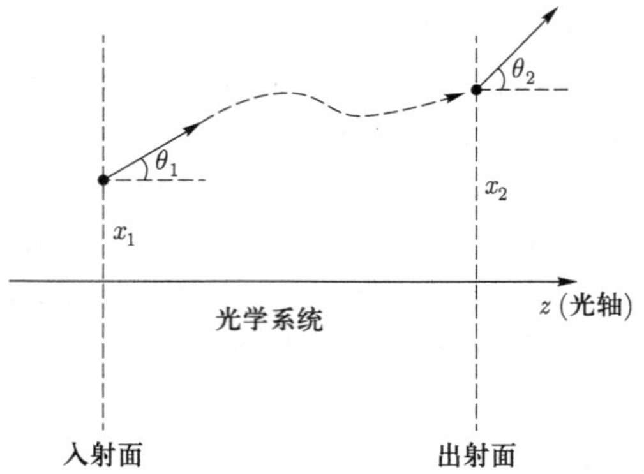

图 1.1-2 光学系统中的光线传播

为说明这点, 让一束光线从入射面上的入射光线坐标 (*input ray coordinates*) $(x_1, v_1 = n_1\theta_1)$ 通过, 当该光线穿过光学系统后, 在其出射面上的出射光线坐标 (*output ray coordinates*) 为 $(x_2, v_2 = n_2\theta_2)$. 在傍轴近似中, 相应的出射量与入射量是线性相关的, 即出射量可以用入射量的加权和 [称为叠加原理 (*principle of superposition*)] 表示为

$$x_2 = Ax_1 + Bv_1,\ \ v_2 = Cx_1 + Dv_1,$$

其中, A, B, C 和 D 分别为权重因子. 可以将上述方程转换为如下矩阵形式

$$\begin{pmatrix} x_2 \\ v_2 \end{pmatrix} = \begin{pmatrix} A & B \\ C & D \end{pmatrix} \begin{pmatrix} x_1 \\ v_1 \end{pmatrix}. \tag{1.1-1}$$

上述 $\begin{pmatrix} A & B \\ C & D \end{pmatrix}$ 矩阵被称为光线变换矩阵 (*ray transfer matrix*). 如果该矩阵是由多个光线变换矩阵相乘所得, 那么该矩阵被称为系统矩阵 (*system matrix*) $\mathcal{S}$. 在下面的内容中, 将导出几个重要的光线变换矩阵.

1.1.1 光线变换矩阵

传输矩阵

如图 1.1-3 所示, 一束光线在折射率为 n 的均匀介质中沿直线传播, 当用光线的坐标来描述其入射面和出射面时, 尝试用一个矩阵来表示该光线在传播了距离 d 以后的入射和出射坐标之间的关系. 因为 $n_1 = n_2 = n$, 且, $\theta_1 = \theta_2$, $v_2 = n_2\theta_2 = n_1\theta_1 = v_1$, 从几何关系可以发现,

Consider the propagation of a paraxial ray through an optical system as shown in Figure 1.1-2. A ray at a given z-plane may be specified by its height x from the optical axis and by its launching angle θ. The convention for the angle is that θ is measured in radians and is anticlockwise positive from the z-axis. The quantities (x, θ) represent the coordinates of the ray for a given z-plane. However, instead of specifying the angle the ray makes with the z-axis, another convention is used. We replace the angle θ by the corresponding $v = n\theta$, where n is the refractive index of the medium in which the ray is traveling. As we will see later, the use of this convention ensures that all the matrices involved are positive unimodular. A *unimodular matrix* is a real square matrix with determinant $+1$ or -1, and a positive unimodular matrix has determinant $+1$.

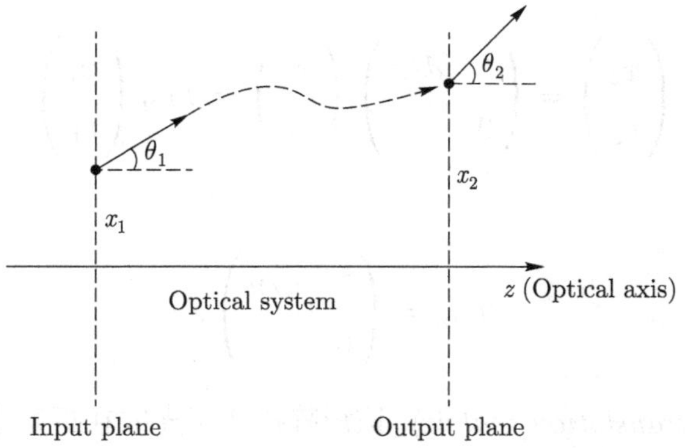

Figure 1.1-2 Ray propagating in an optical system

To clarify the discussion, we let a ray pass through the input plane with the *input ray coordinates* $(x_1, v_1 = n_1\theta_1)$. After the ray passes through the optical system, we denote the *output ray coordinates* $(x_2, v_2 = n_2\theta_2)$ on the output plane. In the paraxial approximation, the corresponding output quantities are linearly dependent on the input quantities. In other words, the output quantities can be expressed in terms of the weighted sum of the input quantities (known as the *principle of superposition*) as follows:

$$x_2 = Ax_1 + Bv_1 \quad \text{and} \quad v_2 = Cx_1 + Dv_1,$$

where A, B, C, and D are the weight factors. We can cast the above equations into a matrix form as

$$\begin{pmatrix} x_2 \\ v_2 \end{pmatrix} = \begin{pmatrix} A & B \\ C & D \end{pmatrix} \begin{pmatrix} x_1 \\ v_1 \end{pmatrix}. \tag{1.1-1}$$

The $\begin{pmatrix} A & B \\ C & D \end{pmatrix}$ matrix in Eq. (1.1-1) is called the *ray transfer matrix*, or the *system matrix* $\mathcal{S}$, if it is represented by the multiplication of ray transfer matrices. In what follows, we shall derive several important ray transfer matrices.

1.1.1 Ray Transfer Matrices

Translation Matrix

A ray travels in a homogenous medium of refractive index n in a straight line (see Figure 1.1-3). Let us denote the input and output planes with the ray's coordinates, and then we try to relate the input and output coordinates with a matrix after the traveling of the distance d. Since $n_1 = n_2 = n$ and $\theta_1 = \theta_2$, $v_2 = n_2\theta_2 = n_1\theta_1 = v_1$. From the geometry, we also find

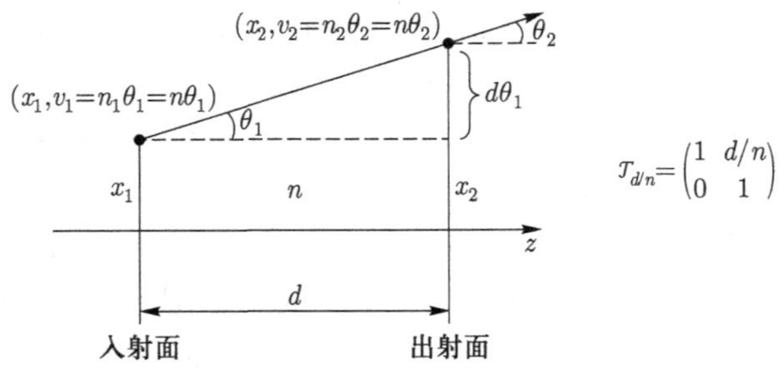

图 1.1-3　均匀介质中入射坐标和出射坐标分别为 $(x_1, v_1 = n_1\theta_1)$ 和 $(x_2, v_2 = n_2\theta_2)$ 的光线传输

$x_2 = x_1 + d\tan\theta_1 \approx x_1 + d\theta_1 = x_1 + dv_1/n$. 因此, 光线的入射坐标和出射坐标之间的关系可以表示为

$$\begin{pmatrix}x_2\\v_2\end{pmatrix} = \begin{pmatrix}1 & d/n\\0 & 1\end{pmatrix}\begin{pmatrix}x_1\\v_1\end{pmatrix} = \mathcal{T}_{d/n}\begin{pmatrix}x_1\\v_1\end{pmatrix}, \tag{1.1-2}$$

这里

$$\mathcal{T}_{d/n} = \begin{pmatrix}1 & d/n\\0 & 1\end{pmatrix}, \tag{1.1-3a}$$

$\mathcal{T}_{d/n}$ 被称为传输矩阵 (*translation matrix*), 该矩阵描述了光线在折射率为 n 的均匀介质中沿着光轴传输了距离 d 以后的一个变换, 其行列式为

$$\left|\mathcal{T}_{d/n}\right| = \begin{vmatrix}1 & d/n\\0 & 1\end{vmatrix} = 1,$$

可以看出, $\mathcal{T}_{d/n}$ 为一个正单位模矩阵. 对于空气中光线的传输, 让 $n = 1$, 那么该传输可以简单地表示为

$$\mathcal{T}_d = \begin{pmatrix}1 & d\\0 & 1\end{pmatrix}. \tag{1.1-3b}$$

折射矩阵

如图 1.1-4, 一个球面将空间分割为折射率为 n_1 和 n_2 的两个区域, 该球面的中心为 C 点, 曲率半径为 R. 曲率半径 (*radius of curvature*) 的符号规定为: 如果曲率中心 C 在球面的右 (左) 侧, 那么该面型的曲率半径取正 (负) 值. 光线到达球面上的 A 点后, 折射进入介质 n_2 中. 注意入射面和出射面是在同一平面内. 因此, 在折射前后, A 点光线的高度是相同的, 即, $x_2 = x_1$. ϕ_i 和 ϕ_t 分别为其入射角和折射角, 各自通过曲面的法线 NAC 与光线之间的夹角来测量. 利用斯涅耳定律和傍轴近似, 可以得到

$$n_1\phi_i = n_2\phi_t. \tag{1.1-4}$$

从图 1.1-4 的几何关系可以知道, $\phi_i = \theta_1 + \phi$, 且, $\phi_t = \theta_2 + \phi$. 因此, 式 (1.1-4) 的左侧可以写为

$$n_1\phi_i = n_1(\theta_1 + \phi) = v_1 + n_1x_1/R, \tag{1.1-5}$$

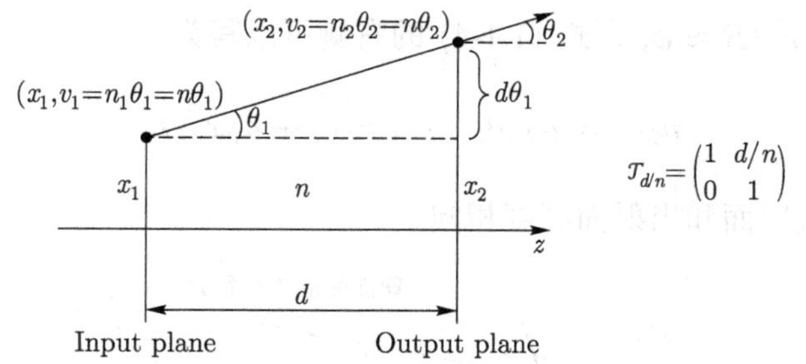

Figure 1.1-3 Ray propagating in a homogeneous medium with the input and output coordinates $(x_1, v_1 = n_1\theta_1)$ and $(x_2, v_2 = n_2\theta_2)$, respectively

$x_2 = x_1 + d\tan\theta_1 \approx x_1 + d\theta_1 = x_1 + dv_1/n$. Therefore, we can relate the output coordinates of the ray with its input coordinates as follows:

$$\begin{pmatrix}x_2\\v_2\end{pmatrix} = \begin{pmatrix}1 & d/n\\0 & 1\end{pmatrix}\begin{pmatrix}x_1\\v_1\end{pmatrix} = \mathcal{T}_{d/n}\begin{pmatrix}x_1\\v_1\end{pmatrix}, \tag{1.1-2}$$

where

$$\mathcal{T}_{d/n} = \begin{pmatrix}1 & d/n\\0 & 1\end{pmatrix}, \tag{1.1-3a}$$

which is called the *translation matrix*. The matrix describes the translation of a ray for a distance d along the optical axis in a homogenous medium of n. The determinant of $\mathcal{T}_{d/n}$ is

$$|\mathcal{T}_{d/n}| = \begin{vmatrix}1 & d/n\\0 & 1\end{vmatrix} = 1,$$

and hence $\mathcal{T}_{d/n}$ is a positive unimodular matrix. For a translation of the ray in air, we have $n = 1$, and the translation can be represented simply by

$$\mathcal{T}_d = \begin{pmatrix}1 & d\\0 & 1\end{pmatrix}. \tag{1.1-3b}$$

Refraction Matrix

We consider a spherical surface separating two regions of refractive indices n_1 and n_2 as shown in Figure 1.1-4. The center of the spherical surface is at C, and its radius of curvature is R. The convention for the *radius of curvature* is as follows. The radius of curvature of the surface is taken to be positive (negative) if the center C of curvature lies to the right (left) of the surface. The ray strikes the surface at the point A and gets refracted into media n_2. Note that the input and output planes are the same. Hence, the height of the ray at A, before and after refraction, is the same, that is, $x_2 = x_1$. ϕ_i and ϕ_t are the angles of incidence and refraction, respectively, which are measured from the normal NAC to the curved surface. Applying Snell's law and using the paraxial approximation, we have

$$n_1\phi_i = n_2\phi_t. \tag{1.1-4}$$

Now, from geometry, we know that $\phi_i = \theta_1 + \phi$ and $\phi_t = \theta_2 + \phi$ (Figure 1.1-4). Hence, the left side of Eq. (1.1-4) becomes

$$n_1\phi_i = n_1(\theta_1 + \phi) = v_1 + n_1 x_1/R, \tag{1.1-5}$$

这里, 利用了 $\sin\phi = x_1/R \approx \phi$, 则式 (1.1-4) 的右侧可以写为

$$n_2\phi_t = n_2(\theta_2 + \phi) = v_2 + n_2x_2/R, \tag{1.1-6}$$

其中, $x_1 = x_2$, 因为入射面和出射面高度相同.

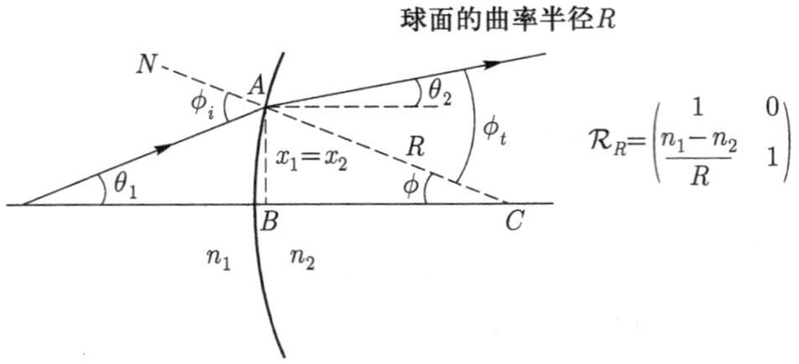

图 1.1-4 被球面分割成折射率分别为 n_1 和 n_2 的两个区域中的光线追迹

最后, 将式 (1.1-5) 和式 (1.1-6) 代入式 (1.1-4), 得

$$v_1 + n_1x_1/R = v_2 + n_2x_2/R,$$

或

$$v_2 = v_1 + (n_1 - n_2)\,x_1/R, \quad 因为\ x_1 = x_2. \tag{1.1-7}$$

将上述公式用系统矩阵方程来表示, 即

$$\begin{pmatrix} x_2 \\ v_2 \end{pmatrix} = \begin{pmatrix} 1 & 0 \\ \dfrac{n_1 - n_2}{R} & 1 \end{pmatrix} \begin{pmatrix} x_1 \\ v_1 \end{pmatrix} = \begin{pmatrix} 1 & 0 \\ -p & 1 \end{pmatrix} \begin{pmatrix} x_1 \\ v_1 \end{pmatrix} = \mathcal{R}_R \begin{pmatrix} x_1 \\ v_1 \end{pmatrix}, \tag{1.1-8}$$

这里

$$\mathcal{R}_R = \begin{pmatrix} 1 & 0 \\ -p & 1 \end{pmatrix}.$$

$\mathcal{R}_R$ 的行列式为

$$|\mathcal{R}_R| = \begin{vmatrix} 1 & 0 \\ -p & 1 \end{vmatrix} = 1.$$

该 2×2 的光线变换矩阵 $\mathcal{R}_p$ 为一正单位模矩阵, 被称为折射矩阵 (*refraction matrix*), 它描述了该球面的折射. 其中 p 的值由下式给出

$$p = \frac{n_2 - n_1}{R},$$

p 被称为球面的折射能力 (屈光能力, 折光能力) (*refracting power*). 当 R 的测量以米为单位时, p 的单位被称为屈光度 (*diopter*). 如果一入射光在经过一个表面后发生会聚 (发散), 那么该 p 值符号为正 (负).

厚透镜矩阵和薄透镜矩阵

如图 1.1-5 所示, 由两个球面组成的厚透镜, 如何求出关联入射光线坐标 (x_1, v_1) 和出射光线坐标 (x_2, v_2) 的系统矩阵? 这里,先求经过球面 R_1 的光线坐标 (x_1, v_1) 与 (x_1', v_1') 的关系,

where we have used $\sin\phi = x_1/R \approx \phi$. Now, the right side of Eq. (1.1-4) is

$$n_2\phi_t = n_2(\theta_2 + \phi) = v_2 + n_2 x_2/R, \tag{1.1-6}$$

where $x_1 = x_2$ as the input and output planes are the same.

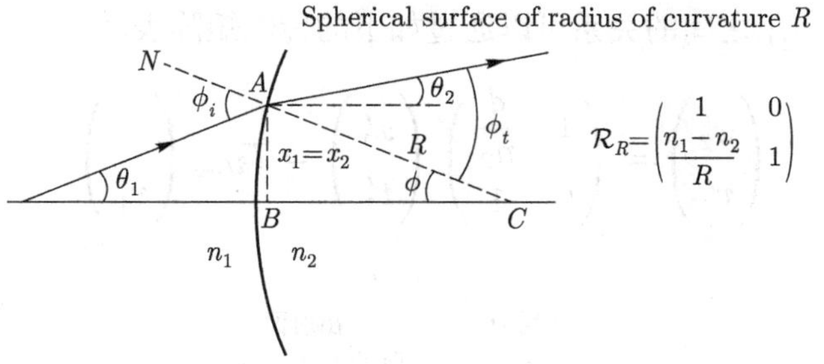

Figure 1.1-4 Ray trajectory during refraction at a spherical surface separating two regions of refractive indices n_1 and n_2

Finally, putting Eqs. (1.1-5) and (1.1-6) into Eq. (1.1-4), we have

$$v_1 + n_1 x_1/R = v_2 + n_2 x_2/R,$$

or

$$v_2 = v_1 + (n_1 - n_2)x_1/R, \quad \text{as} \quad x_1 = x_2. \tag{1.1-7}$$

We can formulate the above equation in terms of a matrix equation as

$$\begin{pmatrix} x_2 \\ v_2 \end{pmatrix} = \begin{pmatrix} 1 & 0 \\ \dfrac{n_1 - n_2}{R} & 1 \end{pmatrix} \begin{pmatrix} x_1 \\ v_1 \end{pmatrix} = \begin{pmatrix} 1 & 0 \\ -p & 1 \end{pmatrix} \begin{pmatrix} x_1 \\ v_1 \end{pmatrix} = \mathcal{R}_R \begin{pmatrix} x_1 \\ v_1 \end{pmatrix}, \tag{1.1-8}$$

where

$$\mathcal{R}_R = \begin{pmatrix} 1 & 0 \\ -p & 1 \end{pmatrix} = \begin{pmatrix} 1 & 0 \\ \dfrac{n_1 - n_2}{R} & 1 \end{pmatrix}.$$

The determinant of $\mathcal{R}_R$ is

$$|\mathcal{R}_R| = \begin{vmatrix} 1 & 0 \\ -p & 1 \end{vmatrix} = 1.$$

The 2×2 ray transfer matrix $\mathcal{R}_R$ is a positive unimodular matrix and is called the *refraction matrix*. The matrix describes refraction for the spherical surface. The quantity p given by

$$p = \frac{n_2 - n_1}{R},$$

is called the *refracting power* of the spherical surface. When R is measured in meters, the unit of p is called *diopters*. If an incident ray is made to converge on (diverge from) by a surface, the power is positive (negative) in sign.

Thick- and Thin-Lens Matrices

A thick lens consists of two spherical surfaces as shown in Figure 1.1-5. We shall find the system matrix that relates the system's input coordinates (x_1, v_1) to system's output ray coordinates (x_2, v_2). Let us first relate (x_1, v_1) to (x_1', v_1') through the spherical surface

其中, (x_1', v_1') 为经过球面 R_1 后的出射光线坐标. 根据式 (1.1-8), 得

$$\begin{pmatrix} x_1' \\ v_1' \end{pmatrix} = \begin{pmatrix} 1 & 0 \\ \dfrac{n_1 - n_2}{R_1} & 1 \end{pmatrix} \begin{pmatrix} x_1 \\ v_1 \end{pmatrix} = \mathcal{R}_{R_1} \begin{pmatrix} x_1 \\ v_1 \end{pmatrix}. \tag{1.1-9}$$

现在, (x_1', v_1') 与 (x_2', v_2') 之间的关系可以通过如下的传输矩阵表示

$$\begin{pmatrix} x_2' \\ v_2' \end{pmatrix} = \begin{pmatrix} 1 & \dfrac{d}{n_2} \\ 0 & 1 \end{pmatrix} \begin{pmatrix} x_1' \\ v_1' \end{pmatrix} = \mathcal{T}_{d/n_2} \begin{pmatrix} x_1' \\ v_1' \end{pmatrix}, \tag{1.1-10}$$

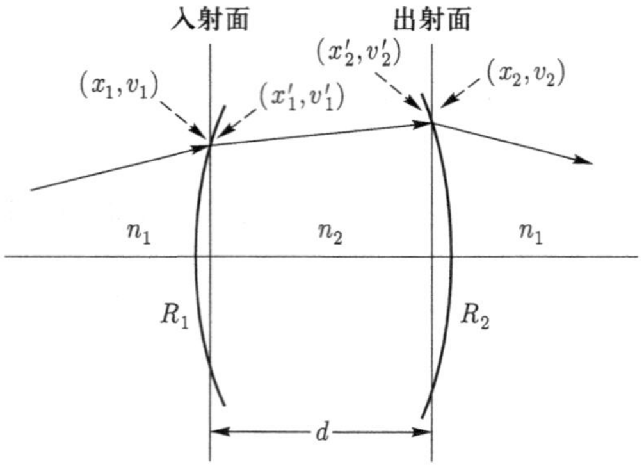

图 1.1-5 厚透镜

其中, (x_2', v_2') 为传输后的出射光线坐标, 同时也是球面 R_2 的入射光线坐标. 最终, 可以将 (x_2', v_2') 与系统的出射光线坐标 (x_2, v_2) 通过以下关系建立起来

$$\begin{pmatrix} x_2 \\ v_2 \end{pmatrix} = \begin{pmatrix} 1 & 0 \\ \dfrac{n_2 - n_1}{R_2} & 1 \end{pmatrix} \begin{pmatrix} x_2' \\ v_2' \end{pmatrix} = \mathcal{R}_{-R_2} \begin{pmatrix} x_2' \\ v_2' \end{pmatrix}. \tag{1.1-11}$$

若将式 (1.1-10) 代入式 (1.1-11), 可得

$$\begin{pmatrix} x_2 \\ v_2 \end{pmatrix} = \begin{pmatrix} 1 & 0 \\ \dfrac{n_2 - n_1}{R_2} & 1 \end{pmatrix} \begin{pmatrix} 1 & \dfrac{d}{n_2} \\ 0 & 1 \end{pmatrix} \begin{pmatrix} x_1' \\ v_1' \end{pmatrix}.$$

接着, 将式 (1.1-9) 代入上式, 即可得到整个系统的系统矩阵方程 (*system matrix equation*) 如下

$$\begin{aligned} \begin{pmatrix} x_2 \\ v_2 \end{pmatrix} &= \begin{pmatrix} 1 & 0 \\ \dfrac{n_2 - n_1}{R_2} & 1 \end{pmatrix} \begin{pmatrix} 1 & \dfrac{d}{n_2} \\ 0 & 1 \end{pmatrix} \begin{pmatrix} 1 & 0 \\ \dfrac{n_1 - n_2}{R_1} & 1 \end{pmatrix} \begin{pmatrix} x_1 \\ v_1 \end{pmatrix} \\ &= \mathcal{R}_{-R_2} \mathcal{T}_{d/n_2} \mathcal{R}_{R_1} \begin{pmatrix} x_1 \\ v_1 \end{pmatrix} = \mathcal{S} \begin{pmatrix} x_1 \\ v_1 \end{pmatrix}. \end{aligned} \tag{1.1-12}$$

至此, 系统的入射光线坐标与出射光线坐标得以关联. 可以发现, 系统矩阵 $\mathcal{S} = \mathcal{R}_{-R_2} \mathcal{T}_{d/n_2} \mathcal{R}_{R_1}$ 为三个光线变换矩阵的乘积. 通常, 系统矩阵是由一系列光线变换矩阵所组成的, 这些光线变换矩阵描述了光线通过该光学系统时所受到的影响. 当光线沿 z 轴正向从左往右传输时, 求解系统矩阵, 其光线变换矩阵要按照从右往左的顺序依次计算. 正是这种使用矩阵形式表示的优势,

defined by R_1. (x_1', v_1') are the output coordinates due to the surface R_1. According to Eq. (1.1-8), we have

$$\begin{pmatrix} x_1' \\ v_1' \end{pmatrix} = \begin{pmatrix} 1 & 0 \\ \dfrac{n_1 - n_2}{R_1} & 1 \end{pmatrix} \begin{pmatrix} x_1 \\ v_1 \end{pmatrix} = \mathcal{R}_{R_1} \begin{pmatrix} x_1 \\ v_1 \end{pmatrix}. \tag{1.1-9}$$

Now, (x_1', v_1') and (x_2', v_2') are related through a translation matrix as follows:

$$\begin{pmatrix} x_2' \\ v_2' \end{pmatrix} = \begin{pmatrix} 1 & \dfrac{d}{n_2} \\ 0 & 1 \end{pmatrix} \begin{pmatrix} x_1' \\ v_1' \end{pmatrix} = \mathcal{T}_{d/n_2} \begin{pmatrix} x_1' \\ v_1' \end{pmatrix}, \tag{1.1-10}$$

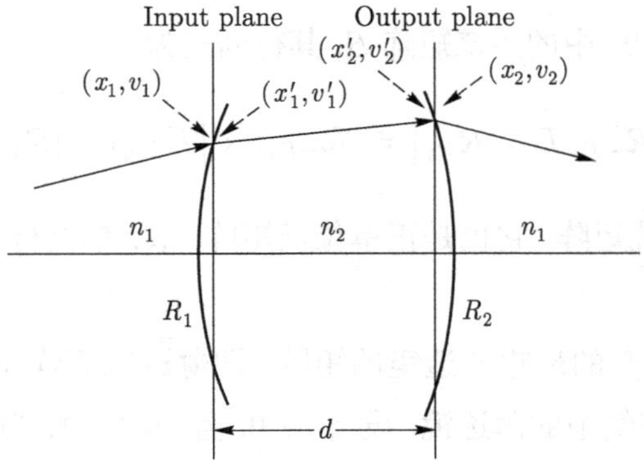

Figure 1.1-5 Thick lens

where (x_2', v_2') are the output coordinates after translation, which are also the coordinates of the input coordinates for the surface R_2. Finally, we relate (x_2', v_2') to the system's output coordinates (x_2, v_2) through

$$\begin{pmatrix} x_2 \\ v_2 \end{pmatrix} = \begin{pmatrix} 1 & 0 \\ \dfrac{n_2 - n_1}{R_2} & 1 \end{pmatrix} \begin{pmatrix} x_2' \\ v_2' \end{pmatrix} = \mathcal{R}_{-R_2} \begin{pmatrix} x_2' \\ v_2' \end{pmatrix}. \tag{1.1-11}$$

If we now substitute Eq. (1.1-10) into Eq. (1.1-11), then we have

$$\begin{pmatrix} x_2 \\ v_2 \end{pmatrix} = \begin{pmatrix} 1 & 0 \\ \dfrac{n_2 - n_1}{R_2} & 1 \end{pmatrix} \begin{pmatrix} 1 & \dfrac{d}{n_2} \\ 0 & 1 \end{pmatrix} \begin{pmatrix} x_1' \\ v_1' \end{pmatrix}.$$

Subsequently, substituting Eq. (1.1-9) into the above equation, we have the *system matrix equation* of the entire system as follows:

$$\begin{aligned} \begin{pmatrix} x_2 \\ v_2 \end{pmatrix} &= \begin{pmatrix} 1 & 0 \\ \dfrac{n_2 - n_1}{R_2} & 1 \end{pmatrix} \begin{pmatrix} 1 & \dfrac{d}{n_2} \\ 0 & 1 \end{pmatrix} \begin{pmatrix} 1 & 0 \\ \dfrac{n_1 - n_2}{R_1} & 1 \end{pmatrix} \begin{pmatrix} x_1 \\ v_1 \end{pmatrix} \\ &= \mathcal{R}_{-R_2} \mathcal{T}_{d/n_2} \mathcal{R}_{R_1} \begin{pmatrix} x_1 \\ v_1 \end{pmatrix} = \mathcal{S} \begin{pmatrix} x_1 \\ v_1 \end{pmatrix}. \end{aligned} \tag{1.1-12}$$

We have now finally related the system's input coordinates to output coordinates. Note that the system matrix, $\mathcal{S} = \mathcal{R}_{-R_2} \mathcal{T}_{d/n_2} \mathcal{R}_{R_1}$, is a product of three ray transfer matrices. In general, the system matrix is made up of a collection of ray transfer matrices to account for the effects of a ray passing through the optical system. As the ray goes from left to right in the positive direction of the z-axis, we obtain the system matrix by writing the ray transfer matrices from right to left. This is precisely the advantage of the matrix formalism in that

使得任意光线在经过光学系统进行传播时, 可以通过依次之间的矩阵相乘进行光线追迹. 让 $\mathcal{A}$ 和 $\mathcal{B}$ 分别为一个 2×2 的矩阵, 即

$$\mathcal{A}=\begin{pmatrix} a & b \\ c & d \end{pmatrix} \quad \text{和} \quad \mathcal{B}=\begin{pmatrix} e & f \\ g & h \end{pmatrix},$$

那么矩阵相乘 (*matrix multiplication*) 的规则为

$$\mathcal{AB}=\begin{pmatrix} a & b \\ c & d \end{pmatrix}\begin{pmatrix} e & f \\ g & h \end{pmatrix}=\begin{pmatrix} ae+bg & af+bh \\ ce+dg & cf+dh \end{pmatrix}.$$

现在, 再回到式 (1.1-12) 中的系统矩阵 $\mathcal{S}$, 其行列式为

$$|\mathcal{S}|=\left|\mathcal{R}_{-R_2}\mathcal{T}_{d/n_2}\mathcal{R}_{R_1}\right|=\left|\mathcal{R}_{-R_2}\right|\times\left|\mathcal{T}_{d/n_2}\right|\times\left|\mathcal{R}_{R_1}\right|=1.$$

可以发现, 即使是系统矩阵, 它也是正单位模矩阵, 故, 单位行列式是系统矩阵的必要而非充分条件.

现在, 推导一焦距为 f 的理想薄透镜的矩阵, 即薄透镜矩阵 (*thin-lens matrix*) $\mathcal{L}_f$. 如图 1.1-5 所示, 对于一个空气中的薄透镜, 设 $d\to 0$, 且, $n_1=1$, 为标注方便, 让 $n_2=n$. 则式 (1.1-12) 中的系统矩阵可写为

$$\begin{aligned}\mathcal{S}&=\begin{pmatrix} 1 & 0 \\ \dfrac{n-1}{R_2} & 1 \end{pmatrix}\begin{pmatrix} 1 & 0 \\ 0 & 1 \end{pmatrix}\begin{pmatrix} 1 & 0 \\ \dfrac{1-n}{R_1} & 1 \end{pmatrix}\\ &=\begin{pmatrix} 1 & 0 \\ \dfrac{n-1}{R_2} & 1 \end{pmatrix}\begin{pmatrix} 1 & 0 \\ \dfrac{1-n}{R_1} & 1 \end{pmatrix}=\begin{pmatrix} 1 & 0 \\ -\dfrac{1}{f} & 1 \end{pmatrix}=\mathcal{L}_f,\end{aligned} \tag{1.1-13}$$

这里, f 为薄透镜的焦距 (*focal length*), 可由下式得到

$$\frac{1}{f}=(n-1)\left(\frac{1}{R_1}-\frac{1}{R_2}\right).$$

对于 $f>0$, 透镜为会聚 (凸) 透镜, 反之, 当 $f<0$ 时, 为发散 (凹) 透镜. 图 1.1-6 所示为理想薄透镜的情况.

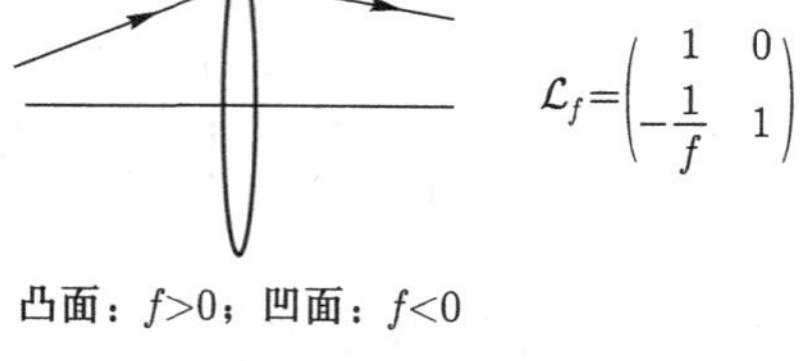

图 1.1-6　焦距为 f 的理想薄透镜及其相应的光线变换矩阵

可以看到, $\mathcal{L}_f$ 的行列式为

$$|\mathcal{L}_f|=\begin{vmatrix} 1 & 0 \\ -\dfrac{1}{f} & 1 \end{vmatrix}=1.$$

any ray, during its propagation through the optical system, can be tracked by successive matrix multiplications. Let $\mathcal{A}$ and $\mathcal{B}$ be the 2×2 matrices as follows:

$$\mathcal{A} = \begin{pmatrix} a & b \\ c & d \end{pmatrix} \quad \text{and} \quad \mathcal{B} = \begin{pmatrix} e & f \\ g & h \end{pmatrix}.$$

Then the rule of matrix multiplication is

$$\mathcal{AB} = \begin{pmatrix} a & b \\ c & d \end{pmatrix} \begin{pmatrix} e & f \\ g & h \end{pmatrix} = \begin{pmatrix} ae+bg & af+bh \\ ce+dg & cf+dh \end{pmatrix}.$$

Now, returning to the system matrix in Eq. (1.1-12), the determinant of the system matrix, $\mathcal{S} = \mathcal{R}_{-R_2}\mathcal{T}_{d/n_2}\mathcal{R}_{R_1}$, is

$$|\mathcal{S}| = \left|\mathcal{R}_{-R_2}\mathcal{T}_{d/n_2}\mathcal{R}_{R_1}\right| = |\mathcal{R}_{-R_2}| \times \left|\mathcal{T}_{d/n_2}\right| \times |\mathcal{R}_{R_1}| = 1.$$

Note that even the system matrix is also positive unimodular. The condition of a unit determinant is a necessary but not a sufficient condition on the system matrix.

We now derive a matrix of an idea thin lens of focal length f, called the *thin-lens matrix*, $\mathcal{L}_f$. For a thin lens in air, we let $d \to 0$ and $n_1 = 1$ in the configuration of Figure 1.1-5. We also write $n_2 = n$ for notational convenience. Then the system matrix in Eq. (1.1-12) becomes

$$\begin{aligned} \mathcal{S} &= \begin{pmatrix} 1 & 0 \\ \dfrac{n-1}{R_2} & 1 \end{pmatrix} \begin{pmatrix} 1 & 0 \\ 0 & 1 \end{pmatrix} \begin{pmatrix} 1 & 0 \\ \dfrac{1-n}{R_1} & 1 \end{pmatrix} \\ &= \begin{pmatrix} 1 & 0 \\ \dfrac{n-1}{R_2} & 1 \end{pmatrix} \begin{pmatrix} 1 & 0 \\ \dfrac{1-n}{R_1} & 1 \end{pmatrix} \\ &= \begin{pmatrix} 1 & 0 \\ -\dfrac{1}{f} & 1 \end{pmatrix} = \mathcal{L}_f, \end{aligned} \tag{1.1-13}$$

where f is the *focal length* of the thin lens and is given by

$$\frac{1}{f} = (n-1)\left(\frac{1}{R_1} - \frac{1}{R_2}\right).$$

For $f > 0$, we have a converging (convex) lens. On the other hand, we have a diverging (concave) lens when $f < 0$. Figure 1.1-6 summarizes the result for the ideal thin lens.

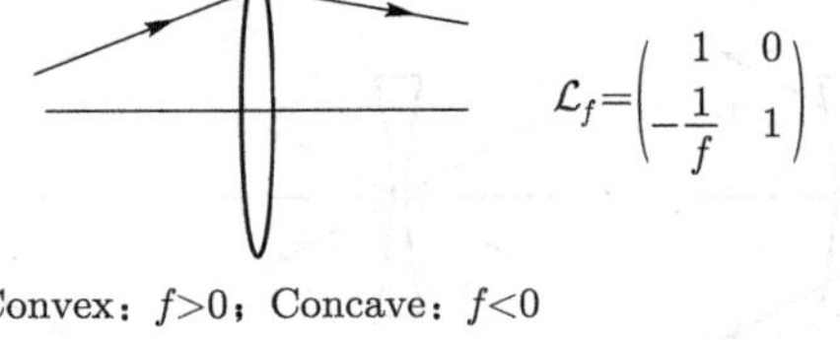

Figure 1.1-6 Ideal thin lens of focal length f and its associated ray transfer matrix

Note that the determinant of $\mathcal{L}_f$ is

$$|\mathcal{L}_f| = \begin{vmatrix} 1 & 0 \\ -\dfrac{1}{f} & 1 \end{vmatrix} = 1.$$

1.1.2　薄透镜的光线追迹

由 1.1.1 节可知, 当考虑一个焦距为 f 的薄透镜时, 根据式 (1.1-13), 其矩阵方程为

$$\begin{pmatrix} x_2 \\ v_2 \end{pmatrix} = \mathcal{L}_f \begin{pmatrix} x_1 \\ v_1 \end{pmatrix} = \begin{pmatrix} 1 & 0 \\ -\dfrac{1}{f} & 1 \end{pmatrix} \begin{pmatrix} x_1 \\ v_1 \end{pmatrix}. \tag{1.1-14}$$

a) 平行于光轴入射的光线

从图 1.1-7a) 可以看出, 对于薄透镜, 入射和出射光线具有同样的高度, 即 $x_1 = x_2$, 根据式 (1.1-14), 有 $v_2 = -x_1/f + v_1$. 则, 当 $v_1 = 0$ 时, 即当入射光线平行于光轴时, $v_2 = -x_1/f$. 对于正的 $x_1, v_2 < 0$, 因为对于一个会聚透镜, 其 $f > 0$. 而对于负的 $x_1, v_2 > 0$. 所有平行于光轴的入射光线在穿过透镜后将会聚于透镜的后焦点 (距离透镜为 f 的位置), 如图 1.1-7a) 所示. 对于一个薄透镜来说, 其前焦点也是在前方距离透镜 f 的位置处.

b) 通过透镜中心入射的光线

对于通过透镜中心的入射光线, 其入射光线坐标为 $(x_1, v_1) = (0, v_1)$, 根据式 (1.1-14), 其出射光线坐标为 $(x_2, v_2) = (0, v_1)$, 如图 1.1-7b) 所示, 可看出, 由于 $v_2 = v_1$, 所有通过透镜中心的入射光线, 其传播方向保持不变.

c) 过透镜前焦点入射的光线

对于这种情况, 如图 1.1-7c) 所示, 当入射光线坐标为 $(x_1, v_1 = x_1/f)$ 时, 由式 (1.1-14), 其出射光线坐标为 $(x_2 = x_1, v_2 = 0)$, 这说明所有出射光线将平行于光轴 $(v_2 = 0)$.

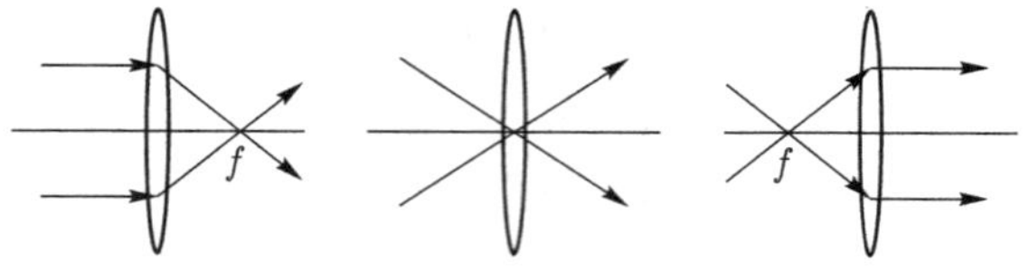

a)　　b)　　c)

图 1.1-7　薄凸透镜的光线追迹

类似地, 对于发散透镜的情况, 可以得到如下结论: 如图 1.1-8a) 所示, 经过折射后偏离光轴发散的光线, 就像该光线来自前方距离透镜为 $|f|$ 的轴上某一点. 如图 1.1-8b) 所示, 通过透镜中心的入射光线, 其传播方向不变. 最后, 如图 1.1-8c) 所示, 对于朝向发散透镜的后焦点而来的入射光线, 在经过透镜后, 其出射光线将平行于光轴传播.

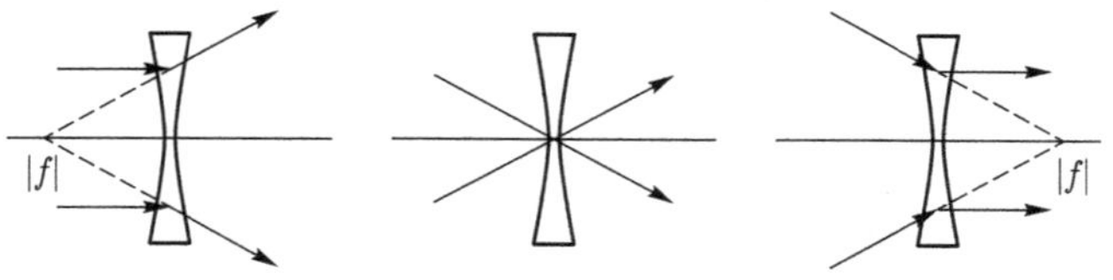

a)　　b)　　c)

图 1.1-8　薄凹透镜的光线追迹

1.1.2 Ray Tracing through a Thin Lens

As we have seen from Section 1.1.1, when a thin lens of focal length f is involved, then the matrix equation, from Eq. (1.1-13), is

$$\begin{pmatrix} x_2 \\ v_2 \end{pmatrix} = \mathcal{L}_f \begin{pmatrix} x_1 \\ v_1 \end{pmatrix} = \begin{pmatrix} 1 & 0 \\ -\dfrac{1}{f} & 1 \end{pmatrix} \begin{pmatrix} x_1 \\ v_1 \end{pmatrix}. \tag{1.1-14}$$

Input Rays Traveling Parallel to the Optical Axis

From Figure 1.1-7a, we recognize that $x_1 = x_2$ as the heights of the input and output rays are the same for the thin lens. Now, according to Eq. (1.1-14), $v_2 = -x_1/f + v_1$. For $v_1 = 0$, that is, the input rays are parallel to the optical axis, $v_2 = -x_1/f$. For positive $x_1, v_2 < 0$ as $f > 0$ for a converging lens. For negative $x_1, v_2 > 0$. All input rays parallel to the optical axis converge behind the lens to the back focal point (a distance of f away from the lens) of the lens as shown in Figure 1.1-7a. Note that for a thin lens, the front focal point is also a distance of f away from the lens.

Input Rays Traveling through the Center of the Lens

For input rays traveling through the center of the lens, their input ray coordinates are $(x_1, v_1) = (0, v_1)$. The output ray coordinates, according to Eq. (1.1-14), are $(x_2, v_2) = (0, v_1)$. Hence, we see as $v_2 = v_1$, all rays traveling through the center of the lens will pass undeviated as shown in Figure 1.1-7b.

Input Rays Passing through the Front Focal Point of the Lens

For this case, the input ray coordinates are $(x_1, v_1 = x_1/f)$, and, according to Eq. (1.1-14), the output ray coordinates are $(x_2 = x_1, v_2 = 0)$, which means that all output rays will be parallel to the optical axis ($v_2 = 0$), as shown in Figure 1.1-7c.

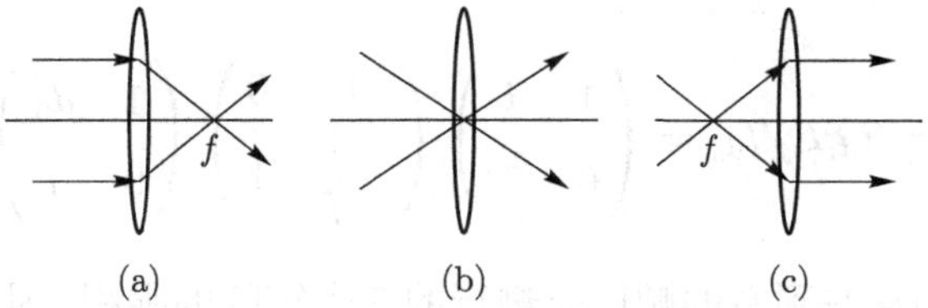

(a) (b) (c)

Figure 1.1-7 Ray tracing through a thin convex lens

Similarly, in the case of a diverging lens, we can draw conclusions as follows. The ray after refraction diverges away from the axis as if it were coming from a point on the axis a distance $|f|$ in front of the lens, as shown in Figure 1.1-8a. The ray traveling through the center of the lens will pass undeviated, as shown in Figure 1.1-8b. Finally, for an input ray appearing to travel toward the back focus point of a diverging lens, the output ray will be parallel to the optical axis, as shown in Figure 1.1-8c.

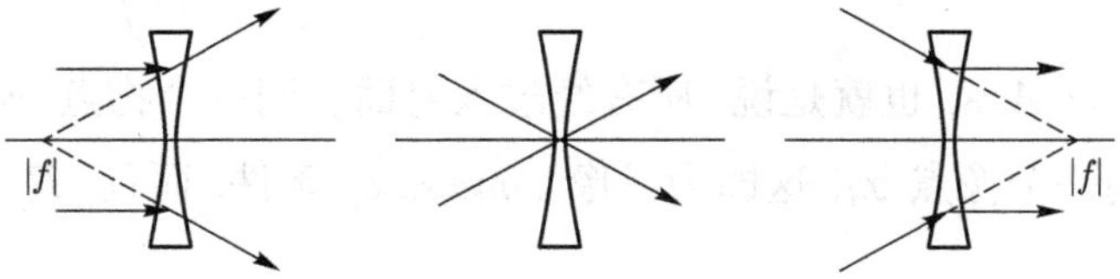

(a) (b) (c)

Figure 1.1-8 Ray tracing through a thin concave lens

举例: 凸透镜成像

如图 1.1-9 所示, 考虑一单透镜成像, 并假定该透镜位于空气中. 先绘出该成像系统的光路图 (*ray diagram*): 一个物体 OO' 位于焦距为 f 的薄透镜前 d_0 位置处, 从物体上一点 O' 处向该透镜发出两束光线. 其中, 第一束光线, 来自 O' 的光线 1, 平行于光轴入射, 如图 1.1-7a) 所示规律, 该平行于光轴的入射光线会聚于透镜的后焦点位置. 第二束光线, 同样从 O' 发出, 即光线 2, 直接穿过透镜的中心传播, 没有改变方向, 如图 1.1-7b) 所示. 这两束光线在透镜的另一侧交会, 形成了 O' 的像点. 该 O' 的像点在图中被记为 I'. 最终的像是实的、倒立的, 被称为实像 (*real image*).

现在, 利用矩阵方法来研究单薄透镜的成像特性. 光学系统的入射面和出射面如图 1.1-9 所示, 设 (x_0, v_0) 和 (x_i, v_i) 分别为光线在 O' 和 I' 处的坐标, 可以看到, 该问题包含三个矩阵, 总的系统矩阵方程为

$$\begin{pmatrix} x_i \\ v_i \end{pmatrix} = \mathcal{T}_{d_i}\mathcal{L}_f\mathcal{T}_{d_0}\begin{pmatrix} x_0 \\ v_0 \end{pmatrix} = \mathcal{S}\begin{pmatrix} x_0 \\ v_0 \end{pmatrix}. \tag{1.1-15}$$

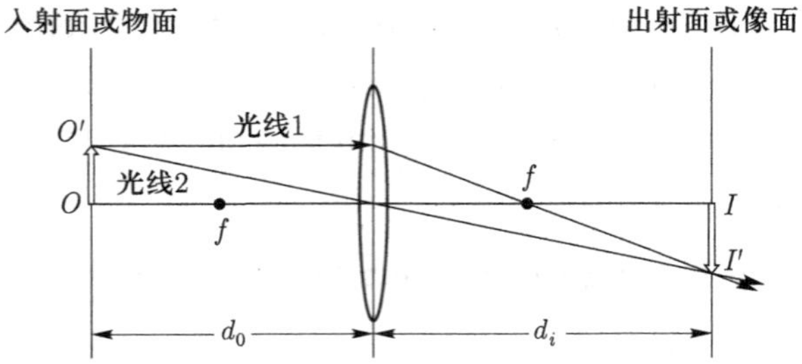

图 1.1-9 单透镜成像的光路图

总的系统矩阵为

$$\mathcal{S} = \mathcal{T}_{d_i}\mathcal{L}_f\mathcal{T}_{d_0} = \begin{pmatrix} 1 & d_i \\ 0 & 1 \end{pmatrix}\begin{pmatrix} 1 & 0 \\ -\dfrac{1}{f} & 1 \end{pmatrix}\begin{pmatrix} 1 & d_0 \\ 0 & 1 \end{pmatrix},$$

如前所述, 被表示为按照从右到左的顺序所进行的三个矩阵的乘积, 因为其光线是沿光轴从左往右传输的. 根据矩阵乘法规则, 式 (1.1-15) 可以被简化为

$$\begin{pmatrix} x_i \\ v_i \end{pmatrix} = \begin{pmatrix} 1-\dfrac{d_i}{f} & d_0+d_i-\dfrac{d_0 d_i}{f} \\ -\dfrac{1}{f} & 1-\dfrac{d_0}{f} \end{pmatrix}\begin{pmatrix} x_0 \\ v_0 \end{pmatrix} = \begin{pmatrix} A & B \\ C & D \end{pmatrix}\begin{pmatrix} x_0 \\ v_0 \end{pmatrix}. \tag{1.1-16}$$

为了讨论成像条件, 这里重点关注上述公式中 $\mathcal{S}$ 的 $\begin{pmatrix} A & B \\ C & D \end{pmatrix}$ 矩阵, 通常, 若 $B = 0$, 则 $x_i = Ax_0 + Bv_0 = Ax_0$, 也就是说, 所有经过入射面上同一个物点 x_0 的光线, 传播后都会经过出射面上的同一个像点 x_i, 这即为成像 (*imaging*) 条件. 而且, $B = 0$ 时, $A = x_i/x_0$,

Example: Imaging by a Convex Lens

We consider a single-lens imaging as shown in Figure 1.1-9, where we assume the lens is immersed in air. We first construct a *ray diagram* for the imaging system. An object OO' is located a distance d_0 in front of a thin lens of focal length f. We send two rays from a point O' towards the lens. Ray 1 from O' is incident parallel to the optical axis, and from Figure 1.1-7a, the input ray parallel to the optical axis converges behind the lens to the back focal point. A second ray, that is, ray 2 also from O', is now drawn through the center of the lens without bending and that is the result from Figure 1.1-7b. The interception of the two rays on the other side of the lens forms an image point of O'. The image point of O' is labeled at I' in the diagram. The final image is real, inverted, and is called a *real image*.

Now we investigate the imaging properties of the single thin lens using the matrix formalism. The input plane and the output plane of the optical system are shown in Figure 1.1-9. We let (x_0, v_0) and (x_i, v_i) represent the coordinates of the ray at O' and I', respectively. We see there are three matrices involved in the problem. The overall system matrix equation becomes

$$\begin{pmatrix} x_i \\ v_i \end{pmatrix} = \mathcal{T}_{d_i}\mathcal{L}_f\mathcal{T}_{d_0}\begin{pmatrix} x_0 \\ v_0 \end{pmatrix} = \mathcal{S}\begin{pmatrix} x_0 \\ v_0 \end{pmatrix}. \tag{1.1-15}$$

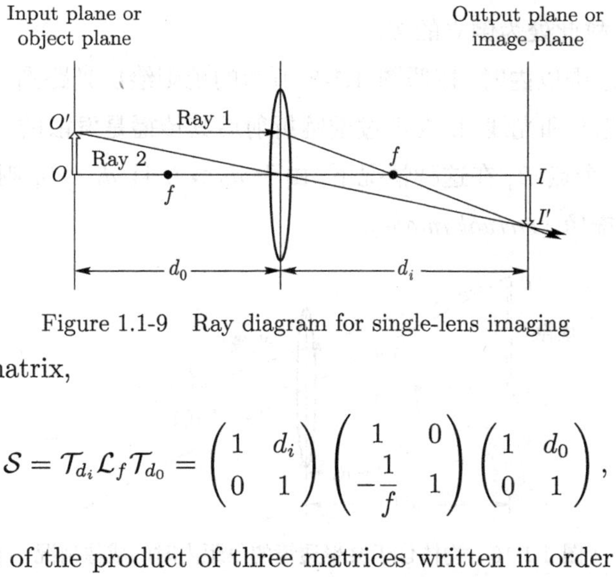

Figure 1.1-9 Ray diagram for single-lens imaging

The overall system matrix,

$$\mathcal{S} = \mathcal{T}_{d_i}\mathcal{L}_f\mathcal{T}_{d_0} = \begin{pmatrix} 1 & d_i \\ 0 & 1 \end{pmatrix}\begin{pmatrix} 1 & 0 \\ -\frac{1}{f} & 1 \end{pmatrix}\begin{pmatrix} 1 & d_0 \\ 0 & 1 \end{pmatrix},$$

is expressed in terms of the product of three matrices written in order from right to left as the ray goes from left to right along the optical axis, as explained earlier. According to the rule of matrix multiplication, Eq. (1.1-15) can be simplified to

$$\begin{pmatrix} x_i \\ v_i \end{pmatrix} = \begin{pmatrix} 1 - \frac{d_i}{f} & d_0 + d_i - \frac{d_0 d_i}{f} \\ -\frac{1}{f} & 1 - \frac{d_0}{f} \end{pmatrix}\begin{pmatrix} x_0 \\ v_0 \end{pmatrix} = \begin{pmatrix} A & B \\ C & D \end{pmatrix}\begin{pmatrix} x_0 \\ v_0 \end{pmatrix}. \tag{1.1-16}$$

To investigate the conditions for imaging, let us concentrate on the $\begin{pmatrix} A & B \\ C & D \end{pmatrix}$ matrix of $\mathcal{S}$ in Eq. (1.1-16). In general, we see that $x_i = Ax_0 + Bv_0 = Ax_0$ if $B = 0$, which means that all rays passing through the input plane at the same object point x_0 will pass through the same image point x_i in the output plane. This is the condition of *imaging*. In addition, for

被定义为成像系统的横向放大率 (*lateral magnification*). 至此, 在薄透镜成像情况下, 当式 (1.1-16) 中的 $B=0$, 会使得 $d_0+d_i-d_0d_i/f=0$, 也就是薄透镜公式 (*thin-lens formula*)

$$\frac{1}{d_0}+\frac{1}{d_i}=\frac{1}{f}. \tag{1.1-17}$$

上述公式里的符号规定为, 若物体位于透镜的左 (右) 侧时, 物距 d_0 为正 (负); 若像位于透镜的右 (左) 侧并且为实 (虚) 的, 则像距 d_i 为正 (负). 在图 1.1-9 中, 可以看出, $d_0>0$, $d_i>0$, 故所成像为实像, 这也表示, 在实际物理空间中, 光线的会聚形成了像点 I'. 因此, 对于成像, 式 (1.1-16) 变为

$$\begin{pmatrix} x_i \\ v_i \end{pmatrix}=\begin{pmatrix} 1-\dfrac{d_i}{f} & 0 \\ -\dfrac{1}{f} & 1-\dfrac{d_0}{f} \end{pmatrix}\begin{pmatrix} x_0 \\ v_0 \end{pmatrix}, \tag{1.1-18}$$

该公式将成像系统中的入射光线坐标和出射光线坐标关联起来. 利用式 (1.1-17), 成像系统的横向放大率 M 为

$$M=A=\frac{x_i}{x_0}=1-\frac{d_i}{f}=-\frac{d_i}{d_0}. \tag{1.1-19}$$

此处符号规定为, 若 $M>0$, 像是正立的; 若 $M<0$, 像是倒立的. 如图 1.1-9 所示, 因为像距 d_i 和物距 d_0 同为正, 故此处为倒立的像.

当物体位于焦距范围以内时, 按照图 1.1-8 所示的规则绘出光路图, 如图 1.1-10 所示. 所不同的是, 此处的光线 1 和光线 2, 在经过透镜折射后其传播是发散的, 不会在透镜的右侧交会. 它们似乎来自同一个点 I', 在这种情况下, 由于 $d_0>0$ 且 $d_i<0$, 则 $M>0$. 最终的像为虚的、正立的, 被称为虚像 (*virtual image*).

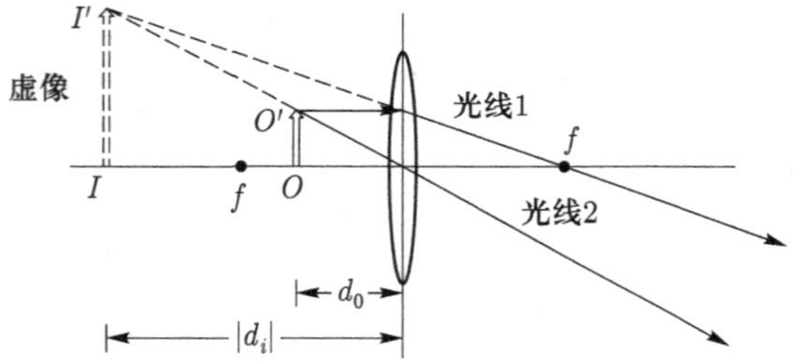

图 1.1-10 物体位于会聚透镜的焦距内时的成像情况

1.2 分辨率、焦深和景深

1.2.1 圆形孔径

透镜的数值孔径 (*numerical aperture*, NA) 可被定义为其物距或像距位于无限远时的情况. 图 1.2-1 给出了一个位于无限远处的物体, 其发出的平行光射向具有圆形孔径 (*circular aperture*) 的透镜, 角度 θ_{im} 用于定义像方的 NA, 则

$$NA_i=n_i\sin(\theta_{im}/2), \tag{1.2-1}$$

$B = 0$, $A = x_i/x_0$ is defined as the *lateral magnification* of the imaging system. Now in our case of thin-lens imaging, $B = 0$ in Eq. (1.1-16) leads to $d_0 + d_i - d_0 d_i/f = 0$, which gives the well-known *thin-lens formula*:

$$\frac{1}{d_0} + \frac{1}{d_i} = \frac{1}{f}. \tag{1.1-17}$$

The sign convention for Eq. (1.1-17) is that the object distance d_0 is positive (negative) if the object is to the left (right) of the lens. If the image distance d_i is positive (negative), the image is to the right (left) of the lens and it is real (virtual). In Figure 1.1-9, we have $d_0 > 0$, $d_i > 0$, and the image is therefore real, which means physically that light rays actually converge to the formed image I'. Hence, for imaging, Eq. (1.11-6) becomes

$$\begin{pmatrix} x_i \\ v_i \end{pmatrix} = \begin{pmatrix} 1 - \dfrac{d_i}{f} & 0 \\ -\dfrac{1}{f} & 1 - \dfrac{d_0}{f} \end{pmatrix} \begin{pmatrix} x_0 \\ v_0 \end{pmatrix}, \tag{1.1-18}$$

which relates the input ray and output ray coordinates in the imaging system. Using Eq. (1.1-17), the *lateral magnification* M of the imaging system is

$$M = A = \frac{x_i}{x_0} = 1 - \frac{d_i}{f} = -\frac{d_i}{d_0}. \tag{1.1-19}$$

The sign convention is that if $M > 0$, the image is erect, and if $M < 0$, the image is inverted. As shown in Figure 1.1-9, we have an inverted image as both d_i and d_0 are positive.

If the object lies within the focal length, as shown in Figure 1.1-10, we follow the rules as given in Figure 1.1-8 to construct a ray diagram. However, now rays 1 and 2, after refraction by the lens, are divergent and do not intercept on the right side of the lens. They seem to come from a point I'. In this case, since $d_0 > 0$ and $d_i < 0$, $M > 0$. The final image is virtual, erect, and is called a *virtual image*.

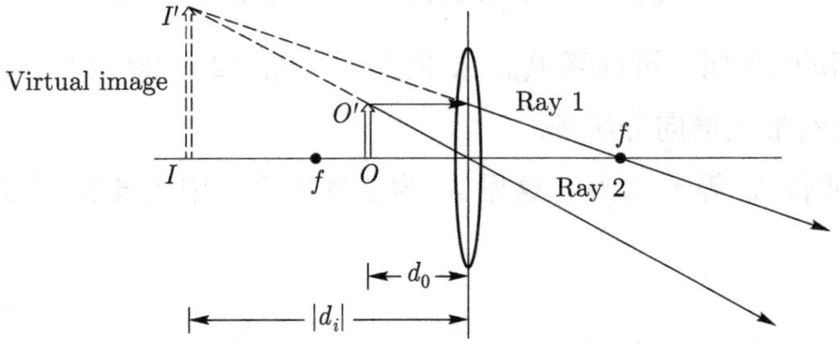

Figure 1.1-10 Imaging of a converging lens with object inside the focal length

1.2 Resolution, Depth of Focus, and Depth of Field

1.2.1 Circular Aperture

The *numerical aperture* (NA) of a lens is defined for an object or image located infinitely far away. Figure 1.2-1 shows an object located at infinity, which sends rays parallel to the lens with a circular aperture. The angle θ_{im} used to define the NA on the image side is

$$NA_i = n_i \sin(\theta_{im}/2), \tag{1.2-1}$$

其中, n_i 为像空间的折射率, 可以发现, 孔径光阑 (*aperture stop*) 限制了光线通过透镜的角度, 这最终影响可以达到的 NA. 接下来求其横向分辨率 Δr.

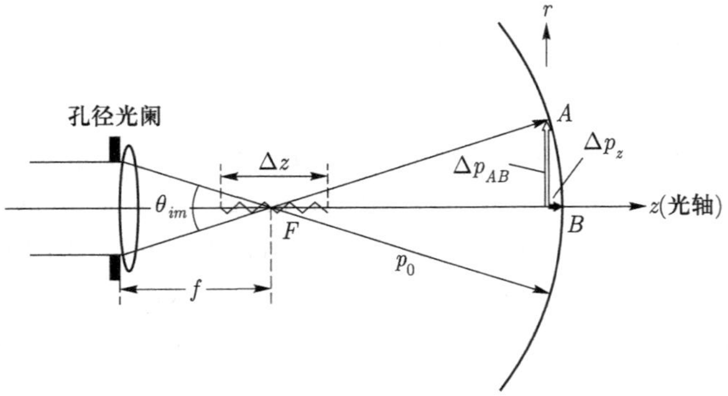

图 1.2-1 求横向分辨率和纵向分辨率的不确定性原理

几何光学中, 由于光被看作粒子, 则每个粒子皆可用其动量 p_0 来表示, 根据量子力学中的不确定性原理 (*uncertainty principle*), 那么粒子位置的最小不确定度 Δr 与其动量的不确定度 Δp_r 之间具有以下公式关系

$$\Delta r \Delta p_r \geqslant h, \tag{1.2-2}$$

其中, h 为普朗克常量 (*Planck's constant*), Δp_r 为光子在 r 方向的动量范围, 光线 FB (主光线, 穿过透镜中心的光线) 沿 r 轴方向的动量为零, 光线 FA (边缘光线, 经过透镜边缘的光线) 沿 r 轴方向的动量为 $\Delta p_{AB} = p_0 \sin(\theta_{im}/2)$, 这里, $p_0 = h/\lambda_0$, λ_0 为像空间介质中的波长. 因此, Δp_r 可以表示为, $\Delta p_r = 2\Delta p_{AB} = 2p_0 \sin(\theta_{im}/2) = \dfrac{2h}{\lambda_0}\sin(\theta_{im}/2)$, 且动量方向上的最大变化范围可通过角度 θ_{im} 来调节, 再将该式代入式 (1.2-2), 可得到其横向分辨率 (*lateral resolution*)

$$\Delta r \geqslant \frac{h}{\Delta p_r} = \frac{h}{2p_0 \sin(\theta_{im}/2)} = \frac{\lambda_0}{2\sin(\theta_{im}/2)}. \tag{1.2-3}$$

该式表明, 为获得高的分辨率, 可提高 $\theta_{im}/2$. 例如, 当 $\theta_{im}/2 = 90^\circ$ 时, Δr 为波长的一半, 即 $\lambda_0/2$, 这即为理论上的最大横向分辨率.

由于像空间的波长 λ_0 等于 λ_v/n_i, 这里 λ_v 为空气或真空中的波长. 由式 (1.2-1), 式 (1.2-3) 变为

$$\Delta r \geqslant \frac{\lambda_v}{2n_i \sin(\theta_{im}/2)} = \frac{\lambda_v}{2NA_i}. \tag{1.2-4}$$

类似地, 可计算出焦深 (*depth of focus*) Δz, 焦深亦称为像空间的轴向分辨率 [纵向分辨率 (*longitudinal resolution*)], 是指像面可以被移动却不影响其清晰成像的轴向距离范围, 为了找到焦深, 利用

$$\Delta z \Delta p_z \geqslant h, \tag{1.2-5}$$

其中, Δp_z 为光线 FB 和 FA 沿 z 方向的动量差, 如图 1.2-1, 由下式给出

$$\Delta p_z = p_0 - p_0 \cos(\theta_{im}/2). \tag{1.2-6}$$

where n_i is refractive index in the image space. Note that the *aperture stop* limits the angle of rays passing through the lens, which affects the achievable NA. Let us now find the lateral resolution, Δr.

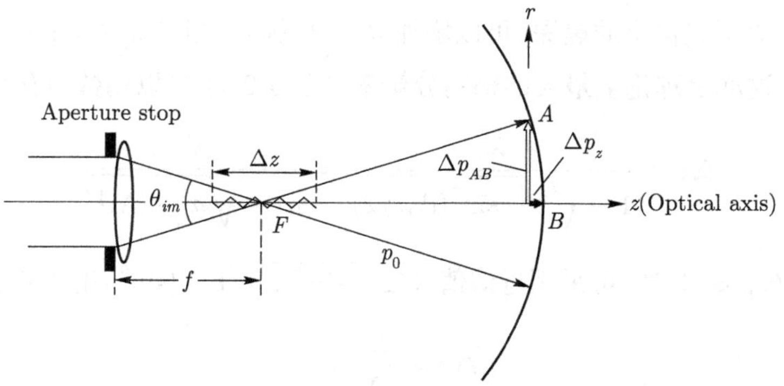

Figure 1.2-1 Uncertainty principle used in finding lateral resolution and depth resolution

Since we treat light as particles in geometrical optics, each particle then can be characterized by its momentum, p_0. According to the *uncertainty principle* in quantum mechanics, we relate the minimum uncertainty in a position of quantum, Δr, to the uncertainty of its momentum, Δp_r, according to the relationship

$$\Delta r \Delta p_r \geqslant h, \tag{1.2-2}$$

where h is *Planck's constant*, and Δp_r is the momentum spread in the r-component of the photons. The momentum of the FB ray (chief ray, a ray passes through the center of the lens) along the r-axis is zero, while the momentum of the FA ray (marginal ray, a ray passes through the edge of the lens) along the r-axis is $\Delta p_{AB} = p_0 \sin(\theta_{im}/2)$, where $p_0 = h/\lambda_0$ with λ_0 being the wavelength in the medium, that is, in the image space. Hence, Δp_r is $\Delta p_r = 2\Delta p_{AB} = 2p_0 \sin(\theta_{im}/2)$ to accommodate a maximum variation (or spread) of the momentum direction by an angle θ_{im}. By substituting this into Eq. (1.2-2), we have *lateral resolution*

$$\Delta r \geqslant \frac{h}{\Delta p_r} = \frac{h}{2p_0 \sin(\theta_{im}/2)} = \frac{\lambda_0}{2\sin(\theta_{im}/2)}. \tag{1.2-3}$$

Note that this equation reveals that in order to achieve high resolution, we can increase $\theta_{im}/2$. For example, when $\theta_{im}/2 = 90°$, Δr is half the wavelength, that is, $\lambda_0/2$, which is the theoretical maximum lateral resolution.

Since the wavelength in the image space, λ_0, is equal to λ_v/n_i, where λ_v is the wavelength in air or in vacuum, Eq. (1.2-3) becomes, using Eq. (1.2-1),

$$\Delta r \geqslant \frac{\lambda_v}{2n_i \sin(\theta_{im}/2)} = \frac{\lambda_v}{2NA_i}. \tag{1.2-4}$$

Similarly, we can calculate the *depth of focus*, Δz. Depth of focus, also called *longitudinal resolution* in the image space, is the axial distance over which the image can be moved without loss of sharpness in the object. To find the depth of focus, we use

$$\Delta z \Delta p_z \geqslant h, \tag{1.2-5}$$

where Δp_z is the momentum difference between rays FB and FA along the z-direction, as shown in Figure 1.2-1, which is given by

$$\Delta p_z = p_0 - p_0 \cos(\theta_{im}/2). \tag{1.2-6}$$

将该式代入式 (1.2-5), 得

$$\Delta z \geqslant \frac{h}{\Delta p_z} = \frac{h}{p_0\left[1-\cos(\theta_{im}/2)\right]} = \frac{\lambda_0}{1-\cos(\theta_{im}/2)}. \tag{1.2-7}$$

从该式可以看出, 为了获得小的焦深, 可以增加 $\theta_{im}/2$. 例如, 当 $\theta_{im}/2 = 90°$ 时, Δz 为一个波长的值, 即为 λ_0, 这亦是理论上最大的轴向分辨率. 式 (1.2-7) 可以用数值孔径表示如下

$$\Delta z \geqslant \frac{\lambda_0}{1-\sqrt{1-\sin^2(\theta_{im}/2)}} = \frac{\lambda_v}{n_i - \sqrt{n_i^2 - NA_i^2}}. \tag{1.2-8}$$

对于小角度, 即, $\theta_{im} \ll 1$ 时, 可采用近似值 $\sqrt{1-\sin^2\beta} \approx 1-(\sin^2\beta)/2$ 来表示

$$\Delta z \geqslant \frac{2n_i\lambda_v}{NA_i^2}. \tag{1.2-9}$$

利用式 (1.2-9) 中的等式, 并结合式 (1.2-4), 可得

$$\frac{(\Delta r)^2}{\Delta z} \approx \frac{\lambda_0}{8}. \tag{1.2-10}$$

在小的数值孔径近似下, 该等式由不确定性原理导出, 并表明, 在成像过程中, 横向分辨率越高, 焦深越短. 例如, 将横向分辨率提高至两倍, 将会导致焦深降低至 1/4.

图 1.2-2 所示为一个具有 θ_{ob} 和 θ_{im} 的成像设备, 其中, θ_{ob} 和 θ_{im} 分别表示物方光线的最大发散角和像方光线的最大会聚角.

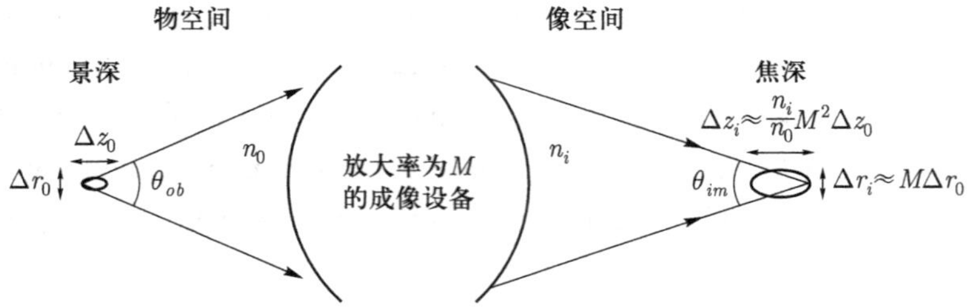

图 1.2-2 成像设备的物空间和像空间分辨率示意图

如果物空间的分辨率由 $\Delta r_0 \approx \lambda_v/2NA_0$ 和 $\Delta z_0 \approx 2n_0\lambda_v/NA_0^2$ 给出, 这里 $NA_0 = n_0\sin(\theta_{ob}/2)$, n_0 为物空间的折射率. 如果设备的横向放大率为 M, 那么像空间的分辨率为 $\Delta r_i \approx M\Delta r_0$. 接下来建立其景深 (*depth of field*) Δz_0 和焦深 (*depth of focus*) Δz_i 之间的关系. 这里, $\Delta z_i \approx 2n_i\lambda_v/NA_i^2$, 其中, $NA_i = n_i\sin(\theta_{im}/2)$. 因为 $\Delta r_i \approx \lambda_v/2NA_i = M\Delta r_0 = M\lambda_v/2NA_0$, 导致 $NA_0 = M \times NA_i$, 因此,

$$\begin{aligned}\Delta z_i &\approx 2n_i\lambda_v/NA_i^2 \\ &= 2n_i\lambda_v/\left(NA_0/M\right)^2 \\ &= M^2\left(2n_i\lambda_v/NA_0^2\right) \\ &= \frac{n_i}{n_0}M^2\Delta z_0.\end{aligned} \tag{1.2-11}$$

By substituting this expression into Eq. (1.2-5), we have

$$\Delta z \geqslant \frac{h}{\Delta p_z} = \frac{h}{p_0\left[1-\cos(\theta_{im}/2)\right]} = \frac{\lambda_0}{1-\cos(\theta_{im}/2)}. \tag{1.2-7}$$

This equation reveals that in order to achieve small depth of focus, we can increase $\theta_{im}/2$. For example, when $\theta_{im}/2 = 90°$, Δz is one wavelength, that is, λ_0, which is the theoretical maximum longitudinal resolution. Equation (1.2-7) can be written in terms of the numerical aperture and is given by

$$\Delta z \geqslant \frac{\lambda_0}{1-\sqrt{1-\sin^2(\theta_{im}/2)}} = \frac{\lambda_v}{n_i - \sqrt{n_i^2 - NA_i^2}}. \tag{1.2-8}$$

For small angles, that is, $\theta_{im} \ll 1$, one can use the approximation $\sqrt{1-\sin^2\beta} \approx 1-(\sin^2\beta)/2$ to get

$$\Delta z \geqslant \frac{2n_i\lambda_v}{NA_i^2}. \tag{1.2-9}$$

Taking the equality in Eq. (1.2-9) and combining with Eq. (1.2-4), we have

$$\frac{(\Delta r)^2}{\Delta z} \approx \frac{\lambda_0}{8}. \tag{1.2-10}$$

under the small-NA approximation. This equality is derived from uncertainty relationship, which says that during imaging the higher the lateral resolution is, the shorter the depth of focus. For example, by increasing the lateral resolution by a factor of 2 will result in the depth of focus decreased by a factor of 4.

Figure 1.2-2 shows an image-forming instrument with θ_{ob} and θ_{im} denoting the ray of maximum divergent angle and maximum convergent angle from the object side and the image side, respectively.

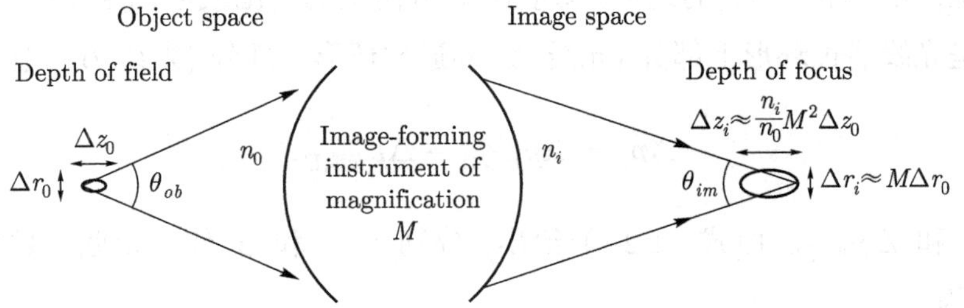

Figure 1.2-2 Image-forming instrument illustrating resolutions in the object space and image space

If the resolutions in the object space are given by $\Delta r_0 \approx \lambda_v/2NA_0$ and $\Delta z_0 \approx 2n_0\lambda_v/NA_0^2$, where $NA_0 = n_0\sin(\theta_{ob}/2)$ and n_0 is the refractive index in the object space, and if the lateral magnification of the instrument is M, the resolution on the image space is then $\Delta r_i \approx M\Delta r_0$. Let us establish the relationship between the *depth of field* Δz_0 and the *depth of focus* Δz_i, where $\Delta z_i \approx 2n_i\lambda_v/NA_i^2$ with $NA_i = n_i\sin(\theta_{im}/2)$. Since $\Delta r_i \approx \lambda_v/2NA_i = M\Delta r_0 = M\lambda_v/2NA_0$, we have $NA_0 = M \times NA_i$. Hence

$$\begin{aligned}\Delta z_i &\approx 2n_i\lambda_v/NA_i^2 \\ &= 2n_i\lambda_v/(NA_0/M)^2 \\ &= M^2(2n_i\lambda_v/NA_0^2) \\ &= \frac{n_i}{n_0}M^2\Delta z_0.\end{aligned} \tag{1.2-11}$$

这一结果表明，其物空间和像空间中的轴向分辨率关系差一个 M^2 因子，以一个倍率为 40×，$NA_0 \approx 0.6$ 的显微物镜为例，当采用波长为 632 nm 的红光时，在空气中 $n_0 = 1$，$\Delta r_0 \approx \lambda_v/2NA_0 \approx 0.5$ μm，景深 $\Delta z_0 \approx 2n_0\lambda_v/NA_0^2 \approx 3.5$ μm，在像空间，其横向放大率为 $\Delta r_i \approx M\Delta r_0 = 40 \times 0.5$ μm $= 20$ μm，且当 $n_i = 1$ 时，其焦深为 $\Delta z_i \approx M^2\Delta z_0 = 40^2 \times 3.5$ μm $=$ 0.56 mm.

1.2.2 环形孔径

显微技术中的三维成像，旨在开发一种方便观察厚的标本的技术，该技术可提供高的横向分辨率，同时保持大的焦深. 但是，计算表明，通过降低透镜的数值孔径可以提高焦深，但这会带来横向分辨率的降低. 下面，考虑采用一个环形孔径 (*annular aperture*)，它具有增加焦深且同时保持横向分辨率的性能. 环形孔径的定义如图 1.2-3 所示，是中心被挡住的通光圆形孔径. 如果一个环形孔径的外半径为 a，内半径为 b，这里定义中心遮挡比 (*central obscuration ratio*) 为 $\varepsilon = b/a$. 当 $\varepsilon = 0$ 时，即为一个清晰的圆形孔径.

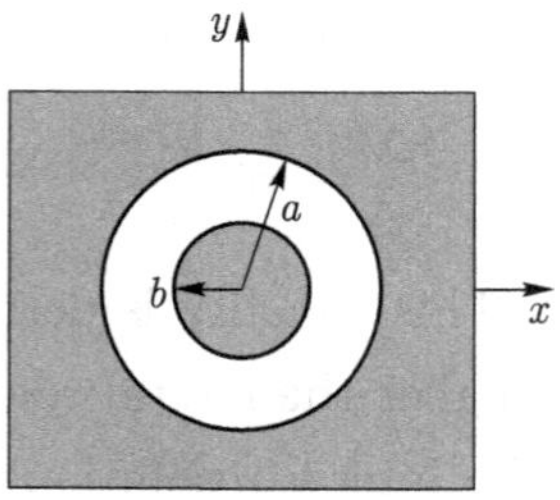

图 1.2-3 环形孔径

如图 1.2-1 所示，由于 $\beta = \theta_{im}$，图 1.2-4 中动量 Δp_r 的 r 分量可能的范围与图 1.2-1 中是相同的，因此，横向分辨率与清晰圆形孔径中的情形是相同的，即当 $\varepsilon = 0$ 时，可由式 (1.2-4) 得到结果. 但是，由于环形中心的遮挡，动量在 z 方向上的范围或不确定度有所不同. 在这种情况下，Δp_z 是光线通过环形上部分 (光线 A) 和通过环形下部分 (光线 B) 之间的动量差，即

$$\Delta p_z = \Delta p_{光线A} - \Delta p_{光线B}, \tag{1.2-12}$$

其中，$\Delta p_{光线A}$ 和 $\Delta p_{光线B}$ 由式 (1.2-6) 给出，分别将 β 和 α 代入余弦函数的辐角. 因此，式 (1.2-12) 变为

$$\Delta p_z = p_0\left[\cos\left(\frac{\alpha}{2}\right) - \cos\left(\frac{\beta}{2}\right)\right],$$

且其焦深为

$$\Delta z_{ann} = \frac{h}{\Delta p_z} = \frac{\lambda_0}{\cos\left(\frac{\alpha}{2}\right) - \cos\left(\frac{\beta}{2}\right)}, \tag{1.2-13}$$

This result indicates that the longitudinal resolutions in the object space and image space are related by a factor of M^2. Take a 40×, $NA_0 \approx 0.6$ microscope objective as an example, we have $\Delta r_0 \approx \lambda_v/2NA_0 \approx 0.5$ μm for red light with wavelength of 632 nm and the depth of field, $\Delta z_0 \approx 2n_0\lambda_v/NA_0^2 \approx 3.5$ μm for $n_0 = 1$ in air. In the image space, the lateral resolution is $\Delta r_i \approx M\Delta r_0 = 40 \times 0.5$ μm $= 20$ μm and the depth of focus is $\Delta z_i \approx M^2\Delta z_0 = 40^2 \times 3.5$ μm $= 0.56$ mm for $n_i = 1$.

1.2.2 Annular Aperture

Three-dimensional imaging in microscopy aims to develop techniques that can provide high lateral resolution, and at the same time maintain a large depth of focus in order to observe a thick specimen. However, calculations have shown that the depth of focus may be increased by reducing the numerical aperture of the lens, but this is achieved at the expense of a decrease in lateral resolution. In what follows, we consider an annular aperture which has the property of increasing the depth of focus and at the same time maintaining the lateral resolution. An *annular aperture* is defined as a clear circular aperture with a central obstruction, as shown in Figure 1.2-3. If the aperture is an annulus with outer radius a and inner radius b, we define a *central obscuration ratio* $\varepsilon = b/a$. For $\varepsilon = 0$, we have a clear circular aperture.

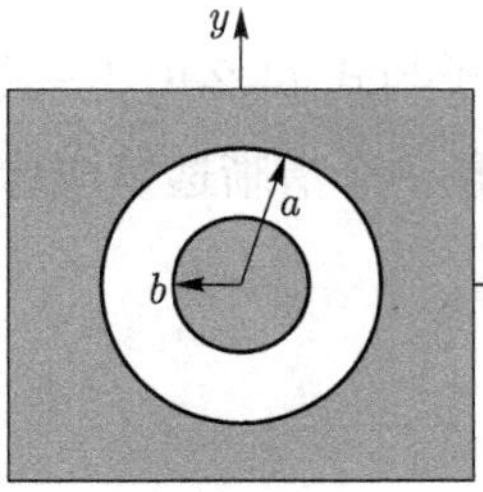

Figure 1.2-3 Annular aperture

Since $\beta = \theta_{im}$ as shown in Figure 1.2-1, the possible spread in the r-component of momentum Δp_r is the same in Figure 1.2-4 as it is in Figure 1.2-1. Hence, the lateral resolution remains the same as that in the case of the clear aperture, that is, $\varepsilon = 0$, given by Eq. (1.2-4). However, the spread or uncertainty in the momentum in the z-direction is different due to the central obscuration of the annulus. Δp_r in this case is the momentum difference between the ray passing the upper part of the annulus (Ray A) and the ray passing the lower part of the annulus (Ray B) and is given by

$$\Delta p_z = \Delta p_{\text{Ray }A} - \Delta p_{\text{Ray }B}, \tag{1.2-12}$$

where $\Delta p_{\text{Ray }A}$ and $\Delta p_{\text{Ray }B}$ are given by Eq. (1.2-6) with β and α substituted into the argument of cosine, respectively. Hence Eq. (1.2-12) becomes

$$\Delta p_z = p_0\left[\cos\left(\frac{\alpha}{2}\right) - \cos\left(\frac{\beta}{2}\right)\right],$$

and the depth of focus is

$$\Delta z_{ann} = \frac{h}{\Delta p_z} = \frac{\lambda_0}{\cos\left(\frac{\alpha}{2}\right) - \cos\left(\frac{\beta}{2}\right)}, \tag{1.2-13}$$

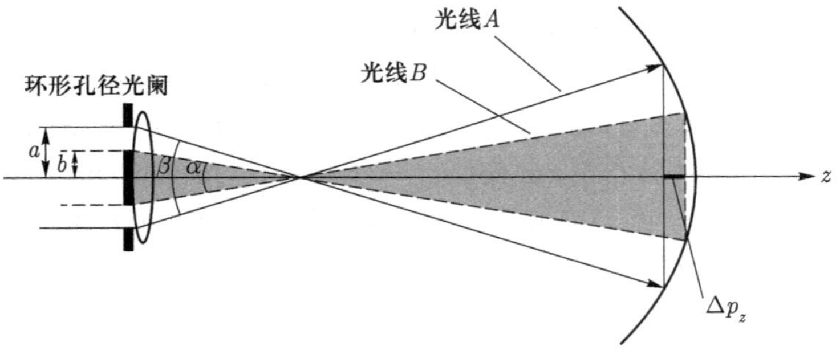

图 1.2-4 对环形孔径光阑的透镜利用不确定性原理求其焦深

此时, 可表示为

$$\Delta z_{ann} = \frac{\lambda_v}{\sqrt{\dfrac{1 - NA^2}{1 + NA^2(\varepsilon^2 - 1)}} - \sqrt{1 - NA^2}}, \tag{1.2-14}$$

其中, $NA = \sin(\beta/2)$, 即, 假定透镜置于空气中, 则, 对于小的 NA, 可由式 (1.2-14) 得出

$$\Delta z_{ann} = \frac{2\lambda_v}{NA^2(1 - \varepsilon^2)} = \frac{\Delta z}{(1 - \varepsilon^2)}. \tag{1.2-15}$$

对于一个清晰的通光孔径, 上述式子与式 (1.2-9) 是一致的, 即为 $\varepsilon = 0$ 的情况; 当有 95% 的遮挡时, 即, $\varepsilon = 0.95$, 此时的焦深会是一个清晰透镜成像焦深的 10 倍.

1.3 举例说明

1.3.1 通过单透镜进行三维成像的举例

图 1.3-1a) 画出了透镜前两个物体进行成像的情况, 可以看出, 两个物体皆位于透镜焦距之外处. 注意, 其放大率是不同的, 因为放大率由物体到透镜的距离决定. 这里考虑纵向放大率 M_z 与之前讨论过的横向放大率 M 之间的关系. 纵向放大率 M_z 是像点沿轴向的移动量 δd_i 与相应的物点沿轴向的移动量 δd_0 之比, 即, $M_z = \delta d_i/\delta d_0$.

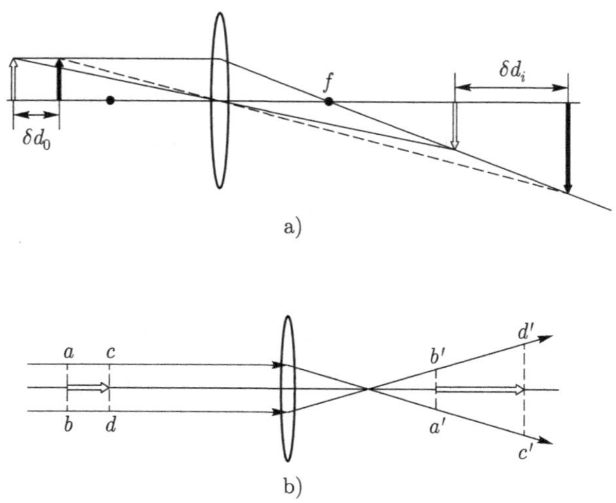

图 1.3-1 a) 不同放大率的示意图; b) 立体成像

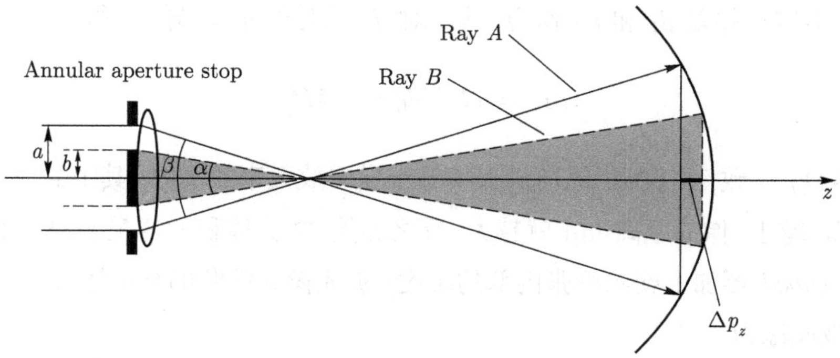

Figure 1.2-4 Uncertainty principle used in finding depth of focus for a lens with annular aperture stop

which can be shown to have the form

$$\Delta z_{ann} = \frac{\lambda_v}{\sqrt{\dfrac{1 - NA^2}{1 + NA^2(\varepsilon^2 - 1)}} - \sqrt{1 - NA^2}}, \tag{1.2-14}$$

where $NA = \sin(\beta/2)$, that is, assuming the lens is being immersed in air. For small NAs, it can be shown that Eq. (1.2-14) is given by

$$\Delta z_{ann} = \frac{2\lambda_v}{NA^2(1 - \varepsilon^2)} = \frac{\Delta z}{(1 - \varepsilon^2)}. \tag{1.2-15}$$

The above equation is consistent with Eq. (1.2-9) for a clear aperture, that is, for the case $\varepsilon = 0$. With 95% obstruction, that is, $\varepsilon = 0.95$, we can increase the depth of focus of a clear lens by more than a factor of 10.

1.3 Illustrative Examples

1.3.1 Three-Dimensional Imaging through a Single-Lens Example

Figure 1.3-1a shows the imaging of two objects in front of the lens, where both of the object lie beyond the focal length of the lens. We note that magnification is different, depending on the object distance to the lens. Let us consider longitudinal magnification, M_z, in addition to lateral magnification, M, considered earlier. Longitudinal magnification M_z is the ratio of an image displacement along the axial direction, δd_i, to the corresponding object displacement, δd_0, that is, $M_z = \delta d_i/\delta d_0$.

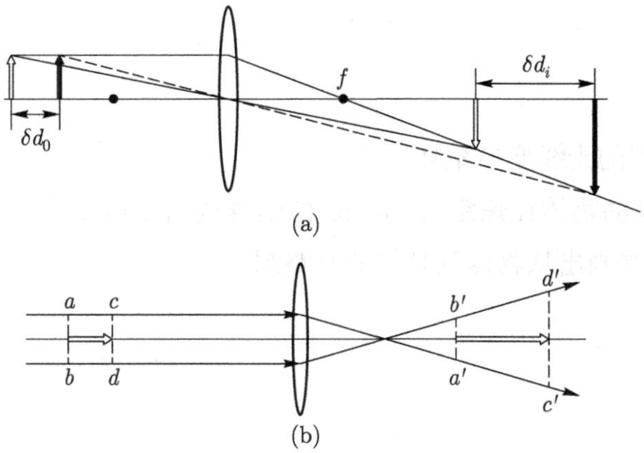

Figure 1.3-1 (a) Illustration of different magnifications and (b) volumetric imaging

利用式 (1.1-17), 并以 d_i 和 d_0 作为变量, 对 d_i 相对于 d_0 求导, 可得

$$M_z = \delta d_i / \delta d_0 = -M^2. \tag{1.3-1}$$

该式与式 (1.2-11) 一致, 并表明, 纵向放大率等于横向放大率的平方, 其中的负号表示物体距透镜的距离 $|d_0|$ 越小, 像的距离 $|d_i|$ 就越大, 反之亦然. 放大体积的情况如图 1.3-1b) 所示, 其中立方体形状 ($abcd$ 再加上朝向纸张内部的维度) 被成像为截断的金字塔形状, 其中 ab 被成像为 $a'b'$, cd 被成像为 $c'd'$.

1.3.2 狭缝的扩散角举例

考虑光从一个宽度为 l_x 的狭缝孔径出射的情况, 如图 1.3-2 所示.

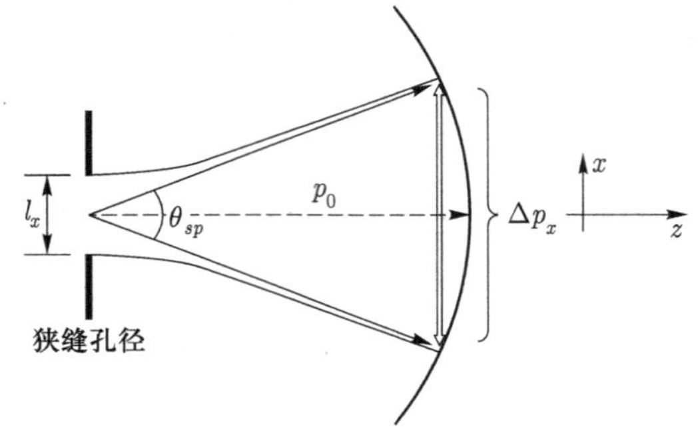

图 1.3-2 狭缝孔径出光情况

将量子的位置最小不确定度 Δx 与其动量不确定度 Δp_x 的关系用以下公式表示

$$\Delta x \Delta p_x \geqslant h.$$

因为光量子可以从孔径的任意一点发出, 且 $\Delta x = l_x$, 故,

$$\Delta p_x \sim \frac{h}{l_x}.$$

定义其扩散角 (*angle of spread*) 为

$$\theta_{sp} \sim \frac{\Delta p_x}{p_0} = \frac{h/l_x}{h/\lambda_0} = \frac{\lambda_0}{l_x}, \tag{1.3-2}$$

这基本上是式 (1.2-3) 对于小角度的结果. 可以看出, 扩散角的大小与孔径的宽度成反比.

习题

1.1 求斯涅耳定律的光线变换矩阵.

1.2 一个 4 cm 高的物体在焦距为 20 cm 的凸透镜前 10 cm 处. 利用矩阵方法, 求其像的位置和横向放大率, 并画出从物体到其像的光路图.

Using Eq. (1.1-17) and treating d_i and d_0 as variables, we take the derivative of d_i with respect to d_0 to obtain

$$M_z = \delta d_i/\delta d_0 = -M^2. \tag{1.3-1}$$

This equation is consistent with Eq. (1.2-11) and states that the longitudinal magnification is equal to the square of the lateral magnification. The minus sign in front of the equation signifies that a decrease in the distance of the object from the lens, $|d_0|$, will result in an increase in the image distance, $|d_i|$, and vice versa. The situation of a magnified volume is shown in Figure 1.3-1b, where a cube volume (*abcd* plus the dimension into the paper) is imaged into a truncated pyramid with *ab* imaged into $a'b'$ and *cd* imaged into $c'd'$.

1.3.2 Angle of Spread from a Slit Example

Consider light emanating from a slit aperture of width l_x, as shown in Figure 1.3-2.

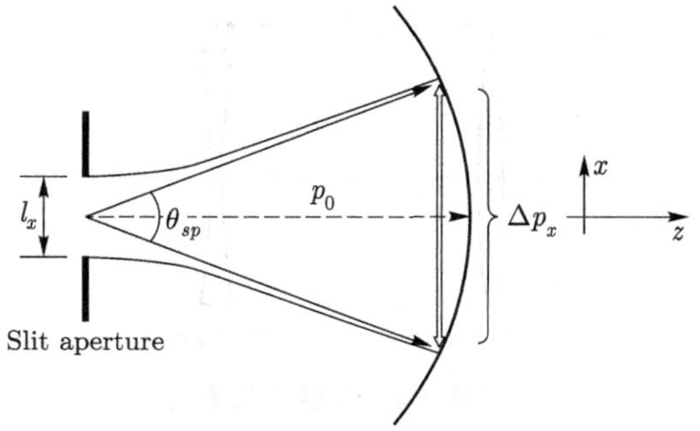

Figure 1.3-2 Spreading from a slit

We relate the minimum uncertainty in position Δx of a quantum to the uncertainty in its momentum Δp_x according to

$$\Delta x \Delta p_x \geqslant h.$$

Because the quantum of light can emerge from any point in the aperture, we have $\Delta x = l_x$. Therefore,

$$\Delta p_x \sim \frac{h}{l_x}.$$

We define the angle of spread as

$$\theta_{sp} \sim \frac{\Delta p_x}{p_0} = \frac{h/l_x}{h/\lambda_0} = \frac{\lambda_0}{l_x}, \tag{1.3-2}$$

which is basically the result in Eq. (1.2-3) for small angles. Note that the spread angle is inversely proportional to the width of the aperture.

Problems

1.1 Find the transfer matrix for Snell's law.

1.2 An object 4 cm tall is 10 cm in front of a convex lens of focal length 20 cm. Using the matrix formalism, find the location and magnification of the image. Draw a ray diagram from the object to the image.

1.3　一个 5 cm 高的幻灯片位于距屏幕 110 cm 的位置处 (见图题 1.3). 求能在屏幕上投射出 50 cm 高的实像的正透镜的焦距是多少? 用矩阵方法来求解.

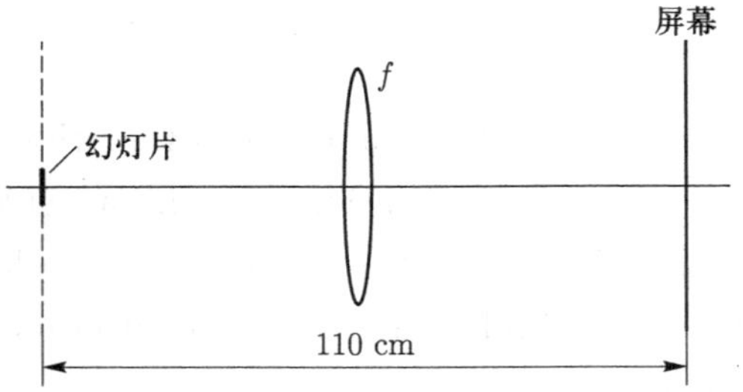

图题 1.3　单透镜系统

1.4　如图题 1.4 所示, 一个 3 cm 高的物体位于凸透镜左侧 24 cm 的光轴上, 使用矩阵方法求出其像的位置和大小, 并绘制从物到像的光路图.

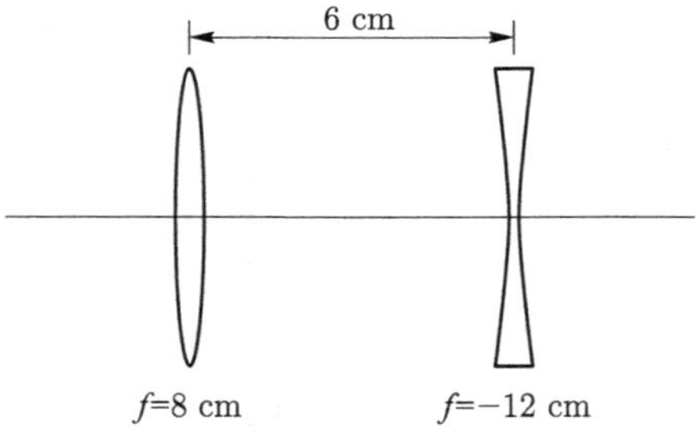

图题 1.4　双透镜系统

1.5　在利用单透镜进行三维成像的示例中, 纵向放大率等于横向放大率的平方, 即 $M_z = \delta d_i/\delta d_0 = -M^2$. 请证明.

1.6　请填写以下几组显微镜物镜的参数 (假定这些透镜置于空气中, 操作时使用的红光的光波波长 $\lambda_v = 632$ nm).

表题 1.6

放大倍数, 数值孔径	像空间分辨率	景深/μm	焦深/mm
10×, 0.1			
20×, 0.4			
40×, 0.6			
60×, 0.8			
100×, 0.95			

1.7　参照图 1.2-4, 从式 (1.2-13) 出发, 证明带有环形孔径光阑的薄透镜的焦深可由下式给出

$$\Delta z_{ann} = \frac{\lambda_v}{\sqrt{\dfrac{1-NA^2}{1+NA^2(\varepsilon^2-1)}} - \sqrt{1-NA^2}},$$

1.3 A slide 5 cm tall is located 110 cm from a screen (see Figure P.1.3). What is the focal length of the positive lens, which will project a real image measuring 50 cm on the screen? Use the matrix formalism to solve the problem.

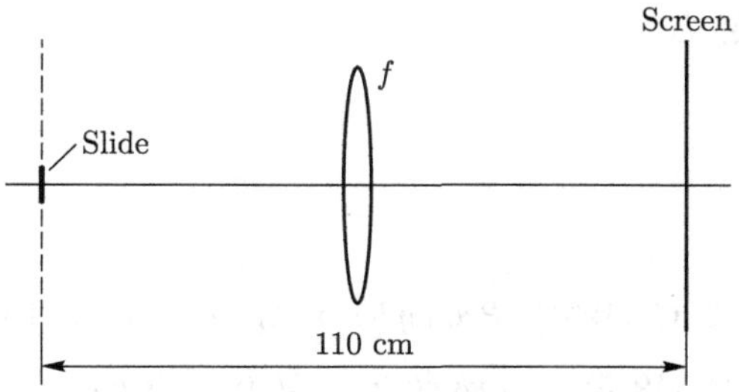

Figure P.1.3 Single-lens system

1.4 If an object 3 cm tall is located on the optical axial 24 cm to the left of the convex lens as shown in Figure P.1.4, find the position and size of the image using the matrix formalism. Also, draw the ray diagram from the object to the image.

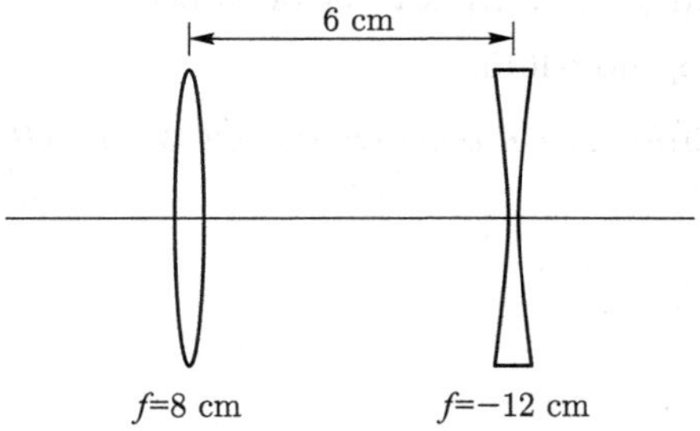

Figure P.1.4 Two-lens system

1.5 In 3-D imaging through a single lens example, it is stated that the longitudinal magnification is equal to the square of the lateral magnification, that is, $M_z = \delta d_i/\delta d_0 = -M^2$. Verify this statement.

1.6 Fill in the blanks below for several microscope objective lenses (assuming the lenses are immersed in air with wavelength of operation using red light of $\lambda_v = 632$ nm).

Table P.1.6

Magnification, NA	Resolution in image space	Depth of field/μm	Depth of focus/mm
10×, 0.1			
20×, 0.4			
40×, 0.6			
60×, 0.8			
100×, 0.95			

1.7 Starting from Eq. (1.2-13) and with reference to Figure 1.2-4, show that the depth of focus of an annular aperture with a thin lens is given by

$$\Delta z_{ann} = \frac{\lambda_v}{\sqrt{\dfrac{1-NA^2}{1+NA^2(\varepsilon^2-1)}} - \sqrt{1-NA^2}},$$

其中, $NA = \sin(\beta/2)$, 且, 对于小的 NA, 可近似为

$$\Delta z_{ann} = \frac{2\lambda_v}{NA^2(1-\varepsilon^2)} = \frac{\Delta z}{(1-\varepsilon^2)},$$

Δz 为清晰圆形孔径的焦深.

参考文献

[1] Banerjee, P. P. and Poon, T.-C. (1991). *Principles of Applied Optics.* Irwin, Illinois.

[2] Gerard, A. and Burch, J. M. (1975). *Introduction to Matrix Methods in Optics.* Wiley, New York.

[3] Hecht, E. (2002). *Optics*, 4th ed.Addison Wesley, California.

[4] Korpel, A. (1970). *United State Patent (#3,614,310)* Electrooptical Apparatus Employing a Hollow Beam for Translating an Image of an Object.

[5] Poon, T.-C. (2007). *Optical Scanning Holography with MATLAB®*. Springer, New York.

[6] Poon, T.-C. and Motamedi, M. (1987). "Optical/digital incoherent image processing for extended depth of field," Applied Optics 26, pp.4612-4615.

[7] Poon, T.-C. and Kim, T. (2018). *Engineering Optics with MATLAB®*, 2nd ed. World Scientific, New Jersey.

where $NA = \sin(\beta/2)$ and for small NAs, it is approximately given by

$$\Delta z_{ann} = \frac{2\lambda_v}{NA^2(1-\varepsilon^2)} = \frac{\Delta z}{(1-\varepsilon^2)},$$

where Δz is the depth of focus for a clear lens.

Bibliography

[1] Banerjee, P. P. and Poon, T.-C. (1991). *Principles of Applied Optics.* Irwin, Illinois.

[2] Gerard, A. and Burch, J. M. (1975). *Introduction to Matrix Methods in Optics.* Wiley, New York.

[3] Hecht, E. (2002). *Optics*, 4th ed.Addison Wesley, California.

[4] Korpel, A. (1970). *United State Patent (#3,614,310)* Electrooptical Apparatus Employing a Hollow Beam for Translating an Image of an Object.

[5] Poon, T.-C. (2007). *Optical Scanning Holography with MATLAB®.* Springer, New York.

[6] Poon, T.-C. and Motamedi, M. (1987). "Optical/digital incoherent image processing for extended depth of field," Applied Optics 26, pp. 4612-4615.

[7] Poon, T.-C. and Kim, T. (2018). *Engineering Optics with MATLAB®*, 2nd ed. World Scientific, New Jersey.

第二章　线性不变系统和傅里叶分析

本章有两部分内容. 第一部分将会对信号与线性系统的基本内容进行阐述, 在后续章节将会遇到衍射以及光学成像系统等情况, 这些可用线性系统进行建模. 第二部分将会介绍傅里叶级数、傅里叶变换的基本性质、卷积 (*convolution*) 和相关 (*correlation*) 的概念等. 事实上, 许多现代光学成像和处理系统, 都可以用傅里叶方法来建模, 而且, 傅里叶方法是分析这类光学系统的主要工具. 这里我们将主要对一维时间信号进行学习, 并扩展到二维空域信号. 许多一维 (1-D) 信号和系统, 其所发展的概念, 亦可应用到二维 (2-D) 系统. 本章同时提供了在后续章节会使用到的重要且基本的数学方法.

2.1　信号与系统

2.1.1　信号的运算

信号 (*signal*) 是一组数据. 对于电气系统来说, 信号是用电压或电流来测量的时间的函数. 对于光学系统来说, 信号是一维、二维甚至三维图像的形式. 本节将介绍一些信号的运算和有用的信号模型. 为简洁起见, 首先主要讨论一维时间信号, 之后将涉及二维信号. 这里用 $x(t)$ 表示一个连续的时间信号 (*continuous-time signal*), 主要有平移 (*shifting*)、缩放 (*scaling*) 和翻转 (*reversal*) 这三种有用的信号运算.

平移

$x(t-t_0)$ 是 $x(t)$ 在时间上的平移形式 (*time-shifted version*), 简称时移. 如果 t_0 为正, 时移向右, 这表示相对于原始信号 $x(t)$, 会有 t_0 的时间延迟. 如果 t_0 为负, 时移向左, 这表示相对于原始信号 $x(t)$, 会有 t_0 的时间超前. 具体如图 2.1-1 所示.

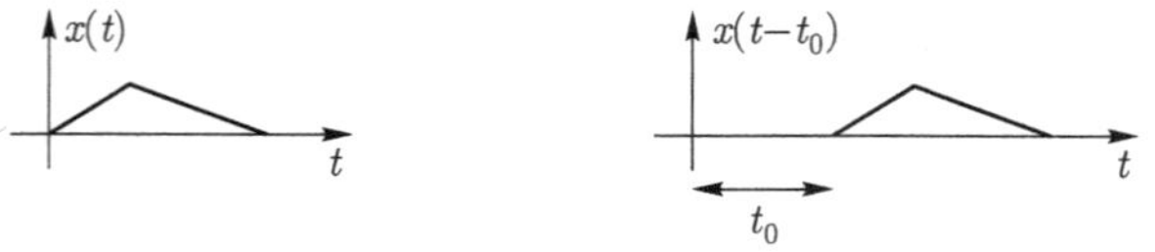

图 2.1-1　t_0 为正时信号的时移

缩放

$x(at)$ 是原始信号 $x(t)$ 的缩放形式 (*scaled version*), 这里 a 是缩放因子, 为一正数. 若 $a>1$, 则会得到压缩了的 $x(t)$ 信号. 若 $a<1$, 则会得到扩展了的 $x(t)$ 信号. 图 2.1-2 分别给出了信号 $x(t)$ 在 $a=2$ 时进行时间压缩和在 $a=1/2$ 时进行时间扩展的举例.

2 Linear Invariant Systems and Fourier Analysis

The purpose of this chapter is twofold. We will first discuss basic aspect of signals and linear systems in the first part. We will see in subsequent chapters that diffraction as well as optical imaging systems can be modelled as linear systems. In the second part, we introduce the basic properties of Fourier series, Fourier transform as well as the concept of convolution and correlation. Indeed, many modern optical imaging and processing systems can be modelled with the Fourier methods, and Fourier analysis is the main tool to analyze such optical systems. We shall study time signals in one dimension and spatial signals in two dimensions will then be covered. Many of the concepts developed for one-dimensional (1-D) signals and systems apply to two-dimensional (2-D) systems. This chapter also serves to provide important and basic mathematical tools to be used in subsequent chapters.

2.1 Signals and Systems

2.1.1 Signal Operations

A *signal* is a set of data. For electrical systems, a signal is a function of time, measured in terms of voltage or current. For optical systems, a signal is of the form of 1-D, 2-D, or even 3-D images. In this section, we introduce some signal operations and useful signal models. We first restrict ourselves to 1-D time signals for brevity. Two-dimensional signals will be covered subsequently. We denote a continuous-time signal by $x(t)$. There are three useful signal operations: shifting, scaling, and reversal.

Shifting

$x(t-t_0)$ is a *time-shifted version* of $x(t)$. If t_0 is positive, the time shift is to the right and it represents a time delay of t_0 seconds to the original signal $x(t)$. If t_0 is negative, the shift is to the left and the original signal $x(t)$ is advanced by t_0. The situation is illustrated in Figure 2.1-1.

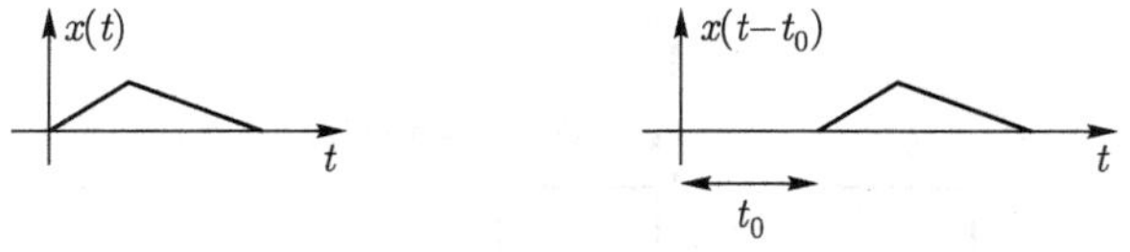

Figure 2.1-1 Time shifting a signal for t_0 positive

Scaling

$x(at)$ is a *scaled version* of the original signal $x(t)$, where a is a scale factor and is a positive number. If $a > 1$, we have a compressed version of $x(t)$. On the other hand, if $a < 1$, we have an expanded version of $x(t)$. Figure 2.1-2 shows a compressed and expanded versions of $x(t)$ for $a = 2$ and $1/2$, respectively.

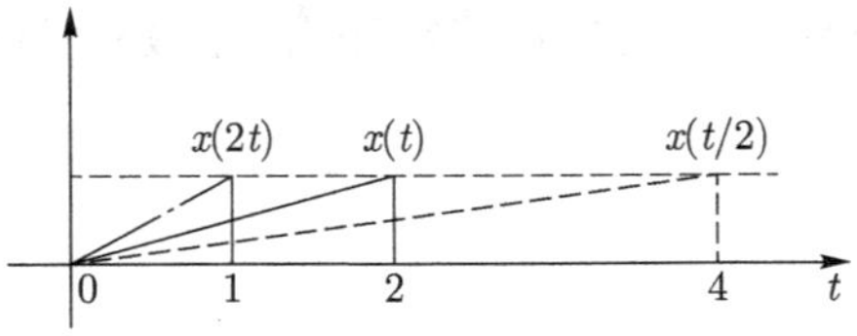

图 2.1-2 信号的时间缩放

翻转

$x(-t)$ 是信号 $x(t)$ 的时间翻转 (*time reversal*). 换句话说, 如果把原始信号 $x(t)$ 中的 t 替换为 $-t$, 则会得到一个关于纵轴 $t=0$ 的镜像信号 $x(t)$, 如图 2.1-3 所示.

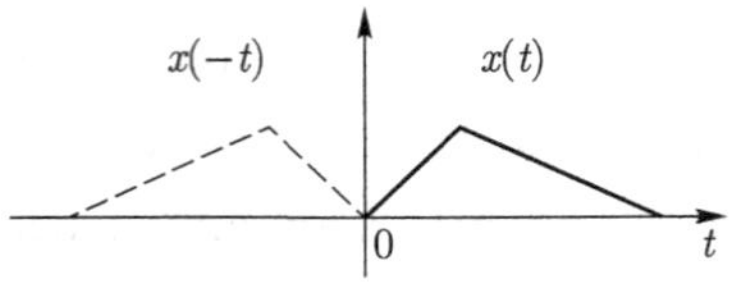

图 2.1-3 信号的时间翻转

举例: 信号的运算

若 $x(t)$ 如图 2.1-4a) 所示, 画出 $x(1-t/2)$.

首先, 将 $x(1-t/2)$ 写为 $x\left(1-\dfrac{t}{2}\right)=x\left(\dfrac{2-t}{2}\right)=x\left[\dfrac{1}{2}(2-t)\right]$, 那么, 可以看出, 其中包含三种信号的操作. 因子 1/2、2 和 −1 分别对应于缩放、平移和翻转. 这里先来进行翻转运算, 由于 $x(t)|_{t\to -t}=x(-t)$, 结果如图 2.1-4b) 所示.接着根据 $x(-t)|_{t\to t/2}=x(-t/2)$ 来进行缩放运算, 如图 2.1-4c) 所示. 最后, 根据 $x\left(-\dfrac{t}{2}\right)\Big|_{t\to t-2}=x\left(-\dfrac{t-2}{2}\right)=x\left(1-\dfrac{t}{2}\right)$, 执行平移运算, 即为最终要绘制出来的信号.

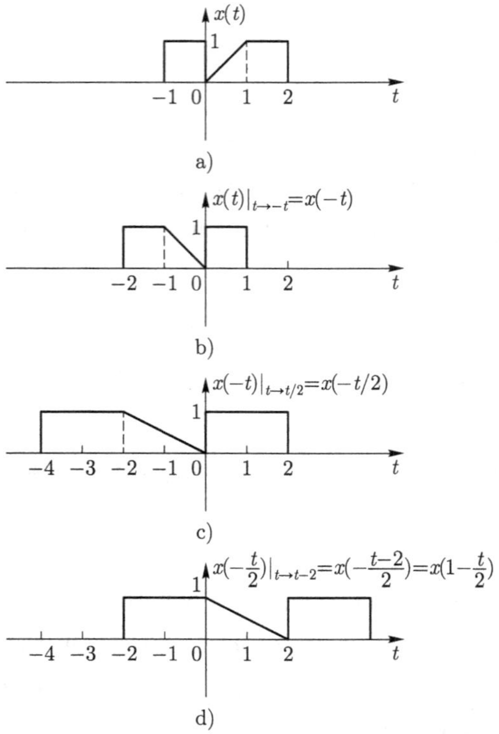

图 2.1-4 信号的运算举例

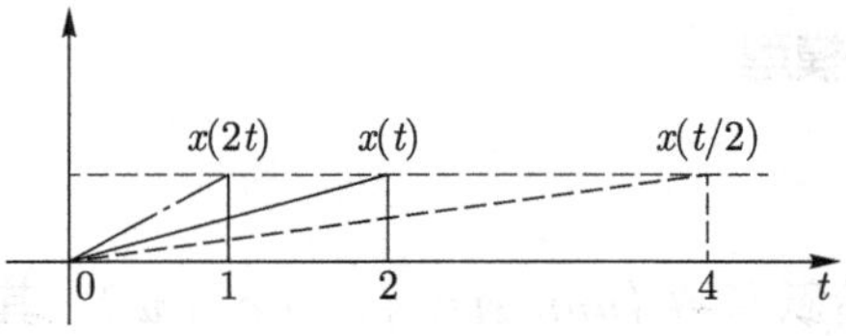

Figure 2.1-2 Time scaling of a signal

Reversal

$x(-t)$ is a *time reversal* of $x(t)$. In other words, if we replace t by $-t$ in the original signal $x(t)$, we have a mirror image of $x(t)$ about the vertical axis $t = 0$. Figure 2.1-3 illustrates the situation.

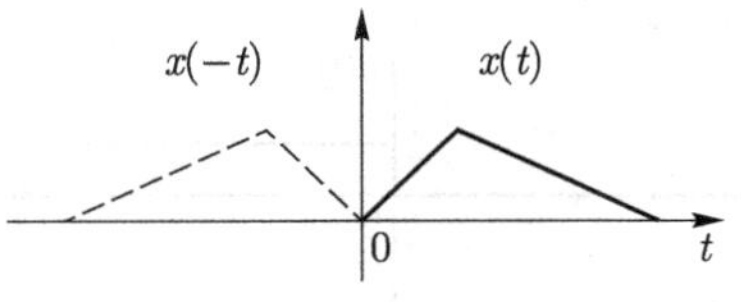

Figure 2.1-3 Time reversal a signal

Example: Signal Operations

Plot $x(1 - t/2)$ if $x(t)$ is shown in Figure 2.1-4a.

Let us first rewrite $x(1 - t/2)$ as $x\left(1 - \frac{t}{2}\right) = x\left(\frac{2-t}{2}\right) = x\left[\frac{1}{2}(2-t)\right]$. We then recognize that there are three signal operations involved. The factors $1/2$, 2, and -1 correspond to scaling, shifting, and reversal, respectively. Let us first perform reversal as $x(t)|_{t\to -t} = x(-t)$ and it is shown in Figure 2.1-4b. We then perform scaling according to $x(-t)|_{t\to t/2} = x(-t/2)$ and the result is shown in Figure 2.1-4c. Finally, we perform shifting according to $x\left(-\frac{t}{2}\right)\Big|_{t\to t-2} = x\left(-\frac{t-2}{2}\right) = x\left(1 - \frac{t}{2}\right)$, which is the desired signal we want to plot.

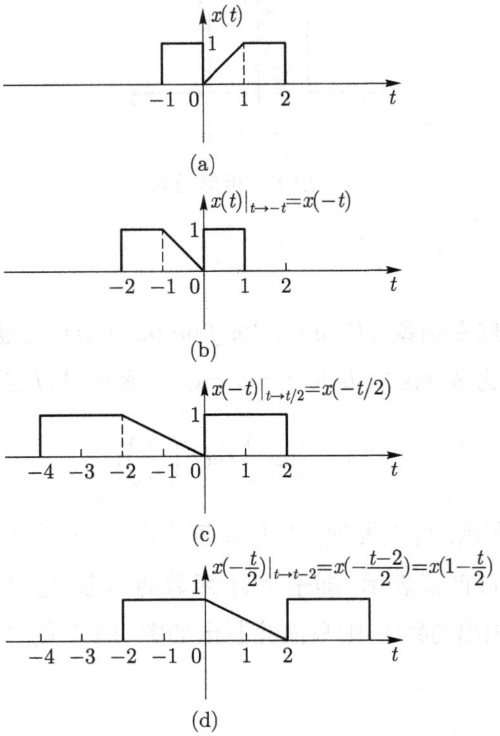

Figure 2.1-4 Signal operations example

2.1.2 一些有用的信号模型

单位阶跃函数, $u(t)$

如图 2.1-5 所示为单位阶跃函数 (*unit step function*) $u(t)$, 其定义为

$$u(t)=\begin{cases}0, & t<0\\ 1, & t>0.\end{cases} \tag{2.1-1}$$

由于 $u(t)$ 在 $t=0$ 处没有唯一值, 且其导数在 $t=0$ 处为无穷大, $u(t)$ 为一个奇异函数 (*singularity function*).

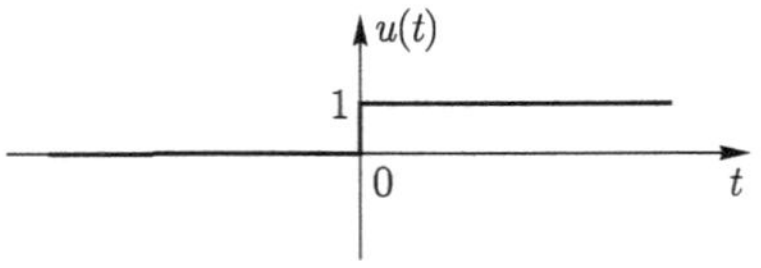

图 2.1-5 单位阶跃函数

矩形函数, $rect(t/a)$

矩形函数 (*rectangular function*) 或简单称为 *rect* 函数 (*rect function*), $rect(t/a)$ 定义如下

$$rect\left(\frac{t}{a}\right)=\begin{cases}1, & |t|<a/2\\ 0, & \text{其他},\end{cases} \tag{2.1-2}$$

其中, a 为函数的宽度. 在电气工程中, *rect* 函数普遍被称为脉冲 (*pulse*), 该函数如图 2.1-6 所示, 由于它的两个边是不连续的, 因此它是一个奇异函数.

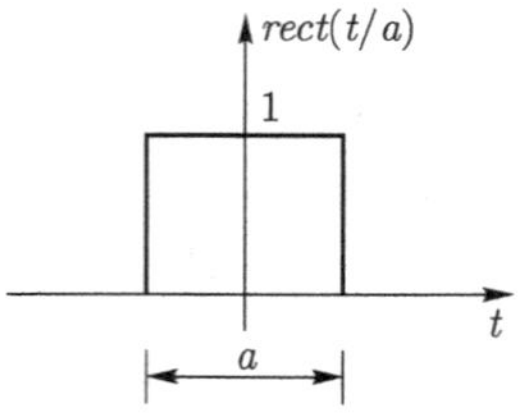

图 2.1-6 矩形函数

狄拉克函数, $\delta(t)$

另一个奇异函数是狄拉克函数 (*Dirac delta function*)$\delta(t)$, 也被称为单位冲激函数 (*unit impulse function*), 或简称为 δ 函数 (*delta function*). δ 函数可以定义如下

$$\delta(t)=\lim_{a\to 0}\frac{1}{a}rect\left(\frac{t}{a}\right).$$

上式情况说明如图 2.1-7 所示. 可以发现, 无论 a 值为多少, 矩形函数下方的面积 $a\times 1/a$ 总是等于 1, 在箭头旁边用 “(1)” 来表示, 即是 $\delta(t)$ 函数的标志. 这个 “1” 有时也被称为是 δ 函数的 “强度” . 因此, 若要画出 $5\delta(t)$, 则只需绘制出如图 2.1-7 的 δ 函数, 并在其箭头旁边处用 “(5)” 表示其强度.

2.1.2 Some Useful Signal Models

Unit Step Function, $u(t)$

The *unit step function*, $u(t)$, is shown in Figure 2.1-5. It is defined by

$$u(t) = \begin{cases} 0, & t < 0 \\ 1, & t > 0. \end{cases} \tag{2.1-1}$$

Because $u(t)$ does not have a unique value at $t = 0$ and its derivative is infinite at $t = 0$, it is a *singularity function.*

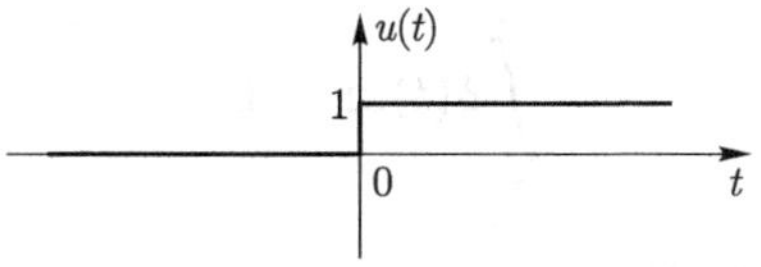

Figure 2.1-5 Unit step function

Rectangular Function, $rect(t/a)$

The *rectangular function* or simply *rect function*, $rect(t/a)$, is given by

$$rect\left(\frac{t}{a}\right) = \begin{cases} 1, & |t| < a/2 \\ 0, & \text{otherwise}, \end{cases} \tag{2.1-2}$$

where a is the width of the function. The rect function is typically called a *pulse* in electrical engineering. The function is shown in Figure 2.1-6. It is a singularity function as it is discontinuous at its two edges.

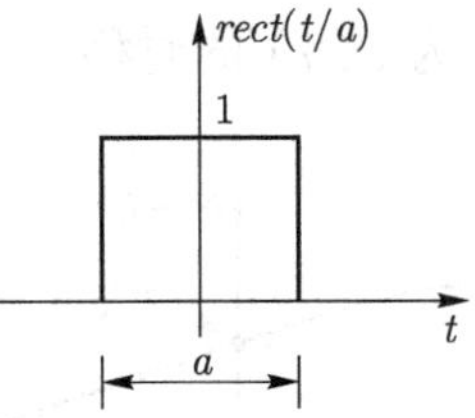

Figure 2.1-6 Rectangular function

Dirac Delta Function, $\delta(t)$

Another singularity function is the *Dirac delta function* $\delta(t)$, which is also known as the *unit impulse function* or just simply as the *delta function.* We can define the delta function as follows:

$$\delta(t) = \lim_{a \to 0} \frac{1}{a} rect\left(\frac{t}{a}\right).$$

The situation of the above equation is illustrated in Figure 2.1-7. Note that the area under the rectangular function is $a \times (1/a)$, which is always equal to 1 regardless the value of a. We denote this unity as "(1)" beside the arrow, which is the symbol of the delta function $\delta(t)$. This unity is sometimes called the "strength" of the delta function. For this reason, to plot $5\delta(t)$, we just plot the delta function as shown in Figure 2.1-7 but denote its strength as "(5)" beside the arrow.

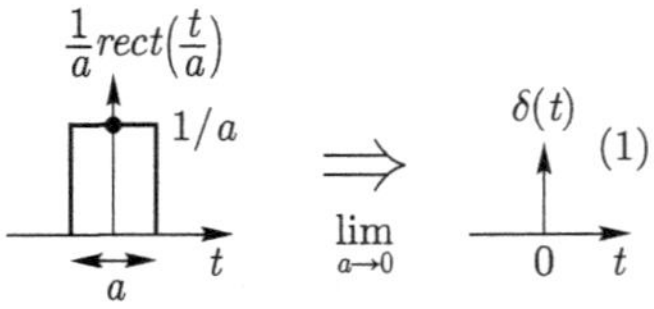

图 2.1-7 利用矩形函数来定义 δ 函数

δ 函数是系统和信号分析中最重要的函数之一, 本书将会讨论其 4 个重要的性质.

性质 #1: 单位面积 (*unit area*)

因为 δ 函数下方的面积为 1, 因此, 可以写出

$$\int_{-\infty}^{\infty} \delta(t)\mathrm{d}t = 1. \tag{2.1-3}$$

实际上, 下面的等式适用于一般情况

$$\int_{a}^{b} \delta\left(t - t_0\right)\mathrm{d}t = 1,$$

若 $a < t_0 < b$, 那么, 除了 $t = t_0$ 以外, $\delta(t - t_0) = 0$.

性质 #2: 乘法性质 (*product property*)

$$x(t)\delta(t - t_0) = x(t_0)\delta(t - t_0). \tag{2.1-4}$$

通过图 2.1-8 所示的图解, 可以很容易地看出该特性. 对于任意一个函数 $x(t)$, 在图中 $t = t_0$ 处, 与具有时移的 δ 函数 $\delta(t - t_0)$ 相交重叠, 那么两个函数的乘积, 显然就等于 $x(t_0)$ 乘以 $\delta(t - t_0)$. 因此, 其结果就是一个强度为 $x(t_0)$ 的时移 δ 函数.

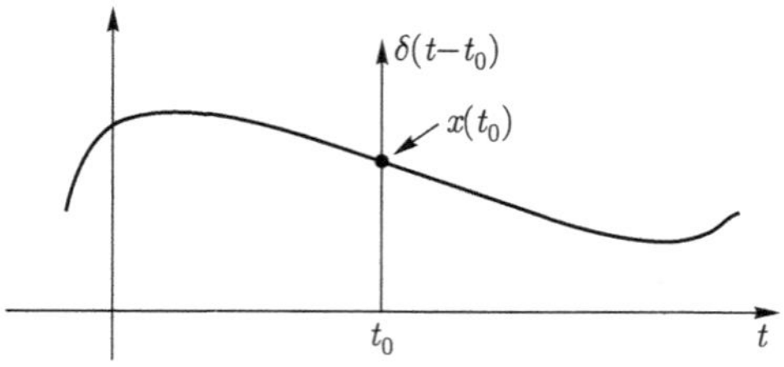

图 2.1-8 乘法性质的说明

性质 #3: 筛选性质 (*sampling property*)

$$\int_{-\infty}^{\infty} x(t)\delta(t - t_0)\mathrm{d}t = x(t_0). \tag{2.1-5}$$

通过性质 #1 和 #2, 可以很容易得到上述结果. 利用性质 #2, 可以写出

$$\begin{aligned}\int_{-\infty}^{\infty} x(t)\delta(t - t_0)\mathrm{d}t &= \int_{-\infty}^{\infty} x(t_0)\delta(t - t_0)\mathrm{d}t \\ &= x(t_0)\int_{-\infty}^{\infty} \delta(t - t_0)\mathrm{d}t = x(t_0),\end{aligned}$$

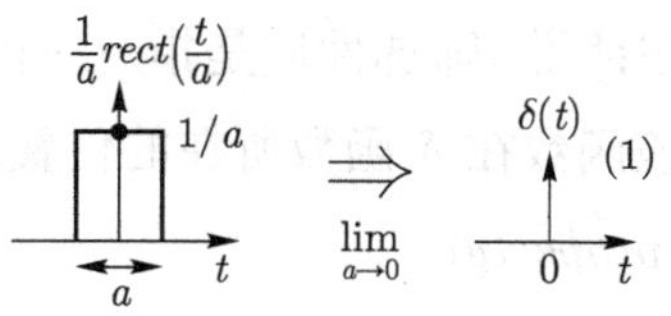

Figure 2.1-7 Dirac delta function defined using a rectangular function

The delta function is one of the most important functions in the study of system and signal analysis. We will discuss four important properties.

Property #1: Unit Area

Since we have found that the area under the delta function is unity, we therefore write

$$\int_{-\infty}^{\infty} \delta(t)dt = 1. \tag{2.1-3}$$

Indeed, the following equality applies in general

$$\int_{a}^{b} \delta(t-t_0)dt = 1,$$

if $a < t_0 < b$ since $\delta(t-t_0) = 0$ except at $t = t_0$.

Property #2: Product Property

$$x(t)\delta(t-t_0) = x(t_0)\delta(t-t_0). \tag{2.1-4}$$

The result of this property can be readily realized by the illustration shown in Figure 2.1-8, where an arbitrary function, $x(t)$, is shown to be overlapped with the time-shifted delta function, $\delta(t-t_0)$, located at $t = t_0$. The product of the two functions is clearly equal to $x(t_0)$ multiplied by $\delta(t-t_0)$. Therefore, the result has become a time-shifted delta function with its strength given by $x(t_0)$.

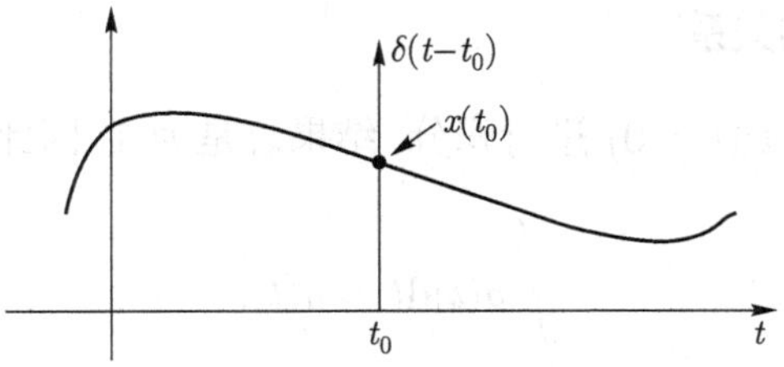

Figure 2.1-8 Illustration of the product property

Property #3: Sampling Property

$$\int_{-\infty}^{\infty} x(t)\delta(t-t_0)dt = x(t_0). \tag{2.1-5}$$

The above result can be obtained readily by the use of Properties #1 and #2. Using Property #2, we write

$$\int_{-\infty}^{\infty} x(t)\delta(t-t_0)dt = \int_{-\infty}^{\infty} x(t_0)\delta(t-t_0)dt$$
$$= x(t_0)\int_{-\infty}^{\infty} \delta(t-t_0)dt = x(t_0),$$

利用性质 #1 得到最后结果. 换句话说, 筛选性质是指, 一个函数 $x(t)$ 和一个时移了 t_0 的 δ 函数 $\delta(t-t_0)$ 的乘积的面积, 与该函数在 δ 函数所在的位置, 即 t_0 处的值相等.

性质 #4: 缩放性质 (*scaling property*)

$$\delta(at)=\frac{1}{|a|}\delta(t),\quad a\neq 0. \tag{2.1-6}$$

为了确定缩放是如何影响 δ 函数的, 这里来计算一下 $\delta(at)$ 在 $a>0$ 时的面积

$$\int_{-\infty}^{\infty}\delta(at)\mathrm{d}t=\int_{-\infty}^{\infty}\delta(t')\frac{\mathrm{d}t'}{a}=\frac{1}{a},$$

其中, 用 $t'=at$ 来简化积分. 通过上式, 得到 $\int_{-\infty}^{\infty}a\delta(at)\mathrm{d}t=1$. 将其与性质 #1 (即 $\int_{-\infty}^{\infty}\delta(t)\mathrm{d}t=1$) 比较, 可以导出 $a\delta(at)=\delta(t)$, 或

$$\delta(at)=\frac{1}{a}\delta(t). \tag{2.1-7a}$$

同样, 对于 $a<0$, 有

$$\int_{-\infty}^{\infty}\delta(at)\mathrm{d}t=\frac{1}{-a},$$

得

$$\delta(at)=\frac{1}{-a}\delta(t), \tag{2.1-7b}$$

对于 a 的所有值, 式 (2.1-7a) 和式 (2.1-7b) 给出了式 (2.1-6) 的结果.

举例: $\boldsymbol{\delta(t)}$ 和 $\boldsymbol{u(t)}$ 的关系

当对 δ 函数从 $-\infty$ 到 t $(t>0)$ 进行积分, 结果总是为 1, 因此, 可得

$$\int_{-\infty}^{t}\delta(t)\mathrm{d}t=u(t). \tag{2.1-8a}$$

对上式进行求导, 得

$$\frac{\mathrm{d}u(t)}{\mathrm{d}t}=\delta(t). \tag{2.1-8b}$$

图 2.1-9 说明了上述关系.

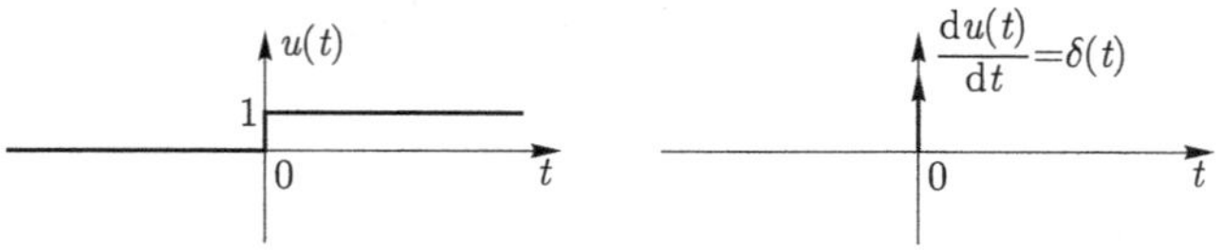

图 2.1-9 $\delta(t)$ 和 $u(t)$ 的关系说明

where we have used Property #1 to obtain the final result. In words, the sampling property says that the area under the product of a function $x(t)$ with a time-shifted delta function, $\delta(t-t_0)$, is equal to the function evaluated at the location of the delta function, which is at t_0.

Property #4: Scaling Property

$$\delta(at)=\frac{1}{|a|}\delta(t) \text{ for } a\neq 0. \tag{2.1-6}$$

To determine how scaling affects the delta function, let us evaluate the area of $\delta(at)$ for $a>0$:

$$\int_{-\infty}^{\infty}\delta(at)\mathrm{d}t=\int_{-\infty}^{\infty}\delta(t')\frac{\mathrm{d}t'}{a}=\frac{1}{a},$$

where we have made the substitution $t'=at$ to simplify the integration. So from the above equation, we have $\int_{-\infty}^{\infty}a\delta(at)\mathrm{d}t=1$. Comparing this with Property #1, that is, $\int_{-\infty}^{\infty}\delta(t)\mathrm{d}t=1$, we deduce that $a\delta(at)=\delta(t)$, or

$$\delta(at)=\frac{1}{a}\delta(t). \tag{2.1-7a}$$

Similarly, for $a<0$, we have

$$\int_{-\infty}^{\infty}\delta(at)\mathrm{d}t=\frac{1}{-a},$$

giving

$$\delta(at)=\frac{1}{-a}\delta(t). \tag{2.1-7b}$$

For all values of a, Eqs. (2.1-7a) and (2.1-7b) give the result in Eq. (2.1-6).

Example: Relationship between $\delta(t)$ and $u(t)$

When we integrate the delta function from $-\infty$ to t, where $t>0$, the result is always unity and hence, we have

$$\int_{-\infty}^{t}\delta(t)\mathrm{d}t=u(t). \tag{2.1-8a}$$

After taking the derivative of Eq. (2.1-8a), we have

$$\frac{\mathrm{d}u(t)}{\mathrm{d}t}=\delta(t). \tag{2.1-8b}$$

Figure 2.1-9 illustrates the above relationship.

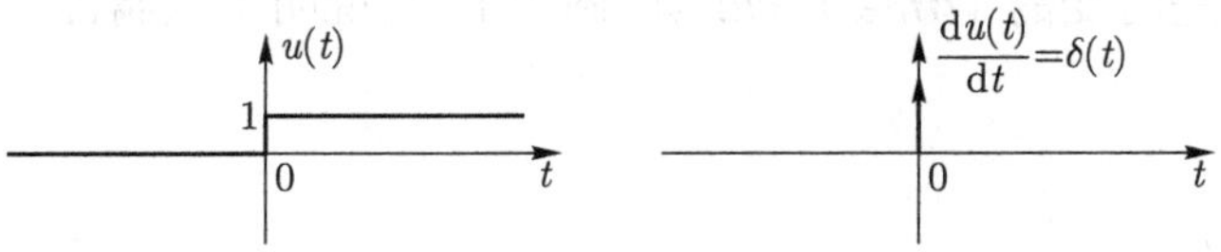

Figure 2.1-9 Illustration of the relationship between $\delta(t)$ and $u(t)$

2.1.3 线性时不变系统 (*linear time-invariant system*)

系统 (*system*) 是用来处理信号的, 它是一种接收输入 $x(t)$ 并产生输出 $y(t)$ 的设备. 图 2.1-10 所示的框图表示出了 $x(t)$ 和 $y(t)$ 之间的一个系统. 可将这种关系象征性地标记为

$$x(t) \to y(t).$$

这种表示可读作 “$x(t)$ 产生 $y(t)$” , 其意义与图 2.1-10 所示相同.

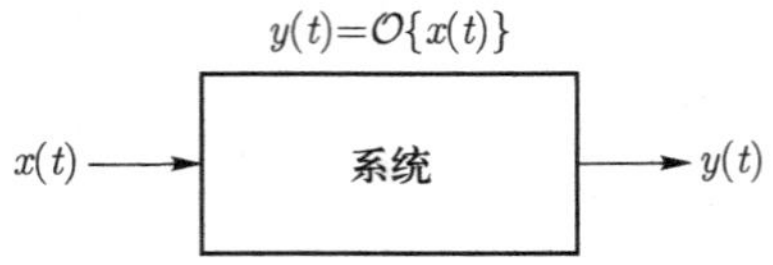

图 2.1-10 一个基本的系统的框图

可用如下所示的一个数学算子来表示该系统

$$y(t) = \mathcal{O}\{x(t)\}, \tag{2.1-9}$$

这里, $\mathcal{O}\{\cdot\}$ 是一个将输入映射到输出的操作算符, 式 (2.1-9) 即关联输入 $x(t)$ 到输出 $y(t)$ 的系统方程 (*system equation*).

如果一个系统具有叠加性 (*superposition*), 那么该系统是线性的 (*linear*), 否则, 就是非线性的 (*nonlinear*). 如果 $x_1 \to y_1$, 且, $x_2 \to y_2$, 那么, 叠加性是指 $x = ax_1+bx_2 \to y = ay_1+by_2$, 这里, a 和 b 为常数. 换言之, 总输出 y 是输入 x_1 与输入 x_2 产生的加权和.

举例: 线性

考虑如下系统方程描述的系统时: $y(t) = \mathcal{O}\{x(t)\} = tx(t)$ 或 $x(t) \to y(t) = tx(t)$. 根据该系统方程, 写出 $x_1 \to y_1 = tx_1$, 且, $x_2 \to y_2 = tx_2$, 该系统是线性的, 因为

$$\begin{aligned} x = ax_1 + bx_2 \to y = tx &= t(ax_1 + bx_2) \\ &= tax_1 + tbx_2 \\ &= ay_1 + by_2. \end{aligned}$$

考虑如下系统方程描述的系统时: $y(t) = \mathcal{O}\{x(t)\} = x^2(t)$ 或 $x(t) \to y(t) = x^2(t)$. 根据该系统方程, 写出 $x_1 \to y_1 = x_1^2$, 且, $x_2 \to y_2 = x_2^2$, 该系统是非线性的, 因为

$$\begin{aligned} x = ax_1 + bx_2 \to y = x^2 &= (ax_1 + bx_2)^2 \\ &= (ax_1)^2 + (bx_2)^2 + 2abx_1x_2 \\ &\neq ay_1 + by_2. \end{aligned}$$

一个系统, 如果其输入的时移导致了输出相同的时移, 则该系统是时不变的 (*time invariant*). 否则, 它是时变的 (*time variant*). 图 2.1-11 说明了该属性, 其中 t_0 是时移或时间延迟.

2.1.3 Linear and Time-Invariant Systems

Systems are used to process signals. A system is a device that accepts an input $x(t)$ and gives out an output $y(t)$. Figure 2.1-10 shows a block diagram representation of a system, relating $x(t)$ and $y(t)$. We show this relation symbolically by

$$x(t) \to y(t).$$

We read this notation as "$x(t)$ produces $y(t)$". It has the same meaning as the block diagram shown in Figure 2.1-10.

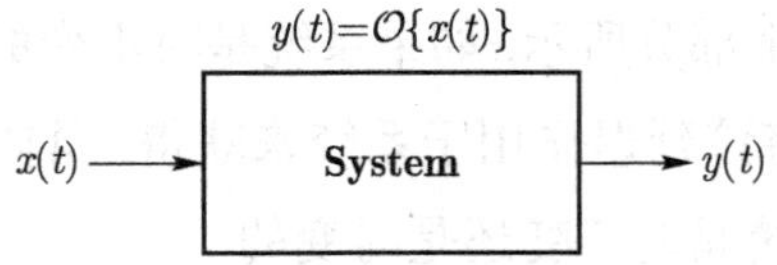

Figure 2.1-10 Block diagram of a general system

We can represent the system by a mathematical operator as follows:

$$y(t) = \mathcal{O}\{x(t)\}, \tag{2.1-9}$$

where $\mathcal{O}\{\cdot\}$ is an operator, mapping the input into the output. Equation (2.1-9) is the *system equation* as it relates the input $x(t)$ to the output $y(t)$.

A system is *linear* if superposition holds. Otherwise, it is *nonlinear.* If $x_1 \to y_1$ and $x_2 \to y_2$; then superposition means $x = ax_1 + bx_2 \to y = ay_1 + by_2$, where a and b are some constants. In words, the overall output y is a weighted sum of the outputs due to inputs x_1 and x_2.

Example: Linearity

Consider the system described by the system equation as follows: $y(t) = \mathcal{O}\{x(t)\} = tx(t)$ or $x(t) \to y(t) = tx(t)$. According to the system equation, we write $x_1 \to y_1 = tx_1$ and $x_2 \to y_2 = tx_2$. The system is linear, since

$$\begin{aligned} x = ax_1 + bx_2 \to y = tx &= t(ax_1 + bx_2) \\ &= tax_1 + tbx_2 \\ &= ay_1 + by_2. \end{aligned}$$

Consider the system described by the system equation as follows: $y(t) = \mathcal{O}\{x(t)\} = x^2(t)$ or $x(t) \to y(t) = x^2(t)$. According to the system equation, we write $x_1 \to y_1 = x_1^2$ and $x_2 \to y_2 = x_2^2$. The system is nonlinear, since

$$\begin{aligned} x = ax_1 + bx_2 \to y = x^2 &= (ax_1 + bx_2)^2 \\ &= (ax_1)^2 + (bx_2)^2 + 2abx_1x_2 \\ &\neq ay_1 + by_2. \end{aligned}$$

A system is *time invariant* if a time shift in the input results in the same time shift at the output. Otherwise, it is *time variant.* Figure 2.1-11 illustrates this property where t_0 is the time shift or the time delay.

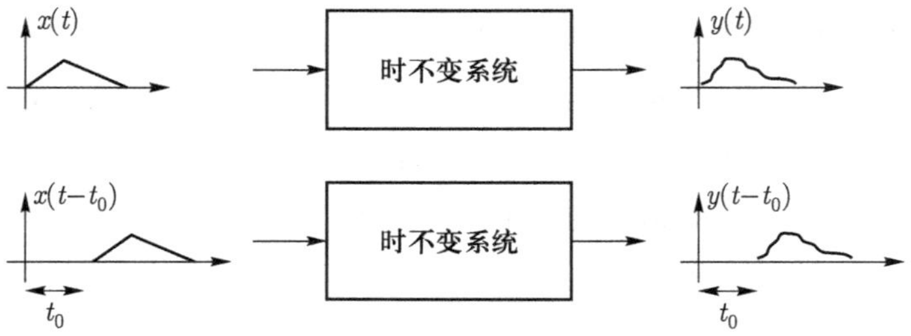

图 2.1-11　时不变系统

另一种说明时不变性质的方法如图 2.1-12 所示. 可通过给输出一个 t_0 的延迟来延迟系统的输出 $y(t)$, 如图 2.1-12 的上半部分所示. 如果系统是时不变的, 那么延迟了的输出 $y(t-t_0)$ 可以通过先给输入延迟 t_0 再将该延迟作用于系统来获得, 具体如图 2.1-12 的下半部分所示. 可利用这个过程来判别一个系统是时不变还是时变的.

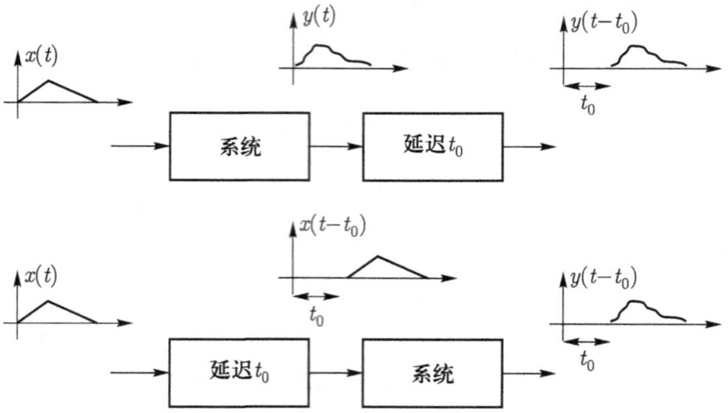

图 2.1-12　时不变性质的说明

举例: 时不变 (*time invariance*)

当一个系统的系统方程如下时: $y(t)=\mathcal{O}\{x(t)\}=\dfrac{\mathrm{d}x}{\mathrm{d}t}$ 或 $x(t)\to y(t)=\dfrac{\mathrm{d}x}{\mathrm{d}t}$, 该系统为一个微分器 (*differentiator*). 根据图 2.1-12 所示的情况确定其时不变性质, 从图 2.1-13 可以看出, 两个最终输出是相同的, 因此所研究的系统是时不变的.

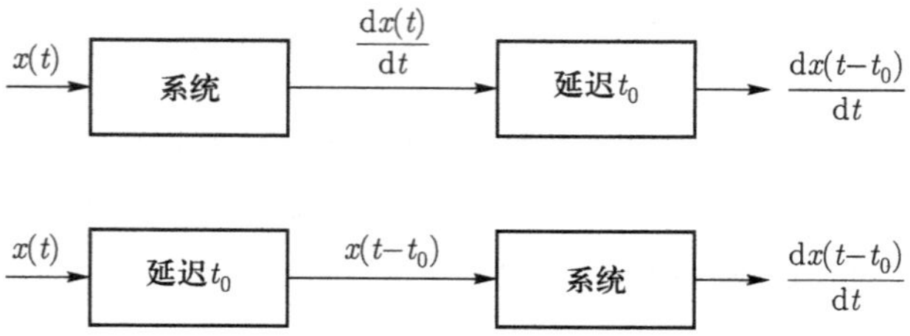

图 2.1-13　时不变系统的判定: 相同的输出

当一个系统的系统方程如下时: $y(t)=\mathcal{O}\{x(t)\}=tx(t)$ 或 $x(t)\to y(t)=tx(t)$. 从图 2.1-14 的结果可以看出, 最终的输出是不相同的, 因此该系统是时变的.

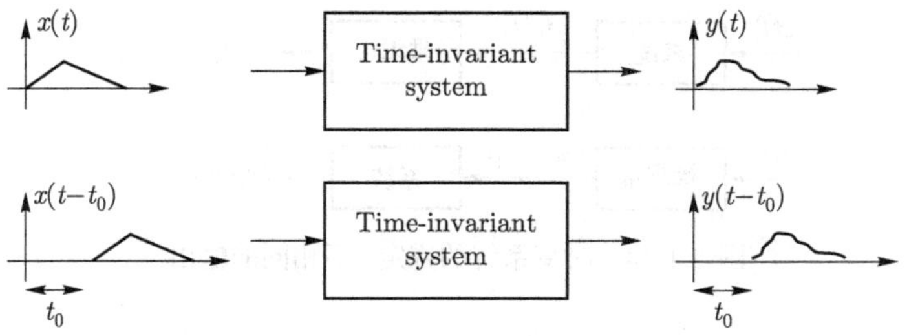

Figure 2.1-11 Time-invariant system

Another way to illustrate the time-invariance property is shown in Figure 2.1-12. We can delay the output $y(t)$ of a system by applying the output to a t_0 delay, as shown in the top part of Figure 2.1-12. If the system is time invariant, then the delayed output $y(t-t_0)$ can also be obtained by first delaying the input by t_0 before applying it to the system, as shown in the bottom half of Figure 2.1-12. This illustration of time-invariance property is constructive as we can use this procedure to test if a system is time invariant or time variant.

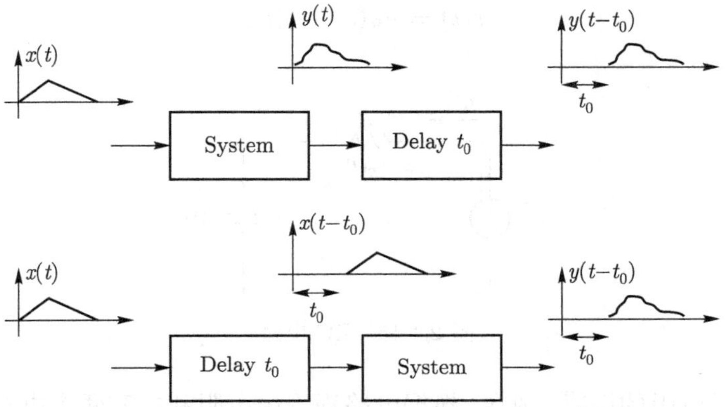

Figure 2.1-12 Illustration of time-invariance property

Example: Time Invariance

Consider the system described by the system equation as follows: $y(t)=\mathcal{O}\{x(t)\}=\dfrac{\mathrm{d}x}{\mathrm{d}t}$ or $x(t)\rightarrow y(t)=\dfrac{\mathrm{d}x}{\mathrm{d}t}$. The system is a *differentiator*. We determine time invariance according to the situations illustrated in Figure 2.1-12. From Figure 2.1-13, we see that the two final outputs are identical, and hence the system under investigation is time invariant.

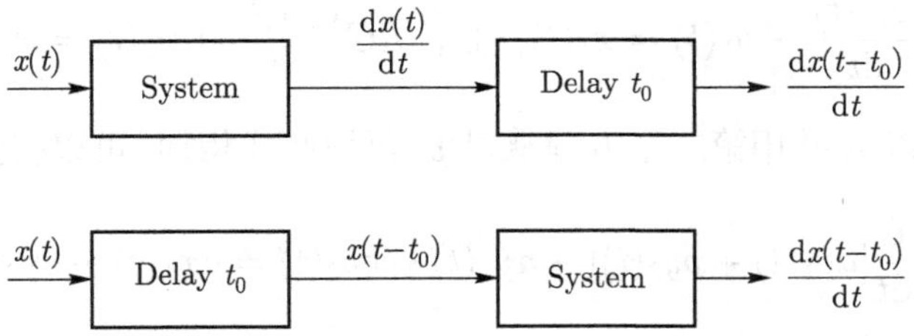

Figure 2.1-13 Time-invariance example: same output

Consider the system described by the system equation as follows: $y(t)=\mathcal{O}\{x(t)\}=tx(t)$ or $x(t)\rightarrow y(t)=tx(t)$. From the results of Figure 2.1-14, we see that the final outputs are not the same, and hence the system is time variant.

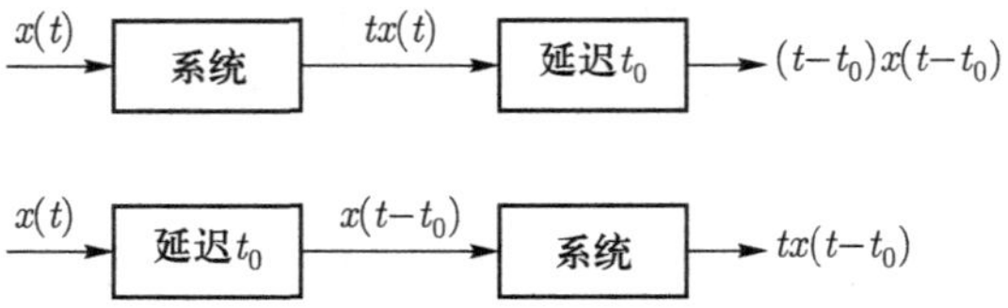

图 2.1-14 时变系统的判定: 不相同的输出

举例: RC 电路为线性时不变系统的情况

一个线性时不变系统是一个线性系统, 同时也是时不变的. 正如后续可以看到的, 线性时不变系统有一个关键的特性, 那就是, 对于一个给定的正弦输入信号, 其输出也是同频率的正弦信号 (但是可能会有不同的振幅和相位). 考虑如图 2.1-15 所示的 RC 电路, 这里先求其系统方程, 假定输入 $x(t)$ 为一个电压源, 输出 $y(t)$ 为通过电容器的电压, 相应地, 根据基本的电路理论, 可以知道

$$x(t) = v_R(t) + y(t), \tag{2.1-10}$$

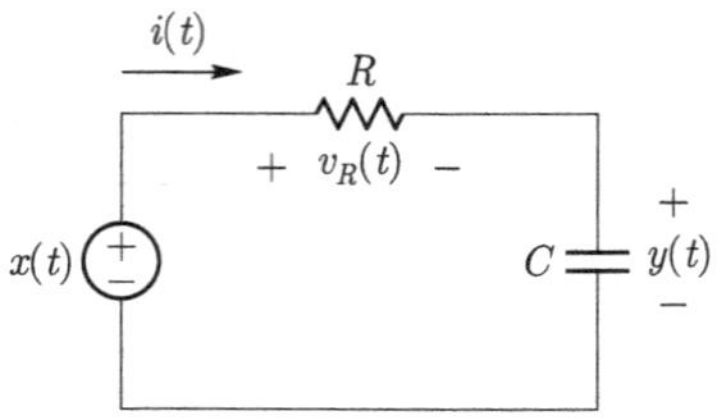

图 2.1-15 RC 电路

其中, $v_R(t)$ 为经过电阻的压降. 现在, 根据电路理论中电阻和电容两个电路元件物理量的符号规定, 即, 当电流进入电压的正极性端时, 电阻和电容有以下关系: $v_R(t) = i(t)R$, 且, $i(t) = C\dfrac{\mathrm{d}y(t)}{\mathrm{d}t}$. 将此关系代入式 (2.1-10), 得

$$RC\frac{\mathrm{d}y(t)}{\mathrm{d}t} + y(t) = x(t). \tag{2.1-11}$$

式 (2.1-11) 为图 2.1-15 所示中电路的系统方程. 若 $x_1 \to y_1$, 且, $x_2 \to y_2$, 那么根据该系统方程, 分别有

$$RC\frac{\mathrm{d}y_1(t)}{\mathrm{d}t} + y_1(t) = x_1(t), \quad 且, \quad RC\frac{\mathrm{d}y_2(t)}{\mathrm{d}t} + y_2(t) = x_2(t),$$

现在用第一个方程乘以 a 并用第二个方程乘以 b, 再把两式相加, 可以得到

$$RC\frac{\mathrm{d}}{\mathrm{d}t}[ay_1(t) + by_2(t)] + ay_1(t) + by_2(t) = ax_1(t) + bx_2(t).$$

此方程即满足叠加原理的 $ax_1(t) + bx_2(t) \to ay_1(t) + by_2(t)$ 的系统方程 [见式 (2.1-11)]. 因此, 该系统为线性的, 同时也是时不变的, 因为其电路元件物理量是常量, 并不是时间的函数. 结果表明, 任何具有常系数的 n 阶线性微分方程都属于线性时不变系统.

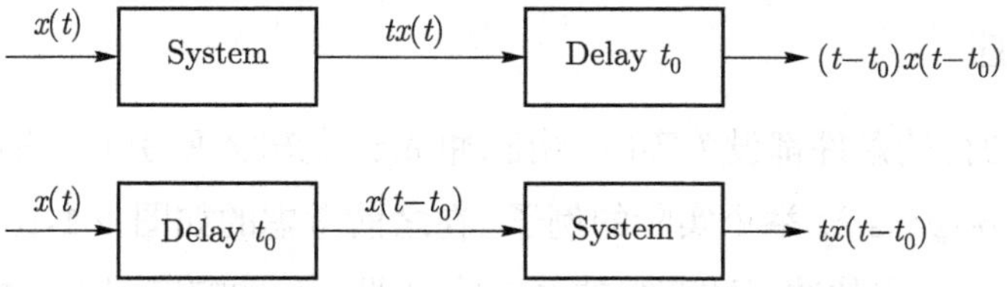

Figure 2.1-14 Time-variance example: different outputs

Example: RC Circuit Being a Linear and Time-Invariant System

A linear and time-invariant system is a linear system that is also time invariant. As we will see later, linear and time-invariant systems have a key property in that, for a given sinusoidal input, the output is also a sinusoidal signal with the same frequency (but possibly with different amplitude and phase). Consider the RC electrical circuit shown in Figure 2.1-15. Let us first find the system equation. We assume that the input $x(t)$ is a voltage source and the output $y(t)$ is the voltage across the capacitor. Accordingly, using basic circuit theory, we get

$$x(t) = v_R(t) + y(t), \tag{2.1-10}$$

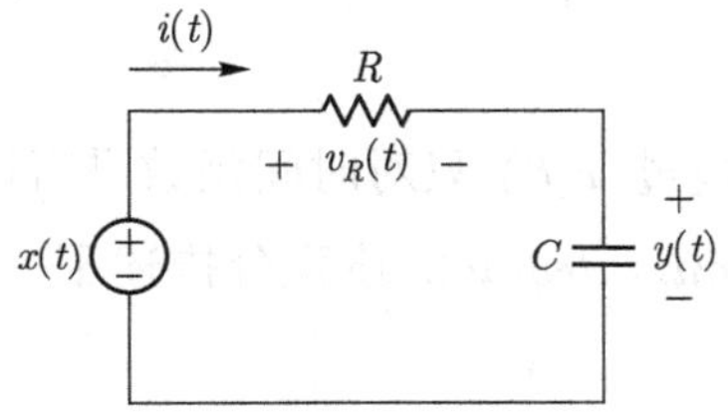

Figure 2.1-15 RC circuit

where $v_R(t)$ is the voltage dropped across the resistor. Now according to the sign convention used in circuit theory for the two circuit elements resistor and capacitor, that is, when a current enters the positive polarity of the voltage, we have the following relationships for the resistor and the capacitor: $v_R(t) = i(t)R$, and $i(t) = C\dfrac{\mathrm{d}y(t)}{\mathrm{d}t}$. Putting these relationships into Eq. (2.1-10), we have

$$RC\frac{\mathrm{d}y(t)}{\mathrm{d}t} + y(t) = x(t). \tag{2.1-11}$$

Equation (2.1-11) is the system equation for the circuit shown in Figure 2.1-15. If $x_1 \to y_1$ and $x_2 \to y_2$ then according to the system equation, we have

$$RC\frac{\mathrm{d}y_1(t)}{\mathrm{d}t} + y_1(t) = x_1(t) \quad \text{and} \quad RC\frac{\mathrm{d}y_2(t)}{\mathrm{d}t} + y_2(t) = x_2(t),$$

respectively. Now multiplying the first equation by a and the second equation by b, and adding them, we have

$$RC\frac{\mathrm{d}}{\mathrm{d}t}[ay_1(t) + by_2(t)] + ay_1(t) + by_2(t) = ax_1(t) + bx_2(t).$$

This equation is the system equation [Eq. (2.1-11)] with

$$ax_1(t) + bx_2(t) \to ay_1(t) + by_2(t),$$

satisfying the principle of superposition. Hence, the system is linear. The system is also time invariant as the circuit elements are constant and not a function of time. As it turns out, any nth order linear differential equation with constant coefficients belongs to the class of linear and time-invariant systems.

2.1.4 脉冲响应

当系统中所有的初始条件都设为零时, 由脉冲 $\delta(t)$ 的输入所引起的输出被称为是系统的脉冲响应 (*impulse response*). 举个实际的例子, 在之前考虑的如图 2.1-15 的 RC 电路中, 当电容器没有电荷存储时, 用脉冲函数获得的系统输出即为脉冲响应. 回到式 (2.1-9), 有 $y(t) = \mathcal{O}\{x(t)\}$, 现在, 对于形式为 $\delta(t-t')$ 的脉冲输入, 其输出 $y(t) = \mathcal{O}\{x(t)\} = \mathcal{O}\{\delta(t-t')\} = h(t;t')$ 被称为线性系统的脉冲响应 (*impulse response*). 利用 δ 函数的筛选性质, 可写出

$$x(t) = \int_{-\infty}^{\infty} x(t')\delta(t-t')\mathrm{d}t',$$

即, $x(t)$ 被看作经加权和平移了的 δ 函数的线性组合. 那么, 线性系统的输出可以写为

$$\begin{aligned} y(t) = \mathcal{O}\{x(t)\} &= \mathcal{O}\left\{\int_{-\infty}^{\infty} x(t')\delta(t-t')\mathrm{d}t'\right\} \\ &= \int_{-\infty}^{\infty} x(t')\mathcal{O}\{\delta(t-t')\}\mathrm{d}t' = \int_{-\infty}^{\infty} x(t')h(t;t')\mathrm{d}t'. \end{aligned} \tag{2.1-12}$$

可以看出, 在上述过程中, 算子绕过 $x(t')$ 只对时间函数进行运算. 上述输入和输出关系的结果被称为叠加积分 (*superposition integral*). 该积分描述了一个线性时变系统 (*linear time-variant system*).

这里, 如果对时移脉冲函数 $\delta(t-t')$ 的响应是 $h(t-t')$, 则该线性系统是时不变的, 即, $\mathcal{O}\{\delta(t-t')\} = h(t-t')$, 那么, 式 (2.1-12) 变为

$$y(t) = \int_{-\infty}^{\infty} x(t')h(t-t')\mathrm{d}t' = x(t) * h(t). \tag{2.1-13}$$

这是一个重要的结果, 脉冲 $h(t;t')$ 仅仅依赖于 $t-t'$, 即, 在线性时不变系统中, $h(t;t') = h(t-t')$, 该积分也可以按照如图 2.1-16 的方式导出.

$\delta(t) \to h(t)$

$\delta(t-t') \to h(t-t')$, 让 $\delta(t)$ 延迟 t' 将会使得输出延迟 t'

$x(t')\delta(t-t') \to x(t')h(t-t'), t'$ 时强度为 $x(t')$ 的脉冲输入情况

$\int_{-\infty}^{\infty} x(t')\delta(t-t')\mathrm{d}t' \to \int_{-\infty}^{\infty} x(t')h(t-t')\mathrm{d}t'$, 对所有不同 t' 的响应进行求和

$x(t) \to y(t) = \int_{-\infty}^{\infty} x(t')h(t-t')\mathrm{d}t'$

图 2.1-16 线性时不变系统中叠加积分的推导

式 (2.1-13) 中的积分即为卷积积分 (*convolution integral*), 其中符号 $*$ 表示 $x(t)$ 和 $h(t)$ 的卷积, $x(t)*h(t)$ 读作 $x(t)$ 卷积 $h(t)$, 对于所有初始条件为零的一个线性时不变系统, 给定任意的输入 $x(t)$, 则, 系统的输出是由该输入与脉冲响应的卷积给出的.

2.1.4 Impulse Responses

The impulse response of a system is the output owing to an impulse $\delta(t)$ input when all the initial conditions in the system have been set to zero. A practical illustrative example is that in our previously considered RC circuit in Figure 2.1-15, the capacitor has no charges stored when we apply an impulse function to obtain the system's output. Returning to Eq. (2.1-9), we have $y(t) = \mathcal{O}\{x(t)\}$. Now, for an impulse input of the form $\delta(t-t')$, the output $y(t) = \mathcal{O}\{x(t)\} = \mathcal{O}\{\delta(t-t')\} = h(t;t')$ is called the *impulse response* of the linear system. Using the sampling property of the delta function, we can write

$$x(t) = \int_{-\infty}^{\infty} x(t')\delta(t-t')\mathrm{d}t',$$

that is, $x(t)$ is regarded as a linear combination of weighted and shifted delta functions. We can then write the output of the linear system as

$$y(t) = \mathcal{O}\{x(t)\} = \mathcal{O}\left\{\int_{-\infty}^{\infty} x(t')\delta(t-t')\mathrm{d}t'\right\}$$

$$= \int_{-\infty}^{\infty} x(t')\mathcal{O}\{\delta(t-t')\}\mathrm{d}t' = \int_{-\infty}^{\infty} x(t')h(t;t')\mathrm{d}t'. \tag{2.1-12}$$

Note that in the above procedure, the operator has moved to pass $x(t')$ and to only operate on a time function. The above result relating the input and output is known as the *superposition integral.* This integral describes a *linear time-variant system.*

Now, a linear system is time invariant if the response to a time-shifted impulse function, $\delta(t-t')$, is given by $h(t-t')$, that is, $\mathcal{O}\{\delta(t-t')\} = h(t-t')$. Equation (2.1-12) then becomes

$$y(t) = \int_{-\infty}^{\infty} x(t')h(t-t')\mathrm{d}t' = x(t) * h(t). \tag{2.1-13}$$

This is an important result. The impulse $h(t;t')$ only depends on $t-t'$, that is, $h(t;t') = h(t-t')$ in linear and time-invariant systems. The integral can also be derived as illustrated in Figure 2.1-16.

$$\delta(t) \to h(t)$$

$\delta(t-t') \to h(t-t')$, delaying $\delta(t)$ by t' will delay the output by t'

$x(t')\delta(t-t') \to x(t')h(t-t')$, when impulse response of amplitude $x(t')$ applied at t'

$\int_{-\infty}^{\infty} x(t')\delta(t-t')\mathrm{d}t' \to \int_{-\infty}^{\infty} x(t')h(t-t')\mathrm{d}t'$, summing up all the responses at different t'

$$x(t) \to y(t) = \int_{-\infty}^{\infty} x(t')h(t-t')\mathrm{d}t'$$

Figure 2.1-16 Derivation of the superposition integral for a linear and time-invariant system

The integral in Eq. (2.1-13) is known as the *convolution integral,* and $*$ is a symbol denoting the convolution of $x(t)$ and $h(t)$. $x(t) * h(t)$ reads as x convolves with h. For a linear and time-invariant system with all the initial conditions being zero, the system output to any input $x(t)$ is given by the convolution between the input and impulse response.

卷积运算实际上是可交换 (*commutative*) 的, 这意味着, 当对两个函数进行卷积时, 其顺序是不会影响卷积结果的. 通过改变式 (2.1-13) 中的变量, 设 $t-t'=t''$, 则, $t'=t-t''$, 且, $\mathrm{d}t'=-\mathrm{d}t''$, 可以得到

$$\begin{aligned} y(t)=x(t)*h(t)&=\int_{-\infty}^{\infty}x(t')h(t-t')\mathrm{d}t' \\ &=\int_{\infty}^{-\infty}x(t-t'')h(t'')(-\mathrm{d}t'') \\ &=\int_{-\infty}^{\infty}x(t-t'')h(t'')\mathrm{d}t''=h(t)*x(t). \end{aligned} \tag{2.1-14}$$

举例: 求卷积的图解法

已知, $x(t)=a\mathrm{e}^{-at}u(t)$, $a>0$, 且, $h(t)=u(t)$, 用图解法来计算其卷积. 根据式 (2.1-13) 中卷积的定义, 有

$$x(t)*h(t)=\int_{-\infty}^{\infty}x(t')h(t-t')\mathrm{d}t'.$$

上面的积分, 简单来说, 就是计算两个函数 $x(t')$ 和 $h(t-t')$ 在不同 t 值时的乘积面积, 因此, 最终结果是 t 的函数. 这里先来考虑如何实现两个函数的乘积 $x(t')h(t-t')$. 根据初始函数 $x(t)$ 和 $h(t)$, 可以画出如图 2.1-17a) 中所示的 $x(t')$ 和 $h(t')$. 现在图中已经有 $x(t')$ 函数, 为得到 $h(t-t')$ 的图形, 先对 $h(t')$ 进行翻转, 得到 $h(-t')$, 如图 2.1-17b) 所示. 然后, 将 $h(-t')$ 平移 t 得到 $h(t-t')$, 如图 2.1-17c) 所示.

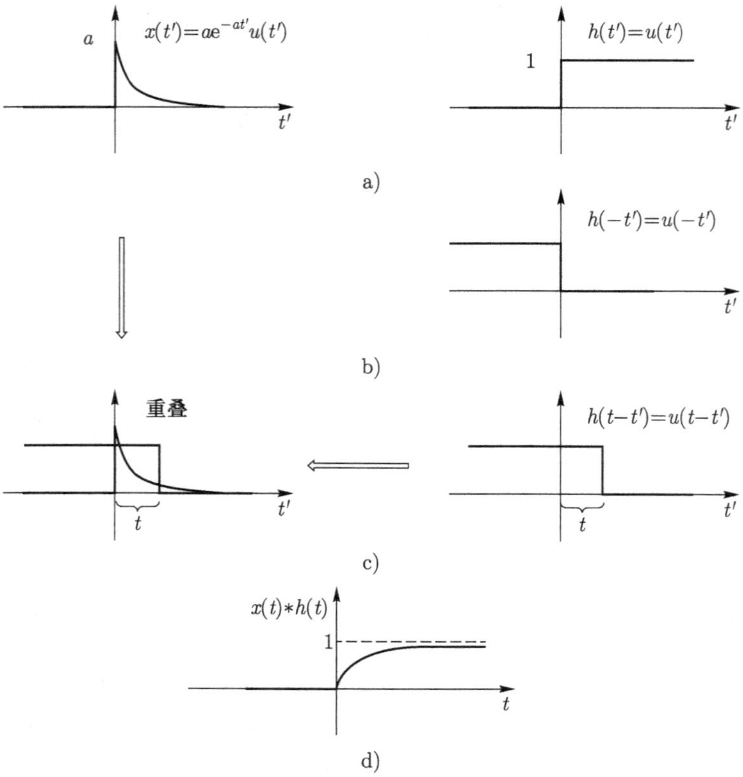

图 2.1-17　卷积的图解法

Convolution operation is actually *commutative*, which means the order in which we perform convolution of the two functions does not change the result. By changing variables in Eq. (2.1-13), we let $t - t' = t''$ so that $t' = t - t''$ and $\mathrm{d}t' = -\mathrm{d}t''$, we have

$$\begin{aligned} y(t) = x(t) * h(t) &= \int_{-\infty}^{\infty} x(t')h(t-t')\mathrm{d}t' \\ &= \int_{\infty}^{-\infty} x(t-t'')h(t'')(-\mathrm{d}t'') \\ &= \int_{-\infty}^{\infty} x(t-t'')h(t'')\mathrm{d}t'' = h(t) * x(t). \end{aligned} \tag{2.1-14}$$

Example: Graphical Method to Find Convolution

Given $x(t) = a\mathrm{e}^{-at}u(t)$, $a > 0$ and $h(t) = u(t)$, we evaluate their convolution by a graphical method. According to the definition of convolution in Eq. (2.1-13), we have

$$x(t) * h(t) = \int_{-\infty}^{\infty} x(t')h(t-t')\mathrm{d}t'.$$

The above integral simply means that we find the areas under the product of two functions, $x(t')$ and $h(t-t')$, for different values of t. Hence, the final answer is a function of t. Let us first visualize how to create the product $x(t')h(t-t')$. From the original functions $x(t)$ and $h(t)$, we plot $x(t')$ and $h(t')$ as shown in the Figure 2.1-17a. Now we have $x(t')$ on a graph. To create the graph of $h(t-t')$, we first take the reversal of $h(t')$ to get $h(-t')$, as shown in Figure 2.1-17b. Next, we shift $h(-t')$ by t to get $h(t-t')$, as shown in Figure 2.1-17c.

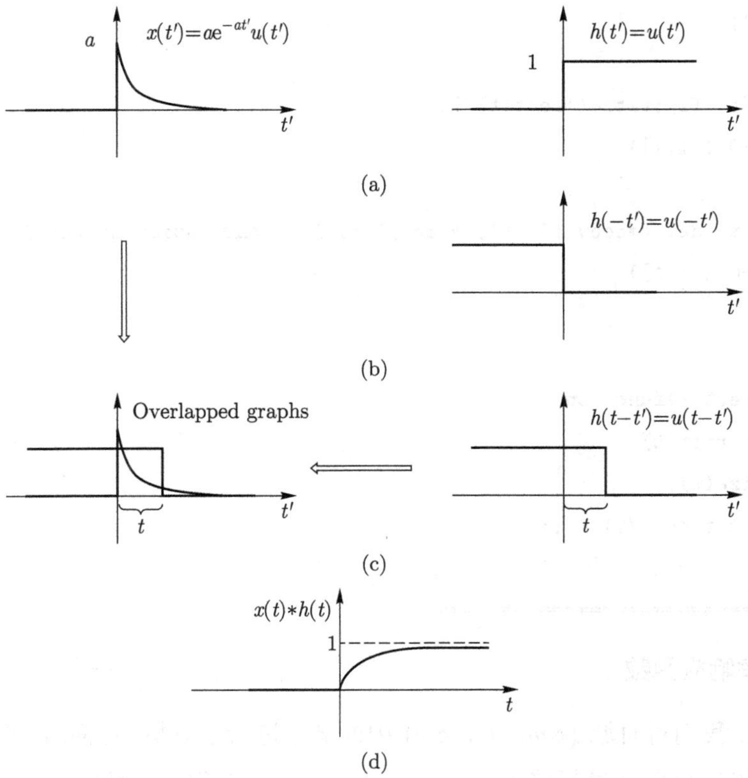

Figure 2.1-17 Graphical understanding of convolution

接下来, 将 $x(t')$ 和 $h(t-t')$ 重叠并绘制在同一个图中, 如图 2.1-17c) 的左侧部分所示. 该图可引导我们写出如下卷积积分时的被积函数及积分限

$$x(t)*h(t)=\int_{-\infty}^{\infty}x(t')h(t-t')\mathrm{d}t'$$

$$=\int_{0}^{t}a\mathrm{e}^{-at'}\mathrm{d}t'=\begin{cases}1-\mathrm{e}^{-at}, & t>0\\ 0, & t<0.\end{cases}$$

上述积分的结果如图 2.1-17d) 所示.

举例: 卷积 MATLAB

计算如下卷积: $rect(t)*rect(t)$. 图 2.1-18a) 绘制了 $rect(t)$, 其卷积结果如图 2.1-18b) 所示.

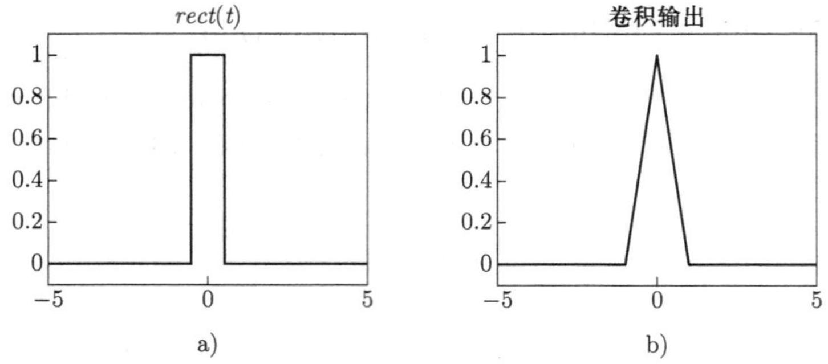

图 2.1-18 a) $rect(t)$ 的绘制; b) $rect(t)*rect(t)$ 的绘制

```
==========================================
% Convolution of rect(t) by itself; convolution_example.m
clear all; close all;
del_t=0.01;
t=-5.0:del_t:5.0;
f1 = rect(t);

figure;plot(t, f1);title('rect(t)');
axis([-5 5 -0.1 1.1])
grid on
figure;plot(t, del_t*conv(f1, f1,'same'));title('convolution output');
axis([-5 5 -0.1 1.1])
grid on

% Defining rect(x)function
function y = rect(t)
y = zeros(size(t));
y(t >= -1/2 & t <= 1/2) = 1;
end
==========================================
```

2.1.5 频率响应函数

到目前为止, 我们对时域 (*time domain*) 中的响应进行了分析, 然而对于线性时不变系统特性的进一步了解, 可以通过频域 (*frequency domain*) 来获得. 这里来看复指数函数 $x(t)=$

We then overlap $x(t')$ and $h(t-t')$ on a single graph, which is shown on the left panel of Figure 2.1-17c. This graph guides us to set up the integrand and the limits for the integral of convolution as follows:

$$x(t) * h(t) = \int_{-\infty}^{\infty} x(t')h(t-t')\mathrm{d}t'$$

$$= \int_{0}^{t} a\mathrm{e}^{-at'}\mathrm{d}t' = \begin{cases} 1-\mathrm{e}^{-at}, & t>0 \\ 0, & t<0. \end{cases}$$

The result of the above integral is plotted in Figure 2.1-17d.

Example: Convolution MATLAB

We perform the following convolution: $rect(t) * rect(t)$. Figure 2.1-18a plots $rect(t)$ and the resulting convolution is plotted in Figure 2.1-18b.

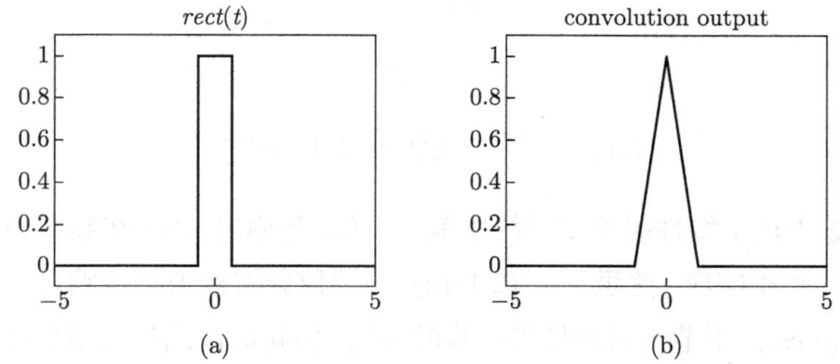

Figure 2.1-18 (a) Plot of $rect(t)$ and (b) plot of $rect(t) * rect(t)$

```
=========================================
% Convolution of rect(t) by itself; convolution_example.m
clear all; close all;
del_t=0.01;
t=-5.0:del_t:5.0;
f1 = rect(t);

figure;plot(t, f1);title('rect(t)');
axis([-5 5 -0.1 1.1])
grid on
figure;plot(t, del_t*conv(f1, f1,'same'));title('convolution output');
axis([-5 5 -0.1 1.1])
grid on

% Defining rect(x)function
function y = rect(t)
y = zeros(size(t));
y(t >= -1/2 & t <= 1/2) = 1;
end
=========================================
```

2.1.5 Frequency Response Functions

While we have analyzed responses in the *time domain* so far, much insight about the behavior of a linear and time-invariant system could be obtained through the *frequency domain*. Let

$e^{j\omega t}$ 的响应, 其中, $j=\sqrt{-1}$, 如果 t 是以 s 为单位来测量的话, 则 ω 是以 rad/s 为单位来测量的弧度频率† (*radian frequency*). 根据式 (2.1-14), 有

$$\begin{aligned} y(t) = x(t) * h(t) &= h(t) * x(t) \\ &= \int_{-\infty}^{\infty} h(t') e^{j\omega(t-t')} dt' \\ &= e^{j\omega t} \int_{-\infty}^{\infty} h(t') e^{-j\omega t'} dt' = H(\omega) e^{j\omega t}, \end{aligned}$$

这里定义了频率响应函数 (*frequency response function*), 也称为传递函数 (*transfer function*) $H(\omega)$, 有

$$H(\omega) = \int_{-\infty}^{\infty} h(t') e^{-j\omega t'} dt'. \tag{2.1-15}$$

因此, 有

$$x(t) = e^{j\omega t} \to y(t) = H(\omega) e^{j\omega t}.$$

频率响应函数表达了线性时不变系统中输入 $x(t)$ 与输出 $y(t)$ 在频域中的关系, 这有利于找出一个系统的频率响应. 这里, 式 (2.1-15) 中的积分被称为脉冲响应 $h(t)$ 的傅里叶变换 (*Fourier transform*), 下节将对此做进一步的讨论. 换句话说, 传递函数 $H(\omega)$ 是线性时不变系统中脉冲响应 $h(t)$ 的傅里叶变换.

举例: RC 电路的频率响应

参考图 2.1-15 所示的 RC 电路, 其系统方程经推导并由下式给出

$$RC\frac{dy(t)}{dt} + y(t) = x(t).$$

通过将 $y(t) = H(\omega)e^{j\omega t}$ 和 $x(t) = e^{j\omega t}$ 代入上式, 可得

$$RCH(\omega)(j\omega)e^{j\omega t} + H(\omega)e^{j\omega t} = e^{j\omega t}.$$

消掉 $e^{j\omega t}$ 项, 并求解 $H(\omega)$, 得

$$H(\omega) = \frac{1}{1+j\omega RC}. \tag{2.1-16a}$$

频率响应函数为一复函数, 利用复数等式 $a+bj=\sqrt{a^2+b^2}e^{j\arctan(b/a)}$, 电路的传递函数可以写为

$$H(\omega) = |H(\omega)|e^{j\theta(\omega)} = \frac{1}{\sqrt{1+(\omega RC)^2}}e^{-j\arctan(\omega RC)}, \tag{2.1-16b}$$

其中, $|H(\omega)| = \dfrac{1}{\sqrt{1+(\omega RC)^2}}$ 和 $\theta(\omega) = -\arctan(\omega RC)$ 分别被称为系统的强度频率谱 (*magnitude frequency spectrum*) 和相位频率谱 (*phase frequency spectrum*). 图 2.1-19 所示为

† 弧度频率 (*radian frequency*) 也被称为角频率 (*angular frequency*).

us examine the response to a complex exponential $x(t) = \mathrm{e}^{\mathrm{j}\omega t}$, where $\mathrm{j} = \sqrt{-1}$ and ω is *radian frequency* measured in radians per second if t is measured in seconds. According to Eq. (2.1-14), we have

$$\begin{aligned} y(t) &= x(t) * h(t) = h(t) * x(t) \\ &= \int_{-\infty}^{\infty} h(t')\mathrm{e}^{\mathrm{j}\omega(t-t')}\mathrm{d}t' \\ &= \mathrm{e}^{\mathrm{j}\omega t}\int_{-\infty}^{\infty} h(t')\mathrm{e}^{-\mathrm{j}\omega t'}\mathrm{d}t' = H(\omega)\mathrm{e}^{\mathrm{j}\omega t}, \end{aligned}$$

where we have defined the *frequency response function*, also called *transfer function* $H(\omega)$ as

$$H(\omega) = \int_{-\infty}^{\infty} h(t')\mathrm{e}^{-\mathrm{j}\omega t'}\mathrm{d}t'. \tag{2.1-15}$$

Hence, we have established that

$$x(t) = \mathrm{e}^{\mathrm{j}\omega t} \to y(t) = H(\omega)\mathrm{e}^{\mathrm{j}\omega t}.$$

A frequency response function expresses the frequency domain relationship between an input $x(t)$ and output $y(t)$ of a linear and time-invariant system and it is useful for finding the frequency response of a system. Note that the integral in Eq. (2.1-15) is called the *Fourier transform* of the impulse response $h(t)$, which we will discuss further in the next subsection. In words, the transfer function $H(\omega)$ is the Fourier transform of the impulse response $h(t)$ in linear and time-invariant systems.

Example: Frequency Response of an RC Circuit

With reference to the RC circuit shown in Figure 2.1-15, its system equation has been derived and given by

$$RC\frac{\mathrm{d}y(t)}{\mathrm{d}t} + y(t) = x(t).$$

By substituting $y(t) = H(\omega)\mathrm{e}^{\mathrm{j}\omega t}$ and $x(t) = \mathrm{e}^{\mathrm{j}\omega t}$ into the above equation, we obtain

$$RCH(\omega)(\mathrm{j}\omega)\mathrm{e}^{\mathrm{j}\omega t} + H(\omega)\mathrm{e}^{\mathrm{j}\omega t} = \mathrm{e}^{\mathrm{j}\omega t}.$$

After canceling the $\mathrm{e}^{\mathrm{j}\omega t}$ term and solving for $H(\omega)$, we have

$$H(\omega) = \frac{1}{1+\mathrm{j}\omega RC}. \tag{2.1-16a}$$

The frequency response function is a complex function. Using the complex number identity $a + b\mathrm{j} = \sqrt{a^2+b^2}\mathrm{e}^{\mathrm{j}\arctan(b/a)}$, we write the transfer function of the circuit as

$$H(\omega) = |H(\omega)|\,\mathrm{e}^{\mathrm{j}\theta(\omega)} = \frac{1}{\sqrt{1+(\omega RC)^2}}\mathrm{e}^{-\mathrm{j}\arctan(\omega RC)}, \tag{2.1-16b}$$

where $|H(\omega)| = \dfrac{1}{\sqrt{1+(\omega RC)^2}}$ and $\theta(\omega) = -\arctan(\omega RC)$ are called the *magnitude frequency spectrum* and *phase frequency spectrum* of the system, respectively. In Figure 2.1-19, we show

$RC = 0.005$ s 的强度频率谱图, 其中, ω 的测量单位为 rad/s. 可见, RC 电路为一简单的信号处理 (*signal processing*) 电路, 确切地说, 该电路对输入 $x(t)$ 进行低通滤波, 因为它允许低频信号通过, 同时令高频信号衰减. 在后续章节中, 将会给出对于光学系统中给定的输入图像要如何进行图像处理 (*image processing*), 以及如何通过求其传递函数来分析该光学系统.

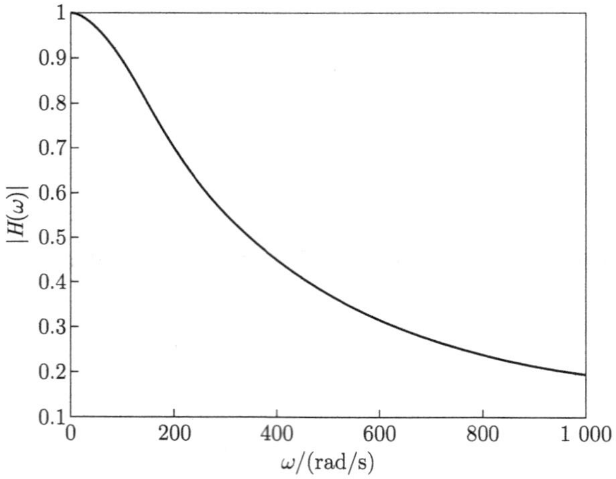

图 2.1-19 RC 滤波器的强度频率谱

运行 MATLAB 程序 RC_filter.m, 得到图 2.1-19, 该程序如下.

```
==========================================
% Plotting of the magnitude frequency response function; RC_filter.m
clear all;
w=0:0.2:1000;
b=1;
H=1./(b+j*w*0.005);
plot(w,abs(H));
xlabel('w in rad/s')
ylabel('|H|')
==========================================
```

举例: 正弦信号输入的响应

在上个例子中, 对于线性时不变系统, 可以得到它的传递函数 $H(\omega)$. 对于复指数输入 $x(t) = \mathrm{e}^{\mathrm{j}\omega t}$, 有输出 $y(t) = H(\omega)\mathrm{e}^{\mathrm{j}\omega t}$. 现在将这一结论扩展到实数信号的正弦信号输入. 由于 $x(t) = \mathrm{e}^{\mathrm{j}\omega t} \to y(t) = H(\omega)\mathrm{e}^{\mathrm{j}\omega t}$, 那么, $x(t) = \mathrm{e}^{-\mathrm{j}\omega t} \to y(t) = H(-\omega)\mathrm{e}^{-\mathrm{j}\omega t}$. 因此, 对于一个正弦信号输入, 根据叠加性, 有, $x(t) = \cos\omega t = \frac{1}{2}\mathrm{e}^{\mathrm{j}\omega t} + \frac{1}{2}\mathrm{e}^{-\mathrm{j}\omega t} \to y(t) = \frac{1}{2}H(\omega)\mathrm{e}^{\mathrm{j}\omega t} + \frac{1}{2}H(-\omega)\mathrm{e}^{-\mathrm{j}\omega t}$. 这里, 从式 (2.1-15), 可知, $H(-\omega) = \int_{-\infty}^{\infty} h(t')\mathrm{e}^{\mathrm{j}\omega t'}\mathrm{d}t'$, 及, $H^*(\omega) = \int_{-\infty}^{\infty} h^*(t')\mathrm{e}^{\mathrm{j}\omega t'}\mathrm{d}t'$. 因此, 可以推得, 当 $h(t)$ 为实数时, 即输入为实数的实际系统情况, 有, $H(-\omega) = H^*(\omega)$, 因此, $x(t) = \cos\omega t = \frac{1}{2}\mathrm{e}^{\mathrm{j}\omega t} + \frac{1}{2}\mathrm{e}^{-\mathrm{j}\omega t} \to$

the plot of the magnitude frequency spectrum for $RC = 0.005$ s, where ω is measured in units of radian/second. Note that the RC circuit is a simple *signal processing* circuit. Indeed, the circuit performs low-pass filtering to the input $x(t)$ as it allows low-frequency signals to pass while attenuating high-frequency signals. In a subsequent chapter, for a given input image presented in an optical system, we will show how *image processing* can be performed and we can analyze an optical system by finding its transfer function.

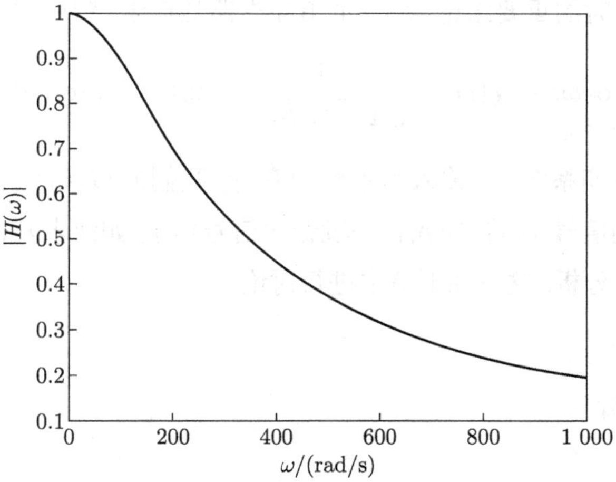

Figure 2.1-19 Magnitude frequency spectrum of an RC filter

Figure 2.1-19 is generated by RC_filter.m shown below.

```
========================================
% Plotting of the magnitude frequency response function; RC_filter.m
clear all;
w=0:0.2:1000;
b=1;
H=1./(b+j*w*0.005);
plot(w,abs(H));
xlabel('w in rad/s')
ylabel('|H|')
========================================
```

Example: Response Due to a Sinusoidal Input

In the last example, for a linear and time-invariant system, we can find its transfer function $H(\omega)$. For a complex exponential input $x(t) = \mathrm{e}^{\mathrm{j}\omega t}$, we have the output $y(t) = H(\omega)\mathrm{e}^{\mathrm{j}\omega t}$. We now extend this result for sinusoidal inputs, which are real signals. Since $x(t) = \mathrm{e}^{\mathrm{j}\omega t} \to y(t) = H(\omega)\mathrm{e}^{\mathrm{j}\omega t}$, then $x(t) = \mathrm{e}^{-\mathrm{j}\omega t} \to y(t) = H(-\omega)\mathrm{e}^{-\mathrm{j}\omega t}$. So for a sinusoidal input, we have $x(t) = \cos\omega t = \frac{1}{2}\mathrm{e}^{\mathrm{j}\omega t} + \frac{1}{2}\mathrm{e}^{-\mathrm{j}\omega t} \to y(t) = \frac{1}{2}H(\omega)\mathrm{e}^{\mathrm{j}\omega t} + \frac{1}{2}H(-\omega)\mathrm{e}^{-\mathrm{j}\omega t}$ according to superposition. Now, from Eq. (2.1-15), we have $H(-\omega) = \int_{-\infty}^{\infty} h(t')\mathrm{e}^{\mathrm{j}\omega t'}\mathrm{d}t'$ and $H^*(\omega) = \int_{-\infty}^{\infty} h^*(t')\mathrm{e}^{\mathrm{j}\omega t'}\mathrm{d}t'$. So we can deduce that for $h(t)$ real, which is the case for physical systems with real inputs, we have $H(-\omega) = H^*(\omega)$. Therefore, $x(t) = \cos\omega t = \frac{1}{2}\mathrm{e}^{\mathrm{j}\omega t} + \frac{1}{2}\mathrm{e}^{-\mathrm{j}\omega t} \to$

$y(t)=\frac{1}{2}H(\omega)\mathrm{e}^{\mathrm{j}\omega t}+\frac{1}{2}H(-\omega)\mathrm{e}^{-\mathrm{j}\omega t}=\frac{1}{2}H(\omega)\mathrm{e}^{\mathrm{j}\omega t}+\frac{1}{2}H^*(\omega)\mathrm{e}^{-\mathrm{j}\omega t}=\mathrm{Re}[H(\omega)\mathrm{e}^{\mathrm{j}\omega t}]$. 由于, $H(\omega)=|H(\omega)|\,\mathrm{e}^{\mathrm{j}\theta(\omega)}$, 最终输出为, $y(t)=\mathrm{Re}\left[H(\omega)\mathrm{e}^{\mathrm{j}\omega t}\right]=\mathrm{Re}\left[|H(\omega)|\,\mathrm{e}^{\mathrm{j}\theta(\omega)}\mathrm{e}^{\mathrm{j}\omega t}\right]=|H(\omega)|\cos(\omega t+\theta(\omega))$. 概括以上结论, 可以写为

$$x(t)=\cos\omega t\to y(t)=|H(\omega)|\cos(\omega t+\theta(\omega)). \tag{2.1-17}$$

这是当输入为正弦信号时的重要结论. 举一个 RC 电路的例子, 利用式 (2.1-16) 可得

$$x(t)=\cos\omega t\to y(t)=\frac{1}{\sqrt{1+(\omega RC)^2}}\cos\left[\omega t-\arctan(\omega RC)\right].$$

现在对于线性时不变系统, 当输入为单个频率时, 我们能够求得其输出, 那么, 当输入为多个频率组成的非周期信号 (如脉冲或者一段音频信号) 时, 如何求其输出? 对这个问题的简短回答就是利用傅里叶分析, 这将在下节中进行讨论.

2.2 傅里叶分析

2.2.1 傅里叶级数

物理系统中出现的任何时间周期函数都可以展开为傅里叶级数 (*Fourier series*). 一个周期为 T_0 的周期信号 $x(t)$, 对所有的 t 都满足 $x(t)=x(t+T_0)$. 其中最小的 T_0 被称为基本周期 (*fundamental period*), 其基频 (*fundamental frequency*) 为 $f_0=\frac{\omega_0}{2\pi}=1/T_0$, 而 ω_0 为基本角频率 (*fundamental radian frequency*). 当 T_0 的单位为秒 (s) 时, f_0 的单位为赫兹 (Hz), ω_0 的单位为弧度/秒 (rad/s). 任何不具有周期性的函数都被称为非周期 (*aperiodic*) 函数. 举一个最简单的周期函数的例子, 如, $\cos(\omega_0 t)$, 因为有 $\cos[\omega_0(t+T_0)]=\cos(\omega_0 t+\omega_0 T_0)=\cos(\omega_0 t+2\pi)=\cos(\omega_0 t)$. 其他周期函数的例子如图 2.2-1 所示. 图 2.2-1a) 为一脉冲序列 (*pulse train*), 这是一个有意思的时间信号, 其光学上的模拟可对应于一个衍射光栅 (*diffraction grating*), 这将在后续章节中看到. 图 2.2-1b) 为一锯齿 (*sawtooth*) 函数.

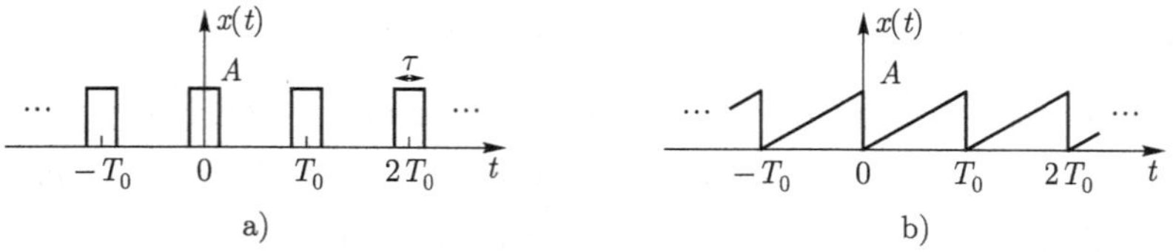

图 2.2-1 周期为 T_0 的周期函数示例: a) 幅值为 A 且脉冲宽度为 τ 的脉冲序列; b) 幅值为 A 的锯齿函数

任一周期为 T_0 的周期信号可以表示为

$$x(t)=\sum_{n=-\infty}^{\infty}X_n\mathrm{e}^{\mathrm{j}n\omega_0 t}, \tag{2.2-1}$$

其中, X_n 为傅里叶系数 (*Fourier coefficient*), ω_0 为 $x(t)$ 的基本角频率或者是一次谐波 (*first harmonic*), 频率 $n\omega_0$ 被称为 n 次谐波 (n*th harmonic*), 其中 n 取从 $-\infty$ 到 ∞ 的整数值. 式 (2.2-1) 的和被称为 $x(t)$ 的傅里叶级数 (*Fourier series*). 对于给定的 $x(t)$, 我们的目标是求 X_n. 为了推导出 X_n,

$y(t) = \frac{1}{2}H(\omega)\mathrm{e}^{\mathrm{j}\omega t} + \frac{1}{2}H(-\omega)\mathrm{e}^{-\mathrm{j}\omega t} = \frac{1}{2}H(\omega)\mathrm{e}^{\mathrm{j}\omega t} + \frac{1}{2}H^*(\omega)\mathrm{e}^{-\mathrm{j}\omega t} = \mathrm{Re}[H(\omega)\mathrm{e}^{\mathrm{j}\omega t}]$. Since $H(\omega) = |H(\omega)|\mathrm{e}^{\mathrm{j}\theta(\omega)}$, we write the final output $y(t) = \mathrm{Re}[H(\omega)\mathrm{e}^{\mathrm{j}\omega t}] = \mathrm{Re}[|H(\omega)|\mathrm{e}^{\mathrm{j}\theta(\omega)}\mathrm{e}^{\mathrm{j}\omega t}] = |H(\omega)|\cos(\omega t + \theta(\omega))$. Summarizing the result, we write

$$x(t) = \cos\omega t \to y(t) = |H(\omega)|\cos(\omega t + \theta(\omega)). \tag{2.1-17}$$

This is an important result for sinusoidal inputs. As an example for the RC circuit, using Eq. (2.1-16), we have

$$x(t) = \cos\omega t \to y(t) = \frac{1}{\sqrt{1+(\omega RC)^2}}\cos\left[\omega t - \arctan(\omega RC)\right].$$

While we have managed to find the output of linear and time-invariant systems for a single frequency input, how do we find the output for aperiodic signals, such as a pulse or a segment of audio signal, which as it turns out consists of many frequencies? A short answer to the question is through the use of Fourier analysis, to be discussed in the next section.

2.2 Fourier Analysis

2.2.1 Fourier Series

Any periodic function in time that appears in physical systems can be expanded into a *Fourier series.* A periodic signal $x(t)$ with period T_0 satisfies $x(t) = x(t + T_0)$ for all t. The smallest such T_0 is called the *fundamental period.* Its *fundamental frequency* is $f_0 = \frac{\omega_0}{2\pi} = 1/T_0$ and ω_0 is the *fundamental radian frequency.* For the units of T_0 in seconds (s), the units of f_0 are hertz (Hz) and of ω_0 are radians per second (rad/s). Any function that is not periodic is called *aperiodic.* One of the simplest periodic functions is the function, for example, $\cos(\omega_0 t)$ as we can see that $\cos[\omega_0(t + T_0)] = \cos(\omega_0 t + \omega_0 T_0) = \cos(\omega_0 t + 2\pi) = \cos\omega_0 t$. Other examples of periodic functions are shown in Figure 2.2-1. Figure 2.2-1a is a *pulse train,* which is an interesting time signal. Its optical analog is a *diffraction grating,* as we will see it in a subsequent chapter. Figure 2.2-1b shows a *sawtooth* function.

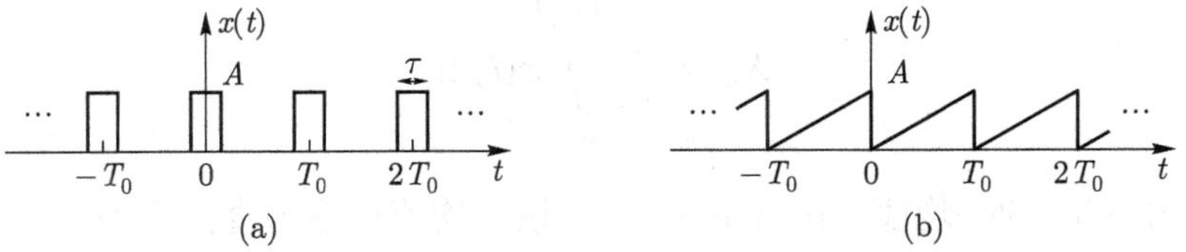

Figure 2.2-1 Examples of periodic functions with period T_0. (a) Pulse train with amplitude A and with pulse width τ and (b) sawtooth with amplitude A

Any periodic signal with T_0 can be represented by

$$x(t) = \sum_{n=-\infty}^{\infty} X_n \mathrm{e}^{\mathrm{j}n\omega_0 t}, \tag{2.2-1}$$

where X_n are called the *Fourier coefficients,* ω_0 is the fundamental radian frequency or the *first harmonic* of $x(t)$, and the frequency $n\omega_0$ is called the nth *harmonic* with n taking on all integer values from $-\infty$ to ∞. The summation in Eq. (2.2-1) is called the *Fourier series* of $x(t)$. For a given $x(t)$, our objective is to find X_n. To derive X_n, we multiply both sides of

在式 (2.2-1) 两端乘以 $\mathrm{e}^{-\mathrm{j}m\omega_0 t}$, 这里, m 为一个整数, 接着, 在一个周期 $T_0 = 2\pi/\omega_0$ 内进行积分. 因此, 有

$$\int_{t_0}^{t_0+T_0} x(t)\mathrm{e}^{-\mathrm{j}m\omega_0 t}\mathrm{d}t = \sum_{n=-\infty}^{\infty} X_n \int_{t_0}^{t_0+T_0} \mathrm{e}^{\mathrm{j}(n-m)\omega_0 t}\mathrm{d}t, \tag{2.2-2}$$

对于任意的实数 t_0. 为了计算等式右边, 可以利用指数的正交性 (*orthogonality of exponents*), 如

$$\int_{t_0}^{t_0+T_0} \mathrm{e}^{\mathrm{j}(n-m)\omega_0 t}\mathrm{d}t = \begin{cases} 0, & m \neq n \\ T_0, & m = n. \end{cases} \tag{2.2-3}$$

因此, 在无限级数中, 只有一项存在, 那就是当 $n = m$ 时, 式 (2.2-2) 变为

$$\int_{t_0}^{t_0+T_0} x(t)\mathrm{e}^{-\mathrm{j}m\omega_0 t}\mathrm{d}t = X_m T_0,$$

或

$$X_m = \frac{1}{T_0}\int_{t_0}^{t_0+T_0} x(t)\mathrm{e}^{-\mathrm{j}m\omega_0 t}\mathrm{d}t.$$

总而言之, 周期为 T_0 的 $x(t)$, 其傅里叶级数被表示为

$$x(t) = \sum_{n=-\infty}^{\infty} X_n \mathrm{e}^{\mathrm{j}n\omega_0 t}, \tag{2.2-4a}$$

其中, $\omega_0 = 2\pi/T_0$, 且

$$X_n = \frac{1}{T_0}\int_{t_0}^{t_0+T_0} x(t)\mathrm{e}^{-\mathrm{j}n\omega_0 t}\mathrm{d}t. \tag{2.2-4b}$$

可知, 在式 (2.2-4b) 中, 当 $n = 0$ 时, 有

$$X_0 = \frac{1}{T_0}\int_{t_0}^{t_0+T_0} x(t)\mathrm{d}t.$$

其中, X_0 为信号 $x(t)$ 的平均值 (*average value*), 该平均值也称为直流值 [*direct current* (DC) *value*], 这个术语源于电路分析, 对于某些信号, 直流值很容易被检测到. 由于傅里叶系数 X_n 是 $\omega = n\omega_0$ 的函数, 且一般来说是复数, 可以写作 $X_n = |X_n|\mathrm{e}^{\mathrm{j}\theta_n(\omega)}$, 这里 $|X_n|$ 和 $\theta_n(\omega)$ 分别被称为信号 $x(t)$ 的强度谱 (*magnitude spectrum*) 和相位谱 (*phase spectrum*).

举例: 正弦函数的傅里叶级数

考虑 $x(t) = \sin(\omega_0 t)$, 求 $\sin(\omega_0 t)$ 的傅里叶系数, 并画出其强度谱和相位谱. 直接的方法就是, 简单地将 $x(t) = \sin(\omega_0 t)$ 代入式 (2.2-4b) 来求 X_n. 然而, 比较简单的方法是, 可通过

Eq. (2.2-1) by $\mathrm{e}^{-\mathrm{j}m\omega_0 t}$, where m is an integer, and then integrate over one period $T_0 = 2\pi/\omega_0$. We, therefore, have

$$\int_{t_0}^{t_0+T_0} x(t)\mathrm{e}^{-\mathrm{j}m\omega_0 t}\mathrm{d}t = \sum_{n=-\infty}^{\infty} X_n \int_{t_0}^{t_0+T_0} \mathrm{e}^{\mathrm{j}(n-m)\omega_0 t}\mathrm{d}t, \tag{2.2-2}$$

for any real number t_0. To evaluate the right side of the equation, we employ the *orthogonality of exponents* stated as follows:

$$\int_{t_0}^{t_0+T_0} \mathrm{e}^{\mathrm{j}(n-m)\omega_0 t}\mathrm{d}t = \begin{cases} 0, & m \neq n \\ T_0, & m = n. \end{cases} \tag{2.2-3}$$

Hence, out of the infinite series, only one term will survive and that is when $n = m$. Equation (2.2-2) becomes

$$\int_{t_0}^{t_0+T_0} x(t)\mathrm{e}^{-\mathrm{j}m\omega_0 t}\mathrm{d}t = X_m T_0,$$

or

$$X_m = \frac{1}{T_0}\int_{t_0}^{t_0+T_0} x(t)\mathrm{e}^{-\mathrm{j}m\omega_0 t}\mathrm{d}t.$$

To summarize, the Fourier series of $x(t)$ with a period T_0 is expressed as

$$x(t) = \sum_{n=-\infty}^{\infty} X_n \mathrm{e}^{\mathrm{j}n\omega_0 t}, \tag{2.2-4a}$$

where $\omega_0 = 2\pi/T_0$ and

$$X_n = \frac{1}{T_0}\int_{t_0}^{t_0+T_0} x(t)\mathrm{e}^{-\mathrm{j}n\omega_0 t}\mathrm{d}t. \tag{2.2-4b}$$

Note that, for $n = 0$ in Eq. (2.2-4b), we have

$$X_0 = \frac{1}{T_0}\int_{t_0}^{t_0+T_0} x(t)\mathrm{d}t.$$

X_0 is the *average value* of the signal $x(t)$. The average value is also called the *Direct Current (DC) value*, a term that is originated from circuit analysis. For some signals, the DC value can be found easily by inspection. Since the Fourier coefficients X_n are functions of $\omega = n\omega_0$ and are complex in general, we can write $X_n = |X_n|\mathrm{e}^{\mathrm{j}\theta_n(\omega)}$, where $|X_n|$ and $\theta_n(\omega)$ are called the *magnitude spectrum* and *phase spectrum* of the signal $x(t)$, respectively.

Example: Fourier Series of a Sine Function

Consider $x(t) = \sin(\omega_0 t)$. Let us find the Fourier coefficients of $\sin(\omega_0 t)$, and also plot its magnitude and phase spectra. In a straightforward way, we simply put $x(t) = \sin(\omega_0 t)$ into Eq. (2.2-4b) to evaluate X_n. However, a simpler approach is to recognize that we can obtain

等式 $\sin(\omega_0 t) = (\mathrm{e}^{\mathrm{j}\omega_0 t} - \mathrm{e}^{-\mathrm{j}\omega_0 t})/(2\mathrm{j})$, 来得到正弦函数的指数形式表达式, 故, 可写为

$$x(t) = \sin(\omega_0 t) = \frac{\mathrm{e}^{\mathrm{j}\omega_0 t} - \mathrm{e}^{-\mathrm{j}\omega_0 t}}{2\mathrm{j}} = \sum_{n=-\infty}^{\infty} X_n \mathrm{e}^{\mathrm{j}n\omega_0 t}.$$

可以看出, n 只能取 $+1$ 和 -1 两个值. 因此, 其傅里叶系数为 $X_1 = |X_1|\mathrm{e}^{\mathrm{j}\theta_1(\omega)} = \dfrac{1}{2\mathrm{j}} = \dfrac{1}{2}\mathrm{e}^{-\mathrm{j}90^\circ}$ 和 $X_{-1} = |X_{-1}|\mathrm{e}^{\mathrm{j}\theta_{-1}(\omega)} = \dfrac{-1}{2\mathrm{j}} = \dfrac{1}{2}\mathrm{e}^{\mathrm{j}90^\circ}$. 图 2.2-2a) 和图 2.2-2b) 分别表示了其强度谱和相位谱, 这些图被称为谱线 (*line spectra*), 因为这些垂直的线条表示了强度和相位. 两幅图中的角频率 ω_0 为正值时所表明的意义容易理解, 但如何解释角频率为负值的情况? 沿着负轴给 ω 定义纯粹是为了数学上的方便, 例如, 用两个复指数 $(\mathrm{e}^{\mathrm{j}\omega_0 t} - \mathrm{e}^{-\mathrm{j}\omega_0 t})/(2\mathrm{j})$ 表示 $\sin(\omega_0 t)$, 负值 ω_0 是没有物理上的意义的.

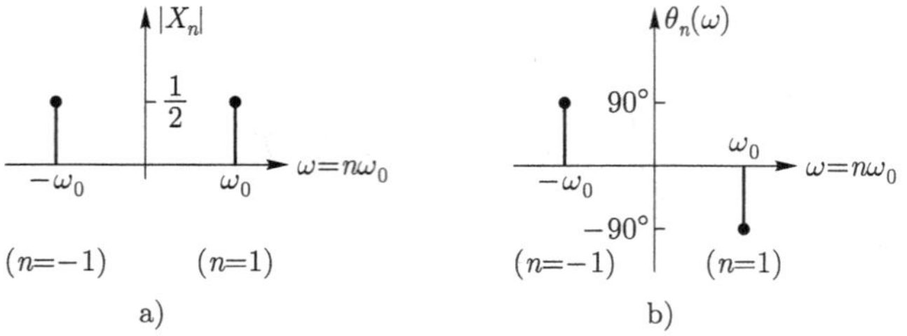

图 2.2-2　正弦信号的 a) 强度谱和 b) 相位谱

举例: 脉冲序列的傅里叶级数

下面来求图 2.2-1a) 中幅度为 A 且脉冲宽度为 τ 的脉冲序列的傅里叶级数. 根据式 (2.2-4b), 有

$$X_n = \frac{1}{T_0}\int_{t_0}^{t_0+T_0} x(t)\mathrm{e}^{-\mathrm{j}n\omega_0 t}\mathrm{d}t = \frac{1}{T_0}\int_{-T_0/2}^{T_0/2} x(t)\mathrm{e}^{-\mathrm{j}n\omega_0 t}\mathrm{d}t = \frac{1}{T_0}\int_{-\tau/2}^{\tau/2} A\mathrm{e}^{-\mathrm{j}n\omega_0 t}\mathrm{d}t$$
$$= \frac{A}{T_0}\frac{\mathrm{e}^{-\mathrm{j}n\omega_0 t}}{-\mathrm{j}n\omega_0}\bigg|_{-\tau/2}^{\tau/2} = \frac{A}{T_0(-\mathrm{j}n\omega_0)}\left(\mathrm{e}^{-\mathrm{j}n\omega_0\tau/2} - \mathrm{e}^{+\mathrm{j}n\omega_0\tau/2}\right).$$

现在, 利用 $2\mathrm{j}\sin\theta = \mathrm{e}^{\mathrm{j}\theta} - \mathrm{e}^{-\mathrm{j}\theta}$, 及 $\omega_0 T_0 = 2\pi$, 上述方程可以写为

$$X_n = \frac{A}{n\pi}\sin\left(\frac{n\omega_0\tau}{2}\right) = \frac{A}{n\pi}\sin\left(\frac{n\pi\tau}{T_0}\right). \tag{2.2-5}$$

可以看出, 脉冲序列的所有信息, 即 A, τ, 以及 T_0, 都包含在傅里叶系数 X_n 中, X_n 可以被表示为 *sinc* 函数 (*sinc function*) 的形式, 该函数在信号处理及傅里叶分析理论中都有着重要的作用.

这里介绍一下 *sinc* 函数 (其发音为 [sink]): $sinc(t) = \dfrac{\sin(\pi t)}{\pi t}$. 图 2.2-3 所示为用 MATLAB 程序绘制的 *sinc* 函数. 可以看出, 在 $t = 0$ 时, $sinc(t)$ 函数的值为 1, 在 $\sin(\pi t)$ 为零的点 (即 $t = \pm 1, \pm 2, \cdots$) 处, $sinc(t)$ 函数的值为零.

an expression for the sine function in terms of exponential functions through the identity $\sin(\omega_0 t) = (e^{j\omega_0 t} - e^{-j\omega_0 t})/(2j)$. So we write

$$x(t) = \sin(\omega_0 t) = \frac{e^{j\omega_0 t} - e^{-j\omega_0 t}}{2j} = \sum_{n=-\infty}^{\infty} X_n e^{jn\omega_0 t}.$$

We can now identify that n only exists for $+1$ and -1. Therefore, the Fourier coefficients are $X_1 = |X_1|e^{j\theta_1(\omega)} = \frac{1}{2j} = \frac{1}{2}e^{-j90^\circ}$ and $X_{-1} = |X_{-1}|e^{j\theta_{-1}(\omega)} = \frac{-1}{2j} = \frac{1}{2}e^{j90^\circ}$. Figure 2.2-2a and b shows the plots for the magnitude spectrum and phase spectrum, respectively. These plots are called *line spectra* because vertical lines indicate the magnitudes and phases. It is easy to understand what a positive value of the angular frequency ω_0 means in both of the plots. How do we interpret a negative frequency? It does not have a physical meaning, as defining ω along the negative axis is purely a mathematical convenience. For example, we represent $\sin(\omega_0 t)$ by two complex exponents $(e^{j\omega_0 t} - e^{-j\omega_0 t})/(2j)$.

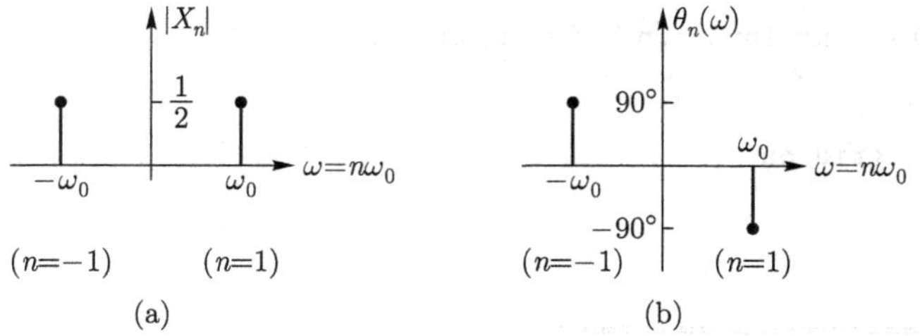

Figure 2.2-2 (a) Magnitude spectrum and (b) phase spectrum of a sine signal

Example: Fourier Series of a Pulse Train

Let us find the Fourier series of a pulse train of an amplitude A and a pulse width of τ shown in Figure 2.2-1a. According to Eq. (2.2-4b), we have

$$X_n = \frac{1}{T_0}\int_{t_0}^{t_0+T_0} x(t)e^{-jn\omega_0 t}dt = \frac{1}{T_0}\int_{-T_0/2}^{T_0/2} x(t)e^{-jn\omega_0 t}dt = \frac{1}{T_0}\int_{-\tau/2}^{\tau/2} Ae^{-jn\omega_0 t}dt$$

$$= \frac{A}{T_0}\frac{e^{-jn\omega_0 t}}{-jn\omega_0}\Bigg|_{-\tau/2}^{\tau/2} = \frac{A}{T_0(-jn\omega_0)}\left(e^{-jn\omega_0\tau/2} - e^{+jn\omega_0\tau/2}\right).$$

Now using $2j\sin\theta = e^{j\theta} - e^{-j\theta}$, and recognizing $\omega_0 T_0 = 2\pi$, the above equation is manipulated to become

$$X_n = \frac{A}{n\pi}\sin\left(\frac{n\omega_0\tau}{2}\right) = \frac{A}{n\pi}\sin\left(\frac{n\pi\tau}{T_0}\right). \tag{2.2-5}$$

Note that all the information of the pulse train, that is, A, τ, and T_0, contain in the Fourier coefficients X_n. X_n can be written in terms of the *sinc function*, which plays an important role in signal processing as well as in the theory of Fourier analysis.

Let us introduce the *sinc* function [pronounced "sink"]: $sinc(t) = \frac{\sin(\pi t)}{\pi t}$. The plot of the sinc function is shown in Figure 2.2-3 along with its MATLAB code. Note that the $sinc(t)$ function has a value of unity at $t = 0$. $sinc(t)$ has zeros at the points at which $\sin(\pi t)$ is zero, which is, at $t = \pm 1, \pm 2, \cdots$.

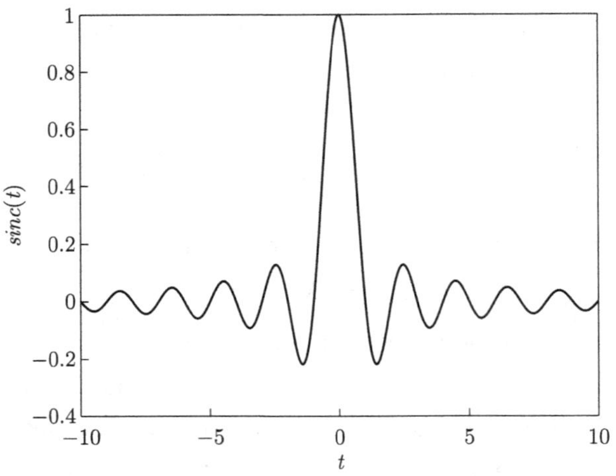

图 2.2-3　*sinc*(*t*) 函数的图示

```
========================================
  % Plotting of the sinc function, sinc_function.m
  t = -10:0.01:10;
  plot(t,sinc(t)), grid on
  xlabel('t')
  ylabel('sinc(t)')
========================================
```

根据 *sinc* 函数的定义, X_n 可改写为

$$X_n = \frac{A}{n\pi}\sin\left(\frac{n\omega_0\tau}{2}\right) = \frac{A\tau}{T_0}sinc\left(\frac{n\omega_0\tau}{2\pi}\right).$$

在图 2.2-4 中, 绘制出 $\tau = 2\pi$, $T_0 = 6\pi$, 且, $A = 3$ 时的 X_n. 可以看出, 由于 X_n 为实数, 可以用一个单图来表示脉冲序列的整个频谱, 这样的图被称为幅度谱 (*amplitude spectrum*).

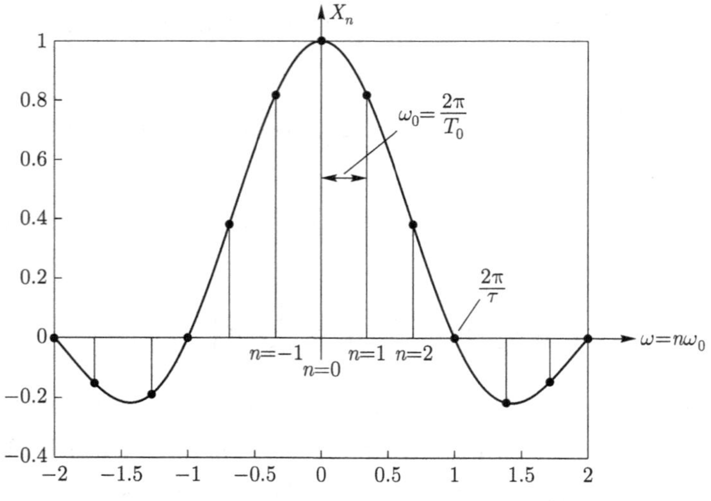

图 2.2-4　$\tau = 2\pi$, $T_0 = 6\pi$ 且 $A = 3$ 时脉冲序列的幅度谱

2.2.2　傅里叶变换

在上一节中, 可以看出, 对于一个给定的周期信号, 可以用傅里叶级数来表示, 且其傅里叶系数用离散 (*discrete*) 的谱线来表示. 本节中, 我们来研究非周期信号, 并介绍傅里叶变换

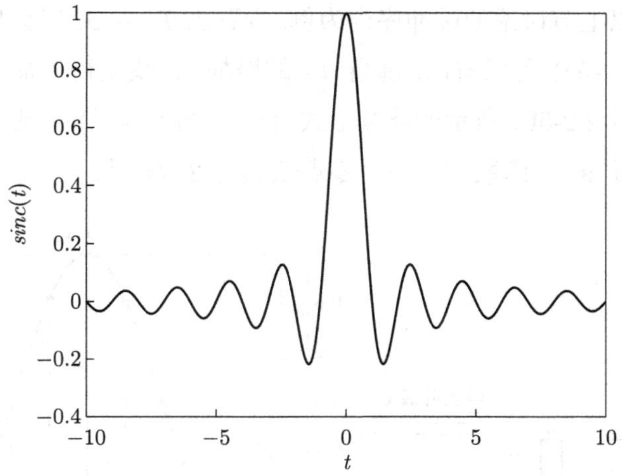

Figure 2.2-3 Plot of $sinc(t)$ function

```
==========================================
% Plotting of the sinc function, sinc_function.m
t = -10:0.01:10;
plot(t,sinc(t)), grid on
xlabel('t')
ylabel('sinc(t)')
==========================================
```

With the definition of the *sinc* function, we can rewrite X_n as

$$X_n = \frac{A}{n\pi}\sin\left(\frac{n\omega_0\tau}{2}\right) = \frac{A\tau}{T_0} sinc\left(\frac{n\omega_0\tau}{2\pi}\right).$$

In Figure 2.2-4, we plot X_n for $\tau = 2\pi$, $T_0 = 6\pi$, and $A = 3$. Note that since X_n is real, we can present the whole spectrum of the pulse train with a single graph, and such a graph is called the *amplitude spectrum.*

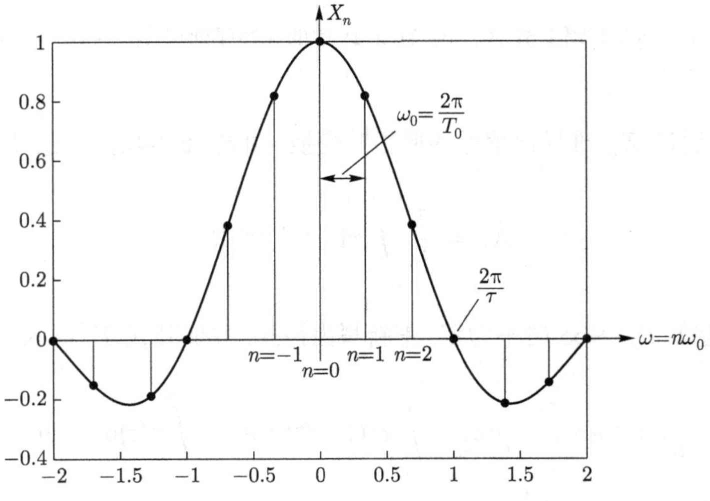

Figure 2.2-4 Amplitude spectrum of a pulse train for $\tau = 2\pi$, $T_0 = 6\pi$, and $A = 3$

2.2.2 Fourier Transform

In the last section, we have seen that we can represent a given periodic signal by the Fourier series and the Fourier coefficients are displayed as discrete line spectra. In this section, we investigate aperiodic signals and introduce the *Fourier transform.* Let us take the pulse train

(*Fourier transform*). 以上节讨论的脉冲序列为例, 看看当 $T_0 \to \infty$ 时会出现什么情况. 图 2.2-5 给出了说明, 从图 2.2-5a) 可以看出, 随着 T_0 的增加, 谱线变得更加密集. 在 $T_0 \to \infty$ 时, $\omega = n\omega_0$ 变成一个如图 2.2-5b) 所示的连续形式, 因此, 图 2.2-5a) 中的离散 *sinc* 函数就变成了 ω 的连续函数, 然而, *sinc* 函数的第一个零点依然在 $2\pi/\tau$ 处.

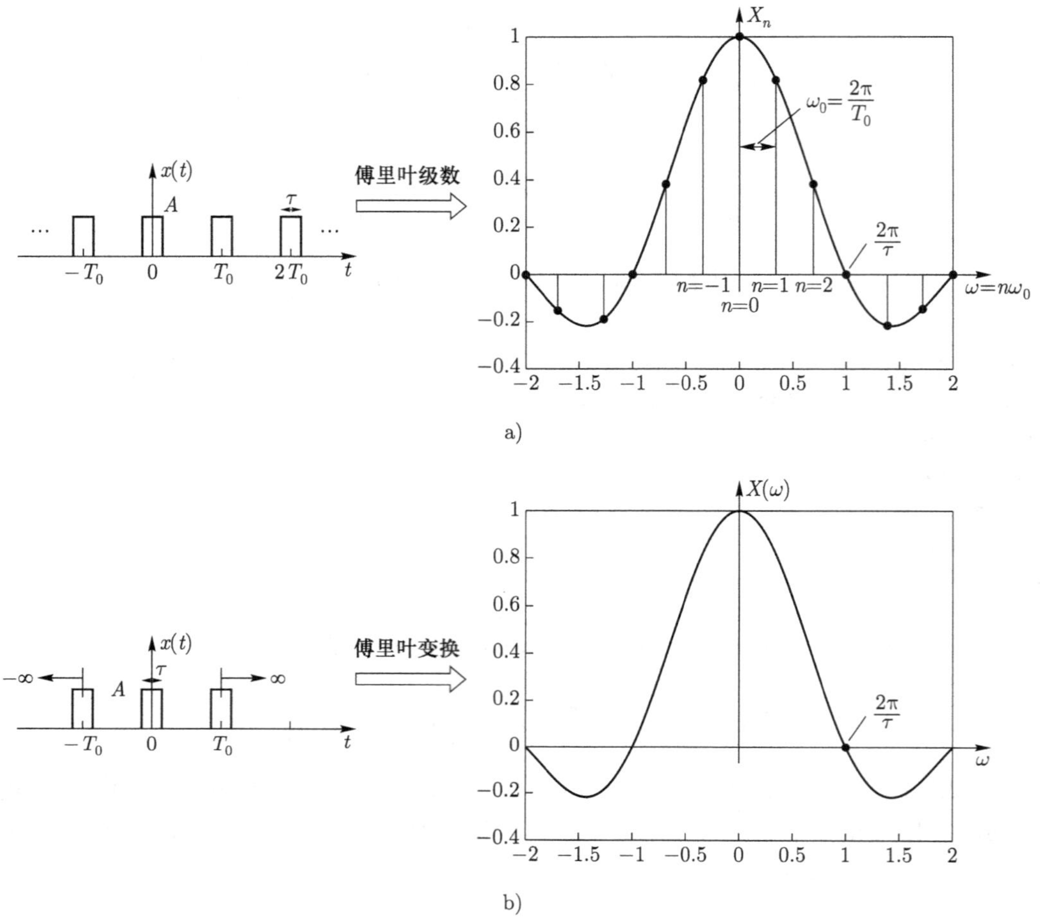

图 2.2-5 a) 周期脉冲序列及其幅度谱 X_n; b) 当图 a) 中的脉冲序列在 $T_0 \to \infty$ 时, 非周期脉冲及其频谱 $X(\omega)$

数学上, 可以通过 X_n 推导出信号的傅里叶变换, 由式 (2.2-4b) 得出, 并重新写为

$$X_n = \frac{1}{T_0}\int_{t_0}^{t_0+T_0} x(t)\mathrm{e}^{-\mathrm{j}n\omega_0 t}\mathrm{d}t.$$

为了让等式的右边在 $T_0 \to \infty$ 时不为零, 重新写出该式, 并取其在 $T_0 \to \infty$ 时的极限如下

$$\lim_{T_0\to\infty} T_0 X_n = \lim_{T_0\to\infty}\int_{-T_0/2}^{T_0/2} x(t)\mathrm{e}^{-\mathrm{j}n\omega_0 t}\mathrm{d}t = \int_{-\infty}^{\infty} x(t)\mathrm{e}^{-\mathrm{j}\omega t}\mathrm{d}t, \tag{2.2-6}$$

在上式的最后一步中, 当 $T_0 \to \infty$ 时, 把 $n\omega_0$ 替换为 ω, 则上述积分就是所谓的信号 $x(t)$ 的正傅里叶变换 (*forward Fourier transform*) $X(\omega)$, 写为

$$\mathcal{F}\{x(t)\} = \int_{-\infty}^{\infty} x(t)\mathrm{e}^{-\mathrm{j}\omega t}\mathrm{d}t = X(\omega), \tag{2.2-7a}$$

discussed from the last section as an example and see what happens when we let $T_0 \to \infty$. Figure 2.2-5 illustrates the situation. From Figure 2.2-5a, we can see that as T_0 increases, the line spectrum becomes denser. At $T_0 \to \infty$, $\omega = n\omega_0$ becomes a continuum as it is illustrated in Figure 2.2-5b. Hence, the discrete *sinc* function in Figure 2.2-5a has become a continuous function in ω. Note that, however, the first zero of the *sinc* function remains at $2\pi/\tau$.

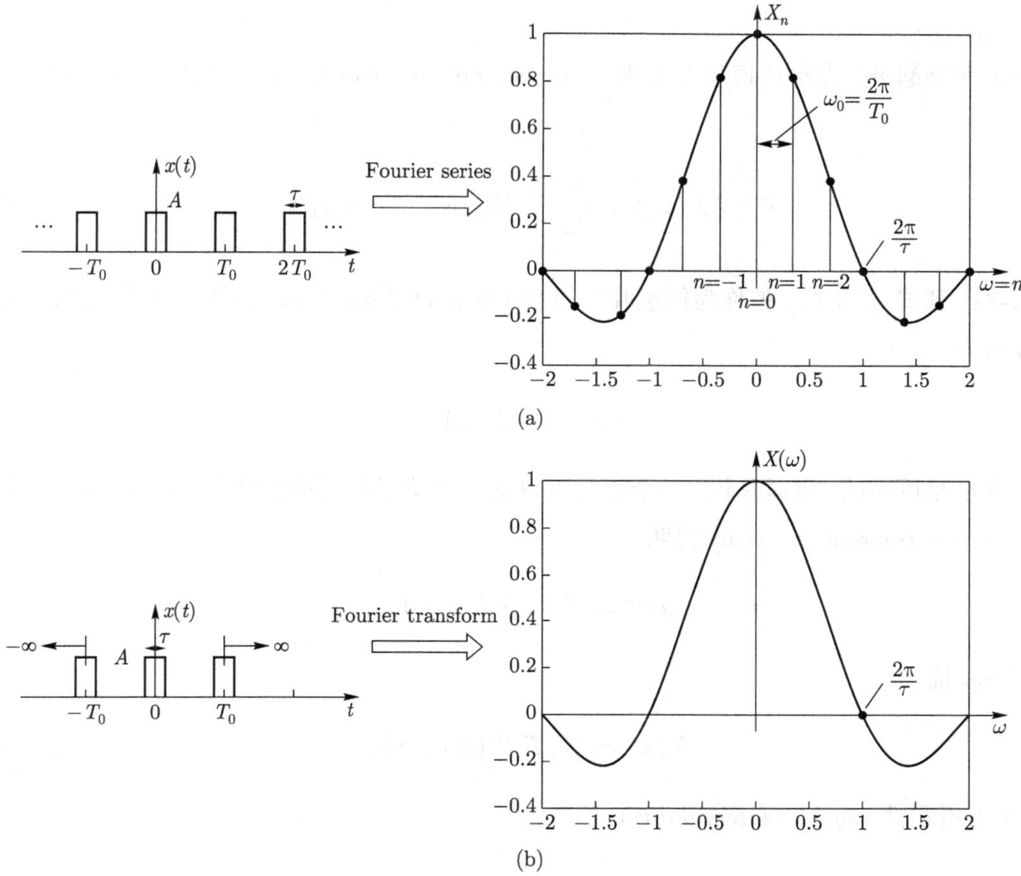

Figure 2.2-5 (a) Periodic pulse train and its amplitude spectrum, X_n and (b) aperiodic pulse and its spectrum, $X(\omega)$ when $T_0 \to \infty$ in the pulse train in (a)

Mathematically, we can derive the Fourier transform of a signal from X_n. From Eq. (2.2-4b) and re-stating it below:

$$X_n = \frac{1}{T_0} \int_{t_0}^{t_0+T_0} x(t) \mathrm{e}^{-\mathrm{j}n\omega_0 t} \mathrm{d}t.$$

In order not to have a zero when $T_0 \to \infty$ on the right-hand side of the equation, we rewrite the equation and take the limit $T_0 \to \infty$ as follows:

$$\lim_{T_0 \to \infty} T_0 X_n = \lim_{T_0 \to \infty} \int_{-T_0/2}^{T_0/2} x(t) \mathrm{e}^{-\mathrm{j}n\omega_0 t} \mathrm{d}t = \int_{-\infty}^{\infty} x(t) \mathrm{e}^{-\mathrm{j}\omega t} \mathrm{d}t, \tag{2.2-6}$$

where we have replaced $n\omega_0$ by ω as $T_0 \to \infty$ in the last step of Eq. (2.2-6). The above integral is known as the *forward Fourier transform* $X(\omega)$ of the signal $x(t)$ and is written as

$$\mathcal{F}\{x(t)\} = \int_{-\infty}^{\infty} x(t) \mathrm{e}^{-\mathrm{j}\omega t} \mathrm{d}t = X(\omega), \tag{2.2-7a}$$

其中, $\mathcal{F}\{x(t)\}$ 是 $x(t)$ 的傅里叶变换的操作算子符号, 在信号处理以及电路分析中, $X(\omega)$ 被称为信号 $x(t)$ 的频谱 (*spectrum*), 由于 $X(\omega)$ 一般为复数, 可以写为

$$X(\omega)=|X(\omega)|\,\mathrm{e}^{\mathrm{j}\theta(\omega)},$$

其中, $|X(\omega)|$ 是信号 $x(t)$ 的强度谱 (*magnitude spectrum*), $\theta(\omega)$ 是信号 $x(t)$ 的相位谱 (*phase spectrum*).

为了完整起见, 表示逆傅里叶变换 (*inverse Fourier transform*) 为 $\mathcal{F}^{-1}\{X(\omega)\}$, 并定义如下

$$\mathcal{F}^{-1}\{X(\omega)\}=\frac{1}{2\pi}\int_{-\infty}^{\infty}X(\omega)\mathrm{e}^{\mathrm{j}\omega t}\mathrm{d}\omega=x(t). \tag{2.2-7b}$$

式 (2.2-7a) 和式 (2.2-7b) 被称为傅里叶变换对 (*Fourier transform pair*), 它们的关系通常用符号表示为

$$x(t)\Longleftrightarrow X(\omega).$$

傅里叶变换是可逆的, 也就是说, 一个函数可以通过其傅里叶变换来恢复. 的确, 傅里叶积分定理 (*Fourier integral theorem*) 说明

$$x(t)=\mathcal{F}^{-1}\{\mathcal{F}\{x(t)\}\}. \tag{2.2-8a}$$

也很容易验证

$$X(\omega)=\mathcal{F}\left\{\mathcal{F}^{-1}\{X(\omega)\}\right\}. \tag{2.2-8b}$$

注意, 平方可积性 (*square integrability*), 即

$$\int_{-\infty}^{\infty}|x(t)|^2\mathrm{d}t<\infty \tag{2.2-9}$$

是傅里叶变换存在的一个充分条件 (*sufficient condition*). 换句话说, 如果满足式 (2.2-9), 则 $x(t)$ 是有傅里叶变换的, 但是, 平方可积性并不是函数具有傅里叶变换的一个必要条件 (*necessary condition*). 一些函数, 如阶跃函数或正弦函数, 并不满足式 (2.2-9), 但仍具有傅里叶变换.

举例: 脉冲的傅里叶变换

当 $x(t)=rect(t)$, 通过式 (2.2-7a) 的积分形式来简单求解, 可以写出

$$\begin{aligned}\mathcal{F}\{rect(t)\}&=\int_{-\infty}^{\infty}rect(t)\mathrm{e}^{-\mathrm{j}\omega t}\mathrm{d}t\\&=\int_{-1/2}^{1/2}1\mathrm{e}^{-\mathrm{j}\omega t}\mathrm{d}t=\left.\frac{\mathrm{e}^{-\mathrm{j}\omega t}}{-\mathrm{j}\omega}\right|_{-1/2}^{1/2}=\frac{\mathrm{e}^{-\mathrm{j}\omega/2}-\mathrm{e}^{\mathrm{j}\omega/2}}{-\mathrm{j}\omega}\\&=2\frac{\mathrm{e}^{\mathrm{j}\omega/2}-\mathrm{e}^{-\mathrm{j}\omega/2}}{2\mathrm{j}\omega}=\frac{\sin(\omega/2)}{\omega/2}=sinc\left(\frac{\omega}{2\pi}\right).\end{aligned}$$

where $\mathcal{F}\{x(t)\}$ is an operator notation for the Fourier transform of $x(t)$. $X(\omega)$ is called the *spectrum* of the signal $x(t)$ in signal processing as well as in circuit analysis. Since $X(\omega)$ is complex in general, we can write

$$X(\omega) = |X(\omega)|\mathrm{e}^{\mathrm{j}\theta(\omega)},$$

where $|X(\omega)|$ is the *magnitude spectrum* and $\theta(\omega)$ is the *phase spectrum* of the signal $x(t)$.

For completeness, we state the *inverse Fourier transform* as $\mathcal{F}^{-1}\{X(\omega)\}$ and defined as

$$\mathcal{F}^{-1}\{X(\omega)\} = \frac{1}{2\pi}\int_{-\infty}^{\infty} X(\omega)\mathrm{e}^{\mathrm{j}\omega t}\mathrm{d}\omega = x(t). \tag{2.2-7b}$$

Equations (2.2-7a) and (2.2-7b) are referred to as the *Fourier transform pair*, and their relationship is often represented symbolically as

$$x(t) \Longleftrightarrow X(\omega).$$

The Fourier transform is *invertible*, that is, it is possible to recover a function from its Fourier transform. Indeed, the *Fourier integral theorem* states that

$$x(t) = \mathcal{F}^{-1}\left\{\mathcal{F}\left\{x(t)\right\}\right\}. \tag{2.2-8a}$$

We can also check easily that

$$X(\omega) = \mathcal{F}\left\{\mathcal{F}^{-1}\left\{X(\omega)\right\}\right\}. \tag{2.2-8b}$$

Note that square integrability, that is,

$$\int_{-\infty}^{\infty} |x(t)|^2\mathrm{d}t < \infty \tag{2.2-9}$$

is a *sufficient condition* for the existence of the Fourier transform. In other words, $x(t)$ is Fourier transformable if Eq. (2.2-9) is satisfied. Square integrability, however, is not a *necessary condition* for the function to be Fourier transformable. Some functions such as a step function or a sinusoidal function do not satisfy Eq. (2.2-9) and yet have Fourier transforms.

Example: Fourier Transform of a Pulse

We simply evaluate the integral given by Eq. (2.2-7a) by recognizing that $x(t) = rect(t)$. Therefore, we write

$$\begin{aligned}
\mathcal{F}\{rect(t)\} &= \int_{-\infty}^{\infty} rect(t)\mathrm{e}^{-\mathrm{j}\omega t}\mathrm{d}t \\
&= \int_{-1/2}^{1/2} 1\mathrm{e}^{-\mathrm{j}\omega t}\mathrm{d}t = \left.\frac{\mathrm{e}^{-\mathrm{j}\omega t}}{-\mathrm{j}\omega}\right|_{-1/2}^{1/2} = \frac{\mathrm{e}^{-\mathrm{j}\omega/2} - \mathrm{e}^{\mathrm{j}\omega/2}}{-\mathrm{j}\omega} \\
&= 2\frac{\mathrm{e}^{\mathrm{j}\omega/2} - \mathrm{e}^{-\mathrm{j}\omega/2}}{2\mathrm{j}\omega} = \frac{\sin(\omega/2)}{\omega/2} = sinc\left(\frac{\omega}{2\pi}\right).
\end{aligned}$$

举例: 用傅里叶变换来分析 RC 电路

之前已经求得 RC 电路的频率响应函数 $H(\omega)$, 并已证明 $H(\omega)$ 的强度对研究电路的滤波特性有帮助. 为方便起见, 这里从电路的系统方程开始, 如下式

$$RC\frac{\mathrm{d}y(t)}{\mathrm{d}t}+y(t)=x(t).$$

在频域 (*frequency domain*) 用傅里叶变换对其分析, 而不是在时域研究其系统的特性. 对上式两边做傅里叶变换, 可得

$$RC\mathcal{F}\left\{\frac{\mathrm{d}y(t)}{\mathrm{d}t}\right\}+\mathcal{F}\{y(t)\}=\mathcal{F}\{x(t)\}. \tag{2.2-10}$$

让 $\mathcal{F}\{x(t)\}=X(\omega)$, 且, $\mathcal{F}\{y(t)\}=Y(\omega)$, 需要求 $\mathcal{F}\left\{\frac{\mathrm{d}y(t)}{\mathrm{d}t}\right\}$. 利用式 (2.2-7b), 写出

$$y(t)=\mathcal{F}^{-1}\{Y(\omega)\}=\frac{1}{2\pi}\int_{-\infty}^{\infty}Y(\omega)\mathrm{e}^{\mathrm{j}\omega t}\mathrm{d}\omega.$$

对上式求导后, 得

$$\frac{\mathrm{d}y(t)}{\mathrm{d}t}=\frac{1}{2\pi}\int_{-\infty}^{\infty}\mathrm{j}\omega Y(\omega)\mathrm{e}^{\mathrm{j}\omega t}\mathrm{d}\omega=\mathcal{F}^{-1}\{\mathrm{j}\omega Y(\omega)\}.$$

现在, 对上式两边做傅里叶变换, 得

$$\mathcal{F}\left\{\frac{\mathrm{d}y(t)}{\mathrm{d}t}\right\}=\mathcal{F}\left\{\mathcal{F}^{-1}\{\mathrm{j}\omega Y(\omega)\}\right\}=\mathrm{j}\omega Y(\omega). \tag{2.2-11}$$

利用这个结果, 式 (2.2-10) 变为

$$RC\mathrm{j}\omega Y(\omega)+Y(\omega)=X(\omega),$$

从而有

$$\frac{Y(\omega)}{X(\omega)}=\frac{1}{1+\mathrm{j}\omega RC}=H(\omega). \tag{2.2-12}$$

此结果与式 (2.1-16a) 相同. 然而, 由以上结果可知, 频率响应函数 $H(\omega)$ 就是输出谱 $Y(\omega)$ 与输入谱 $X(\omega)$ 的简单比值. 因此, 可以看到, 傅里叶变换在确定系统的频率响应 (或传递函数) 方面是非常有用的. 在接下来的章节中, 我们还将利用傅里叶变换来介绍光学系统的相干传递函数 (*coherent transfer function*) 和光学传递函数 (*optical transfer function*). 作为这个例子的最后一点, 参考式 (2.1-15), 重新写出

$$H(\omega)=\int_{-\infty}^{\infty}h(t')\mathrm{e}^{-\mathrm{j}\omega t'}\mathrm{d}t',$$

从傅里叶变换的定义可以看出, 有 $H(\omega)=\mathcal{F}\{h(t)\}$. 也就是说, 系统的频率响应函数是脉冲响应 $h(t)$ 的傅里叶变换.

Example: Use of Fourier Transform for RC-Circuit Analysis

We have found the frequency response function $H(\omega)$ of an RC circuit earlier and shown that the magnitude of $H(\omega)$ allows us to investigate the filtering characteristic of the circuit. Let us start with the system equation of the circuit, stated below for convenience:

$$RC\frac{\mathrm{d}y(t)}{\mathrm{d}t} + y(t) = x(t).$$

Instead of investigating the behavior of the system in the time domain, we could analyze it in the *frequency domain* with the Fourier transform. Taking the Fourier transform of both sides of the above equation, we obtain

$$RC\mathcal{F}\left\{\frac{\mathrm{d}y(t)}{\mathrm{d}t}\right\} + \mathcal{F}\{y(t)\} = \mathcal{F}\{x(t)\}. \tag{2.2-10}$$

Let $\mathcal{F}\{x(t)\} = X(\omega)$ and $\mathcal{F}\{y(t)\} = Y(\omega)$. We now need to find $\mathcal{F}\left\{\frac{\mathrm{d}y(t)}{\mathrm{d}t}\right\}$. Using Eq. (2.2-7b), we write

$$y(t) = \mathcal{F}^{-1}\{Y(\omega)\} = \frac{1}{2\pi}\int_{-\infty}^{\infty} Y(\omega)\mathrm{e}^{\mathrm{j}\omega t}\mathrm{d}\omega.$$

After differentiating the above equation, we get

$$\frac{\mathrm{d}y(t)}{\mathrm{d}t} = \frac{1}{2\pi}\int_{-\infty}^{\infty} \mathrm{j}\omega Y(\omega)\mathrm{e}^{\mathrm{j}\omega t}\mathrm{d}\omega = \mathcal{F}^{-1}\{\mathrm{j}\omega Y(\omega)\}.$$

Now, taking the Fourier transform on both sides of the above equation, we have

$$\mathcal{F}\left\{\frac{\mathrm{d}y(t)}{\mathrm{d}t}\right\} = \mathcal{F}\left\{\mathcal{F}^{-1}\{\mathrm{j}\omega Y(\omega)\}\right\} = \mathrm{j}\omega Y(\omega). \tag{2.2-11}$$

Using this result, Eq. (2.2-10) becomes

$$RC\mathrm{j}\omega Y(\omega) + Y(\omega) = X(\omega),$$

which gives

$$\frac{Y(\omega)}{X(\omega)} = \frac{1}{1+\mathrm{j}\omega RC} = H(\omega). \tag{2.2-12}$$

This result is identical to that shown in Eq. (2.1-16a). However, from the above result, we see that the frequency response function $H(\omega)$ is simply the ratio of the output spectrum $Y(\omega)$ to the input spectrum $X(\omega)$. Hence, we can see that the Fourier transform is very useful in determining the frequency response (or transfer function) of the system. In a subsequent chapter, we will also take advantage of the Fourier transform to introduce the *coherent transfer function* and *optical transfer function* for optical systems. As a final note to this example, and with reference to Eq. (2.1-15) re-stated below

$$H(\omega) = \int_{-\infty}^{\infty} h(t')\mathrm{e}^{-\mathrm{j}\omega t'}\mathrm{d}t',$$

we see that from the definition of the Fourier transform, we have $H(\omega) = \mathcal{F}\{h(t)\}$. In words, it says the frequency response function of the system is the Fourier transform of the impulse response $h(t)$.

举例: 非周期信号的响应

前面讨论过线性时不变系统中输入为正弦信号的响应情况. 现在来求任意非周期信号的响应. 由式 (2.2-12), 写出

$$Y(\omega) = X(\omega)H(\omega). \tag{2.2-13a}$$

在时域里, 有

$$y(t) = \mathcal{F}^{-1}\{Y(\omega)\} = \mathcal{F}^{-1}\{X(\omega)H(\omega)\}, \tag{2.2-13b}$$

这是线性时不变系统的一个重要结果. 它说的是, 当输入信号通过系统时, 其输入谱 $X(\omega)$ 被系统的传递函数 $H(\omega)$ 修改或滤波, 经过滤波后的输入将作为系统的输出. 事实上, 我们已经看到了一个简单的 RC 电路 (见图 2.1-15) 作为低通滤波器. 系统的传递函数决定滤波特性.

2.3 二维傅里叶分析

二维 (2-D) 傅里叶变换的研究紧接着上节对一维 (1-D) 傅里叶变换的讨论. 因此, 之前对一维傅里叶变换的研究是很有意义的. 许多工程问题中, 如数字图像处理和光学系统分析, 输入和输出通常是二维的, 在某些情况下是高维的. 对于平面图像 $f(x, y)$ 的情况, 我们处理以 x 和 y 标记的两个空间维度. 本节中, 将主要讨论二维信号.

2.3.1 二维傅里叶变换

含有两个空间变量 x 和 y 的函数 $f(x, y)$ 被称为二维信号或二维空间函数. 一个常见的例子就是摄影图像或计算机生成的图像, 其中变量 x 和 y 是图像的空间坐标, $f(x, y)$ 是灰度值 (*gray value*) 的幅度. 然而, 函数 $f(x, y)$ 通常为复数. 在这种情况下, 一个二维复函数可以被表示为实部和虚部的形式, 或强度和相位的形式, 由于复数 $z = a + b\mathrm{j} = \sqrt{a^2 + b^2}\mathrm{e}^{\mathrm{j}\arctan(b/a)}$, 其中, a 和 b 是 z 的实部和虚部, $\sqrt{a^2 + b^2}$ 和 $\arctan(b/a)$ 是 z 的强度和相位部分. 当处理一维时间信号时, 有两个自变量 $(t; \omega)$, 其中, t 以 s 为单位, 时间角频率 (*radian temporal frequency*) ω 以 rad/s 为单位. 对于 $f(x, y)$, 有四个自变量 $(x, y; k_x, k_y)$, 其中, x 和 y 的单位为 m, k_x 和 k_y 被称为空间角频率 (*radian spatial frequency*), 单位为 rad/m.

空间信号 $f(x, y)$ 的二维正傅里叶变换 (*2-D forward Fourier transform*) 被定义为

$$\mathcal{F}\{f(x, y)\} = F(k_x, k_y) = \iint_{-\infty}^{\infty} f(x, y)\mathrm{e}^{\mathrm{j}k_x x + \mathrm{j}k_y y}\mathrm{d}x\mathrm{d}y, \tag{2.3-1a}$$

而二维逆傅里叶变换 (*2-D inverse Fourier transform*) 为

$$\mathcal{F}^{-1}\left\{F(k_x, k_y)\right\} = f(x, y) = \frac{1}{4\pi^2}\iint_{-\infty}^{\infty} F(k_x, k_y)\mathrm{e}^{-\mathrm{j}k_x x - \mathrm{j}k_y y}\mathrm{d}k_x\mathrm{d}k_y. \tag{2.3-1b}$$

Example: Response Due to an Aperiodic Signal

We have discussed a response in a linear and time-invariant system due to a sinusoidal input. Now, we can find the response due to any aperiodic signal. Using Eq. (2.2-12), we write

$$Y(\omega) = X(\omega)H(\omega). \tag{2.2-13a}$$

In the time domain, we have

$$y(t) = \mathcal{F}^{-1}\{Y(\omega)\} = \mathcal{F}^{-1}\{X(\omega)H(\omega)\}, \tag{2.2-13b}$$

This is an important result for linear and time-invariant systems. It says that the input spectrum $X(\omega)$ is modified or filtered by the system's transfer function $H(\omega)$ as the input signal goes through the system to give a filtered version of the input as an output. Indeed, we have seen a simple RC circuit (see Figure 2.1-15) as a low-pass filter. Filtering is characterized by the transfer function of the system.

2.3 Fourier Analysis in Two Dimensions

The study of two-dimensional (2-D) Fourier transform follows closely the study of the one-dimensional (1-D) Fourier transform discussed in the last section. Hence, our devotion to the 1-D Fourier transform will prove to be useful. Many engineering problems such as in digital image processing and optical system analysis, the inputs and outputs are typically 2-D and, in some cases, higher dimensional. For the case of planar images, $f(x, y)$, we deal with two spatial dimensions, which we label x and y. In this section, we devote our time to 2-D signals.

2.3.1 The Two-Dimensional Fourier Transform

A function, $f(x, y)$, of two spatial variables x and y is called a 2-D signal or a 2-D spatial function. A common example is a photographic image or a computer-generated image, wherein the variables x and y are the spatial coordinates of the image, and $f(x, y)$ is the amplitude of gray values. However, in general, the function $f(x, y)$ takes on complex values. In that case, depiction of a 2-D complex function can be represented in terms of the real and imaginary parts, or the magnitude and phase parts since a complex number $z = a + b\mathrm{j} = \sqrt{a^2 + b^2}\mathrm{e}^{\mathrm{j}\arctan(b/a)}$, where a and b are the real and imaginary parts of z, and $\sqrt{a^2 + b^2}$ and $\arctan(b/a)$ are the magnitude and phase parts of z. When dealing with 1-D time signals, we have two independent transform variables $(t; \omega)$, where t is measured in seconds, and radian temporal frequency ω is measured in rad/s. With $f(x, y)$, we have four independent transform variables $(x, y; k_x, k_y)$, where x and y are measured in meters, and k_x and k_y are called *radian spatial frequencies* in rad/m.

The *two-dimensional (2-D) forward Fourier transform* of a spatial signal $f(x, y)$ is defined as

$$\mathcal{F}\{f(x, y)\} = F(k_x, k_y) = \iint_{-\infty}^{\infty} f(x, y)\mathrm{e}^{\mathrm{j}k_x x + \mathrm{j}k_y y}\mathrm{d}x\mathrm{d}y, \tag{2.3-1a}$$

and the *two-dimensional (2-D) inverse Fourier transform* is

$$\mathcal{F}^{-1}\{F(k_x, k_y)\} = f(x, y) = \frac{1}{4\pi^2}\iint_{-\infty}^{\infty} F(k_x, k_y)\mathrm{e}^{-\mathrm{j}k_x x - \mathrm{j}k_y y}\mathrm{d}k_x\mathrm{d}k_y. \tag{2.3-1b}$$

$f(x,y)$ 和 $F(k_x,k_y)$ 是一个傅里叶变换对, 用符号表示为

$$f(x,y) \Longleftrightarrow F(k_x,k_y).$$

这里备注二维 “空间” 傅里叶变换及其逆变换的定义不同于 “时间” 傅里叶变换及其逆变换的定义, 对于前者, 分别用 $\mathrm{j}k_x x+\mathrm{j}k_y y$ 和 $-\mathrm{j}k_x x-\mathrm{j}k_y y$ 作为其正变换和逆变换的指数, 而对于后者, 分别用 $-\mathrm{j}\omega t$ 和 $\mathrm{j}\omega t$ 作为其指数. 这样做是为了与行波的工程惯例一致, 在此规定中, $\mathrm{Re}\left[\mathrm{e}^{\mathrm{j}\omega t-\mathrm{j}kz}\right]$ 表示一个时间频率为 ω、传播常数为 k、沿 $+z$ 方向传播的波, 其中, $\mathrm{Re}[\cdot]$ 表示取括号内复值的实部.

2.3.2　一些二维傅里叶变换的算例

二维 δ 函数及其另一个定义

二维 δ 函数被定义为

$$\delta(x,y)=\delta(x)\delta(y).$$

该函数用于模拟光学系统中一理想点源或针孔孔径. 根据式 (2.3-1a) 中傅里叶变换的定义, 有

$$\mathcal{F}\{\delta(x,y)\}=\iint_{-\infty}^{\infty}\delta(x,y)\mathrm{e}^{\mathrm{j}k_x x+\mathrm{j}k_y y}\mathrm{d}x\mathrm{d}y,$$

由于 $\delta(x,y)$ 是两个自变量的函数, 且为可分离函数 (*separable function*), 将上述方程重新写为两个函数的乘积, 其中每个函数只有一个变量, 如下所示

$$\begin{aligned}\mathcal{F}\{\delta(x,y)\}=\mathcal{F}\{\delta(x)\delta(y)\}&=\int_{-\infty}^{\infty}\delta(x)\mathrm{e}^{\mathrm{j}k_x x}\mathrm{d}x\times\int_{-\infty}^{\infty}\delta(y)\mathrm{e}^{\mathrm{j}k_y y}\mathrm{d}y\\&=\left.\mathrm{e}^{\mathrm{j}k_x x}\right|_{x=0}\times\left.\mathrm{e}^{\mathrm{j}k_y y}\right|_{y=0}=1,\end{aligned}$$

这里, 利用式 (2.1-5) 中 δ 函数的筛选性质, 即, 用 $\int_{-\infty}^{\infty}x(t)\delta(t-t_0)\mathrm{d}t=x(t_0)$ 来求得最后的结果, 可以象征性地将这个结果表示如下

$$\delta(x,y)\Longleftrightarrow 1. \tag{2.3-2}$$

傅里叶变换对如图 2.3-1 所示.

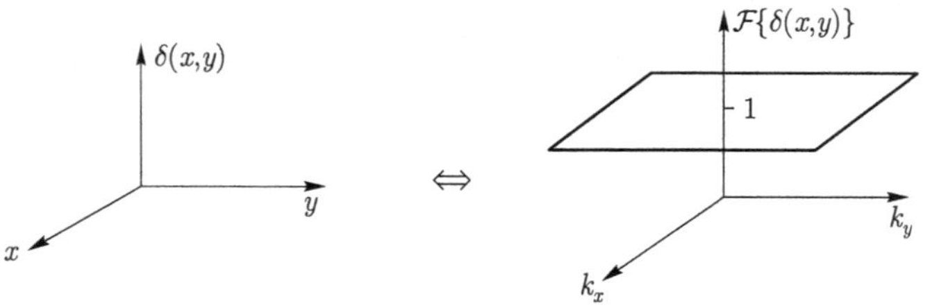

图 2.3-1　二维 δ 函数及其傅里叶变换

$f(x,y)$ and $F(k_x,k_y)$ is a Fourier transform pair, and symbolically we write

$$f(x,y) \Longleftrightarrow F(k_x,k_y).$$

We remark that the definitions of 2-D "spatial" Fourier and inverse transforms are different from the definitions of "temporal" Fourier and inverse transforms in that, for the former, we use $\mathrm{j}k_x x + \mathrm{j}k_y y$ and $-\mathrm{j}k_x x - \mathrm{j}k_y y$ as the exponents for the forward and inverse transforms, respectively, whereas $-\mathrm{j}\omega t$ and $\mathrm{j}\omega t$ serve as the exponents in the latter. This is done to be consistent with the engineering convention for a travelling wave. In the convention, $\mathrm{Re}[\mathrm{e}^{\mathrm{j}\omega t-\mathrm{j}kz}]$ denotes a wave travelling in the $+z$ direction with temporal frequency ω and a propagation constant k, where $\mathrm{Re}[\cdot]$ denotes taking the real part of the bracketed complex quantity.

2.3.2 Calculation Examples of Some 2-D Fourier Transforms

2-D Delta Function and Its Alternative Definition

The 2-D delta function is defined as

$$\delta(x,y) = \delta(x)\delta(y).$$

The function is used to model an ideal point source or a pin hole aperture in optical systems. From the definition of the Fourier transform in Eq. (2.3-1a), we have

$$\mathcal{F}\{\delta(x,y)\} = \iint_{-\infty}^{\infty} \delta(x,y)\mathrm{e}^{\mathrm{j}k_x x+\mathrm{j}k_y y}\mathrm{d}x\mathrm{d}y,$$

Since $\delta(x,y)$ is a function of two independent variables and is a *separable function*, we rewrite the above equation as a product of two functions, each of them depending on only one variable as follows:

$$\begin{aligned}\mathcal{F}\{\delta(x,y)\} = \mathcal{F}\{\delta(x)\delta(y)\} &= \int_{-\infty}^{\infty} \delta(x)\mathrm{e}^{\mathrm{j}k_x x}\mathrm{d}x \times \int_{-\infty}^{\infty} \delta(y)\mathrm{e}^{\mathrm{j}k_y y}\mathrm{d}y \\ &= \mathrm{e}^{\mathrm{j}k_x x}\Big|_{x=0} \times \mathrm{e}^{\mathrm{j}k_y y}\Big|_{y=0} = 1,\end{aligned}$$

where we have used the sampling property of a delta function from Eq. (2.1-5), that is, $\int_{-\infty}^{\infty} x(t)\delta(t-t_0)\mathrm{d}t = x(t_0)$ to obtain the final result. We can state the result symbolically as

$$\delta(x,y) \Longleftrightarrow 1. \tag{2.3-2}$$

The Fourier transform pair is illustrated in Figure 2.3-1.

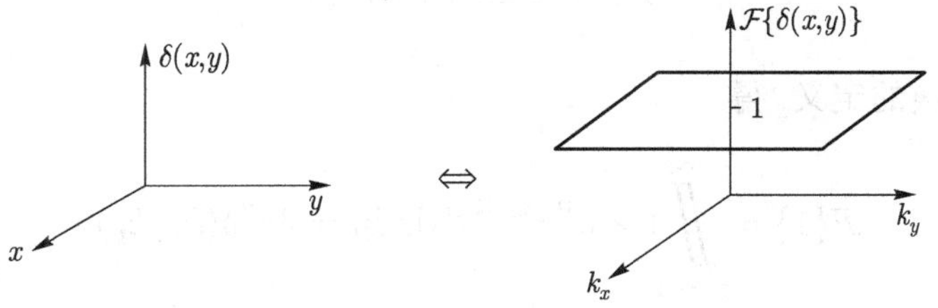

Figure 2.3-1 A 2-D delta function and its Fourier transform

前面已经用矩形函数定义过 δ 函数 (见图 2.1-7). 接下来求 δ 函数的积分定义, 这将利于简化一些积分. 这里来求 1 的二维傅里叶变换. 根据傅里叶变换的定义, 写出

$$\mathcal{F}\{1\}=\iint_{-\infty}^{\infty} 1\times \mathrm{e}^{\mathrm{j}k_x x+\mathrm{j}k_y y}\mathrm{d}x\mathrm{d}y=\int_{-\infty}^{\infty} 1\times \mathrm{e}^{\mathrm{j}k_x x}\mathrm{d}x\times\int_{-\infty}^{\infty} 1\times \mathrm{e}^{\mathrm{j}k_y y}\mathrm{d}y.$$

现在, 利用微积分的标准方法

$$\int_{-\infty}^{\infty} 1\times \mathrm{e}^{\mathrm{j}k_x x}\mathrm{d}x=\left.\frac{\mathrm{e}^{\mathrm{j}k_x x}}{\mathrm{j}k_x}\right|_{-\infty}^{\infty},$$

当求其极限时, 值是不确定的, 可以发现 $\left.\mathrm{e}^{\mathrm{j}\theta}\right|_{\theta\to\infty}=\left.(\cos\theta+\mathrm{j}\sin\theta)\right|_{\theta\to\infty}$ 是不确定的. 为了绕过这个问题, 对 δ 函数进行如下的逆变换

$$\begin{aligned}\mathcal{F}^{-1}\left\{\delta(k_x,k_y)\right\}&=\frac{1}{4\pi^2}\iint_{-\infty}^{\infty}\delta(k_x,k_y)\mathrm{e}^{-\mathrm{j}k_x x-\mathrm{j}k_y y}\mathrm{d}k_x\mathrm{d}k_y\\&=\frac{1}{4\pi^2}\left.\mathrm{e}^{-\mathrm{j}k_x x-\mathrm{j}k_y y}\right|_{k_x=k_y=0}=\frac{1}{4\pi^2},\end{aligned}$$

这里用到了 δ 函数的筛选性质来求积分. 现在, 对上式两边同时做变换, 可得

$$\mathcal{F}^{-1}\left\{4\pi^2\delta(k_x,k_y)\right\}=1.$$

最后, 对上式进行傅里叶变换, 并由傅里叶积分定理可得

$$\mathcal{F}\left\{\mathcal{F}^{-1}\left\{4\pi^2\delta(k_x,k_y)\right\}\right\}=4\pi^2\delta(k_x,k_y)=\mathcal{F}\{1\},\tag{2.3-3a}$$

或

$$1\Longleftrightarrow 4\pi^2\delta(k_x,k_y).\tag{2.3-3b}$$

傅里叶变换对如图 2.3-2 所示.

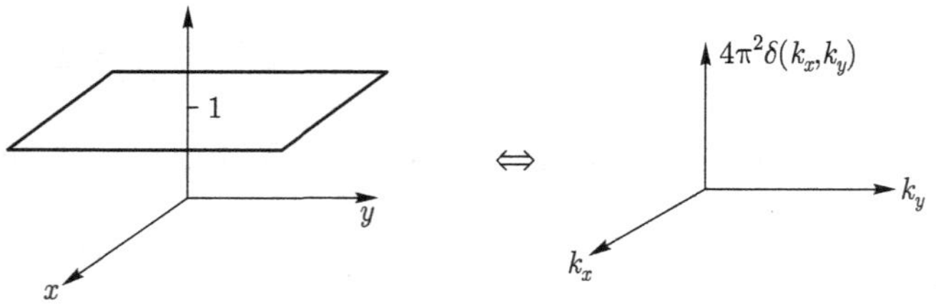

图 2.3-2 一个常数和它的傅里叶变换

由式 (2.3-3a) 的结果, 写出

$$\mathcal{F}\{1\}=4\pi^2\delta\left(k_x,k_y\right),$$

或者利用傅里叶变换的定义, 得

$$\mathcal{F}\{1\}=\iint_{-\infty}^{\infty} 1\times \mathrm{e}^{\mathrm{j}k_x x+\mathrm{j}k_y y}\mathrm{d}x\mathrm{d}y=4\pi^2\delta(k_x,k_y).$$

We have seen the definition of the delta function with a rectangular function (see Figure 2.1-7). In what follows, we obtain an integral definition of the delta function, which is useful for simplifying some integrals. Let us find the 2-D Fourier transform of 1. According to the definition of the Fourier transform, we write

$$\mathcal{F}\{1\}=\iint_{-\infty}^{\infty} 1\times \mathrm{e}^{\mathrm{j}k_x x+\mathrm{j}k_y y}\mathrm{d}x\mathrm{d}y=\int_{-\infty}^{\infty} 1\times \mathrm{e}^{\mathrm{j}k_x x}\mathrm{d}x\times\int_{-\infty}^{\infty} 1\times \mathrm{e}^{\mathrm{j}k_y y}\mathrm{d}y.$$

Now, employing standard techniques of calculus,

$$\int_{-\infty}^{\infty} 1\times \mathrm{e}^{\mathrm{j}k_x x}\mathrm{d}x=\left.\frac{\mathrm{e}^{\mathrm{j}k_x x}}{\mathrm{j}k_x}\right|_{-\infty}^{\infty},$$

which is indeterminate when we evaluate the limits. As we can see $\mathrm{e}^{\mathrm{j}\theta}\big|_{\theta\to\infty}=(\cos\theta+\mathrm{j}\sin\theta)\big|_{\theta\to\infty}$ indeterminate. To bypass such an issue, let us take the inverse transform of the delta function as follows:

$$\begin{aligned}\mathcal{F}^{-1}\{\delta(k_x,k_y)\}&=\frac{1}{4\pi^2}\iint_{-\infty}^{\infty}\delta(k_x,k_y)\mathrm{e}^{-\mathrm{j}k_x x-\mathrm{j}k_y y}\mathrm{d}k_x\mathrm{d}k_y\\&=\frac{1}{4\pi^2}\left.\mathrm{e}^{-\mathrm{j}k_x x-\mathrm{j}k_y y}\right|_{k_x=k_y=0}=\frac{1}{4\pi^2},\end{aligned}$$

where we have used the sampling property of the delta function to evaluate the integral. Now, let us take the transform on both sides of the above equation, we then have

$$\mathcal{F}^{-1}\{4\pi^2\delta(k_x,k_y)\}=1.$$

Finally, taking the transform of the above and using the Fourier integral theorem,

$$\mathcal{F}\left\{\mathcal{F}^{-1}\left\{4\pi^2\delta(k_x,k_y)\right\}\right\}=4\pi^2\delta(k_x,k_y)=\mathcal{F}\{1\}, \tag{2.3-3a}$$

or

$$1\Longleftrightarrow 4\pi^2\delta(k_x,k_y). \tag{2.3-3b}$$

The Fourier transform pair is illustrated in Figure 2.3-2.

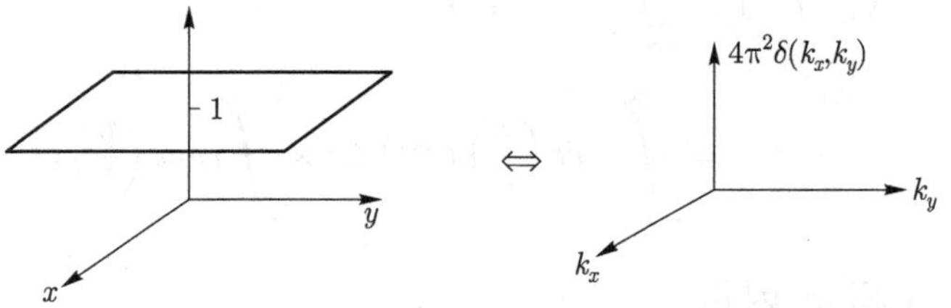

Figure 2.3-2 A constant and its Fourier transform

From the result of Eq. (2.3-3a), we write

$$\mathcal{F}\{1\}=4\pi^2\delta(k_x,k_y),$$

or using the definition of the Fourier transform, we obtain

$$\mathcal{F}\{1\}=\iint_{-\infty}^{\infty} 1\times \mathrm{e}^{\mathrm{j}k_x x+\mathrm{j}k_y y}\mathrm{d}x\mathrm{d}y=4\pi^2\delta(k_x,k_y).$$

根据上面的等式, 有

$$\delta(k_x,k_y)=\frac{1}{4\pi^2}\iint\limits_{-\infty}^{\infty}\mathrm{e}^{\mathrm{j}k_x x+\mathrm{j}k_y y}\mathrm{d}x\mathrm{d}y.$$

如果在积分中做如下替换: (k_x,k_y) 换为 (x,y), 以及 (x,y) 换为 (x',y'), 则积分变为

$$\delta(x,y)=\frac{1}{4\pi^2}\iint\limits_{-\infty}^{\infty}\mathrm{e}^{\mathrm{j}xx'+\mathrm{j}yy'}\mathrm{d}x'\mathrm{d}y'.$$

由于 δ 函数为偶函数, 这由图 2.1-7 及定义可知, 有

$$\delta(-x,-y)=\frac{1}{4\pi^2}\iint\limits_{-\infty}^{\infty}\mathrm{e}^{-\mathrm{j}xx'-\mathrm{j}yy'}\mathrm{d}x'\mathrm{d}y'=\delta(x,y).$$

故, 可以把 δ 函数一般的积分定义写为

$$\delta(x,y)=\frac{1}{4\pi^2}\iint\limits_{-\infty}^{\infty}\mathrm{e}^{\pm\mathrm{j}xx'\pm\mathrm{j}yy'}\mathrm{d}x'\mathrm{d}y'. \tag{2.3-4}$$

二维矩形函数

二维矩形函数是一个可分离函数, 被定义为

$$rect\left(\frac{x}{a}\right)rect\left(\frac{y}{b}\right)=rect\left(\frac{x}{a},\frac{y}{b}\right). \tag{2.3-5}$$

在图 2.3-3a) 和图 2.3-3b) 中, 分别画出在 $a=b$ 时函数的三维图和灰度图. 在灰度图中, 假定幅度为 1 表示为白色, 幅度为 0 表示为黑色. 该函数也可以有效地表示光学系统中的一个矩形孔径, 其中幅度为 1 表示孔径的开口或透明区域, 幅度为 0 表示孔径区域是不透明的. 在后续章节中讨论衍射时, 会再用这种方式表示孔径.

求矩形函数的傅里叶变换, 根据定义

$$\begin{aligned}\mathcal{F}\left\{rect\left(\frac{x}{a},\frac{y}{b}\right)\right\}&=\iint\limits_{-\infty}^{\infty}rect\left(\frac{x}{a},\frac{y}{b}\right)\mathrm{e}^{\mathrm{j}k_x x+\mathrm{j}k_y y}\mathrm{d}x\mathrm{d}y\\&=\int_{-\infty}^{\infty}rect\left(\frac{x}{a}\right)\mathrm{e}^{\mathrm{j}k_x x}\mathrm{d}x\times\int_{-\infty}^{\infty}rect\left(\frac{y}{b}\right)\mathrm{e}^{\mathrm{j}k_y y}\mathrm{d}y.\end{aligned}$$

现在可以分开计算每个积分. 因此

$$\begin{aligned}\int_{-\infty}^{\infty}rect\left(\frac{x}{a}\right)\mathrm{e}^{\mathrm{j}k_x x}\mathrm{d}x&=\int_{-a/2}^{a/2}1\times\mathrm{e}^{\mathrm{j}k_x x}\mathrm{d}x\\&=\left.\frac{\mathrm{e}^{\mathrm{j}k_x x}}{\mathrm{j}k_x}\right|_{-a/2}^{a/2}=\frac{1}{\mathrm{j}k_x}\left(\mathrm{e}^{\mathrm{j}k_x a/2}-\mathrm{e}^{-\mathrm{j}k_x a/2}\right)=\frac{1}{\mathrm{j}k_x}\left[2\mathrm{j}\sin\left(k_x a/2\right)\right]\end{aligned}$$

From the above equality, we write

$$\delta(k_x, k_y) = \frac{1}{4\pi^2} \iint_{-\infty}^{\infty} \mathrm{e}^{\mathrm{j}k_x x + \mathrm{j}k_y y} \mathrm{d}x \mathrm{d}y.$$

If we make the following substitutions in the integral: (k_x, k_y) by (x, y), and (x, y) by (x', y'), the integral becomes

$$\delta(x, y) = \frac{1}{4\pi^2} \iint_{-\infty}^{\infty} \mathrm{e}^{\mathrm{j}xx' + \mathrm{j}yy'} \mathrm{d}x' \mathrm{d}y'.$$

Since the delta function is an even function, which follows clearly from the definition according to Figure (2.1-7), we have

$$\delta(-x, -y) = \frac{1}{4\pi^2} \iint_{-\infty}^{\infty} \mathrm{e}^{-\mathrm{j}xx' - \mathrm{j}yy'} \mathrm{d}x' \mathrm{d}y' = \delta(x, y).$$

So we can write a general integral definition of the delta function as

$$\delta(x, y) = \frac{1}{4\pi^2} \iint_{-\infty}^{\infty} \mathrm{e}^{\pm\mathrm{j}xx' \pm\mathrm{j}yy'} \mathrm{d}x' \mathrm{d}y'. \tag{2.3-4}$$

2-D Rectangular Function

The 2-D rectangular function is a separable function and is defined as

$$rect\left(\frac{x}{a}\right) rect\left(\frac{y}{b}\right) = rect\left(\frac{x}{a}, \frac{y}{b}\right). \tag{2.3-5}$$

In Figure 2.3-3a and b, we have shown the 3-D plot and the gray scale plot of the function for $a = b$, respectively. In the gray scale plot, we have assumed that an amplitude of 1 translates to "white" and an amplitude of 0 to "black." The function could also be effective in representing a rectangular aperture within an optical system, where an amplitude of 1 means the opening or clear area of the aperture, and an amplitude of 0 means the area of the aperture is opaque. We will return to this aspect of the representation of an aperture in a subsequent chapter when we deal with diffraction.

Let us find the Fourier transform of the rectangular function. By definition,

$$\begin{aligned}\mathcal{F}\left\{rect\left(\frac{x}{a}, \frac{y}{b}\right)\right\} &= \iint_{-\infty}^{\infty} rect\left(\frac{x}{a}, \frac{y}{b}\right) \mathrm{e}^{\mathrm{j}k_x x + \mathrm{j}k_y y} \mathrm{d}x \mathrm{d}y \\ &= \int_{-\infty}^{\infty} rect\left(\frac{x}{a}\right) \mathrm{e}^{\mathrm{j}k_x x} \mathrm{d}x \times \int_{-\infty}^{\infty} rect\left(\frac{y}{b}\right) \mathrm{e}^{\mathrm{j}k_y y} \mathrm{d}y.\end{aligned}$$

We can now evaluate each integral separately. Therefore,

$$\begin{aligned}\int_{-\infty}^{\infty} rect\left(\frac{x}{a}\right) \mathrm{e}^{\mathrm{j}k_x x} \mathrm{d}x &= \int_{-a/2}^{a/2} 1 \times \mathrm{e}^{\mathrm{j}k_x x} \mathrm{d}x \\ &= \left.\frac{\mathrm{e}^{\mathrm{j}k_x x}}{\mathrm{j}k_x}\right|_{-a/2}^{a/2} = \frac{1}{\mathrm{j}k_x}\left(\mathrm{e}^{\mathrm{j}k_x a/2} - \mathrm{e}^{-\mathrm{j}k_x a/2}\right) = \frac{1}{\mathrm{j}k_x}\left[2\mathrm{j} \sin\left(k_x a/2\right)\right]\end{aligned}$$

$$= \frac{\sin(k_x a/2)}{k_x/2} = \frac{a\sin(k_x a/2)}{k_x a/2} = \frac{a\sin(\pi k_x a/2\pi)}{\pi k_x a/2\pi} = a\ sinc\left(\frac{k_x a}{2\pi}\right).$$

类似地, 有

$$\int_{-\infty}^{\infty} rect\left(\frac{y}{b}\right) \mathrm{e}^{\mathrm{j}k_y y}\mathrm{d}y = b\ sinc\left(\frac{k_y b}{2\pi}\right),$$

最后, 有

$$\mathcal{F}\left\{rect\left(\frac{x}{a},\frac{y}{b}\right)\right\} = ab\, sinc\left(\frac{k_x a}{2\pi}\right) sinc\left(\frac{k_y b}{2\pi}\right) = ab\, sinc\left(\frac{k_x a}{2\pi},\frac{k_y b}{2\pi}\right),$$

或

$$rect\left(\frac{x}{a},\frac{y}{b}\right) \Longleftrightarrow ab\, sinc\left(\frac{k_x a}{2\pi},\frac{k_y b}{2\pi}\right). \tag{2.3-6}$$

可以看出, 二维 *sinc* 函数也是一个可分离函数. 图 2.3-3c) 给出了当 $a=b$ 时矩形函数的二维傅里叶变换的强度图. 图 2.3-3b) 和图 2.3-3c) 为下面 m-文件运行后生成的结果.

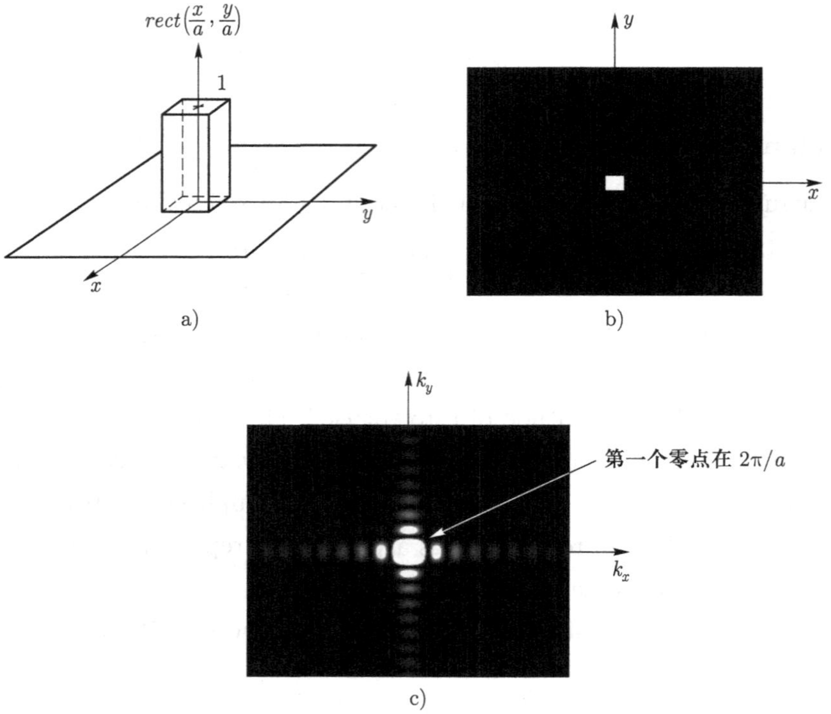

图 2.3-3　二维矩形函数在 $a=b$ 时的 a) 三维图; b) 灰度图; c) 傅里叶变换的强度图

```
========================================
    %Fourier transform of a square function; FT2D_rect.m
    clear

    L=1; %length of display area
    N=256; % number of sampling points
    dx=L/(N-1); % dx : step size

    % Create square image, M by M square, rect(x/a), M=odd number
```

$$= \frac{\sin(k_x a/2)}{k_x/2} = \frac{a\sin(k_x a/2)}{k_x a/2} = \frac{a\sin(\pi k_x a/2\pi)}{\pi k_x a/2\pi} = a\ sinc\left(\frac{k_x a}{2\pi}\right).$$

Similarly,

$$\int_{-\infty}^{\infty} rect\left(\frac{y}{b}\right) e^{jk_y y} dy = b\ sinc\left(\frac{k_y b}{2\pi}\right),$$

and finally, we have

$$\mathcal{F}\left\{rect\left(\frac{x}{a},\frac{y}{b}\right)\right\} = ab\,sinc\left(\frac{k_x a}{2\pi}\right) sinc\left(\frac{k_y b}{2\pi}\right) = ab\,sinc\left(\frac{k_x a}{2\pi},\frac{k_y b}{2\pi}\right),$$

or

$$rect\left(\frac{x}{a},\frac{y}{b}\right) \Longleftrightarrow ab\,sinc\left(\frac{k_x a}{2\pi},\frac{k_y b}{2\pi}\right). \tag{2.3-6}$$

Note that the 2-D sinc function is also a separable function. Figure 2.3-3c shows the magnitude of the 2-D Fourier transform of a rectangular function for $a = b$. Figure 2.3-3b and c is generated using the m-file shown below.

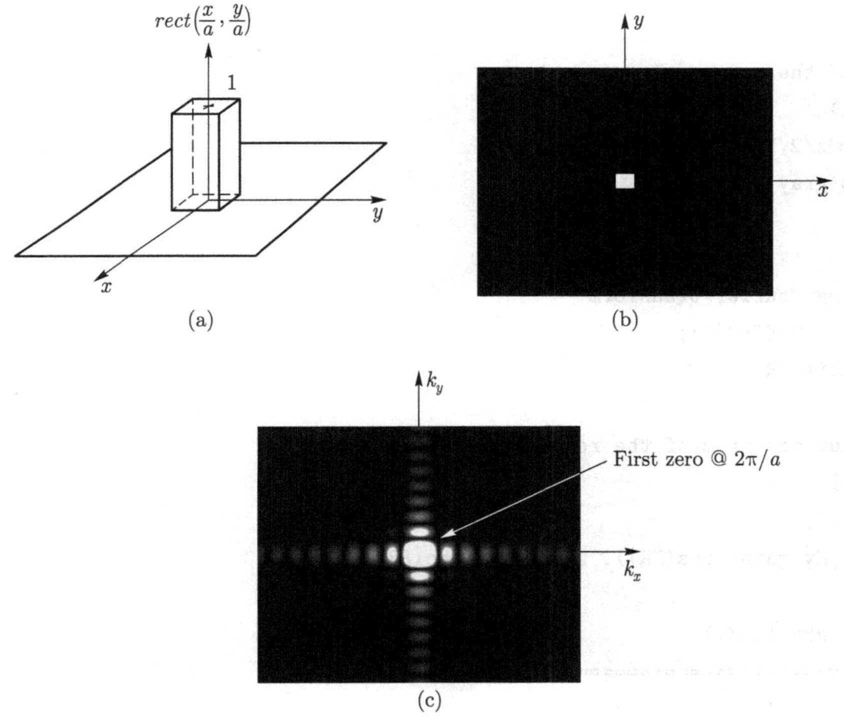

Figure 2.3-3 (a) 3-D plot, (b) gray scale plot, and (c) magnitude of the Fourier transform, of the 2-D rectangular function for $a = b$

```
===========================================
    %Fourier transform of a square function; FT2D_rect.m
    clear

    L=1; %length of display area
    N=256; % number of sampling points
    dx=L/(N-1); % dx : step size

    % Create square image, M by M square, rect(x/a), M=odd number
```

```
M=17;
a=M/256
R=zeros(256); %assign a matrix (256x256) of zeros
r=ones(M); % assign a matrix (MxM) of ones
n=(M-1)/2;
R(128-n:128+n,128-n:128+n)=r;
%End of creating square input image

%Axis Scaling
for k=1:256
   X(k)=1/255*(k-1)-L/2;
   Y(k)=1/255*(k-1)-L/2;

   %Kx=(2*pi*k)/((N-1)*dx)
   %in our case, N=256, dx=1/255

   Kx(k)=(2*pi*(k-1))/((N-1)*dx)-((2*pi*(256-1))/((N-1)*dx))/2;
   Ky(k)=(2*pi*(k-1))/((N-1)*dx)-((2*pi*(256-1))/((N-1)*dx))/2;
 end

%Image of the rectangular fucntion
figure(1)
image(X+dx/2,Y+dx/2,255*R);
colormap(gray(256));
axis off

%Computing Fourier transform
FR=(1/256)^2*fft2(R);
FR=fftshift(FR);

%Magnitude spectrum of the rectangular function
figure(2);
gain=10000;
image(Kx,Ky,gain*(abs(FR)).^2/max(max(abs(FR))).^2)
axis off
colormap(gray(256))
```
==

圆域函数

圆域函数 (*circular function*) 定义为

$$circ\left(\frac{r}{r_0}\right)=\begin{cases}1, & r<r_0\\ 0, & \text{其他},\end{cases} \tag{2.3-7}$$

其中, $r=\sqrt{x^2+y^2}$, 由于透镜和孔径光阑通常都是圆形的, 因此该函数在光学领域有着特殊的作用. 因为所关心的问题具有圆对称性, 这里先在极坐标中表示其傅里叶变换, 在 (x,y) 和 (k_x,k_y) 平面上做直角坐标到极坐标的变换如下

$$x=r\cos\theta,\ \ y=r\sin\theta,\ \ k_x=k_r\cos\phi,\ \ k_y=k_r\sin\phi. \tag{2.3-8}$$

```
M=17;
a=M/256
R=zeros(256); %assign a matrix (256x256) of zeros
r=ones(M); % assign a matrix (MxM) of ones
n=(M-1)/2;
R(128-n:128+n,128-n:128+n)=r;
%End of creating square input image

%Axis Scaling
for k=1:256
   X(k)=1/255*(k-1)-L/2;
   Y(k)=1/255*(k-1)-L/2;

   %Kx=(2*pi*k)/((N-1)*dx)
   %in our case, N=256, dx=1/255

   Kx(k)=(2*pi*(k-1))/((N-1)*dx)-((2*pi*(256-1))/((N-1)*dx))/2;
   Ky(k)=(2*pi*(k-1))/((N-1)*dx)-((2*pi*(256-1))/((N-1)*dx))/2;
 end

%Image of the rectangular fucntion
figure(1)
image(X+dx/2,Y+dx/2,255*R);
colormap(gray(256));
axis off

%Computing Fourier transform
FR=(1/256)^2*fft2(R);
FR=fftshift(FR);

%Magnitude spectrum of the rectangular function
figure(2);
gain=10000;
image(Kx,Ky,gain*(abs(FR)).^2/max(max(abs(FR))).^2)
axis off
colormap(gray(256))
========================================
```

Circular Function

The circular function is defined as

$$circ\left(\frac{r}{r_0}\right)=\begin{cases}1, & r<r_0\\ 0, & \text{otherwise,}\end{cases} \tag{2.3-7}$$

where $r=\sqrt{x^2+y^2}$. This function is particularly useful in optics where lenses and aperture stops are often circular in shape. Since the problem of interest has circular symmetry, let us first express the Fourier transform in polar coordinates. We make a transformation from rectangular coordinates to polar coordinates in both the (x,y) and the (k_x,k_y) planes as follows:

$$x=r\cos\theta,\ \ y=r\sin\theta,\ \ k_x=k_r\cos\phi,\ \ k_y=k_r\sin\phi. \tag{2.3-8}$$

图 2.3-4 给出了空域和频域中两个坐标之间的关系.

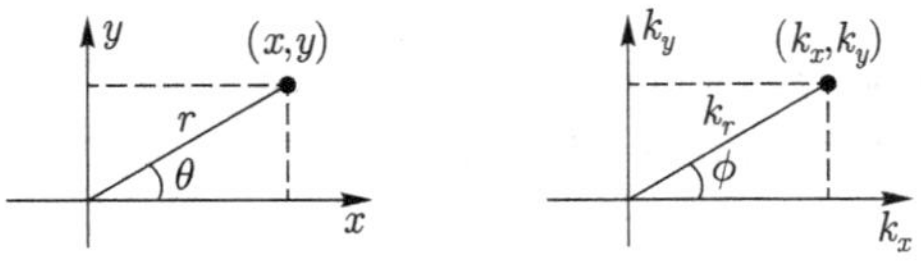

图 2.3-4 直角坐标和极坐标之间的关系

因此, $f(x,y)$ 在直角坐标系下的二维傅里叶变换, 即

$$\mathcal{F}\{f(x,y)\}=\iint\limits_{-\infty}^{\infty} f(x,y)\mathrm{e}^{\mathrm{j}k_x x+\mathrm{j}k_y y}\mathrm{d}x\mathrm{d}y=F(k_x,k_y),$$

变为

$$\begin{aligned}\mathcal{F}\{\bar{f}(r,\theta)\}&=\int_0^{\infty}\int_0^{2\pi}\bar{f}(r,\theta)\mathrm{e}^{\mathrm{j}k_r r(\cos\theta\cos\phi+\sin\theta\sin\phi)}r\mathrm{d}\theta\mathrm{d}r\\&=\int_0^{\infty}\int_0^{2\pi}\bar{f}(r,\theta)\mathrm{e}^{\mathrm{j}k_r r\cos(\theta-\phi)}r\mathrm{d}\theta\mathrm{d}r=\bar{F}(k_r,\phi),\end{aligned}\tag{2.3-9}$$

这里用了式 (2.3-8) 对积分进行变换, 且有微分面积元 $\mathrm{d}x\mathrm{d}y=r\mathrm{d}\theta\mathrm{d}r$. 可以知道, $f(x,y)$ 在极坐标中表示为 $\bar{f}(r,\theta), F(k_x,k_y)$ 在极坐标中表示为 $\bar{F}\,(k_r,\phi)$.

如果函数 $\bar{f}(r,\theta)=\bar{f}_r(r)\bar{f}_\theta(\theta)$, 则该函数 $\bar{f}(r,\theta)$ 在极坐标下是可分离的. 当 $\bar{f}_\theta(\theta)=1, \bar{f}(r,\theta)$ 为圆对称 (*circularly symmetric*) 的, 即, $\bar{f}(r,\theta)=\bar{f}_r(r)$, 式 (2.3-9) 变为

$$\mathcal{F}\left\{\bar{f}_r(r)\right\}=\int_0^{\infty}r\bar{f}_r(r)\left[\int_0^{2\pi}\mathrm{e}^{\mathrm{j}k_r r\cos(\theta-\phi)}\mathrm{d}\theta\right]\mathrm{d}r.\tag{2.3-10}$$

由于

$$J_0(\beta)=\frac{1}{2\pi}\int_0^{2\pi}\mathrm{e}^{\mathrm{j}\beta\cos(\theta-\phi)}\mathrm{d}\theta,\tag{2.3-11}$$

其中, $J_0(\beta)$ 为第一类零阶贝塞尔函数 (*zero-order Bessel function of the first kind*), 式 (2.3-10) 可被写作

$$\mathcal{F}\left\{\bar{f}_r(r)\right\}=2\pi\int_0^{\infty}r\bar{f}_r(r)J_0(k_r r)\mathrm{d}r=\mathcal{B}\left\{\bar{f}_r(r)\right\}.\tag{2.3-12}$$

式 (2.3-12) 定义了圆对称问题中会出现的傅里叶–贝塞尔变换 (*Fourier-Bessel transform*). 在推导了傅里叶–贝塞尔变换之后, 即可求出圆域函数的傅里叶变换. 根据式 (2.3-12), 对于 $\bar{f}_r(r)=circ\left(\dfrac{r}{r_0}\right)$, 有

$$\mathcal{B}\left\{circ\left(\frac{r}{r_0}\right)\right\}=2\pi\int_0^{\infty}r\,circ\left(\frac{r}{r_0}\right)J_0(k_r r)\mathrm{d}r=2\pi\int_0^{r_0}rJ_0(k_r r)\mathrm{d}r.\tag{2.3-13}$$

现在, 利用以下贝塞尔函数的等式

$$J_1(\alpha)=\frac{1}{\alpha}\int_0^{\alpha}\beta J_0(\beta)\mathrm{d}\beta,\tag{2.3-14}$$

Figure 2.3-4 shows the relations between the two coordinates in the spatial domain as well as frequency domain.

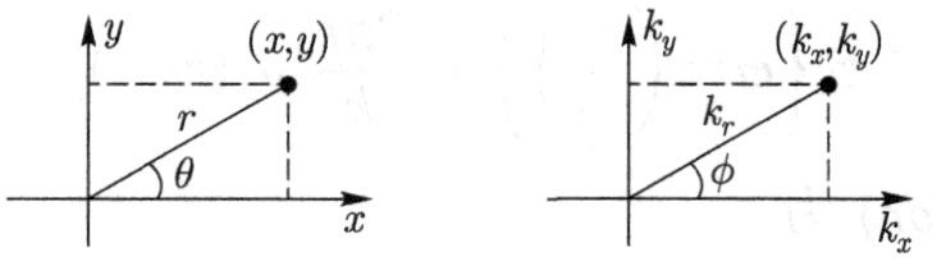

Figure 2.3-4 Relations between rectangular and polar coordinates

Hence, the 2-D Fourier transform in rectangular coordinates of $f(x,y)$, that is,

$$\mathcal{F}\{f(x,y)\} = \iint_{-\infty}^{\infty} f(x,y)\mathrm{e}^{\mathrm{j}k_x x+\mathrm{j}k_y y}\mathrm{d}x\mathrm{d}y = F(k_x,k_y),$$

transforms to

$$\begin{aligned}\mathcal{F}\{\bar{f}(r,\theta)\} &= \int_0^{\infty}\int_0^{2\pi} \bar{f}(r,\theta)\mathrm{e}^{\mathrm{j}k_r r(\cos\theta\cos\phi+\sin\theta\sin\phi)}r\mathrm{d}\theta\mathrm{d}r \\ &= \int_0^{\infty}\int_0^{2\pi} \bar{f}(r,\theta)\mathrm{e}^{\mathrm{j}k_r r\cos(\theta-\phi)}r\mathrm{d}\theta\mathrm{d}r = \bar{F}(k_r,\phi),\end{aligned} \tag{2.3-9}$$

where we have used Eq. (2.3-8) to transform the integral and recognized that the differential area element $\mathrm{d}x\mathrm{d}y = r\mathrm{d}\theta\mathrm{d}r$. Note that $f(x,y)$ in polar coordinates is $\bar{f}(r,\theta)$ and $F(k_x,k_y)$ in polar coordinates is $\bar{F}(k_r,\phi)$.

A function $\bar{f}(r,\theta)$ is separable in polar coordinates if $\bar{f}(r,\theta)=\bar{f}_r(r)\bar{f}_\theta(\theta)$. When $\bar{f}_\theta(\theta) = 1$, $\bar{f}(r,\theta)$ is *circularly symmetric*, that is, $\bar{f}(r,\theta)=\bar{f}_r(r)$, Eq. (2.3-9) becomes

$$\mathcal{F}\{\bar{f}_r(r)\} = \int_0^{\infty} r\bar{f}_r(r)\left[\int_0^{2\pi}\mathrm{e}^{\mathrm{j}k_r r\cos(\theta-\phi)}\mathrm{d}\theta\right]\mathrm{d}r. \tag{2.3-10}$$

Since

$$J_0(\beta) = \frac{1}{2\pi}\int_0^{2\pi}\mathrm{e}^{\mathrm{j}\beta\cos(\theta-\phi)}\mathrm{d}\theta, \tag{2.3-11}$$

where $J_0(\beta)$ is the *zero-order Bessel function of the first kind*, Eq. (2.3-10) can be written as

$$\mathcal{F}\left\{\bar{f}_r(r)\right\} = 2\pi\int_0^{\infty} r\bar{f}_r(r)J_0(k_r r)\mathrm{d}r = \mathcal{B}\left\{\bar{f}_r(r)\right\}. \tag{2.3-12}$$

Equation (2.3-12) defines the *Fourier-Bessel transform* and arises in circularly symmetric problems. After developing the Fourier-Bessel transform, we can now find the Fourier transform of the circular function. According to Eq. (2.3-12), for $\bar{f}_r(r) = circ\left(\frac{r}{r_0}\right)$, we have

$$\mathcal{B}\left\{circ\left(\frac{r}{r_0}\right)\right\} = 2\pi\int_0^{\infty} r\,circ\left(\frac{r}{r_0}\right)J_0(k_r r)\mathrm{d}r = 2\pi\int_0^{r_0} rJ_0(k_r r)\mathrm{d}r. \tag{2.3-13}$$

Now, using the following identity of the Bessel functions:

$$J_1(\alpha) = \frac{1}{\alpha}\int_0^{\alpha}\beta J_0(\beta)\mathrm{d}\beta, \tag{2.3-14}$$

其中, $J_1(\alpha)$ 为第一类一阶贝塞尔函数 (*first-order Bessel function of the first kind*), 式 (2.3-13) 可简化为

$$\mathcal{B}\left\{circ\left(\frac{r}{r_0}\right)\right\}=\frac{2\pi r_0}{k_r}J_1(r_0k_r). \tag{2.3-15}$$

定义 *jinc* 函数 (*jinc function*) 为

$$jinc(r)=\frac{J_1(r)}{r}, \tag{2.3-16}$$

式 (2.3-15) 变为

$$\mathcal{B}\left\{circ\left(\frac{r}{r_0}\right)\right\}=2\pi {r_0}^2 jinc(r_0k_r). \tag{2.3-17}$$

图 2.3-5 给出了傅里叶变换对, 及利用下面的 m-文件生成的结果.

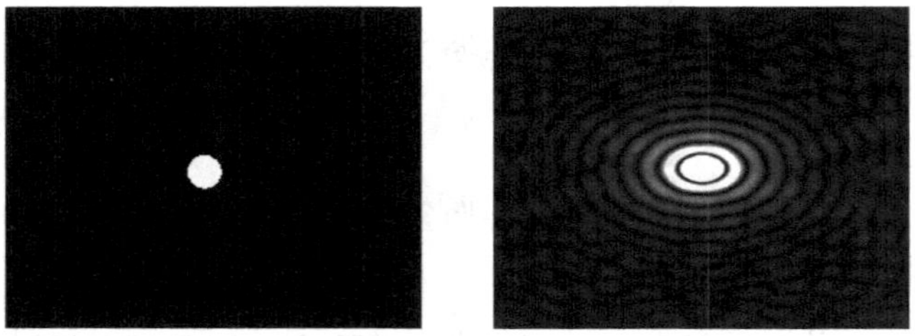

$circ\left(\frac{r}{r_0}\right)$ $\Leftrightarrow$ $2\pi r_0^2 jinc(r_0k_r)$

图 2.3-5 一个圆域函数及其傅里叶变换的强度

```
=========================================
%fft_circular_function.m
%Simulation of Fourier transformation of a circular function
clear

I=imread('smallcircle.bmp','bmp'); %Input image of 256 by 256
I=I(:,:,1);
figure(1) %displaying input
colormap(gray(255));
image(I)
axis off

FI=fft2(I);
FI=fftshift(FI);
max1=max(FI);
max2=max(max1);
scale=1.0/max2;
FI=FI.*scale;

figure(2) %Gray scale image of the absolute value of the transform
colormap(gray(255));
image(10*(abs(256*FI)));
axis off
=========================================
```

where $J_1(\alpha)$ is the *first-order Bessel function of the first kind*, Eq. (2.3-13) can be simplified to give

$$\mathcal{B}\left\{circ\left(\frac{r}{r_0}\right)\right\} = \frac{2\pi r_0}{k_r}J_1(r_0k_r). \tag{2.3-15}$$

Defining the *jinc function* as

$$jinc(r) = \frac{J_1(r)}{r}, \tag{2.3-16}$$

Eq. (2.3-15) becomes

$$\mathcal{B}\left\{circ\left(\frac{r}{r_0}\right)\right\} = 2\pi r_0^2 jinc(r_0k_r). \tag{2.3-17}$$

Figure 2.3-5 illustrates the Fourier transform pair and it is generated using the m-file shown below.

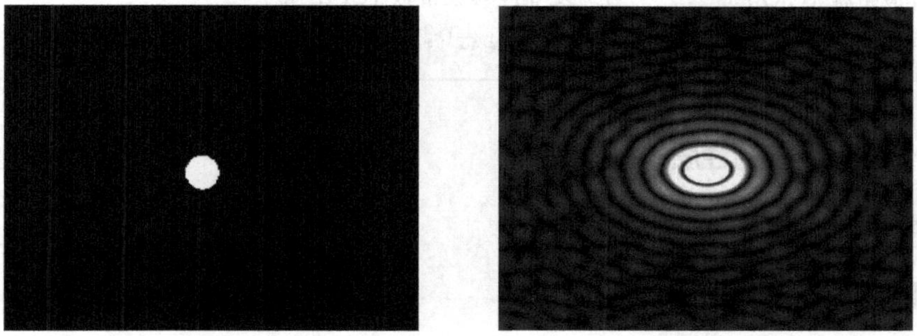

$circ\left(\frac{r}{r_0}\right) \Leftrightarrow 2\pi r_0^2 jinc(r_0k_r)$

Figure 2.3-5 A circular function and the magnitude of its Fourier transform

```
==========================================
%fft_circular_function.m
%Simulation of Fourier transformation of a circular function
clear

I=imread('smallcircle.bmp','bmp'); %Input image of 256 by 256
I=I(:,:,1);
figure(1) %displaying input
colormap(gray(255));
image(I)
axis off

FI=fft2(I);
FI=fftshift(FI);
max1=max(FI);
max2=max(max1);
scale=1.0/max2;
FI=FI.*scale;

figure(2) %Gray scale image of the absolute value of the transform
colormap(gray(255));
image(10*(abs(256*FI)));
axis off
==========================================
```

表 2.1 总结了本节的结果, 并列出了一些有用的傅里叶变换对.

表 2.1 一些常用的傅里叶变换对

$f(x,y)$	$F(k_x,k_y)$
1. δ 函数: $\delta(x,y)$	常数: 1
2. 常数: 1	δ 函数: $4\pi^2\delta(k_x,k_y)$
3. 矩形函数: $rect\left(\frac{x}{a}\right)rect\left(\frac{y}{b}\right)=rect\left(\frac{x}{a},\frac{y}{b}\right)$	$sinc$ 函数: $a\ sinc\left(\frac{k_xa}{2\pi}\right)b\ sinc\left(\frac{k_yb}{2\pi}\right)=ab\ sinc\left(\frac{k_xa}{2\pi},\frac{k_yb}{2\pi}\right)$
4. 圆域函数: $circ\left(\frac{r}{r_0}\right)$	$jinc$ 函数: $2\pi{r_0}^2jinc(r_0k_r)$
5. 高斯函数: $\mathrm{e}^{-\alpha(x^2+y^2)}$	高斯函数: $\frac{\pi}{\alpha}\mathrm{e}^{-\frac{1}{4\alpha}(k_x^2+k_y^2)}$
6. 复菲涅耳波带板 (CFZP): $\mathrm{e}^{-\mathrm{j}\alpha(x^2+y^2)}$	复菲涅耳波带板 (CFZP): $\frac{-\mathrm{j}\pi}{\alpha}\mathrm{e}^{\frac{\mathrm{j}}{4\alpha}(k_x^2+k_y^2)}$

2.3.3 傅里叶变换的性质

本节讨论傅里叶变换的一些有用的性质.

1. 线性性质 (*linearity property*)

由于傅里叶变换是一个线性运算, 具有可叠加性.

若

$$f_1(x,y)\Longleftrightarrow F_1(k_x,k_y),\ \ f_2(x,y)\Longleftrightarrow F_2(k_x,k_y),$$

那么

$$af_1(x,y)+bf_2(x,y)\Longleftrightarrow aF_1(k_x,k_y)+bF_2(k_x,k_y), \tag{2.3-18}$$

其中, a 和 b 一般是任意复常数. 该性质的推导来自傅里叶变换的积分定义. 作为一个例子, 来求一个由下式给出的 "条纹" 图案的傅里叶变换

$$f(x,y)=1+\cos(ax+by). \tag{2.3-19}$$

对其进行傅里叶变换, 有

$$\begin{aligned}\mathcal{F}\{f(x,y)\}&=\mathcal{F}\{1+\cos(ax+by)\}\\&=\mathcal{F}\left\{1+\frac{1}{2}\left[\mathrm{e}^{\mathrm{j}(ax+by)}+\mathrm{e}^{-\mathrm{j}(ax+by)}\right]\right\}\\&=\iint_{-\infty}^{\infty}1\mathrm{e}^{\mathrm{j}k_xx+\mathrm{j}k_yy}\mathrm{d}x\mathrm{d}y+\iint_{-\infty}^{\infty}\frac{1}{2}\mathrm{e}^{\mathrm{j}(ax+by)}\mathrm{e}^{\mathrm{j}k_xx+\mathrm{j}k_yy}\mathrm{d}x\mathrm{d}y\\&\quad+\iint_{-\infty}^{\infty}\frac{1}{2}\mathrm{e}^{-\mathrm{j}(ax+by)}\mathrm{e}^{\mathrm{j}k_xx+\mathrm{j}k_yy}\mathrm{d}x\mathrm{d}y.\end{aligned}$$

Table 2.1 summarizes our results in the section and lists some of the useful Fourier transform pairs.

Table 2.1 Select Fourier transform pairs

$f(x,y)$	$F(k_x,k_y)$
1. Delta function: $\delta(x,y)$	Constant of unity: 1
2. Constant of unity: 1	Delta function: $4\pi^2\delta(k_x,k_y)$
3. Rectangular function: $rect\left(\frac{x}{a}\right) rect\left(\frac{y}{b}\right) = rect\left(\frac{x}{a},\frac{y}{b}\right)$	*sinc* function: $a\ sinc\left(\frac{k_x a}{2\pi}\right) b\ sinc\left(\frac{k_y b}{2\pi}\right) = ab\ sinc\left(\frac{k_x a}{2\pi},\frac{k_y b}{2\pi}\right)$
4. Circular function: $circ\left(\frac{r}{r_0}\right)$	*jinc* function: $2\pi {r_0}^2 jinc(r_0 k_r)$
5. Gaussian function: $\mathrm{e}^{-\alpha(x^2+y^2)}$	Gaussian function: $\frac{\pi}{\alpha}\mathrm{e}^{-\frac{1}{4\alpha}(k_x^2+k_y^2)}$
6. Complex Fresnel zone plate (CFZP): $\mathrm{e}^{-\mathrm{j}\alpha(x^2+y^2)}$	Complex Fresnel zone plate (CFZP): $\frac{-\mathrm{j}\pi}{\alpha}\mathrm{e}^{\frac{\mathrm{j}}{4\alpha}(k_x^2+k_y^2)}$

2.3.3 Properties of the Fourier Transform

In this section, we will discuss some of the useful properties of the Fourier transform.

1. Linearity Property

Since the Fourier transform is a linear operation, superposition applies.

If

$$f_1(x,y) \Longleftrightarrow F_1(k_x,k_y),\ \ f_2(x,y) \Longleftrightarrow F_2(k_x,k_y),$$

then

$$af_1(x,y)+bf_2(x,y) \Longleftrightarrow aF_1(k_x,k_y)+bF_2(k_x,k_y), \tag{2.3-18}$$

where a and b are any arbitrary complex constants in general. The derivation of this property follows from the integral definition of the Fourier transform. As an example, let us find the Fourier transform of a "fringe" pattern given by

$$f(x,y) = 1+\cos(ax+by). \tag{2.3-19}$$

Taking the Fourier transform, we have

$$\begin{aligned}
\mathcal{F}\{f(x,y)\} &= \mathcal{F}\{1+\cos(ax+by)\} \\
&= \mathcal{F}\left\{1+\frac{1}{2}\left[\mathrm{e}^{\mathrm{j}(ax+by)}+\mathrm{e}^{-\mathrm{j}(ax+by)}\right]\right\} \\
&= \iint_{-\infty}^{\infty} 1\mathrm{e}^{\mathrm{j}k_x x+\mathrm{j}k_y y}\mathrm{d}x\mathrm{d}y + \iint_{-\infty}^{\infty} \frac{1}{2}\mathrm{e}^{\mathrm{j}(ax+by)}\mathrm{e}^{\mathrm{j}k_x x+\mathrm{j}k_y y}\mathrm{d}x\mathrm{d}y \\
&\quad + \iint_{-\infty}^{\infty} \frac{1}{2}\mathrm{e}^{-\mathrm{j}(ax+by)}\mathrm{e}^{\mathrm{j}k_x x+\mathrm{j}k_y y}\mathrm{d}x\mathrm{d}y.
\end{aligned}$$

把方程右边第二项和第三项的指数中相似的项进行组合, 得

$$\mathcal{F}\{1+\cos(ax+by)\} = \iint_{-\infty}^{\infty} 1\mathrm{e}^{\mathrm{j}k_x x+\mathrm{j}k_y y}\mathrm{d}x\mathrm{d}y + \iint_{-\infty}^{\infty} \frac{1}{2}\mathrm{e}^{\mathrm{j}(k_x+a)x+\mathrm{j}(k_y+b)y}\mathrm{d}x\mathrm{d}y + \iint_{-\infty}^{\infty} \frac{1}{2}\mathrm{e}^{\mathrm{j}(k_x-a)x+\mathrm{j}(k_y-b)y}\mathrm{d}x\mathrm{d}y.$$

由式 (2.3-4) 中的 δ 函数的积分定义, 可以得到

$$\mathcal{F}\{1+\cos(ax+by)\} = 4\pi^2\delta(k_x,k_y)+2\pi^2\delta(k_x+a,k_y+b)+2\pi^2\delta(k_x-a,k_y-b). \tag{2.3-20}$$

在图 2.3-6a) 中, 画出了 $f(x,y)=1+\cos(ax+by)$ 在 $a=b$ 时的灰度图像. 图 2.3-6b) 用三个 δ 函数的三维图画出了相应的傅里叶变换. 图 2.3-6 是用以下 m-文件生成的结果.

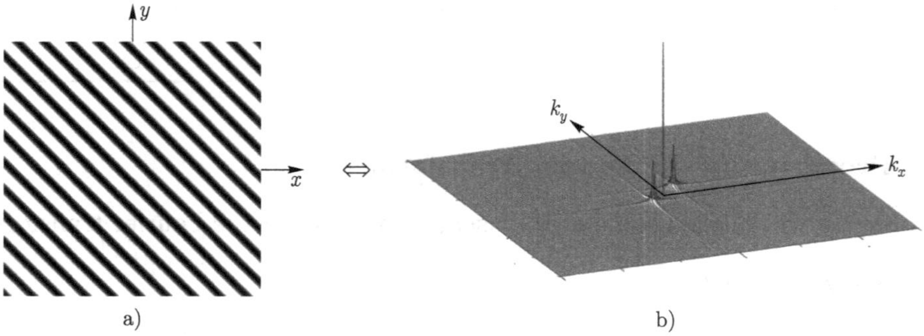

图 2.3-6 a) 条纹图案; b) 相应的傅里叶变换

```
=========================================
   % plot2d.m
   clear
   close all
   x=-15:0.15:15;
   y=-x;
   [x,y]=meshgrid(x,y);
   z=1 + cos(2*x+2*y);
   figure(1)
   imshow(z);
   Fz=fftshift(fft2(z));
   Fzmax=max(max(Fz));
   Fz1=Fz./Fzmax;
   figure(2)
   imshow(abs(Fz1));
   figure(3)
   plot3(x,y,abs(Fz1));
=========================================
```

2. 平移性质 (*shifting property*)

如果

$$f(x,y) \Longleftrightarrow F(k_x,k_y),$$

By grouping similar terms in the exponents of the second and the third terms on the right side of the equation, we have

$$\mathcal{F}\{1+\cos(ax+by)\} = \iint\limits_{-\infty}^{\infty} 1\mathrm{e}^{\mathrm{j}k_x x+\mathrm{j}k_y y}\mathrm{d}x\mathrm{d}y + \iint\limits_{-\infty}^{\infty} \frac{1}{2}\mathrm{e}^{\mathrm{j}(k_x+a)x+\mathrm{j}(k_y+b)y}\mathrm{d}x\mathrm{d}y + \iint\limits_{-\infty}^{\infty} \frac{1}{2}\mathrm{e}^{\mathrm{j}(k_x-a)x+\mathrm{j}(k_y-b)y}\mathrm{d}x\mathrm{d}y.$$

Now, by recognizing the integral definition of the delta function from Eq. (2.3-4), we finally obtain

$$\mathcal{F}\{1+\cos(ax+by)\} = 4\pi^2\delta(k_x,k_y) + 2\pi^2\delta(k_x+a,k_y+b) + 2\pi^2\delta(k_x-a,k_y-b). \quad (2.3\text{-}20)$$

In Figure 2.3-6a, we illustrate the grayscale image of $f(x,y) = 1+\cos(ax+by)$ with $a=b$. Its corresponding Fourier transform is illustrated with a 3-D plot with three delta functions in Figure 2.3-6b. Figure 2.3-6 is generated using the m-file shown below.

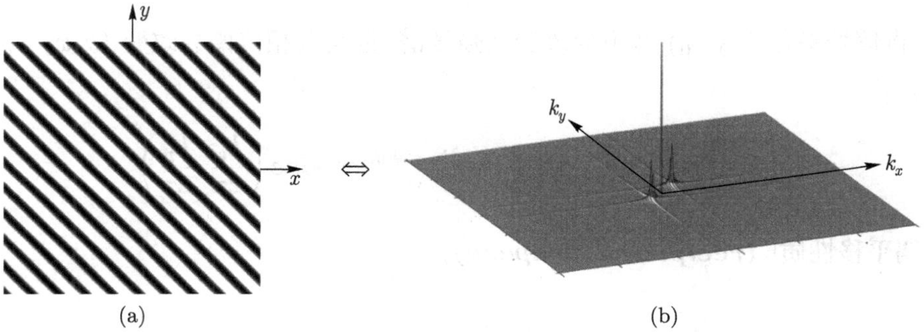

Figure 2.3-6 (a) The fringe pattern and (b) its corresponding Fourier transform

```
==========================================
% plot2d.m
clear
close all
x=-15:0.15:15;
y=-x;
[x,y]=meshgrid(x,y);
z=1 + cos(2*x+2*y);
figure(1)
imshow(z);
Fz=fftshift(fft2(z));
Fzmax=max(max(Fz));
Fz1=Fz./Fzmax;
figure(2)
imshow(abs(Fz1));
figure(3)
plot3(x,y,abs(Fz1));
==========================================
```

2. Shifting Property

If

$$f(x,y) \Longleftrightarrow F(k_x,k_y),$$

那么

$$f(x-x_0,y-y_0) \Longleftrightarrow \mathrm{e}^{\mathrm{j}k_x x_0+\mathrm{j}k_y y_0}F(k_x,k_y). \tag{2.3-21}$$

由定义, 有

$$\mathcal{F}\{f(x-x_0,y-y_0)\}=\iint_{-\infty}^{\infty} f(x-x_0,y-y_0)\mathrm{e}^{\mathrm{j}k_x x+\mathrm{j}k_y y}\mathrm{d}x\mathrm{d}y.$$

让 $x-x_0=x'$ 及 $y-y_0=y'$, 有

$$\begin{aligned}\mathcal{F}\{f(x-x_0,y-y_0)\}&=\iint_{-\infty}^{\infty} f(x',y')\mathrm{e}^{\mathrm{j}k_x(x'+x_0)+\mathrm{j}k_y(y'+y_0)}\mathrm{d}x'\mathrm{d}y'\\&=\mathrm{e}^{\mathrm{j}k_x x_0+\mathrm{j}k_y y_0}\iint_{-\infty}^{\infty} f(x',y')\mathrm{e}^{\mathrm{j}k_x x'+\mathrm{j}k_y y'}\mathrm{d}x'\mathrm{d}y'\\&=\mathrm{e}^{\mathrm{j}k_x x_0+\mathrm{j}k_y y_0}F(k_x,k_y).\end{aligned}$$

结果表明, 将信号平移 (x_0,y_0) 并不会改变其强度谱, 但是其相位谱会改变 $(k_x x_0,k_y y_0)$. 举一个简单的例子

$$\mathcal{F}\{rect(x-x_0,y-y_0)\}=\mathrm{e}^{\mathrm{j}k_x x_0+\mathrm{j}k_y y_0}\ sinc\left(\frac{k_x}{2\pi},\frac{k_y}{2\pi}\right).$$

3. 对偶平移性质† (*reciprocal property*)

如果

$$f(x,y) \Longleftrightarrow F(k_x,k_y),$$

那么

$$\mathrm{e}^{\mathrm{j}\alpha x+\mathrm{j}\beta y}f(x,y) \Longleftrightarrow F(k_x+\alpha,k_y+\beta). \tag{2.3-22}$$

由定义, 有

$$\begin{aligned}\mathcal{F}\left\{\mathrm{e}^{\mathrm{j}\alpha x+\mathrm{j}\beta y}f(x,y)\right\}&=\iint_{-\infty}^{\infty}\mathrm{e}^{\mathrm{j}\alpha x+\mathrm{j}\beta y}f(x,y)\mathrm{e}^{\mathrm{j}k_x x+\mathrm{j}k_y y}\mathrm{d}x\mathrm{d}y\\&=\iint_{-\infty}^{\infty} f(x,y)\mathrm{e}^{\mathrm{j}(k_x+\alpha)x+\mathrm{j}(k_y+\beta)y}\mathrm{d}x\mathrm{d}y\\&=F(k_x+\alpha,k_y+\beta).\end{aligned}$$

该结果表明, 线性相位因子 $\mathrm{e}^{\mathrm{j}\alpha x+\mathrm{j}\beta y}$ 的乘积会使信号的频谱发生 $k_x=-\alpha$ 和 $k_y=-\beta$ 的平移. 举一个简单的例子

$$\mathcal{F}\left\{\mathrm{e}^{\mathrm{j}\alpha x+\mathrm{j}\beta y}rect(x,y)\right\}=sinc\left(\frac{k_x+\alpha}{2\pi},\frac{k_y+\beta}{2\pi}\right).$$

图 2.3-7 为原函数 $rect(x,y)$ 乘以线性相位因子 $\mathrm{e}^{\mathrm{j}\alpha x+\mathrm{j}\beta y}$ 后的原强度谱及其平移后的强度谱.

† 又名频移性质 (*frequency shifting property*).

then

$$f(x-x_0, y-y_0) \Longleftrightarrow \mathrm{e}^{\mathrm{j}k_x x_0+\mathrm{j}k_y y_0} F(k_x, k_y). \tag{2.3-21}$$

By definition,

$$\mathcal{F}\{f(x-x_0, y-y_0)\} = \iint_{-\infty}^{\infty} f(x-x_0, y-y_0)\mathrm{e}^{\mathrm{j}k_x x+\mathrm{j}k_y y}\mathrm{d}x\mathrm{d}y.$$

Letting $x - x_0 = x'$ and $y - y_0 = y'$, we have

$$\begin{aligned}\mathcal{F}\{f(x-x_0, y-y_0)\} &= \iint_{-\infty}^{\infty} f(x', y')\mathrm{e}^{\mathrm{j}k_x(x'+x_0)+\mathrm{j}k_y(y'+y_0)}\mathrm{d}x'\mathrm{d}y' \\ &= \mathrm{e}^{\mathrm{j}k_x x_0+\mathrm{j}k_y y_0}\iint_{-\infty}^{\infty} f(x', y')\mathrm{e}^{\mathrm{j}k_x x'+\mathrm{j}k_y y'}\mathrm{d}x'\mathrm{d}y' \\ &= \mathrm{e}^{\mathrm{j}k_x x_0+\mathrm{j}k_y y_0} F(k_x, k_y).\end{aligned}$$

The result shows that shifting a signal by (x_0, y_0) does not change its magnitude spectrum. The phase spectrum, however, is changed by $(k_x x_0, k_y y_0)$. A simple example is

$$\mathcal{F}\{rect(x-x_0, y-y_0)\} = \mathrm{e}^{\mathrm{j}k_x x_0+\mathrm{j}k_y y_0}\ sinc\left(\frac{k_x}{2\pi}, \frac{k_y}{2\pi}\right).$$

3. Reciprocal Shifting Property

If

$$f(x, y) \Longleftrightarrow F(k_x, k_y),$$

then

$$\mathrm{e}^{\mathrm{j}\alpha x+\mathrm{j}\beta y} f(x, y) \Longleftrightarrow F(k_x+\alpha, k_y+\beta). \tag{2.3-22}$$

By definition,

$$\begin{aligned}\mathcal{F}\left\{\mathrm{e}^{\mathrm{j}\alpha x+\mathrm{j}\beta y} f(x, y)\right\} &= \iint_{-\infty}^{\infty} \mathrm{e}^{\mathrm{j}\alpha x+\mathrm{j}\beta y} f(x, y)\mathrm{e}^{\mathrm{j}k_x x+\mathrm{j}k_y y}\mathrm{d}x\mathrm{d}y \\ &= \iint_{-\infty}^{\infty} f(x, y)\mathrm{e}^{\mathrm{j}(k_x+\alpha)x+\mathrm{j}(k_y+\beta)y}\mathrm{d}x\mathrm{d}y \\ &= F(k_x+\alpha, k_y+\beta).\end{aligned}$$

The result indicates that the multiplication of a linear phase factor $\mathrm{e}^{\mathrm{j}\alpha x+\mathrm{j}\beta y}$ shifts the spectrum of that signal by $k_x = -\alpha$ and $k_y = -\beta$. A simple exalmple is

$$\mathcal{F}\left\{\mathrm{e}^{\mathrm{j}\alpha x+\mathrm{j}\beta y} rect(x, y)\right\} = sinc\left(\frac{k_x+\alpha}{2\pi}, \frac{k_y+\beta}{2\pi}\right).$$

Figure 2.3-7 shows the original magnitude spectrum and its shifted magnitude spectrum when the original function rect (x, y) is multiplied by the linear phase factor $\mathrm{e}^{\mathrm{j}\alpha x+\mathrm{j}\beta y}$.

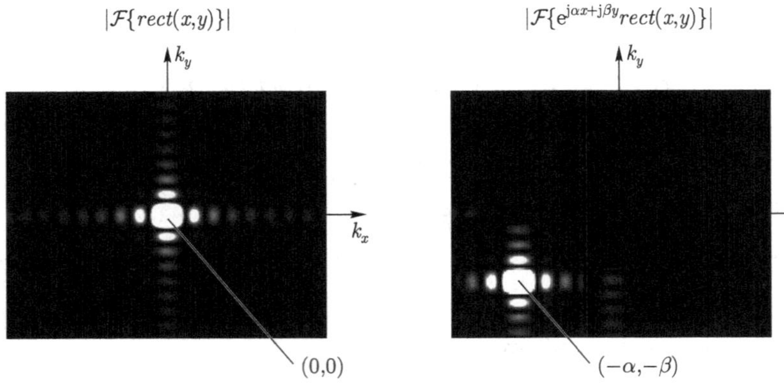

图 2.3-7 对偶平移性质的说明

4. 缩放性质 (*scaling property*)

如果

$$f(x,y)\Longleftrightarrow F(k_x,k_y),$$

则

$$f(ax,by)\Longleftrightarrow \frac{1}{|ab|}F\left(\frac{k_x}{a},\frac{k_y}{b}\right),\tag{2.3-23}$$

其中, a 和 b 为任意复常数. 为简便起见, 考虑正的实常数 $a>0$ 和 $b>0$ 的情况, 根据定义, 有

$$\mathcal{F}\{f(ax,by)\}=\iint\limits_{-\infty}^{\infty} f(ax,by)\mathrm{e}^{\mathrm{j}k_x x+\mathrm{j}k_y y}\mathrm{d}x\mathrm{d}y.$$

让 $ax=x'$ 及 $ay=y'$, 有

$$\begin{aligned}\mathcal{F}\{f(ax,by)\}&=\iint\limits_{-\infty}^{\infty} f(x',y')\mathrm{e}^{\frac{\mathrm{j}k_x}{a}x'+\frac{\mathrm{j}k_y}{b}y'}(\mathrm{d}x'/a)(\mathrm{d}y'/b)\\&=\frac{1}{ab}\iint\limits_{-\infty}^{\infty} f(x',y')\mathrm{e}^{\mathrm{j}\frac{k_x}{a}x'+\mathrm{j}\frac{k_y}{b}y'}\mathrm{d}x'\mathrm{d}y'=\frac{1}{ab}F\left(\frac{k_x}{a},\frac{k_y}{b}\right).\end{aligned}$$

当 a 为负时, x' 的积分限会随着积分变量的改变而反向改变, 于是对于 $a<0$, 有

$$\mathcal{F}\{f(ax,by)\}=-\frac{1}{ab}F\left(\frac{k_x}{a},\frac{k_y}{b}\right).$$

这两种结果可以合并成如下更为紧凑的形式

$$\mathcal{F}\{f(ax,by)\}=\frac{1}{|ab|}F\left(\frac{k_x}{a},\frac{k_y}{b}\right).$$

缩放性质如图 2.3-8 所示. 图 2.3-8a) 表示了一个 $x_0\times x_0$ 白色区域所示大小的矩形函数, 即, $rect\left(\frac{x}{x_0},\frac{y}{x_0}\right)$. 根据式 (2.3-6), 其对应的变换 $x_0^2 sinc\left(\frac{x_0k_x}{2\pi},\frac{x_0k_y}{2\pi}\right)$ 也是知道的, 由于 $sinc$ 函数的第一个零点位置在 $2\pi/x_0$ 处, 另一个类似的矩形函数如图 2.3-8b) 所示, 只是现在其

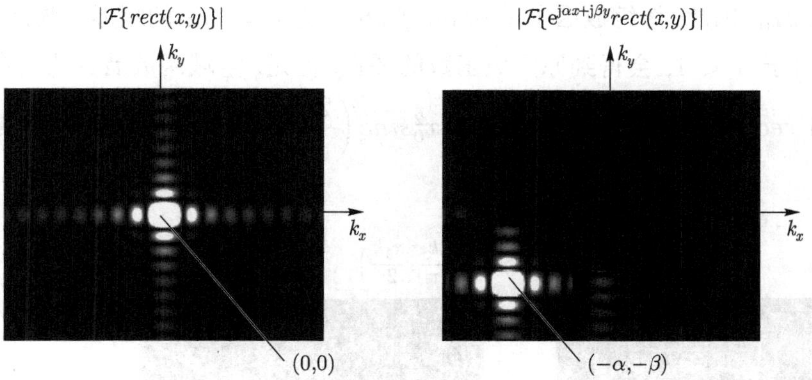

Figure 2.3-7 Illustration of the reciprocal shifting property

4. Scaling Property

If

$$f(x,y) \Longleftrightarrow F(k_x,k_y),$$

then

$$f(ax,by) \Longleftrightarrow \frac{1}{|ab|}F\left(\frac{k_x}{a},\frac{k_y}{b}\right), \tag{2.3-23}$$

where a and b are any arbitrary complex constants. For brevity, we consider positive real constants $a>0$ and $b>0$. By definition, we have

$$\mathcal{F}\{f(ax,by)\} = \iint\limits_{-\infty}^{\infty} f(ax,by)e^{jk_x x+jk_y y}dxdy.$$

Letting $ax=x'$ and $ay=y'$, we have

$$\mathcal{F}\{f(ax,by)\} = \iint\limits_{-\infty}^{\infty} f(x',y')e^{\frac{jk_x}{a}x'+\frac{jk_y}{b}y'}(dx'/a)(dy'/b)$$

$$= \frac{1}{ab}\iint\limits_{-\infty}^{\infty} f(x',y')e^{j\frac{k_x}{a}x'+j\frac{k_y}{b}y'}dx'dy' = \frac{1}{ab}F\left(\frac{k_x}{a},\frac{k_y}{b}\right).$$

When a is negative, the limits on the x' integral are reversed as the variable of integration is changed so that

$$\mathcal{F}\{f(ax,by)\} = -\frac{1}{ab}F\left(\frac{k_x}{a},\frac{k_y}{b}\right).$$

for $a<0$. These two results can be combined into the more compact form as follows:

$$\mathcal{F}\{f(ax,by)\} = \frac{1}{|ab|}F\left(\frac{k_x}{a},\frac{k_y}{b}\right).$$

The scaling property is illustrated graphically by Figure 2.3-8. Figure 2.3-8a displays a square function with "white" area of $x_0\times x_0$, that is, $rect\left(\frac{x}{x_0},\frac{y}{x_0}\right)$. According to Eq. (2.3-6), its corresponding transform, $x_0^2 sinc\left(\frac{x_0k_x}{2\pi},\frac{x_0k_y}{2\pi}\right)$, is also shown. Note that the first zero of the *sinc* function is at $2\pi/x_0$. A similar square function is shown in Figure 2.3-8b, except

面积为 $2x_0 \times 2x_0$, 因此, 该缩放因子 (*scaling factor*) 为 $a = 1/2$ 的矩形函数是图 2.3-8a) 的展宽. 因此, 对于 $a < 1$, 会得到原空域函数的展宽. 但是, 空域函数展宽两倍会带来频谱一半的压缩, 如 $rect\left(\dfrac{x}{2x_0}, \dfrac{y}{2x_0}\right)$ 的频谱为 $4x_0^2 sinc\left(\dfrac{2x_0 k_x}{2\pi}, \dfrac{2x_0 k_y}{2\pi}\right)$, 其第一个零点位置会在 π/x_0 处.

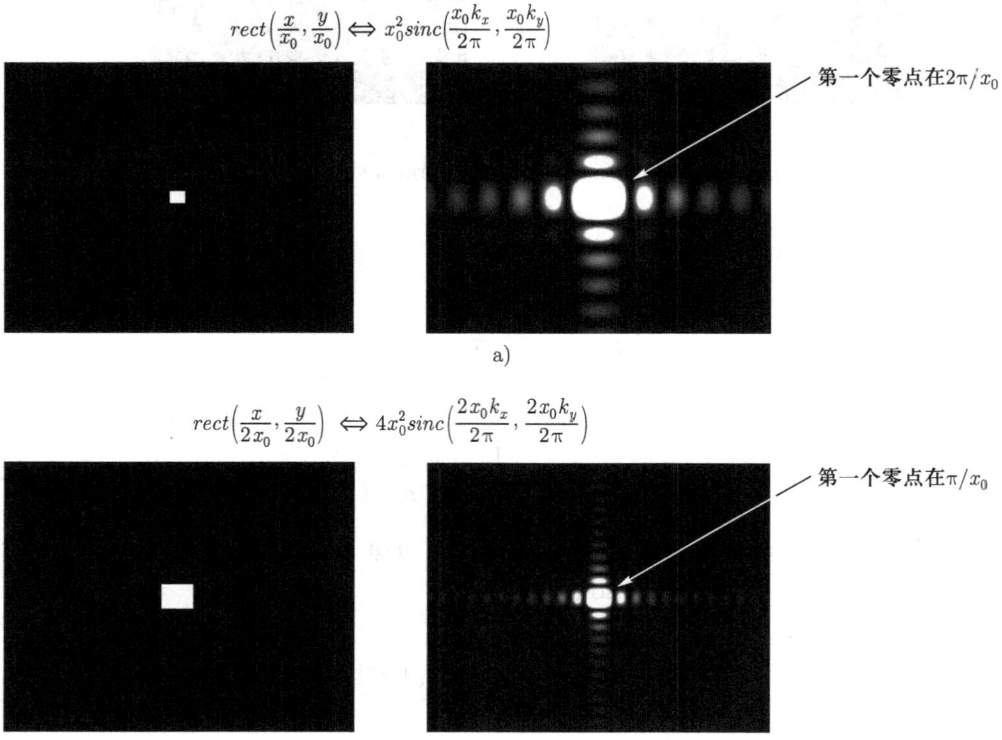

图 2.3-8　缩放性质的说明

这些结论说明, 函数 $f(ax, by)$ 代表了函数 $f(x, y)$ 在 x 和 y 方向分别以因子 a 和 b 进行缩放. 同样, 函数 $F\left(\dfrac{k_x}{a}, \dfrac{k_y}{b}\right)$ 代表了函数 $F(k_x, k_y)$ 在 k_x 和 k_y 方向的频率分别以因子 $1/a$ 和 $1/b$ 进行缩放. 对于 $a < 1$ 且 $b < 1$, 会对原信号在空域进行展宽, 这会导致频域里相应因子的频谱压缩. 另一方面, 对于 $a > 1$ 且 $b > 1$, 会对原信号在空域进行压缩, 这会导致频域中相应因子的频谱展宽.

5. 微分性质 (*differentiation property*)

如果

$$f(x, y) \Longleftrightarrow F(k_x, k_y),$$

那么

$$\frac{\partial f(x, y)}{\partial x} \Longleftrightarrow -\mathrm{j}k_x F(k_x, k_y). \tag{2.3-24}$$

上述性质的推导类似于前述一维傅里叶变换例子中对一个 RC 电路进行傅里叶变换的情况,

that its area is now $2x_0 \times 2x_0$. Hence, the square function is a stretched version of the first one with a scaling factor $a = 1/2$. Hence, for $a < 1$, we have an expansion of the original spatial function. However, stretching the spatial function by a factor of 2 leads to compression of the spectrum by the same factor as we can see that the spectrum of $rect\left(\frac{x}{2x_0}, \frac{y}{2x_0}\right)$ is $4x_0^2 sinc\left(\frac{2x_0 k_x}{2\pi}, \frac{2x_0 k_y}{2\pi}\right)$ with its first zero at π/x_0.

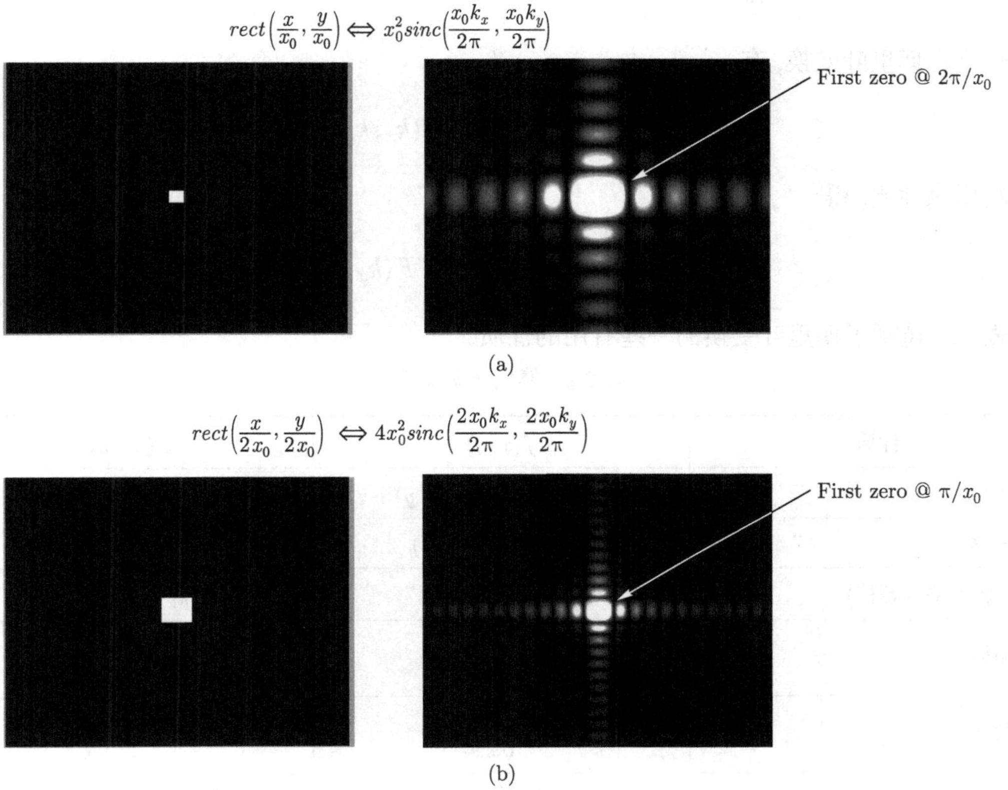

Figure 2.3-8 Illustration of the scaling property

These conclusions lead us to state that the function $f(ax, by)$ represents the function $f(x, y)$ scaled by factors a and b along the x- and y-directions, respectively. Similarly, the function $F\left(\frac{k_x}{a}, \frac{k_y}{b}\right)$ represents the function $F(k_x, k_y)$ scaled in frequencies by $1/a$ and $1/b$ along the k_x and k_y, respectively. For $a < 1$ and $b < 1$, we have spatial expansion of the original signal, which leads to spectral compression by the same factor. On the other hand, for $a > 1$ and $b > 1$, we have spatial compression of the original signal, which leads to spectral expansion by the same factor.

5. Differentiation Property

If

$$f(x, y) \Longleftrightarrow F(k_x, k_y),$$

then

$$\frac{\partial f(x, y)}{\partial x} \Longleftrightarrow -\mathrm{j}k_x F(k_x, k_y). \tag{2.3-24}$$

The derivation of the above property is similar to that in the 1-D Fourier transform in the earlier example on the use of the Fourier Transform for an RC circuit. Starting from Eq.

从式 (2.3-1b), 即二维逆傅里叶变换的定义, 可知

$$f(x,y)=\frac{1}{4\pi^2}\iint\limits_{-\infty}^{\infty}F(k_x,k_y)\mathrm{e}^{-\mathrm{j}k_x x-\mathrm{j}k_y y}\mathrm{d}k_x\mathrm{d}k_y.$$

上式两边对 x 变量求微分

$$\frac{\partial f(x,y)}{\partial x}=\frac{1}{4\pi^2}\iint\limits_{-\infty}^{\infty}-\mathrm{j}k_xF(k_x,k_y)\mathrm{e}^{-\mathrm{j}k_x x-\mathrm{j}k_y y}\mathrm{d}k_x\mathrm{d}k_y=\mathcal{F}^{-1}\left\{-\mathrm{j}k_xF(k_x,k_y)\right\}.$$

对两边进行傅里叶变换, 有

$$\frac{\partial f(x,y)}{\partial x}\Longleftrightarrow -\mathrm{j}k_xF(k_x,k_y).$$

重复利用该性质, 得

$$\frac{\partial^n f(x,y)}{\partial x^n}\Longleftrightarrow (-\mathrm{j}k_x)^n F(k_x,k_y). \tag{2.3-25}$$

表 2.2 概括了傅里叶变换的一些有用的性质.

表 2.2 傅里叶变换性质

性质	$f(x,y)$	$F(k_x,k_y)$
1. 线性	$af_1(x,y)+bf_2(x,y)$	$aF_1(k_x,k_y)+bF_2(k_x,k_y)$
2. 平移	$f(x-x_0,y-y_0)$	$\mathrm{e}^{\mathrm{j}k_x x_0+\mathrm{j}k_y y_0}F(k_x,k_y)$
3. 对偶平移 (频移)	$\mathrm{e}^{\mathrm{j}\alpha x+\mathrm{j}\beta y}f(x,y)$	$F(k_x+\alpha,k_y+\beta)$
4. 缩放	$f(ax,by)$	$\frac{1}{\lvert ab\rvert}F\left(\frac{k_x}{a},\frac{k_y}{b}\right)$
5. 微分	$\frac{\partial f(x,y)}{\partial x}$	$-\mathrm{j}k_xF(k_x,k_y)$
6. 高阶微分	$\frac{\partial^n f(x,y)}{\partial x^n}$	$(-\mathrm{j}k_x)^n F(k_x,k_y)$
7. 混合微分	$\frac{\partial^2 f(x,y)}{\partial x\partial y}$	$(-\mathrm{j}k_x)(-\mathrm{j}k_y)F(k_x,k_y)$
8. 共轭	$f^*(x,y)$	$F^*(-k_x,-k_y)$

2.3.4 二维卷积、相关和匹配滤波

2-D 卷积

在线性时不变系统中, 我们已经讨论过卷积积分. 本节将讨论二维卷积 (*2-D convolution*) 的概念. 随后, 我们将讨论另一个被称为相关 (*correlation*) 的重要运算.

在光学中, 可以将线性时不变系统的概念扩展到所谓的线性空不变 (*linear space-invariant*, LSI) 系统. 因此, 二维卷积积分可写为

$$g(x,y)=\iint\limits_{-\infty}^{\infty}f(x',y')h(x-x',y-y')\mathrm{d}x'\mathrm{d}y'=f(x,y)*h(x,y), \tag{2.3-26}$$

其中, $f(x,y)$ 是线性空不变系统的二维输入, $h(x,y)$ 和 $g(x,y)$ 分别为系统相应的脉冲响应和

(2.3-1b), that is, the definition of the 2-D inverse Fourier transform, we have

$$f(x,y) = \frac{1}{4\pi^2}\iint_{-\infty}^{\infty} F(k_x,k_y)\mathrm{e}^{-\mathrm{j}k_x x-\mathrm{j}k_y y}\mathrm{d}k_x\mathrm{d}k_y.$$

Differentiation of both sides of the above equation with respect to the x variable yields

$$\frac{\partial f(x,y)}{\partial x} = \frac{1}{4\pi^2}\iint_{-\infty}^{\infty} -\mathrm{j}k_x F(k_x,k_y)\mathrm{e}^{-\mathrm{j}k_x x-\mathrm{j}k_y y}\mathrm{d}k_x\mathrm{d}k_y = \mathcal{F}^{-1}\left\{-\mathrm{j}k_x F(k_x,k_y)\right\}.$$

Taking the Fourier transform on both sides, we have

$$\frac{\partial f(x,y)}{\partial x} \Longleftrightarrow -\mathrm{j}k_x F(k_x,k_y).$$

Repeated application of this property yields

$$\frac{\partial^n f(x,y)}{\partial x^n} \Longleftrightarrow (-\mathrm{j}k_x)^n F(k_x,k_y). \tag{2.3-25}$$

Table 2.2 summarizes some of the useful properties of the Fourier transform.

Table 2.2 Fourier transform properties

Property	$f(x,y)$	$F(k_x,k_y)$
1. Linearity	$af_1(x,y)+bf_2(x,y)$	$aF_1(k_x,k_y)+bF_2(k_x,k_y)$
2. Shifting	$f(x-x_0,y-y_0)$	$\mathrm{e}^{\mathrm{j}k_x x_0+\mathrm{j}k_y y_0}F(k_x,k_y)$
3. Reciprocal shifting	$\mathrm{e}^{\mathrm{j}\alpha x+\mathrm{j}\beta y}f(x,y)$	$F(k_x+\alpha,k_y+\beta)$
4. Scaling	$f(ax,by)$	$\frac{1}{\lvert ab\rvert}F\left(\frac{k_x}{a},\frac{k_y}{b}\right)$
5. Differentiation	$\frac{\partial f(x,y)}{\partial x}$	$-\mathrm{j}k_x F(k_x,k_y)$
6. Higher-order differentiation	$\frac{\partial^n f(x,y)}{\partial x^n}$	$(-\mathrm{j}k_x)^n F(k_x,k_y)$
7. Mixed differentiation	$\frac{\partial^2 f(x,y)}{\partial x\partial y}$	$(-\mathrm{j}k_x)(-\mathrm{j}k_y)F(k_x,k_y)$
8. Conjugation	$f^*(x,y)$	$F^*(-k_x,-k_y)$

2.3.4 2-D Convolution, Correlation, and Matched Filtering

2-D Convolution

We have discussed that in a linear and time-invariant system, the convolution integral is involved. In this section, we will discuss the concept of 2-D convolution. Subsequently, we will discuss another important operation called *correlation*.

In optics, we can extend the concept of a linear and time-invariant system to the so-called *linear space-invariant* system. Hence, we can write the 2-D convolution integral as follows:

$$g(x,y) = \iint_{-\infty}^{\infty} f(x',y')h(x-x',y-y')\mathrm{d}x'\mathrm{d}y' = f(x,y)*h(x,y), \tag{2.3-26}$$

where $f(x,y)$ is the 2-D input to a linear space-invariant system. $h(x,y)$ and $g(x,y)$ are the corresponding impulse response and the output of the system, respectively. In optical

输出. 在光学系统中, $h(x,y)$ 通常被称为点扩散函数 (*point spread function*, PSF), 它描述了光学系统对一个点源的响应. 图 2.1-11 描述了时不变的概念. 将输入信号 $x(t)$ 延迟任意常数 t_0, 会得到同样被延迟了 t_0 的输出. 那么光学系统中的空不变性意味着什么呢? 图 2.3-9 澄清了这一概念. 从该图的下半部分可以看到, 当输入图像 $f(x,y)$ 移动到一个新的原点 (x_0,y_0) 时, 它的输出在输出平面上也会有相应移动, 但其函数形式仍与 $g(x,y)$ 相同. 在光学系统的输入平面上, 将遵循空间不变性的平面称为等晕面 (*isoplanatic patch*), 我们将在后续章节中遇到线性空不变光学系统.

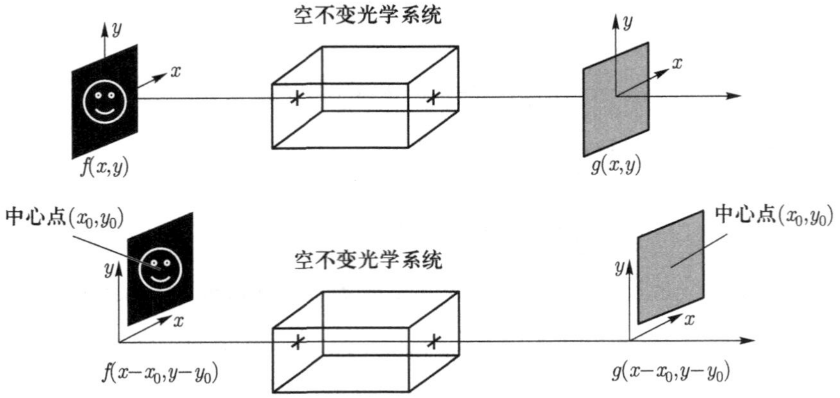

图 2.3-9 空间不变性的概念

由于二维卷积运算概念的理解有点复杂, 这里尝试提供一些清晰的解释. 图 2.3-10 用 $f(x,y)$ 和 $h(x,y)$ 两个函数说明了二维卷积的概念, 该讨论主要基于图 2.1-17 中对一维信号的说明. 根据卷积的定义, 两个函数的卷积基本上是计算两个函数 $f(x',y')$ 和 $h(x-x',y-y')$ 在不同位移 (x,y) 下乘积的不同面积. 因此, 首先需要计算两个原函数的乘积 $f(x',y')h(x-x',y-y')$, 在如图 2.3-10 中的第一行给出了两个原函数. 在第二行, 把它们放在 x',y' 轴上,

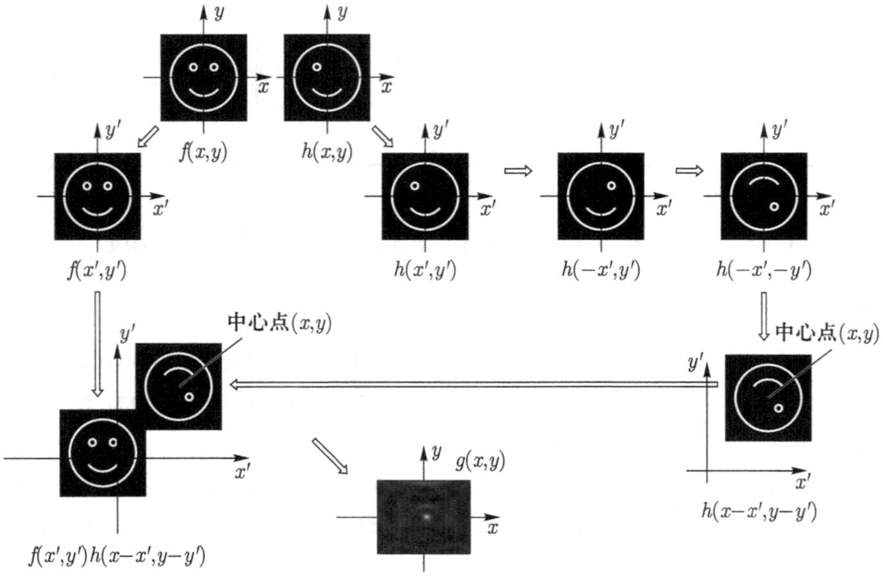

图 2.3-10 二维卷积的概念

systems, $h(x, y)$ is often called the *point spread function* (PSF), which describes the response of an optical system to a point source. The concept of time invariance has been described in Figure 2.1-11. Delaying the input signal $x(t)$ by any constant t_0 gives the same output but delayed by exactly t_0. What does it mean by space invariance in optical systems? In Figure 2.3-9, we clarify this concept. From the bottom half of the figure, we see that as the input image $f(x, y)$ is shifted to a new origin (x_0, y_0), its output, remaining the same functional form as $g(x, y)$, is shifted accordingly on the output plane. The *isoplanatic patch* is the name given to the input plane over which the optical system obeys space invariance. We will encounter linear and space-invariant optical systems later in the coming chapters.

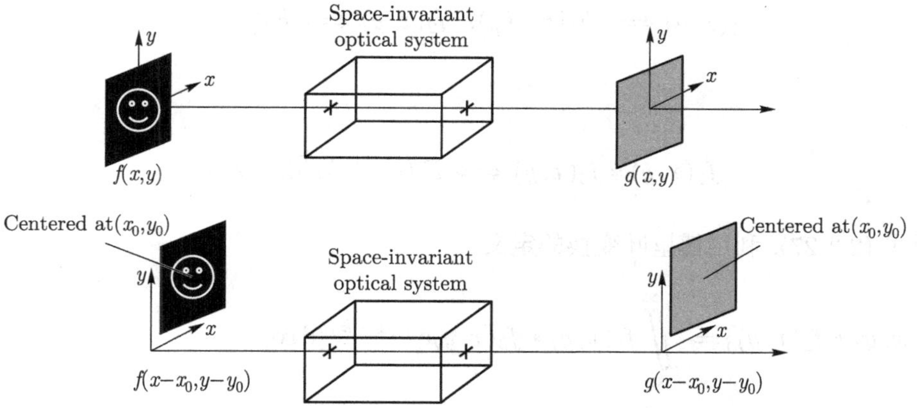

Figure 2.3-9 Space-invariance concept

As the convolution operation in two dimensions is a bit complicated concept to grasp, we try to provide some clarity in the process. In Figure 2.3-10, we illustrate the concept of convolution involving two 2-D functions, $f(x, y)$ and $h(x, y)$. The discussion on the illustration follows closely to that in Figure 2.1-17 for 1-D signals. According to the definition of convolution, the convolution of the two functions basically involves the calculation of the different areas under the product of two functions, $f(x', y')$ and $h(x - x', y - y')$, for different shifts, (x, y). So we need first to create the product $f(x', y')h(x - x', y - y')$ from the two original functions. The two original functions are shown in the first row in Figure 2.3-10. On

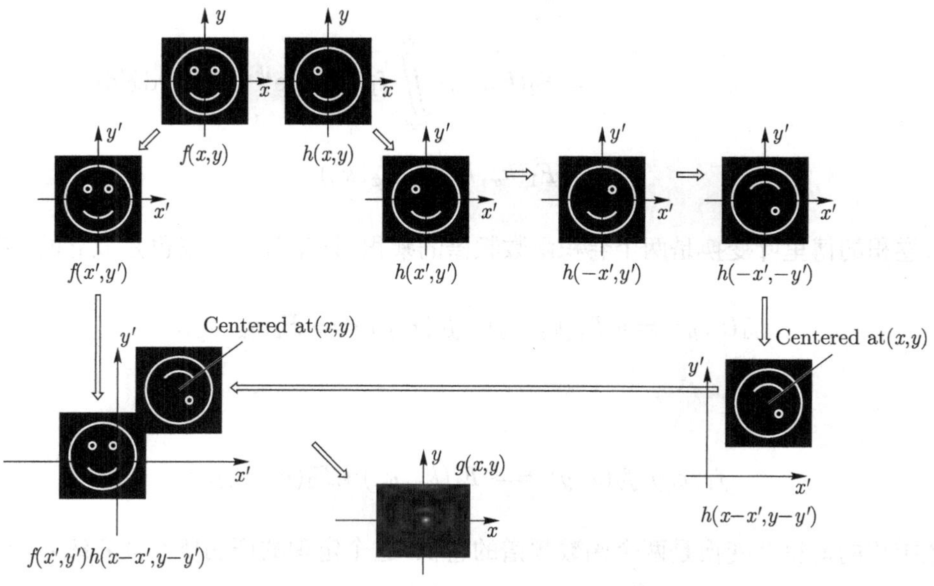

Figure 2.3-10 Concept of 2-D convolution

就得到了 $f(x',y')$ 图像. 为了得到 $h(x-x',y-y')$ 图像, 首先对 $h(x',y')$ 在 x' 轴上进行翻转, 再在 y' 轴上进行翻转, 得到 $h(-x',-y')$, 如第二行所示. 一旦得到 $h(-x',-y')$ 图像, 就将它在 x'-y' 平面上沿 x',y' 轴分别平移 x 和 y 得到 $h(x-x',y-y')$. 现在如图中第三行左图中 x'-y' 平面中所示, 让 $f(x',y')$ 和 $h(x-x',y-y')$ 进行重叠, 最后, 计算在不同位移 (x,y) 下乘积的面积, 得到 $g(x,y)$ 的二维图, 如图中的灰度图所示.

卷积定理 (*convolution theorem*)

如果

$$f_1(x,y) \Longleftrightarrow F_1(k_x,k_y),\quad f_2(x,y) \Longleftrightarrow F_2(k_x,k_y),$$

那么

$$f_1(x,y) * f_2(x,y) \Longleftrightarrow F_1(k_x,k_y)F_2(k_x,k_y). \tag{2.3-27}$$

为了证明式 (2.3-27), 利用傅里叶变换的定义

$$\begin{aligned}\mathcal{F}\{f_1(x,y) * f_2(x,y)\} &= \iint\limits_{-\infty}^{\infty} f_1(x,y) * f_2(x,y)\mathrm{e}^{\mathrm{j}k_x x+\mathrm{j}k_y y}\mathrm{d}x\mathrm{d}y \\ &= \iint\limits_{-\infty}^{\infty}\iint\limits_{-\infty}^{\infty} f_1(x',y')f_2(x-x',y-y')\mathrm{d}x'\mathrm{d}y'\mathrm{e}^{\mathrm{j}k_x x+\mathrm{j}k_y y}\mathrm{d}x\mathrm{d}y. \end{aligned}\tag{2.3-28}$$

先对 x 和 y 积分

$$\iint\limits_{-\infty}^{\infty} f_2(x-x',y-y')\mathrm{e}^{\mathrm{j}k_x x+\mathrm{j}k_y y}\mathrm{d}x\mathrm{d}y = \mathrm{e}^{\mathrm{j}k_x x'+\mathrm{j}k_y y'}F_2(k_x,k_y),$$

上式是根据表 2.2 中傅里叶变换的平移性质所得, 将上述结果代入式 (2.3-28), 得

$$\begin{aligned}\mathcal{F}\{f_1(x,y) * f_2(x,y)\} &= \iint\limits_{-\infty}^{\infty} f_1(x',y')\mathrm{e}^{\mathrm{j}k_x x'+\mathrm{j}k_y y'}F_2(k_x,k_y)\mathrm{d}x'\mathrm{d}y' \\ &= F_2(k_x,k_y)\iint\limits_{-\infty}^{\infty} f_1(x',y')\mathrm{e}^{\mathrm{j}k_x x'+\mathrm{j}k_y y'}\mathrm{d}x'\mathrm{d}y' \\ &= F_1(k_x,k_y)F_2(k_x,k_y).\end{aligned}$$

两个函数卷积的傅里叶变换是两个卷积函数频谱的乘积. 该定理反过来也是成立的. 若

$$f_1(x,y) \Longleftrightarrow F_1(k_x,k_y),\quad f_2(x,y) \Longleftrightarrow F_2(k_x,k_y),$$

那么

$$f_1(x,y)f_2(x,y) \Longleftrightarrow F_1(k_x,k_y) * F_2(k_x,k_y). \tag{2.3-29}$$

两个函数乘积的傅里叶变换是两个函数频谱的卷积, 这个定理的证明留给读者练习. 在后续章节的光学系统中, 我们会有机会体会到卷积定理的作用.

the second row, we put them on the x', y'-axis. Now we have $f(x', y')$ on a graph. To create the graph of $h(x-x', y-y')$, we first take the reversal of $h(x', y')$ upon the x'-axis and then upon the y'-axis to get $h(-x', -y')$, as shown in the second row. Once we have $h(-x', -y')$ we shift it by x and y to get $h(x-x', y-y')$ on the x', y'-plane. We can now overlap $f(x', y')$ and $h(x-x', y-y')$ as shown in the left-hand side of the third row on the x', y'-plane. Finally, we calculate the areas of the product for different shifts (x, y) to obtain a 2-D plot of $g(x, y)$, shown as a gray-scale plot in the figure.

Convolution Theorems

If

$$f_1(x,y) \Longleftrightarrow F_1(k_x,k_y),\quad f_2(x,y) \Longleftrightarrow F_2(k_x,k_y),$$

then

$$f_1(x,y) * f_2(x,y) \Longleftrightarrow F_1(k_x,k_y)F_2(k_x,k_y). \tag{2.3-27}$$

To prove Eq. (2.3-27), we use the definition of the Fourier transform:

$$\begin{aligned}\mathcal{F}\{f_1(x,y) * f_2(x,y)\} &= \iint_{-\infty}^{\infty} f_1(x,y) * f_2(x,y)\mathrm{e}^{\mathrm{j}k_x x+\mathrm{j}k_y y}\mathrm{d}x\mathrm{d}y \\ &= \iint_{-\infty}^{\infty}\iint_{-\infty}^{\infty} f_1(x',y')f_2(x-x',y-y')\mathrm{d}x'\mathrm{d}y'\mathrm{e}^{\mathrm{j}k_x x+\mathrm{j}k_y y}\mathrm{d}x\mathrm{d}y. \end{aligned} \tag{2.3-28}$$

Let us integrate over x and y first, and the integral over x and y is

$$\iint_{-\infty}^{\infty} f_2(x-x',y-y')\mathrm{e}^{\mathrm{j}k_x x+\mathrm{j}k_y y}\mathrm{d}x\mathrm{d}y = \mathrm{e}^{\mathrm{j}k_x x'+\mathrm{j}k_y y'}F_2(k_x,k_y),$$

which is according to the shifting property of the Fourier transform shown in Table 2.2. Substituting the above result into Eq. (2.3-28), we have

$$\begin{aligned}\mathcal{F}\{f_1(x,y) * f_2(x,y)\} &= \iint_{-\infty}^{\infty} f_1(x',y')\mathrm{e}^{\mathrm{j}k_x x'+\mathrm{j}k_y y'}F_2(k_x,k_y)\mathrm{d}x'\mathrm{d}y' \\ &= F_2(k_x,k_y)\iint_{-\infty}^{\infty} f_1(x',y')\mathrm{e}^{\mathrm{j}k_x x'+\mathrm{j}k_y y'}\mathrm{d}x'\mathrm{d}y' \\ &= F_1(k_x,k_y)F_2(k_x,k_y).\end{aligned}$$

The Fourier transform of the convolution of two functions is the product of the spectra of the two convolving functions. The converse of this theorem also holds. If

$$f_1(x,y) \Longleftrightarrow F_1(k_x,k_y),\quad f_2(x,y) \Longleftrightarrow F_2(k_x,k_y),$$

then

$$f_1(x,y)f_2(x,y) \Longleftrightarrow F_1(k_x,k_y) * F_2(k_x,k_y). \tag{2.3-29}$$

The Fourier transform of the product of two functions is the convolution of the spectra of the two functions. The proof of this theorem is left as an exercise for the reader. We will have the opportunity to appreciate the power of the convolution theorems in subsequent chapters in optical systems.

相关 (*correlation*)

现在来讨论另一个重要的运算, 即相关 (*correlation*). 两个函数 $f(x,y)$ 和 $h(x,y)$ 的相关 $C_{fh}(x,y)$, 被定义为

$$C_{fh}(x,y)=\iint\limits_{-\infty}^{\infty} f^*(x',y')h(x+x',y+y')\mathrm{d}x'\mathrm{d}y'=f(x,y)\otimes h(x,y), \tag{2.3-30}$$

其中, f^* 是 f 的复共轭, $\otimes$ 是表示 $f(x,y)$ 与 $h(x,y)$ 相关的符号. $f(x,y)\otimes h(x,y)$ 读作 f 与 h 的相关. 现在来解释 $f(x,y)$ 和 $h(x,y)$ 这两个函数的相关. 与两个函数的卷积类似, 相关运算也关系到计算两个函数 $f^*(x',y')$ 和 $h(x+x',y+y')$ 在不同位移 (x,y) 时乘积的面积. 为简单起见, 假定 f 为实数, 即, 为解释相关运算, 让 $f^*=f$. 如图 2.3-11 中第一行给出两个原函数. 在第二行, 将其放在 x',y' 轴上, 得到 $f(x',y')$ 图像. 为得到 $h(x+x',y+y')$ 图像, 将 $h(x',y')$ 在 x',y' 平面上沿 x',y' 轴分别平移 x 和 y, 得到 $h(x+x',y+y')$. 现在如第三行左边图中所示, 让 $f(x',y')$ 和 $h(x+x',y+y')$ 在 x'-y' 平面上重叠. 最终, 求得不同位移 (x,y) 时乘积的面积, 并绘出 $C_{fh}(x,y)$ 的二维图, 如图中灰度图所示. $C_{fh}(x,y)$ 被称为 $f(x,y)$ 和 $h(x,y)$ 的互相关 (*cross-correlation*). 而 $C_{ff}(x,y)$ 是 $f(x,y)$ 的自相关 (*auto-correlation*).

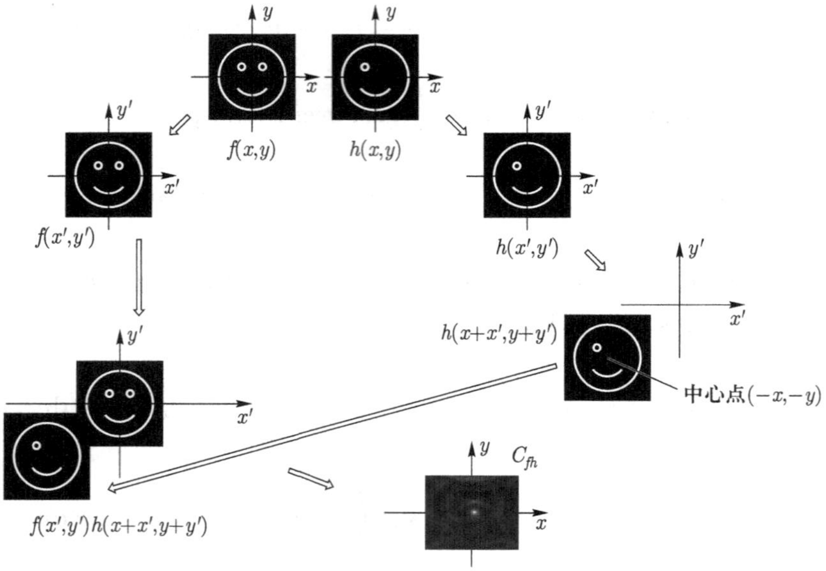

图 2.3-11　二维相关的概念

举例: 用图解法求解相关

给定 $x(t)=a\mathrm{e}^{-at}u(t), a>0$ 及 $h(t)=u(t)$, 用图解法 (*graphical method*) 来求其相关. 根据式 (2.3-30) 中对一维信号卷积的定义, 有

$$C_{xh}(t)=x(t)\otimes h(t)=\int\limits_{-\infty}^{\infty} x^*(t')h(t+t')\mathrm{d}t'.$$

需要求出两个函数 $x^*(t')$ 和 $h(t+t')$ 在不同 t 值下乘积的面积. 我们先来想想如何生成

Correlation

We will now discuss another important operation called *correlation*. The correlation, $C_{fh}(x,y)$, of two functions $f(x,y)$ and $h(x,y)$, is defined as

$$C_{fh}(x,y)=\iint\limits_{-\infty}^{\infty} f^*(x',y')h(x+x',y+y')\mathrm{d}x'\mathrm{d}y' = f(x,y)\otimes h(x,y), \tag{2.3-30}$$

where f^* is the complex conjugate of f, and $\otimes$ is a symbol denoting the correlation of $f(x,y)$ and $h(x,y)\cdot f(x,y)\otimes h(x,y)$ reads as f correlates with h. Let us now illustrate the correlation of the two functions, $f(x,y)$ and $h(x,y)$. Similar to the convolution of the two functions, the correlation operation involves the calculation of the areas under the product of the two functions, $f^*(x',y')$ and $h(x+x',y+y')$, for different shifts (x,y). For simplicity, we assume f is real, that is, $f^*=f$, for the illustration of the correlation operation. The two original functions are shown in the first row in Figure 2.3-11. On the second row, we put them on the x',y'-axis. Now we have $f(x',y')$ on a graph. To create the graph of $h(x+x',y+y')$, we shift $h(x',y')$ by x and y to get $h(x+x',y+y')$ on the x',y'-plane. We can now overlap $f(x',y')$ and $h(x+x',y+y')$ as shown in the left-hand side of the third row on the x',y'-plane. Finally, we calculate the areas of the product for different shifts (x,y) to obtain a 2-D plot of $C_{fh}(x,y)$, shown as a gray-scale plot in the figure. $C_{fh}(x,y)$ is called the *cross-correlation* of $f(x,y)$ and $h(x,y)$, and $C_{ff}(x,y)$ is the *auto-correlation* of $f(x,y)$.

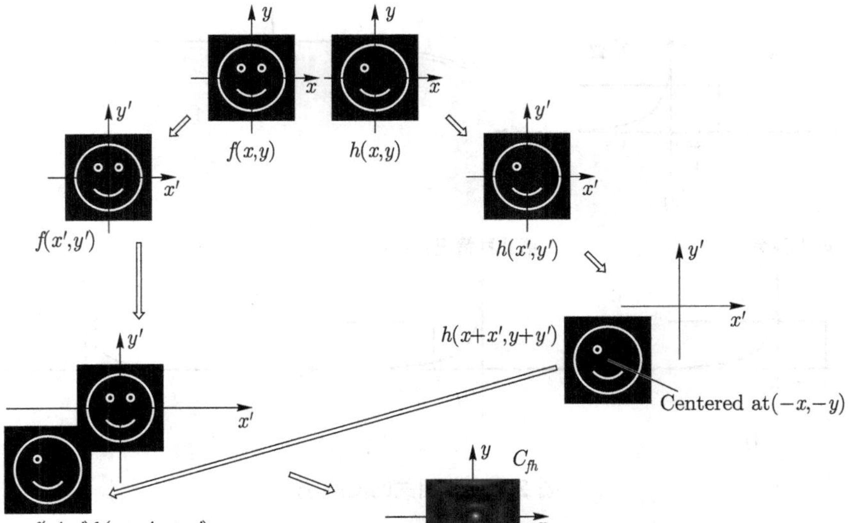

Figure 2.3-11 Concept of 2-D correlation

Example: Graphical Method to Find Correlation

Given $x(t)=a\mathrm{e}^{-at}u(t), a>0$ and $h(t)=u(t)$, we evaluate their correlation by a graphical method. According to the definition of convolution in Eq. (2.3-30) for 1-D signals, we have

$$C_{xh}(t)=x(t)\otimes h(t)=\int\limits_{-\infty}^{\infty} x^*(t')h(t+t')\mathrm{d}t'.$$

We need to find the areas under the product of two functions, $x^*(t')$ and $h(t+t')$, for different values of t. Let us visualize how to create the product $x^*(t')h(t+t')$. From the

$x^*(t')h(t+t')$. 由原函数 $x(t)$ 和 $h(t)$, 画出如图 2.3-12a) 中所示的 $x(t')$ 和 $h(t')$. 现在图中有了 $x(t')$, 为了得到 $h(t+t')$ 的图形, 如图 2.3-12b) 中所示, 将 $h(t')$ 平移 t 得到 $h(t+t')$, 然后如图 2.3-12c) 所示, 将 $x(t')$ 和 $h(t+t')$ 在一个图上重叠. 这个过程引导我们确定被积函数和相关积分的积分限. 有两种情况关系到相关积分时积分限的确定. 第 I 种情况是当 $t>0$ 时, 利用图 2.3-12d) 来进行相关积分

$$C_{xh}(t)=x(t)\otimes h(t)=\int_{-\infty}^{\infty}x(t')h(t+t')\mathrm{d}t'=\int_{0}^{\infty}a\mathrm{e}^{-at'}\mathrm{d}t'=1.$$

第Ⅱ种情况是当 $t<0$ 时, 利用图 2.3-12e) 来进行相关积分

$$C_{xh}(t)=x(t)\otimes h(t)=\int_{-\infty}^{\infty}x(t')h(t+t')\mathrm{d}t'=\int_{|t|}^{\infty}a\mathrm{e}^{-at'}\mathrm{d}t'=\mathrm{e}^{-a|t|}.$$

$C_{xh}(t)$ 的最终图示如图 2.3-12f) 所示.

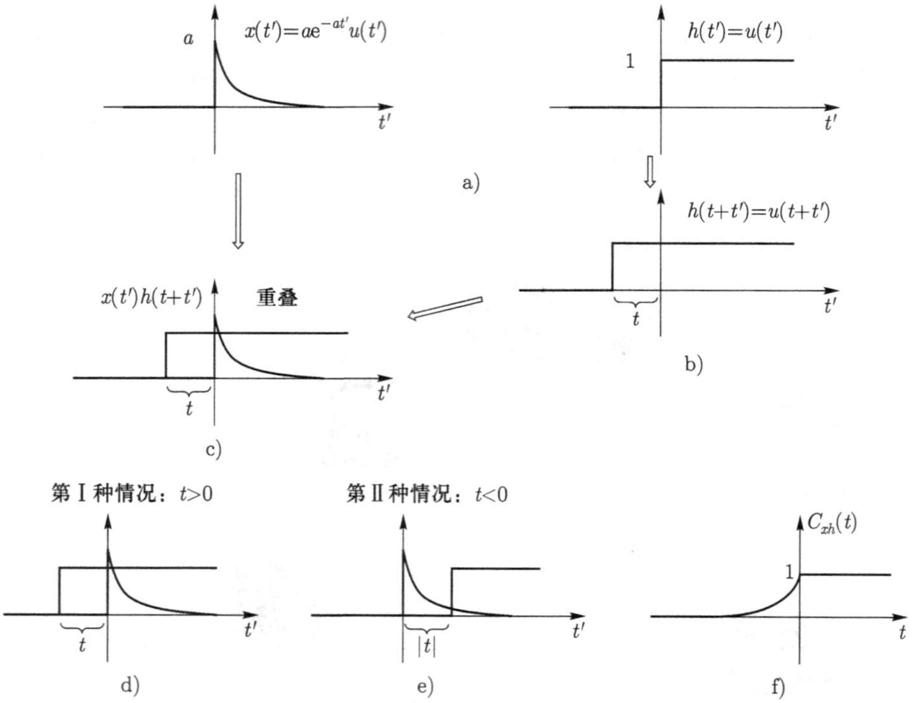

图 2.3-12　相关的图解法

匹配滤波 (*matched filtering*)

在线性空不变 (LSI) 系统中, 匹配滤波器 (*matched filter*) 是指, 如果输入信号 (图像) $s(x,y)$ 的脉冲响应 $h(x,y)$ 由 $s^*(-x,-y)$ 给定, 则称滤波器与该特定的输入信号 (图像) $s(x,y)$ 相匹配. 在 LSI 系统中, 通过将脉冲响应与其输入信号 $s(x,y)$ 进行卷积, 即可得到输出

$$o(x,y)=s(x,y)*h(x,y)=s(x,y)*s^*(-x,-y).$$

将上式重新写为

$$o(x,y)=s^*(-x,-y)*s(x,y)=\iint_{-\infty}^{\infty}s^*(-x',-y')s(x-x',y-y')\mathrm{d}x'\mathrm{d}y'.$$

original functions $x(t)$ and $h(t)$, we plot $x(t')$ and $h(t')$ as shown in Figure 2.3-12a. Now we have $x(t')$ on a graph. To create the graph of $h(t+t')$, we shift $h(t')$ by t to get $h(t+t')$, as shown in Figure 2.3-12b. We then overlap $x(t')$ and $h(t+t')$ on a single graph, which is shown in Figure 2.3-12c. This graph guides us to set up the integrand and the limits for the integral of correlation. There are two cases involved in terms of setting up the limits of the correlation integral. In case I, when $t>0$, we use Figure 2.3-12d to help setting up the correlation integral:

$$C_{xh}(t) = x(t) \otimes h(t) = \int_{-\infty}^{\infty} x(t')h(t+t')\mathrm{d}t' = \int_{0}^{\infty} a\mathrm{e}^{-at'}\mathrm{d}t' = 1.$$

In case Ⅱ, when $t<0$, we use Figure 2.3-12e to set up the correlation integral:

$$C_{xh}(t) = x(t) \otimes h(t) = \int_{-\infty}^{\infty} x(t')h(t+t')\mathrm{d}t' = \int_{|t|}^{\infty} a\mathrm{e}^{-at'}\mathrm{d}t' = \mathrm{e}^{-a|t|}.$$

The final plot of $C_{xh}(t)$ is shown in Figure 2.3-12f.

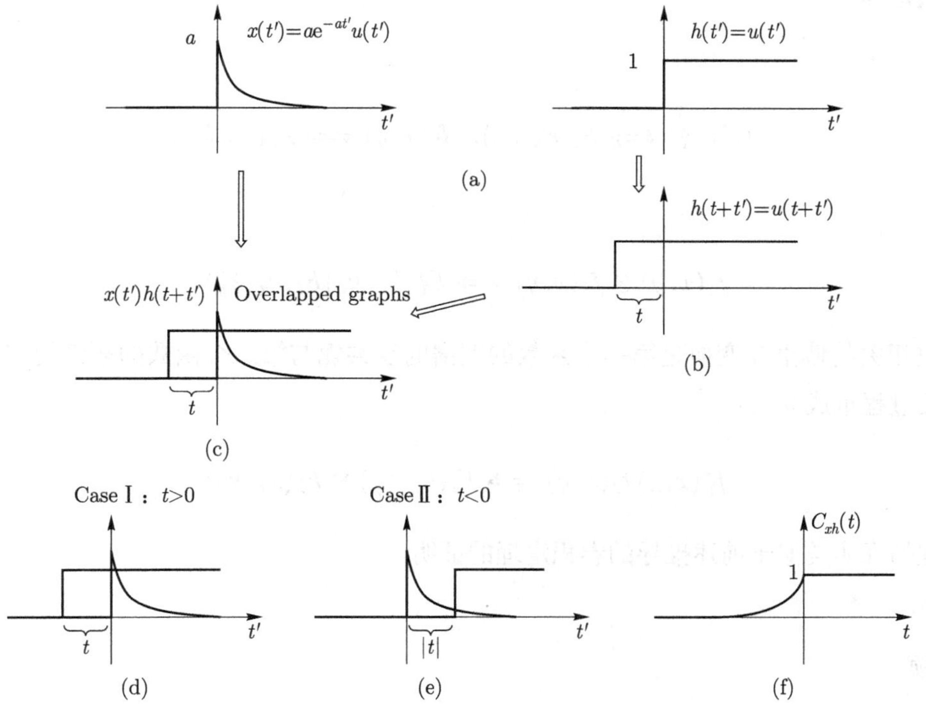

Figure 2.3-12 Graphical method for correlation

Matched Filtering

In a linear space-invariant (LSI) system, a *matched filter* is said to be matched to the particular signal (image) $s(x,y)$ if its impulse response $h(x,y)$ is given by $s^*(-x,-y)$. In LSI systems, we find the output $o(x,y)$ by convolving the input $s(x,y)$ with its impulse response:

$$o(x,y) = s(x,y) * h(x,y) = s(x,y) * s^*(-x,-y).$$

Let us rewrite the above as follows:

$$o(x,y) = s^*(-x,-y) * s(x,y) = \iint_{-\infty}^{\infty} s^*(-x',-y')s(x-x',y-y')\mathrm{d}x'\mathrm{d}y'.$$

让 $-x'=x''$ 及 $-y'=y''$, 上述积分变为

$$o(x,y)=\iint\limits_{-\infty}^{\infty} s^*(x'',y'')s(x+x'',y+y'')\mathrm{d}x''\mathrm{d}y''$$
$$=s(x,y)\otimes s(x,y)=C_{ss}(x,y).$$

可以发现, 如果输入为 $g(x,y)$, 则

$$o(x,y)=g(x,y)*h(x,y)=g(x,y)*s^*(-x,-y)$$
$$=s(x,y)\otimes g(x,y)=C_{sg}(x,y).$$

因此, 在匹配滤波中进行两个函数的相关, 可以证明, $|C_{ss}(0,0)|\geqslant|C_{ss}(x,y)|$, 这意味着自相关的强度总是中央最大, 实际中可利用自相关来作为比较两个函数相似性的一种手段. 因此, 匹配滤波的概念在光学模式识别 (*optical pattern recognition*) 中具有一定意义, 第五章讨论光学中的复数滤波 (*complex filtering*) 时我们会再回到这个话题.

相关定理

如果

$$f_1(x,y)\Longleftrightarrow F_1(k_x,k_y),\quad f_2(x,y)\Longleftrightarrow F_2(k_x,k_y),$$

那么

$$f_1(x,y)\otimes f_2(x,y)\Longleftrightarrow F_1^*(k_x,k_y)F_2(k_x,k_y). \tag{2.3-31}$$

两个函数相关的傅里叶变换是第一个函数的频谱的复共轭与第二个函数的频谱的乘积. 这个定理的反过程也成立, 即

$$f_1^*(x,y)f_2(x,y)\Longleftrightarrow F_1(k_x,k_y)\otimes F_2(k_x,k_y). \tag{2.3-32}$$

上述定理的证明类似于前述推导的卷积定理的证明.

习题

2.1 对于图题 2.1 所示的信号 $x(t)$, 画出: a) $x(2t-4)$, b) $x(2-4t)$.

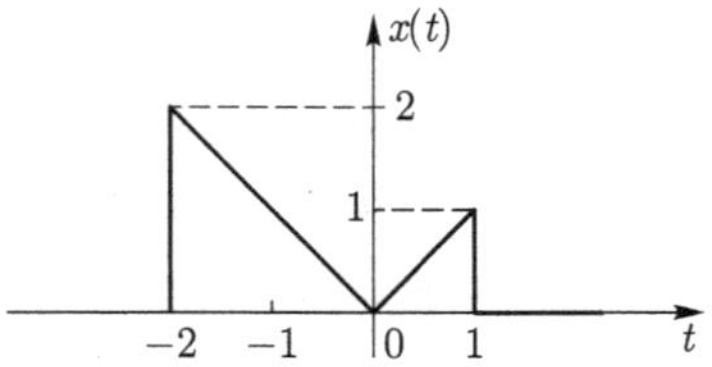

图题 2.1

2.2 求以下积分:

a) $\displaystyle\int_{-\infty}^{\infty}\delta(t+2)\mathrm{e}^{-2(t-1)}\mathrm{d}t.$

We let $-x' = x''$ and $-y' = y''$, the above integral becomes

$$o(x,y) = \iint_{-\infty}^{\infty} s^*(x'', y'')s(x+x'', y+y'')\mathrm{d}x''\mathrm{d}y''$$
$$= s(x,y) \otimes s(x,y) = C_{ss}(x,y).$$

Note that if the input is $g(x,y)$, we have

$$o(x,y) = g(x,y) * h(x,y) = g(x,y) * s^*(-x,-y)$$
$$= s(x,y) \otimes g(x,y) = C_{sg}(x,y).$$

Therefore, in matched filtering, we perform correlation of two functions. As it turns out, $|C_{ss}(0,0)| \geqslant |C_{ss}(x,y)|$, which means the magnitude of the autocorrelation always has a central maximum, facilitating the use of autocorrelation as a means for comparing the similarity of two functions. *Matched filtering* is, therefore, a useful concept in *optical pattern recognition*. We will return to this topic in Chapter 5 when we discuss *complex filtering* in optics.

Correlation Theorems

If

$$f_1(x,y) \Longleftrightarrow F_1(k_x,k_y), \quad f_2(x,y) \Longleftrightarrow F_2(k_x,k_y),$$

then

$$f_1(x,y) \otimes f_2(x,y) \Longleftrightarrow F_1^*(k_x,k_y)F_2(k_x,k_y). \tag{2.3-31}$$

The Fourier transform of the correlation of two functions is the product of the complex conjugate of the spectrum of the first function with the spectrum of the second function. The converse of this theorem also holds, that is,

$$f_1^*(x,y)f_2(x,y) \Longleftrightarrow F_1(k_x,k_y) \otimes F_2(k_x,k_y). \tag{2.3-32}$$

The proof of the above theorems follows similarly to that of the convolution theorem derived earlier.

Problems

2.1 For the signal $x(t)$ shown in Figure P.2.1, sketch (a) $x(2t-4)$ and (b) $x(2-4t)$.

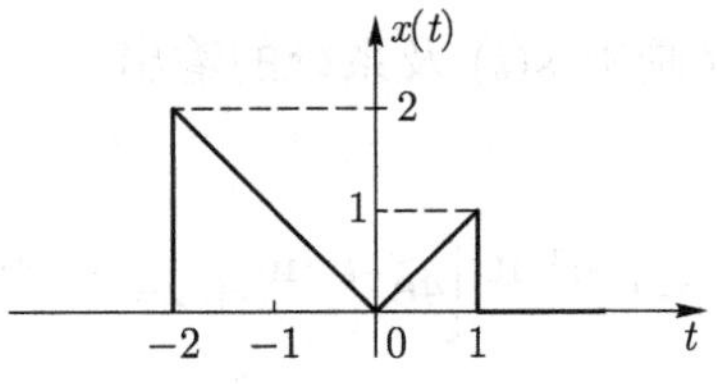

Figure P.2.1

2.2 Evaluate the following integrals:

(a) $\displaystyle\int_{-\infty}^{\infty} \delta(t+2)\mathrm{e}^{-2(t-1)}\mathrm{d}t.$

b) $\int_{-\infty}^{\infty}\delta(t+2)\mathrm{e}^{-2(t-1)}u(t)\mathrm{d}t$.

c) $\int_{-\infty}^{\infty}\delta(at)\mathrm{d}t$, a 是正实数.

2.3 判断由 $y(t)=t^2x(t)$ 描述的系统是否为: a) 线性的, b) 时不变的.

2.4 判断以下式

$$\left[\frac{\mathrm{d}y(t)}{\mathrm{d}t}\right]^2+2y(t)=x(t)$$

描述的系统是线性还是非线性的.

2.5 对于图题 2.5 所示电路:

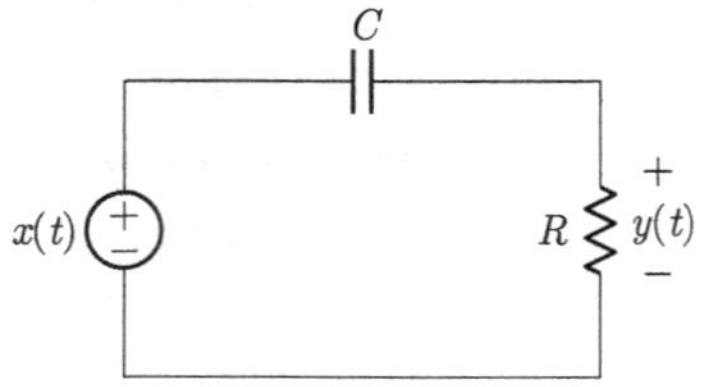

图题 2.5

a) 求输出 $y(t)$ 与输入 $x(t)$ 之间的微分方程.

b) 用两种方法求出电路的频率响应函数 $H(\omega)$, 并确定该电路是进行低通滤波还是高通滤波.

2.6 给定信号 $x(t)=10rect\left(\frac{t}{6}\right)\mathrm{e}^{-t^2}$, 噪声源 $n(t)=\sin(10t)+\frac{1}{2}\sin(12t)$, 系统 A 的脉冲响应 $h_{\mathrm{A}}(t)=rect(10t)$, 系统 B 的脉冲响应为 $h_{\mathrm{B}}(t)=rect\left(\frac{t}{2}\right)$, 编写 MATLAB 代码完成以下任务:

a) 画出信号 $x(t)$.

b) 画出噪声信号 $x_n(t)=x(t)+n(t)$.

c) 画出 $x_n(t)*h_{\mathrm{A}}(t)$.

d) 画出 $x_n(t)*h_{\mathrm{B}}(t)$.

e) 从以上结果分析, 哪个系统除噪的效果更好? 并给出解释.

2.7 假设信号 $s(t)=\mathrm{e}^{-t^2/10}\left[4\mathrm{e}^{-t^2/10}+2\mathrm{e}^{-(t-1)^2}\right]$ 被输入到脉冲响应为 $h(t)$ 的线性时不变系统:

a) 设计 $h(t)$ 使该系统与此输入 $s(t)$ 相匹配.

b) 编写一个 MTALAB 程序画出 $s(t)$ 及系统的输出.

c) 另一个信号 $g(t)$ 为

$$g(t)=\mathrm{e}^{-t^2/10}\left[4\mathrm{e}^{-t^2/10}+2\mathrm{e}^{-(t-2)^2}\right],$$

被输入给同 a) 中定义的 $h(t)$ 一样的系统, 画出 $g(t)$ 和该系统的输出.

d) 根据 a)、b) 和 c) 的结果, 给出一些结论.

2.8 求 $rect\left(\frac{x}{x_0}\right)*rect\left(\frac{x}{x_0}\right)$:

a) 用卷积的图解法.

(b) $\int_{-\infty}^{\infty} \delta(t+2)\mathrm{e}^{-2(t-1)}u(t)\mathrm{d}t$.

(c) $\int_{-\infty}^{\infty} \delta(at)\mathrm{d}t$, where a is real and positive.

2.3 Determine if the system described by $y(t) = t^2x(t)$ is (a) linear, (b) time invariant.

2.4 Determine if the system described by

$$\left[\frac{\mathrm{d}y(t)}{\mathrm{d}t}\right]^2 + 2y(t) = x(t)$$

is linear or nonlinear.

2.5 For the circuit shown in Figure P.2.5:

(a) Find the differential equation relating the output $y(t)$ to the input $x(t)$.

(b) Find the frequency response function of the circuit $H(\omega)$ using two methods and determine if the circuit performs low-pass filtering or high-pass filtering (Figure P.2.5).

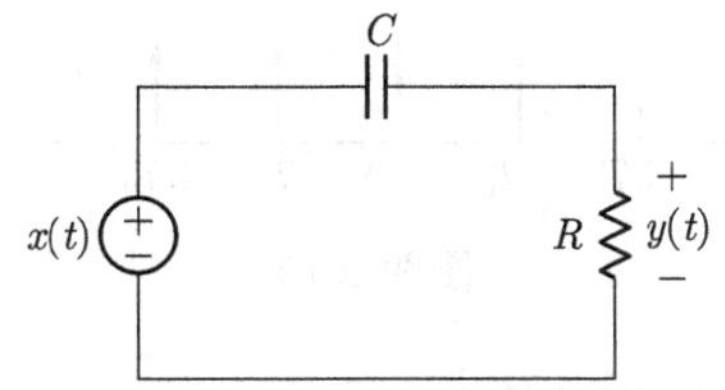

Figure P.2.5

2.6 Given a signal $x(t) = 10rect\left(\frac{t}{6}\right)\mathrm{e}^{-t^2}$, a noise source $n(t) = \sin(10t) + \frac{1}{2}\sin(12t)$, an impulse response of system A being $h_{\mathrm{A}}(t) = rect(10t)$, and an impulse response of system B being $h_{\mathrm{B}}(t) = rect\left(\frac{t}{2}\right)$, write MATLAB codes to perform the following tasks:

(a) Plot $x(t)$.

(b) Plot noisy signal $x_n(t) = x(t) + n(t)$.

(c) Plot $x_n(t) * h_{\mathrm{A}}(t)$.

(d) Plot $x_n(t) * h_{\mathrm{B}}(t)$.

(e) From the above results, which system does a better job of cleaning up the noise? Please explain.

2.7 Suppose a signal is described by $s(t) = \mathrm{e}^{-t^2/10}\left[4\mathrm{e}^{-t^2/10} + 2\mathrm{e}^{-(t-1)^2}\right]$ and this signal is fed to a linear and time-invariant system with its impulse response given by $h(t)$:

(a) Design $h(t)$ such that the system is matched to the input $s(t)$.

(b) Write a MTALAB program to plot $s(t)$ and the system's output.

(c) Another signal $g(t)$ described by

$$g(t) = \mathrm{e}^{-t^2/10}\left[4\mathrm{e}^{-t^2/10} + 2\mathrm{e}^{-(t-2)^2}\right]$$

is fed to the same system with $h(t)$ defined in part (a). Plot $g(t)$ and the system's output.

(d) Make some concluding remarks from the results of (a), (b), and (c).

2.8 Find $rect\left(\frac{x}{x_0}\right) * rect\left(\frac{x}{x_0}\right)$:

(a) Using the graphical method of convolution.

b) 用傅里叶变换.

2.9 对图题 2.9 中所示的 $x_1(t)$ 和 $x_2(t)$, 用图解法求出在 $-1 < t < 1$ 时的 $x_1(t) * x_2(t)$.

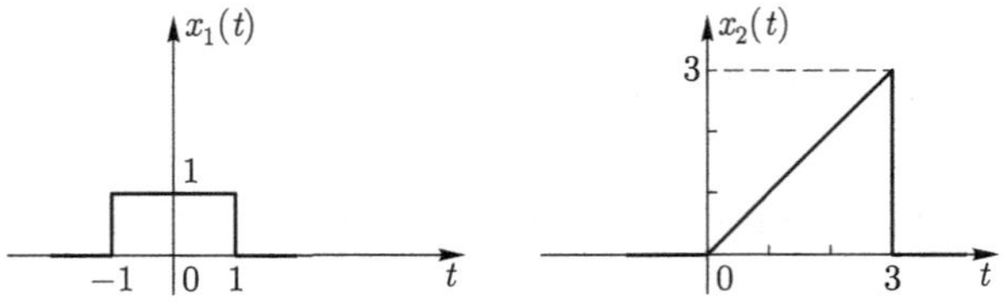

图题 2.9

2.10 求周期为 T_0 的脉冲序列的傅里叶级数并画出相应的谱. 该脉冲序列如图题 2.10 所示, 其数学定义如下

$$x(t)=\sum_{n=-\infty}^{\infty}\delta(t-nT_0)=\delta_{T_0}(t).$$

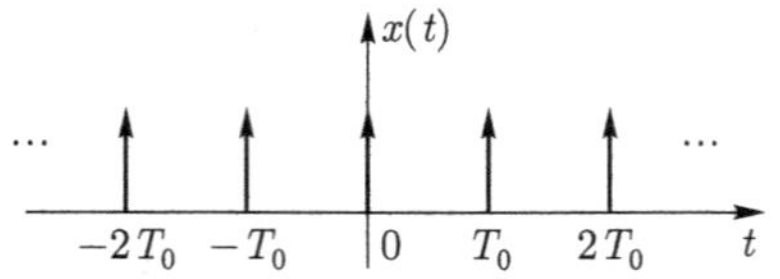

图题 2.10

2.11 求脉冲序列 $\delta_{T_0}(t)$ 的傅里叶变换.

2.12 验证表 2.1 中第 5 项的傅里叶变换对.

2.13 a) 利用等式

$$J_1(\alpha)=\frac{1}{\alpha}\int_0^{\alpha}\beta J_0(\beta)\mathrm{d}\beta,$$

证明

$$\mathcal{B}\left\{circ\left(\frac{r}{r_0}\right)\right\}=\frac{2\pi r_0}{k_r}J_1(r_0k_r)=2\pi r_0^2 jinc(r_0k_r).$$

b) 利用 a) 的结果求 $jinc(0)$.

2.14 定义一维空域函数的傅里叶变换的形式如下

$$\mathcal{F}\{f(x)\}=F\left(k_x\right)=\int_{-\infty}^{\infty}f(x)\mathrm{e}^{\mathrm{j}k_x x}\mathrm{d}x.$$

求 $\mathcal{F}\left\{\mathrm{e}^{\mathrm{j}\pi x}f\left(x-\frac{1}{2}\right)\right\}$.

2.15 验证表 2.2 中傅里叶变换的共轭性质.

2.16 如果

$$f_1(x,y)\Longleftrightarrow F_1(k_x,k_y)\text{ 且 }f_2(x,y)\Longleftrightarrow F_2(k_x,k_y),$$

证明:

a) $f_1(x,y)\otimes f_2(x,y)\Longleftrightarrow F_1^*(k_x,k_y)F_2(k_x,k_y)$.

b) $f_1^*(x,y)f_2(x,y)\Longleftrightarrow F_1(k_x,k_y)\otimes F_2(k_x,k_y)$.

(b) Using the Fourier transform.

2.9 For $x_1(t)$ and $x_2(t)$ shown in Figure P.2.9, find $x_1(t) * x_2(t)$ for $-1 < t < 1$ using the graphical method.

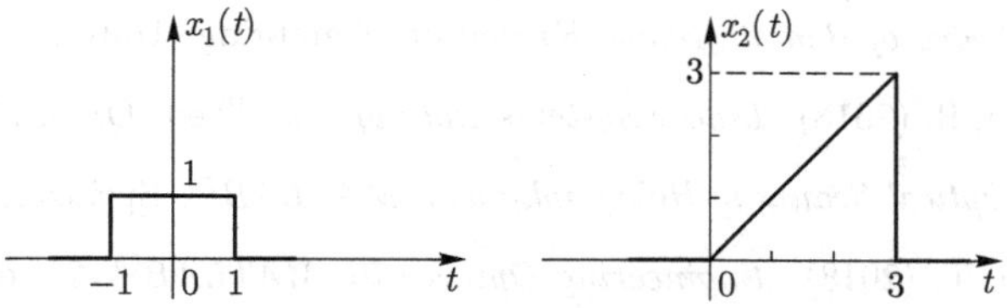

Figure P.2.9

2.10 Find the Fourier series and sketch the corresponding spectrum for the impulse train with period T_0. The impulse train is depicted in Figure P.2.10 and defined mathematically as follows:

$$x(t) = \sum_{n=-\infty}^{\infty} \delta(t - nT_0) = \delta_{T_0}(t).$$

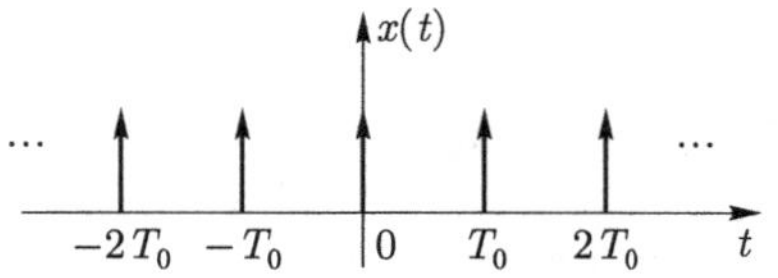

Figure P.2.10

2.11 Find the Fourier transform of the impulse train $\delta_{T_0}(t)$.

2.12 Verify the Fourier transform pair of item #5 in Table 2.1.

2.13 (a) Using the identity

$$J_1(\alpha) = \frac{1}{\alpha}\int_0^{\alpha} \beta J_0(\beta)\mathrm{d}\beta,$$

show that

$$\mathcal{B}\left\{circ\left(\frac{r}{r_0}\right)\right\} = \frac{2\pi r_0}{k_r}J_1(r_0 k_r) = 2\pi r_0^2 jinc(r_0 k_r).$$

(b) Using the result in part (a), find $jinc(0)$.

2.14 Let us define a 1-D version of the Fourier transform of a spatial function as follows:

$$\mathcal{F}\{f(x)\} = F(k_x) = \int_{-\infty}^{\infty} f(x)\mathrm{e}^{\mathrm{j}k_x x}\mathrm{d}x.$$

Find $\mathcal{F}\left\{\mathrm{e}^{\mathrm{j}\pi x} f\left(x - \frac{1}{2}\right)\right\}$.

2.15 Verify the Fourier transform property of conjugation in Table 2.2.

2.16 If

$$f_1(x, y) \Longleftrightarrow F_1(k_x, k_y) \text{ and } f_2(x, y) \Longleftrightarrow F_2(k_x, k_y),$$

show that:

(a) $f_1(x, y) \otimes f_2(x, y) \Longleftrightarrow F_1^*(k_x, k_y)F_2(k_x, k_y)$.

(b) $f_1^*(x, y)f_2(x, y) \Longleftrightarrow F_1(k_x, k_y) \otimes F_2(k_x, k_y)$.

参考文献

[1] Banerjee, P. P. and Poon, T.-C. (1991). *Principles of Applied Optics.* Irwin, Illinois.

[2] Blahut, R. (2004). *Theory of Remote Image Formation.* Cambridge University Press, Cambridge.

[3] Lathi, B. P. and Green, R. (2018). *Linear Systems and Signals*, 3rd ed. Oxford University Press, Oxford.

[4] Poon, T.-C. (2007). *Optical Scanning Holography with MATLAB®*. Springer, New York.

[5] Poon, T.-C. and Kim, T. (2018). *Engineering Optics with MATLAB®*, 2nd ed. World Scientific, New Jersey.

[6] Ulaby, F. T. and Yagle, A.E. (2016). *Engineering Signals and Systems in Continuous and Discrete Time*, 2nd ed. National Technology & Science Press, Austin.

Bibliography

[1] Banerjee, P. P. and Poon, T.-C. (1991). *Principles of Applied Optics.* Irwin, Illinois.

[2] Blahut, R. (2004). *Theory of Remote Image Formation.* Cambridge University Press, Cambridge.

[3] Lathi, B. P. and Green, R. (2018). *Linear Systems and Signals*, 3rd ed. Oxford University Press, Oxford.

[4] Poon, T.-C. (2007). *Optical Scanning Holography with MATLAB®*. Springer, New York.

[5] Poon, T.-C. and Kim, T. (2018). *Engineering Optics with MATLAB®*, 2nd ed. World Scientific, New Jersey.

[6] Ulaby, F. T. and Yagle, A.E. (2016). *Engineering Signals and Systems in Continuous and Discrete Time*, 2nd ed. National Technology & Science Press, Austin.

第三章　波的传播与波动光学

在第一章中, 我们介绍了高斯光学, 并使用矩阵形式来描述通过光学系统的光线. 光线是基于光的粒子特性. 因为光有双重性, 所以光也是波.1924 年, 德布罗意 (de Borglie) 提出了 de Borglie 假说 (*de Borglie hypothesis*), 该假说将波长和动量之间的关系表述为 $\lambda_0 = h/p$, 其中, p 为粒子 (称为光子) 运动时的动量 (单位为 kg·m/s), λ_0 是波的波长 (单位为 m), $h = 6.626\times10^{-34}$ J·s 是普朗克常量. 在本章, 我们将探索光的波动特性, 主要是为了解释如光的干涉和衍射等现象的波动效应.

3.1　麦克斯韦方程

在电磁学中, 我们关心电磁场 [*electromagnetic* (EM) *field*] 的四个矢量: 电场强度 (*electric field*) $\boldsymbol{E}$ (V/m), 电感强度 (*electric flux density*) $\boldsymbol{D}$ (C/m^2), 磁场强度 (*magnetic field*) $\boldsymbol{H}$ (A/m) 和磁感强度 (*magnetic flux density*) $\boldsymbol{B}$ (Wb/m^2). 电磁场是空间中的位置如 x, y, z 和时间 t 的函数, 这些量由四个麦克斯韦方程 (*Maxwell's equations*) 所决定, 如下所示.

电场高斯定律

电场高斯定律 (*electric Gauss's law*) 描述了静电场和产生静电场的电荷之间的关系. 如下式所述

$$\boldsymbol{\nabla}\cdot\boldsymbol{D} = \rho_v, \tag{3.1-1}$$

其中, ρ_v 是电荷密度 (*electric charge density*) (C/m^3). 符号 $\boldsymbol{\nabla}$ 是矢量偏微分算子 (*vector partial-differentiation operator*), 通常称为 $\boldsymbol{\nabla}$ (del) 算子, 为

$$\boldsymbol{\nabla} = \frac{\partial}{\partial x}\hat{\boldsymbol{x}} + \frac{\partial}{\partial y}\hat{\boldsymbol{y}} + \frac{\partial}{\partial z}\hat{\boldsymbol{z}}, \tag{3.1-2}$$

其中, $\hat{\boldsymbol{x}}$, $\hat{\boldsymbol{y}}$ 和 $\hat{\boldsymbol{z}}$ 分别为 x, y 和 z 方向上的单位矢量 (*unit vector*). 对于分量为 A_x, A_y 和 A_z 的矢量 $\boldsymbol{A}$, 可以写出 $\boldsymbol{A} = A_x\hat{\boldsymbol{x}} + A_y\hat{\boldsymbol{y}} + A_z\hat{\boldsymbol{z}}$. $\boldsymbol{\nabla}\cdot\boldsymbol{A}$ 被称为 $\boldsymbol{A}$ 的散度 (*divergence*) 或者简称 "del dot $\boldsymbol{A}$", 其中, 点表示两个矢量的点乘 (*dot product*) 运算或标量积 (*scalar product*) 运算. 另一个名称 "标量积" 强调点乘的结果是一个标量, 而非矢量. $\boldsymbol{\nabla}\cdot\boldsymbol{A}$ 为矢量 $\boldsymbol{\nabla}$ 和 $\boldsymbol{A}$ 的点乘, 可写为

$$\begin{aligned}\boldsymbol{\nabla}\cdot\boldsymbol{A} &= \left(\frac{\partial}{\partial x}\hat{\boldsymbol{x}} + \frac{\partial}{\partial y}\hat{\boldsymbol{y}} + \frac{\partial}{\partial z}\hat{\boldsymbol{z}}\right)\cdot(A_x\hat{\boldsymbol{x}} + A_y\hat{\boldsymbol{y}} + A_z\hat{\boldsymbol{z}})\\ &= \frac{\partial A_x}{\partial x} + \frac{\partial A_y}{\partial y} + \frac{\partial A_z}{\partial z}.\end{aligned} \tag{3.1-3}$$

给定两个矢量 $\boldsymbol{A}$ 和 $\boldsymbol{B}$, 其点乘的定义为

$$\boldsymbol{A}\cdot\boldsymbol{B} = |\boldsymbol{A}|\,|\boldsymbol{B}|\cos\theta_{AB} = AB\cos\theta_{AB},$$

3 Wave Propagation and Wave Optics

We have introduced Gaussian optics and used a matrix formalism to describe light rays through optical systems in Chapter 1. Light rays are based on the particle nature of light. Since light has a dual nature, light is waves as well. In 1924, de Broglie formulated the *de Broglie hypothesis*, which relates wavelength and momentum as $\lambda_0 = h/p$, where p is the momentum (in kg·m/s) of the particle (called photon) when the particle moves, λ_0 is the wavelength (in m) of the wave, and $h = 6.626 \times 10^{-34}$ J·s is Planck's constant. In this chapter, we explore the wave nature of light, which accounts for wave effects such as interference and diffraction.

3.1 Maxwell's Equations

In electromagnetics, we are concerned with four vector quantities called *electromagnetic (EM) fields:* the electric field $\boldsymbol{E}$ (V/m), the electric flux density $\boldsymbol{D}$ (C/m^2), the magnetic field $\boldsymbol{H}$ (A/m), and the magnetic flux density $\boldsymbol{B}$ (Wb/m^2). These EM fields are functions of position in space, that is, x, y, z, and time, t, and they are governed by four Maxwell's equations as follows.

Electric Gauss's Law

Electric Gauss's law describes the relationship between a static electric field and the electric charges that generate it. The law is stated as

$$\boldsymbol{\nabla} \cdot \boldsymbol{D} = \rho_v, \tag{3.1-1}$$

where ρ_v is the electric charge density (C/m^3). The symbol $\boldsymbol{\nabla}$ is a vector partial-differentiation operator, often called the $\boldsymbol{\nabla}$ (del) operator and is given by

$$\boldsymbol{\nabla} = \frac{\partial}{\partial x}\hat{\boldsymbol{x}} + \frac{\partial}{\partial y}\hat{\boldsymbol{y}} + \frac{\partial}{\partial z}\hat{\boldsymbol{z}}, \tag{3.1-2}$$

where $\hat{\boldsymbol{x}}$, $\hat{\boldsymbol{y}}$, and $\hat{\boldsymbol{z}}$ are the *unit vectors* in the x, y, and z-directions, respectively. For a vector $\boldsymbol{A}$ with components A_x, A_y and A_z, we write $\boldsymbol{A} = A_x\hat{\boldsymbol{x}} + A_y\hat{\boldsymbol{y}} + A_z\hat{\boldsymbol{z}}$. $\boldsymbol{\nabla} \cdot \boldsymbol{A}$ is called the *divergence* of $\boldsymbol{A}$ or simply called "del dot $\boldsymbol{A}$," where the dot denotes an operation equivalent to the *dot product* or *scalar product* of two vectors. The alternative name "scalar product" emphasizes that the result of the product is a scalar, rather than a vector. $\boldsymbol{\nabla} \cdot \boldsymbol{A}$ is the dot product of the vectors $\boldsymbol{\nabla}$ and $\boldsymbol{A}$ given by

$$\begin{aligned}\boldsymbol{\nabla} \cdot \boldsymbol{A} &= \left(\frac{\partial}{\partial x}\hat{\boldsymbol{x}} + \frac{\partial}{\partial y}\hat{\boldsymbol{y}} + \frac{\partial}{\partial z}\hat{\boldsymbol{z}}\right) \cdot (A_x\hat{\boldsymbol{x}} + A_y\hat{\boldsymbol{y}} + A_z\hat{\boldsymbol{z}}) \\ &= \frac{\partial A_x}{\partial x} + \frac{\partial A_y}{\partial y} + \frac{\partial A_z}{\partial z}.\end{aligned} \tag{3.1-3}$$

Given two vectors $\boldsymbol{A}$ and $\boldsymbol{B}$, the dot product is defined as

$$\boldsymbol{A} \cdot \boldsymbol{B} = |\boldsymbol{A}|\,|\boldsymbol{B}| \cos\theta_{AB} = AB\cos\theta_{AB},$$

其中, $|\boldsymbol{A}| = A$ 且 $|\boldsymbol{B}| = B$ 分别为 $\boldsymbol{A}$ 和 $\boldsymbol{B}$ 的强度 (*magnitude*), θ_{AB} 是 $\boldsymbol{A}$ 和 $\boldsymbol{B}$ 之间较小的角度. 可以看出, 点乘运算具有交换律 (*commutative property*) 和分配律 (*distributive property*), 即

$$\boldsymbol{A}\cdot\boldsymbol{B} = \boldsymbol{B}\cdot\boldsymbol{A} \text{ (交换律)},$$
$$\boldsymbol{A}\cdot(\boldsymbol{B}+\boldsymbol{C}) = \boldsymbol{A}\cdot\boldsymbol{B}+\boldsymbol{A}\cdot\boldsymbol{C} \text{ (分配律)}.$$

举例: 给定方向上某矢量的分量

可用点乘来求给定方向上某矢量的分量. 标示如图 3.1-1所示的 x-y 平面上的 $\boldsymbol{B}$, $\boldsymbol{B}$ 沿单位矢量 $\hat{\boldsymbol{x}}$ 的分量为

$$B_x = \boldsymbol{B}\cdot\hat{\boldsymbol{x}} = |\boldsymbol{B}||\hat{\boldsymbol{x}}|\cos\theta_{Bx} = B\cos\theta_{Bx}.$$

类似地, $\boldsymbol{B}$ 沿单位矢量 $\hat{\boldsymbol{y}}$ 的分量为

$$\begin{aligned} B_y = \boldsymbol{B}\cdot\hat{\boldsymbol{y}} &= |\boldsymbol{B}||\hat{\boldsymbol{y}}|\cos\theta_{By} = B\cos\theta_{By} \\ &= B\cos(90^\circ-\theta_{Bx}) = B\sin\theta_{Bx}. \end{aligned}$$

所以, 可以写出

$$\boldsymbol{B} = B_x\hat{\boldsymbol{x}} + B_y\hat{\boldsymbol{y}} = (\boldsymbol{B}\cdot\hat{\boldsymbol{x}})\hat{\boldsymbol{x}} + (\boldsymbol{B}\cdot\hat{\boldsymbol{y}})\hat{\boldsymbol{y}}.$$

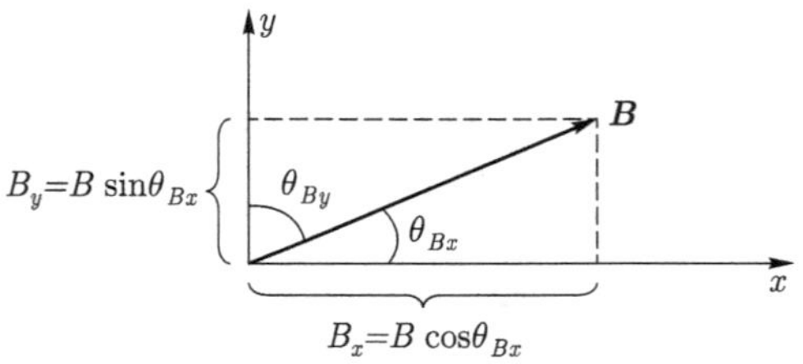

图 3.1-1 矢量分量示意图

磁场高斯定律

磁场高斯定律 (*magnetic Gauss's law*) 表明不存在 "磁单极子 (*magnetic monopole*)". 换句话说, 并不存在与电荷相类似的磁单极子. 例如, 条形磁铁的两端被称为北极和南极, 若这两个磁极同时存在, 则被称为磁偶极子 (*magnetic dipole*). 磁场高斯定律可以表述如下

$$\nabla\cdot\boldsymbol{B} = 0. \tag{3.1-4}$$

法拉第定律

法拉第定律 (*Faraday's law*) 描述了时变磁场如何产生电场, 定律用以下公式描述

$$\nabla\times\boldsymbol{E} = -\frac{\partial\boldsymbol{B}}{\partial t} \tag{3.1-5}$$

式 (3.1-2) 中定义过矢量算子 ∇, $\nabla\times\boldsymbol{A}$ 被称为 $\boldsymbol{A}$ 的旋度 (*curl*) 或简称为 "叉乘 (*del cross*)$\boldsymbol{A}$", 其中, 叉表示两个矢量的叉乘 (*cross product*) 或矢量积 (*vector product*) 运算. 另一个名称

where $|\boldsymbol{A}| = A$ and $|\boldsymbol{B}| = B$ are the magnitudes of $\boldsymbol{A}$ and $\boldsymbol{B}$, respectively. θ_{AB} is the smaller angle between $\boldsymbol{A}$ and $\boldsymbol{B}$. Note that the dot product is commutative and distributive, that is,

$$\boldsymbol{A}\cdot\boldsymbol{B} = \boldsymbol{B}\cdot\boldsymbol{A} \text{ (commutative property)},$$
$$\boldsymbol{A}\cdot(\boldsymbol{B}+\boldsymbol{C}) = \boldsymbol{A}\cdot\boldsymbol{B}+\boldsymbol{A}\cdot\boldsymbol{C} \text{ (distributive property)}.$$

Example: Component of a Vector in a Given Direction

We can use the dot product to find the component of a vector in a given direction. Let us denote $\boldsymbol{B}$ on the x-y plane as shown in Figure 3.1-1. The component of $\boldsymbol{B}$ along the unit vector $\hat{\boldsymbol{x}}$ is

$$B_x = \boldsymbol{B}\cdot\hat{\boldsymbol{x}} = |\boldsymbol{B}||\hat{\boldsymbol{x}}|\cos\theta_{Bx} = B\cos\theta_{Bx}.$$

Similarly, the component of $\boldsymbol{B}$ along the unit vector $\hat{\boldsymbol{y}}$ is

$$\begin{aligned} B_y &= \boldsymbol{B}\cdot\hat{\boldsymbol{y}} = |\boldsymbol{B}||\hat{\boldsymbol{y}}|\cos\theta_{By} = B\cos\theta_{By} \\ &= B\cos(90^\circ - \theta_{Bx}) = B\sin\theta_{Bx}. \end{aligned}$$

So we can write

$$\boldsymbol{B} = B_x\hat{\boldsymbol{x}} + B_y\hat{\boldsymbol{y}} = (\boldsymbol{B}\cdot\hat{\boldsymbol{x}})\hat{\boldsymbol{x}} + (\boldsymbol{B}\cdot\hat{\boldsymbol{y}})\hat{\boldsymbol{y}}.$$

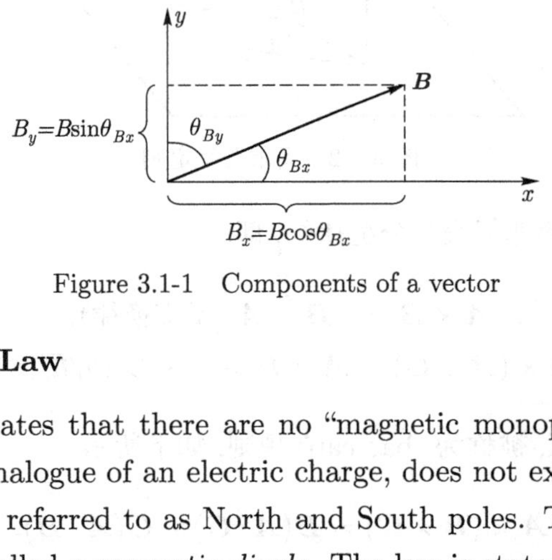

Figure 3.1-1 Components of a vector

Magnetic Gauss's Law

Magnetic Gauss's Law states that there are no "magnetic monopoles". In other words, a magnetic monopole, an analogue of an electric charge, does not exist. For example, the two ends of a bar magnet are referred to as North and South poles. The two poles exist at the same time and they are called a *magnetic dipole*. The law is stated as

$$\boldsymbol{\nabla}\cdot\boldsymbol{B} = 0. \tag{3.1-4}$$

Faraday's Law

Faraday's law describes how a time-varying magnetic field generates an electric field. The law is stated as

$$\boldsymbol{\nabla}\times\boldsymbol{E} = -\frac{\partial\boldsymbol{B}}{\partial t} \tag{3.1-5}$$

With the vector operator $\boldsymbol{\nabla}$ defined in Eq. (3.1-2), $\boldsymbol{\nabla}\times\boldsymbol{A}$ is called the *curl* of $\boldsymbol{A}$ or simply called "del cross $\boldsymbol{A}$," where the cross denotes an operation equivalent to the *cross product* or *vector product* of two vectors. The alternative name "vector product" emphasizes that the

“矢量积” 强调的是叉乘的结果为一个矢量. $\boldsymbol{\nabla}\times\boldsymbol{A}$ 为矢量 $\boldsymbol{\nabla}$ 和 $\boldsymbol{A}$ 的叉乘运算, 用下式给出

$$\begin{aligned}\boldsymbol{\nabla}\times\boldsymbol{A}&=\begin{vmatrix}\hat{\boldsymbol{x}}&\hat{\boldsymbol{y}}&\hat{\boldsymbol{z}}\\ \dfrac{\partial}{\partial x}&\dfrac{\partial}{\partial y}&\dfrac{\partial}{\partial z}\\ A_x&A_y&A_z\end{vmatrix}\\ &=\hat{\boldsymbol{x}}\begin{vmatrix}\dfrac{\partial}{\partial y}&\dfrac{\partial}{\partial z}\\ A_y&A_z\end{vmatrix}-\hat{\boldsymbol{y}}\begin{vmatrix}\dfrac{\partial}{\partial x}&\dfrac{\partial}{\partial z}\\ A_x&A_z\end{vmatrix}+\hat{\boldsymbol{z}}\begin{vmatrix}\dfrac{\partial}{\partial x}&\dfrac{\partial}{\partial y}\\ A_x&A_y\end{vmatrix}\\ &=\hat{\boldsymbol{x}}\left(\frac{\partial A_z}{\partial y}-\frac{\partial A_y}{\partial z}\right)-\hat{\boldsymbol{y}}\left(\frac{\partial A_z}{\partial x}-\frac{\partial A_x}{\partial z}\right)+\hat{\boldsymbol{z}}\left(\frac{\partial A_y}{\partial x}-\frac{\partial A_x}{\partial y}\right).\end{aligned}\tag{3.1-6}$$

举例: 叉乘运算

给定两个矢量 $\boldsymbol{A}$ 和 $\boldsymbol{B}$, 其叉乘运算被定义为

$$\boldsymbol{A}\times\boldsymbol{B}=AB\sin(\theta_{AB})\hat{\boldsymbol{n}},$$

其中, $\hat{\boldsymbol{n}}$ 是垂直于 $\boldsymbol{A}$ 和 $\boldsymbol{B}$ 所定义平面的单位向量, θ_{AB} 是 $\boldsymbol{A}$ 和 $\boldsymbol{B}$ 之间较小的角度, 如图 3.1-2 所示. $\hat{\boldsymbol{n}}$ 的方向用右手定则确定, 右手手指从 A 沿 θ_{AB} 向 B 弯曲四指, 此时大拇指指向为 $\hat{\boldsymbol{n}}$ 的方向.

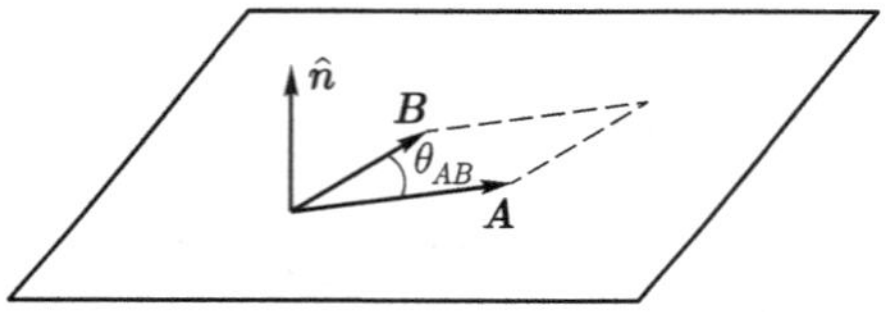

图 3.1-2 叉乘右手定则

叉乘运算是不服从交换律但服从分配律的, 即

$$\boldsymbol{A}\times\boldsymbol{B}=-\boldsymbol{B}\times\boldsymbol{A}\ (\text{反交换律}),$$
$$\boldsymbol{A}\times(\boldsymbol{B}+\boldsymbol{C})=\boldsymbol{A}\times\boldsymbol{B}+\boldsymbol{A}\times\boldsymbol{C}\ (\text{分配律}).$$

最后, 还有向量三重积, 被称为 “bac-cab” 规则, 如下所示

$$\boldsymbol{A}\times(\boldsymbol{B}\times\boldsymbol{C})=\boldsymbol{B}(\boldsymbol{A}\cdot\boldsymbol{C})-\boldsymbol{C}(\boldsymbol{A}\cdot\boldsymbol{B}).$$

安培定律

安培定律 (*Ampere's law*) 指出, 磁场可以通过两种方式产生: 一种是电流 ($\boldsymbol{J_C}$), 另一种是随时间变化的电场 ($\partial\boldsymbol{D}/\partial t$). 该定律可以表述如下

$$\boldsymbol{\nabla}\times\boldsymbol{H}=\boldsymbol{J}=\boldsymbol{J_C}+\frac{\partial\boldsymbol{D}}{\partial t}.\tag{3.1-7}$$

其中, 矢量 $\boldsymbol{J_C}$ 是传导电流密度 (*current density*) ($\mathrm{A/m^2}$), 它来自电荷 (如在导体中存在的电子) 的移动.

至此, 可以将麦克斯韦的四个方程总结如下

$$\boldsymbol{\nabla}\cdot\boldsymbol{D}=\rho_v\ (\text{电场高斯定律}),\tag{3.1-8a}$$

result of the product is a vector. $\nabla \times \boldsymbol{A}$ is the cross product of the vectors ∇ and $\boldsymbol{A}$ given by

$$\begin{aligned}\nabla \times \boldsymbol{A} &= \begin{vmatrix} \hat{\boldsymbol{x}} & \hat{\boldsymbol{y}} & \hat{\boldsymbol{z}} \\ \dfrac{\partial}{\partial x} & \dfrac{\partial}{\partial y} & \dfrac{\partial}{\partial z} \\ A_x & A_y & A_z \end{vmatrix} \\ &= \hat{\boldsymbol{x}} \begin{vmatrix} \dfrac{\partial}{\partial y} & \dfrac{\partial}{\partial z} \\ A_y & A_z \end{vmatrix} - \hat{\boldsymbol{y}} \begin{vmatrix} \dfrac{\partial}{\partial x} & \dfrac{\partial}{\partial z} \\ A_x & A_z \end{vmatrix} + \hat{\boldsymbol{z}} \begin{vmatrix} \dfrac{\partial}{\partial x} & \dfrac{\partial}{\partial y} \\ A_x & A_y \end{vmatrix} \\ &= \hat{\boldsymbol{x}} \left(\frac{\partial A_z}{\partial y} - \frac{\partial A_y}{\partial z} \right) - \hat{\boldsymbol{y}} \left(\frac{\partial A_z}{\partial x} - \frac{\partial A_x}{\partial z} \right) + \hat{\boldsymbol{z}} \left(\frac{\partial A_y}{\partial x} - \frac{\partial A_x}{\partial y} \right). \end{aligned} \tag{3.1-6}$$

Example: Cross Product

Given two vectors $\boldsymbol{A}$ and $\boldsymbol{B}$, the cross product is defined as

$$\boldsymbol{A} \times \boldsymbol{B} = AB \sin(\theta_{AB}) \hat{\boldsymbol{n}},$$

where $\hat{\boldsymbol{n}}$ is a unit vector normal to the plane defined by $\boldsymbol{A}$ and $\boldsymbol{B}$, and θ_{AB} is the smaller angle between $\boldsymbol{A}$ and $\boldsymbol{B}$. The situation is shown in Figure 3.1-2. The direction of $\hat{\boldsymbol{n}}$ is found using the right-hand rule by rotating $\boldsymbol{A}$ towards $\boldsymbol{B}$ through the angle θ_{AB} with the fingers, and the thumb points to the $\hat{\boldsymbol{n}}$ direction.

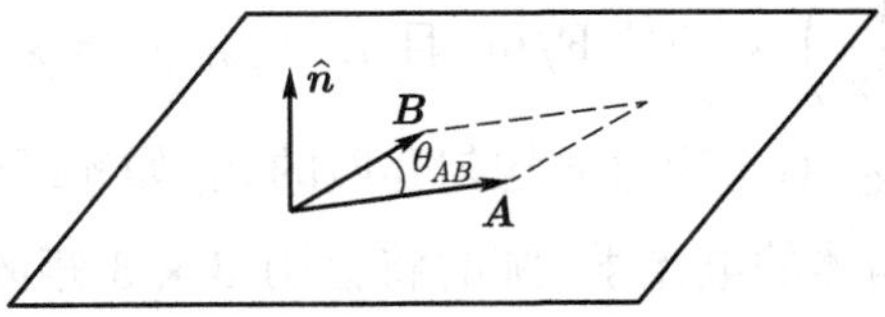

Figure 3.1-2 The cross product by the right-hand rule

The cross product is anti-commutative and distributive, that is,

$$\boldsymbol{A} \times \boldsymbol{B} = -\boldsymbol{B} \times \boldsymbol{A} \text{ (anti-commutative property)},$$
$$\boldsymbol{A} \times (\boldsymbol{B} + \boldsymbol{C}) = \boldsymbol{A} \times \boldsymbol{B} + \boldsymbol{A} \times \boldsymbol{C} \text{ (distributive property)}.$$

Finally, there is the vector triple product known as the "bac-cab" rule given by

$$\boldsymbol{A} \times (\boldsymbol{B} \times \boldsymbol{C}) = \boldsymbol{B}(\boldsymbol{A} \cdot \boldsymbol{C}) - \boldsymbol{C}(\boldsymbol{A} \cdot \boldsymbol{B}).$$

Ampere's Law

Ampere's law states that magnetic fields can be generated in two ways: by electric current ($\boldsymbol{J_C}$) and by time-varying electric fields ($\partial \boldsymbol{D}/\partial t$). The law is stated as

$$\nabla \times \boldsymbol{H} = \boldsymbol{J} = \boldsymbol{J_C} + \frac{\partial \boldsymbol{D}}{\partial t}. \tag{3.1-7}$$

The vector quantity $\boldsymbol{J_C}$ is the conduction current density (A/m^2) and it comes from the movement of electric charges such as electrons being present in conductors.

We summarize Maxwell's four equations as follows:

$$\nabla \cdot \boldsymbol{D} = \rho_v \text{ (electric Gauss's law)}, \tag{3.1-8a}$$

$$\nabla \cdot \boldsymbol{B} = 0 \text{ (磁场高斯定律)}, \tag{3.1-8b}$$

$$\nabla \times \boldsymbol{E} = -\frac{\partial \boldsymbol{B}}{\partial t} \text{ (法拉第定律)}, \tag{3.1-8c}$$

$$\nabla \times \boldsymbol{H} = \boldsymbol{J} = \boldsymbol{J_C} + \frac{\partial \boldsymbol{D}}{\partial t} \text{ (安培定律)}. \tag{3.1-8d}$$

其中, ρ_v 和 $\boldsymbol{J_C}$ 是产生电磁场的来源. 为了完全确定麦克斯韦方程中的四个场量, 还需要本构关系 (*constitutive relations*)

$$\boldsymbol{D} = \varepsilon \boldsymbol{E}, \tag{3.1-9}$$

以及

$$\boldsymbol{B} = \mu \boldsymbol{H}, \tag{3.1-10}$$

其中, ε 和 μ 分别为介质或材料的介电常数 (*electrical permittivity*) (F/m) 和磁导率 (*magnetic permeability*) (H/m), 介电材料 (*dielectric material*) 用 ε 表征, 磁性材料 (*magnetic material*) 用 μ 表征, 其本构关系并不简单. 一般来说, ε 和 μ 由场强决定, 且是空间 (位置) 和时间的函数. 如果它们的值不依赖于位置, 则为均匀材料处理 (*homogeneous*). 如果它们的值不依赖于场在该介质中的振幅, 则该介质是线性的 (*linear*). 如果 ε 和 μ 在任意给定点的所有方向上都相同, 则该介质为各向同性介质 (*isotropic material*). 对于线性、均匀、各向同性的介质, 如真空或自由空间, $\varepsilon = \varepsilon_0 \approx \left(\frac{1}{36\pi}\right) \times 10^{-9}$ F/m, 且 $\mu = \mu_0 \approx 4\pi \times 10^{-7}$ H/m. 本书中讨论的都是非磁性材料, 因此取 $\mu = \mu_0$. 对于穿过非均匀介质的光, 如光纤的光, 则处理 $\varepsilon(x, y, z)$. 在第七章中, 我们将讨论光通过晶体的电光学, 那时将会用 3×3 矩阵的 ε 来分析各向异性介质, 此外, 我们还将讨论 ε 是时间函数的声光学 (*acousto-optics*).

3.2 矢量波动方程

上一节阐述了麦克斯韦方程组及其本构关系. 实际上, 对于给定的 ρ_v 和 $\boldsymbol{J_C}$, 可以求解其电磁场. 现在来推导 ε 和 μ 均为常数时这一重要特殊情况下的波动方程, 此时介质是线性、均匀和各向同性的. 空气就是这类介质中一个很好的例子. 接下来用波动方程来描述电磁场的传播.

为导出 $\boldsymbol{E}$ 的波动方程, 对 $\boldsymbol{E}$ 的旋度方程再进行旋度运算. $\boldsymbol{E}$ 的旋度方程由式 (3.1-8c) 中的法拉第定律可得. 因此, 有

$$\nabla \times (\nabla \times \boldsymbol{E}) = \nabla \times \left(-\frac{\partial \boldsymbol{B}}{\partial t}\right) = -\frac{\partial}{\partial t}(\nabla \times \boldsymbol{B}),$$

这里, 交换了 ∇ 和 $\partial/\partial t$ 的运算. 同时在上式右边, 利用式 (3.1-8d) 以及式 (3.1-9) 和式 (3.1-10) 两个本构方程, 有

$$\nabla \cdot \boldsymbol{B} = 0 \text{ (magnetic Gauss's law)}, \tag{3.1-8b}$$

$$\nabla \times \boldsymbol{E} = -\frac{\partial \boldsymbol{B}}{\partial t} \text{ (Faraday's law)}, \tag{3.1-8c}$$

$$\nabla \times \boldsymbol{H} = \boldsymbol{J} = \boldsymbol{J}_C + \frac{\partial \boldsymbol{D}}{\partial t} \text{ (Ampere's Law)}. \tag{3.1-8d}$$

ρ_v and $\boldsymbol{J}_C$ are the sources responsible for generating the electromagnetic fields. In order to completely determine the four field quantities in Maxwell's equations, we need also the *constitutive relations*:

$$\boldsymbol{D} = \varepsilon \boldsymbol{E}, \tag{3.1-9}$$

and

$$\boldsymbol{B} = \mu \boldsymbol{H}, \tag{3.1-10}$$

where ε and μ are the *electrical permittivity* (F/m) and *magnetic permeability* (H/m) of the medium or material, respectively. While *dielectric materials* are characterized by ε, μ is used to characterize *magnetic materials*. The constitutive relations are rarely simple. In general, ε and μ depend on the field strength and they are functions of space (position) and time. If they do not depend on position, we are dealing with *homogeneous* materials. If they do not depend on the amplitude of the fields in the medium, the materials are *linear*. We have *isotropic materials* if ε and μ are the same in all directions from any given point. In the case for a linear, homogenous, and isotropic medium such as in vacuum or free space, $\varepsilon = \varepsilon_0 \approx \left(\frac{1}{36\pi}\right) \times 10^{-9}$ F/m, and $\mu = \mu_0 \approx 4\pi \times 10^{-7}$ H/m. Throughout this book, we deal with nonmagnetic materials, and hence we take $\mu = \mu_0$. For light travelling though inhomogeneous media such as optical fibers, we deal with $\varepsilon(x, y, z)$. In Chapter 7, we deal with electro-optics in which light traveling through crystals is analyzed by ε of a 3×3 matrix for anisotropic media. Also in the same chapter, we discuss acousto-optics when ε is a function of time.

3.2 Vector Wave Equations

In the last section, we have enunciated Maxwell's equations and the constitutive relations. Indeed, we could solve for the electromagnetic fields for a given ρ_v and $\boldsymbol{J}_C$. We shall now derive the wave equation for one of the important special cases in that the medium is characterized by ε and μ both being constant, that is, the medium is linear, homogeneous and isotropic. A good example of such a medium is air. We will see the wave equations describe the propagation of the electromagnetic fields.

To derive the wave equation in $\boldsymbol{E}$, we take the curl of the curl equation for $\boldsymbol{E}$. The curl equation for $\boldsymbol{E}$ is Faraday's law given in Eq. (3.1-8c). We, therefore, have

$$\nabla \times (\nabla \times \boldsymbol{E}) = \nabla \times \left(-\frac{\partial \boldsymbol{B}}{\partial t}\right) = -\frac{\partial}{\partial t}(\nabla \times \boldsymbol{B}),$$

where we have interchanged ∇ and $\partial/\partial t$ operations. Using Eq. (3.1-8d), and the two constitutive equations in Eqs. (3.1-9) and (3.1-10) in the right-hand side of the above equation,

$$\nabla\times(\nabla\times\boldsymbol{E})=-\mu\varepsilon\frac{\partial^2\boldsymbol{E}}{\partial t^2}-\mu\frac{\partial\boldsymbol{J_C}}{\partial t}. \tag{3.2-1}$$

使用含有 del 运算的矢量等式

$$\nabla\times(\nabla\times\boldsymbol{A})=\nabla(\nabla\cdot\boldsymbol{A})-\nabla^2\boldsymbol{A},$$

其中, $\nabla^2\boldsymbol{A}$ 是 $\boldsymbol{A}$ 的拉普拉斯算子 (*Laplacian*), 在笛卡儿坐标系 (*Cartesian coordinates*) 中表示为

$$\nabla^2\boldsymbol{A}=\frac{\partial^2\boldsymbol{A}}{\partial x^2}+\frac{\partial^2\boldsymbol{A}}{\partial y^2}+\frac{\partial^2\boldsymbol{A}}{\partial z^2},$$

式 (3.2-1) 变为

$$\nabla(\nabla\cdot\boldsymbol{E})-\nabla^2\boldsymbol{E}=-\mu\varepsilon\frac{\partial^2\boldsymbol{E}}{\partial t^2}-\mu\frac{\partial\boldsymbol{J_C}}{\partial t}. \tag{3.2-2}$$

由式 (3.1-8a) 和式 (3.1-9), 可以得出 $\nabla(\nabla\cdot\boldsymbol{E})=\dfrac{1}{\varepsilon}\nabla\rho_v$. 因此, 式 (3.2-2) 最终变为

$$\nabla^2\boldsymbol{E}-\mu\varepsilon\frac{\partial^2\boldsymbol{E}}{\partial t^2}=\mu\frac{\partial\boldsymbol{J_C}}{\partial t}+\frac{1}{\varepsilon}\nabla\rho_v, \tag{3.2-3}$$

这是 $\boldsymbol{E}$ 在线性、均匀、各向同性介质中, 同时在源 $\boldsymbol{J_C}$ 和 ρ_v 存在的情况下的非齐次矢量波动方程 (*inhomogeneous vector wave equation*). 为了推导 $\boldsymbol{H}$ 的波动方程, 我们只需对 $\boldsymbol{H}$ 的旋度方程再取旋度, 然后用类似的方法即可得到 $\boldsymbol{H}$ 的矢量波动方程

$$\nabla^2\boldsymbol{H}-\mu\varepsilon\frac{\partial^2\boldsymbol{H}}{\partial t^2}=-\nabla\times\boldsymbol{J_C}. \tag{3.2-4}$$

对于无源介质, 即, $\boldsymbol{J_C}=\rho_v=0$, 上述两个波动方程可以变为

$$\nabla^2\boldsymbol{E}-\mu\varepsilon\frac{\partial^2\boldsymbol{E}}{\partial t^2}=0, \tag{3.2-5a}$$

和

$$\nabla^2\boldsymbol{H}-\mu\varepsilon\frac{\partial^2\boldsymbol{H}}{\partial t^2}=0, \tag{3.2-5b}$$

这两个方程为齐次矢量波动方程 (*homogeneous vector wave equation*). 图 3.2-1 总结了这种情况. 在图 3.2-1 中, 假设源 $\boldsymbol{J_C}$ 和 ρ_v 置于一个以 ε 和 μ 为特征的限定区域 V′ 内, 根据式 (3.2-3) 和式 (3.2-4) 可以求得该区域所产生的场. 然而, 一旦所产生的行波场到达无源区域 V, 则该场必须满足齐次矢量波动方程.

需要引起注意的是, 式 (3.2-5a) 和式 (3.2-5b) 中的波动方程在科学和工程的许多分支中都会遇到, 它们的解会以后面我们看到的行波的形式出现. 式 (3.2-5a) 和式 (3.2-5b) 并不是独立的两个方程, 因为它们都是由式 (3.1-8c) 和式 (3.1-8d) 得到的. 通过波动方程可以得到一个重要结果, 那就是光速的表达式, v 在介质中可以表示为

$$v=\frac{1}{\sqrt{\varepsilon\mu}}=\frac{1}{\sqrt{\varepsilon_0\mu_0}}=c\approx 3\times10^8\ \text{m/s},\ \text{在自由空间 (真空中)}, \tag{3.2-6}$$

光速的理论值最早是由麦克斯韦计算出来的, 并且事实上, 这也让麦克斯韦将光归结为一种以波的形式存在的电磁扰动.

we have

$$\nabla \times (\nabla \times \boldsymbol{E}) = -\mu\varepsilon \frac{\partial^2 \boldsymbol{E}}{\partial t^2} - \mu \frac{\partial \boldsymbol{J_C}}{\partial t}. \tag{3.2-1}$$

Using the vector identity involving the del operator:

$$\nabla \times (\nabla \times \boldsymbol{A}) = \nabla(\nabla \cdot \boldsymbol{A}) - \nabla^2 \boldsymbol{A},$$

where $\nabla^2 \boldsymbol{A}$ is the *Laplacian* of $\boldsymbol{A}$, which in *Cartesian coordinates* is given by

$$\nabla^2 \boldsymbol{A} = \frac{\partial^2 \boldsymbol{A}}{\partial x^2} + \frac{\partial^2 \boldsymbol{A}}{\partial y^2} + \frac{\partial^2 \boldsymbol{A}}{\partial z^2},$$

Eq. (3.2-1) becomes

$$\nabla(\nabla \cdot \boldsymbol{E}) - \nabla^2 \boldsymbol{E} = -\mu\varepsilon \frac{\partial^2 \boldsymbol{E}}{\partial t^2} - \mu \frac{\partial \boldsymbol{J_C}}{\partial t}. \tag{3.2-2}$$

Using Eqs. (3.1-8a) and (3.1-9), we have $\nabla(\nabla \cdot \boldsymbol{E}) = \frac{1}{\varepsilon}\nabla\rho_v$. Therefore, Eq. (3.2-2) finally becomes

$$\nabla^2 \boldsymbol{E} - \mu\varepsilon \frac{\partial^2 \boldsymbol{E}}{\partial t^2} = \mu \frac{\partial \boldsymbol{J_C}}{\partial t} + \frac{1}{\varepsilon}\nabla\rho_v, \tag{3.2-3}$$

which is the *inhomogeneous vector wave equation* for $\boldsymbol{E}$ in a linear, homogenous, and isotropic medium in the presence of sources $\boldsymbol{J_C}$ and ρ_v. To derive the wave equation for $\boldsymbol{H}$, we simply take the curl of the curl equation for $\boldsymbol{H}$ and proceed with similar procedures to obtain the vector wave equation for $\boldsymbol{H}$:

$$\nabla^2 \boldsymbol{H} - \mu\varepsilon \frac{\partial^2 \boldsymbol{H}}{\partial t^2} = -\nabla \times \boldsymbol{J_C}. \tag{3.2-4}$$

For a source-free medium, that is, $\boldsymbol{J_C} = \rho_v = 0$, the two wave equations become

$$\nabla^2 \boldsymbol{E} - \mu\varepsilon \frac{\partial^2 \boldsymbol{E}}{\partial t^2} = 0, \tag{3.2-5a}$$

and

$$\nabla^2 \boldsymbol{H} - \mu\varepsilon \frac{\partial^2 \boldsymbol{H}}{\partial t^2} = 0, \tag{3.2-5b}$$

which are the *homogeneous vector wave equations.* Figure 3.2-1 summarizes the situation. In Figure 3.2-1, we assume that the sources $\boldsymbol{J_C}$ and ρ_v are localized in a limited region V′ characterized by ε and μ. We can solve for the generated fields in the region according to Eqs. (3.2-3) and (3.2-4). However, once the generated traveling fields reach the source-free region V, the fields must satisfy the homogeneous vector wave equations.

It is important to note that the wave equations in Eqs. (3.2-5a) and (3.2-5b) are encountered in many branches of science and engineering and have solutions in the form of traveling waves as we will see later. Equations (3.2-5a) and (3.2-5b) are not independent since they are both obtained from Eqs. (3.1-8c) and (3.1-8d). One important result from the wave equations is that we can obtain an expression of the speed of light, v, in the medium as

$$v = \frac{1}{\sqrt{\varepsilon\mu}} = \frac{1}{\sqrt{\varepsilon_0\mu_0}} = c \approx 3\times10^8 \text{ m/s, in free space (vacuum).} \tag{3.2-6}$$

This theoretical value of the speed of light was first calculated by Maxwell and indeed this led Maxwell to conclude that light is an electromagnetic disturbance in the form of waves.

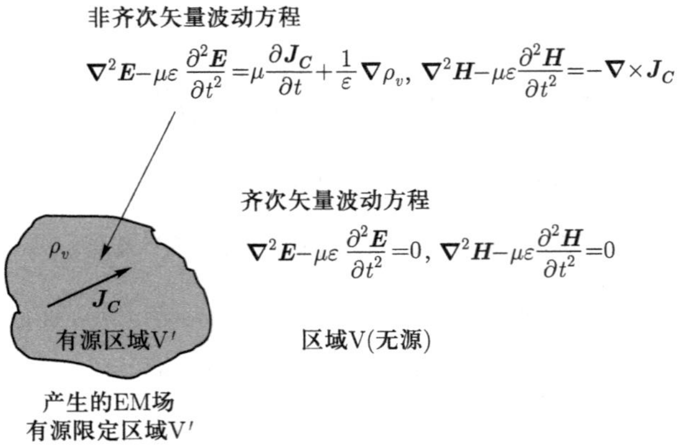

图 3.2-1 有源区域 V′ 和无源区域 V

3.3 行波解与坡印亭矢量

这里, 我们来求波动方程的一些最简单的解. 式 (3.2-5a) 中的波动方程等价于三个标量方程, $\boldsymbol{E}$ 的每个分量对应一个标量方程. 令 $\boldsymbol{E}=E_x\hat{\boldsymbol{x}}+E_y\hat{\boldsymbol{y}}+E_z\hat{\boldsymbol{z}}$, 并代入式 (3.2-5a) 中, 可得

$$\begin{aligned}\nabla^2\boldsymbol{E}&=\frac{\partial^2\boldsymbol{E}}{\partial x^2}+\frac{\partial^2\boldsymbol{E}}{\partial y^2}+\frac{\partial^2\boldsymbol{E}}{\partial z^2}=\left(\frac{\partial^2}{\partial x^2}+\frac{\partial^2}{\partial y^2}+\frac{\partial^2}{\partial z^2}\right)(E_x\hat{\boldsymbol{x}}+E_y\hat{\boldsymbol{y}}+E_z\hat{\boldsymbol{z}})\\&=\mu\varepsilon\frac{\partial^2\boldsymbol{E}}{\partial t^2}=\mu\varepsilon\frac{\partial^2}{\partial t^2}(E_x\hat{\boldsymbol{x}}+E_y\hat{\boldsymbol{y}}+E_z\hat{\boldsymbol{z}}).\end{aligned}$$

通过比较方程两边的 $\hat{\boldsymbol{x}}$ 分量, 得到

$$\frac{\partial^2E_x}{\partial x^2}+\frac{\partial^2E_x}{\partial y^2}+\frac{\partial^2E_x}{\partial z^2}=\mu\varepsilon\frac{\partial^2E_x}{\partial t^2},$$

或者

$$\nabla^2E_x=\mu\varepsilon\frac{\partial^2E_x}{\partial t^2}.$$

同样地, 通过比较其他分量, 可以得到与上式相同的 E_y 或 E_z 分量方程. 因此, 可以写出

$$\nabla^2\psi=\mu\varepsilon\frac{\partial^2\psi}{\partial t^2}=\frac{1}{v^2}\frac{\partial^2\psi}{\partial t^2},\tag{3.3-1}$$

其中, $\psi(x,y,z;t)$ 可以表示电场 $\boldsymbol{E}$ 的任一分量 E_x, E_y 或 E_z, 式 (3.3-1) 称为三维标量波动方程 (*3-D scalar wave equation*). 波动光学 (*wave optics*) 的起点是基于标量衍射理论 (*scalar diffraction theory*) 的标量波动方程, 这点将在后续章节中看到.

对于振荡的频率为 ω_0 的单色光场, 式 (3.3-1) 的通解为

$$\psi(x,y,z;t)=c_1f^+(\omega_0t-\boldsymbol{k}_0\cdot\boldsymbol{R})+c_2f^-(\omega_0t+\boldsymbol{k}_0\cdot\boldsymbol{R}),\tag{3.3-2}$$

当条件为

$$\frac{\omega_0^2}{k_{0x}^2+k_{0y}^2+k_{0z}^2}=\frac{\omega_0^2}{k_0^2}=v^2,$$

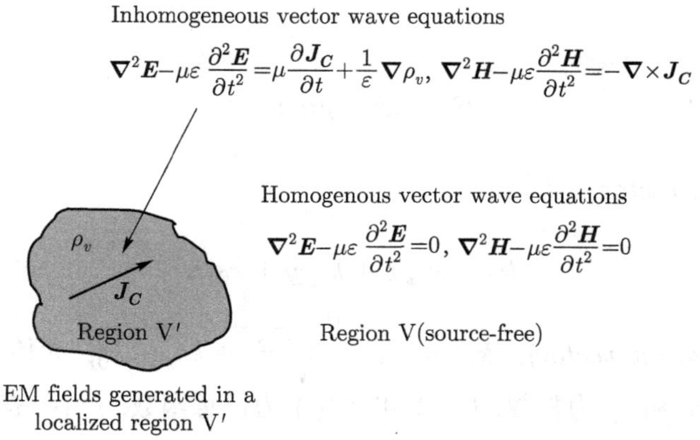

Figure 3.2-1 Region V′ contains sources, whereas region V is source-free

3.3 Traveling-Wave Solutions and the Poynting Vector

Let us find some of the simplest solutions in the wave equation. The wave equation in Eq. (3.2-5a) is equivalent to three scalar equations – one for each component of $\boldsymbol{E}$. We let $\boldsymbol{E} = E_x\hat{\boldsymbol{x}} + E_y\hat{\boldsymbol{y}} + E_z\hat{\boldsymbol{z}}$ and substitute this into Eq. (3.2-5a) to obtain

$$\nabla^2\boldsymbol{E} = \frac{\partial^2\boldsymbol{E}}{\partial x^2} + \frac{\partial^2\boldsymbol{E}}{\partial y^2} + \frac{\partial^2\boldsymbol{E}}{\partial z^2} = \left(\frac{\partial^2}{\partial x^2} + \frac{\partial^2}{\partial y^2} + \frac{\partial^2}{\partial z^2}\right)(E_x\hat{\boldsymbol{x}} + E_y\hat{\boldsymbol{y}} + E_z\hat{\boldsymbol{z}})$$

$$= \mu\varepsilon\frac{\partial^2\boldsymbol{E}}{\partial t^2} = \mu\varepsilon\frac{\partial^2}{\partial t^2}(E_x\hat{\boldsymbol{x}} + E_y\hat{\boldsymbol{y}} + E_z\hat{\boldsymbol{z}}).$$

By comparing the $\hat{\boldsymbol{x}}$-component on both sides of the equation, we obtain

$$\frac{\partial^2 E_x}{\partial x^2} + \frac{\partial^2 E_x}{\partial y^2} + \frac{\partial^2 E_x}{\partial z^2} = \mu\varepsilon\frac{\partial^2 E_x}{\partial t^2},$$

or

$$\nabla^2 E_x = \mu\varepsilon\frac{\partial^2 E_x}{\partial t^2}.$$

Similarly, by comparing other components, we have the same type of equation shown above for the E_y and E_z components. Therefore, we can write

$$\nabla^2\psi = \mu\varepsilon\frac{\partial^2\psi}{\partial t^2} = \frac{1}{v^2}\frac{\partial^2\psi}{\partial t^2}, \tag{3.3-1}$$

where $\psi(x, y, z; t)$ may represent any component, E_x, E_y or E_z, of the electric field $\boldsymbol{E}$. Equation (3.3-1) is called the *3-D scalar wave equation*. The starting point for *wave optics* is the scalar wave equation in which *scalar diffraction theory* is based, as we will see in subsequent subsections.

For monochromatic light fields oscillating at the frequency ω_0, the general solution to Eq. (3.3-1) is

$$\psi(x, y, z; t) = c_1 f^+(\omega_0 t - \boldsymbol{k}_0\cdot\boldsymbol{R}) + c_2 f^-(\omega_0 t + \boldsymbol{k}_0\cdot\boldsymbol{R}), \tag{3.3-2}$$

with the condition that

$$\frac{\omega_0^2}{k_{0x}^2 + k_{0y}^2 + k_{0z}^2} = \frac{\omega_0^2}{k_0^2} = v^2,$$

其中

$$\boldsymbol{R} = x\hat{\boldsymbol{x}} + y\hat{\boldsymbol{y}} + z\hat{\boldsymbol{z}}$$

是位置矢量 (*position vector*), 且

$$\boldsymbol{k}_0 = k_{0x}\hat{\boldsymbol{x}} + k_{0y}\hat{\boldsymbol{y}} + k_{0z}\hat{\boldsymbol{z}}$$

是传播矢量 (*propagation vector*), $|\boldsymbol{k}_0| = k_0 = \sqrt{k_{0x}^2 + k_{0y}^2 + k_{0z}^2}$ 为传播常数 (*propagation constant*) (rad/m), c_1 和 c_2 为常数, $f^+(\cdot)$ 和 $f^-(\cdot)$ 为任意函数, $f^+(\cdot)$ 表示沿 $\boldsymbol{k}_0$ 方向传播的波, $f^-(\cdot)$ 表示沿 $-\boldsymbol{k}_0$ 方向传播的波. 图 3.3-1 表示 $\boldsymbol{k}_0$ 及其分量.

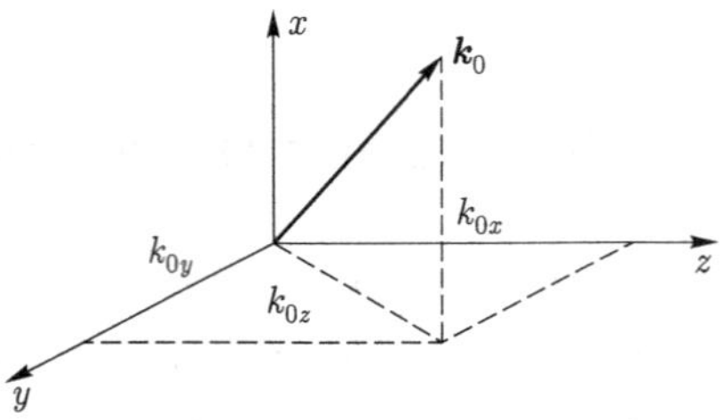

图 3.3-1 传播矢量

举例: 行波解 (*travelling–wave solution*)

举个简单的例子, 式 (3.3-2) 中, 令 $c_2 = 0$, $c_1 = 1$, 且有变量 $\xi = \omega_0 t - k_0 z$ 的 $f^+(\cdot)$. 在这个例子中, 我们证明 $\psi(z;t) = f^+(\omega_0 t - k_0 z)$ 是标量波动方程的一个解. 首先, 有

$$\frac{\partial \psi}{\partial z} = \frac{\partial f^+}{\partial \xi}\frac{\partial \xi}{\partial z} = \frac{\partial f^+}{\partial \xi}(-k_0),$$

和

$$\frac{\partial^2 \psi}{\partial z^2} = (-k_0)\frac{\partial^2 f^+}{\partial \xi^2}\frac{\partial \xi}{\partial z} = (-k_0)^2\frac{\partial^2 f^+}{\partial \xi^2}. \tag{3.3-3}$$

此时

$$\frac{\partial \psi}{\partial t} = \frac{\partial f^+}{\partial \xi}\frac{\partial \xi}{\partial t} = \frac{\partial f^+}{\partial \xi}\omega_0,$$

且

$$\frac{\partial^2 \psi}{\partial t^2} = \omega_0\frac{\partial^2 f^+}{\partial \xi^2}\frac{\partial \xi}{\partial t} = (\omega_0)^2\frac{\partial^2 f^+}{\partial \xi^2}. \tag{3.3-4}$$

将式 (3.3-4) 代入式 (3.3-3), 得

$$\frac{\partial^2 \psi}{\partial z^2} = \frac{(k_0)^2}{(\omega_0)^2}\frac{\partial^2 \psi}{\partial t^2}.$$

因为 $\dfrac{\omega_0^2}{k_0^2} = v^2$, 上式变为

$$\frac{\partial^2 \psi}{\partial z^2} = \frac{1}{v^2}\frac{\partial^2 \psi}{\partial t^2}, \tag{3.3-5}$$

where

$$\boldsymbol{R} = x\hat{\boldsymbol{x}} + y\hat{\boldsymbol{y}} + z\hat{\boldsymbol{z}}$$

is the *position vector*, and

$$\boldsymbol{k}_0 = k_{0x}\hat{\boldsymbol{x}} + k_{0y}\hat{\boldsymbol{y}} + k_{0z}\hat{\boldsymbol{z}}$$

is the *propagation vector* with $|\boldsymbol{k}_0| = k_0 = \sqrt{k_{0x}^2 + k_{0y}^2 + k_{0z}^2}$ being the *propagation constant* (rad/m). c_1 and c_2 are some constants. $f^+(\cdot)$ and $f^-(\cdot)$ are arbitrary functions, where $f^+(\cdot)$ depicts a wave propagating along the $\boldsymbol{k}_0$ direction, and $f^-(\cdot)$ means a wave is propagating along the $-\boldsymbol{k}_0$ direction. Figure 3.3-1 illustrates $\boldsymbol{k}_0$ with its components.

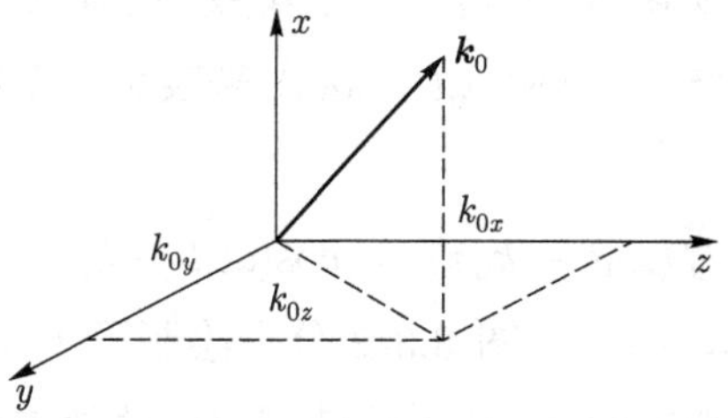

Figure 3.3-1 Propagation vector

Example: Travelling-Wave Solution

With reference to Eq. (3.3-2), let us take $c_2 = 0$, $c_1 = 1$, and $f^+(\cdot)$ of the variable $\xi = \omega_0 t - k_0 z$ as a simple example. In this example, we show that $\psi(z;t) = f^+(\omega_0 t - k_0 z)$ is a solution of the scalar wave equation. To start with, we have

$$\frac{\partial \psi}{\partial z} = \frac{\partial f^+}{\partial \xi}\frac{\partial \xi}{\partial z} = \frac{\partial f^+}{\partial \xi}(-k_0),$$

and

$$\frac{\partial^2 \psi}{\partial z^2} = (-k_0)\frac{\partial^2 f^+}{\partial \xi^2}\frac{\partial \xi}{\partial z} = (-k_0)^2\frac{\partial^2 f^+}{\partial \xi^2}. \tag{3.3-3}$$

Now,

$$\frac{\partial \psi}{\partial t} = \frac{\partial f^+}{\partial \xi}\frac{\partial \xi}{\partial t} = \frac{\partial f^+}{\partial \xi}\omega_0,$$

and

$$\frac{\partial^2 \psi}{\partial t^2} = \omega_0\frac{\partial^2 f^+}{\partial \xi^2}\frac{\partial \xi}{\partial t} = (\omega_0)^2\frac{\partial^2 f^+}{\partial \xi^2}. \tag{3.3-4}$$

Substituting Eq. (3.3-4) into Eq. (3.3-3), we obtain

$$\frac{\partial^2 \psi}{\partial z^2} = \frac{(k_0)^2}{(\omega_0)^2}\frac{\partial^2 \psi}{\partial t^2}.$$

Since $\dfrac{\omega_0^2}{k_0^2} = v^2$, the above equation becomes

$$\frac{\partial^2 \psi}{\partial z^2} = \frac{1}{v^2}\frac{\partial^2 \psi}{\partial t^2}, \tag{3.3-5}$$

该式是式 (3.3-1) 所示的三维波动方程的一维形式. 因此, $\psi(z;t)=f^+(\omega_0 t-k_0 z)$ 是标量波动方程的一个解. 实际上, 式 (3.3-5) 的通解为

$$\psi(z;t)=c_1 f^+(\omega_0 t-k_0 z)+c_2 f^-(\omega_0 t+k_0 z). \tag{3.3-6}$$

令 $f^+(\cdot)=\mathrm{e}^{\mathrm{j}(\cdot)}$ 作为例子, 可以得到

$$\psi(z;t)=\mathrm{e}^{\mathrm{j}(\omega_0 t-k_0 z)} \tag{3.3-7}$$

作为一个简单的解. 这个解被称为波动方程的平面波解 (*plane-wave solution*). 尽管用复数形式表示, 但这种波被称为单位振幅的平面波 (*plane wave*). 由于 ψ 表示电磁场的一个分量, 所以它一定是关于空间和时间的实函数. 要得到一个实数, 只需取该复函数的实部即可. 因此, 可以用 $\mathrm{Re}[\psi(z;t)]=\mathrm{Re}[\mathrm{e}^{\mathrm{j}(\omega_0 t-k_0 z)}]=\cos(\omega_0 t-k_0 z)$ 来表示一个物理量, 其中, $\mathrm{Re}[\cdot]$ 表示取括号内复值的实部.

以平面波解为例, 来说明 $f(\omega_0 t-k_0 z)=\cos(\omega_0 t-k_0 z)$ 是一个行波, 为简单起见, 取 $\omega_0=k_0=1$, 因此 $v=\omega_0/k_0=1$ m/s. 图 3.3-2 分别绘制了 $t=0$ 和 $t=2$ s 时的 $\cos(t-z)$. 可以看出, 如果画出波上一个给定的点, 比如峰值点 P, 并随着时间的推移跟踪它, 则可以测量出所谓的波的相速度 (*phase velocity*), 在这个特例当中, 相速度为 1 m/s. 同样地, 可以追踪并画出以相同相速度传播的波的波谷 P'.

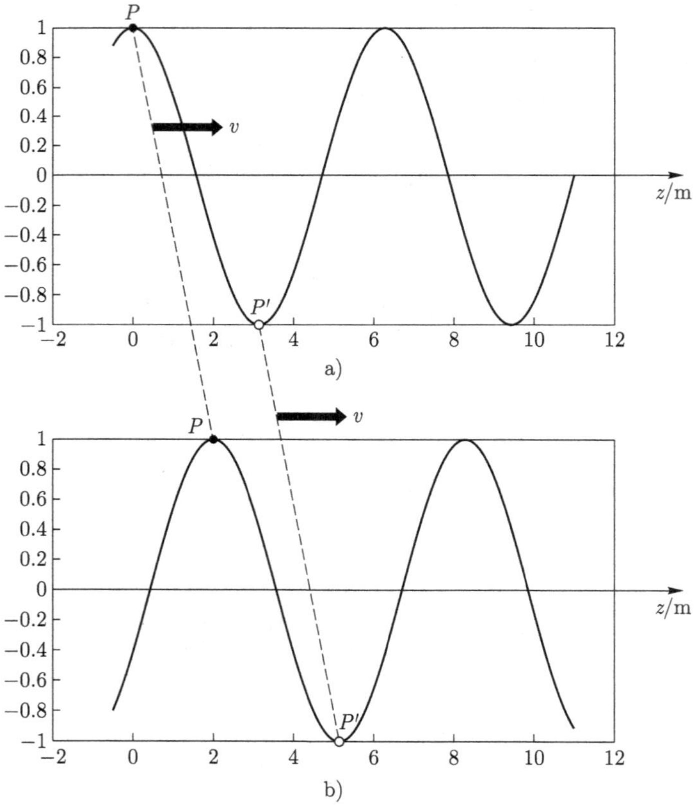

图 3.3-2 a) $t=0$ 和 b) $t=2$ s 时 z 的函数 $f(t-z)=\cos(t-z)$

which is a 1-D version of the 3-D wave equation shown in Eq. (3.3-1). Hence, $\psi(z;t) = f^+(\omega_0 t - k_0 z)$ is a solution of the scalar wave equation. Indeed, the general solution to Eq. (3.3-5) is

$$\psi(z;t) = c_1 f^+(\omega_0 t - k_0 z) + c_2 f^-(\omega_0 t + k_0 z). \tag{3.3-6}$$

Let us take $f^+(\cdot) = \mathrm{e}^{\mathrm{j}(\cdot)}$ as an example. We then have

$$\psi(z;t) = \mathrm{e}^{\mathrm{j}(\omega_0 t - k_0 z)} \tag{3.3-7}$$

as a simple solution. This solution is called a *plane-wave solution* to the wave equation. The wave is called a *plane wave* of unit amplitude, albeit written in a complex notation. Since ψ represents a component of the electromagnetic field, it must be a real function of space and time. To obtain a real quantity, we simply take the real part of the complex function. We, therefore, have $\mathrm{Re}[\psi(z;t)] = \mathrm{Re}[\mathrm{e}^{\mathrm{j}(\omega_0 t - k_0 z)}] = \cos(\omega_0 t - k_0 z)$ to represent a physical quantity, where $\mathrm{Re}[\cdot]$ means taking the real part of the quantity being bracketed.

Taking the plane wave solution as an example, let us now illustrate that $f(\omega_0 t - k_0 z) = \cos(\omega_0 t - k_0 z)$ is a travelling wave. For simplicity, we take $\omega_0 = k_0 = 1$, and hence $v = \omega_0/k_0 = 1$ m/s. In Figure 3.3-2a and b, we plot $\cos(t - z)$ for $t = 0$ and $t = 2$ s, respectively. Note that if we track a given point on the wave, such as the peak P, and follow it as time advances, we can measure the so-called *phase velocity* of the wave, which is 1 m/s in this particular example. Similarly, we can track the trough P' of the wave travelling at the same phase velocity.

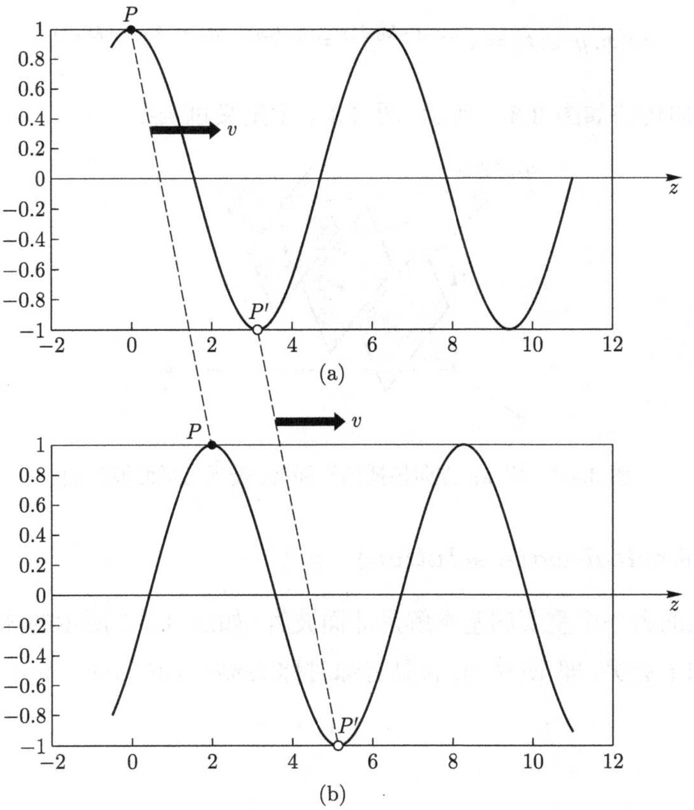

Figure 3.3-2 Plots of $f(t - z) = \cos(t - z)$ as a function of z at (a) $t = 0$ and (b) $t = 2$ s. z is measured in meters

平面波解 (*plane-wave solution*)

在上面的例子中, 我们已经简要地讨论了平面波解. $\psi(z;t)=\mathrm{e}^{\mathrm{j}(\omega_0 t-k_0 z)}$ 是式 (3.3-5) 的平面波解, 其中平面波沿 z 方向传播, 写为

$$\psi(z;t)=\mathrm{e}^{\mathrm{j}(\omega_0 t-k_0 z)}=\mathrm{e}^{\mathrm{j}\omega_0 t}\mathrm{e}^{-\mathrm{j}\theta(z)}, \tag{3.3-8}$$

其中, $\theta(z)=k_0 z=\dfrac{2\pi}{\lambda_0}z$ 被称为波长为 λ_0 的波的相位. 设坐标原点为零相位参考位置, 即 $\theta(z=0)=0$, 这就意味着在整个 $z=0$ 的平面上, 相位为 0. 在 $z=\lambda_0$ 的位置处, $\theta(z=\lambda_0)=\dfrac{2\pi}{\lambda_0}\lambda_0=2\pi$. 因此, 每传播一个波长的距离, 波的相位就会增加 2π. 因此, 可以画出沿 z 方向传播的平面波前 (*planar wavefront*), 如图 3.3-3 所示.

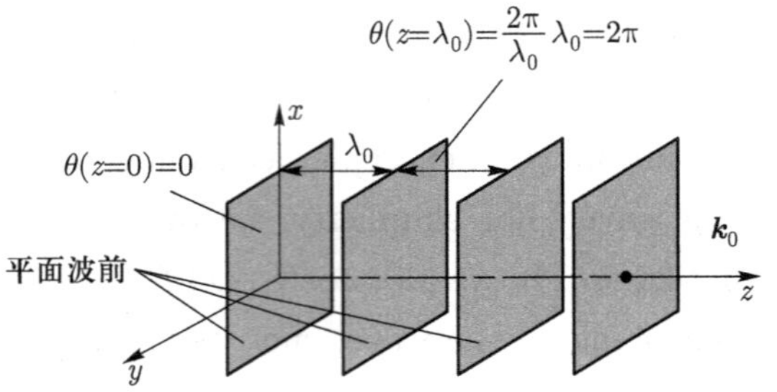

图 3.3-3 沿 z 方向传播的平面波前

对于平面波沿 $\boldsymbol{k}_0$ 传播的总方向, 根据式 (3.3-2), 有

$$\psi(x,y,z;t)=\mathrm{e}^{\mathrm{j}(\omega_0 t-\boldsymbol{k}_0\cdot\boldsymbol{R})}=\mathrm{e}^{\mathrm{j}(\omega_0 t-k_{0x}x-k_{0y}y-k_{0z}z)}. \tag{3.3-9}$$

平面波前沿 $\boldsymbol{k}_0$ 方向传播如图 3.3-4 所示, 图 3.3-1 中定义过 $\boldsymbol{k}_0$.

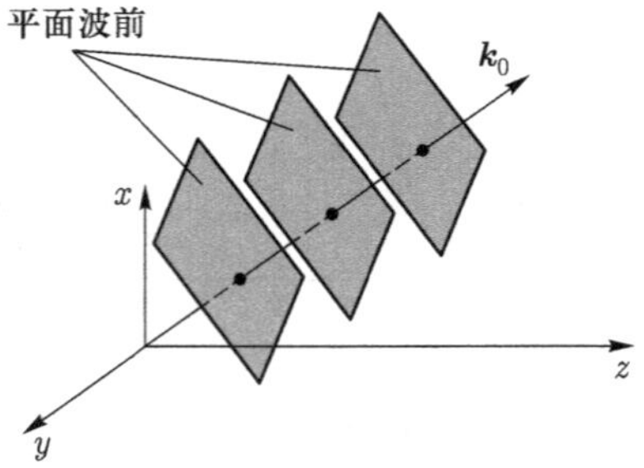

图 3.3-4 沿 $\boldsymbol{k}_0$ 方向传播的平面波, 表现为平面波阵面

球面波解 (*spherical-wave solution*)

标量波动方程的另一个重要的基本解是球面波解. 如图 3.3-5 所示的球坐标系中, 对于球面波, ψ 只与 R 和 t 有关, 即 $\psi(R,t)$, 且具有球对称 (*spherical symmetry*). 式 (3.3-1) 波动方

Plane-Wave Solution

We have briefly discussed the plane-wave solution in the above example. $\psi(z;t) = \mathrm{e}^{\mathrm{j}(\omega_0 t - k_0 z)}$ is the plane-wave solution for Eq. (3.3-5), where the plane wave is propagating along the z-direction. Let us write

$$\psi(z;t) = \mathrm{e}^{\mathrm{j}(\omega_0 t - k_0 z)} = \mathrm{e}^{\mathrm{j}\omega_0 t}\mathrm{e}^{-\mathrm{j}\theta(z)}, \tag{3.3-8}$$

where $\theta(z) = k_0 z = \dfrac{2\pi}{\lambda_0}z$ is called the phase of the wave with λ_0 being the wavelength of the wave. Let us take the origin of the coordinates as a zero-phase reference position, that is, $\theta(z=0) = 0$. What it means is that over the plane $z = 0$, we have zero phase. At $z = \lambda_0$, $\theta(z=\lambda_0) = \dfrac{2\pi}{\lambda_0}\lambda_0 = 2\pi$. So for every distance of propagation of a wavelength, the phase of the wave gains 2π. We, therefore, have what is known as the *planar wavefronts* along the z-direction. The situation is illustrated in Figure 3.3-3.

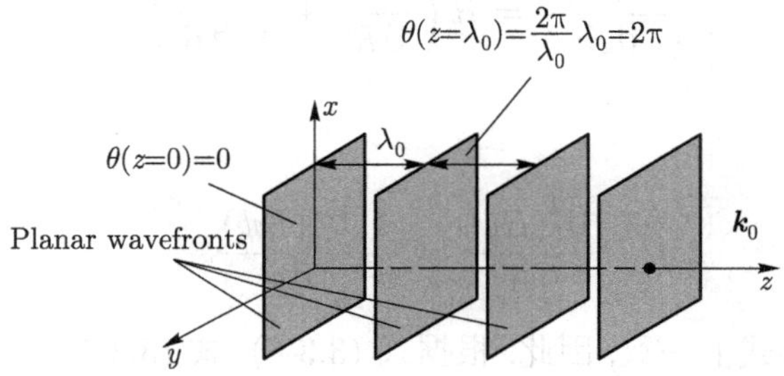

Figure 3.3-3 Plane wave propagating along the z-direction exhibiting planar wavefronts

For a general direction of plane-wave propagation along $\boldsymbol{k}_0$, according to Eq. (3.3-2), we write

$$\psi(x,y,z;t) = \mathrm{e}^{\mathrm{j}(\omega_0 t - \boldsymbol{k}_0 \cdot \boldsymbol{R})} = \mathrm{e}^{\mathrm{j}(\omega_0 t - k_{0x}x - k_{0y}y - k_{0z}z)}. \tag{3.3-9}$$

and the planar wavefronts propagate along the direction of $\boldsymbol{k}_0$ as shown in Figure 3.3-4, where $\boldsymbol{k}_0$ has been defined in Figure 3.3-1.

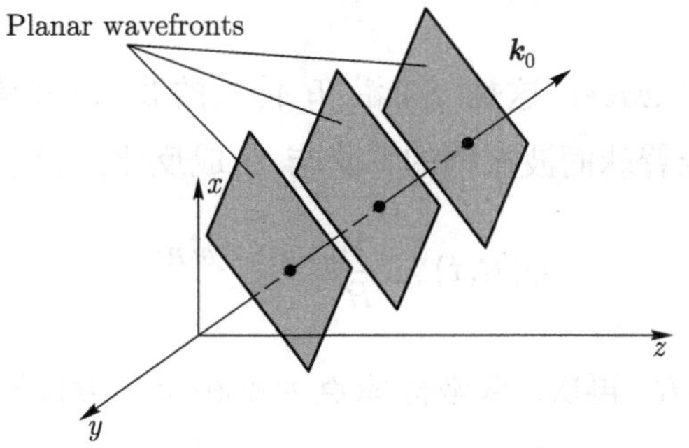

Figure 3.3-4 Plane wave propagating along the $\boldsymbol{k}_0$-direction exhibiting planar wavefronts

Spherical-Wave Solution

Another important basic solution to the scalar wave equation is the *spherical-wave solution.* We consider the spherical coordinates shown in Figure 3.3-5. For a spherical wave, ψ depends only on R and t, that is, $\psi(R,t)$, and we have *spherical symmetry.* $\boldsymbol{\nabla}^2\psi$ in the wave equation

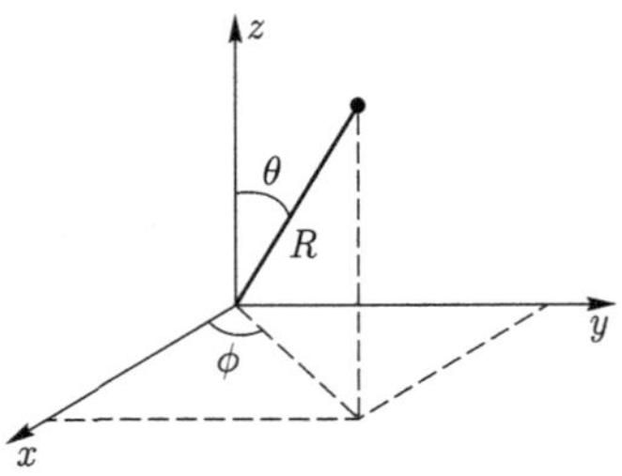

图 3.3-5 球坐标系

程中的 $\nabla^2\psi$ 被简化为 $\nabla^2\psi=\dfrac{\partial^2\psi}{\partial R^2}+\dfrac{2}{R}\dfrac{\partial\psi}{\partial R}$, 因此, 波动方程变为

$$\nabla^2\psi=\frac{\partial^2\psi}{\partial R^2}+\frac{2}{R}\frac{\partial\psi}{\partial R}=\frac{1}{v^2}\frac{\partial^2\psi}{\partial t^2}. \tag{3.3-10}$$

根据微积分中的链式法则, 有

$$\frac{\partial^2(R\psi)}{\partial R^2}=R\left(\frac{\partial^2\psi}{\partial R^2}+\frac{2}{R}\frac{\partial\psi}{\partial R}\right),$$

式 (3.3-10) 变为

$$\frac{\partial^2(R\psi)}{\partial R^2}=\frac{1}{v^2}\frac{\partial^2(R\psi)}{\partial t^2}. \tag{3.3-11}$$

该等式与式 (3.3-5) 在形式上一致, 因此, 根据式 (3.3-6), 式 (3.3-11) 的通解为

$$R\psi=c_1f^+(\omega_0t-k_0R)+c_2f^-(\omega_0t+k_0R).$$

因此, 式 (3.3-10) 的最终通解为

$$\psi(R;t)=c_1\frac{f^+(\omega_0t-k_0R)}{R}+c_2\frac{f^-(\omega_0t+k_0R)}{R}. \tag{3.3-12}$$

同样, 对于 $c_2=0$, $c_1=1$, 且 $f^+(\cdot)=\mathrm{e}^{\mathrm{j}(\cdot)}$, 其简单解为

$$\psi(R;t)=\frac{1}{R}\mathrm{e}^{\mathrm{j}(\omega_0t-k_0R)}, \tag{3.3-13}$$

这被称为球面波 (*spherical wave*). 这是一个沿 R 传播的波, 也就是, 远离坐标原点的球面波, 并带有因子 $1/R$, 该项意味着球面波振幅的下降与 R 成反比, 可以写为

$$\psi(R;t)=\frac{1}{R}\mathrm{e}^{\mathrm{j}\omega_0t}\mathrm{e}^{-\mathrm{j}\theta(R)},$$

其中, $\theta(R)=k_0R=\dfrac{2\pi}{\lambda_0}R$. 再次, 取坐标原点为零相位参考位置, 即, $\theta(R=0)=0$, 那么, $\theta(R=\lambda_0)=\dfrac{2\pi}{\lambda_0}\lambda_0=2\pi$, 因此, 对于每一个传播波长距离的球面波, 波的相位增加 2π. 图 3.3-6 说明了在 y-z 平面上横截面的情况.

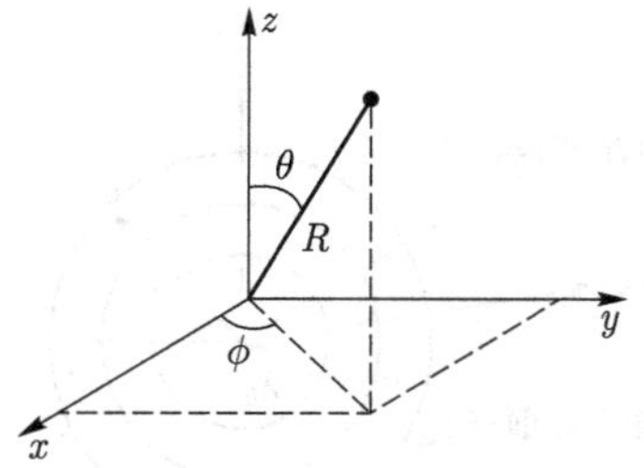

Figure 3.3-5 Spherical coordinates system

of Eq. 3.3-1 is reduced to $\nabla^2\psi = \dfrac{\partial^2\psi}{\partial R^2} + \dfrac{2}{R}\dfrac{\partial\psi}{\partial R}$. Hence, the wave equation becomes

$$\nabla^2\psi = \frac{\partial^2\psi}{\partial R^2} + \frac{2}{R}\frac{\partial\psi}{\partial R} = \frac{1}{v^2}\frac{\partial^2\psi}{\partial t^2}. \tag{3.3-10}$$

Using the chain rule in calculus, we have

$$\frac{\partial^2(R\psi)}{\partial R^2} = R\left(\frac{\partial^2\psi}{\partial R^2} + \frac{2}{R}\frac{\partial\psi}{\partial R}\right),$$

and Eq. (3.3-10) becomes

$$\frac{\partial^2(R\psi)}{\partial R^2} = \frac{1}{v^2}\frac{\partial^2(R\psi)}{\partial t^2}. \tag{3.3-11}$$

This equation is of the same form as that of Eq. (3.3-5), and therefore, the general solution of Eq. (3.3-11) is, according to Eq. (3.3-6),

$$R\psi = c_1 f^+(\omega_0 t - k_0 R) + c_2 f^-(\omega_0 t + k_0 R).$$

Therefore, the final general solution to Eq. (3.3-10) is

$$\psi(R;t) = c_1\frac{f^+(\omega_0 t - k_0 R)}{R} + c_2\frac{f^-(\omega_0 t + k_0 R)}{R}. \tag{3.3-12}$$

Again, for $c_2 = 0$, $c_1 = 1$, and $f^+(\cdot) = \mathrm{e}^{\mathrm{j}(\cdot)}$, it has a simple solution given by

$$\psi(R;t) = \frac{1}{R}\mathrm{e}^{\mathrm{j}(\omega_0 t - k_0 R)}, \tag{3.3-13}$$

which is called a *spherical wave.* This is a wave propagating along R, that is, away from the origin of the coordinates and with the factor $1/R$ term implying that the amplitude of a spherical wave decreases inversely with R. We can write

$$\psi(R;t) = \frac{1}{R}\mathrm{e}^{\mathrm{j}\omega_0 t}\mathrm{e}^{-\mathrm{j}\theta(R)},$$

where $\theta(R) = k_0 R = \dfrac{2\pi}{\lambda_0}R$. Again we take the origin of the coordinates as a zero-phase reference position, that is, $\theta(R=0) = 0$, and then $\theta(R=\lambda_0) = \dfrac{2\pi}{\lambda_0}\lambda_0 = 2\pi$. Therefore, for every distance of propagation of a wavelength, the phase of the wave gains 2π. Figure 3.3-6 illustrates the situation with a cross-section on the y-z plane.

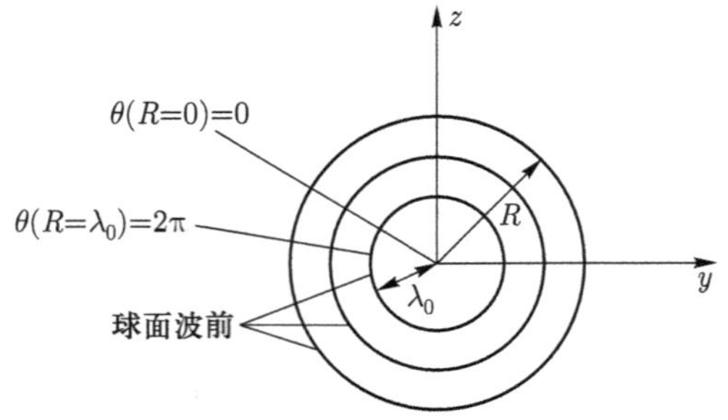

图 3.3-6 球面波前横截面示意图

坡印亭矢量

现在考虑在自由空间中存在的电场和磁场之间的关系, 在此基础上, 引入电磁传播中功率流的概念, 并定义坡印亭矢量 (*Poynting vector*).

由上节可知, 对于式 (3.2-5a) 给定的波动方程, $\psi(x,y,z;t)=\mathrm{e}^{\mathrm{j}(\omega_0 t-k_0 z)}$ 可以表示电场 $\boldsymbol{E}$ 的任意分量 E_x, E_y 或 E_z, 因此, 可以写出式 (3.2-5a) 和式 (3.2-5b) 所给出的波动方程的通解, 分别为

$$\begin{aligned}\boldsymbol{E}&=E_x\hat{\boldsymbol{x}}+E_y\hat{\boldsymbol{y}}+E_z\hat{\boldsymbol{z}}\\&=E_{0x}\mathrm{e}^{\mathrm{j}(\omega_0 t-k_0 z)}\hat{\boldsymbol{x}}+E_{0y}\mathrm{e}^{\mathrm{j}(\omega_0 t-k_0 z)}\hat{\boldsymbol{y}}+E_{0z}\mathrm{e}^{\mathrm{j}(\omega_0 t-k_0 z)}\hat{\boldsymbol{z}},\end{aligned}\tag{3.3-14}$$

和

$$\begin{aligned}\boldsymbol{H}&=H_x\hat{\boldsymbol{x}}+H_y\hat{\boldsymbol{y}}+H_z\hat{\boldsymbol{z}}\\&=H_{0x}\mathrm{e}^{\mathrm{j}(\omega_0 t-k_0 z)}\hat{\boldsymbol{x}}+H_{0y}\mathrm{e}^{\mathrm{j}(\omega_0 t-k_0 z)}\hat{\boldsymbol{y}}+H_{0z}\mathrm{e}^{\mathrm{j}(\omega_0 t-k_0 z)}\hat{\boldsymbol{z}}.\end{aligned}\tag{3.3-15}$$

其中, E_{0x}, E_{0y}, E_{0z} 和 H_{0x}, H_{0y}, H_{0z} 一般是复常数. 同样, 为了表示一个物理量, 我们只需取 $\boldsymbol{E}$ 或 $\boldsymbol{H}$ 的实部. 然而, 这些通解必须满足麦克斯韦方程. 对于无源情况, 即 $\rho_v=0$ 时的电场高斯定律 [见式 (3.1-1)], 有

$$\nabla\cdot\boldsymbol{E}=0,$$

该式给出

$$\left(\frac{\partial}{\partial x}\hat{\boldsymbol{x}}+\frac{\partial}{\partial y}\hat{\boldsymbol{y}}+\frac{\partial}{\partial z}\hat{\boldsymbol{z}}\right)\cdot\left(E_{0x}\mathrm{e}^{\mathrm{j}(\omega_0 t-k_0 z)}\hat{\boldsymbol{x}}+E_{0y}\mathrm{e}^{\mathrm{j}(\omega_0 t-k_0 z)}\hat{\boldsymbol{y}}+E_{0z}\mathrm{e}^{\mathrm{j}(\omega_0 t-k_0 z)}\hat{\boldsymbol{z}}\right)=0,$$

或

$$\frac{\partial}{\partial z}E_{0z}\mathrm{e}^{\mathrm{j}(\omega_0 t-k_0 z)}=-\mathrm{j}k_0E_{0z}\mathrm{e}^{\mathrm{j}(\omega_0 t-k_0 z)}=0,$$

说明

$$E_{0z}=0,\tag{3.3-16}$$

因为 $-\mathrm{j}k_0\mathrm{e}^{\mathrm{j}(\omega_0 t-k_0 z)}$ 一般不为零. 同样, 将 $\boldsymbol{H}$ 代入磁场高斯定律 [见式 (3.1-4)], 得到

$$H_{0z}=0.\tag{3.3-17}$$

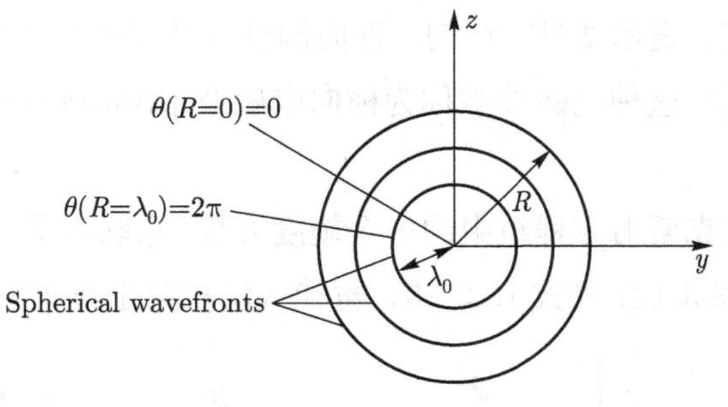

Figure 3.3-6 Cross section of wavefronts of a spherical wave

The Poynting Vector

Let us now consider the relationship between existing electric and magnetic fields in free space. In this connection, we introduce the concept of power flow during electromagnetic propagation and define the *Poynting vector.*

We have seen from the last section that $\psi(x, y, z; t) = \mathrm{e}^{\mathrm{j}(\omega_0 t - k_0 z)}$ may represent any component, E_x, E_y, or E_z, of the electric field $\boldsymbol{E}$ for the wave equation given by Eq. (3.2-5a). Hence, we can write the general solution of the wave equations given by Eqs. (3.2-5a) and (3.2-5b) as

$$\begin{aligned}\boldsymbol{E} &= E_x\hat{\boldsymbol{x}} + E_y\hat{\boldsymbol{y}} + E_z\hat{\boldsymbol{z}}\\ &= E_{0x}\mathrm{e}^{\mathrm{j}(\omega_0 t - k_0 z)}\hat{\boldsymbol{x}} + E_{0y}\mathrm{e}^{\mathrm{j}(\omega_0 t - k_0 z)}\hat{\boldsymbol{y}} + E_{0z}\mathrm{e}^{\mathrm{j}(\omega_0 t - k_0 z)}\hat{\boldsymbol{z}},\end{aligned} \tag{3.3-14}$$

and

$$\begin{aligned}\boldsymbol{H} &= H_x\hat{\boldsymbol{x}} + H_y\hat{\boldsymbol{y}} + H_z\hat{\boldsymbol{z}}\\ &= H_{0x}\mathrm{e}^{\mathrm{j}(\omega_0 t - k_0 z)}\hat{\boldsymbol{x}} + H_{0y}\mathrm{e}^{\mathrm{j}(\omega_0 t - k_0 z)}\hat{\boldsymbol{y}} + H_{0z}\mathrm{e}^{\mathrm{j}(\omega_0 t - k_0 z)}\hat{\boldsymbol{z}},\end{aligned} \tag{3.3-15}$$

respectively. E_{0x}, E_{0y}, E_{0z} and H_{0x}, H_{0y}, H_{0z} are complex constants in general. Again, to represent a physical quantity, we simply take the real part of $\boldsymbol{E}$ or $\boldsymbol{H}$. However, these general solutions must satisfy the Maxwell's equations. From electric Gauss's law [see Eq. (3.1-1)] along with a source-free situation, that is, $\rho_v = 0$, we have

$$\boldsymbol{\nabla}\cdot\boldsymbol{E} = 0,$$

which gives

$$\left(\frac{\partial}{\partial x}\hat{\boldsymbol{x}} + \frac{\partial}{\partial y}\hat{\boldsymbol{y}} + \frac{\partial}{\partial z}\hat{\boldsymbol{z}}\right)\cdot\left(E_{0x}\mathrm{e}^{\mathrm{j}(\omega_0 t - k_0 z)}\hat{\boldsymbol{x}} + E_{0y}\mathrm{e}^{\mathrm{j}(\omega_0 t - k_0 z)}\hat{\boldsymbol{y}} + E_{0z}\mathrm{e}^{\mathrm{j}(\omega_0 t - k_0 z)}\hat{\boldsymbol{z}}\right) = 0,$$

or

$$\frac{\partial}{\partial z}E_{0z}\mathrm{e}^{\mathrm{j}(\omega_0 t - k_0 z)} = -\mathrm{j}k_0 E_{0z}\mathrm{e}^{\mathrm{j}(\omega_0 t - k_0 z)} = 0,$$

implying

$$E_{0z} = 0, \tag{3.3-16}$$

as $-\mathrm{j}k_0\mathrm{e}^{\mathrm{j}(\omega_0 t - k_0 z)}$ in general is not zero. Similarly, by substituting $\boldsymbol{H}$ into magnetic Gauss's Law [see Eq. (3.1-4)], we get

$$H_{0z} = 0. \tag{3.3-17}$$

式 (3.3-16) 和式 (3.3-17) 表示线性、均匀、各向同性介质 (如自由空间) 中, 电场和磁场在传播方向上没有传播分量. 这种电磁波被称为横向电磁波 [*transverse electromagnetic* (TEM) *wave*].

现在, 我们来研究麦克斯韦方程组中的一个旋度方程, 它联系了 $\boldsymbol{E}$ 和 $\boldsymbol{H}$ 之间的关系. 将 $E_{0z}=H_{0z}=0$ 代入式 (3.3-14) 和式 (3.3-15), 再代入法拉第定律 [见式 (3.1-5)], 得

$$\begin{aligned}\nabla\times\boldsymbol{E}&=\begin{vmatrix}\hat{\boldsymbol{x}}&\hat{\boldsymbol{y}}&\hat{\boldsymbol{z}}\\\dfrac{\partial}{\partial x}&\dfrac{\partial}{\partial y}&\dfrac{\partial}{\partial z}\\E_{0x}\mathrm{e}^{\mathrm{j}(\omega_0t-k_0z)}&E_{0y}\mathrm{e}^{\mathrm{j}(\omega_0t-k_0z)}&0\end{vmatrix}\\&=\mathrm{j}k_0E_{0y}\mathrm{e}^{\mathrm{j}(\omega_0t-k_0z)}\hat{\boldsymbol{x}}-\mathrm{j}k_0E_{0x}\mathrm{e}^{\mathrm{j}(\omega_0t-k_0z)}\hat{\boldsymbol{y}},\end{aligned}$$

和

$$\begin{aligned}-\frac{\partial\boldsymbol{B}}{\partial t}=-\frac{\partial\mu\boldsymbol{H}}{\partial t}&=-\mu\frac{\partial}{\partial t}\left[H_{0x}\mathrm{e}^{\mathrm{j}(\omega_0t-k_0z)}\hat{\boldsymbol{x}}+H_{0y}\mathrm{e}^{\mathrm{j}(\omega_0t-k_0z)}\hat{\boldsymbol{y}}\right]\\&=-\mathrm{j}\mu\omega_0H_{0x}\mathrm{e}^{\mathrm{j}(\omega_0t-k_0z)}\hat{\boldsymbol{x}}-\mathrm{j}\mu\omega_0H_{0y}\mathrm{e}^{\mathrm{j}(\omega_0t-k_0z)}\hat{\boldsymbol{y}}.\end{aligned}$$

因此

$$\nabla\times\boldsymbol{E}=-\frac{\partial\boldsymbol{B}}{\partial t}$$

给出

$$\mathrm{j}k_0E_{0y}\mathrm{e}^{\mathrm{j}(\omega_0t-k_0z)}\hat{\boldsymbol{x}}-\mathrm{j}k_0E_{0x}\mathrm{e}^{\mathrm{j}(\omega_0t-k_0z)}\hat{\boldsymbol{y}}=-\mathrm{j}\mu\omega_0H_{0x}\mathrm{e}^{\mathrm{j}(\omega_0t-k_0z)}\hat{\boldsymbol{x}}-\mathrm{j}\mu\omega_0H_{0y}\mathrm{e}^{\mathrm{j}(\omega_0t-k_0z)}\hat{\boldsymbol{y}}.$$

通过比较上式两边的 $\hat{\boldsymbol{x}}$ 和 $\hat{\boldsymbol{y}}$ 分量, 有

$$H_{0x}=-\frac{1}{\eta}E_{0y},\quad H_{0y}=\frac{1}{\eta}E_{0x},\tag{3.3-18}$$

其中, η 被称为介质的特性阻抗 (*characteristic impedance*), 由下式给出

$$\eta=\frac{\omega_0}{k_0}\mu=v\mu=\sqrt{\frac{\mu}{\varepsilon}}.\tag{3.3-19}$$

特征阻抗的单位为欧姆, 它在自由空间的值为 $\eta=\sqrt{\mu_0/\varepsilon_0}=377\ \Omega$.

利用式 (3.3-18), 可以写出

$$\boldsymbol{E}=E_{0x}\mathrm{e}^{\mathrm{j}(\omega_0t-k_0z)}\hat{\boldsymbol{x}}+E_{0y}\mathrm{e}^{\mathrm{j}(\omega_0t-k_0z)}\hat{\boldsymbol{y}},\tag{3.3-20a}$$

和

$$\begin{aligned}\boldsymbol{H}&=H_{0x}\mathrm{e}^{\mathrm{j}(\omega_0t-k_0z)}\hat{\boldsymbol{x}}+H_{0y}\mathrm{e}^{\mathrm{j}(\omega_0t-k_0z)}\hat{\boldsymbol{y}}\\&=-\frac{1}{\eta}E_{0y}\mathrm{e}^{\mathrm{j}(\omega_0t-k_0z)}\hat{\boldsymbol{x}}+\frac{1}{\eta}E_{0x}\mathrm{e}^{\mathrm{j}(\omega_0t-k_0z)}\hat{\boldsymbol{y}}.\end{aligned}\tag{3.3-20b}$$

一旦知道了 $\boldsymbol{E}$ 场的分量 E_{0x} 和 E_{0y}, 就可以直接求出 $\boldsymbol{H}$ 场, 因为对于给定的介质, η 是已知的. 此外, 可以看出

$$\boldsymbol{E}\cdot\boldsymbol{H}=-E_{0x}\mathrm{e}^{\mathrm{j}(\omega_0t-k_0z)}\frac{1}{\eta}E_{0y}\mathrm{e}^{\mathrm{j}(\omega_0t-k_0z)}+E_{0y}\mathrm{e}^{\mathrm{j}(\omega_0t-k_0z)}\frac{1}{\eta}E_{0x}\mathrm{e}^{\mathrm{j}(\omega_0t-k_0z)}=0,$$

Equations (3.3-16) and (3.3-17) mean that there is no component of the electric and magnetic fields in the direction of propagation in a linear, homogeneous, and isotropic medium (such as free space). Such electromagnetic (EM) waves are called *transverse electromagnetic (TEM) waves.*

Now, let us work with one of the curl equations from Maxwell's equations, which relates $\boldsymbol{E}$ to $\boldsymbol{H}$. Substituting Eqs. (3.3-14) and (3.3-15) with $E_{0z} = H_{0z} = 0$ into Faraday's law [see Eq. (3.1-5)], we have

$$\begin{aligned}\nabla \times \boldsymbol{E} &= \begin{vmatrix} \hat{\boldsymbol{x}} & \hat{\boldsymbol{y}} & \hat{\boldsymbol{z}} \\ \dfrac{\partial}{\partial x} & \dfrac{\partial}{\partial y} & \dfrac{\partial}{\partial z} \\ E_{0x}\mathrm{e}^{\mathrm{j}(\omega_0 t - k_0 z)} & E_{0y}\mathrm{e}^{\mathrm{j}(\omega_0 t - k_0 z)} & 0 \end{vmatrix} \\ &= \mathrm{j}k_0 E_{0y}\mathrm{e}^{\mathrm{j}(\omega_0 t - k_0 z)}\hat{\boldsymbol{x}} - \mathrm{j}k_0 E_{0x}\mathrm{e}^{\mathrm{j}(\omega_0 t - k_0 z)}\hat{\boldsymbol{y}},\end{aligned}$$

and

$$\begin{aligned}-\frac{\partial \boldsymbol{B}}{\partial t} = -\frac{\partial \mu \boldsymbol{H}}{\partial t} &= -\mu\frac{\partial}{\partial t}\left[H_{0x}\mathrm{e}^{\mathrm{j}(\omega_0 t - k_0 z)}\hat{\boldsymbol{x}} + H_{0y}\mathrm{e}^{\mathrm{j}(\omega_0 t - k_0 z)}\hat{\boldsymbol{y}}\right] \\ &= -\mathrm{j}\mu\omega_0 H_{0x}\mathrm{e}^{\mathrm{j}(\omega_0 t - k_0 z)}\hat{\boldsymbol{x}} - \mathrm{j}\mu\omega_0 H_{0y}\mathrm{e}^{\mathrm{j}(\omega_0 t - k_0 z)}\hat{\boldsymbol{y}}.\end{aligned}$$

Hence,

$$\nabla \times \boldsymbol{E} = -\frac{\partial \boldsymbol{B}}{\partial t}$$

gives

$$\mathrm{j}k_0 E_{0y}\mathrm{e}^{\mathrm{j}(\omega_0 t - k_0 z)}\hat{\boldsymbol{x}} - \mathrm{j}k_0 E_{0x}\mathrm{e}^{\mathrm{j}(\omega_0 t - k_0 z)}\hat{\boldsymbol{y}} = -\mathrm{j}\mu\omega_0 H_{0x}\mathrm{e}^{\mathrm{j}(\omega_0 t - k_0 z)}\hat{\boldsymbol{x}} - \mathrm{j}\mu\omega_0 H_{0y}\mathrm{e}^{\mathrm{j}(\omega_0 t - k_0 z)}\hat{\boldsymbol{y}}.$$

By comparing the $\hat{\boldsymbol{x}}$ and $\hat{\boldsymbol{y}}$ components on both sides of the above equation, we have

$$H_{0x} = -\frac{1}{\eta}E_{0y}, \quad H_{0y} = \frac{1}{\eta}E_{0x}, \tag{3.3-18}$$

where η is called the *characteristic impedance* of the medium and given by

$$\eta = \frac{\omega_0}{k_0}\mu = v\mu = \sqrt{\frac{\mu}{\varepsilon}}. \tag{3.3-19}$$

The characteristic impendence has the units of ohm and its value for free space is $\eta = \sqrt{\mu_0/\varepsilon_0} = 377\ \Omega$.

Using Eq. (3.3-18), we can now write down the $\boldsymbol{E}$ and $\boldsymbol{H}$ fields as follows:

$$\boldsymbol{E} = E_{0x}\mathrm{e}^{\mathrm{j}(\omega_0 t - k_0 z)}\hat{\boldsymbol{x}} + E_{0y}\mathrm{e}^{\mathrm{j}(\omega_0 t - k_0 z)}\hat{\boldsymbol{y}}, \tag{3.3-20a}$$

and

$$\begin{aligned}\boldsymbol{H} &= H_{0x}\mathrm{e}^{\mathrm{j}(\omega_0 t - k_0 z)}\hat{\boldsymbol{x}} + H_{0y}\mathrm{e}^{\mathrm{j}(\omega_0 t - k_0 z)}\hat{\boldsymbol{y}} \\ &= -\frac{1}{\eta}E_{0y}\mathrm{e}^{\mathrm{j}(\omega_0 t - k_0 z)}\hat{\boldsymbol{x}} + \frac{1}{\eta}E_{0x}\mathrm{e}^{\mathrm{j}(\omega_0 t - k_0 z)}\hat{\boldsymbol{y}}.\end{aligned} \tag{3.3-20b}$$

Once we know the components of the $\boldsymbol{E}$ field, that is, E_{0x} and E_{0y}, we can find the $\boldsymbol{H}$ field directly as η is known for a given medium. Additionally, we note that

$$\boldsymbol{E}\cdot\boldsymbol{H} = -E_{0x}\mathrm{e}^{\mathrm{j}(\omega_0 t - k_0 z)}\frac{1}{\eta}E_{0y}\mathrm{e}^{\mathrm{j}(\omega_0 t - k_0 z)} + E_{0y}\mathrm{e}^{\mathrm{j}(\omega_0 t - k_0 z)}\frac{1}{\eta}E_{0x}\mathrm{e}^{\mathrm{j}(\omega_0 t - k_0 z)} = 0,$$

这意味着, 电场和磁场是相互正交的.

$$\boldsymbol{S}=\boldsymbol{E}\times\boldsymbol{H}, \tag{3.3-21}$$

量 $\boldsymbol{S}$ 被称为坡印亭矢量, 其单位为瓦特每平方米 ($\mathrm{W/m^2}$). 计算 $\boldsymbol{S}$, 有

$$\boldsymbol{S}=\begin{vmatrix}\hat{\boldsymbol{x}} & \hat{\boldsymbol{y}} & \hat{\boldsymbol{z}}\\ E_{0x}\mathrm{e}^{\mathrm{j}(\omega_0 t-k_0 z)} & E_{0y}\mathrm{e}^{\mathrm{j}(\omega_0 t-k_0 z)} & 0\\ -\dfrac{1}{\eta}E_{0y}\mathrm{e}^{\mathrm{j}(\omega_0 t-k_0 z)} & \dfrac{1}{\eta}E_{0x}\mathrm{e}^{\mathrm{j}(\omega_0 t-k_0 z)} & 0\end{vmatrix}=\frac{1}{\eta}(E_{0x}^2+E_{0y}^2)\mathrm{e}^{2\mathrm{j}(\omega_0 t-k_0 z)}\hat{\boldsymbol{z}}, \tag{3.3-22}$$

为了简单起见, 假设 E_{0x} 和 E_{0y} 都为实数. 注意, 电磁波携带能量, 且其能量流在波传播的方向上进行. 图 3.3-7 表示了当电场在 $\hat{\boldsymbol{x}}$ 方向时的坡印亭矢量的方向.

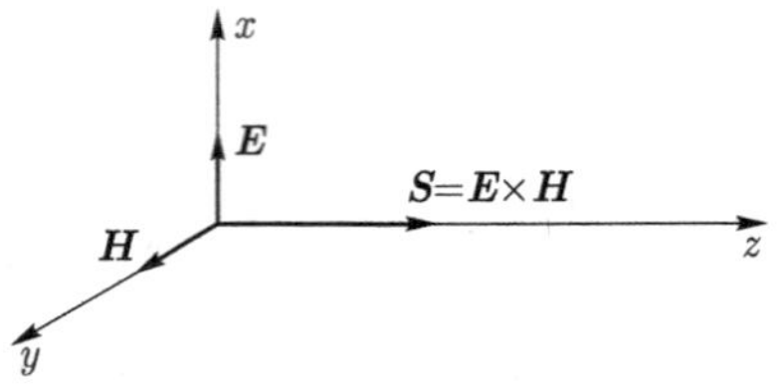

图 3.3-7 在 $\hat{\boldsymbol{z}}$ 方向传播的一种横向电磁 (TEM) 波

举例: 平均功率计算及强度

在以 μ 和 ε 为特征的线性、均匀、各向同性介质中, 考虑一个电场在 $\hat{\boldsymbol{x}}$ 方向, 并沿 $\hat{\boldsymbol{z}}$ 方向传播的简单例子. 如图 3.3-7 所示. 因此, 可以写出

$$\boldsymbol{E}=E_0\mathrm{e}^{\mathrm{j}(\omega_0 t-k_0 z)}\hat{\boldsymbol{x}}=|E_0|\mathrm{e}^{-\mathrm{j}\phi}\mathrm{e}^{\mathrm{j}(\omega_0 t-k_0 z)}\hat{\boldsymbol{x}},$$

考虑到 E_0 可能为复数, 由于电场必须是空间和时间的实函数, 因此只需对上式取实部, 得到一个实函数的表达式: $\mathrm{Re}[\boldsymbol{E}]=|E_0|\cos(\omega_0 t-k_0 z-\phi)\hat{\boldsymbol{x}}$. 根据式 (3.3-21), 磁场为

$$\boldsymbol{H}=\frac{1}{\eta}|E_0|\mathrm{e}^{\mathrm{j}(\omega_0 t-k_0 z-\phi)}\hat{\boldsymbol{y}},$$

或

$$\mathrm{Re}[\boldsymbol{H}]=\frac{1}{\eta}|E_0|\cos(\omega_0 t-k_0 z-\phi)\hat{\boldsymbol{y}},$$

因为 $\boldsymbol{S}\propto\hat{\boldsymbol{x}}\times\hat{\boldsymbol{y}}=\hat{\boldsymbol{z}}$. 可以看出, $\boldsymbol{S}$ 是时间的函数, 这对于定义时间平均功率密度更为方便

$$\begin{aligned}\boldsymbol{S}_{\boldsymbol{ave}}&=\frac{1}{T_0}\int_0^{T_0}\mathrm{Re}[\boldsymbol{E}]\times\mathrm{Re}[\boldsymbol{H}]\mathrm{d}t\\&=\frac{1}{T_0}\int_0^{T_0}|E_0|\cos(\omega_0 t-k_0 z-\phi)\hat{\boldsymbol{x}}\times\frac{1}{\eta}|E_0|\cos(\omega_0 t-k_0 z-\phi)\hat{\boldsymbol{y}}\mathrm{d}t\end{aligned}$$

meaning that the electric and magnetic fields are orthogonal to each other.

The quantity

$$\boldsymbol{S} = \boldsymbol{E} \times \boldsymbol{H}, \tag{3.3-21}$$

is called the Poynting vector and is measured in watts per square meter (W/m^2). Evaluating $\boldsymbol{S}$, we have

$$\boldsymbol{S} = \begin{vmatrix} \hat{\boldsymbol{x}} & \hat{\boldsymbol{y}} & \hat{\boldsymbol{z}} \\ E_{0x}\mathrm{e}^{\mathrm{j}(\omega_0 t - k_0 z)} & E_{0y}\mathrm{e}^{\mathrm{j}(\omega_0 t - k_0 z)} & 0 \\ -\dfrac{1}{\eta}E_{0y}\mathrm{e}^{\mathrm{j}(\omega_0 t - k_0 z)} & \dfrac{1}{\eta}E_{0x}\mathrm{e}^{\mathrm{j}(\omega_0 t - k_0 z)} & 0 \end{vmatrix} = \frac{1}{\eta}(E_{0x}^2 + E_{0y}^2)\mathrm{e}^{2\mathrm{j}(\omega_0 t - k_0 z)}\hat{\boldsymbol{z}}, \tag{3.3-22}$$

where have assumed both E_{0x} and E_{0y} are real for simplicity. We notice that the EM waves carry energy and the energy flow occurs in the direction of propagation of the wave. Figure 3.3-7 illustrates the direction of the Poynting vector if the electrical field is in the $\hat{\boldsymbol{x}}$-direction.

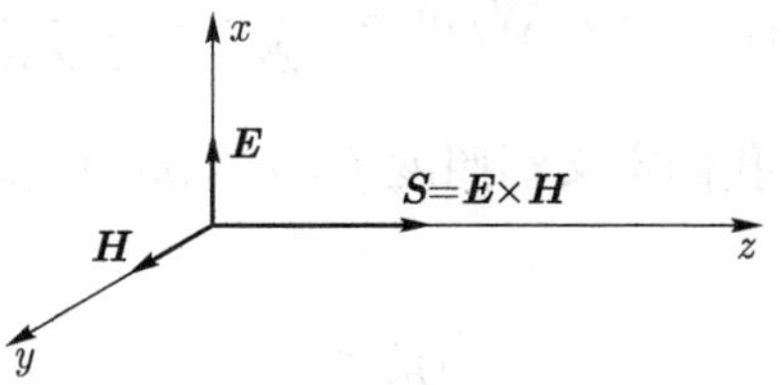

Figure 3.3-7 A transverse electromagnetic (TEM) wave propagating in the $\hat{\boldsymbol{z}}$-direction

Example: Average Power Calculation and Intensity

In a linear, homogenous, and isotropic medium characterized by ε and μ, we consider a simple case in that the electric field is along the $\hat{\boldsymbol{x}}$-direction and propagates along the $\hat{\boldsymbol{z}}$-direction. The situation is shown in Figure 3.3-7. Hence, we write

$$\boldsymbol{E} = E_0\mathrm{e}^{\mathrm{j}(\omega_0 t - k_0 z)}\hat{\boldsymbol{x}} = |E_0|\mathrm{e}^{-\mathrm{j}\phi}\mathrm{e}^{\mathrm{j}(\omega_0 t - k_0 z)}\hat{\boldsymbol{x}},$$

taking into account that E_0 could be complex. Since the electrical field must be a real function of space and time, we simply take the real part of the above to have a real function expression: $\mathrm{Re}[\boldsymbol{E}] = |E_0|\cos(\omega_0 t - k_0 z - \phi)\hat{\boldsymbol{x}}$. The magnetic field, according to Eq. (3.3-21), is

$$\boldsymbol{H} = \frac{1}{\eta}|E_0|\mathrm{e}^{\mathrm{j}(\omega_0 t - k_0 z - \phi)}\hat{\boldsymbol{y}},$$

or

$$\mathrm{Re}[\boldsymbol{H}] = \frac{1}{\eta}|E_0|\cos(\omega_0 t - k_0 z - \phi)\hat{\boldsymbol{y}}$$

as $\boldsymbol{S} \propto \hat{\boldsymbol{x}} \times \hat{\boldsymbol{y}} = \hat{\boldsymbol{z}}$. Note that $\boldsymbol{S}$ is a function of time and it is more convenient to define the time-averaged power density:

$$\begin{aligned} \boldsymbol{S_{ave}} &= \frac{1}{T_0}\int_0^{T_0} \mathrm{Re}[\boldsymbol{E}] \times \mathrm{Re}[\boldsymbol{H}]\mathrm{d}t \\ &= \frac{1}{T_0}\int_0^{T_0} |E_0|\cos(\omega_0 t - k_0 z - \phi)\hat{\boldsymbol{x}} \times \frac{1}{\eta}|E_0|\cos(\omega_0 t - k_0 z - \phi)\hat{\boldsymbol{y}}\mathrm{d}t \end{aligned}$$

$$= \frac{1}{T_0}\int_0^{T_0} \frac{|E_0|^2}{\eta}\cos^2(\omega_0 t - k_0 z - \phi)\mathrm{d}t\hat{\boldsymbol{z}} = \frac{1}{2}\frac{|E_0|^2}{\eta}\hat{\boldsymbol{z}},$$

其中

$$\frac{1}{T_0}\int_0^{T_0}\cos^2(\omega_0 t - \theta)\mathrm{d}t = \frac{1}{2},$$

这里 θ 为某常数, 且 $T_0 = 2\pi/\omega_0$ 是正弦函数的周期. 可以看出, 因为平均功率是实数, 在积分前取 $\boldsymbol{E}$ 和 $\boldsymbol{H}$ 的实部.

对于将 1 mW 均匀传输到自由空间 1 mm^2 区域的激光器, 时间平均功率密度为 1 mW/1 mm^2=10^3 W/m^2=1 kW/m^2, 这约为太阳在地球表面辐射的功率密度. 因此, 激光束中电场的峰值大小可根据

$$|\boldsymbol{S_{ave}}| = 10^3\ \mathrm{W/m} = \frac{|E_0|^2}{2\eta} = \frac{|E_0|^2}{2\times 377},$$

给定 $|E_0| = 868.3$ V/m, 通常, 我们定义辐照度 (*irradiance*) I_{ir} 为 $\boldsymbol{S_{ave}}$ 的强度, 即, $|\boldsymbol{S_{ave}}| = S_{ave}$,

$$I_{ir} = S_{ave} = \frac{|E_0|^2}{2\eta} \propto I = |E_0|^2. \tag{3.3-23}$$

在信息光学中, 我们通常将强度 (*intensity*) I 表示为场的强度的平方, 即, $I = |E_0|^2$.

3.4 基于傅里叶变换的标量衍射理论

上文讨论了波在线性、均匀和各向同性介质中的传播, 波在这些介质中的传播是没有阻碍的. 衍射 (*diffraction*) 是指当波遇到孔径时所发生的现象. 孔径 (*aperture*) 是光通过时的开口. 如果我们想处理衍射的矢量性质, 就要运用电磁光学 (*electromagnetic optics*), 并需要在边界条件下求解矢量麦克斯韦方程组 [见式 (3.1-8)]. 在这种情况下, 就有了电磁波的矢量衍射理论 (*vector diffraction theory*). 在本节中, 我们讨论标量衍射 (*scalar diffraction*). 在标量衍射理论中, 求解边界条件下的三维标量波动方程 [见式 (3.3-1)]. 标量理论仅在衍射结构 (*diffracting structures*) 大于光的波长时才有效. 结果表明, 三维标量波动方程满足叠加原理. 因此, 所要介绍的理论对于第二章所讨论的线性系统是有效的. 叠加原理为我们后续研究的衍射和干涉 (*interference*) 提供了基础.

标量衍射理论通常从亥姆霍兹方程 (*Helmholtz equation*) 开始, 并用高等数学的格林定理 (*Green's theorem*) 来转换方程, 因此, 一些物理意义很容易被掩盖. 这里将通过傅里叶变换和三维标量波动方程来展开标量衍射理论. 事实上, 正如我们将看到的, 衍射的概念可用一种极其简单的方法推导出来. 我们称这种方法为基于傅里叶变换的标量衍射理论 [*Fourier transform* (FT)-*based scalar diffraction theory*]. 通常来说, 所发展的理论是针对单色波的,

$$= \frac{1}{T_0} \int_0^{T_0} \frac{|E_0|^2}{\eta} \cos^2(\omega_0 t - k_0 z - \phi) \mathrm{d}t \hat{\boldsymbol{z}} = \frac{1}{2} \frac{|E_0|^2}{\eta} \hat{\boldsymbol{z}},$$

as

$$\frac{1}{T_0} \int_0^{T_0} \cos^2(\omega_0 t - \theta) \mathrm{d}t = \frac{1}{2},$$

where θ is some constant, and $T_0 = 2\pi/\omega_0$ is the period of the sinusoidal function. Note that we are using the real part of $\boldsymbol{E}$ and $\boldsymbol{H}$ before the integration because the average power is a real quantity.

For a laser designed to deliver 1 mW uniformly onto an area of 1 mm^2 in free space, the time-averaged power density is 1 mW/1 mm^2=10^3 W/m^2=1 kW/m^2, which is about the power density of the sun's radiation on the surface of the Earth. The peak magnitude of the electric field in the laser beam is, therefore, calculated according to

$$|\boldsymbol{S}_{\boldsymbol{ave}}| = 10^3 \text{ W/m} = \frac{|E_0|^2}{2\eta} = \frac{|E_0|^2}{2 \times 377},$$

giving $|E_0| = 868.3$ V/m. Typically, we define *irradiance* I_{ir} as the magnitude of $\boldsymbol{S}_{\boldsymbol{ave}}$, that is, $|\boldsymbol{S}_{\boldsymbol{ave}}| = S_{ave}$,

$$I_{ir} = S_{ave} = \frac{|E_0|^2}{2\eta} \propto I = |E_0|^2. \tag{3.3-23}$$

In information optics, we usually speak of the *intensity* I being taken to be the square magnitude of the field, that is, $I = |E_0|^2$.

3.4 Fourier Transform-Based Scalar Diffraction Theory

We have discussed wave propagation in a linear, homogeneous, and isotropic medium in which waves travel with no obstruction. *Diffraction* refers to phenomena that occur when a wave encounters an aperture. An *aperture* is an opening through which light travels. If we want to deal with the vectorial nature of diffraction, we employ *electromagnetic optics* in which vectorial Maxwell's equations [see Eq. (3.1-8)] are needed to be solved under boundary conditions. In that case, we have *vector diffraction theory* for electromagnetic waves. In this section, we deal with *scalar diffraction.* In scalar diffraction theory, we solve the 3-D scalar wave equation [see Eq. (3.3-1)] subject to boundary conditions. The scalar theory is only valid when the diffracting structures are large compared with the wavelength of light. As it turns out, the 3-D scalar wave equation satisfies the principle of superposition. Hence, the theory to be developed is valid for linear systems as discussed in Chapter 2. The principle of superposition provides the basis for our subsequent study of diffraction and *interference.*

Scalar diffraction theory typically is developed by starting with the Helmholtz equation and converting the equation using highly mathematical Green's theorem and as such, some of the physical insight is easily obscured. We will develop scalar diffraction theory through the use of Fourier transform and the 3-D scalar wave equation. Indeed, as we shall see, the concept of diffraction can be derived in a much simpler way. We call such an approach as *Fourier transform-based (FT-based) scalar diffraction theory.* As usual, the theory developed is for

即仅有单一波长的波. 然而, 这个结果可以通过使用傅里叶分析推广到非单色波. 这种推广的情况类似于在第二章中讨论的电路脉冲的输入和输出的关系.

图 3.4-1 给出了衍射几何图样. 孔径 $t(x,y)$ 位于 $z=0$. 在 $z=0^-$ 处时间频率为 ω_0 的入射波为

$$\psi(x,y,z=0^-;t)=\psi_p(x,y;z=0^-)\mathrm{e}^{\mathrm{j}\omega_0 t}. \tag{3.4-1}$$

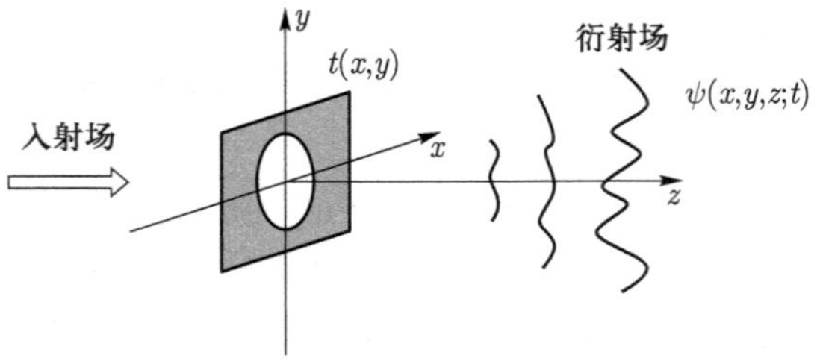

图 3.4-1 衍射几何图样

$\psi_p(x,y;z=0^-)$ 在光学上被称为复振幅 (*complex amplitude*). $\psi_p(x,y;z=0^-)$ 是一个包含振幅和相位信息的复函数, 在电气工程中被称为相位 (*phasor*). 现在定义孔径的振幅透射函数 (*amplitude transmittance function*), 也简称为透过率函数 (*transparency function*), 即在 $z=0$ 平面上, $\psi_p(x,y;z=0^+)$ 的透射复振幅 (复光场) 与入射复振幅 (复光场) $\psi_p(x,y;z=0^-)$ 的比值

$$t(x,y)=\frac{\psi_p(x,y;z=0^+)}{\psi_p(x,y;z=0^-)}. \tag{3.4-2a}$$

那么

$$\psi_p(x,y;z=0^+)=\psi_p(x,y;z=0^-)t(x,y)=\psi_{p0}(x,y). \tag{3.4-2b}$$

这个复振幅是该情形的初始条件或边界条件. 对于给定的入射场和透过率函数, 我们来求其衍射场 $\psi(x,y,z;t)$.

为了求距离孔径 z 处的场分布, 将解的形式建模为以下形式

$$\psi(x,y,z;t)=\psi_p(x,y;z)\mathrm{e}^{\mathrm{j}\omega_0 t}, \tag{3.4-3}$$

其中, $\psi_p(x,y;z)$ 为在初始条件 $\psi_{p0}(x,y)$ 下将要求解的未知复振幅. 将 $\psi(x,y,\ z;t)=\psi_p(x,y;z)\mathrm{e}^{\mathrm{j}\omega_0 t}$ 代入三维标量波动方程式 (3.3-1), 得

$$\frac{\partial^2\psi_p}{\partial x^2}\mathrm{e}^{\mathrm{j}\omega_0 t}+\frac{\partial^2\psi_p}{\partial y^2}\mathrm{e}^{\mathrm{j}\omega_0 t}+\frac{\partial^2\psi_p}{\partial z^2}\mathrm{e}^{\mathrm{j}\omega_0 t}=\frac{1}{v^2}(\mathrm{j}\omega_0)^2\psi_p\mathrm{e}^{\mathrm{j}\omega_0 t},$$

得到 $\psi_p(x,y;z)$ 的亥姆霍兹方程 (*Helmholtz equation*)

$$\frac{\partial^2\psi_p}{\partial x^2}+\frac{\partial^2\psi_p}{\partial y^2}+\frac{\partial^2\psi_p}{\partial z^2}+k_0^2\psi_p=0. \tag{3.4-4}$$

可以看出, 该亥姆霍兹方程与时间无关. 在这个过程中, 我们完成了降维.

monochromatic waves, that is, waves with a single wavelength. The result, nevertheless, can be generalized to non-monochromatic waves simply by using Fourier analysis. The situation of such generalization is analogous to finding the output due to a pulse input to a circuit discussed in Chapter 2.

Figure 3.4-1 illustrates the diffraction geometry. The aperture $t(x, y)$ is located at $z = 0$. The incident wave of temporal frequency of ω_0 at $z = 0^-$ is given by

$$\psi(x, y, z = 0^-; t) = \psi_p(x, y; z = 0^-)\mathrm{e}^{\mathrm{j}\omega_0 t}. \tag{3.4-1}$$

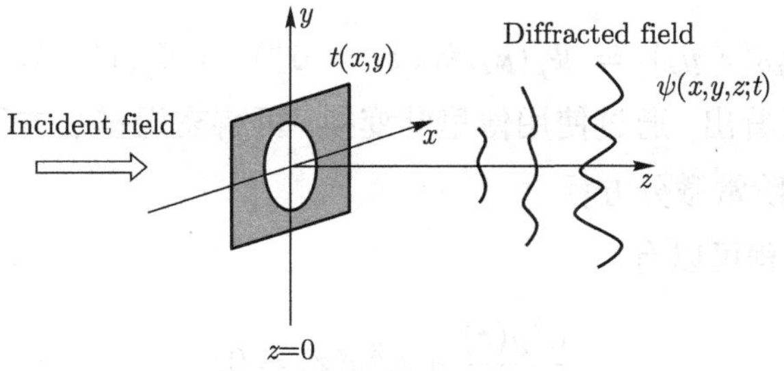

Figure 3.4-1 Diffraction geometry

The quantity $\psi_p(x, y; z = 0^-)$ is referred to as the *complex amplitude* in optics. $\psi_p(x, y; z = 0^-)$ is a complex function containing the amplitude and phase information and is called a *phasor* in electrical engineering. Let us definite the *amplitude transmittance function*, also simply called *transparency function*, of the aperture as the ratio of the transmitted complex field $\psi_p(x, y; z = 0^+)$ to the incident complex field $\psi_p(x, y; z = 0^-)$ in the $z = 0$ plane,

$$t(x, y) = \frac{\psi_p(x, y; z = 0^+)}{\psi_p(x, y; z = 0^-)}. \tag{3.4-2a}$$

Then

$$\psi_p(x, y; z = 0^+) = \psi_p(x, y; z = 0^-)t(x, y) = \psi_{p0}(x, y). \tag{3.4-2b}$$

This complex amplitude is the initial condition or the boundary condition of the situation.

For a given incident field and the transparency function, we seek to find the diffracted field $\psi(x, y, z; t)$.

To find the field distribution at z away from the aperture, we model the solution in the form of

$$\psi(x, y, z; t) = \psi_p(x, y; z)\mathrm{e}^{\mathrm{j}\omega_0 t}, \tag{3.4-3}$$

where $\psi_p(x, y; z)$ is the unknown complex field to be found with the given initial condition $\psi_{p0}(x, y)$. Substituting $\psi(x, y, z; t) = \psi_p(x, y; z)\mathrm{e}^{\mathrm{j}\omega_0 t}$ into the 3-D scalar wave in Eq. (3.3-1), we obtain

$$\frac{\partial^2 \psi_p}{\partial x^2}\mathrm{e}^{\mathrm{j}\omega_0 t} + \frac{\partial^2 \psi_p}{\partial y^2}\mathrm{e}^{\mathrm{j}\omega_0 t} + \frac{\partial^2 \psi_p}{\partial z^2}\mathrm{e}^{\mathrm{j}\omega_0 t} = \frac{1}{v^2}(\mathrm{j}\omega_0)^2\psi_p\mathrm{e}^{\mathrm{j}\omega_0 t},$$

which becomes the *Helmholtz equation* for $\psi_p(x, y; z)$,

$$\frac{\partial^2 \psi_p}{\partial x^2} + \frac{\partial^2 \psi_p}{\partial y^2} + \frac{\partial^2 \psi_p}{\partial z^2} + k_0^2\psi_p = 0. \tag{3.4-4}$$

Note that the Helmholtz equation contains no time dependence. We have accomplished dimensionality reduction in the process.

现在用傅里叶变换方法来求上述方程的解.通过对式 (3.4-4) 进行二维傅里叶变换, 并使用表 2.2 的变换对 $\dfrac{\partial^n f(x,y)}{\partial x^n} \Longleftrightarrow (-\mathrm{j}k_x)^n F(k_x,k_y)$, 得

$$(-\mathrm{j}k_x)^2\Psi_p(k_x,k_y;z)+(-\mathrm{j}k_y)^2\Psi_p(k_x,k_y;z)+\frac{\mathrm{d}^2\Psi_p(k_x,k_y;z)}{\mathrm{d}z^2}+k_0^2\Psi_p(k_x,k_y;z)=0,$$

这里 $\mathcal{F}\{\psi_p(x,y;z)\}=\Psi_p(k_x,k_y;z)$. 上式简化为

$$\frac{\mathrm{d}^2\Psi_p}{\mathrm{d}z^2}+(k_0^2-k_x^2-k_y^2)\Psi_p=0, \tag{3.4-5}$$

可通过初始条件 $\mathcal{F}\{\psi_{p0}(x,y)\}=\Psi_p(k_x,k_y;z=0^+)=\Psi_{p0}(k_x,k_y)$ 来求解, $\psi_{p0}(x,y)$ 由式 (3.4-2b) 得出. 可以看出, 通过使用傅里叶变换, 可将亥姆霍兹方程的二阶偏微分方程简化为 $\Psi_p(k_x,k_y;z)$ 的二阶常微分方程.

这种形式的微分方程可以有

$$\frac{\mathrm{d}^2y(z)}{\mathrm{d}z^2}+\alpha^2y(z)=0,$$

以下解的形式

$$y(z)=y(0)\exp(-\mathrm{j}\alpha z),$$

写出式 (3.4-5) 的解如下

$$\Psi_p(k_x,k_y;z)=\Psi_{p0}(k_x,k_y)\mathrm{e}^{-\mathrm{j}k_0\sqrt{\left(1-k_x^2/k_0^2-k_y^2/k_0^2\right)}z}, \tag{3.4-6}$$

这对应了沿 z 方向传播的波. 可以看出, 在第二章中利用传递函数 $H(\omega)$ 描述线性时不变系统时, 该方程在概念上与式 (2.2-13a) 是等同的. 因此, 我们可以定义一些衍射的传递函数, 其所涉及的系统为线性空不变系统. 因此, 我们定义传播的空间传递函数 (*spatial transfer function of propagation*) $\mathcal{H}(k_x,k_y;z)$ 为

$$\mathcal{H}(k_x,k_y;z)=\Psi_p(k_x,k_y;z)/\Psi_{p0}(k_x,k_y)=\mathrm{e}^{-\mathrm{j}k_0\sqrt{\left(1-k_x^2/k_0^2-k_y^2/k_0^2\right)}z}. \tag{3.4-7}$$

这个传递函数将 $z=0^+$ 处的输入谱 $\Psi_{p0}(k_x,k_y)$ 和 z 处的输出谱 $\Psi_p(k_x,k_y;z)$ 联系起来, 为了求复振幅 $\psi_p(x,y;z)$, 只需对式 (3.4-7) 进行逆傅里叶变换即可得到

$$\psi_p(x,y;z)=\mathcal{F}^{-1}\{\Psi_p(k_x,k_y;z)\}=\mathcal{F}^{-1}\{\Psi_{p0}(k_x,k_y)\mathcal{H}(k_x,k_y;z)\},$$

或

$$\begin{aligned}\psi_p(x,y;z)&=\mathcal{F}^{-1}\left\{\Psi_{p0}(k_x,k_y)\mathrm{e}^{-\mathrm{j}k_0\sqrt{\left(1-k_x^2/k_0^2-k_y^2/k_0^2\right)}z}\right\}\\&=\frac{1}{4\pi^2}\iint_{-\infty}^{\infty}\Psi_{p0}(k_x,k_y)\mathrm{e}^{-\mathrm{j}k_0\sqrt{\left(1-k_x^2/k_0^2-k_y^2/k_0^2\right)}z}\mathrm{e}^{-\mathrm{j}k_xx-\mathrm{j}k_yy}\mathrm{d}k_x\mathrm{d}k_y.\end{aligned} \tag{3.4-8}$$

这一结果非常重要, 因为它是接下来将要讨论的菲涅耳衍射 (*Fresnel diffraction*) 和夫琅禾费衍射 (*Fraunhofer diffraction*) 的基础. 现在给出式 (3.4-8) 的物理解释.

We now use the Fourier transform technique to find the solution to Eq. (3.4-4). By taking the 2-D Fourier transform of Eq. (3.4-4) and using the transform pair $\dfrac{\partial^n f(x,y)}{\partial x^n} \Longleftrightarrow (-\mathrm{j}k_x)^n F(k_x,k_y)$ from Table 2.2, we obtain

$$(-\mathrm{j}k_x)^2\Psi_p(k_x,k_y;z)+(-\mathrm{j}k_y)^2\Psi_p(k_x,k_y;z)+\frac{\mathrm{d}^2\Psi_p(k_x,k_y;z)}{\mathrm{d}z^2}+k_0^2\Psi_p(k_x,k_y;z)=0,$$

where $\mathcal{F}\{\psi_p(x,y;z)\}=\Psi_p(k_x,k_y;z)$. The above equation is simplified to

$$\frac{\mathrm{d}^2\Psi_p}{\mathrm{d}z^2}+(k_0^2-k_x^2-k_y^2)\Psi_p=0, \tag{3.4-5}$$

which is to be solved subject to the initial condition $\mathcal{F}\{\psi_{p0}(x,y)\}=\Psi_p(k_x,k_y;z=0^+)=\Psi_{p0}(k_x,k_y)$. $\psi_{p0}(x,y)$ is given by Eq. (3.4-2b). Note that by using the Fourier transform, we have managed to reduce the second-order partial differential equation of the Helmholtz equation into a second-order ordinary differential equation in $\Psi_p(k_x,k_y;z)$.

Recognizing the differential equation of the form

$$\frac{\mathrm{d}^2y(z)}{\mathrm{d}z^2}+\alpha^2y(z)=0$$

has the solution given by

$$y(z)=y(0)\exp(-\mathrm{j}\alpha z),$$

we can write the solution to Eq. (3.4-5) as

$$\Psi_p(k_x,k_y;z)=\Psi_{p0}(k_x,k_y)\mathrm{e}^{-\mathrm{j}k_0\sqrt{\left(1-k_x^2/k_0^2-k_y^2/k_0^2\right)}z} \tag{3.4-6}$$

for waves propagating along the z-direction. Note that conceptually this equation is identical to Eq. (2.2-13a) when we describe linear time-invariant systems in terms of the transfer function $H(\omega)$ in Chapter 2. Hence, we can define some kind of transfer function for diffraction and the system involved is being a linear space-invariant system. We, therefore, define the *spatial transfer function of propagation* $\mathcal{H}(k_x,k_y;z)$ as

$$\mathcal{H}(k_x,k_y;z)=\Psi_p(k_x,k_y;z)/\Psi_{p0}(k_x,k_y)=\mathrm{e}^{-\mathrm{j}k_0\sqrt{\left(1-k_x^2/k_0^2-k_y^2/k_0^2\right)}z}. \tag{3.4-7}$$

This transfer function relates the input spectrum $\Psi_{p0}(k_x,k_y)$ at $z=0^+$ and output spectrum $\Psi_p(k_x,k_y;z)$ at z. To find the complex field $\psi_p(x,y;z)$, we simply take the inverse Fourier transform of Eq. (3.4-7) to give

$$\psi_p(x,y;z)=\mathcal{F}^{-1}\{\Psi_p(k_x,k_y;z)\}=\mathcal{F}^{-1}\{\Psi_{p0}(k_x,k_y)\mathcal{H}(k_x,k_y;z)\},$$

or

$$\psi_p(x,y;z)=\mathcal{F}^{-1}\left\{\Psi_{p0}(k_x,k_y)\mathrm{e}^{-\mathrm{j}k_0\sqrt{\left(1-k_x^2/k_0^2-k_y^2/k_0^2\right)}z}\right\}$$

$$=\frac{1}{4\pi^2}\iint_{-\infty}^{\infty}\Psi_{p0}(k_x,k_y)\mathrm{e}^{-\mathrm{j}k_0\sqrt{\left(1-k_x^2/k_0^2-k_y^2/k_0^2\right)}z}\mathrm{e}^{-\mathrm{j}k_xx-\mathrm{j}k_yy}\mathrm{d}k_x\mathrm{d}k_y. \tag{3.4-8}$$

This is a very important result as it is the foundation to *Fresnel diffraction* and *Fraunhofer diffraction*, to be discussed next. Let us now bring out the physical interpretation of Eq. (3.4-8).

对于给定的 $z=0^+$ 平面上的场分布, 即 $\psi_p(x,y;z=0^+)=\psi_{p0}(x,y)$, 通过计算式 (3.4-8), 可以求出距离这个场 z 处且平行于 x-y 平面的场分布. 式 (3.4-8) 中的 $\Psi_{p0}(k_x,k_y)$ 项是 $\psi_{p0}(x,y)$ 的傅里叶变换, 根据式 (2.3-1b), 有

$$\psi_{p0}(x,y)=\mathcal{F}^{-1}\left\{\Psi_{p0}(k_x,k_y)\right\}=\frac{1}{4\pi^2}\iint\limits_{-\infty}^{\infty}\Psi_{p0}(k_x,k_y)\mathrm{e}^{-\mathrm{j}k_x x-\mathrm{j}k_y y}\mathrm{d}k_x\mathrm{d}k_y.$$

上述积分的物理意义是, 我们首先认识到一个传播矢量为 $\boldsymbol{k}_0$ 的单位振幅平面波可以表示为 $\mathrm{e}^{\mathrm{j}(\omega_0 t-k_{0x}x-k_{0y}y-k_{0z}z)}$ [见式 (3.3-9)]. 根据式 (3.4-3), 平面波的复振幅为

$$\mathrm{e}^{-\mathrm{j}k_{0x}x-\mathrm{j}k_{0y}y-\mathrm{j}k_{0z}z}.$$

则在 $z=0$ 处的场分布或平面波分量为

$$\mathrm{e}^{-\mathrm{j}k_{0x}x-\mathrm{j}k_{0y}y}.$$

将此式与上述 $\psi_{p0}(x,y)$ 给出的方程进行比较, 可以发现, 场分布 $\psi_{p0}(x,y)$ 的空间角频率变量 k_x 和 k_y, 可分别被看作平面波的 k_{0x} 和 k_{0y} [即所谓的传播矢量 (波矢) 分量]. 因此, $\Psi_{p0}(k_x,k_y)\mathrm{e}^{-\mathrm{j}k_x x-\mathrm{j}k_y y}$ 是振幅为 $\Psi_{p0}(k_x,k_y)$ 的平面波在各个不同方向 k_x 和 k_y 的求和, 可以得到 $z=0^+$ 处的场分布 $\psi_{p0}(x,y)$, 因此, $\Psi_{p0}(k_x,k_y)$ 也称为场分布 $\psi_{p0}(x,y)$ 的平面波角谱 (*angular plane wave spectrum*).

为了求出距离 z 处的场分布, 我们只需让不同的平面波传播 z 的距离, 这意味着, 只需将平面波的变量 k_z 看作 k_{0z}, 即会产生 $\mathrm{e}^{-\mathrm{j}k_z z}$ 或 $\mathrm{e}^{-\mathrm{j}k_{0z}z}$ 的相移

$$\Psi_{p0}(k_x,k_y)\mathrm{e}^{-\mathrm{j}k_x x-\mathrm{j}k_y y}\mathrm{e}^{-\mathrm{j}k_{0z}z}. \tag{3.4-9}$$

现在通过对不同方向的 k_x 和 k_y 进行求和, 有

$$\begin{aligned}\psi_p(x,y;z)&=\frac{1}{4\pi^2}\iint\limits_{-\infty}^{\infty}\Psi_{p0}(k_x,k_y)\mathrm{e}^{-\mathrm{j}k_x x-\mathrm{j}k_y y-\mathrm{j}k_z z}\mathrm{d}k_x\mathrm{d}k_y\\&=\mathcal{F}^{-1}\left\{\Psi_{p0}(k_x,k_y)\mathrm{e}^{-\mathrm{j}k_{0z}z}\right\}.\end{aligned}$$

因 $k_0=\sqrt{k_{0x}^2+k_{0y}^2+k_{0z}^2}$, 故 $k_z=k_{0z}=\pm k_0\sqrt{1-k_x^2/k_0^2-k_y^2/k_0^2}$. 上述关系的 + 号表示波沿 z 轴正向传播. 因此, 有

$$\begin{aligned}\psi_p(x,y;z)&=\mathcal{F}^{-1}\left\{\Psi_{p0}(k_x,k_y)\mathrm{e}^{-\mathrm{j}k_{0z}z}\right\}\\&=\mathcal{F}^{-1}\left\{\Psi_{p0}(k_x,k_y)\mathrm{e}^{-\mathrm{j}k_0\sqrt{\left(1-k_x^2/k_0^2-k_y^2/k_0^2\right)}\,z}\right\},\end{aligned}$$

得到式 (3.4-8) 并给出该式的物理意义. 可以看出, 场分布 $\psi_p(x,y;z)$ 的平面波角谱由下式给出

$$\Psi_{p0}(k_x,k_y)\mathrm{e}^{-\mathrm{j}k_{0z}z},$$

For a given field distribution over the $z = 0^+$ plane, that is, $\psi_p(x, y; z = 0^+) = \psi_{p0}(x, y)$, we can find the field distribution across a plane parallel to the x-y plane but at a distance z from it by calculating Eq. (3.4-8). The term $\Psi_{p0}(k_x, k_y)$ in Eq. (3.4-8) is a Fourier transform of $\psi_{p0}(x, y)$, and according to Eq. (2.3-1b), we have

$$\psi_{p0}(x, y) = \mathcal{F}^{-1}\{\Psi_{p0}(k_x, k_y)\} = \frac{1}{4\pi^2}\iint_{-\infty}^{\infty} \Psi_{p0}(k_x, k_y)\mathrm{e}^{-\mathrm{j}k_x x - \mathrm{j}k_y y}\mathrm{d}k_x \mathrm{d}k_y.$$

The physical meaning of the above integral is that we first recognize that a unit-amplitude plane wave propagating with propagation vector $\boldsymbol{k}_0$ is given by $\mathrm{e}^{\mathrm{j}(\omega_0 t - k_{0x}x - k_{0y}y - k_{0z}z)}$ [see Eq. (3.3-9)]. The complex field of the plane wave, according to Eq. (3.4-3), is

$$\mathrm{e}^{-\mathrm{j}k_{0x}x - \mathrm{j}k_{0y}y - \mathrm{j}k_{0z}z}.$$

The field distribution or the plane-wave component at $z = 0$ is then

$$\mathrm{e}^{-\mathrm{j}k_{0x}x - \mathrm{j}k_{0y}y}.$$

By comparing this equation with the one given by $\psi_{p0}(x, y)$ above and recognizing that the spatial radian frequency variables k_x and k_y of the field distribution $\psi_{p0}(x, y)$ can be regarded as k_{0x} and k_{0y} of the plane wave, respectively. Therefore, $\Psi_{p0}(k_x, k_y)\mathrm{e}^{-\mathrm{j}k_x x - \mathrm{j}k_y y}$ is the plane wave with amplitude $\Psi_{p0}(k_x, k_y)$ and by summing over various directions of k_x and k_y, we have the field distribution $\psi_{p0}(x, y)$ at $z = 0^+$. For this reason, $\Psi_{p0}(k_x, k_y)$ is also called the *angular plane wave spectrum* of the field distribution $\psi_{p0}(x, y)$.

To find the field distribution a distance of z away, we simply let the various plane waves to propagate over a distance z, which means acquiring a phase shift of $\mathrm{e}^{-\mathrm{j}k_z z}$ or $\mathrm{e}^{-\mathrm{j}k_{0z}z}$ by noting that the variable k_z is regarded as k_{0z} of the plane wave:

$$\Psi_{p0}(k_x, k_y)\mathrm{e}^{-\mathrm{j}k_x x - \mathrm{j}k_y y}\mathrm{e}^{-\mathrm{j}k_{0z}z}. \tag{3.4-9}$$

Now by summing over various directions of k_x and k_y, we have

$$\begin{aligned}\psi_p(x, y; z) &= \frac{1}{4\pi^2}\iint_{-\infty}^{\infty} \Psi_{p0}(k_x, k_y)\mathrm{e}^{-\mathrm{j}k_x x - \mathrm{j}k_y y - \mathrm{j}k_z z}\mathrm{d}k_x \mathrm{d}k_y \\ &= \mathcal{F}^{-1}\left\{\Psi_{p0}(k_x, k_y)\mathrm{e}^{-\mathrm{j}k_{0z}z}\right\}.\end{aligned}$$

Recall that $k_0 = \sqrt{k_{0x}^2 + k_{0y}^2 + k_{0z}^2}$ and hence $k_z = k_{0z} = \pm k_0\sqrt{1 - k_x^2/k_0^2 - k_y^2/k_0^2}$. We keep the + sign in the above relation to represent waves traveling in the positive z-direction. Hence, we have

$$\begin{aligned}\psi_p(x, y; z) &= \mathcal{F}^{-1}\left\{\Psi_{p0}(k_x, k_y)\mathrm{e}^{-\mathrm{j}k_{0z}z}\right\} \\ &= \mathcal{F}^{-1}\left\{\Psi_{p0}(k_x, k_y)\mathrm{e}^{-\mathrm{j}k_0\sqrt{(1 - k_x^2/k_0^2 - k_y^2/k_0^2)}z}\right\},\end{aligned}$$

and we immediately recover Eq. (3.4-8) and provide physical meaning to the equation. Note that the angular plane wave spectrum of the field distribution $\psi_p(x, y; z)$ is given by

$$\Psi_{p0}(k_x, k_y)\mathrm{e}^{-\mathrm{j}k_{0z}z},$$

或

$$\Psi_{p0}(k_x,k_y)\mathrm{e}^{-\mathrm{j}k_0\sqrt{\left(1-k_x^2/k_0^2-k_y^2/k_0^2\right)}\,z},$$

该式表示平面波的传播, 也就是, 必须有 $1-k_x^2/k_0^2-k_y^2/k_0^2\geqslant 0$ 或 $k_0^2\geqslant k_x^2+k_y^2$.

当 $1-k_x^2/k_0^2-k_y^2/k_0^2<0$ 或 $k_x^2+k_y^2\geqslant k_0^2$ 时, 超高空间频率 k_x,k_y 所对应的谱范围为倏逝波 (*evanescent wave*), 此时, $k_{0z}=\pm k_0\sqrt{1-k_x^2/k_0^2-k_y^2/k_0^2}=\pm\mathrm{j}k_0\sqrt{k_x^2/k_0^2+k_y^2/k_0^2-1}$. 物理上, 当上式取负号时, 波的振幅会随着离开原点的距离的增大而按指数衰减, 因此

$$\begin{aligned}\Psi_{p0}(k_x,k_y)\mathrm{e}^{-\mathrm{j}k_{0z}z}&=\Psi_{p0}(k_x,k_y)\mathrm{e}^{-\mathrm{j}\left(-\mathrm{j}k_0\sqrt{k_x^2/k_0^2+k_y^2/k_0^2-1}\right)z}\\&=\Psi_{p0}(k_x,k_y)\mathrm{e}^{-k_0z\sqrt{k_x^2/k_0^2+k_y^2/k_0^2-1}}.\end{aligned}$$

上式表示在 x-y 平面上传播的场, 其振幅在 z 方向上呈指数衰减. 也就是说, 由于 z 方向上的阻尼效应 (*damping effect*), 当 $z\gg\lambda_0$ 时, 倏逝波的特性会消失或观察不到. 对倏逝波的观测和测量是光学中超分辨率成像的重要问题之一. 需要指出的是, 式 (3.4-8) 的使用通常被称为角谱法 (*angular spectrum method*, ASM), 该方法利用快速傅里叶变换 (*fast Fourier transform*, FFT) 来数值计算光场通过距离 z 的衍射.

由于 $\mathcal{H}(k_x,k_y;z)$ 为传播的空间传递函数, 其逆变换 $G(x,y;z)$ 被称为在线性、均匀、各向同性介质中传播的空间脉冲响应 (*spatial impulse response of propagation*)

$$\begin{aligned}G(x,y;z)&=\mathcal{F}^{-1}\left\{\mathcal{H}(k_x,k_y;z)\right\}\\&=\frac{1}{4\pi^2}\iint\limits_{-\infty}^{\infty}\mathrm{e}^{-\mathrm{j}k_0\sqrt{\left(1-k_x^2/k_0^2-k_y^2/k_0^2\right)}\,z}\mathrm{e}^{-\mathrm{j}k_xx-\mathrm{j}k_yy}\mathrm{d}k_x\mathrm{d}k_y.\end{aligned}\tag{3.4-10a}$$

上式的解比较复杂, Stark (1982) 给出了其解的形式

$$G(x,y;z)=\frac{\mathrm{j}k_0\mathrm{e}^{-\mathrm{j}k_0\sqrt{x^2+y^2+z^2}}}{2\pi\sqrt{x^2+y^2+z^2}}\frac{z}{\sqrt{x^2+y^2+z^2}}\left(1+\frac{1}{\mathrm{j}k_0\sqrt{x^2+y^2+z^2}}\right).\tag{3.4-10b}$$

式 (3.4-8) 允许通过傅里叶变换求其衍射场, 我们也可以将空域中 z 处的场分布写成

$$\psi_p(x,y;z)=\psi_{p0}(x,y)*G(x,y;z).\tag{3.4-11}$$

一旦得到 $\psi_p(x,y;z)$, 最终的解即可由式 (3.4-3) 给出, 对于实的物理量, 只需取复函数的实部, 即, $\mathrm{Re}\left[\psi_p(x,y;z)\mathrm{e}^{\mathrm{j}\omega_0t}\right]$.

3.4.1 惠更斯原理

式 (3.4-11) 中 $G(x,y;z)$ 的卷积积分非常复杂, 通常使用一些实际的物理假设来简化

or

$$\Psi_{p0}(k_x, k_y)\mathrm{e}^{-\mathrm{j}k_0\sqrt{\left(1-k_x^2/k_0^2-k_y^2/k_0^2\right)}\,z},$$

which is for the propagation of plane waves, that is, we must have $1 - k_x^2/k_0^2 - k_y^2/k_0^2 \geqslant 0$ or $k_0^2 \geqslant k_x^2 + k_y^2$.

The range of super high spatial frequencies k_x, k_y covers the *evanescent waves* when $1 - k_x^2/k_0^2 - k_y^2/k_0^2 < 0$ or $k_x^2 + k_y^2 \geqslant k_0^2$. It follows that $k_{0z} = \pm k_0\sqrt{1 - k_x^2/k_0^2 - k_y^2/k_0^2} = \pm \mathrm{j}k_0\sqrt{k_x^2/k_0^2 + k_y^2/k_0^2 - 1}$. Physically, we simply choose the negative sign so that the wave amplitude decays with distance from the source rather than growing exponentially. Hence,

$$\begin{aligned}\Psi_{p0}(k_x, k_y)\mathrm{e}^{-\mathrm{j}k_{0z}z} &= \Psi_{p0}(k_x, k_y)\mathrm{e}^{-\mathrm{j}\left(-\mathrm{j}k_0\sqrt{k_x^2/k_0^2+k_y^2/k_0^2-1}\right)z} \\ &= \Psi_{p0}(k_x, k_y)\mathrm{e}^{-k_0 z\sqrt{k_x^2/k_0^2+k_y^2/k_0^2-1}}.\end{aligned}$$

This represents a field propagating on the x-y plane with an exponentially decreasing amplitude in the z-direction. In other words, the feature of the evanescent waves is absent or not observable at a distance $z \gg \lambda_0$ due to the damping effect in the z-direction. The ability to observe or measure evanescent waves is one of the important problems in optics for super-resolution imaging. We want to point out that the use of Eq. (3.4-8) is commonly referred to as the *angular spectrum method* (ASM) to calculate numerically the diffraction of a wave field though a distance z using the *fast Fourier transform* (FFT).

Since $\mathcal{H}(k_x, k_y; z)$ is the spatial transfer function of propagation, its inverse transform $G(x, y; z)$ is called the *spatial impulse response of propagation* in the linear, homogeneous, and isotropic medium:

$$\begin{aligned}G(x, y; z) &= \mathcal{F}^{-1}\left\{\mathcal{H}(k_x, k_y; z)\right\} \\ &= \frac{1}{4\pi^2}\iint_{-\infty}^{\infty} \mathrm{e}^{-\mathrm{j}k_0\sqrt{\left(1-k_x^2/k_0^2-k_y^2/k_0^2\right)}\,z}\mathrm{e}^{-\mathrm{j}k_x x-\mathrm{j}k_y y}\mathrm{d}k_x\mathrm{d}k_y.\end{aligned} \tag{3.4-10a}$$

The solution to the above equation is formidable and its solution has been provided in the literature [Stark (1982)]:

$$G(x, y; z) = \frac{\mathrm{j}k_0\mathrm{e}^{-\mathrm{j}k_0\sqrt{x^2+y^2+z^2}}}{2\pi\sqrt{x^2+y^2+z^2}}\frac{z}{\sqrt{x^2+y^2+z^2}}\left(1 + \frac{1}{\mathrm{j}k_0\sqrt{x^2+y^2+z^2}}\right). \tag{3.4-10b}$$

While Eq. (3.4-8) allows us to find the diffracted field through the Fourier transform, we can also write the field distribution at z in the spatial domain as

$$\psi_p(x, y; z) = \psi_{p0}(x, y) * G(x, y; z). \tag{3.4-11}$$

Once we have $\psi_p(x, y; z)$, the final solution is then given by Eq. (3.4-3), and for physical real quantity, we simple take the real part of the complex function, that is, $\mathrm{Re}\left[\psi_p(x, y; z)\mathrm{e}^{\mathrm{j}\omega_0 t}\right]$.

3.4.1 Huygens' Principle

The convolution integral involving $G(x, y; z)$ in Eq. (3.4-11) is very complicated and, often, we make some realistic physical assumptions to simplify the functional form of $G(x, y; z)$.

$G(x,y;z)$ 的函数形式. 假设所观测的光场分布在衍射孔径多个波长之外, 即, $z \gg \lambda_0 = 2\pi/k_0$, 则

$$1+\frac{1}{\mathrm{j}k_0\sqrt{x^2+y^2+z^2}}=1+\frac{1}{\mathrm{j}\dfrac{2\pi}{\lambda_0}z\sqrt{1+\dfrac{x^2+y^2}{z^2}}}\approx 1.$$

现在, 对于小角度 [傍轴近似 (*paraxial approximation*)], 即 $z \gg \sqrt{x^2+y^2}$, 有 $\sqrt{x^2+y^2+z^2} = z\sqrt{1+(x^2+y^2)/z^2} \approx z$. 在物理上, 这意味着只考虑沿 z 轴传播的近轴区域, 那么, $G(x,y;z)$ 中的因子 $\dfrac{z}{\sqrt{x^2+y^2+z^2}}$ 可以被 1 替换, 则传播的空间脉冲响应变为

$$G(x,y;z)=\frac{\mathrm{j}k_0\mathrm{e}^{-\mathrm{j}k_0\sqrt{x^2+y^2+z^2}}}{2\pi\sqrt{x^2+y^2+z^2}}. \tag{3.4-12}$$

将式 (3.3-13) 中给出的球面波表达式写出, 并将其复振幅项改写如下

$$\psi(R;t)=\frac{1}{R}\mathrm{e}^{\mathrm{j}(\omega_0 t-k_0 R)}=\psi_p(R)\mathrm{e}^{\mathrm{j}\omega_0 t}.$$

可以得到球面波的复振幅为

$$\psi_p(R)=\frac{1}{R}\mathrm{e}^{-\mathrm{j}k_0 R}. \tag{3.4-13}$$

由于 $\sqrt{x^2+y^2+z^2}=R$ (见图 3.3-5), 同时通过比较式 (3.4-13), 可观察到式 (3.4-12) 中的 $G(x,y;z)$ 基本上是振幅为 $\dfrac{\mathrm{j}k_0}{2\pi}$ 的球面波, 现在, 利用式 (3.4-12) 写出式 (3.4-11) 中的卷积积分, 得到众所周知的惠更斯–菲涅耳衍射积分 (*Huygens-Fresnel diffraction integral*)

$$\begin{aligned}\psi_p(x,y;z)&=\psi_{p0}(x,y)*G(x,y;z)=\psi_{p0}(x,y)*\frac{\mathrm{j}k_0\mathrm{e}^{-\mathrm{j}k_0 R}}{2\pi R}\\&=\frac{\mathrm{j}k_0}{2\pi}\iint_{-\infty}^{\infty}\psi_{p0}(x',y')\frac{\mathrm{e}^{-\mathrm{j}k_0 R'}}{R'}\mathrm{d}x'\mathrm{d}y',\end{aligned} \tag{3.4-14}$$

其中, $R'=\sqrt{(x-x')^2+(y-y')^2+z^2}$. 惠更斯–菲涅耳衍射积分基本上是惠更斯原理 (*Huygens' principle*) 的数学表述, 它表明经过衍射孔径的每一个波前无阻碍的点都是一个球面波源. 在任意一点的复振幅是所有这些球面波的叠加, 情况如图 3.4-2 所示. 当我们想求出距离光源 (孔径) 平面 x'-y' 为 z 的观测平面 x-y 上 P 点的复振幅时, 首先考虑光源平面上的微元面积 $\mathrm{d}x'\mathrm{d}y'$. 由微元面积处发出的球面波在 P 点处的场为

$$\frac{\mathrm{j}k_0}{2\pi}\psi_{p0}(x',y')\frac{\mathrm{e}^{-\mathrm{j}k_0 R'}}{R'}\mathrm{d}x'\mathrm{d}y',$$

其中, $\dfrac{\mathrm{j}k_0}{2\pi}\psi_{p0}(x',y')$ 为球面波的振幅, 因为 $\psi_{p0}(x,y)=\psi_p(x,y;z=0^-)t(x,y)$ [见式 (3.4-2)],

With the assumption that we observe the field distribution many wavelengths away from the diffracting aperture, that is, $z \gg \lambda_0 = 2\pi/k_0$, the term

$$1 + \frac{1}{\mathrm{j}k_0\sqrt{x^2+y^2+z^2}} = 1 + \frac{1}{\mathrm{j}\dfrac{2\pi}{\lambda_0}z\sqrt{1+\dfrac{x^2+y^2}{z^2}}} \approx 1.$$

Now, for small angles (*paraxial approximation*), that is, $z \gg \sqrt{x^2+y^2}$, we have $\sqrt{x^2+y^2+z^2} = z\sqrt{1+(x^2+y^2)/z^2} \approx z$. Physically, it means we only consider small excursions around the axis of propagation along z. The factor $\dfrac{z}{\sqrt{x^2+y^2+z^2}}$ in $G(x,y;z)$ can then be replaced by 1. The spatial impulse response of propagation now becomes

$$G(x,y;z) = \frac{\mathrm{j}k_0 \mathrm{e}^{-\mathrm{j}k_0\sqrt{x^2+y^2+z^2}}}{2\pi\sqrt{x^2+y^2+z^2}}. \tag{3.4-12}$$

Let us state the spherical wave given in Eq. (3.3-13) and rewrite it terms of its complex amplitude as follows:

$$\psi(R;t) = \frac{1}{R}\mathrm{e}^{\mathrm{j}(\omega_0 t - k_0 R)} = \psi_p(R)\mathrm{e}^{\mathrm{j}\omega_0 t}.$$

We then notice that the complex amplitude of a spherical wave is

$$\psi_p(R) = \frac{1}{R}\mathrm{e}^{-\mathrm{j}k_0 R}. \tag{3.4-13}$$

Since $\sqrt{x^2+y^2+z^2} = R$ (see Figure 3.3-5), we observe that $G(x,y;z)$ in Eq. (3.4-12) is basically a spherical wave with amplitude of $\dfrac{\mathrm{j}k_0}{2\pi}$ upon comparing with Eq. (3.4-13). We now write out the convolution integral in Eq. (3.4-11) by using Eq. (3.4-12), and we have the well-known *Huygens-Fresnel diffraction integral*:

$$\begin{aligned}\psi_p(x,y;z) &= \psi_{p0}(x,y) * G(x,y;z) = \psi_{p0}(x,y) * \frac{\mathrm{j}k_0 \mathrm{e}^{-\mathrm{j}k_0 R}}{2\pi R} \\ &= \frac{\mathrm{j}k_0}{2\pi}\iint_{-\infty}^{\infty} \psi_{p0}(x',y')\frac{\mathrm{e}^{-\mathrm{j}k_0 R'}}{R'}\mathrm{d}x'\mathrm{d}y',\end{aligned} \tag{3.4-14}$$

where $R' = \sqrt{(x-x')^2+(y-y')^2+z^2}$. The Huygens-Fresnel diffraction integral is basically a mathematical statement of the *Huygens' principle*, which states that every unobstructed point of a wavefront from the diffracting aperture serves as a source of spherical waves. The complex amplitude at any point beyond is the superposition of all these spherical waves. The situation is depicted in Figure 3.4-2. As we want to find the complex field at point P on the observation x-y plane that is at a distance z from the source (aperture) x'-y' plane, we first consider a differential area $\mathrm{d}x'\mathrm{d}y'$ on the source plane. The field at point P due to spherical waves emanating from the differential area is

$$\frac{\mathrm{j}k_0}{2\pi}\psi_{p0}(x',y')\frac{\mathrm{e}^{-\mathrm{j}k_0 R'}}{R'}\mathrm{d}x'\mathrm{d}y',$$

where the term $\dfrac{\mathrm{j}k_0}{2\pi}\psi_{p0}(x',y')$ is the amplitude of the spherical wave, which is a weighting factor due to the aperture and light illumination on the aperture as $\psi_{p0}(x,y) = \psi_p(x,y;z =$

所以球面波的振幅是受孔径及孔径上光照影响的加权因子. 为了求 P 点处的总光场, 我们只需对孔径内的所有微元面积求和即可

$$\iint_{-\infty}^{\infty} \frac{\mathrm{j}k_0}{2\pi}\psi_{p0}(x',y')\frac{\mathrm{e}^{-\mathrm{j}k_0 R'}}{R'}\mathrm{d}x'\mathrm{d}y',$$

这即是式 (3.4-14) 的结果.

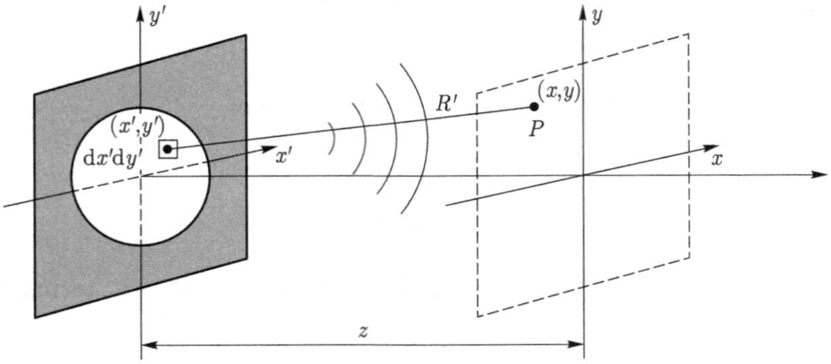

图 3.4-2 光源和观测平面的衍射图示

3.4.2 菲涅耳衍射和夫琅禾费衍射

菲涅耳衍射 (*Fresnel diffraction*)

惠更斯原理是由式 (3.4-14) 和式 (3.4-12) 推导出来的, 对于实际应用来说比较复杂. 但是, 可以通过进一步简化 $G(x,y;z)$ 来得到更为简单可用的表达式. 使用傍轴近似, 即, $\sqrt{x^2+y^2+z^2}=z\sqrt{1+(x^2+y^2)/z^2}\approx z$, 可以用 z 代替分母, 得

$$G(x,y;z)=\frac{\mathrm{j}k_0\mathrm{e}^{-\mathrm{j}k_0\sqrt{x^2+y^2+z^2}}}{2\pi\sqrt{x^2+y^2+z^2}}\approx\frac{\mathrm{j}k_0\mathrm{e}^{-\mathrm{j}k_0\sqrt{x^2+y^2+z^2}}}{2\pi z}. \tag{3.4-15}$$

然而, 相位项对于近似来说更加敏感, 因为相位项 $k_0\sqrt{x^2+y^2+z^2}$ 乘了一个很大的波数 $k_0=2\pi/\lambda_0$. 对于可见光谱区中的典型光谱 $\lambda_0=0.6\times10^{-6}$ m, $k_0\approx10^7$ rad/m. 波的传播会引起相位的改变, 即使小到几个波长的波传播也会引起相位值显著的改变. 例如, 对于 $z=5\lambda_0$, 有 $k_0\sqrt{x^2+y^2+z^2}\approx k_0 z=10^7\times5\times0.6\times10^{-6}$ rad $=30$ rad. 为此, 要为该相位项寻找一个更佳的近似. 当 $\varepsilon<1$ 时, 指数幂可以被展开为级数, $\sqrt{1+\varepsilon}\approx1+\varepsilon/2-\varepsilon^2/8+\cdots$. 因此, 如果用 x 和 y 的二次项 (*quadratic*) 来近似, 有

$$\sqrt{x^2+y^2+z^2}=z\sqrt{1+\frac{x^2+y^2}{z^2}}\approx z+\frac{x^2+y^2}{2z}.$$

这种近似被称为菲涅耳近似 (*Fresnel approximation*), 式 (3.4-15) 即所谓的傅里叶光学中的空间脉冲响应 (*spatial impulse response in Fourier optics*) $h(x,y;z)$, 即

$$G(x,y;z)=\frac{\mathrm{j}k_0\mathrm{e}^{-\mathrm{j}k_0\sqrt{x^2+y^2+z^2}}}{2\pi z}\approx\frac{\mathrm{j}k_0}{2\pi z}\mathrm{e}^{-\mathrm{j}k_0 z}\mathrm{e}^{\frac{-\mathrm{j}k_0(x^2+y^2)}{2z}}=h(x,y;z). \tag{3.4-16}$$

利用 $h(x,y;z)$, 式 (3.4-11) 变为

$$\psi_p(x,y;z)=\psi_{p0}(x,y)*G(x,y;z)\approx\psi_{p0}(x,y)*h(x,y;z)$$

$0^-)t(x, y)$ [see Eq. (3.4-2)]. To find the total field at point P, we simply sum over all the differential areas in the aperture to obtain

$$\iint_{-\infty}^{\infty} \frac{\mathrm{j}k_0}{2\pi} \psi_{p0}(x', y') \frac{\mathrm{e}^{-\mathrm{j}k_0 R'}}{R'} \mathrm{d}x'\mathrm{d}y',$$

which is Eq. (3.4-14).

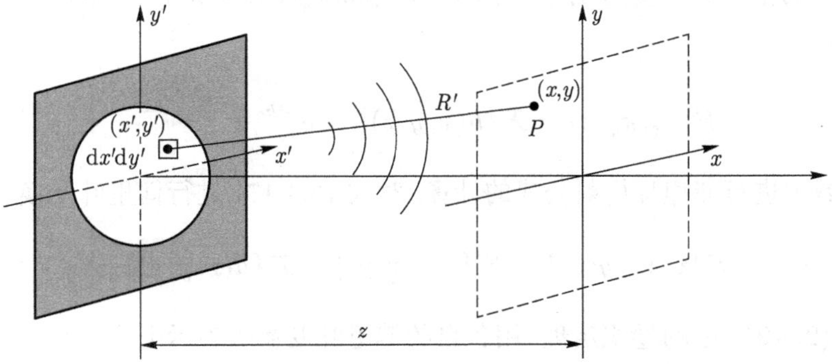

Figure 3.4-2 Diffraction geometry illustrating the source and observation planes

3.4.2 Fresnel Diffraction and Fraunhofer Diffraction

Fresnel Diffraction

The Huygens' principle, derived from Eq. (3.4-14) along with Eq. (3.4-12), is fairly complicated for practical applications. However, we can further reduce $G(x, y; z)$ to obtain a more simple and usable expression. Using the paraxial approximation, that is, $\sqrt{x^2 + y^2 + z^2} = z\sqrt{1 + (x^2 + y^2)/z^2} \approx z$, we can replace the denominator factor by z to get

$$G(x, y; z) = \frac{\mathrm{j}k_0 \mathrm{e}^{-\mathrm{j}k_0\sqrt{x^2+y^2+z^2}}}{2\pi\sqrt{x^2 + y^2 + z^2}} \approx \frac{\mathrm{j}k_0 \mathrm{e}^{-\mathrm{j}k_0\sqrt{x^2+y^2+z^2}}}{2\pi z}. \tag{3.4-15}$$

However, the phase term is more sensitive to the approximation. The reason is that the phase term $k_0\sqrt{x^2 + y^2 + z^2}$ is multiplied by a very large wavenumber $k_0 = 2\pi/\lambda_0$. For a typical value of $\lambda_0 = 0.6\times10^{-6}$ m in the visible region of the light spectrum, $k_0 \approx 10^7$ rad/m. Phase changes due to wave propagation of as little as several wavelengths change the value of the phase significantly. For example, for $z = 5\lambda_0$, $k_0\sqrt{x^2 + y^2 + z^2} \approx k_0 z = 10^7 \times 5 \times 0.6 \times 10^{-6}$ rad $= 30$ rad. In light of this, we seek to have a better approximation for the phase term. The exponent can be expanded as the series $\sqrt{1+\varepsilon} \approx 1 + \varepsilon/2 - \varepsilon^2/8 + \cdots$ for $\varepsilon < 1$. Therefore, if approximated by terms up to quadratic in x and y, we have

$$\sqrt{x^2 + y^2 + z^2} = z\sqrt{1 + \frac{x^2 + y^2}{z^2}} \approx z + \frac{x^2 + y^2}{2z}.$$

This approximation is called the *Fresnel approximation*, and Eq. (3.4-15) becomes the so-called *spatial impulse response in Fourier optics*, $h(x, y; z)$, as

$$G(x, y; z) = \frac{\mathrm{j}k_0 \mathrm{e}^{-\mathrm{j}k_0\sqrt{x^2+y^2+z^2}}}{2\pi z} \approx \frac{\mathrm{j}k_0}{2\pi z} \mathrm{e}^{-\mathrm{j}k_0 z} \mathrm{e}^{\frac{-\mathrm{j}k_0(x^2+y^2)}{2z}} = h(x, y; z). \tag{3.4-16}$$

Using $h(x, y; z)$, Eq. (3.4-11) becomes

$$\psi_p(x, y; z) = \psi_{p0}(x, y) * G(x, y; z) \approx \psi_{p0}(x, y) * h(x, y; z)$$

$$= e^{-jk_0 z}\frac{jk_0}{2\pi z}\iint_{-\infty}^{\infty}\psi_{p0}(x',y')\exp\left\{\frac{-jk_0}{2z}\left[(x-x')^2+(y-y')^2\right]\right\}dx'dy'. \tag{3.4-17}$$

式 (3.4-17) 被称为菲涅耳衍射公式 (*Fresnel diffraction formula*), 描述的是光束在传播过程中具有初始复振幅 (*initial complex amplitude*) $\psi_{p0}(x,y)$ 的菲涅耳衍射 (*Fresnel diffraction*), 通常也被称为近场衍射 (*near field diffraction*). 对脉冲响应 $h(x,y;z)$ 进行傅里叶变换, 得到傅里叶光学中的空间频率传递函数 (*spatial frequency transfer function in Fourier optics*) $H(k_x,k_y;z)$ 为

$$H(k_x,k_y;z)=\mathcal{F}\{h(x,y;z)\}=e^{-jk_0 z}e^{\frac{j(k_x^2+k_y^2)z}{2k_0}}. \tag{3.4-18}$$

为了在频域中进行菲涅耳衍射公式的计算, 对式 (3.4-17) 进行傅里叶变换

$$\mathcal{F}\{\psi_p(x,y;z)\}=\mathcal{F}\{\psi_{p0}(x,y)\}\times\mathcal{F}\{h(x,y;z)\},$$

式中使用了式 (2.3-27) 中的卷积定理. 用各自的傅里叶变换来表示上式, 有

$$\Psi_p(k_x,k_y;z)=\Psi_{p0}(k_x,k_y)H(k_x,k_y;z),$$

最终在空域, 就有

$$\psi_p(x,y;z)=\mathcal{F}^{-1}\{\Psi_p(k_x,k_y;z)\}=\mathcal{F}^{-1}\{\Psi_{p0}(k_x,k_y)H(k_x,k_y;z)\}, \tag{3.4-19}$$

这是用傅里叶变换表示的菲涅耳衍射公式.

另外, 衍射场的计算可以用一次傅里叶变换表示. 对于式 (3.4-17), 将其积分中指数里的二次项展开, 并将所有带撇号的变量在积分中重组

$$\begin{aligned}\psi_p(x,y;z)&=e^{-jk_0 z}\frac{jk_0}{2\pi z}\iint_{-\infty}^{\infty}\psi_{p0}(x',y')e^{\frac{-jk_0}{2z}\left[(x^2+y^2)+(x'^2+y'^2)-2xx'-2yy'\right]}dx'dy'\\&=e^{-jk_0 z}\frac{jk_0}{2\pi z}e^{\frac{-jk_0}{2z}(x^2+y^2)}\iint_{-\infty}^{\infty}\psi_{p0}(x',y')e^{\frac{-jk_0}{2z}(x'^2+y'^2)}e^{\frac{jk_0}{z}(xx'+yy')}dx'dy'.\end{aligned} \tag{3.4-20}$$

将该式与式 (2.3-1a) 的傅里叶变换的定义进行比较, 为方便起见, 重写式子如下

$$\mathcal{F}\{f(x,y)\}=F(k_x,k_y)=\iint_{-\infty}^{\infty}f(x,y)e^{jk_x x+jk_y y}dxdy,$$

用一次傅里叶变换运算写出菲涅耳衍射公式, 如下所示

$$\psi_p(x,y;z)=e^{-jk_0 z}\frac{jk_0}{2\pi z}e^{\frac{-jk_0}{2z}(x^2+y^2)}\mathcal{F}\left\{\psi_{p0}(x,y)e^{\frac{-jk_0}{2z}(x^2+y^2)}\right\}\bigg|_{k_x=\frac{k_0x}{z},\,k_y=\frac{k_0y}{z}}. \tag{3.4-21}$$

夫琅禾费衍射 (*Fraunhofer diffraction*)

由式 (3.4-20) 到式 (3.4-21) 可知, 菲涅耳衍射可以通过源场 $\psi_{p0}(x,y)$ 与二次相位因子 $e^{\frac{-jk_0}{2z}(x^2+y^2)}$ 的乘积的傅里叶变换得到. 如果采用所谓的夫琅禾费近似 (*Fraunhofer approximation*)

$$\frac{k_0}{2}\left(x'^2+y'^2\right)\Big|_{\max}=z_R\ll z, \tag{3.4-22}$$

$$= \mathrm{e}^{-\mathrm{j}k_0 z}\frac{\mathrm{j}k_0}{2\pi z}\iint_{-\infty}^{\infty}\psi_{p0}(x',y')\exp\left\{\frac{-\mathrm{j}k_0}{2z}\left[(x-x')^2+(y-y')^2\right]\right\}\mathrm{d}x'\mathrm{d}y'. \quad (3.4\text{-}17)$$

Equation (3.4-17) is called the *Fresnel diffraction formula* and describes *Fresnel diffraction*, often known as *near field diffraction*, of a beam during propagation and having the initial complex profile $\psi_{p0}(x, y)$. By taking the Fourier transform of the impulse response $h(x, y; z)$, we have the *spatial frequency transfer function in Fourier optics* $H(k_x, k_y; z)$:

$$H(k_x,k_y;z) = \mathcal{F}\{h(x,y;z)\} = \mathrm{e}^{-\mathrm{j}k_0 z}\mathrm{e}^{\frac{\mathrm{j}(k_x^2+k_y^2)z}{2k_0}}. \quad (3.4\text{-}18)$$

To perform the calculation of the Fresnel diffraction formula in the frequency domain, we take the Fourier transform of Eq. (3.4-17):

$$\mathcal{F}\{\psi_p(x,y;z)\} = \mathcal{F}\{\psi_{p0}(x,y)\}\times\mathcal{F}\{h(x,y;z)\},$$

where we have used the convolution theorem in Eq. (2.3-27). Expressing the above using their transforms, we have

$$\Psi_p(k_x,k_y;z) = \Psi_{p0}(k_x,k_y)H(k_x,k_y;z),$$

and finally in the spatial domain, we have

$$\psi_p(x,y;z) = \mathcal{F}^{-1}\{\Psi_p(k_x,k_y;z)\} = \mathcal{F}^{-1}\{\Psi_{p0}(k_x,k_y)H(k_x,k_y;z)\}, \quad (3.4\text{-}19)$$

which is the Fresnel diffraction formulas written in terms of Fourier transform.

Alternatively, the calculation of the diffracted field can be performed using a single Fourier transform. We rewrite Eq. (3.4-17) by expanding the quadratic terms in the exponent inside the integral and grouping all the primed variables within the integral:

$$\psi_p(x,y;z) = \mathrm{e}^{-\mathrm{j}k_0 z}\frac{\mathrm{j}k_0}{2\pi z}\iint_{-\infty}^{\infty}\psi_{p0}(x',y')\mathrm{e}^{\frac{-\mathrm{j}k_0}{2z}\left[(x^2+y^2)+(x'^2+y'^2)-2xx'-2yy'\right]}\mathrm{d}x'\mathrm{d}y'$$

$$= \mathrm{e}^{-\mathrm{j}k_0 z}\frac{\mathrm{j}k_0}{2\pi z}\mathrm{e}^{\frac{-\mathrm{j}k_0}{2z}(x^2+y^2)}\iint_{-\infty}^{\infty}\psi_{p0}(x',y')\mathrm{e}^{\frac{-\mathrm{j}k_0}{2z}(x'^2+y'^2)}\mathrm{e}^{\frac{\mathrm{j}k_0}{z}(xx'+yy')}\mathrm{d}x'\mathrm{d}y'. \quad (3.4\text{-}20)$$

By comparing this equation with the definition of the Fourier transform from Eq. (2.3-1a) re-stated below for convenience:

$$\mathcal{F}\{f(x,y)\} = F(k_x,k_y) = \iint_{-\infty}^{\infty} f(x,y)\mathrm{e}^{\mathrm{j}k_x x+\mathrm{j}k_y y}\mathrm{d}x\mathrm{d}y,$$

we can write the Fresnel diffraction formulas using a single Fourier transform operation as follows:

$$\psi_p(x,y;z) = \mathrm{e}^{-\mathrm{j}k_0 z}\frac{\mathrm{j}k_0}{2\pi z}\mathrm{e}^{\frac{-\mathrm{j}k_0}{2z}(x^2+y^2)}\mathcal{F}\left\{\psi_{p0}(x,y)\mathrm{e}^{\frac{-\mathrm{j}k_0}{2z}(x^2+y^2)}\right\}\Bigg|_{k_x=\frac{k_0 x}{z},\,k_y=\frac{k_0 y}{z}}. \quad (3.4\text{-}21)$$

Fraunhofer Diffraction

From Eq. (3.4-20) to Eq. (3.4-21), we see that Fresnel diffraction can be obtained from a Fourier transform of the product of the source distribution $\psi_{p0}(x, y)$ and a quadratic phase factor $\mathrm{e}^{\frac{-\mathrm{j}k_0}{2z}(x^2+y^2)}$. If we now impose the so-called *Fraunhofer approximation*

$$\frac{k_0}{2}\left(x'^2+y'^2\right)\Big|_{\max} = z_R \ll z, \quad (3.4\text{-}22)$$

式 (3.4-20) 内积分中的二次相位因子 (*quadratic phase factor*) 在整个孔径上近似为 1, 使式 (3.4-21) 变为

$$\psi_p(x,y;z)=\mathrm{e}^{-\mathrm{j}k_0z}\frac{\mathrm{j}k_0}{2\pi z}\mathrm{e}^{\frac{-\mathrm{j}k_0}{2z}(x^2+y^2)}\mathcal{F}\left\{\psi_{p0}(x,y)\right\}\Big|_{k_x=\frac{k_0x}{z},\,k_y=\frac{k_0y}{z}}. \tag{3.4-23}$$

这种情况发生时, 处于夫琅禾费衍射或远场衍射 (*far field diffraction*) 区域. z_R 被称为瑞利长度 (*Rayleigh length*), 如果用波长 λ_0 表示, 有

$$z_R=\frac{\pi}{\lambda_0}(x'^2+y'^2)\big|_{\max}.$$

$\pi(x'^2+y'^2)\big|_{\max}$ 定义了源的最大面积, 若该面积除以波长的值远小于观测距离 z, 则观察的衍射场为夫琅禾费衍射. 式 (3.4-23) 为夫琅禾费衍射公式 (*Fraunhofer diffraction formula*).

式 (3.4-23) 说明, 光学上可以计算出 $\psi_{p0}(x,y)$ 的谱, 即 $\mathcal{F}\left\{\psi_{p0}(x,y)\right\}$, 这只需在远大于瑞利长度的地方观察衍射场即可得到. 光学上可通过光学元件直接对图像进行处理, 这点对于光学图像处理 (*optical image processing*) 来说非常有利. 光学图像处理的概念来自之前在第二章讨论的信号处理. 由式 (2.2-13b), 得

$$y(t)=\mathcal{F}^{-1}\left\{Y(\omega)\right\}=\mathcal{F}^{-1}\left\{X(\omega)H(\omega)\right\}.$$

需要计算 $x(t)$ 的输入谱, 即 $X(\omega)$ 时, 该频谱被系统的传递函数 $H(\omega)$ 进行 "处理", 得到处理后的输出 $y(t)$. 为了探索这种处理输入的方法, 我们需要从输入谱开始. 由式 (3.4-23), 我们的确可以得到输入谱 $\psi_{p0}(x,y)$, 然而, 该谱受二次相位因子 $\mathrm{e}^{\frac{-\mathrm{j}k_0}{2z}(x^2+y^2)}$ 影响引入了误差. 因此需要以某种方式来消除该二次相位误差. 在下一节中, 我们将介绍光学系统中的透镜. 事实证明, 一个理想的透镜具有相位变换特性, 这使得这个过程可实现精确的傅里叶变换.

举例: 方形孔径的夫琅禾费衍射

由单位振幅平面波照射的方形孔径, 在孔径出口处的复振幅为 $\psi_{p0}(x,y)=rect\left(\frac{x}{l_x},\frac{y}{l_x}\right)$, $l_x\times l_x$ 为开口面积. 根据夫琅禾费衍射公式 (*Fraunhofer diffraction formula*), 即式 (3.4-23), 距离孔径为 z 处的复振幅为

$$\begin{aligned}\psi_p(x,y;z)&=\mathrm{e}^{-\mathrm{j}k_0z}\frac{\mathrm{j}k_0}{2\pi z}\mathrm{e}^{\frac{-\mathrm{j}k_0}{2z}(x^2+y^2)}\mathcal{F}\left\{\psi_{p0}(x,y)\right\}\Big|_{k_x=\frac{k_0x}{z},\,k_y=\frac{k_0y}{z}}\\&=\mathrm{e}^{-\mathrm{j}k_0z}\frac{\mathrm{j}k_0}{2\pi z}\mathrm{e}^{\frac{-\mathrm{j}k_0}{2z}(x^2+y^2)}\mathcal{F}\left\{rect\left(\frac{x}{l_x},\frac{y}{l_x}\right)\right\}\Bigg|_{k_x=\frac{k_0x}{z},k_y=\frac{k_0y}{z}}\end{aligned}$$

利用式 (2.3-6) 的结果或利用表 2.1, 上式变为

$$\begin{aligned}\psi_p(x,y;z)&=\mathrm{e}^{-\mathrm{j}k_0z}\frac{\mathrm{j}k_0}{2\pi z}\mathrm{e}^{\frac{-\mathrm{j}k_0}{2z}(x^2+y^2)}l_xl_x sinc\left(\frac{k_xl_x}{2\pi},\frac{k_yl_x}{2\pi}\right)\Bigg|_{k_x=\frac{k_0x}{z},k_y=\frac{k_0y}{z}}\\&=\mathrm{e}^{-\mathrm{j}k_0z}\frac{\mathrm{j}k_0}{2\pi z}\mathrm{e}^{\frac{-\mathrm{j}k_0}{2z}(x^2+y^2)}l_xl_x sinc\left(\frac{k_0x}{z}\frac{l_x}{2\pi},\frac{k_0y}{z}\frac{l_x}{2\pi}\right).\end{aligned}$$

the quadratic phase factor within the integral of Eq. (3.4-20) is approximated unity over the entire aperture, giving Eq. (3.4-21) to become

$$\psi_p(x, y; z) = \mathrm{e}^{-\mathrm{j}k_0 z} \frac{\mathrm{j}k_0}{2\pi z} \mathrm{e}^{\frac{-\mathrm{j}k_0}{2z}(x^2+y^2)} \mathcal{F}\{\psi_{p0}(x, y)\}\Big|_{k_x=\frac{k_0 x}{z},\, k_y=\frac{k_0 y}{z}}. \tag{3.4-23}$$

When this happens, we are in the region of *Fraunhofer diffraction* or *far field diffraction.* z_R is called the *Rayleigh length* and if we express it in terms of the wavelength λ_0, we have

$$z_R = \frac{\pi}{\lambda_0}(x'^2 + y'^2)\big|_{\max}.$$

The term $\pi(x'^2 + y'^2)\big|_{\max}$ defines the maximum area of the source and if this area divided by the wavelength is much less than the observation distance z, we observe diffracted field under Fraunhofer diffraction. Equation (3.4-23) is the *Fraunhofer diffraction formula.*

Equation (3.4-23) says that we can compute optically the spectrum of $\psi_{p0}(x, y)$, that is, $\mathcal{F}\{\psi_{p0}(x, y)\}$, by simply observing the diffracted field much greater than the Rayleigh distance. That is good news for *optical image processing* as we could perform image processing optically. The notion of optical image processing comes from our previous experiences with signal processing discussed in Chapter 2. From Eq. (2.2-13b), we have

$$y(t) = \mathcal{F}^{-1}\{Y(\omega)\} = \mathcal{F}^{-1}\{X(\omega)H(\omega)\}.$$

We need to compute the input spectrum of $x(t)$, that is, $X(\omega)$, so that the spectrum is "processed" by the transfer function, $H(\omega)$, of the system to obtain the processed output $y(t)$. In order to explore such way of processing the input, we need the spectrum of the input as a start. From Eq. (3.4-23), we do have the spectrum of the input $\psi_{p0}(x, y)$. However, the spectrum is corrupted by the quadratic phase term, $\mathrm{e}^{\frac{-\mathrm{j}k_0}{2z}(x^2+y^2)}$, and that introduces errors into the spectrum. We need, somehow, to eliminate the quadratic phase errors. In the next section, we introduce a lens in an optical system. As it turns out, an ideal lens carries a phase transformation property that will allow us to achieve an exact Fourier transformation in the process.

Example: Fraunhofer Diffraction of a Square Aperture

The complex amplitude of a square aperture illuminated by a plane wave of unity amplitude is represented by $\psi_{p0}(x, y) = rect\left(\frac{x}{l_x}, \frac{y}{l_x}\right)$ at the exit of the aperture, where $l_x \times l_x$ is the area of the opening. According to the *Fraunhofer diffraction formula*, that is, Eq. (3.4-23), the complex field at a distance of z away from the aperture is given by

$$\begin{aligned}\psi_p(x, y; z) &= \mathrm{e}^{-\mathrm{j}k_0 z} \frac{\mathrm{j}k_0}{2\pi z} \mathrm{e}^{\frac{-\mathrm{j}k_0}{2z}(x^2+y^2)} \mathcal{F}\{\psi_{p0}(x, y)\}\Big|_{k_x=\frac{k_0 x}{z},\, k_y=\frac{k_0 y}{z}} \\ &= \mathrm{e}^{-\mathrm{j}k_0 z} \frac{\mathrm{j}k_0}{2\pi z} \mathrm{e}^{\frac{-\mathrm{j}k_0}{2z}(x^2+y^2)} \mathcal{F}\left\{rect\left(\frac{x}{l_x}, \frac{y}{l_x}\right)\right\}\Bigg|_{k_x=\frac{k_0 x}{z}, k_y=\frac{k_0 y}{z}}.\end{aligned}$$

Using the result of Eq. (2.3-6) or through the use of Table 2.1, the above equation becomes

$$\begin{aligned}\psi_p(x, y; z) &= \mathrm{e}^{-\mathrm{j}k_0 z} \frac{\mathrm{j}k_0}{2\pi z} \mathrm{e}^{\frac{-\mathrm{j}k_0}{2z}(x^2+y^2)} l_x l_x sinc\left(\frac{k_x l_x}{2\pi}, \frac{k_y l_x}{2\pi}\right)\Bigg|_{k_x=\frac{k_0 x}{z}, k_y=\frac{k_0 y}{z}} \\ &= \mathrm{e}^{-\mathrm{j}k_0 z} \frac{\mathrm{j}k_0}{2\pi z} \mathrm{e}^{\frac{-\mathrm{j}k_0}{2z}(x^2+y^2)} l_x l_x sinc\left(\frac{k_0 x}{z}\frac{l_x}{2\pi}, \frac{k_0 y}{z}\frac{l_x}{2\pi}\right).\end{aligned}$$

根据式 (3.3-22), 强度为

$$I = \psi_p \psi_p^* = |\psi_p|^2 \propto sinc^2\left(\frac{x}{z}\frac{l_x}{\lambda_0}, \frac{y}{z}\frac{l_x}{\lambda_0}\right).$$

sinc 函数沿 x 轴方向的第一个零点出现在 $x = \pm z\lambda_0/l_x$ 处, 扩散角 (*angle of spread* 或 *spread angle*) 被定义为 $\theta_{sp} = \lambda_0/l_x$, 具体如图 3.4-3 所示. 可以发现, 从衍射中得到的扩散角 θ_{sp}, 与第一章讨论狭缝孔径的由不确定性原理得到的扩散角是一致的 (见 1.3 节).

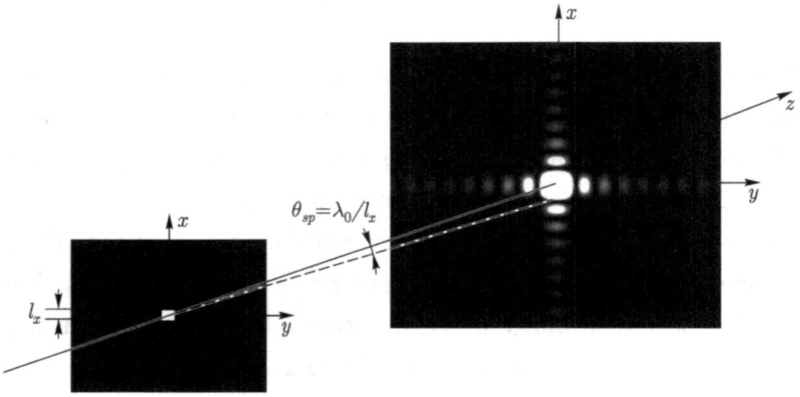

图 3.4-3 夫琅禾费衍射中强度图样的扩散角示意图

3.4.3 理想透镜的相变特性

前面的章节讨论了孔径的衍射. 本节讨论光通过理想透镜 (*ideal lens*) 的情况. 理想透镜是无厚度、无限大的纯相位光学元件. 因此, 理想透镜只改变入射波前的相位. 这里给出理想透镜的透过率函数

$$t_f(x, y) = \mathrm{e}^{\mathrm{j}\frac{k_0}{2f}(x^2+y^2)}, \tag{3.4-24}$$

其中, f 是透镜的焦距. 对于会聚 (或正) 透镜, $f > 0$, 而当 $f < 0$, 为发散 (或负) 透镜.

对于入射到透镜上的振幅为 A 的平面波, 求透镜后焦面上的复振幅分布. 如图 3.4-4 所示, 假设透镜位于 $z = 0$ 的平面上. 根据式 (3.4-2b), 透镜后的场分布为

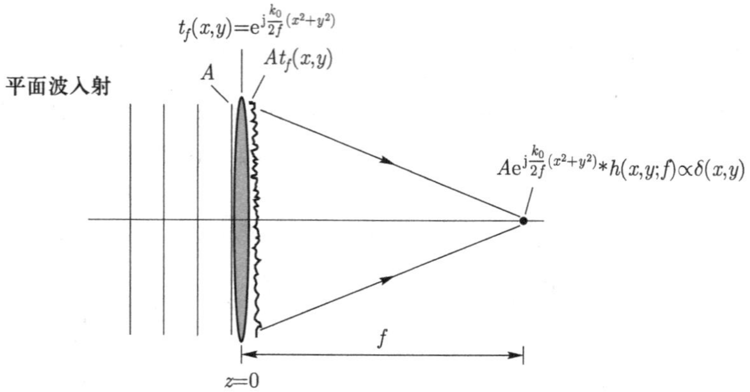

图 3.4-4 振幅为 A 的平面波入射到焦距为 f 的理想透镜上

The intensity pattern, according to Eq. (3.3-22), is

$$I = \psi_p \psi_p^* = |\psi_p|^2 \propto sinc^2\left(\frac{x}{z}\frac{l_x}{\lambda_0}, \frac{y}{z}\frac{l_x}{\lambda_0}\right).$$

The first zeros of the *sinc* function along the x-direction occur at $x = \pm z\lambda_0/l_x$, and the *angle of spread* or *spread angle* can be defined as $\theta_{sp} = \lambda_0/l_x$. The situation is shown in Figure 3.4-3. Note that the spread angle θ_{sp} obtained from diffraction is consistent with that derived from the uncertainty principle discussed in Chapter 1 for a slit aperture (see Section 1.3).

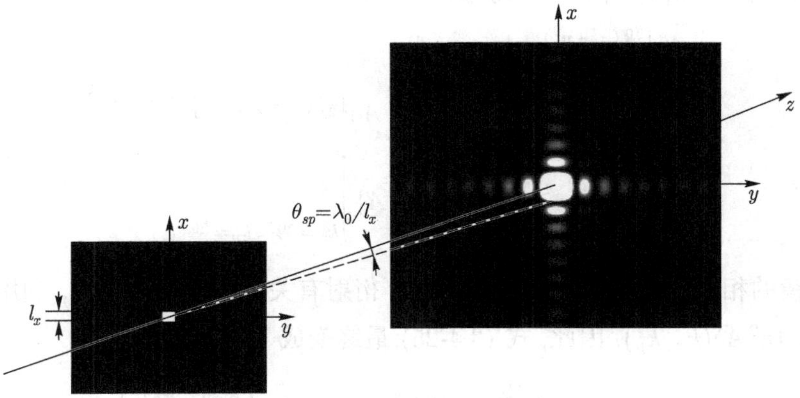

Figure 3.4-3 Fraunhofer diffraction of a square

3.4.3 Phase Transforming Property of an Ideal Lens

We have discussed light diffraction by aperture in the previous section. In this section, we will discuss the passage of light through an *ideal lens*. An ideal lens is a pure phase optical element of infinite extent and has no thickness. Hence, the ideal lens only modifies the phase of the incoming wavefronts. The transparency function of the ideal lens is given by

$$t_f(x, y) = e^{j\frac{k_0}{2f}(x^2+y^2)}, \tag{3.4-24}$$

where f is the focal length of the lens. For a converging (or positive) lens, $f > 0$ and $f < 0$ is for a diverging (or negative) lens.

For a plane wave of amplitude A incident on the lens, let us find the complex field distribution in the back focal plane of the lens. We assume that the lens is located on the plane $z = 0$. The situation is shown in Figure 3.4-4. According to Eq. (3.4-2b), the field distribution after the lens is

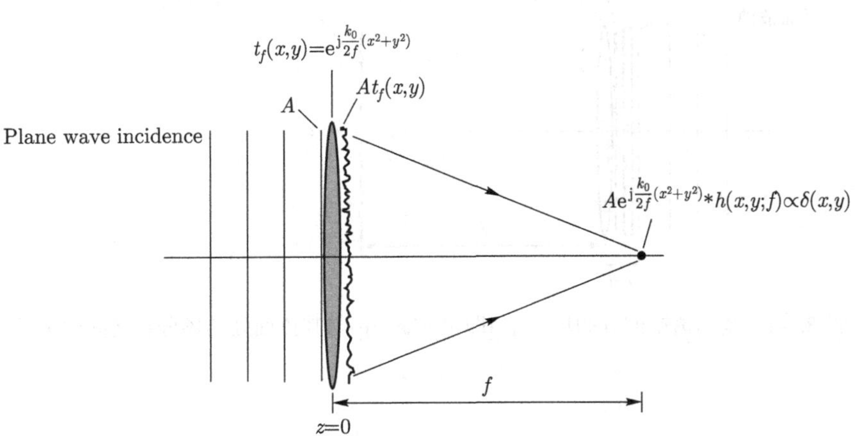

Figure 3.4-4 Plane wave of amplitude A incident on the ideal lens of focal length f

$$\psi_{p0}(x,y)=\psi_p(x,y;z=0^+)=\psi_p(x,y;z=0^-)t_f(x,y).$$

振幅为 A 的入射平面波可由式 (3.3-8) 给出为 $\psi(z;t)=A\mathrm{e}^{\mathrm{j}(\omega_0 t-k_0 z)}=\psi_p\mathrm{e}^{\mathrm{j}\omega_0 t}$, 因此, $\psi_p(x,y;z=0^-)=\psi_p=A\mathrm{e}^{-\mathrm{j}k_0 z}\big|_{z=0^-}=A$ 为紧贴透镜前方的振幅, $\psi_{p0}(x,y)=At_f(x,y)=A\mathrm{e}^{\mathrm{j}\frac{k_0}{2f}(x^2+y^2)}$. 现在使用菲涅耳衍射公式 [见式 (3.4-17) 或式 (3.4-21)] 来计算透镜后焦面, 即, $z=f$ 处的场分布

$$\begin{aligned}\psi_p(x,y;z=f)&=\psi_{p0}(x,y)*h(x,y;f)\\&=A\mathrm{e}^{\mathrm{j}\frac{k_0}{2f}(x^2+y^2)}*h(x,y;f)\\&=\mathrm{e}^{-\mathrm{j}k_0 f}\frac{\mathrm{j}k_0}{2\pi f}\mathrm{e}^{\frac{-\mathrm{j}k_0}{2f}(x^2+y^2)}\mathcal{F}\left\{A\mathrm{e}^{\mathrm{j}\frac{k_0}{2f}(x^2+y^2)}\mathrm{e}^{\frac{-\mathrm{j}k_0}{2f}(x^2+y^2)}\right\}\Bigg|_{k_x=\frac{k_0x}{f},\,k_y=\frac{k_0y}{f}}\\&=\mathrm{e}^{-\mathrm{j}k_0 f}\frac{\mathrm{j}k_0}{2\pi f}\mathrm{e}^{\frac{-\mathrm{j}k_0}{2f}(x^2+y^2)}\mathcal{F}\{A\}\Bigg|_{k_x=\frac{k_0x}{f},\,k_y=\frac{k_0y}{f}}.\end{aligned}\tag{3.4-25}$$

可观察到, 透镜的相位函数正好消掉了与菲涅耳衍射有关的二次相位函数, 利用表 2.1, A 的傅里叶变换为 $4\pi^2A\delta(k_x,k_y)$, 因此, 式 (3.4-25) 最终变成

$$\psi_p(x,y;z=f)=\mathrm{e}^{-\mathrm{j}k_0 f}\frac{\mathrm{j}k_0}{2\pi f}\mathrm{e}^{\frac{-\mathrm{j}k_0}{2f}(x^2+y^2)}A4\pi^2\delta\left(\frac{k_0x}{f},\frac{k_0y}{f}\right)\propto\delta(x,y).$$

这个结果非常有意义, 因为它与几何光学一致, 即所有平行于光轴的入射光线在透镜后都会聚到一个被称为后焦点的点上. 由式 (3.4-24) 给出的相位函数的函数形式对于理想透镜确实是正确的, 这一结论得以证明.

从式 (3.4-25) 可以看出, A 为紧贴透镜前方的场分布, 则, 广义上, 式 (3.4-25) 可以写为

$$\psi_p(x,y;z=f)=\mathrm{e}^{-\mathrm{j}k_0 f}\frac{\mathrm{j}k_0}{2\pi f}\mathrm{e}^{\frac{-\mathrm{j}k_0}{2f}(x^2+y^2)}\mathcal{F}\{\psi_{fl}(x,y)\}\Bigg|_{k_x=\frac{k_0x}{f},\,k_y=\frac{k_0y}{f}},\tag{3.4-26}$$

其中, $\psi_{fl}(x,y)$ 表示紧贴透镜前方的场分布. 例如, 如图 3.4-5 所示, 如果物理情况改变为在紧贴透镜前方放置一透过率函数为 $t(x,y)$ 的透明片, 则紧贴透镜前方的场分布变为 $\psi_{fl}(x,y)=A\,t(x,y)$. 所讨论的透过率函数可以是镜头前的某种孔径, 也可以仅仅是一个输入图像.

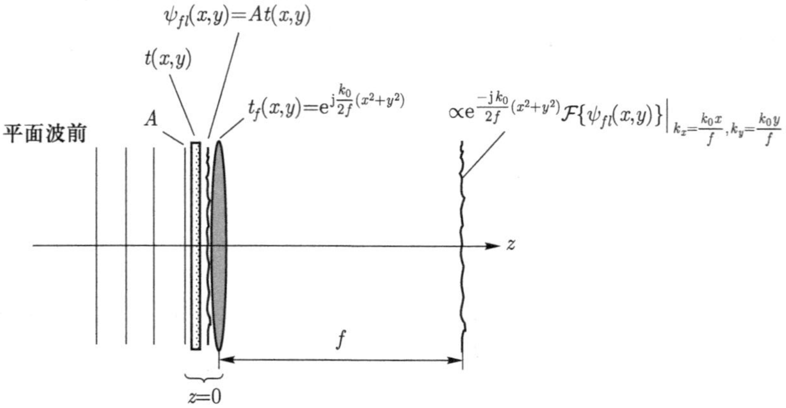

图 3.4-5 紧贴透镜前方的场分布 [即 $\psi_{fl}(x,y)$] 与其焦面处的场分布之间的关系

$$\psi_{p0}(x,y)=\psi_p(x,y;z=0^+)=\psi_p(x,y;z=0^-)t_f(x,y).$$

Since the incident plane wave of amplitude A is given by Eq. (3.3-8) as $\psi(z;t)=Ae^{\mathrm{j}(\omega_0 t-k_0 z)}=\psi_p e^{\mathrm{j}\omega_0 t}$. Hence, $\psi_p(x,y;z=0^-)=\psi_p=Ae^{-\mathrm{j}k_0 z}\big|_{z=0^-}=A$ is the amplitude just immediately before the lens and $\psi_{p0}(x,y)=At_f(x,y)=Ae^{\mathrm{j}\frac{k_0}{2f}(x^2+y^2)}$. We can now use the Fresnel diffraction formula [Eq. (3.4-17) or Eq. (3.4-21)] to calculate the field distribution in the back focal plane of the lens for $z=f$:

$$\begin{aligned}\psi_p(x,y;z=f)&=\psi_{p0}(x,y)*h(x,y;f)\\&=Ae^{\mathrm{j}\frac{k_0}{2f}(x^2+y^2)}*h(x,y;f)\\&=e^{-\mathrm{j}k_0 f}\frac{\mathrm{j}k_0}{2\pi f}e^{\frac{-\mathrm{j}k_0}{2f}(x^2+y^2)}\mathcal{F}\left\{Ae^{\mathrm{j}\frac{k_0}{2f}(x^2+y^2)}e^{\frac{-\mathrm{j}k_0}{2f}(x^2+y^2)}\right\}\Bigg|_{k_x=\frac{k_0x}{f},\,k_y=\frac{k_0y}{f}}\\&=e^{-\mathrm{j}k_0 f}\frac{\mathrm{j}k_0}{2\pi f}e^{\frac{-\mathrm{j}k_0}{2f}(x^2+y^2)}\mathcal{F}\{A\}\Bigg|_{k_x=\frac{k_0x}{f},\,k_y=\frac{k_0y}{f}}.\end{aligned}\tag{3.4-25}$$

We observe that the lens phase function cancels out exactly the quadratic phase function associated with Fresnel diffraction and the Fourier transform of A, using Table 2.1, is $4\pi^2A\delta(k_x,k_y)$. Therefore, Eq. (3.4-25) finally becomes

$$\psi_p(x,y;z=f)=e^{-\mathrm{j}k_0 f}\frac{\mathrm{j}k_0}{2\pi f}e^{\frac{-\mathrm{j}k_0}{2f}(x^2+y^2)}A4\pi^2\delta\left(\frac{k_0x}{f},\frac{k_0y}{f}\right)\propto\delta(x,y).$$

The result makes perfect sense as it is consistent with geometrical optics in that all input rays parallel to the optical axis converges behind the lens to a point called the back focal point. What we have just demonstrated is that the functional form of the phase function given by Eq. (3.4-24) is indeed correct for the ideal lens.

From Eq. (3.4-25), we recognize that A is the field distribution just in front of the lens. Therefore, we can generalize Eq. (3.4-25) to

$$\psi_p(x,y;z=f)=e^{-\mathrm{j}k_0 f}\frac{\mathrm{j}k_0}{2\pi f}e^{\frac{-\mathrm{j}k_0}{2f}(x^2+y^2)}\mathcal{F}\{\psi_{fl}(x,y)\}\Bigg|_{k_x=\frac{k_0x}{f},\,k_y=\frac{k_0y}{f}},\tag{3.4-26}$$

where $\psi_{fl}(x,y)$ denotes the field distribution just in front of the lens. For example, if the physical situation is changed to include a transparency function $t(x,y)$ just in front of the lens as shown in Figure 3.4-5, the field distribution just in front of the lens becomes $\psi_{fl}(x,y)=At(x,y)$. The transparency function under discussion could represent some kind of aperture in front of the lens or simply as an input image.

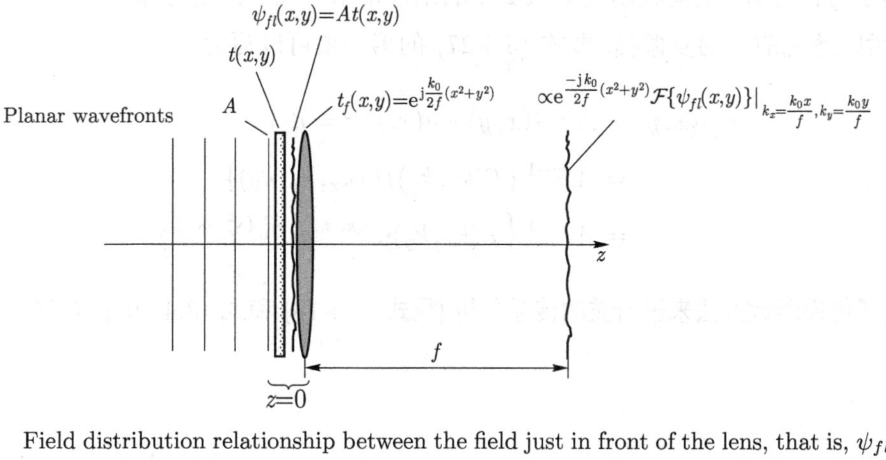

Figure 3.4-5 Field distribution relationship between the field just in front of the lens, that is, $\psi_{fl}(x,y)$, and the field on the focal plane of the lens

举例: 透过率函数作为输入置于理想透镜前的情况

这里考虑一个更一般的情况, 其中透过率函数 $t(x,y)$ 被放置在理想透镜前距离 d_0 处, 如图 3.4-6a) 所示, 该透明片被一振幅为 A 的平面波照射, 故, $\psi_p(x,y;z=0^-)=A$ 为紧贴 $t(x,y)$ 的透明片前方的复振幅, $A\times t(x,y)$ 为经过透明片后的场分布. 通常来说, 如果该透明片被某个总的复振幅 $\psi_p^i(x,y)$ 照射, 即, $\psi_p(x,y;z=0^-)=\psi_p^i(x,y)$, 那么, 紧贴透明片后方的场分布由 $\psi_p^i(x,y)\times t(x,y)$ 给出. 对于目前的情况, 仅有 $\psi_p^i(x,y)=A$.

首先利用菲涅耳衍射公式 [见式 (3.4-17)] 计算透镜前 $z=d_0$ 处的场分布, 可由 $[A\times t(x,y)]*h(x,y;z=d_0)$ 给出. 故, 透镜后的光场为 $\{[A\times t(x,y)]*h(x,y;z=d_0)\}\times t_f(x,y)$, 最后, 在后焦面处的光场, 即经过距离为 f 的菲涅耳衍射后, 场分布为

$$\psi_p(x,y;z=d_0+f)=\{\{[A\times t(x,y)]*h(x,y;z=d_0)\}\times t_f(x,y)\}*h(x,y;f). \tag{3.4-27}$$

这个方程可以用图 3.4-6b) 所示的系统框图来表示。

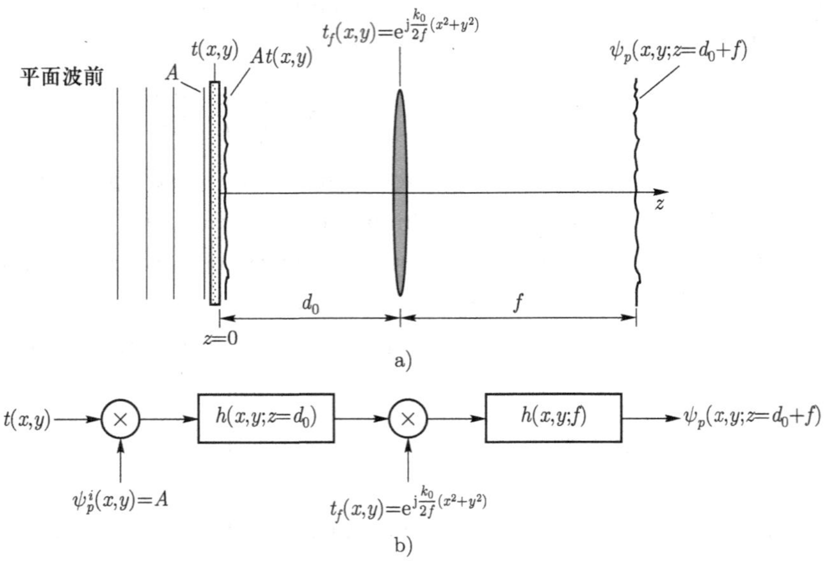

图 3.4-6 平面波照射位于焦距是 f 的理想透镜前距离为 d_0 处透明片 $t(x,y)$ 的情况: a) 实际情况; b) 系统框图

式 (3.4-27) 是图 3.4-6a) 中物理情况的一种准确的表示, 但由于卷积运算的重复, 计算比较复杂。为了方便计算, 这里利用式 (3.4-26) 的结果来求焦面上的场分布。为了做到这一点, 必须先求得紧贴透镜前方的复振幅, 即式 (3.4-27) 的第一项可以写为

$$\begin{aligned}\psi_{fl}(x,y)&=A\times t(x,y)*h(x,y;z=d_0)\\&=A\mathcal{F}^{-1}\{T(k_x,k_y)H(k_x,k_y;d_0)\}\\&=A\mathcal{F}^{-1}\left\{T(k_x,k_y)e^{-jk_0d_0}e^{j\frac{d_0}{2k_0}(k_x^2+k_y^2)}\right\},\end{aligned} \tag{3.4-28}$$

其中, 使用了传递函数方法来进行波的传播分析 [见式 (3.4-18) 和式 (3.4-19)], 同时, $\mathcal{F}\{t(x,y)\}=$

Example: Transparency Function as an Input Placed in Front of the Ideal Lens

Let us consider a more general situation where a transparency function $t(x,y)$ is placed a distance d_0 in front of the idea lens. The transparency is illuminated by a plane wave with amplitude A, as shown in Figure 3.4-6a. Therefore, $\psi_p(x,y;z=0^-)=A$ is the complex amplitude just immediately before $t(x,y)$ and $A\times t(x,y)$ is the field distribution after the transparency. In general, if the transparency is illuminated by some general complex field $\psi_p^i(x,y)$, that is, $\psi_p(x,y;z=0^-)=\psi_p^i(x,y)$, the field distribution immediately behind the transparency is then given by $\psi_p^i(x,y)\times t(x,y)$. For the current situation, we simply have $\psi_p^i(x,y)=A$.

We first use the Fresnel diffraction formula [Eq. (3.4-17)] to calculate the field distribution just before the lens at $z=d_0$, which is given by $[A\times t(x,y)]*h(x,y;z=d_0)$. The field after the lens is, therefore, given by $\{[A\times t(x,y)]*h(x,y;z=d_0)\}\times t_f(x,y)$. Finally, the field at the back focal plane, after Fresnel diffraction of a distance of f, is given by

$$\psi_p\left(x,y;z=d_0+f\right)=\{\{[A\times t(x,y)]*h\left(x,y;z=d_0\right)\}\times t_f(x,y)\}*h(x,y;f). \qquad (3.4\text{-}27)$$

This equation can be represented by a system block diagram shown in Figure 3.4-6b.

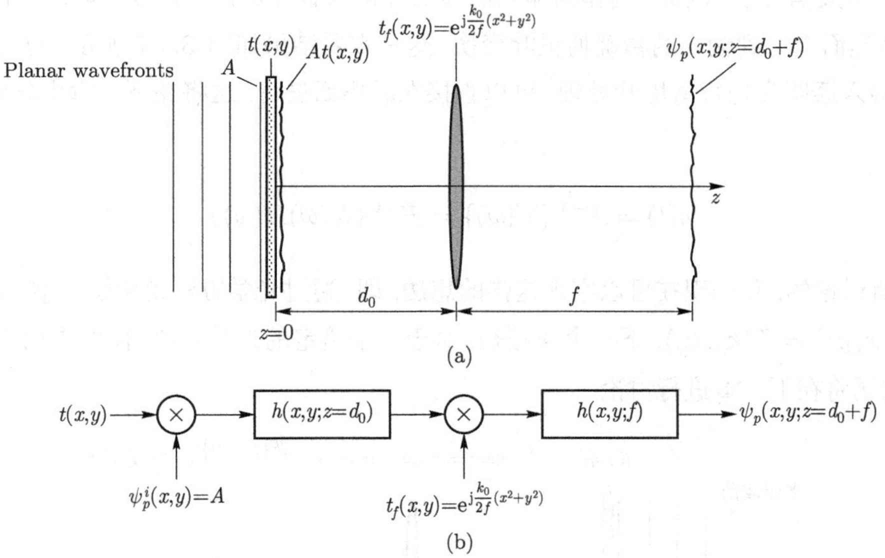

Figure 3.4-6 Plane wave illumination of a transparency $t(x,y)$ located a distance d_0 in front of an ideal lens of focal length: (a) physical situation and (b) system block diagram

Equation (3.4-27) is a precise way to represent the physical situation in Figure 3.4-6a, but it is complicated to evaluate owing to the repeated convolution operations. Let us employ the result of Eq. (3.4-26) to facilitate the calculation to obtain the field distribution on the focal plane. In order to do it, we must first find the complex field just in front of the lens, which is the first step in Eq. (3.4-27):

$$\begin{aligned}\psi_{fl}(x,y)&=A\times t(x,y)*h(x,y;z=d_0)\\&=A\mathcal{F}^{-1}\left\{T(k_x,k_y)H\left(k_x,k_y;d_0\right)\right\}\\&=A\mathcal{F}^{-1}\left\{T(k_x,k_y)\mathrm{e}^{-\mathrm{j}k_0d_0}\mathrm{e}^{\mathrm{j}\frac{d_0}{2k_0}(k_x^2+k_y^2)}\right\},\end{aligned} \qquad (3.4\text{-}28)$$

where we have used the transfer function approach to wave propagation [see Eqs. (3.4-18) and (3.4-19)] and $\mathcal{F}\{t(x,y)\}=T(k_x,k_y)$ is the spectrum of the input. Now, the complex

$T(k_x, k_y)$ 为输入谱. 将式 (3.4-28) 代入式 (3.4-26), 得到焦面上的复振幅, 有

$$
\begin{aligned}
\psi_p(x,y;z=f+d_0) &= \mathrm{e}^{-\mathrm{j}k_0 f}\frac{\mathrm{j}k_0}{2\pi f}\mathrm{e}^{\frac{-\mathrm{j}k_0}{2f}(x^2+y^2)}\mathcal{F}\left\{\psi_{fl}(x,y)\right\}\Big|_{k_x=\frac{k_0x}{f},\,k_y=\frac{k_0y}{f}}\\
&= \mathrm{e}^{-\mathrm{j}k_0 f}\frac{\mathrm{j}k_0}{2\pi f}A\mathrm{e}^{\frac{-\mathrm{j}k_0}{2f}(x^2+y^2)}\mathcal{F}\left\{\mathcal{F}^{-1}\left\{T(k_x,k_y)\mathrm{e}^{-\mathrm{j}k_0d_0}\mathrm{e}^{\frac{\mathrm{j}d_0}{2k_0}(k_x^2+k_y^2)}\right\}\right\}\Big|_{k_x=\frac{k_0x}{f},\,k_y=\frac{k_0y}{f}}\\
&= \mathrm{e}^{-\mathrm{j}k_0 f}\frac{\mathrm{j}k_0}{2\pi f}A\mathrm{e}^{\frac{-\mathrm{j}k_0}{2f}(x^2+y^2)}T(k_x,k_y)\mathrm{e}^{-\mathrm{j}k_0d_0}\mathrm{e}^{\frac{\mathrm{j}d_0}{2k_0}\left(k_x^2+k_y^2\right)}\Big|_{k_x=\frac{k_0x}{f},\,k_y=\frac{k_0y}{f}}\\
&= \mathrm{e}^{-\mathrm{j}k_0(f+d_0)}\frac{\mathrm{j}k_0}{2\pi f}A\mathrm{e}^{\frac{-\mathrm{j}k_0}{2f}(x^2+y^2)}T(k_x,k_y)\Big|_{k_x=\frac{k_0x}{f},\,k_y=\frac{k_0y}{f}}\times\mathrm{e}^{\mathrm{j}\frac{d_0}{2k_0}\left[\left(\frac{k_0x}{f}\right)^2+\left(\frac{k_0y}{f}\right)^2\right]}.
\end{aligned}
$$

最后

$$
\psi_p(x,y;z=f+d_0)=\mathrm{e}^{-\mathrm{j}k_0(f+d_0)}\frac{\mathrm{j}k_0}{2\pi f}A\mathrm{e}^{-\mathrm{j}\frac{k_0}{2f}\left(1-\frac{d_0}{f}\right)(x^2+y^2)}\mathcal{F}\left\{t(x,y)\right\}\Big|_{k_x=\frac{k_0x}{f},\,k_y=\frac{k_0y}{f}},\tag{3.4-29}
$$

其中, 为了得到上述结果, 通过重组指数函数中的 (x^2+y^2) 项来简化公式. 可以看出, 如在式 (3.4-23) 中, 一个相位曲率因子 (*phase curvature factor*) 再次出现在傅里叶变换之前, 但当 $d_0=f$ 时又消失了, 因此, 当作为输入的透明片被放置在透镜的前焦面时, 除了一些常数外, 会在后焦面上得到输入的精确傅里叶变换. 这一重要结果如图 3.4-7 所示. 对透镜前焦面处放置的输入透明片进行傅里叶处理, 可以直接在后焦面进行, 这将在下一章中看到. 根据信号处理

$$
y(t)=\mathcal{F}^{-1}\left\{Y(\omega)\right\}=\mathcal{F}^{-1}\left\{X(\omega)H(\omega)\right\},
$$

如第二章所讨论的, 我们现在可以实现这样的想法, 即, 通过光学方法得到输入谱 $X(\omega)$, 它等同于 $\mathcal{F}\{t(x,y)\}=T(k_x,k_y)$. 下一个问题是, 对于一个给定的光学系统, 需要求出它的传递函数, 这个问题将在下一章进行讨论.

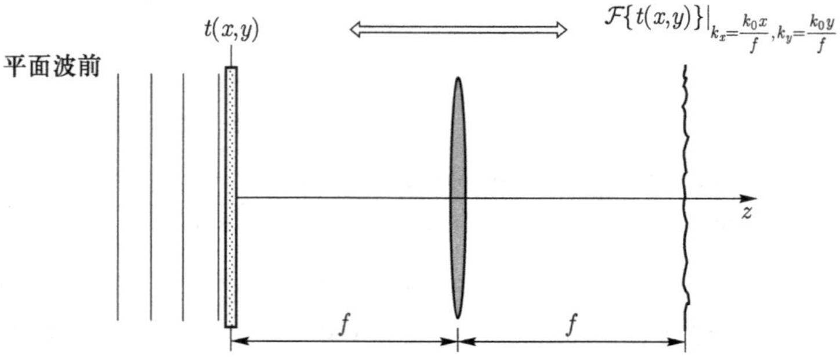

图 3.4-7　理想透镜作为光学傅里叶变换器

举例: 通过圆形孔径聚焦

将一个圆形孔径紧贴在透过率函数为 $t_f(x,y)=\mathrm{e}^{\mathrm{j}\frac{k_0}{2f}(x^2+y^2)}$ 的理想透镜前方, 在透镜后表面, 单位振幅平面波照射圆形孔径的复振幅可以表示为 $\psi_{p0}(x,y)=circ\left(\frac{r}{r_0}\right)t_f(x,y)=circ\left(\frac{r}{r_0}\right)\mathrm{e}^{\mathrm{j}\frac{k_0}{2f}(x^2+y^2)}$, 其中, r_0 是孔径半径. 根据菲涅耳衍射公式, 即式 (3.4-17) 或式 (3.4-21),

field at the focal plane is given by substituting Eq. (3.4-28) into Eq. (3.4-26), and we have

$$\begin{aligned}\psi_p(x,y;z=f+d_0)&=\mathrm{e}^{-\mathrm{j}k_0 f}\frac{\mathrm{j}k_0}{2\pi f}\mathrm{e}^{\frac{-\mathrm{j}k_0}{2f}(x^2+y^2)}\mathcal{F}\left\{\psi_{fl}(x,y)\right\}\Big|_{k_x=\frac{k_0x}{f},\,k_y=\frac{k_0y}{f}}\\&=\mathrm{e}^{-\mathrm{j}k_0 f}\frac{\mathrm{j}k_0}{2\pi f}A\mathrm{e}^{\frac{-\mathrm{j}k_0}{2f}(x^2+y^2)}\mathcal{F}\Big\{\mathcal{F}^{-1}\Big\{T(k_x,k_y)\mathrm{e}^{-\mathrm{j}k_0d_0}\mathrm{e}^{\frac{\mathrm{j}d_0}{2k_0}(k_x^2+k_y^2)}\Big\}\Big\}\Big|_{k_x=\frac{k_0x}{f},\,k_y=\frac{k_0y}{f}}\\&=\mathrm{e}^{-\mathrm{j}k_0 f}\frac{\mathrm{j}k_0}{2\pi f}A\mathrm{e}^{\frac{-\mathrm{j}k_0}{2f}(x^2+y^2)}T(k_x,k_y)\mathrm{e}^{-\mathrm{j}k_0d_0}\mathrm{e}^{\frac{\mathrm{j}d_0}{2k_0}(k_x^2+k_y^2)}\Big|_{k_x=\frac{k_0x}{f},\,k_y=\frac{k_0y}{f}}\\&=\mathrm{e}^{-\mathrm{j}k_0(f+d_0)}\frac{\mathrm{j}k_0}{2\pi f}A\mathrm{e}^{\frac{-\mathrm{j}k_0}{2f}(x^2+y^2)}T(k_x,k_y)\Big|_{k_x=\frac{k_0x}{f},\,k_y=\frac{k_0y}{f}}\times\mathrm{e}^{\mathrm{j}\frac{d_0}{2k_0}\left[\left(\frac{k_0x}{f}\right)^2+\left(\frac{k_0y}{f}\right)^2\right]},\end{aligned}$$

or finally,

$$\psi_p(x,y;z=f+d_0)=\mathrm{e}^{-\mathrm{j}k_0(f+d_0)}\frac{\mathrm{j}k_0}{2\pi f}A\mathrm{e}^{-\mathrm{j}\frac{k_0}{2f}\left(1-\frac{d_0}{f}\right)(x^2+y^2)}\mathcal{F}\left\{t(x,y)\right\}\Big|_{k_x=\frac{k_0x}{f},\,k_y=\frac{k_0y}{f}},\tag{3.4-29}$$

where we have grouped the (x^2+y^2) terms in the exponents to simplify in order to obtain the above result. Note that, as in Eq. (3.4-23), a phase curvature factor again precedes the Fourier transform but vanished when $d_0=f$. Thus, when the input transparency is placed in the front focal plane of the lens, aside from some constants we have the exact Fourier transform of the input on the back focal plane. This important result is illustrated in Figure 3.4-7. Fourier processing on an input transparency located on the front focal plane may now be performed on the back focal plane, as will be seen in the next chapter. With reference to signal processing where

$$y(t)=\mathcal{F}^{-1}\left\{Y(\omega)\right\}=\mathcal{F}^{-1}\left\{X(\omega)H(\omega)\right\},$$

as discussed in Chapter 2, we now can implement such an idea as the spectrum of the input $X(\omega)$, which is equivalent to $\mathcal{F}\left\{t(x,y)\right\}=T(k_x,k_y)$, has been obtained optically. The next question is, for a given optical system, we need to find its transfer function. We will address this in the next chapter.

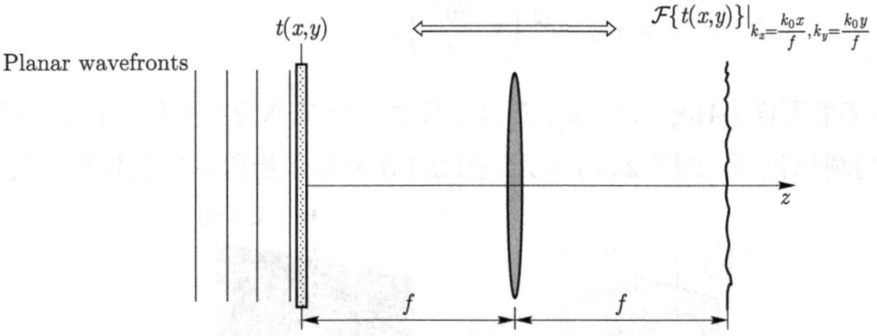

Figure 3.4-7 Ideal lens as an optical Fourier transformer

Example: Focusing through a Circular Aperture

We place a circular aperture against an ideal lens of the transparency function given by $t_f(x,y)=\mathrm{e}^{\mathrm{j}\frac{k_0}{2f}(x^2+y^2)}$. The complex amplitude of a circular aperture illuminated by a plane wave of unity amplitude is then represented by $\psi_{p0}(x,y)=circ\left(\frac{r}{r_0}\right)t_f(x,y)=circ\left(\frac{r}{r_0}\right)\mathrm{e}^{\mathrm{j}\frac{k_0}{2f}(x^2+y^2)}$ at the exit of the lens, where r_0 is radius of the aperture. Accord-

将距孔径 $f+z$ 处的复振幅写成

$$\begin{aligned}\psi_p(x,y;f+z)&=\mathrm{e}^{-\mathrm{j}k_0(f+z)}\frac{\mathrm{j}k_0}{2\pi(f+z)}\mathrm{e}^{\frac{-\mathrm{j}k_0}{2(f+z)}(x^2+y^2)}\mathcal{F}\left\{\psi_{p0}(x,y)\mathrm{e}^{\frac{-\mathrm{j}k_0}{2(f+z)}(x^2+y^2)}\right\}\bigg|_{k_x=\frac{k_0x}{f+z},\,k_y=\frac{k_0y}{f+z}}\\&=\mathrm{e}^{-\mathrm{j}k_0(f+z)}\frac{\mathrm{j}k_0}{2\pi(f+z)}\mathrm{e}^{\frac{-\mathrm{j}k_0}{2(f+z)}(x^2+y^2)}\\&\quad\cdot\mathcal{F}\left\{circ\left(\frac{r}{r_0}\right)\mathrm{e}^{\mathrm{j}\frac{k_0}{2f}(x^2+y^2)}\mathrm{e}^{\frac{-\mathrm{j}k_0}{2(f+z)}(x^2+y^2)}\right\}\bigg|_{k_x=\frac{k_0x}{f+z},\,k_y=\frac{k_0y}{f+z}}.\end{aligned}$$

由于这个问题具有圆对称性, 可以用傅里叶–贝塞尔变换将上述方程重写为

$$\psi_p(r;f+z)=\mathrm{e}^{-\mathrm{j}k_0(f+z)}\frac{\mathrm{j}k_0}{2\pi(f+z)}\mathrm{e}^{\frac{-\mathrm{j}k_0}{2(f+z)}r^2}\mathcal{B}\left\{circ\left(\frac{r}{r_0}\right)\mathrm{e}^{\mathrm{j}\frac{k_0}{2f}r^2}\mathrm{e}^{\frac{-\mathrm{j}k_0}{2(f+z)}r^2}\right\}\bigg|_{k_r=\frac{k_0r}{f+z}}.\tag{3.4-30}$$

A) 会聚的圆形光束: 艾里斑

为了求焦面上的复振幅, 令上式中 $z=0$, 有

$$\psi_p(r;f)=\mathrm{e}^{-\mathrm{j}k_0f}\frac{\mathrm{j}k_0}{2\pi f}\mathrm{e}^{\frac{-\mathrm{j}k_0}{2f}r^2}\mathcal{B}\left\{circ\left(\frac{r}{r_0}\right)\right\}\bigg|_{k_r=\frac{k_0r}{f}}.$$

由于圆域函数的傅里叶–贝塞尔变换由式 (2.3-17) 给出

$$\mathcal{B}\left\{circ\left(\frac{r}{r_0}\right)\right\}=2\pi r_0^2jinc(r_0k_r),$$

焦面上的复振幅为

$$\psi_p(r;f)=\mathrm{e}^{-\mathrm{j}k_0f}\frac{\mathrm{j}k_0}{2\pi f}\mathrm{e}^{\frac{-\mathrm{j}k_0}{2f}r^2}2\pi r_0^2jinc\left(r_0\frac{k_0r}{f}\right),$$

焦面上的光强为

$$\begin{aligned}I(r,f)=|\psi_p(r;f)|^2&=\left(\frac{k_0r_0^2}{f}\right)^2jinc^2\left(r_0\frac{k_0r}{f}\right)\\&=I_0jinc^2\left(r_0\frac{k_0r}{f}\right),\end{aligned}\tag{3.4-31}$$

这就是所谓的艾里图样 (*Airy pattern*). 艾里图样由一个被称为艾里斑 (*Airy disk*) 的中央圆盘构成, 被一系列圆环包围, 如图 2.3-5 所示. 图 3.4-8 表示了当前示例的物理情况.

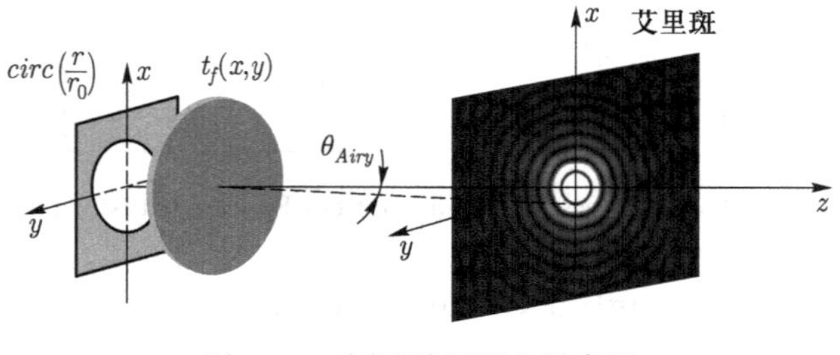

图 3.4-8 圆形孔径聚焦示意图

现在来求中央圆盘的半径. 由于 $jinc(x)=\dfrac{J_1(x)}{x}$, 故 J_1 的第一个零点决定了其半径.

ing to the Fresnel diffraction formula, that is, Eq. (3.4-17) or Eq. (3.4-21), we write the complex field at a distance of $f+z$ away from the aperture as

$$\psi_p(x,y;f+z) = \mathrm{e}^{-\mathrm{j}k_0(f+z)}\frac{\mathrm{j}k_0}{2\pi(f+z)}\mathrm{e}^{\frac{-\mathrm{j}k_0}{2(f+z)}(x^2+y^2)}\mathcal{F}\left\{\psi_{p0}(x,y)\mathrm{e}^{\frac{-\mathrm{j}k_0}{2(f+z)}(x^2+y^2)}\right\}\Bigg|_{k_x=\frac{k_0x}{f+z},\,k_y=\frac{k_0y}{f+z}}$$

$$= \mathrm{e}^{-\mathrm{j}k_0(f+z)}\frac{\mathrm{j}k_0}{2\pi(f+z)}\mathrm{e}^{\frac{-\mathrm{j}k_0}{2(f+z)}(x^2+y^2)} \cdot\mathcal{F}\left\{circ\left(\frac{r}{r_0}\right)\mathrm{e}^{\mathrm{j}\frac{k_0}{2f}(x^2+y^2)}\mathrm{e}^{\frac{-\mathrm{j}k_0}{2(f+z)}(x^2+y^2)}\right\}\Bigg|_{k_x=\frac{k_0x}{f+z},\,k_y=\frac{k_0y}{f+z}}.$$

Since the problem has circular symmetry, we can rewrite the above equation in terms of the Fourier-Bessel transform as

$$\psi_p(r;f+z) = \mathrm{e}^{-\mathrm{j}k_0(f+z)}\frac{\mathrm{j}k_0}{2\pi(f+z)}\mathrm{e}^{\frac{-\mathrm{j}k_0}{2(f+z)}r^2}\mathcal{B}\left\{circ\left(\frac{r}{r_0}\right)\mathrm{e}^{\mathrm{j}\frac{k_0}{2f}r^2}\mathrm{e}^{\frac{-\mathrm{j}k_0}{2(f+z)}r^2}\right\}\Bigg|_{k_r=\frac{k_0r}{f+z}}. \quad (3.4\text{-}30)$$

A) Focused Circular Beam: The Airy Pattern

To find the complex field at the focal plane, we let $z=0$ in the above equation:

$$\psi_p(r;f) = \mathrm{e}^{-\mathrm{j}k_0f}\frac{\mathrm{j}k_0}{2\pi f}\mathrm{e}^{\frac{-\mathrm{j}k_0}{2f}r^2}\mathcal{B}\left\{circ\left(\frac{r}{r_0}\right)\right\}\Bigg|_{k_r=\frac{k_0r}{f}}.$$

Since the Fourier-Bessel transform of a circular function is given by Eq. (2.3-17) as

$$\mathcal{B}\left\{circ\left(\frac{r}{r_0}\right)\right\} = 2\pi r_0^2 jinc(r_0k_r),$$

the complex field in the focal plane is then

$$\psi_p(r;f) = \mathrm{e}^{-\mathrm{j}k_0f}\frac{\mathrm{j}k_0}{2\pi f}\mathrm{e}^{\frac{-\mathrm{j}k_0}{2f}r^2}2\pi r_0^2 jinc\left(r_0\frac{k_0r}{f}\right),$$

and the intensity on the focal plane is

$$I(r,f) = |\psi_p(r;f)|^2 = \left(\frac{k_0r_0^2}{f}\right)^2 jinc^2\left(r_0\frac{k_0r}{f}\right)$$
$$= I_0 jinc^2\left(r_0\frac{k_0r}{f}\right), \quad (3.4\text{-}31)$$

which is known as the *Airy pattern*. The Airy pattern consists of a central disc, known as the *Airy disk*, surrounded by rings, as illustrated in Figure 2.3-5. Figure 3.4-8 shows the physical situation of the current example.

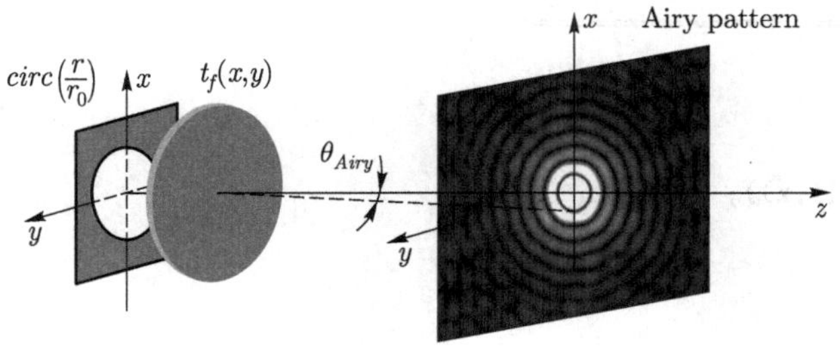

Figure 3.4-8 Focusing through a circular aperture

Let us find the radius of the central disc. Since $jinc(x) = \dfrac{J_1(x)}{x}$, the first zero of J_1,

图 3.4-9 绘制了 $J_1(x)$ 并估计它的第一个零点出现在 $x \approx 3.835$ 处. 让式 (3.4-31) 中的 $jinc$ 函数的自变量等于 3.835, 即可得到圆盘的半径 $r = r_{Airy}$, 有

$$r_0 \frac{k_0 r_{Airy}}{f} \approx 3.835,$$

或

$$r_{Airy} \approx 1.22 f \frac{\lambda_0}{2r_0} = f\theta_{Airy}, \tag{3.4-32}$$

这里

$$\theta_{Airy} = 1.22 \frac{\lambda_0}{2r_0},$$

其为扩散角, 是艾里斑扩开的半角. 可以发现, 中心峰 r_{Airy} 的半径可以用来度量横向分辨率. 根据第一章讨论的测不准原理, 由式 (1.2-3), 圆形孔径的横向分辨率为

$$\Delta r \approx \frac{\lambda_0}{2\sin(\theta_{im}/2)}, \tag{3.4-33}$$

对于小角度, 这里 $\sin(\theta_{im}/2)$ 由 r_0/f 给出, 由式 (3.4-33), Δr 变为

$$\Delta r \approx f \frac{\lambda_0}{2r_0},$$

这与式 (3.4-32) 给出的结果一致.

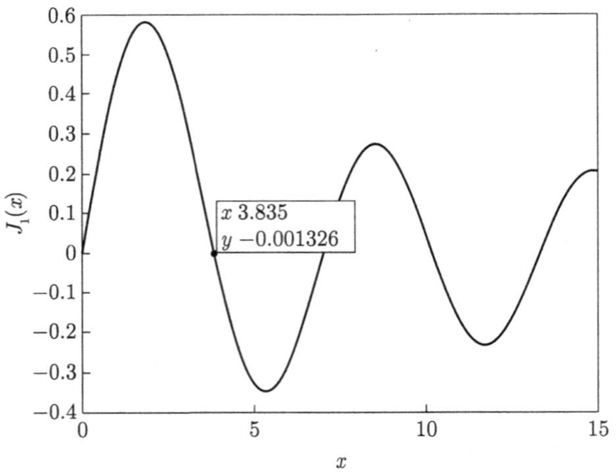

图 3.4-9 第一类一阶贝塞尔函数

图 3.4-9 是根据下面的 m-文件生成的.

```
========================================
%besselroot.m
x=0:0.001:15;
plot (x, besselj(1,x)), grid on
j1=besselj(1,x);
xlabel ('x')
ylabel('J1(x)')
========================================
```

therefore, determines the radius. In Figure 3.4-9, we plot $J_1(x)$ and estimate its first zero, which happens at $x \approx 3.835$. By letting the argument of the *jinc* function in Eq. (3.4-31) equal to 3.835, we have the radius of the disk $r = r_{Airy}$:

$$r_0 \frac{k_0 r_{Airy}}{f} \approx 3.835,$$

or

$$r_{Airy} \approx 1.22 f \frac{\lambda_0}{2r_0} = f\theta_{Airy}, \tag{3.4-32}$$

where

$$\theta_{Airy} = 1.22 \frac{\lambda_0}{2r_0},$$

is the spread angle, which is the half angle subtended by the Airy disk. Note that the radius of the central peak r_{Airy} is a measure of lateral resolution. From the uncertainty principle discussed in Chapter 1, the lateral resolution of a circular aperture, according to Eq. (1.2-3), is

$$\Delta r \approx \frac{\lambda_0}{2\sin(\theta_{im}/2)}, \tag{3.4-33}$$

where $\sin(\theta_{im}/2)$ is given by r_0/f for small angles. Δr from Eq. (3.4-33) then becomes

$$\Delta r \approx f \frac{\lambda_0}{2r_0},$$

which is consistent to that given by Eq. (3.4-32).

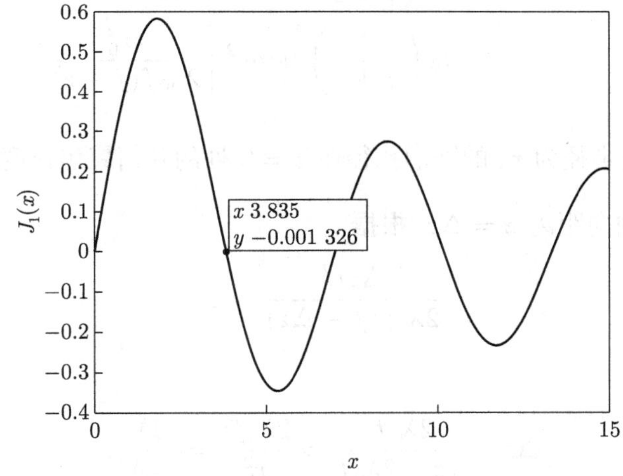

Figure 3.4-9 Plot of the first-order Bessel function of the first kind

Figure 3.4-9 is generated using the m-file below.

```
==========================================
%besselroot.m
x=0:0.001:15;
plot (x, besselj(1,x)), grid on
j1=besselj(1,x);
xlabel ('x')
ylabel('J1(x)')
==========================================
```

B) 焦深

求轴上强度. 从式 (3.4-30), 有

$$\begin{aligned}\psi_p(r; f+z) &= \mathrm{e}^{-\mathrm{j}k_0(f+z)}\frac{\mathrm{j}k_0}{2\pi(f+z)}\mathrm{e}^{\frac{-\mathrm{j}k_0}{2(f+z)}r^2}\mathcal{B}\left\{circ\left(\frac{r}{r_0}\right)\mathrm{e}^{\mathrm{j}\frac{k_0}{2f}r^2}\mathrm{e}^{\frac{-\mathrm{j}k_0}{2(f+z)}r^2}\right\}\Bigg|_{k_r=\frac{k_0 r}{f+z}} \\ &= \mathrm{e}^{-\mathrm{j}k_0(f+z)}\frac{\mathrm{j}k_0}{2\pi(f+z)}\mathrm{e}^{\frac{-\mathrm{j}k_0}{2(f+z)}r^2}2\pi\int_0^{r_0} r'\mathrm{e}^{\mathrm{j}\frac{k_0}{2f}r'^2}\mathrm{e}^{\frac{-\mathrm{j}k_0}{2(f+z)}r'^2}J_0\left(k_r r'\right)\mathrm{d}r'\Bigg|_{k_r=\frac{k_0 r}{f+z}} \\ &= \mathrm{e}^{-\mathrm{j}k_0(f+z)}\frac{\mathrm{j}k_0}{2\pi\left(f+z\right)}\mathrm{e}^{\frac{-\mathrm{j}k_0}{2(f+z)}r^2}2\pi\int_0^{r_0} r'\mathrm{e}^{\mathrm{j}\frac{k_0}{2}\varepsilon r'^2}J_0\left(\frac{k_0 r}{f+z}r'\right)\mathrm{d}r'\end{aligned}$$

其中, $\varepsilon = z/[f(f+z)]$, 故, 轴上或轴向的复振幅为

$$\psi_p(0; f+z) = \mathrm{e}^{-\mathrm{j}k_0(f+z)}\frac{\mathrm{j}k_0}{f+z}\int_0^{r_0} r'\mathrm{e}^{\mathrm{j}\frac{k_0}{2}\varepsilon r'^2}\mathrm{d}r'.$$

积分之后, 有

$$\psi_p(0; f+z) = \mathrm{e}^{-\mathrm{j}k_0(f+z)}\frac{\mathrm{j}k_0 r_0^2}{2(f+z)}\frac{\sin(k_0\varepsilon r_0^2/4)}{k_0\varepsilon r_0^2/4}\mathrm{e}^{\mathrm{j}k_0\varepsilon r_0^2/4}.$$

轴向强度为

$$\begin{aligned}I(0, f+z) &= \psi_p(0; f+z)\psi_p^*(0; f+z) \\ &= I_0\left(\frac{f}{f+z}\right)^2 sinc^2\left(\frac{\varepsilon r_0^2}{2\lambda_0}\right) \\ &= I_0\left(\frac{f}{f+z}\right)^2 sinc^2\left[\frac{z r_0^2}{2\lambda_0 f(f+z)}\right],\end{aligned} \tag{3.4-34}$$

其中, $I_0 = \left(\dfrac{\pi r_0^2}{\lambda_0 f}\right)^2$ 是半径为 r_0 的圆形孔径在 $z=0$ 处的几何聚焦强度. 现在来求 *sinc* 函数的第一个零点对应的轴向距离 $z = \Delta z$, 根据

$$\frac{\Delta z r_0^2}{2\lambda_0 f(f+\Delta z)} = 1,$$

求出

$$\Delta z = \frac{2\lambda_0 f^2}{r_0^2 - 2\lambda_0 f} \approx \frac{2\lambda_0 f^2}{r_0^2} = \frac{2\lambda_0}{NA^2}. \tag{3.4-35}$$

因 $r_0 \gg \lambda_0$, $r_0 \sim f$, 且, 对于小角度, $r_0/f \approx NA$, Δz 为焦深 [见式 (1.2-9)], 在第一章中, 通过使用测不准原理已导出.

3.5 高斯光束光学

激光器发出的光束通常具有高斯振幅分布. 因此, 研究高斯光束 (*Gaussian beam*) 的衍射非常重要. 考虑一个 $z=0$ 处为平面波前 (*planar wavefront*) 且沿 z 方向传播的高斯光束

$$\psi_{p0}(x, y) = \mathrm{e}^{-(x^2+y^2)/\omega_0^2}, \tag{3.5-1}$$

B) Depth of Focus

Let us find the on-axis intensity. From Eq. (3.4-30), we have

$$\begin{aligned}
\psi_p(r;f+z) &= \mathrm{e}^{-\mathrm{j}k_0(f+z)}\frac{\mathrm{j}k_0}{2\pi(f+z)}\mathrm{e}^{\frac{-\mathrm{j}k_0}{2(f+z)}r^2}\mathcal{B}\left\{circ\left(\frac{r}{r_0}\right)\mathrm{e}^{\mathrm{j}\frac{k_0}{2f}r^2}\mathrm{e}^{\frac{-\mathrm{j}k_0}{2(f+z)}r^2}\right\}\Bigg|_{k_r=\frac{k_0r}{f+z}}\\
&= \mathrm{e}^{-\mathrm{j}k_0(f+z)}\frac{\mathrm{j}k_0}{2\pi(f+z)}\mathrm{e}^{\frac{-\mathrm{j}k_0}{2(f+z)}r^2}2\pi\int_0^{r_0} r'\mathrm{e}^{\mathrm{j}\frac{k_0}{2f}r'^2}\mathrm{e}^{\frac{-\mathrm{j}k_0}{2(f+z)}r'^2}J_0\left(k_r r'\right)\mathrm{d}r'\Bigg|_{k_r=\frac{k_0r}{f+z}}\\
&= \mathrm{e}^{-\mathrm{j}k_0(f+z)}\frac{\mathrm{j}k_0}{2\pi\left(f+z\right)}\mathrm{e}^{\frac{-\mathrm{j}k_0}{2(f+z)}r^2}2\pi\int_0^{r_0} r'\mathrm{e}^{\mathrm{j}\frac{k_0}{2}\varepsilon r'^2}J_0\left(\frac{k_0r}{f+z}r'\right)\mathrm{d}r'
\end{aligned}$$

where $\varepsilon = z/[f(f+z)]$. Hence, the on-axis or the axial complex field is

$$\psi_p(0;f+z) = \mathrm{e}^{-\mathrm{j}k_0(f+z)}\frac{\mathrm{j}k_0}{f+z}\int_0^{r_0} r'\mathrm{e}^{\mathrm{j}\frac{k_0}{2}\varepsilon r'^2}\mathrm{d}r'.$$

After integrating, we have

$$\psi_p(0;f+z) = \mathrm{e}^{-\mathrm{j}k_0(f+z)}\frac{\mathrm{j}k_0r_0^2}{2(f+z)}\frac{\sin(k_0\varepsilon r_0^2/4)}{k_0\varepsilon r_0^2/4}\mathrm{e}^{\mathrm{j}k_0\varepsilon r_0^2/4}.$$

The axial intensity is then given by

$$\begin{aligned}
I(0,f+z) &= \psi_p(0;f+z)\psi_p^*(0;f+z)\\
&= I_0\left(\frac{f}{f+z}\right)^2 sinc^2\left(\frac{\varepsilon r_0^2}{2\lambda_0}\right)\\
&= I_0\left(\frac{f}{f+z}\right)^2 sinc^2\left[\frac{zr_0^2}{2\lambda_0 f(f+z)}\right],
\end{aligned}\tag{3.4-34}$$

where $I_0 = \left(\frac{\pi r_0^2}{\lambda_0 f}\right)^2$ is the intensity of the geometrical focus at $z=0$ due to the circular aperture of radius r_0. Let us find the axial distance $z=\Delta z$ at which the *sinc* function finds its first zero:

$$\frac{\Delta z r_0^2}{2\lambda_0 f(f+\Delta z)} = 1,$$

From this, we solve for

$$\Delta z = \frac{2\lambda_0 f^2}{r_0^2 - 2\lambda_0 f} \approx \frac{2\lambda_0 f^2}{r_0^2} = \frac{2\lambda_0}{NA^2}. \tag{3.4-35}$$

as $r_0 \gg \lambda_0$, $r_0 \sim f$ and $r_0/f \approx NA$ for small angles. Note that Δz is the depth of focus [see Eq. (1.2-9)], derived in Chapter 1 through the use of the uncertainty principle.

3.5 Gaussian Beam Optics

Optical beams from lasers typically have a Gaussian amplitude distribution. Therefore, the study of the diffraction of a *Gaussian beam* is of great importance. Let us consider a Gaussian beam propagating along the z-direction with planar wavefronts at $z=0$. Hence,

$$\psi_{p0}(x,y) = \mathrm{e}^{-(x^2+y^2)/\omega_0^2}, \tag{3.5-1}$$

其中, ω_0 是高斯光束的束腰 (*beam waist*), 因为在距离 z 轴 ω_0 远的位置处, 光束的振幅会下降 $1/\mathrm{e}$ 的系数. 现在可以由式 (3.4-19) 的频域方程求出光束的非涅耳衍射

$$\psi_p(x,y;z)=\mathcal{F}^{-1}\left\{\Psi_p(k_x,k_y;z)\right\}=\mathcal{F}^{-1}\left\{\Psi_{p0}(k_x,k_y)H(k_x,k_y;z)\right\}. \tag{3.5-2}$$

高斯光束的傅里叶变换可由表 2.1 中的第 5 项求得, 其结果为

$$\Psi_{p0}(k_x,k_y)=\mathcal{F}\left\{\mathrm{e}^{-(x^2+y^2)/\omega_0^2}\right\}=\pi\omega_0^2\mathrm{e}^{\frac{-(k_x^2+k_y^2)\omega_0^2}{4}}. \tag{3.5-3}$$

利用傅里叶光学中的空间频率传递函数式 (3.4-18), 式 (3.5-2) 变为

$$\begin{aligned}\psi_p(x,y;z)&=\mathcal{F}^{-1}\left\{\pi\omega_0^2\mathrm{e}^{\frac{-(k_x^2+k_y^2)\omega_0^2}{4}}\times\mathrm{e}^{-\mathrm{j}k_0z}\mathrm{e}^{\frac{\mathrm{j}(k_x^2+k_y^2)z}{2k_0}}\right\}\\&=\mathrm{e}^{-\mathrm{j}k_0z}\mathcal{F}^{-1}\left\{\pi\omega_0^2\mathrm{e}^{\mathrm{j}\frac{(k_x^2+k_y^2)}{2k_0}q}\right\},\end{aligned} \tag{3.5-4}$$

其中, q 为高斯光束的 q 参数 (*q-parameter of the Gaussian beam*), 由下式给出

$$q(z)=z+\mathrm{j}z_{RG}, \tag{3.5-5}$$

其中, z_{RG} 为高斯光束的瑞利范围 (*Raleigh range of the Gaussian beam*)

$$z_{RG}=\frac{k_0\omega_0^2}{2}. \tag{3.5-6}$$

式 (3.5-4) 中复高斯函数的逆变换可通过再次使用表 2.1 中的第 5 项得到

$$\psi_p(x,y;z)=\mathrm{e}^{-\mathrm{j}k_0z}\frac{\mathrm{j}k_0\omega_0^2}{2q}\mathrm{e}^{-\mathrm{j}\frac{k_0}{2q}(x^2+y^2)}. \tag{3.5-7}$$

这是用 q 参数表示的衍射高斯光束的一个重要结果.

将式 (3.5-5) 代入式 (3.5-7), 经过推导, 有

$$\psi_p(x,y;z)=\frac{\omega_0}{\omega(z)}\mathrm{e}^{-\frac{x^2+y^2}{\omega^2(z)}}\mathrm{e}^{-\frac{\mathrm{j}k_0}{2R(z)}(x^2+y^2)}\mathrm{e}^{-\mathrm{j}\phi(z)}\mathrm{e}^{-\mathrm{j}k_0z}, \tag{3.5-8a}$$

这里

$$\omega(z)=\omega_0\sqrt{\left[1+\left(\frac{z}{z_{RG}}\right)^2\right]}, \tag{3.5-8b}$$

是高斯光束在距离 z 处的束腰, 且

$$R(z)=z\left[1+\left(\frac{z_{RG}}{z}\right)^2\right], \tag{3.5-8c}$$

是在距离 z 处的高斯光束的曲率半径 (*radius of curvature of the Gaussian beam*), 最终

$$\phi(z)=-\arctan\left(\frac{z}{z_{RG}}\right). \tag{3.5-8d}$$

图 3.5-1 描述了式 (3.5-1) 所给出的具有初始轮廓的高斯光束的传播情况. 下面将参考图 3.5-1 讨论一些高斯光束的衍射特性.

where ω_0 is called the *beam waist* of the Gaussian beam as it is at this distance from the z-axis, that the amplitude of the beam falls by a factor 1/e. We can now find Fresnel diffraction of the beam using the frequency domain equation from Eq. (3.4-19):

$$\psi_p(x, y; z) = \mathcal{F}^{-1}\{\Psi_p(k_x, k_y; z)\} = \mathcal{F}^{-1}\{\Psi_{p0}(k_x, k_y)H(k_x, k_y; z)\}. \tag{3.5-2}$$

The Fourier transform of the Gaussian beam can be found by using item #5 of Table 2.1, and the result is

$$\Psi_{p0}(k_x, k_y) = \mathcal{F}\left\{\mathrm{e}^{-(x^2+y^2)/\omega_0^2}\right\} = \pi\omega_0^2 \mathrm{e}^{\frac{-(k_x^2+k_y^2)\omega_0^2}{4}}. \tag{3.5-3}$$

Using the spatial frequency transfer function in Fourier optics from Eq. (3.4-18), Eq. (3.5-2) becomes

$$\begin{aligned}\psi_p(x, y; z) &= \mathcal{F}^{-1}\left\{\pi\omega_0^2 \mathrm{e}^{\frac{-(k_x^2+k_y^2)\omega_0^2}{4}} \times \mathrm{e}^{-\mathrm{j}k_0 z}\mathrm{e}^{\frac{\mathrm{j}(k_x^2+k_y^2)z}{2k_0}}\right\} \\ &= \mathrm{e}^{-\mathrm{j}k_0 z}\mathcal{F}^{-1}\left\{\pi\omega_0^2 \mathrm{e}^{\mathrm{j}\frac{(k_x^2+k_y^2)}{2k_0}q}\right\},\end{aligned} \tag{3.5-4}$$

where q is the *q-parameter of the Gaussian beam*, given by

$$q(z) = z + \mathrm{j}z_{RG}, \tag{3.5-5}$$

with z_{RG} defined as the *Raleigh range of the Gaussian beam*:

$$z_{RG} = \frac{k_0\omega_0^2}{2}. \tag{3.5-6}$$

The inverse transform of the complex Gaussian in Eq. (3.5-4) can be found by using item #5 of Table 2.1 again to give

$$\psi_p(x, y; z) = \mathrm{e}^{-\mathrm{j}k_0 z}\frac{\mathrm{j}k_0\omega_0^2}{2q}\mathrm{e}^{-\mathrm{j}\frac{k_0}{2q}(x^2+y^2)}. \tag{3.5-7}$$

This is an important result for a diffracted Gaussian beam expressed in terms of the q-parameter.

Substituting Eq. (3.5-5) into Eq. (3.5-7), we have, after some manipulations,

$$\psi_p(x, y; z) = \frac{\omega_0}{\omega(z)}\mathrm{e}^{-\frac{x^2+y^2}{\omega^2(z)}}\mathrm{e}^{-\frac{\mathrm{j}k_0}{2R(z)}(x^2+y^2)}\mathrm{e}^{-\mathrm{j}\phi(z)}\mathrm{e}^{-\mathrm{j}k_0 z}, \tag{3.5-8a}$$

where

$$\omega(z) = \omega_0\sqrt{\left[1 + \left(\frac{z}{z_{RG}}\right)^2\right]}, \tag{3.5-8b}$$

is the waist of the Gaussian beam at a distance z, and

$$R(z) = z\left[1 + \left(\frac{z_{RG}}{z}\right)^2\right], \tag{3.5-8c}$$

is the *radius of curvature of the Gaussian beam* at a distance z, and finally

$$\phi(z) = -\arctan\left(\frac{z}{z_{RG}}\right). \tag{3.5-8d}$$

Figure 3.5-1 illustrates the propagation of a Gaussian beam with initial profile given by Eq. (3.5-1). In what follows, we will discuss some Gaussian beam properties upon diffraction with reference to Figure 3.5-1.

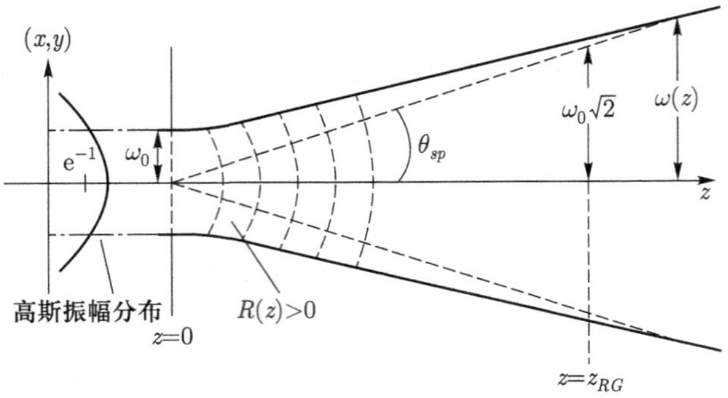

图 3.5-1 高斯光束在衍射时的扩散

光束束腰 $\omega(z)$

由式 (3.5-8b) 可以发现, 当光波在瑞利范围 z_{RG} 内衍射时, 束腰 $\omega(z)$ 位置近似恒定, 当 $z=z_{RG}$ 时, 束腰增加到 $\omega_0\sqrt{2}$, 在远场范围 $z\gg z_{RG}$ 时, 根据 $\omega(z)\approx\omega_0 z/z_{RG}$, 束腰将呈线性增长.

曲率半径 $R(z)$

由式 (3.5-8c) 可知, 曲率半径在初始时是无限大的, 即 $R(z=0)=\infty$, 这与假设高斯光束在初始 $z=0$ 处其波前为平面时的曲率半径一致. 当 $z>0$ 且 $R(z)>0$ 时, 波前发生弯曲. 因此, 我们预期其等相面 (*equiphase surface*) (即波前) 在 $z>0$ 时是曲率半径为 $R(z)$ 的球面, 即 $R(z)>0$. 的确, 当 $z\gg z_{RG}$ 且 $R(z\gg z_{RG})\approx z$ 时, 高斯光束与从点源发出的球面波前相似. 可以看出, 该符号规定与第一章讨论折射矩阵时所定义的球面曲率半径 R 相同.

发散角或扩散角 θ_{sp}

从图 3.5-1 的几何图中可以看出, 高斯光束的发散角 (*angular of divergence*) 或扩散角 (*angular of spread*) θ_{sp} 由 $\tan\theta_{sp}=\omega(z)/z$ 得到. 在远场, 即 $z\gg z_{RG}$ 时, 扩散角为

$$\theta_{sp}\approx\frac{\omega(z)}{z}=\frac{\lambda_0}{\pi\omega_0}, \tag{3.5-9}$$

其中, $\omega(z)\approx 2z/(k_0\omega_0)$, λ_0 为高斯光束在折射率为 n 的介质中传播时的波长.

古依相位 $\phi(z)$

式 (3.5-8a) 中的 $\mathrm{e}^{-\mathrm{j}\phi(z)}\mathrm{e}^{-\mathrm{j}k_0z}$ 项表示光束沿 z 轴传输时所获得的相位, 其中, $\phi(z)=-\arctan\left(\dfrac{z}{z_{RG}}\right)$ 为古依相位 (*Gouy phase*), 高斯光束产生的相位相对于平面波的相位有一个小的偏差, 在式 (3.3-8) 中曾讨论过平面波的相位由 k_0z 给出. 产生这一偏差的原因是, 高斯光束可以看作一系列在不同方向传播的平面波的叠加, 所有这些平面波的叠加使得高斯光束相对于沿光轴方向传输平面波产生了相移, 这称为古依相位.

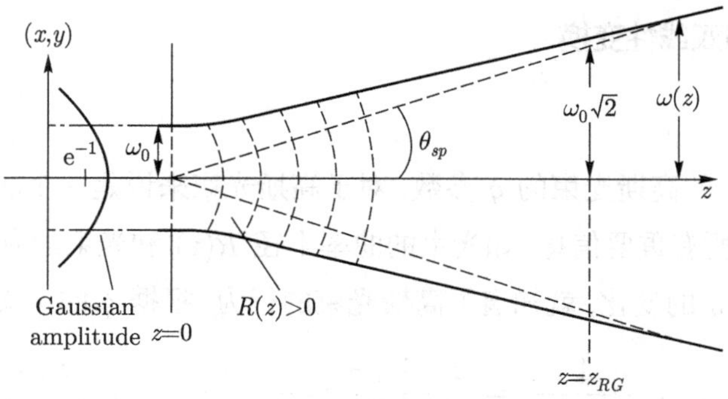

Figure 3.5-1 Spreading of a Gaussian beam upon diffraction

Beam Waist $\omega(z)$

From Eq. (3.5-8b), we observe that the beam waist $\omega(z)$ stays approximately constant when it has been diffracted well within the Raleigh range z_{RG}. At $z = z_{RG}$ the beam waist increases to $\omega_0\sqrt{2}$. In the far field when $z \gg z_{RG}$, the beam waist increases linearly according to $\omega(z) \approx \omega_0 z/z_{RG}$.

Radius of Curvature $R(z)$

From Eq. (3.5-8c), we see that the radius of curvature is initially infinite as $R(z = 0) = \infty$, which is consistent with that of the initial Gaussian beam when we assume that it is with planar wavefronts at $z = 0$. The wavefronts are curved when $z > 0$ as $R(z) > 0$. Therefore, we anticipate that the *equiphase surfaces* (i.e., wavefronts) are spherical with a radius of curvature given by $R(z)$ for $z > 0$ as $R(z) > 0$. Indeed, the Gaussian beam resembles a spherical wavefront from a point source when $z \gg z_{RG}$ as $R(z \gg z_{RG}) \approx z$. Note that the sign convention is the same as the radius of curvature defined for a spherical surface when we discussed refraction matrix in Chapter 1.

Divergence or Angular Spread, θ_{sp}

The *divergence* or the *angular spread* of the Gaussian beam, θ_{sp}, is obtained by $\tan\theta_{sp} = \omega(z)/z$, as it is found from the geometry shown in Figure 3.5-1. In the far field, that is, $z \gg z_{RG}$, the divergence can be determined by

$$\theta_{sp} \approx \frac{\omega(z)}{z} = \frac{\lambda_0}{\pi\omega_0}, \tag{3.5-9}$$

as $\omega(z) \approx 2z/(k_0\omega_0)$ and λ_0 is the wavelength in the medium of refractive index n in which the Gaussian beam propagates.

Gouy Phase $\phi(z)$

The term $e^{-j\phi(z)}e^{-jk_0 z}$ in Eq. (3.5-8a) represents the phase acquired as the beam propagates along the z-axis, where $\phi(z) = -\arctan\left(\dfrac{z}{z_{RG}}\right)$ is called the *Gouy phase*. This phase gives a small deviation from that of a plane wave, which is simply given by $k_0 z$ as we have discussed this in Eq. (3.3-8). The reason is that we have a beam, which consists of superposition of plane waves propagating in different directions. The overall phase shift is the superposition of all these plane waves, giving rise to the Gouy phase.

3.5.1 q 变换和双线性变换

q 变换

由式 (3.5-5) 引入的高斯光束的 q 参数, 对于高斯光束来说是一个非常有用的特性, 因为它包含了关于光束的所有重要信息, 如光束的曲率半径 $R(z)$ 和光束的束腰 $\omega(z)$. 一旦知道了光束通过光学系统时 q 的变化, 就知道了高斯光束的行为. 根据 q 的定义, 可以写出

$$\begin{aligned}\frac{1}{q(z)} &= \frac{1}{z+\mathrm{j}z_{RG}} = \frac{z-\mathrm{j}z_{RG}}{z^2+z_{RG}^2} = \frac{z}{z^2+z_{RG}^2} - \mathrm{j}\frac{z_{RG}}{z^2+z_{RG}^2} \\ &= \frac{1}{R(z)} - \mathrm{j}\frac{2}{k_0\omega^2(z)},\end{aligned} \tag{3.5-10}$$

其中, 用式 (3.5-8b) 和式 (3.5-8c) 来简化上述结果. 显然, q 参数包含了高斯光束 $R(z)$ 和 $\omega(z)$ 的信息. 例如, q 的实部, 即 $\mathrm{Re}[1/q]$ 等于 $1/R$, 因此, q 常被称为高斯光束的复曲率 (*complex curvature*). q 的虚部, 即 $\mathrm{Im}[1/q]$, 是对 $1/\omega^2(z)$ 的度量.

由于 q 的重要性, 接下来将研究当高斯光束在光学系统中传播时 q 是如何变化的.

双线性变换 (*bilinear transformation*)

从第一章讨论的高斯光学开始, 我们已经利用傍轴近似下的光线变换矩阵的形式描述了光学系统中光线的行为. 在本章中, 我们用菲涅耳衍射公式的形式描述傍轴近似下波的传播. 事实证明, 这两种形式的关联在于衍射公式可以写成光线变换矩阵中各元素的形式, 这提供了高斯光学和衍射理论之间的联系. 这种联系由下面的积分给出, 通常被称为柯林斯积分 (*Collins' integral*)

$$\psi_p(x,y;z) = \mathrm{e}^{-\mathrm{j}k_0 z}\frac{\mathrm{j}k_0}{2\pi B}\iint_{-\infty}^{\infty}\psi_{p0}(x',y')\mathrm{e}^{-\mathrm{j}\frac{k_0}{2B}\left[A(x'^2+y'^2)+D(x^2+y^2)-2(xx'+yy')\right]}\mathrm{d}x'\mathrm{d}y', \tag{3.5-11}$$

其中, 输入平面 x'-y' 和输出平面 x-y 根据图 3.4-2 来定义, z 为两个平面之间的距离, 参数 A, B, C, D 为第一章讨论的系统矩阵的矩阵元素.

举例: 空气中的传输矩阵

根据式 (1.1-3b), 空气中距离为 z 时的传输矩阵为

$$\mathcal{T}_z = \begin{pmatrix} A & B \\ C & D \end{pmatrix} = \begin{pmatrix} 1 & z \\ 0 & 1 \end{pmatrix}.$$

将 A, B, C, D 参数相应地代入式 (3.5-11), 有

$$\psi_p(x,y;z) = \mathrm{e}^{-\mathrm{j}k_0 z}\frac{\mathrm{j}k_0}{2\pi z}\iint_{-\infty}^{\infty}\psi_{p0}(x',y')\mathrm{e}^{-\mathrm{j}\frac{k_0}{2z}\left[(x'^2+y'^2)+(x^2+y^2)-2(xx'+yy')\right]}\mathrm{d}x'\mathrm{d}y'$$

3.5.1 q-Transformation and Bilinear Transformation

q-Transformation

The q-parameter of the Gaussian beam, introduced by Eq. (3.5-5), is a very useful quality to know for a Gaussian beam as it contains all the important information, namely its curvature $R(z)$ and its waist $\omega(z)$, about the beam. Once we know how the q transforms as the beam going through an optical system, we know how the Gaussian beam behaves. From the definition of q, we write

$$\begin{aligned}\frac{1}{q(z)} &= \frac{1}{z+\mathrm{j}z_{RG}} = \frac{z-\mathrm{j}z_{RG}}{z^2+z_{RG}^2} = \frac{z}{z^2+z_{RG}^2} - \mathrm{j}\frac{z_{RG}}{z^2+z_{RG}^2} \\ &= \frac{1}{R(z)} - \mathrm{j}\frac{2}{k_0\omega^2(z)},\end{aligned} \tag{3.5-10}$$

where we have used Eqs. (3.5-8b) and (3.5-8c) to simplify the above results. Clearly, we see that the q-parameter contains the information on $R(z)$ and $\omega(z)$ of the Gaussian beam. For example, the real part of q, that is, $\mathrm{Re}[1/q]$, is equal to $1/R$, and for this reason q is often called the *complex curvature* of a Gaussian beam. The imaginary part of q, that is, $\mathrm{Im}[1/q]$, is a measure of $1/\omega^2(z)$.

Because of the importance of q, we shall develop how q evolves as a Gaussian beam is propagating along an optical system.

Bilinear Transformation

From Gaussian optics discussed in Chapter 1, we have described the behavior of light rays in optical systems using the formalism of the ray transfer matrix under the paraxial approximation. In this chapter, we have described wave propagation using the formalism of the Fresnel diffraction formula, which is also developed under the paraxial approximation. As it turns out, the two formalisms are related in that the diffraction formula can be written in terms of the elements of the ray transfer matrix, providing a connection between Gaussian optics and diffraction theory. The connection is given by the following integral, often known as *Collins' integral*:

$$\psi_p(x,y;z) = \mathrm{e}^{-\mathrm{j}k_0 z}\frac{\mathrm{j}k_0}{2\pi B}\iint_{-\infty}^{\infty}\psi_{p0}(x',y')\mathrm{e}^{-\mathrm{j}\frac{k_0}{2B}\left[A(x'^2+y'^2)+D(x^2+y^2)-2(xx'+yy')\right]}\mathrm{d}x'\mathrm{d}y', \tag{3.5-11}$$

where the input x'-y' plane and the output x-y plane are defined according to Figure 3.4-2 with z being the distance between the two planes, and the parameters A, B, C, D are the matrix elements of the system matrix discussed in Chapter 1.

Example: Translation Matrix in Air

According to Eq. (1.1-3b), the translation matrix in air for a distance z is given by

$$\mathcal{T}_z = \begin{pmatrix} A & B \\ C & D \end{pmatrix} = \begin{pmatrix} 1 & z \\ 0 & 1 \end{pmatrix}.$$

Substituting the A, B, C, D parameters into Eq. (3.5-11) accordingly, we have

$$\psi_p(x,y;z) = \mathrm{e}^{-\mathrm{j}k_0 z}\frac{\mathrm{j}k_0}{2\pi z}\iint_{-\infty}^{\infty}\psi_{p0}(x',y')\mathrm{e}^{-\mathrm{j}\frac{k_0}{2z}\left[(x'^2+y'^2)+(x^2+y^2)-2(xx'+yy')\right]}\mathrm{d}x'\mathrm{d}y'$$

$$= \mathrm{e}^{-\mathrm{j}k_0 z} \frac{\mathrm{j}k_0}{2\pi z} \iint_{-\infty}^{\infty} \psi_{p0}(x', y') \mathrm{e}^{-\mathrm{j}\frac{k_0}{2z}\left[(x-x')^2+(y-y')^2\right]} \mathrm{d}x' \mathrm{d}y',$$

这与式 (3.4-17) 中 $\psi_{p0}(x, y)$ 产生距离为 z 的菲涅耳衍射相同.

举例: 理想透镜前放置一透过率函数 (透明片) 时的情况

参考图 3.4-6a) 中的物理情况, 输入平面在透镜前方的距离为 d_0, 透镜后的输出平面在后焦面上, 根据式 (1.1-12) 给出的过程, 得到其系统矩阵为

$$\mathcal{S} = \begin{pmatrix} A & B \\ C & D \end{pmatrix} = \begin{pmatrix} 1 & f \\ 0 & 1 \end{pmatrix} \begin{pmatrix} 1 & 0 \\ -\dfrac{1}{f} & 1 \end{pmatrix} \begin{pmatrix} 1 & d_0 \\ 0 & 1 \end{pmatrix} = \begin{pmatrix} 0 & f \\ -\dfrac{1}{f} & 1-\dfrac{d_0}{f} \end{pmatrix}.$$

将 A, B, C, D 参数代入式 (3.5-11), 相应地, 由 $z = f + d_0$, 且对于单位振幅的平面波入射, 有 $\psi_{p0}(x, y) = t(x, y)$, 则

$$\begin{aligned}
\psi_p(x, y; z = f + d_0) &= \mathrm{e}^{-\mathrm{j}k_0(f+d_0)} \frac{\mathrm{j}k_0}{2\pi f} \iint_{-\infty}^{\infty} t(x', y') \mathrm{e}^{-\mathrm{j}\frac{k_0}{2f}\left[\left(1-\frac{d_0}{f}\right)(x^2+y^2)-2(xx'+yy')\right]} \mathrm{d}x' \mathrm{d}y' \\
&= \mathrm{e}^{-\mathrm{j}k_0(f+d_0)} \frac{\mathrm{j}k_0}{2\pi f} \mathrm{e}^{-\mathrm{j}\frac{k_0}{2f}\left(1-\frac{d_0}{f}\right)(x^2+y^2)} \iint_{-\infty}^{\infty} t(x', y') \mathrm{e}^{\frac{\mathrm{j}k_0}{f}(xx'+yy')} \mathrm{d}x' \mathrm{d}y' \\
&= \mathrm{e}^{-\mathrm{j}k_0(f+d_0)} \frac{\mathrm{j}k_0}{2\pi f} \mathrm{e}^{-\mathrm{j}\frac{k_0}{2f}\left(1-\frac{d_0}{f}\right)(x^2+y^2)} \mathcal{F}\left\{t(x, y)\right\}\Big|_{k_x=\frac{k_0 x}{f},\, k_y=\frac{k_0 y}{f}},
\end{aligned}$$

这与式 (3.4-29) 相同. 可以看出, 利用柯林斯积分和系统矩阵的知识, 可以很容易地得到沿光学系统的复振幅.

既然我们已经讨论过柯林斯积分, 这里就用它来推导高斯光束通过以系统矩阵为特征的光学系统的双线性变换 (*bilinear transformation*). 双线性变换是变量的数学映射 (*mathematical mapping*), 在我们的例子中, 所讨论的变量是 q 参数.

假设在 $z = 0$ 处有一个初始高斯光束 $\mathrm{e}^{-(x^2+y^2)/\omega_0^2}$, 由式 (3.5-5) 可知, $q(z) = z + \mathrm{j}z_{RG}$. 因此, 在输入平面 $z = 0$ 处, 有 $q(z = 0) = q_0 = \mathrm{j}z_{RG} = \mathrm{j}\dfrac{k_0\omega_0^2}{2}$, 并且, $\psi_p(x, y; z = 0) = \psi_{p0}(x, y) = \mathrm{e}^{-(x^2+y^2)/\omega_0^2} = \mathrm{e}^{-\mathrm{j}\frac{k_0}{2q_0}(x^2+y^2)}$ 可以用 q 参数来表示.

应用柯林斯积分, z 是输入平面和输出平面之间的距离, 写出

$$\begin{aligned}
\psi_p(x, y; z) &= \mathrm{e}^{-\mathrm{j}k_0 z} \frac{\mathrm{j}k_0}{2\pi B} \iint_{-\infty}^{\infty} \psi_{p0}(x', y') \mathrm{e}^{-\mathrm{j}\frac{k_0}{2B}\left[A(x'^2+y'^2)+D(x^2+y^2)-2(xx'+yy')\right]} \mathrm{d}x' \mathrm{d}y' \\
&= \mathrm{e}^{-\mathrm{j}k_0 z} \frac{\mathrm{j}k_0}{2\pi B} \iint_{-\infty}^{\infty} \mathrm{e}^{-\mathrm{j}\frac{k_0}{2q_0}(x'^2+y'^2)} \mathrm{e}^{-\mathrm{j}\frac{k_0}{2B}\left[A(x'^2+y'^2)+D(x^2+y^2)-2(xx'+yy')\right]} \mathrm{d}x' \mathrm{d}y' \\
&= \mathrm{e}^{-\mathrm{j}k_0 z} \frac{\mathrm{j}k_0}{2\pi B} \mathrm{e}^{-\mathrm{j}\frac{k_0}{2B}\left[D(x^2+y^2)\right]} \iint_{-\infty}^{\infty} \mathrm{e}^{-\mathrm{j}\frac{k_0}{2q_0}(x'^2+y'^2)} \mathrm{e}^{-\mathrm{j}\frac{k_0}{2B}\left[A(x'^2+y'^2)\right]} \mathrm{e}^{\frac{\mathrm{j}k_0}{B}(xx'+yy')} \mathrm{d}x' \mathrm{d}y'
\end{aligned}$$

$$= \mathrm{e}^{-\mathrm{j}k_0 z}\frac{\mathrm{j}k_0}{2\pi z}\iint_{-\infty}^{\infty}\psi_{p0}(x',y')\mathrm{e}^{-\mathrm{j}\frac{k_0}{2z}\left[(x-x')^2+(y-y')^2\right]}\mathrm{d}x'\mathrm{d}y',$$

which is identical to Eq. (3.4-17) for $\psi_{p0}(x, y)$ undergoing Fresnel diffraction for a distance z.

Example: Input Transparency Function in Front of the Ideal Lens

With reference to the physical situation in Figure 3.4-6a, from the input plane, which is a distance d_0 in front of the lens to the output plane on the back focal plane, we have the system matrix as, according to the process given by Eq. (1.1-12),

$$\mathcal{S} = \begin{pmatrix} A & B \\ C & D \end{pmatrix} = \begin{pmatrix} 1 & f \\ 0 & 1 \end{pmatrix}\begin{pmatrix} 1 & 0 \\ -\dfrac{1}{f} & 1 \end{pmatrix}\begin{pmatrix} 1 & d_0 \\ 0 & 1 \end{pmatrix} = \begin{pmatrix} 0 & f \\ -\dfrac{1}{f} & 1-\dfrac{d_0}{f} \end{pmatrix}.$$

Substituting the A, B, C, D parameters into Eq. (3.5-11) accordingly with $z = f + d_0$ and $\psi_{p0}(x, y) = t(x, y)$ for unit amplitude plane wave incidence, we have

$$\psi_p(x,y;z=f+d_0) = \mathrm{e}^{-\mathrm{j}k_0(f+d_0)}\frac{\mathrm{j}k_0}{2\pi f}\iint_{-\infty}^{\infty}t(x',y')\mathrm{e}^{-\mathrm{j}\frac{k_0}{2f}\left[\left(1-\frac{d_0}{f}\right)(x^2+y^2)-2(xx'+yy')\right]}\mathrm{d}x'\mathrm{d}y'$$

$$= \mathrm{e}^{-\mathrm{j}k_0(f+d_0)}\frac{\mathrm{j}k_0}{2\pi f}\mathrm{e}^{-\mathrm{j}\frac{k_0}{2f}\left(1-\frac{d_0}{f}\right)(x^2+y^2)}\iint_{-\infty}^{\infty}t(x',y')\mathrm{e}^{\frac{\mathrm{j}k_0}{f}(xx'+yy')}\mathrm{d}x'\mathrm{d}y'$$

$$= \mathrm{e}^{-\mathrm{j}k_0(f+d_0)}\frac{\mathrm{j}k_0}{2\pi f}\mathrm{e}^{-\mathrm{j}\frac{k_0}{2f}\left(1-\frac{d_0}{f}\right)(x^2+y^2)}\mathcal{F}\left\{t(x,y)\right\}\Big|_{k_x=\frac{k_0x}{f},\,k_y=\frac{k_0y}{f}},$$

which is identical to Eq. (3.4-29). As we can see, using Collins' integral along with the knowledge of the system matrix, we can obtain the complex field along the optical system easily.

Now that we have discussed Collins' integral, we use it to derive the *bilinear transformation* of a Gaussian beam propagating through an optical system characterized by its system matrix. The bilinear transformation is a *mathematical mapping* of variables and in our case, the variable in question is the q-parameter.

We assume we have an initial Gaussian beam $\mathrm{e}^{-(x^2+y^2)/\omega_0^2}$ at $z = 0$. From Eq. (3.5-5), $q(z) = z + \mathrm{j}z_{RG}$. Therefore, at input plane $z = 0$, we have $q(z=0) = q_0 = \mathrm{j}z_{RG} = \mathrm{j}\dfrac{k_0\omega_0^2}{2}$, and we can express $\psi_p(x, y; z = 0) = \psi_{p0}(x, y) = \mathrm{e}^{-(x^2+y^2)/\omega_0^2} = \mathrm{e}^{-\mathrm{j}\frac{k_0}{2q_0}(x^2+y^2)}$ in terms of the q-parameter.

Applying Collins' integral with z being the distance between the input plane and output plane, we write

$$\psi_p(x,y;z) = \mathrm{e}^{-\mathrm{j}k_0 z}\frac{\mathrm{j}k_0}{2\pi B}\iint_{-\infty}^{\infty}\psi_{p0}(x',y')\mathrm{e}^{-\mathrm{j}\frac{k_0}{2B}\left[A(x'^2+y'^2)+D(x^2+y^2)-2(xx'+yy')\right]}\mathrm{d}x'\mathrm{d}y'$$

$$= \mathrm{e}^{-\mathrm{j}k_0 z}\frac{\mathrm{j}k_0}{2\pi B}\iint_{-\infty}^{\infty}\mathrm{e}^{-\mathrm{j}\frac{k_0}{2q_0}(x'^2+y'^2)}\mathrm{e}^{-\mathrm{j}\frac{k_0}{2B}\left[A(x'^2+y'^2)+D(x^2+y^2)-2(xx'+yy')\right]}\mathrm{d}x'\mathrm{d}y'$$

$$= \mathrm{e}^{-\mathrm{j}k_0 z}\frac{\mathrm{j}k_0}{2\pi B}\mathrm{e}^{-\mathrm{j}\frac{k_0}{2B}\left[D(x^2+y^2)\right]}\iint_{-\infty}^{\infty}\mathrm{e}^{-\mathrm{j}\frac{k_0}{2q_0}(x'^2+y'^2)}\mathrm{e}^{-\mathrm{j}\frac{k_0}{2B}\left[A(x'^2+y'^2)\right]}\mathrm{e}^{\frac{\mathrm{j}k_0}{B}(xx'+yy')}\mathrm{d}x'\mathrm{d}y'$$

$$= \mathrm{e}^{-\mathrm{j}k_0 z}\frac{\mathrm{j}k_0}{2\pi B}\mathrm{e}^{-\frac{\mathrm{j}k_0}{2B}\left[D(x^2+y^2)\right]}\iint\limits_{-\infty}^{\infty}\mathrm{e}^{-\mathrm{j}\left(\frac{k_0}{2q_0}+\frac{k_0 A}{2B}\right)(x'^2+y'^2)}\mathrm{e}^{\frac{\mathrm{j}k_0}{B}(xx'+yy')}\mathrm{d}x'\mathrm{d}y',$$

根据傅里叶变换可以写成

$$\psi_p(x,y;z)=\mathrm{e}^{-\mathrm{j}k_0 z}\frac{\mathrm{j}k_0}{2\pi B}\mathrm{e}^{-\mathrm{j}\frac{k_0}{2B}\left[D(x^2+y^2)\right]}\mathcal{F}\left\{\mathrm{e}^{-\mathrm{j}k_0\left(\frac{1}{2q_0}+\frac{A}{2B}\right)(x^2+y^2)}\right\}\Bigg|_{k_x=\frac{k_0 x}{B},k_y=\frac{k_0 y}{B}}.$$

复高斯函数可以进行傅里叶变换, 根据表 2.1 给出

$$\psi_p(x,y;z)=\mathrm{e}^{-\mathrm{j}k_0 z}\frac{q_0}{Aq_0+B}\mathrm{e}^{-\mathrm{j}k_0\frac{Cq_0+D}{2(Aq_0+B)}(x^2+y^2)}=\mathrm{e}^{-\mathrm{j}k_0 z}\frac{q_0}{Aq_0+B}\mathrm{e}^{-\mathrm{j}\frac{k_0}{2q_1}(x^2+y^2)}. \quad (3.5\text{-}12\mathrm{a})$$

可以发现, $\mathrm{e}^{-\mathrm{j}\frac{k_0}{2q_1}(x^2+y^2)}$ 为一高斯光束, 其中

$$q_1=\frac{Aq_0+B}{Cq_0+D}. \quad (3.5\text{-}12\mathrm{b})$$

这被称为 q 参数的双线性变换 (*bilinear transformation*), 因为对于给定的系统矩阵, 它将 q_0 与 q_1 联系了起来.

3.5.2 双线性变换的应用实例

高斯光束在空气 (自由空间) 中传播距离 z 的情况

空气中距离 z 的传输矩阵表达式为

$$\mathcal{T}_z=\begin{pmatrix}A & B\\ C & D\end{pmatrix}=\begin{pmatrix}1 & z\\ 0 & 1\end{pmatrix}.$$

假定 $\psi_p(x,y;z=0)=\mathrm{e}^{-\mathrm{j}\frac{k_0}{2q_0}(x^2+y^2)}$, 根据式 (3.5-12b) 及 $\mathcal{T}_z$, 有

$$q_1(z)=z+q_0, \quad (3.5\text{-}13)$$

这与式 (3.5-5) 一致. 式 (3.5-13) 描述高斯光束的 q 参数在空气 (自由空间) 中的传输规律 (*translation law for a Gaussian beam*). 其复振幅可由式 (3.5-12a) 给出 (对于 $A=1$, $B=z$)

$$\psi_p(x,y;z)=\mathrm{e}^{-\mathrm{j}k_0 z}\frac{q_0}{q_0+z}\mathrm{e}^{-\mathrm{j}\frac{k_0(x^2+y^2)}{2q_1}}=\mathrm{e}^{-\mathrm{j}k_0 z}\frac{q_0}{q_1}\mathrm{e}^{-\mathrm{j}\frac{k_0(x^2+y^2)}{2q_1}}=\mathrm{e}^{-\mathrm{j}k_0 z}\frac{\mathrm{j}k_0\omega_0^2}{2q}\mathrm{e}^{-\mathrm{j}\frac{k_0}{2q}(x^2+y^2)},$$

这与式 (3.5-7) 中的结果相同, 因为 q_1 为 q 的输出.

高斯光束的透镜成像定律

为了分析焦距为 f 的理想透镜对高斯光束的影响, 我们回忆一下薄透镜矩阵 $\mathcal{L}_f$ [见式 (1.1-13)]

$$\mathcal{L}_f=\begin{pmatrix}1 & 0\\ -\dfrac{1}{f} & 1\end{pmatrix}.$$

利用式 (3.5-12b) 的双线性变换及 $\mathcal{L}_f$, 得

$$q_1=\frac{q_0}{-\dfrac{1}{f}q_0+1}=\frac{fq_0}{f-q_0},$$

或

$$\frac{1}{q_1}=\frac{1}{q_0}-\frac{1}{f}. \quad (3.5\text{-}14)$$

这是高斯光束的透镜成像定律 (*lens law for a Gaussian beam*), 它将透镜前后的 q 参数联系了起来.

$$= e^{-jk_0 z}\frac{jk_0}{2\pi B}e^{-\frac{jk_0}{2B}[D(x^2+y^2)]}\iint_{-\infty}^{\infty} e^{-j\left(\frac{k_0}{2q_0}+\frac{k_0 A}{2B}\right)(x'^2+y'^2)}e^{\frac{jk_0}{B}(xx'+yy')}dx'dy',$$

which can be written in terms of the Fourier transform as

$$\psi_p(x,y;z)=e^{-jk_0 z}\frac{jk_0}{2\pi B}e^{-j\frac{k_0}{2B}[D(x^2+y^2)]}\mathcal{F}\left\{e^{-jk_0\left(\frac{1}{2q_0}+\frac{A}{2B}\right)(x^2+y^2)}\right\}\Bigg|_{k_x=\frac{k_0 x}{B},k_y=\frac{k_0 y}{B}}.$$

The complex Gaussian function can be Fourier transformed, according to Table 2.1, to give

$$\psi_p(x,y;z) = e^{-jk_0 z}\frac{q_0}{Aq_0+B}e^{-jk_0\frac{Cq_0+D}{2(Aq_0+B)}(x^2+y^2)} = e^{-jk_0 z}\frac{q_0}{Aq_0+B}e^{-j\frac{k_0}{2q_1}(x^2+y^2)}. \tag{3.5-12a}$$

where we recognize that the term $e^{-j\frac{k_0}{2q_1}(x^2+y^2)}$ is a Gaussian beam with

$$q_1 = \frac{Aq_0+B}{Cq_0+D}. \tag{3.5-12b}$$

This is called the *bilinear transformation* for the q-parameter as it relates q_0 to q_1 for a given system matrix.

3.5.2 Examples on the Use of the Bilinear Transformation

Propagation of a Gaussian Beam with a Distance z in Air

The translation matrix in air for a distance z is given by

$$\mathcal{T}_z = \begin{pmatrix} A & B \\ C & D \end{pmatrix} = \begin{pmatrix} 1 & z \\ 0 & 1 \end{pmatrix}.$$

Assume $\psi_p(x,y;z=0) = e^{-j\frac{k_0}{2q_0}(x^2+y^2)}$. According to Eq. (3.5-12b) along with $\mathcal{T}_z$,

$$q_1(z) = z + q_0, \tag{3.5-13}$$

which is consistent with Eq. (3.5-5). Equation (3.5-13) is the *translation law for a Gaussian beam* propagating in air. Now, the complex field is given by Eq. (3.5-12a) (for $A=1$, $B=z$):

$$\psi_p(x,y;z) = e^{-jk_0 z}\frac{q_0}{q_0+z}e^{-j\frac{k_0}{2q_1}(x^2+y^2)}$$

$$= e^{-jk_0 z}\frac{q_0}{q_1}e^{-j\frac{k_0}{2q_1}(x^2+y^2)} = e^{-jk_0 z}\frac{jk_0 w_0^2}{2q}e^{-j\frac{k_0}{2q}(x^2+y^2)},$$

which is the same result as in Eq. (3.5-7) as q_1 is the output q.

Lens Law for a Gaussian Beam

To analyze the effects of an ideal lens of focal length f on a Gaussian beam, we recall the thin-lens matrix, $\mathcal{L}_f$ [see Eq. (1.1-13)]:

$$\mathcal{L}_f = \begin{pmatrix} 1 & 0 \\ -\frac{1}{f} & 1 \end{pmatrix}.$$

Using the bilinear transformation from Eq. (3.5-12b) along with $\mathcal{L}_f$, we obtain

$$q_1 = \frac{q_0}{-\frac{1}{f}q_0+1} = \frac{fq_0}{f-q_0},$$

or

$$\frac{1}{q_1} = \frac{1}{q_0} - \frac{1}{f}. \tag{3.5-14}$$

This is the *lens law for a Gaussian beam*, which relates the qs before and after the lens.

高斯光束的聚焦

考虑一高斯光束通过由两个变换矩阵组成的光学系统的传输, 例如高斯光束通过透镜的聚焦. 这一般会涉及两个矩阵, 如图 3.5-2 所示, 将输入光束参数 q_1 与输出光束参数 q_3 联系起来.

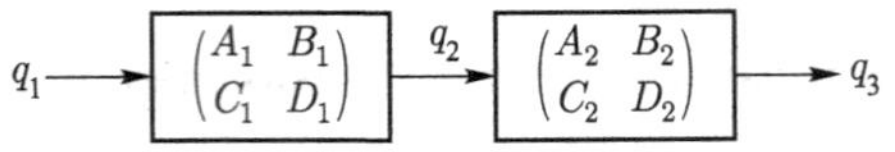

图 3.5-2 涉及两个或多个变换矩阵的 q 变换

利用式 (3.5-12b), 可以把不同 q 之间的关系写成

$$q_2 = \frac{A_1 q_1 + B_1}{C_1 q_1 + D_1},$$

和

$$q_3 = \frac{A_2 q_2 + B_2}{C_2 q_2 + D_2}.$$

结合这两个等式, 有

$$\begin{aligned} q_3 &= \frac{A_2 \dfrac{A_1 q_1 + B_1}{C_1 q_1 + D_1} + B_2}{C_2 \dfrac{A_1 q_1 + B_1}{C_1 q_1 + D_1} + D_2} = \frac{A_2(A_1 q_1 + B_1) + B_2(C_1 q_1 + D_1)}{C_2(A_1 q_1 + B_1) + D_2(C_1 q_1 + D_1)} \\ &= \frac{(A_2 A_1 + B_2 C_1) q_1 + A_2 B_1 + B_2 D_1}{(C_2 A_1 + D_2 C_1) q_1 + C_2 B_1 + D_2 D_1} = \frac{A_T q_1 + B_T}{C_T q_1 + D_T}, \end{aligned} \tag{3.5-15}$$

其中, A_T, B_T, C_T, D_T 是关联输出平面上的 q_3 和输入平面上的 q_1 的系统矩阵的元素, 可由下式给出

$$\begin{pmatrix} A_T & B_T \\ C_T & D_T \end{pmatrix} = \begin{pmatrix} A_2 & B_2 \\ C_2 & D_2 \end{pmatrix} \begin{pmatrix} A_1 & B_1 \\ C_1 & D_1 \end{pmatrix}.$$

现在考虑 $z = 0$ 处为平面波前的高斯光束, 即束腰为 ω_0 时的高斯光束入射到焦距为 f 的理想透镜上, 如图 3.5-3 所示. 我们将在透镜后找到输出光束的束腰位置. 其输入平面紧贴着透镜的左侧, 而输出平面位于透镜 z 距离远处. 根据式 (3.5-10), 有

$$\frac{1}{q(z)} = \frac{1}{R(z)} - \mathrm{j}\frac{2}{k_0 \omega^2(z)}.$$

注意, 当 $R(z=0)=\infty$ 时, 输入光束的最小束腰为 ω_0. 因此, 输入 q 为 $q_0 = \left[-\mathrm{j}\dfrac{2}{k_0\omega^2(z=0)}\right]^{-1} = \dfrac{\mathrm{j}k_0\omega_0^2}{2} = \mathrm{j}z_{RG}$, 且它为纯虚数.

根据式 (3.5-15), 输出 q 为

$$q_1 = \frac{A_T q_0 + B_T}{C_T q_0 + D_T},$$

这里

$$\begin{pmatrix} A_T & B_T \\ C_T & D_T \end{pmatrix} = \mathcal{T}_z \mathcal{L}_f = \begin{pmatrix} 1 & z \\ 0 & 1 \end{pmatrix} \begin{pmatrix} 1 & 0 \\ -\dfrac{1}{f} & 1 \end{pmatrix} = \begin{pmatrix} 1 - \dfrac{z}{f} & z \\ -\dfrac{1}{f} & 1 \end{pmatrix}.$$

Gaussian Beam Focusing

We consider the propagation of a Gaussian beam through an optical system that consists of two transfer matrices, such as the focusing of a Gaussian beam through a lens. The general situation involving two matrices is illustrated in Figure 3.5-2. Let us relate the input beam parameter, q_1, to the output beam parameter, q_3.

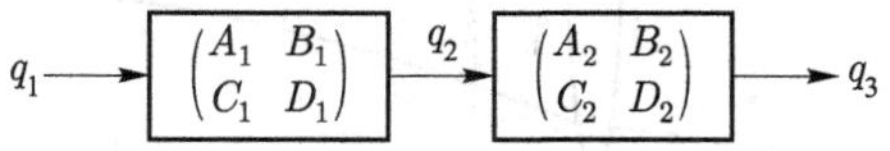

Figure 3.5-2 q-transformation involving more than one transfer matrix

Using Eq. (3.5-12b), we can write the relations between the various qs as follows:

$$q_2 = \frac{A_1 q_1 + B_1}{C_1 q_1 + D_1} \text{ and } q_3 = \frac{A_2 q_2 + B_2}{C_2 q_2 + D_2}.$$

Combining the two equations, we have

$$\begin{aligned} q_3 &= \frac{A_2 \dfrac{A_1 q_1 + B_1}{C_1 q_1 + D_1} + B_2}{C_2 \dfrac{A_1 q_1 + B_1}{C_1 q_1 + D_1} + D_2} = \frac{A_2(A_1 q_1 + B_1) + B_2(C_1 q_1 + D_1)}{C_2(A_1 q_1 + B_1) + D_2(C_1 q_1 + D_1)} \\ &= \frac{(A_2 A_1 + B_2 C_1) q_1 + A_2 B_1 + B_2 D_1}{(C_2 A_1 + D_2 C_1) q_1 + C_2 B_1 + D_2 D_1} = \frac{A_T q_1 + B_T}{C_T q_1 + D_T}, \end{aligned} \tag{3.5-15}$$

where A_T, B_T, C_T, D_T are the elements of the system matrix relating q_3 on the output plane to q_1 on the input plane, which is given conveniently by

$$\begin{pmatrix} A_T & B_T \\ C_T & D_T \end{pmatrix} = \begin{pmatrix} A_2 & B_2 \\ C_2 & D_2 \end{pmatrix} \begin{pmatrix} A_1 & B_1 \\ C_1 & D_1 \end{pmatrix}.$$

Let us now consider a Gaussian beam of planar wavefronts at $z = 0$, that is incident at its waist ω_0 on an ideal lens of focal length f, as shown in Figure 3.5-3. We will find the location of the waist of the output beam after the lens. The input plane is the immediate left of the lens and the output plane is at a distance z away from the lens. With reference to Eq. (3.5-10), we have

$$\frac{1}{q(z)} = \frac{1}{R(z)} - \mathrm{j}\frac{2}{k_0 \omega^2(z)}.$$

Note that the minimum waist of the input beam, ω_0, is when $R(z = 0) = \infty$. Therefore, the input q is $q_0 = \left[-\mathrm{j}\dfrac{2}{k_0 \omega^2(z=0)}\right]^{-1} = \dfrac{\mathrm{j} k_0 \omega_0^2}{2} = \mathrm{j} z_{RG}$ and it is purely imaginary.

The output q, according to Eq. (3.5-15), is

$$q_1 = \frac{A_T q_0 + B_T}{C_T q_0 + D_T},$$

where

$$\begin{pmatrix} A_T & B_T \\ C_T & D_T \end{pmatrix} = \mathcal{T}_z \mathcal{L}_f = \begin{pmatrix} 1 & z \\ 0 & 1 \end{pmatrix} \begin{pmatrix} 1 & 0 \\ -\dfrac{1}{f} & 1 \end{pmatrix} = \begin{pmatrix} 1 - \dfrac{z}{f} & z \\ -\dfrac{1}{f} & 1 \end{pmatrix}.$$

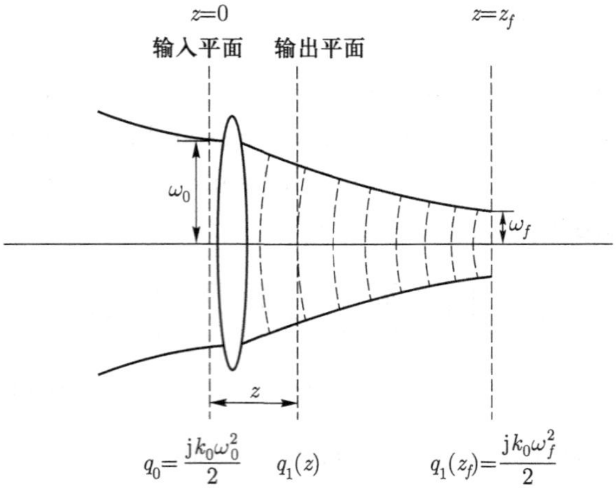

图 3.5-3 高斯光束的聚焦

因此

$$q_1(z)=\frac{\left(1-\frac{z}{f}\right)q_0+z}{-\frac{1}{f}q_0+1}=\frac{fq_0}{f-q_0}+z.$$

当 q 变成纯虚数时, 即当光束具有平面波前时, 该高斯光束被定义在 $z=z_f$ 处聚焦 (最小束腰). 设 ω_f 为聚焦时的束腰, 因此, 可以写出

$$q_1(z_f)=\frac{fq_0}{f-q_0}+z_f=\frac{\mathrm{j}k_0\omega_f^2}{2},$$

将 $q_0=\mathrm{j}z_{RG}$ 代入上式, 经过一些运算, 得

$$\left(z_f-\frac{fz_{RG}^2}{f^2+z_{RG}^2}\right)+\mathrm{j}\frac{f^2z_{RG}}{f^2+z_{RG}^2}=\frac{\mathrm{j}k_0\omega_f^2}{2}. \tag{3.5-16}$$

由上式实部的等式, 有

$$z_f-\frac{fz_{RG}^2}{f^2+z_{RG}^2}=0,$$

或者同样地

$$z_f=\frac{fz_{RG}^2}{f^2+z_{RG}^2}=\frac{f}{1+\left(\frac{f}{z_{RG}}\right)^2}\leqslant f. \tag{3.5-17}$$

由于 $z_{RG}=\dfrac{k_0\omega_0^2}{2}$ 永远是正数, 故 $z_f\leqslant f$. 换句话说, 高斯光束并不精确地聚焦在透镜的几何后焦点处. 相反, 焦点会移到靠近透镜的位置. 这种现象被称为焦移 (*focal shift*) [Poon (1988)], 这点在高功率激光应用中极为重要, 尤其是对精确聚焦至关重要. 可以看出, 对于平面波入射, 即 $\omega_0\to\infty$, $z_{RG}\gg f$, 且 $z_f=f$.

现在, 写出式 (3.5-16) 虚部的等式, 有

$$\frac{f^2z_{RG}}{f^2+z_{RG}^2}=\frac{k_0\omega_f^2}{2}.$$

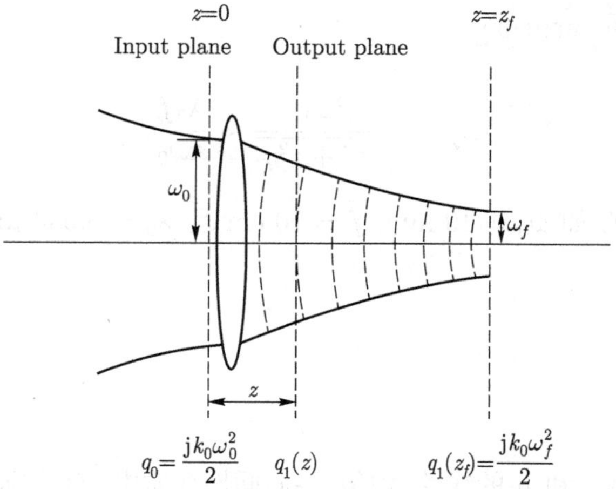

Figure 3.5-3 Focusing of a Gaussian beam

Therefore,

$$q_1(z) = \frac{\left(1-\frac{z}{f}\right)q_0 + z}{-\frac{1}{f}q_0 + 1} = \frac{fq_0}{f-q_0} + z.$$

The Gaussian beam is defined to be at focus (minimum waist) at $z = z_f$ when its q becomes purely imaginary, that is, when the beam having planar wavefronts. Let ω_f be the waist at focus; we, therefore, can write

$$q_1(z_f) = \frac{fq_0}{f-q_0} + z_f = \frac{\mathrm{j}k_0\omega_f^2}{2},$$

Substituting $q_0 = \mathrm{j}z_{RG}$ into the above equation, and after some manipulations, we obtain

$$\left(z_f - \frac{fz_{RG}^2}{f^2+z_{RG}^2}\right) + \mathrm{j}\frac{f^2 z_{RG}}{f^2+z_{RG}^2} = \frac{\mathrm{j}k_0\omega_f^2}{2}. \tag{3.5-16}$$

Equating the real part of Eq. (3.5-16), we have

$$z_f - \frac{fz_{RG}^2}{f^2+z_{RG}^2} = 0,$$

or equivalently

$$z_f = \frac{fz_{RG}^2}{f^2+z_{RG}^2} = \frac{f}{1+\left(\frac{f}{z_{RG}}\right)^2} \leqslant f. \tag{3.5-17}$$

Since $z_{RG} = \dfrac{k_0\omega_0^2}{2}$ is always a positive number, $z_f \leqslant f$. In other words, the Gaussian beam does not focus exactly at the geometrical back focus of the lens. Instead, the focus is shifted close to the lens. This phenomenon is called the *focal shift* [Poon (1988)] and it is crucial for high laser power applications in which precise focus is of paramount importance. Note that for a plane wave incidence, that is, $\omega_0 \to \infty$, $z_{RG} \gg f$, and $z_f = f$.

Now, equating the imaginary part of Eq. (3.5-16), we have

$$\frac{f^2 z_{RG}}{f^2+z_{RG}^2} = \frac{k_0\omega_f^2}{2}.$$

在分母上有 $z_{RG}=\dfrac{k_0\omega_0^2}{2}$, 可以求出

$$\omega_f=\frac{f\omega_0}{\sqrt{f^2+z_{RG}^2}}\approx\frac{\lambda_0 f}{\pi\omega_0}. \tag{3.5-18}$$

当 ω_0 值大时, $z_{RG}\gg f$, 如 $\omega_0=10\ \text{mm}$, $f=10\ \text{cm}$ 且 $\lambda_0=0.633\ \mu\text{m}$ (红光) 时, 聚焦光斑尺寸 $\omega_f\approx 2\ \mu\text{m}$.

习题

3.1 证明在特性为 ε 和 μ 的线性、均匀、各向同性介质中, $\boldsymbol{H}$ 的波动方程为

$$\nabla^2\boldsymbol{H}-\mu\varepsilon\frac{\partial^2\boldsymbol{H}}{\partial t^2}=-\nabla\times\boldsymbol{J_C}.$$

3.2 证明三维标量波动方程

$$\nabla^2\psi=\mu\varepsilon\frac{\partial^2\psi}{\partial t^2}=\frac{1}{v^2}\frac{\partial^2\psi}{\partial t^2},$$

是一个线性微分方程. 如果一个微分方程是线性的, 则叠加原理成立, 这为干涉和衍射提供了基础.

3.3 确定下列哪个函数描述了行波:

a) $\psi(z,t)=\sin[(3z-4t)(3z+4z)]$.

b) $\psi(z,t)=\cos^3\left[\left(\dfrac{z}{3}+\dfrac{t}{b}\right)^2\right]$.

c) $\psi(z,t)=\tan(3t-4z)$.

同时, 求出这些行波的相速度和方向 ($+z$ 或 $-z$ 方向).

3.4 利用非涅耳衍射公式, 求:

a) 距离点源为 z 处的复振幅 [提示: 可以将点源建模为 $\delta(x,y)$].

b) 距离平面波 z 处的复振幅 (提示: 可利用傅里叶变换来求解).

3.5 利用习题 3.4a) 的结果, 证明当点物成点像时, 对应焦距为 f 的理想透镜的透过率函数为 $t_f(x,y)=\mathrm{e}^{\mathrm{j}\frac{k_0}{2f}(x^2+y^2)}$, 如图题 3.5 所示.

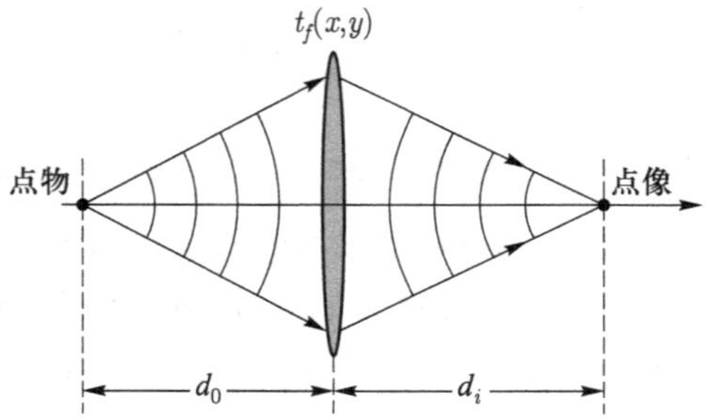

图题 3.5

3.6 单位振幅平面波照射到 $t(x,y)=circ\left(\dfrac{r}{r_0}\right)$ 的圆形孔径上, 证明其扩散角为

With $z_{RG} = \frac{k_0\omega_0^2}{2}$ in the nominator, we solve for ω_f:

$$\omega_f = \frac{f\omega_0}{\sqrt{f^2 + z_{RG}^2}} \approx \frac{\lambda_0 f}{\pi\omega_0}. \tag{3.5-18}$$

for a large ω_0 as $z_{RG} \gg f$. For $\omega_0 = 10$ mm, $f = 10$ cm, and $\lambda_0 = 0.633$ μm (red light), the focused spot size $\omega_f \approx 2$ μm.

Problems

3.1 Show that the wave equation for $\boldsymbol{H}$ in a linear, homogeneous, and isotopic medium characterized by ε and μ is given by

$$\boldsymbol{\nabla}^2\boldsymbol{H} - \mu\varepsilon\frac{\partial^2\boldsymbol{H}}{\partial t^2} = -\boldsymbol{\nabla}\times\boldsymbol{J}_C.$$

3.2 Show that the 3-D scalar wave equation

$$\boldsymbol{\nabla}^2\psi = \mu\varepsilon\frac{\partial^2\psi}{\partial t^2} = \frac{1}{v^2}\frac{\partial^2\psi}{\partial t^2},$$

is a linear differential equation. If a differential equation is linear, the principle of superposition holds, which provides the basis for interference and diffraction.

3.3 Determine which of the following functions describe travelling waves:

(a) $\psi(z,t) = \sin[(3z - 4t)(3z + 4z)]$.

(b) $\psi(z,t) = \cos^3\left[\left(\frac{z}{3} + \frac{t}{b}\right)^2\right]$.

(c) $\psi(z,t) = \tan(3t - 4z)$.

Also, find the phase velocity and direction ($+z$ or $-z$ direction) for those travelling waves.

3.4 Employ the Fresnel diffraction formula,

(a) find the complex field at a distance z away from a point source [Hint: you can model a point source by $\delta(x,y)$], and

(b) find the complex field at a distance z away from a plane wave. (Hint: You might want to take advantage of the Fourier transform to find your final answer.)

3.5 Using the result of P.3.4(a), show that the transparency function of the ideal lens of focal length f is given by $t_f(x,y) = \mathrm{e}^{\mathrm{j}\frac{k_0}{2f}(x^2+y^2)}$ when a point object is imaged to give an image point, as shown in Figure P.3.5.

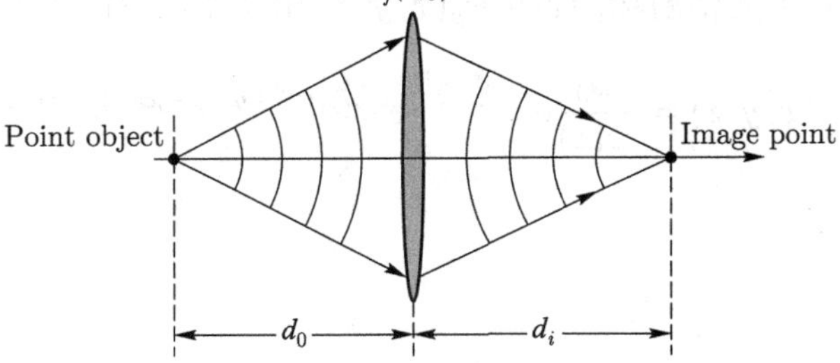

Figure P.3.5

3.6 For a plane wave of unit amplitude incident on a circular aperture $t(x,y) = circ\left(\frac{r}{r_0}\right)$,

$$\theta_{sp}=1.22\frac{\lambda_0}{2r_0}.$$

3.7 参考图题 3.7, 利用柯林斯积分, 证明焦面上的复振幅由式 (3.4-29) 给出, 即

$$\psi_p(x,y;z=f+d_0)=\mathrm{e}^{-\mathrm{j}k_0(f+d_0)}\frac{\mathrm{j}k_0}{2\pi f}A\mathrm{e}^{-\mathrm{j}\frac{k_0}{2f}\left(1-\frac{d_0}{f}\right)(x^2+y^2)}\mathcal{F}\left\{t(x,y)\right\}\Big|_{k_x=\frac{k_0x}{f},\,k_y=\frac{k_0y}{f}}.$$

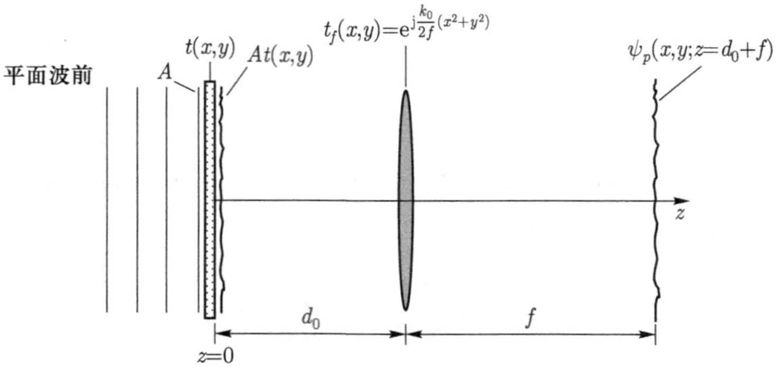

图题 3.7

3.8 图题 3.8 为一 4-f 成像系统, 输入 $t(x,y)$, 被一振幅为 A 的平面波照射.

a) 对于像面上的复振幅, 写出类似于式 (3.4-27) 含有卷积和乘法的表达式.

b) 根据 a) 的结果, 画出光学系统的框图.

c) 证明在像面上, 复振幅分布与 $t(-x,-y)$ 成正比.

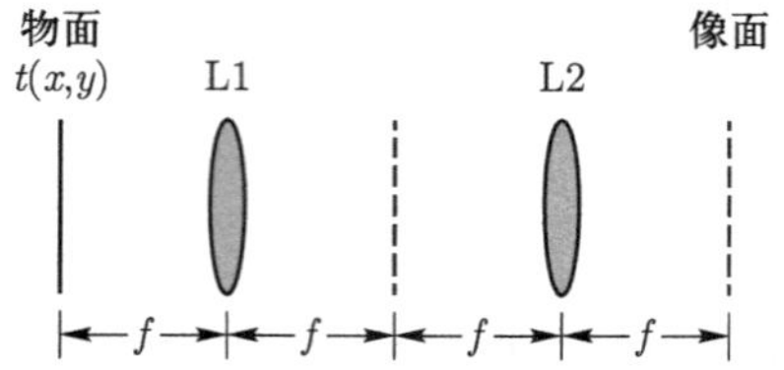

图题 3.8

3.9 对于初始轮廓 (*initial profile*) 由下式给出的高斯光束

$$\psi_p(x,y;z=0)=\psi_{p0}(x,y)=\mathrm{e}^{-\mathrm{j}\frac{k_0}{2q_0}(x^2+y^2)},\quad q_0=\mathrm{j}z_{RG}=\mathrm{j}\frac{k_0\omega_0^2}{2},$$

已证得它到距离 z 处的衍射复振幅为

$$\psi_p(x,y;z)=\mathrm{e}^{-\mathrm{j}k_0z}\frac{\mathrm{j}k_0\omega_0^2}{2q}\mathrm{e}^{-\mathrm{j}\frac{k_0}{2q}(x^2+y^2)}.$$

将 $q(z)=z+\mathrm{j}z_{RG}$ 代入上述衍射场, 证明 $\psi_p(x,y;z)$ 由式 (3.5-8) 给出, 即

$$\psi_p(x,y;z)=\frac{\omega_0}{\omega(z)}\mathrm{e}^{\frac{-(x^2+y^2)}{\omega^2(z)}}\mathrm{e}^{\frac{-\mathrm{j}k_0}{2R(z)}(x^2+y^2)}\mathrm{e}^{-\mathrm{j}\phi(z)}\mathrm{e}^{-\mathrm{j}k_0z}.$$

show that the spread angle is given by

$$\theta_{sp} = 1.22\frac{\lambda_0}{2r_0}.$$

3.7 With reference to Figure P.3.7, employ Collins' integral and show that the complex field on the focal plane is given by Eq. (3.4-29), that is,

$$\psi_p(x, y; z = f + d_0) = \mathrm{e}^{-\mathrm{j}k_0(f+d_0)}\frac{\mathrm{j}k_0}{2\pi f}A\mathrm{e}^{-\mathrm{j}\frac{k_0}{2f}\left(1-\frac{d_0}{f}\right)(x^2+y^2)}\mathcal{F}\{t(x, y)\}\Big|_{k_x=\frac{k_0 x}{f},\, k_y=\frac{k_0 y}{f}}.$$

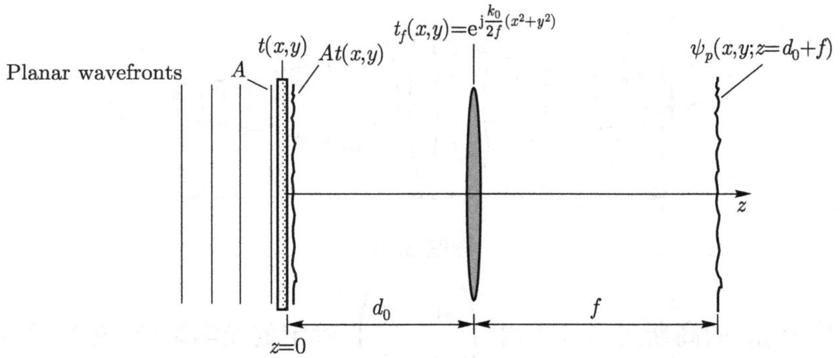

Figure P.3.7

3.8 Figure P.3.8 shows a 4-f imaging system. The input, $t(x, y)$, is illuminated by a plane wave of amplitude A.

(a) Give an expression involving convolution and multiplications similar to that in Eq. (3.4-27) for the complex field on the image plane.

(b) From the result in part (a), draw a block diagram of the optical system.

(c) Show that on the image plane, the complex field distribution is proportional to $t(-x, -y)$ (Figure P.3.8).

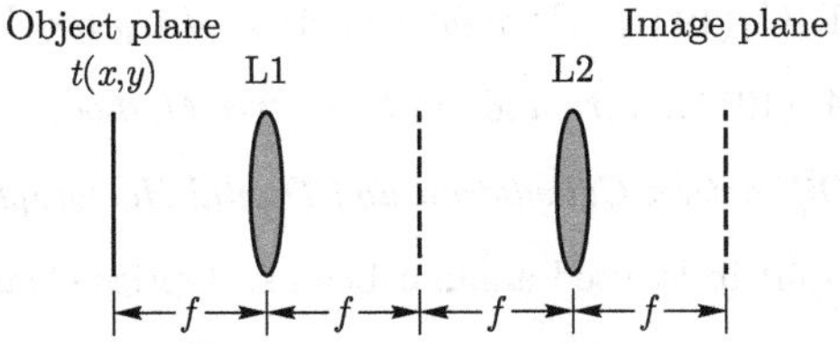

Figure P.3.8

3.9 For a Gaussian beam with initial profile given by

$$\psi_p(x, y; z = 0) = \psi_{p0}(x, y) = \mathrm{e}^{-\mathrm{j}\frac{k_0}{2q_0}(x^2+y^2)} \text{ with } q_0 = \mathrm{j}z_{RG} = \mathrm{j}\frac{k_0\omega_0^2}{2},$$

we have shown that its diffracted complex field at a distance z is

$$\psi_p(x, y; z) = \mathrm{e}^{-\mathrm{j}k_0 z}\frac{\mathrm{j}k_0\omega_0^2}{2q}\mathrm{e}^{-\mathrm{j}\frac{k_0}{2q}(x^2+y^2)}.$$

By substituting $q(z) = z + \mathrm{j}z_{RG}$ into the above diffracted field, show that $\psi_p(x, y; z)$ is given by Eq. (3.5-8), which is given below:

$$\psi_p(x, y; z) = \frac{\omega_0}{\omega(z)}\mathrm{e}^{\frac{-(x^2+y^2)}{\omega^2(z)}}\mathrm{e}^{\frac{-\mathrm{j}k_0}{2R(z)}(x^2+y^2)}\mathrm{e}^{-\mathrm{j}\phi(z)}\mathrm{e}^{-\mathrm{j}k_0 z}.$$

3.10 参考图题 3.10, 假设一个束腰为 ω_0 的高斯光束

$$\psi_{p0}(x,y)=\mathrm{e}^{-(x^2+y^2)/\omega_0^2},$$

经过 d_0 的距离传播到焦距为 f 的理想透镜, 到距离 d_f 处时, 该高斯光束转换为束腰 ω_f 的平面波.

a) 求其 ω_f 和 d_f.

b) 当 $\omega_0\to\infty$, 即, 高斯光波变为平面波时, 通过 a) 的结果, 找出一个能将 d_0, d_f 和 f 联系起来的公式.

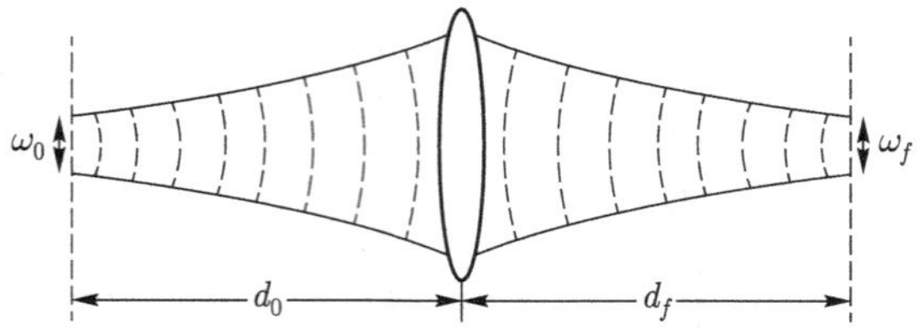

图题 3.10

3.11 参数为 q_1 的高斯光束进入由 $\begin{pmatrix}A & B\\ C & D\end{pmatrix}$ 确定的光学系统. 在光学系统的出口处, 该高斯光束的参数变为 q_2, 证明两个束腰的比值为

$$\frac{\omega_2}{\omega_1}=\left|A+\frac{B}{q_1}\right|,$$

其中, ω_1 和 ω_2 分别为输入和输出光束的束腰.

参考文献

[1] Banerjee, P. P. and Poon, T.-C. (1991). *Principles of Applied Optics.* Irwin, Illinois.

[2] Gerard, A. and Burch, J. M. (1975). *Introduction to Matrix Methods in Optics.* Wiley, New York.

[3] Li, J. and Wu, Y. (2016). *Diffraction Calculation and Digital Holography I.* Science Press, Beijing.

[4] Poon, T.-C. (1988). "Focal shift in focused annular beams," Optics Communications 6, pp.401-406.

[5] Poon, T.-C. (2007). *Optical Scanning Holography with MATLAB®.* Springer, New York.

[6] Poon, T.-C. and Kim, T. (2018). *Engineering Optics with MATLAB®*, 2nd ed. World Scientific, New Jersey.

[7] Stark, H. (1982). *Application of Optical Transforms.* Academic Press, Florida.

3.10 With reference to Figure P.3.10, we assume that a Gaussian beam with a waist ω_0,

$$\psi_{p0}(x,y) = \mathrm{e}^{-(x^2+y^2)/\omega_0^2},$$

travels at a distance d_0 to an ideal lens of focal length f. At a distance d_f, the Gaussian beam is transformed to a beam of planar wavefronts with waist ω_f.

(a) Find ω_f and d_f.

(b) When $\omega_0 \to \infty$, that is, the Gaussian has become a plane wave, find an equation relating d_0, d_f, and f from the result you have obtained in part (a).

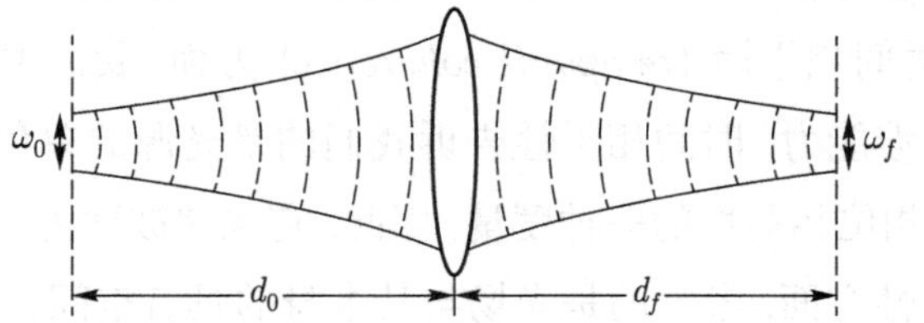

Figure P.3.10

3.11 A Gaussian beam of q_1 enters an optical system specified by $\begin{pmatrix} A & B \\ C & D \end{pmatrix}$. At the exit of the optical system, the Gaussian beam has q_2. Show that the ratio of the two waists is

$$\frac{\omega_2}{\omega_1} = \left| A + \frac{B}{q_1} \right|,$$

where ω_1 and ω_2 are the waists of the input and output beams, respectively.

Bibliography

[1] Banerjee, P. P. and Poon, T.-C. (1991). *Principles of Applied Optics.* Irwin, Illinois.

[2] Gerard, A. and Burch, J. M. (1975). *Introduction to Matrix Methods in Optics.* Wiley, New York.

[3] Li, J. and Wu, Y. (2016). *Diffraction Calculation and Digital Holography I.* Science Press, Beijing.

[4] Poon, T.-C. (1988). "Focal shift in focused annular beams," Optics Communications 6, pp. 401-406.

[5] Poon, T.-C. (2007). *Optical Scanning Holography with MATLAB®.* Springer, New York.

[6] Poon, T.-C. and Kim, T. (2018). *Engineering Optics with MATLAB®*, 2nd ed. World Scientific, New Jersey.

[7] Stark, H. (1982). *Application of Optical Transforms.* Academic Press, Florida.

第四章　空间相干和非相干光学系统

4.1　时间相干性和空间相干性

光的相干性在形式处理上是一个非常复杂的问题. 本节将定性地描述光学相干性以便得到一些基本的概念. 在时间相干性 (*temporal coherence*) 方面, 我们关心的是光场与其具有时间延迟的光场之间的干涉能力. 时间相干性告诉我们的是光源的单色性如何, 它是光波在其传播方向上不同点处的相位相关性的一种度量, 因此, 它也常被称为纵向相干性 (*longitudinal coherence*). 在空间相干性方面, 考虑的是光场与其本身的具有空间位移的光场之间的干涉能力. 空间相干性 (*spatial coherence*) 是垂直于光波传播方向的波面上不同点之间的相位相关性的一种度量, 它通常被称为横向相干性 (*transverse coherence*). 空间相干性告诉我们波前相位的均匀性如何. 将这两种影响完全分离是不可能的, 但是指出其意义非常重要.

时间相干性

第三章中关于波动光学的讨论是不完整的, 因为其没有考虑光波干涉 (*interference*) 存在的条件. 干涉是一种现象, 是两束光波相叠加的结果. 光源可以被分为激光光源和热源. 热源的一个典型例子就是气体放电灯, 如灯泡. 在这样的光源中, 光是通过受激的原子发出的. 根据现代物理学, 在原子和分子中的电子能通过发射或吸收光子来实现能级间跃迁, 该光子的能量必须刚好等于两个能级间的能量差. 光源中的原子并不是连续地发出波, 而是发出特定频率的 "波列 (*wave train*)". 根据 $\hbar\omega_0 = E_2 - E_1$, 频率 ω_0 与参与跃迁的原子能级之间的能量差有关, 其中 E_2 与 E_1 为能级, $\hbar = h/(2\pi)$, h 为普朗克常量.

光波的时间相干性的度量方法是, 通过将光波分成两列不同传播距离的光波, 然后重新合成它们以形成干涉条纹图案. 图 4.1-1 是可度量时间相干性的迈克耳孙干涉仪 (*Michelson interferometer*).

在镀了半透半反银膜的镜片上, 来自光源的入射波列被分为两列相等振幅的波. 一列沿路径 1 传播, 在 M1 镜上被反射. 另一列沿路径 2 传播稍远的距离后在 M2 镜处被反射. 这两列反射波列以一定位移从干涉仪中出来, 相应的位移等于路径长度的差值 $2d$, 其中 d 如图中所示定义. 如果 $2d$ 远远小于如图 4.1-1a) 中所示波列的长度, 则这两个波列几乎叠加且能够干涉. 形成的干涉条纹会很清晰, 这种情况即具有时间相干性. 当通过移动 M2 使其离开 M1$'$ 来

4 Spatial Coherent and Incoherent Optical Systems

4.1 Temporal Coherence and Spatial Coherence

The formal treatment of coherence of light is a rather complicated topic. In this section, we shall describe optical coherence qualitatively to have some basic understanding. In *temporal coherence*, we are concerned with the ability of a light field to interfere with a time delayed version of itself. Temporal coherence tells us how monochromatic a source is, and it is a measure of the correlation of light wave's phase at different points along the direction of propagation. Therefore, it is often referred to as *longitudinal coherence.* In spatial coherence, the ability of a light field to interfere with a spatially shifted version of itself is considered. *Spatial coherence* is a measure of the correlation of a light wave's phase at different points transverse to the direction of propagation. It is often referred to as *transverse coherence.* Spatial coherence tells us how uniform the phase of a wavefront is. It is never possible to separate completely the two effects, but it is important to point out their significance.

Temporal Coherence

A discussion of wave optics from the last chapter is incomplete without considering the conditions that light waves exist for *interference.* Interference is a phenomenon as a result of summing two light waves. Light sources can be categorized into laser sources and thermal sources. A good example of a thermal source is a gas-discharge lamp such as a light bulb. In such a light source, light is emitted by excited atoms. According to modern physics, electrons in atoms and molecules can make transitions in energy levels by emitting or absorbing a photon, whose energy must be exactly equal to the energy difference between the two energy levels. The atoms of a light source do not continuously send out waves but rather they send out "*wave trains*" of a particular frequency. The frequency ω_0 is related to the energy difference between the atomic levels involved in the transition according to $\hbar\omega_0 = E_2 - E_1$, where E_2 and E_1 are energy levels and $\hbar = h/(2\pi)$ with h being Plank's constant.

We can measure the temporal coherence of the light wave by splitting the light wave into two paths of different distances and then recombine them to form an interference fringe pattern. Figure 4.1-1 shows the *Michelson interferometer* for the measure of temporal coherence.

At the half-silvered mirror, the incoming wave train from the source splits into two waves of equal amplitude, one travelling along path 1 to be reflected by mirror M1. The other wave taking path 2 travels a slightly larger distance and reflected by mirror M2. The two reflected wave trains are shown emerging from the interferometer with some displacement. The displacement is equal to the difference in path length of $2d$, where d is defined in the figure. If $2d$ is much smaller than the length of the wave trains as shown in Figure 4.1-1a, the two wave trains are almost superposed and are able to interfere. The created interference pattern will be sharp. We have temporal coherence. As we increase the path difference $2d$

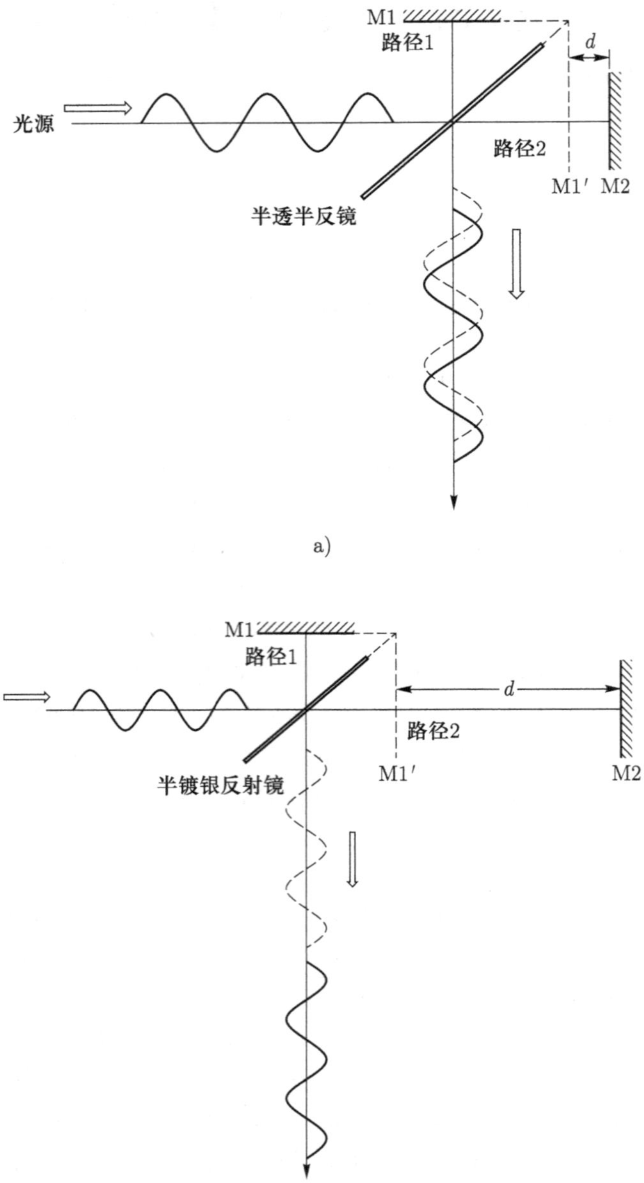

图 4.1-1 迈克耳孙干涉仪: a) 路径差 $2d$ 比光源发出的波列长度小时, 干涉出现; b) 路径差 $2d$ 大于光源发出的波列长度时, 没有干涉发生

增加路径差 $2d$ 时, 从该干涉仪中出现的两个波列将减少重叠并且干涉条纹的清晰度会下降. 当来自同一光源初始波列所分成的两个波列的路径差 $2d$ 大于波列长度时, 两列波列不会再重叠且干涉现象不再发生, 如图 4.1-1b) 所示. 对于这一简单情况, 可以将波列的持续时间 τ_c 定义为相干时间 (*coherence time*), 则相干长度 (*coherence length*) l_c 等于 $c\tau_c$, 其中 c 为真空中的光速. 然而, 波列从干涉仪中出来时仍然有可能重叠, 原因在于可能有其他原子会随机地发出频率或波长略微不同的任意长度的波列 (例如, 由于原子的热运动). 实际上, 有大量的波列会从光源中不同的原子发射出来.

假设光源发出波长范围为 λ 到 $\lambda+\Delta\lambda$ 的光波, $\Delta\lambda$ 是光源的光谱宽度 (*spectral width*).

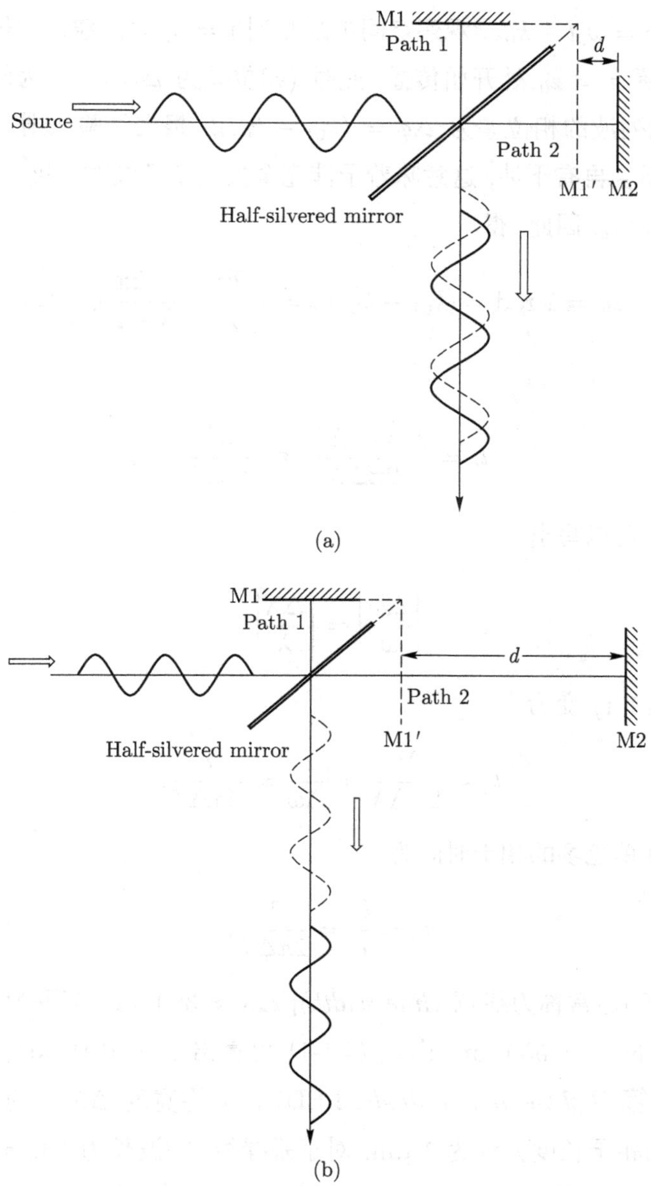

Figure 4.1-1 Michelson interferometer: (a) The path difference $2d$ is less than the length of the wave train emitted by the source. Interference occurs. (b) The path difference $2d$ is greater than the length of the wave train emitted by the source. Interference cannot occur

by translating M2 away from M1′, the two wave trains emerging from the interferometer will overlap less and created interference pattern will be less sharp. When $2d$ is larger than the length of the wave trains, which come from the same initial wave train from the source, the two wave trains can no longer overlap and no interference can occur. The situation is shown in Figure 4.1-1b. For this simple case, we can define the *coherence time* as the duration of the wave train, τ_c. The *coherence length* l_c, therefore, equals $c\tau_c$, where c is the speed of light in vacuum. However, it is still possible to have wave trains overlap as they emerge from the interferometer. The reason is that there might be another atom emitting a wave train of random length and at random times with slightly different frequencies or different wavelengths (e.g., due to the thermal motion of the atom). In practice, there are enormous number of wave trains emitted from different atoms within the source.

Let us assume the source emits waves with wavelengths ranging from λ to $\lambda + \Delta\lambda$, and $\Delta\lambda$ is the *spectral width* of the source. Remember that the phase of a propagating wave is

记波传播的相位为 $\phi = \omega_0 t - k_0 z$, 观察空间中某时刻 $t = t_0$ 时的波形. 假定有波数为 k_{01} 和 k_{02} 的两列波, 从位置 $z = z_0$ 处开始传播, 此时 (相位差为 $\Delta\phi = 0$) 为同相位. 在传播到一定的距离 l 处时, 两列波的相位差为 $\Delta\phi = (k_{01} - k_{02})l$. 根据经验法则 (*rule of thumb*), 当 $\Delta\phi \approx 1$ rad 时, 光不会再有干涉, 这意味着干涉完全失去了条纹对比度 (*fringe contrast*). 对于 $\Delta\phi = 1$ rad, l 变为 l_c, 因此, 得

$$\Delta\phi = 1\ \text{rad} = (k_{01} - k_{02})l_c = \left(\frac{2\pi}{\lambda} - \frac{2\pi}{\lambda + \Delta\lambda}\right) l_c,$$

或

$$l_c = \frac{\lambda(\lambda + \Delta\lambda)}{2\pi\Delta\lambda} \approx \frac{\lambda^2}{2\pi\Delta\lambda}. \tag{4.1-1}$$

由于 $\lambda f = \lambda\dfrac{\omega}{2\pi} = c$, 可以写出

$$\left|\frac{\Delta\omega}{\omega}\right| = \left|\frac{\Delta\lambda}{\lambda}\right|.$$

利用这个结果, 式 (4.1-1) 变为

$$l_c \approx \frac{\lambda^2}{2\pi\Delta\lambda} = \frac{c}{\Delta\omega} = \frac{c}{2\pi\Delta f},$$

或者带宽为 Δf 的准单色场的相干时间为

$$\tau_c = \frac{l_c}{c} = \frac{1}{2\pi\Delta f}. \tag{4.1-2}$$

白光的光谱宽度 [通常称为线宽 (*line width*)] $\Delta\lambda \approx 300$ nm, 范围为 400 nm 到 700 nm 之间. 如果取平均波长 $\lambda = 550$ nm, 由式 (4.1-1) 可算出 $l_c \sim 0.16$ μm, 这是一个非常短的相干长度. 发光二极管 (*light-emitting diode*, LED) 的光谱宽度 $\Delta\lambda$ 大约为 50 nm 且对波长为 632 nm 的红光的相干长度大约为 1 μm. 对于汞绿线, 在波长为 546 nm 处其线宽 $\Delta\lambda$ 约为 0.025 nm, 其相干长度约为 0.2 cm. 激光通常有较长的相干长度. 氦氖激光器可以发出相干长度超过 5 m 的相干光, 但最常见的相干长度为 20 cm. 一些工业用的线宽 $\Delta\lambda$ 约为 10^{-5} nm 的二氧化碳 (CO_2) 激光器, 能发出波长 $\lambda = 10.6$ μm 的红外光, 其相干长度大概有几千米. 在后续章节中讨论全息术时, 相干长度在物光波和参考光波之间产生干涉的可允许的路径差上有个上限. 如果要干涉条纹可以被观察到, 则路径分离的距离差必须小于所用光源的相干长度.

空间相干性

首先考虑众所周知的杨氏干涉实验 (*Young's interference experiment*), 如图 4.1-2 所示. 间距为 $2x_0$ 的两个点光源 S1 和 S2 产生的球面波在距离光源平面 l 远处的屏上干涉.

根据菲涅耳衍射 [见式 (3.4-17)], 振幅为 A_1、相位为 θ_1 的点源 S1 在屏幕上的复振幅为

$$\begin{aligned}\psi_{S1} &= A_1 e^{-j\theta_1}\delta(x - x_0, y) * h(x, y; z = l) = A_1 e^{-j\theta_1} h(x - x_0, y; z = l) \\ &= A_1 e^{-j\theta_1}\frac{jk_0}{2\pi l} e^{-jk_0 l} e^{\frac{-jk_0[(x-x_0)^2 + y^2]}{2l}}.\end{aligned}$$

$\phi = \omega_0 t - k_0 z$. We monitor the wave pattern in space at some time $t = t_0$. At some distance l, the phase difference between two waves with wavenumbers k_{01} and k_{02} which are in phase at some $z = z_0$ becomes out of phase by $\Delta\phi = (k_{01} - k_{02})l$. As a *rule of thumb*, when $\Delta\phi \approx 1$ rad, we say that the light is no longer considered coherent, which means interference loose its *fringe contrast* completely. For $\Delta\phi = 1$ rad, l becomes l_c. We, therefore, have

$$\Delta\phi = 1 \text{ rad} = (k_{01} - k_{02})l_c = \left(\frac{2\pi}{\lambda} - \frac{2\pi}{\lambda + \Delta\lambda}\right) l_c,$$

or

$$l_c = \frac{\lambda(\lambda + \Delta\lambda)}{2\pi\Delta\lambda} \approx \frac{\lambda^2}{2\pi\Delta\lambda}. \tag{4.1-1}$$

Since $\lambda f = \lambda \dfrac{\omega}{2\pi} = c$, we can establish that

$$\left|\frac{\Delta\omega}{\omega}\right| = \left|\frac{\Delta\lambda}{\lambda}\right|.$$

Using this result, Eq. (4.1-1) becomes

$$l_c \approx \frac{\lambda^2}{2\pi\Delta\lambda} = \frac{c}{\Delta\omega} = \frac{c}{2\pi\Delta f},$$

or the coherence time of a quasi-monochromatic field of bandwidth Δf is

$$\tau_c = \frac{l_c}{c} = \frac{1}{2\pi\Delta f}. \tag{4.1-2}$$

White light has a spectral width (often called *line width*) $\Delta\lambda \approx 300$ nm, ranging from 400 nm to 700 nm. If we take the average wavelength at $\lambda = 550$ nm, Eq. (4.1-1) gives $l_c \sim 0.16$ μm, a very short coherence length. Light-emitting diodes (LEDs) have a spectral width $\Delta\lambda$ of about 50 nm and have a coherence length of about 1 μm for red color with wavelength of 632 nm. As for the green line of mercury at λ =546 nm having a line width $\Delta\lambda$ about 0.025 nm, its coherence length is about 0.2 cm. Lasers typically have a long coherence length. Helium-neon lasers can produce light with coherence lengths greater than 5 m but 20 cm is typical. Some industrial CO_2 lasers of line width $\Delta\lambda$ of around 10^{-5} nm producing emission at the infrared wavelength of $\lambda = 10.6$ μm would give a coherence length of around a couple of kilometers. When we discuss holography in subsequent chapters, the coherence length places an upper limit on the allowable path difference between the object and reference wave for interference. If interference fringes are to be observed, the path separation must be less than the coherence length of the source used.

Spatial Coherence

We first consider the well-known Young's interference experiment shown in Figure 4.1-2. Two point sources S1 and S2 separated by a distance $2x_0$ generate spherical waves that are interfered on a screen at a distance l.

According to Fresnel diffraction [see Eq. (3.4-17)], the complex field of point source S1 of amplitude A_1 with phase θ_1 on the screen is given by

$$\begin{aligned}\psi_{S1} &= A_1 e^{-j\theta_1}\delta(x - x_0, y) * h(x, y; z = l) = A_1 e^{-j\theta_1} h(x - x_0, y; z = l) \\ &= A_1 e^{-j\theta_1}\frac{jk_0}{2\pi l} e^{-jk_0 l} e^{\frac{-jk_0[(x-x_0)^2+y^2]}{2l}}.\end{aligned}$$

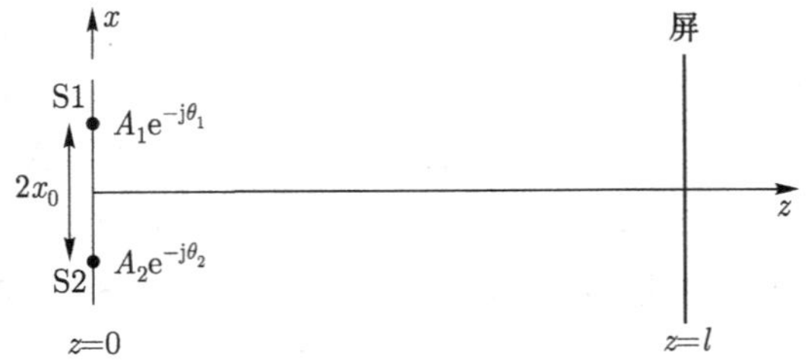

图 4.1-2 杨氏干涉实验

相似地, 对振幅为 A_2、相位为 θ_2 的点源 S2, 其在屏幕上的复振幅为

$$\psi_{S2} = A_2\mathrm{e}^{-\mathrm{j}\theta_2}\delta(x+x_0,y)*h(x,y;z=l) = A_2\mathrm{e}^{-\mathrm{j}\theta_2}\frac{\mathrm{j}k_0}{2\pi l}\mathrm{e}^{-\mathrm{j}k_0 l}\mathrm{e}^{\frac{-\mathrm{j}k_0[(x+x_0)^2+y^2]}{2l}}.$$

则对于屏幕上的总光场 ψ_t, 根据叠加原理有

$$\psi_t = \psi_{S1} + \psi_{S2},$$

相应的强度 $I(x,y)$ 为总光场的振幅的平方 (*square magnitude*) [见式 (3.3-23)]

$$\begin{aligned}
I(x,y) &= |\psi_{S1}+\psi_{S2}|^2 = |\psi_{S1}|^2 + |\psi_{S2}|^2 + \psi_{S1}\psi_{S2}^* + \psi_{S1}^*\psi_{S2} \\
&= \left(A_1\frac{k_0}{2\pi l}\right)^2 + \left(A_2\frac{k_0}{2\pi l}\right)^2 + A_1\mathrm{e}^{-\mathrm{j}\theta_1}\frac{\mathrm{j}k_0}{2\pi l}\mathrm{e}^{-\mathrm{j}k_0 l}\mathrm{e}^{\frac{-\mathrm{j}k_0[(x-x_0)^2+y^2]}{2l}} \\
&\quad \cdot A_2\mathrm{e}^{+\mathrm{j}\theta_2}\frac{-\mathrm{j}k_0}{2\pi l}\mathrm{e}^{+\mathrm{j}k_0 l}\mathrm{e}^{\frac{\mathrm{j}k_0[(x+x_0)^2+y^2]}{2l}} + c.c. \\
&= \left(A_1\frac{k_0}{2\pi l}\right)^2 + \left(A_2\frac{k_0}{2\pi l}\right)^2 + A_1A_2\mathrm{e}^{-\mathrm{j}(\theta_1-\theta_2)} \\
&\quad \cdot \left(\frac{k_0}{2\pi l}\right)^2\mathrm{e}^{\frac{-\mathrm{j}k_0[(x-x_0)^2+y^2]}{2l}}\mathrm{e}^{\frac{\mathrm{j}k_0[(x+x_0)^2+y^2]}{2l}} + c.c. \\
&= |\psi_{S1}|^2 + |\psi_{S2}|^2 + A_1A_2\mathrm{e}^{-\mathrm{j}(\theta_1-\theta_2)}\left(\frac{k_0}{2\pi l}\right)^2\mathrm{e}^{\frac{-\mathrm{j}k_0(-4xx_0)}{2l}} + c.c.
\end{aligned}$$

其中, *c.c.* 表示复共轭. 上式可用实函数表示为

$$\begin{aligned}
I(x,y) &= |\psi_{S1}|^2 + |\psi_{S2}|^2 + 2A_1A_2\left(\frac{k_0}{2\pi l}\right)^2\cos\left[\frac{k_0}{l}2x_0x + (\theta_2-\theta_1)\right] \\
&= |\psi_{S1}|^2 + |\psi_{S2}|^2 + 2A_1A_2\left(\frac{k_0}{2\pi l}\right)^2\cos\left[\frac{2\pi}{\Lambda}x + (\theta_2-\theta_1)\right],
\end{aligned} \tag{4.1-3}$$

屏幕上沿 x 方向的条纹, 其条纹周期$\Lambda = \dfrac{\lambda_0 l}{2x_0}$. 根据两个点源之间的相位差 $\theta_1-\theta_2$, 近光轴 (z) 的第一个峰值所移动的位置 x_p 为

$$x_p = \frac{\Lambda}{2\pi}(\theta_1-\theta_2). \tag{4.1-4}$$

式 (4.1-3) 中的最后一项为干涉项. 若场是空间相干 (*spatially coherent*) 的, 则在 θ_1 和 θ_2 之间就有一个固定的相位关系, 根据式 (4.1-3) 可以获得屏上的条纹图案, 其强度分布图可由下式求得

$$I(x,y)|_{coh} = |\psi_{S1}+\psi_{S2}|^2, \tag{4.1-5}$$

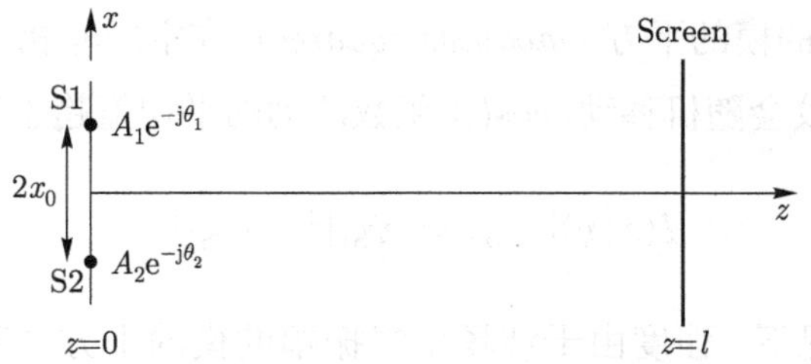

Figure 4.1-2 Young's interference experiment

Similarly, for point source S2 of amplitude A_2 with phase θ_2, the complex field on the screen is given by

$$\psi_{S2} = A_2 e^{-j\theta_2}\delta(x + x_0, y) * h(x, y; z = l) = A_2 e^{-j\theta_2}\frac{jk_0}{2\pi l}e^{-jk_0 l}e^{\frac{-jk_0[(x+x_0)^2+y^2]}{2l}}.$$

The total field ψ_t on the screen, according to the principle of superposition, is

$$\psi_t = \psi_{S1} + \psi_{S2},$$

and the corresponding intensity $I(x, y)$ is the *square magnitude* of the total field [see Eq. (3.3-23)]:

$$\begin{aligned}
I(x,y) &= |\psi_{S1} + \psi_{S2}|^2 = |\psi_{S1}|^2 + |\psi_{S2}|^2 + \psi_{S1}\psi_{S2}^* + \psi_{S1}^*\psi_{S2} \\
&= \left(A_1\frac{k_0}{2\pi l}\right)^2 + \left(A_2\frac{k_0}{2\pi l}\right)^2 + A_1 e^{-j\theta_1}\frac{jk_0}{2\pi l}e^{-jk_0 l}e^{\frac{-jk_0[(x-x_0)^2+y^2]}{2l}} \\
&\quad \cdot A_2 e^{+j\theta_2}\frac{-jk_0}{2\pi l}e^{+jk_0 l}e^{\frac{jk_0[(x+x_0)^2+y^2]}{2l}} + c.c. \\
&= \left(A_1\frac{k_0}{2\pi l}\right)^2 + \left(A_2\frac{k_0}{2\pi l}\right)^2 + A_1A_2 e^{-j(\theta_1-\theta_2)} \\
&\quad \cdot \left(\frac{k_0}{2\pi l}\right)^2 e^{\frac{-jk_0[(x-x_0)^2+y^2]}{2l}} e^{\frac{jk_0[(x+x_0)^2+y^2]}{2l}} + c.c. \\
&= |\psi_{S1}|^2 + |\psi_{S2}|^2 + A_1A_2 e^{-j(\theta_1-\theta_2)}\left(\frac{k_0}{2\pi l}\right)^2 e^{\frac{-jk_0(-4xx_0)}{2l}} + c.c.
\end{aligned}$$

where *c.c.* denotes the complex conjugate. We can express the above equation in terms of a real function:

$$\begin{aligned}
I(x,y) &= |\psi_{S1}|^2 + |\psi_{S2}|^2 + 2A_1A_2\left(\frac{k_0}{2\pi l}\right)^2 \cos\left[\frac{k_0}{l}2x_0 x + (\theta_2 - \theta_1)\right] \\
&= |\psi_{S1}|^2 + |\psi_{S2}|^2 + 2A_1A_2\left(\frac{k_0}{2\pi l}\right)^2 \cos\left[\frac{2\pi}{\Lambda}x + (\theta_2 - \theta_1)\right],
\end{aligned} \tag{4.1-3}$$

There is a fringe pattern along the x-direction on the screen with the fringe period $\Lambda = \frac{\lambda_0 l}{2x_0}$. The first peak x_p near the optical axis (z) is shifted according to the phase difference $\theta_1 - \theta_2$ between the two point sources:

$$x_p = \frac{\Lambda}{2\pi}(\theta_1 - \theta_2). \tag{4.1-4}$$

The last term of Eq. (4.1-3) is the interference term. If the fields are *spatially coherent*, there is a fixed phase relationship between θ_1 and θ_2, and we have the fringe pattern on the screen given by Eq. (4.1-3) as the intensity pattern is calculated according to

$$I(x,y)|_{coh} = |\psi_{S1} + \psi_{S2}|^2, \tag{4.1-5}$$

即, 强度是两个复振幅和的模的平方 (*modulus-squared*). 在空间非相干情况中, $\theta_1-\theta_2$ 是随机波动的, 因此 x_p 或者条纹会随机移动. $\cos(\cdot)$ 函数波动的平均值趋于零. 因此, 式 (4.1-3) 变为

$$I(x,y)|_{incoh}=|\psi_{\mathrm{S1}}|^2+|\psi_{\mathrm{S2}}|^2. \tag{4.1-6}$$

因此, 在完全非相干情况下, 强度由干涉场中复振幅的模的平方之和获得. 式 (4.1-5) 和式 (4.1-6) 是后续章节中考虑相干成像和非相干成像的基础. 在相干成像中, 将所有的复振幅叠加即可求得全部的强度, 而在非相干成像中, 则直接对强度进行处理.

4.2　空间相干图像处理

基于第二章对信号滤波的讨论, 由式 (2.2-13b), 可知

$$y(t)=\mathcal{F}^{-1}\{Y(\omega)\}=\mathcal{F}^{-1}\{X(\omega)H(\omega)\}.$$

当输入信号 $x(t)$ 经过系统后得到输出信号 $y(t)$, 则其输入谱 $X(\omega)$ 被系统传递函数 $H(\omega)$ 修改或滤波, $y(t)$ 是 $x(t)$ 经过滤波的信号. 为了实现与上式类似的 "图像" 滤波, 下面分析一个如图 4.2-1 所示标准的双透镜相干图像处理.

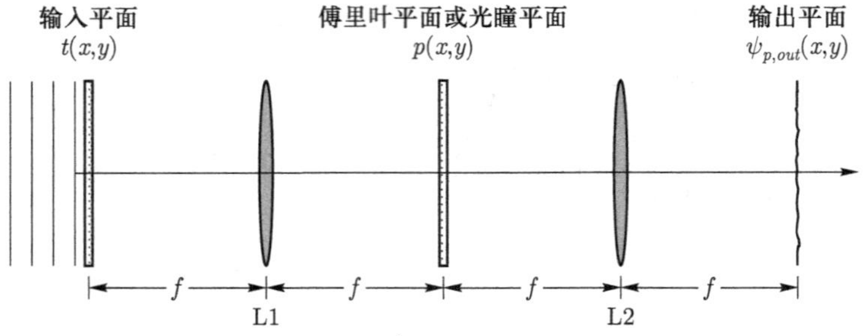

图 4.2-1　平面波照明下标准的 4-f 光学图像处理系统

4.2.1　瞳函数, 相干点扩散函数, 相干传递函数

图 4.2-1 所示的双透镜系统为所谓的 4-f 系统, 其中透镜 L1 与透镜 L2 有相同的焦距 f. 透明片 $t(x,y)$ 形式作为输入被一单位振幅平面波 (为相干光波) 照射, 透过率函数 $p(x,y)$ 被置于光学系统的共焦面上. 共焦面也经常被称为处理系统的傅里叶 (平) 面 (*Fourier plane*). 原因在于 $t(x,y)$ 的频谱或者是傅里叶变换恰好出现在透镜 L1 的后焦面上, 即 $T\left(\dfrac{k_0x}{f},\dfrac{k_0y}{f}\right)$, 而 T 是 $t(x,y)$ 的傅里叶变换. 傅里叶平面也被称为光学系统的光瞳 (平) 面 (*pupil plane*). 因此, $p(x,y)$ 被称为瞳函数 (*pupil function*), 可以看出, 输入的频谱被 $p(x,y)$ 修改, 在瞳函数后的光场立刻变为 $T\left(\dfrac{k_0x}{f},\dfrac{k_0y}{f}\right)p(x,y)$, 该光场再一次被透镜 L2 进行傅里叶变换, 得到输出

that is, the intensity is computed as the modulus-squared of the sum of the two complex fields. In the spatially incoherent case, $\theta_1 - \theta_2$ is randomly fluctuating and, hence, x_p, or the fringes, moves randomly. The $\cos(\cdot)$ fluctuations average out to zero. Hence, Eq. (4.1-3) becomes

$$I(x,y)|_{incoh} = |\psi_{S1}|^2 + |\psi_{S2}|^2. \tag{4.1-6}$$

Therefore, in the perfectly incoherent case, the intensity is computed as the sum of the moduli-squared of the complex fields of the interfering fields. Equations (4.1-5) and (4.1-6) are the bases when we consider coherent imaging and incoherent imaging in subsequent sections. In coherent imaging, we sum up all the complex fields and then find the overall intensity, whereas in incoherent imaging, we manipulate the intensities directly.

4.2 Spatial Coherent Image Processing

With reference to our earlier discussion on signal filtering in Chapter 2, from Eq. (2.2-13b), we have

$$y(t) = \mathcal{F}^{-1}\{Y(\omega)\} = \mathcal{F}^{-1}\{X(\omega)H(\omega)\}.$$

The input spectrum $X(\omega)$ is modified or filtered by the system's transfer function $H(\omega)$ as the input signal, $x(t)$, goes through the system to give an output $y(t)$, which is the filtered version of the input. To effectuate "image" filtering similar to that in the above equation, we analyze a standard two-lens coherent image processing shown in Figure 4.2-1.

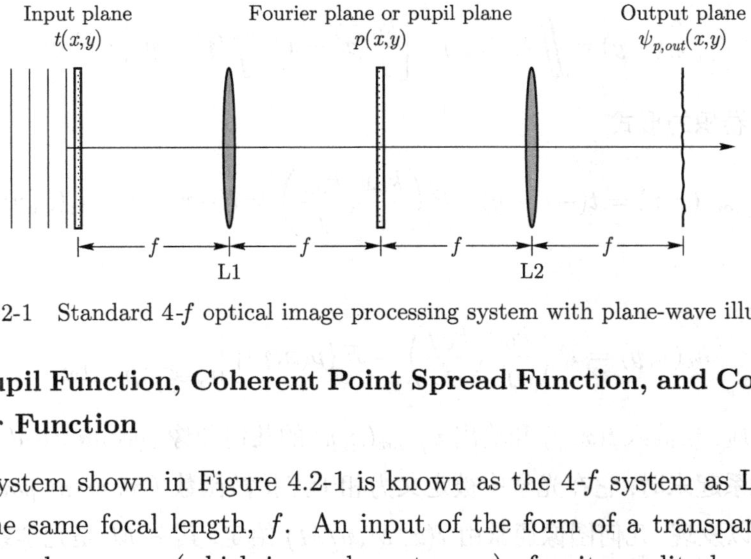

Figure 4.2-1 Standard 4-f optical image processing system with plane-wave illumination

4.2.1 Pupil Function, Coherent Point Spread Function, and Coherent Transfer Function

The two-lens system shown in Figure 4.2-1 is known as the 4-f system as Lens 1 and Lens 2 both have the same focal length, f. An input of the form of a transparency, $t(x,y)$, is illuminated by a plane wave (which is a coherent wave) of unit amplitude. A transparency function, $p(x,y)$, is placed on the confocal plane of the optical system. The confocal plane is often called the *Fourier plane* of the processing system. The reason is that the back focal plane of Lens 1 is the plane where the spectrum or the Fourier transform of $t(x,y)$ appears (see Figure 3.4-7), which is $T\left(\frac{k_0x}{f}, \frac{k_0y}{f}\right)$ with T being the Fourier transform of $t(x,y)$. The Fourier plane is also called the *pupil plane* of the processing system. For this reason, $p(x,y)$ is called the *pupil function.* We can see that the spectrum of the input is now modified by $p(x,y)$ and the field immediately after the pupil function is $T\left(\frac{k_0x}{f}, \frac{k_0y}{f}\right)p(x,y)$. This field

平面上的复振幅为

$$\psi_{p,out}(x,y)=\mathcal{F}\left\{T\left(\frac{k_0x}{f},\frac{k_0y}{f}\right)p(x,y)\right\}\Bigg|_{k_x=\frac{k_0x}{f},\,k_y=\frac{k_0y}{f}}. \tag{4.2-1}$$

由于

$$T\left(\frac{k_0x}{f},\frac{k_0y}{f}\right)=\mathcal{F}\left\{t(x,y)\right\}\Big|_{k_x=\frac{k_0x}{f},\,k_y=\frac{k_0y}{f}}=\iint t(x',y')\mathrm{e}^{\frac{\mathrm{j}k_0}{f}(xx'+yy')}\mathrm{d}x'\mathrm{d}y', \tag{4.2-2}$$

式 (4.2-1) 重写为

$$\begin{aligned}\psi_{p,out}(x,y)&=\mathcal{F}\left\{\iint t(x',y')\mathrm{e}^{\frac{\mathrm{j}k_0}{f}(xx'+yy')}\mathrm{d}x'\mathrm{d}y'p(x,y)\right\}\Bigg|_{k_x=\frac{k_0x}{f},\,k_y=\frac{k_0y}{f}}\\&=\iint\left[\iint t(x',y')\mathrm{e}^{\frac{\mathrm{j}k_0}{f}(x''x'+y''y')}\mathrm{d}x'\mathrm{d}y'p(x'',y'')\right]\mathrm{e}^{\frac{\mathrm{j}k_0}{f}(xx''+yy'')}\mathrm{d}x''\mathrm{d}y''.\end{aligned} \tag{4.2-3}$$

我们把其中带有两撇的坐标变量重组并先在这些坐标上对其进行积分, 有

$$\begin{aligned}\iint p(x'',y'')\mathrm{e}^{\frac{\mathrm{j}k_0}{f}(x''x'+y''y')}\mathrm{e}^{\frac{\mathrm{j}k_0}{f}(xx''+yy'')}\mathrm{d}x''\mathrm{d}y''&=\iint p(x'',y'')\mathrm{e}^{\frac{\mathrm{j}k_0}{f}[x''(x'+x)+y''(y'+y)]}\mathrm{d}x''\mathrm{d}y''\\&=\mathcal{F}\left\{p(x,y)\right\}\Big|_{k_x=\frac{k_0}{f}(x'+x),\,k_y=\frac{k_0}{f}(y'+y)}\\&=P\left[\frac{k_0}{f}(x'+x),\frac{k_0}{f}(y'+y)\right].\end{aligned}$$

将这个结果代入式 (4.2-3), 有

$$\psi_{p,out}(x,y)=\iint t(x',y')P\left[\frac{k_0}{f}(x'+x),\frac{k_0}{f}(y'+y)\right]\mathrm{d}x'\mathrm{d}y'.$$

该式亦可表示为卷积的形式

$$\psi_{p,out}(x,y)=t(-x,-y)*P\left(\frac{k_0x}{f},\frac{k_0y}{f}\right)=t(-x,-y)*h_c(x,y), \tag{4.2-4}$$

其中

$$h_c(x,y)=P\left(\frac{k_0x}{f},\frac{k_0y}{f}\right)=\mathcal{F}\left\{p(x,y)\right\}\Big|_{k_x=\frac{k_0x}{f},\,k_y=\frac{k_0y}{f}}$$

是系统的脉冲响应, 当输入 $t(x,y)$ 和输出 $\psi_{p,out}(x,y)$ 的几何图像 (*geometrical image*) $t(-x,-y)$ 通过卷积积分联系起来时, 它在光学中被定义为相干点扩散函数 (*coherent point spread function*, CPSF). 可以发现, 几何图像通常由 $t(x/M,y/M)$ 给定, 其中 M 是光学系统中的横向放大率. 目前考虑的光学系统 (见图 4.2-1), $M=-1$. 因此, 图 4.2-1 所示的光学系统为线性空不变相干系统 (*linear space-invariant coherent system*). 通过仔细设计瞳函数 $p(x,y)$, 可以改变图像处理的能力, 同时, 瞳函数的设计也可以用于特定的应用, 比如长景深的三维成像或是提高分辨率以进行精确测量等, 被称为瞳函数工程 (*pupil engineering*). 由于瞳函数在输入的傅里叶平面上, 它实际上改变了输入的频谱. 因此, 瞳函数也被称为空间滤波器 (*spatial filter*) ——能够改变输入信号的空间频率的物理滤波器. 在文献中, 这样的处理通常被称为空间滤波 (*spatial filtering*).

again will be Fourier transformed by Lens 2 to give the complex field on the output plane as

$$\psi_{p,out}(x,y) = \mathcal{F}\left\{T\left(\frac{k_0 x}{f}, \frac{k_0 y}{f}\right) p(x,y)\right\}\bigg|_{k_x=\frac{k_0 x}{f},\, k_y=\frac{k_0 y}{f}}. \tag{4.2-1}$$

Since

$$T\left(\frac{k_0 x}{f}, \frac{k_0 y}{f}\right) = \mathcal{F}\{t(x,y)\}\Big|_{k_x=\frac{k_0 x}{f},\, k_y=\frac{k_0 y}{f}} = \iint t(x',y')\mathrm{e}^{\frac{\mathrm{j}k_0}{f}(xx'+yy')}\mathrm{d}x'\mathrm{d}y', \tag{4.2-2}$$

we rewrite Eq. (4.2-1) as

$$\begin{aligned}\psi_{p,out}(x,y) &= \mathcal{F}\left\{\iint t(x',y')\mathrm{e}^{\frac{\mathrm{j}k_0}{f}(xx'+yy')}\mathrm{d}x'\mathrm{d}y' p(x,y)\right\}\bigg|_{k_x=\frac{k_0 x}{f},\, k_y=\frac{k_0 y}{f}} \\ &= \iint\left[\iint t(x',y')\mathrm{e}^{\frac{\mathrm{j}k_0}{f}(x''x'+y''y')}\mathrm{d}x'\mathrm{d}y' p(x'',y'')\right]\mathrm{e}^{\frac{\mathrm{j}k_0}{f}(xx''+yy'')}\mathrm{d}x''\mathrm{d}y''. \end{aligned} \tag{4.2-3}$$

Let us group the double-primed coordinates and integrate over these coordinates first, and we have

$$\begin{aligned}\iint p(x'',y'')\mathrm{e}^{\frac{\mathrm{j}k_0}{f}(x''x'+y''y')}\mathrm{e}^{\frac{\mathrm{j}k_0}{f}(xx''+yy'')}\mathrm{d}x''\mathrm{d}y'' &= \iint p(x'',y'')\mathrm{e}^{\frac{\mathrm{j}k_0}{f}[x''(x'+x)+y''(y'+y)]}\mathrm{d}x''\mathrm{d}y'' \\ &= \mathcal{F}\{p(x,y)\}\Big|_{k_x=\frac{k_0}{f}(x'+x),\, k_y=\frac{k_0}{f}(y'+y)} \\ &= P\left[\frac{k_0}{f}(x'+x), \frac{k_0}{f}(y'+y)\right].\end{aligned}$$

Substituting this result back into Eq. (4.2-3), we have

$$\psi_{p,out}(x,y) = \iint t(x',y')P\left[\frac{k_0}{f}(x'+x), \frac{k_0}{f}(y'+y)\right]\mathrm{d}x'\mathrm{d}y'.$$

which can be expressed in terms of convolution as

$$\psi_{p,out}(x,y) = t(-x,-y) * P\left(\frac{k_0 x}{f}, \frac{k_0 y}{f}\right) = t(-x,-y) * h_c(x,y), \tag{4.2-4}$$

where

$$h_c(x,y) = P\left(\frac{k_0 x}{f}, \frac{k_0 y}{f}\right) = \mathcal{F}\{p(x,y)\}\Big|_{k_x=\frac{k_0 x}{f},\, k_y=\frac{k_0 y}{f}}$$

is the impulse response of the system and defined as the *coherent point spread function* (CPSF) in optics as the *geometrical image* $t(-x,-y)$ of the input $t(x,y)$ and the output $\psi_{p,out}(x,y)$ are related through the convolution integral. Note that the geometrical image in general is given by $t(x/M, y/M)$, where M is the lateral magnification of the optical system. In the optical system presently under consideration (see Figure 4.2-1), we have $M=-1$. Therefore, the optical system shown in Figure 4.2-1 is a *linear space-invariant coherent system.* Image processing capabilities can be varied by carefully designing the pupil function, $p(x,y)$, and the design of pupils to accommodate certain intended applications such as long depth of field for 3-D imaging or enhanced resolution for precision measurement, etc., is called *pupil engineering.* Since the pupil function is on the Fourier plane of the input, it actually modifies the spectrum of the input. Hence, the pupil function is also called a *spatial filter*−physical filter capable of modifying the spatial frequencies of the input. Such a process is often called *spatial filtering* in the literature.

根据定义, 脉冲响应的傅里叶变换即为传递函数. 相干点扩散函数的傅里叶变换即被称为相干传递函数 (*coherent transfer function*, CTF)

$$\begin{aligned} H_c(k_x,k_y) &= \mathcal{F}\{h_c(x,y)\} = \mathcal{F}\left\{P\left(\frac{k_0x}{f},\frac{k_0y}{f}\right)\right\} \\ &= p(x,y)\Big|_{x=\frac{-fk_x}{k_0},\,y=\frac{-fk_y}{k_0}} = p\left(\frac{-fk_x}{k_0},\frac{-fk_y}{k_0}\right). \end{aligned} \tag{4.2-5}$$

对式 (4.2-4) 进行傅里叶变换, 有

$$\begin{aligned} \mathcal{F}\{\psi_{p,out}(x,y)\} &= \mathcal{F}\left\{t(-x,-y)*P\left(\frac{k_0x}{f},\frac{k_0y}{f}\right)\right\} \\ &= \mathcal{F}\{t(-x,-y)\}\,\mathcal{F}\left\{P\left(\frac{k_0x}{f},\frac{k_0y}{f}\right)\right\} \\ &= \mathcal{F}\{t(-x,-y)\}\,H_c(k_x,k_y) \\ &= T(-k_x,-k_y)H_c(k_x,k_y), \end{aligned}$$

或

$$\psi_{p,out}(x,y) = \mathcal{F}^{-1}\{T(-k_x,-k_y)H_c(k_x,k_y)\}. \tag{4.2-6}$$

与下式相比 [见式 (2.2-13b)], 这个等式在滤波方面的含义与本书第二章中讨论的相同

$$y(t) = \mathcal{F}^{-1}\{Y(\omega)\} = \mathcal{F}^{-1}\{X(\omega)H(\omega)\}.$$

然而, 在相干图像处理方面, 因为透镜仅仅处理正傅里叶变换, 又由于双透镜系统中自然会产生两次傅里叶变换, 因此我们处理的是 $t(x,y)$ 的翻转和倒转的频谱, 即 $T(-k_x,-k_y)$. 最终, 一旦求出输出的复振幅, 相应的图像强度即为

$$I(x,y) = \psi_{p,out}(x,y)\psi^*_{p,out}(x,y) = |t(-x,-y)*h_c(x,y)|^2, \tag{4.2-7}$$

这是相干图像处理 (*coherent image processing*) 的基础.

4.2.2 相干图像处理实例

举例: 全通滤波, $p(x,y)=1$

由式 (4.2-4), 相干点扩散函数为

$$h_c(x,y) = \mathcal{F}\{1\}\Big|_{k_x=\frac{k_0x}{f},\,k_y=\frac{k_0y}{f}} = 4\pi^2\delta\left(\frac{k_0x}{f},\frac{k_0y}{f}\right) = 4\pi^2\left(\frac{f}{k_0}\right)^2\delta(x,y),$$

其中, 用 δ 函数的缩放性质求得上式最后一步 [见式 (2.1-6)]. 根据式 (4.2-4), 输出为

$$\begin{aligned} \psi_{p,out}(x,y) &= t(-x,-y)*h_c(x,y) \\ &= 4\pi^2\left(\frac{f}{k_0}\right)^2 t(-x,-y)*\delta(x,y) \propto t(-x,-y). \end{aligned} \tag{4.2-8}$$

上式表示的成像系统会在输出平面上输出翻转和倒转的像, 这与如图 4.2-2 所示的由射线光学图所表示的意思一致.

By definition, the Fourier transform of the impulse response is the transfer function. The transform of the CPSF is called the *coherent transfer function* (CTF):

$$H_c(k_x, k_y) = \mathcal{F}\{h_c(x,y)\} = \mathcal{F}\left\{P\left(\frac{k_0 x}{f}, \frac{k_0 y}{f}\right)\right\}$$
$$= p(x,y)\Big|_{x=\frac{-fk_x}{k_0},\, y=\frac{-fk_y}{k_0}} = p\left(\frac{-fk_x}{k_0}, \frac{-fk_y}{k_0}\right). \quad (4.2\text{-}5)$$

If we take the Fourier transform of Eq. (4.2-4), we have

$$\mathcal{F}\{\psi_{p,out}(x,y)\} = \mathcal{F}\left\{t(-x,-y) * P\left(\frac{k_0 x}{f}, \frac{k_0 y}{f}\right)\right\}$$
$$= \mathcal{F}\{t(-x,-y)\}\,\mathcal{F}\left\{P\left(\frac{k_0 x}{f}, \frac{k_0 y}{f}\right)\right\}$$
$$= \mathcal{F}\{t(-x,-y)\}\,H_c(k_x, k_y)$$
$$= T(-k_x, -k_y) H_c(k_x, k_y),$$

or

$$\psi_{p,out}(x,y) = \mathcal{F}^{-1}\left\{T(-k_x,-k_y)H_c(k_x,k_y)\right\}. \quad (4.2\text{-}6)$$

This equation has the same connotation in terms of filtering as we have discussed earlier in Chapter 2 when comparing to the equation below [see Eq. (2.2-13b)]:

$$y(t) = \mathcal{F}^{-1}\{Y(\omega)\} = \mathcal{F}^{-1}\{X(\omega)H(\omega)\}.$$

However, in coherent image processing, we process the flipped and inverted spectrum of $t(x,y)$, that is, $T(-k_x,-k_y)$, owing to the double Fourier transformation naturally occurring in the two-lens system as the lenses only perform forward Fourier transform. Finally, once the output complex field is found, the corresponding image intensity is

$$I(x,y) = \psi_{p,out}(x,y)\psi^*_{p,out}(x,y) = |t(-x,-y) * h_c(x,y)|^2, \quad (4.2\text{-}7)$$

which is the basis for *coherent image processing*.

4.2.2 Coherent Image Processing Examples

Example: All-Pass Filtering, $p(x,y) = 1$

From Eq. (4.2-4), the CPSF is then given by

$$h_c(x,y) = \mathcal{F}\{1\}\Big|_{k_x=\frac{k_0 x}{f},\, k_y=\frac{k_0 y}{f}} = 4\pi^2\delta\left(\frac{k_0 x}{f}, \frac{k_0 y}{f}\right) = 4\pi^2\left(\frac{f}{k_0}\right)^2 \delta(x,y),$$

where we have used the scaling property of the delta function to obtain the last step of the equation [see Eq. (2.1-6)]. The output, according to Eq. (4.2-4), is

$$\psi_{p,out}(x,y) = t(-x,-y) * h_c(x,y)$$
$$= 4\pi^2\left(\frac{f}{k_0}\right)^2 t(-x,-y) * \delta(x,y) \propto t(-x,-y). \quad (4.2\text{-}8)$$

This represents an imaging system with the output being flipped and inverted on the output plane, which is consistent with the ray optics diagram shown in Figure 4.2-2.

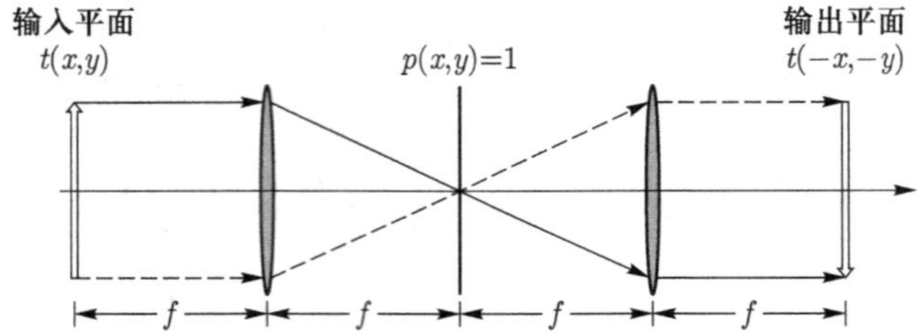

图 4.2-2 双透镜成像系统

由式 (4.2-5), 相干传递函数为

$$H_c(k_x,k_y)=p(x,y)\Big|_{x=\frac{-fk_x}{k_0},\,y=\frac{-fk_y}{k_0}}=1,$$

表示全通滤波 (*all-pass filtering*).

另一个分析该问题的方法是通过式 (4.2-1), 其中输入会进行两次傅里叶变换, 因此, 有

$$\begin{aligned}\psi_{p,out}(x,y)&=\mathcal{F}\left\{T\left(\frac{k_0x}{f},\frac{k_0y}{f}\right)p(x,y)\right\}\Big|_{k_x=\frac{k_0x}{f},\,k_y=\frac{k_0y}{f}}\\&=\mathcal{F}\left\{T\left(\frac{k_0x}{f},\frac{k_0y}{f}\right)\right\}\Big|_{k_x=\frac{k_0x}{f},\,k_y=\frac{k_0y}{f}}.\end{aligned}$$

利用式 (4.2-2), 上式变为

$$\begin{aligned}\psi_{p,out}(x,y)&=\mathcal{F}\left\{\iint t(x',y')\mathrm{e}^{\frac{\mathrm{j}k_0}{f}(xx'+yy')}\mathrm{d}x'\mathrm{d}y'\right\}\Big|_{k_x=\frac{k_0x}{f},\,k_y=\frac{k_0y}{f}}\\&=\iint\left[\iint t(x',y')\mathrm{e}^{\frac{\mathrm{j}k_0}{f}(x''x'+y''y')}\mathrm{d}x'\mathrm{d}y'\right]\mathrm{e}^{\mathrm{j}(k_xx''+k_yy'')}\mathrm{d}x''\mathrm{d}y''\\&=\iint\left[\iint t(x',y')\mathrm{e}^{\frac{\mathrm{j}k_0}{f}(x''x'+y''y')}\mathrm{d}x'\mathrm{d}y'\right]\mathrm{e}^{\mathrm{j}\left(\frac{k_0x}{f}x''+\frac{k_0y}{f}y''\right)}\mathrm{d}x''\mathrm{d}y'',\end{aligned}\tag{4.2-9}$$

其中, 用傅里叶变换的定义 [见式 (2.3-1a)] 写出上述积分. 现在来处理带有两撇的坐标变量的积分

$$\begin{aligned}&\iint\mathrm{e}^{\frac{\mathrm{j}k_0}{f}(x''x'+y''y')}\mathrm{e}^{\mathrm{j}\left(\frac{k_0x}{f}x''+\frac{k_0y}{f}y''\right)}\mathrm{d}x''\mathrm{d}y''\\=&\iint\mathrm{e}^{\frac{\mathrm{j}k_0}{f}\left[x''\left(x'+x\right)+y''(y'+y)\right]}\mathrm{d}x''\mathrm{d}y''\\=&\,4\pi^2\delta\left(\frac{k_0}{f}(x'+x),\frac{k_0}{f}(y'+y)\right)=4\pi^2\left(\frac{f}{k_0}\right)^2\delta(x'+x,y'+y),\end{aligned}\tag{4.2-10}$$

这里, 我们来回顾 δ 函数的积分定义 [见式 (2.3-4)]

$$\delta(x,y)=\frac{1}{4\pi^2}\iint\limits_{-\infty}^{\infty}\mathrm{e}^{\pm\mathrm{j}xx'\pm\mathrm{j}yy'}\mathrm{d}x'\mathrm{d}y'.$$

将式 (4.2-10) 代入式 (4.2-9), 得

$$\begin{aligned}\psi_{p,out}(x,y)&=4\pi^2\left(\frac{f}{k_0}\right)^2\iint t(x',y')\delta(x'+x,y'+y)\mathrm{d}x'\mathrm{d}y'\\&=4\pi^2\left(\frac{f}{k_0}\right)^2t(-x,-y)\propto t(-x,-y)\end{aligned}$$

这与式 (4.2-8) 的结果一样.

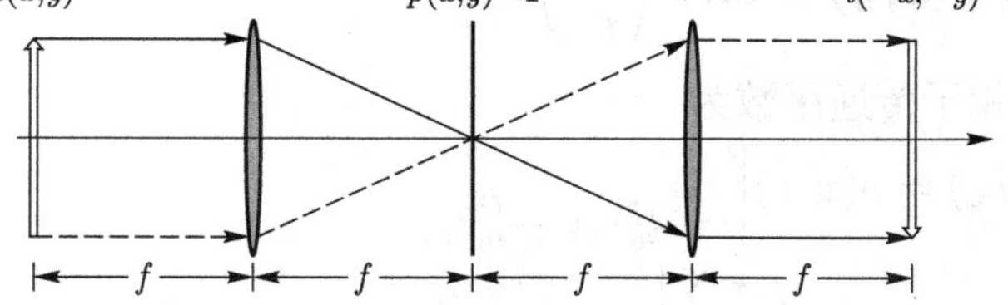

Figure 4.2-2 Two-lens imaging system

According to Eq. (4.2-5), the coherent transfer function is

$$H_c(k_x, k_y) = p(x, y)\Big|_{x=\frac{-fk_x}{k_0},\, y=\frac{-fk_y}{k_0}} = 1,$$

representing *all-pass filtering*.

An alternative way to analyze the problem is through Eq. (4.2-1), where the input has gone through double Fourier transforms; we, therefore, have

$$\begin{aligned}\psi_{p,out}(x, y) &= \mathcal{F}\left\{T\left(\frac{k_0 x}{f}, \frac{k_0 y}{f}\right) p(x, y)\right\}\Big|_{k_x=\frac{k_0 x}{f},\, k_y=\frac{k_0 y}{f}} \\ &= \mathcal{F}\left\{T\left(\frac{k_0 x}{f}, \frac{k_0 y}{f}\right)\right\}\Big|_{k_x=\frac{k_0 x}{f},\, k_y=\frac{k_0 y}{f}}.\end{aligned}$$

Using Eq. (4.2-2), the equation above becomes

$$\begin{aligned}\psi_{p,out}(x, y) &= \mathcal{F}\left\{\iint t(x', y')\mathrm{e}^{\frac{\mathrm{j}k_0}{f}(xx'+yy')}\mathrm{d}x'\mathrm{d}y'\right\}\Big|_{k_x=\frac{k_0 x}{f},\, k_y=\frac{k_0 y}{f}} \\ &= \iint\left[\iint t(x', y')\mathrm{e}^{\frac{\mathrm{j}k_0}{f}(x''x'+y''y')}\mathrm{d}x'\mathrm{d}y'\right]\mathrm{e}^{\mathrm{j}(k_x x''+k_y y'')}\mathrm{d}x''\mathrm{d}y'' \\ &= \iint\left[\iint t(x', y')\mathrm{e}^{\frac{\mathrm{j}k_0}{f}(x''x'+y''y')}\mathrm{d}x'\mathrm{d}y'\right]\mathrm{e}^{\mathrm{j}\left(\frac{k_0 x}{f}x''+\frac{k_0 y}{f}y''\right)}\mathrm{d}x''\mathrm{d}y'',\end{aligned} \tag{4.2-9}$$

where we have used the definition of Fourier transform [see Eq. (2.3-1a)] to set up the above integral. Let us now perform the integration involving the double primed coordinates first:

$$\begin{aligned}&\iint \mathrm{e}^{\frac{\mathrm{j}k_0}{f}(x''x'+y''y')}\mathrm{e}^{\mathrm{j}\left(\frac{k_0 x}{f}x''+\frac{k_0 y}{f}y''\right)}\mathrm{d}x''\mathrm{d}y'' \\ &= \iint \mathrm{e}^{\frac{\mathrm{j}k_0}{f}[x''(x'+x)+y''(y'+y)]}\mathrm{d}x''\mathrm{d}y'' \\ &= 4\pi^2\delta\left(\frac{k_0}{f}(x'+x), \frac{k_0}{f}(y'+y)\right) = 4\pi^2\left(\frac{f}{k_0}\right)^2\delta(x'+x, y'+y),\end{aligned} \tag{4.2-10}$$

where we have recalled the integral definition of the delta function as [see Eq (2.3-4)]

$$\delta(x, y) = \frac{1}{4\pi^2}\iint_{-\infty}^{\infty}\mathrm{e}^{\pm\mathrm{j}xx'\pm\mathrm{j}yy'}\mathrm{d}x'\mathrm{d}y'.$$

Putting Eq. (4.2-10) into Eq. (4.2-9), we have

$$\begin{aligned}\psi_{p,out}(x, y) &= 4\pi^2\left(\frac{f}{k_0}\right)^2\iint t(x', y')\delta(x'+x, y'+y)\mathrm{d}x'\mathrm{d}y' \\ &= 4\pi^2\left(\frac{f}{k_0}\right)^2 t(-x, -y) \propto t(-x, -y)\end{aligned}$$

which is the same result in Eq. (4.2-8).

举例: 低通滤波, $p(x,y) = circ\left(\dfrac{r}{r_0}\right)$

根据式 (4.2-5), 相干传递函数为

$$\begin{aligned} H_c(k_x,k_y) &= p(x,y)\Big|_{x=\frac{-fk_x}{k_0},\,y=\frac{-fk_y}{k_0}} \\ &= circ\left(\frac{r}{r_0}\right)\Big|_{x=\frac{-fk_x}{k_0},\,y=\frac{-fk_y}{k_0}} \\ &= circ\left(\frac{\sqrt{x^2+y^2}}{r_0}\right)\Bigg|_{x=\frac{-fk_x}{k_0},\,y=\frac{-fk_y}{k_0}} = circ\left(\frac{k_r}{r_0k_0/f}\right), \end{aligned}$$

其中, $k_r = \sqrt{k_x^2+k_y^2}$. 这时为低通滤波 (*low-pass filtering*), 因为光瞳平面的圆孔开口大小仅允许低于 r_0k_0/f 空间频率的光通过. 图 4.2-3 给出了低通滤波的例子. 在图 4.2-3a) 和图 4.2-3b) 中, 分别给出了原始图像及其频谱. 在图 4.2-3c) 和图 4.2-3e) 中, 给出了 r_0 取不同值时低通滤波后的频谱, 并在图 4.2-3d) 和图 4.2-3f) 中分别给出其相应的滤波后的图像.当 "白色" 圆的半径为 r_0k_0/f 时, 经过滤波后的频谱由原始图像的频谱与 $circ\left(\dfrac{k_r}{r_0k_0/f}\right)$ 相乘而得 [见式 (4.2-6)]. 可以看到, 图 4.2-3c) 中的半径比图 4.2-3e) 中的半径大, 这表示更多的空间频率通过了该光学系统. 一般来说, 经低通滤波后的图像与原始图像相比是模糊的或平滑的.

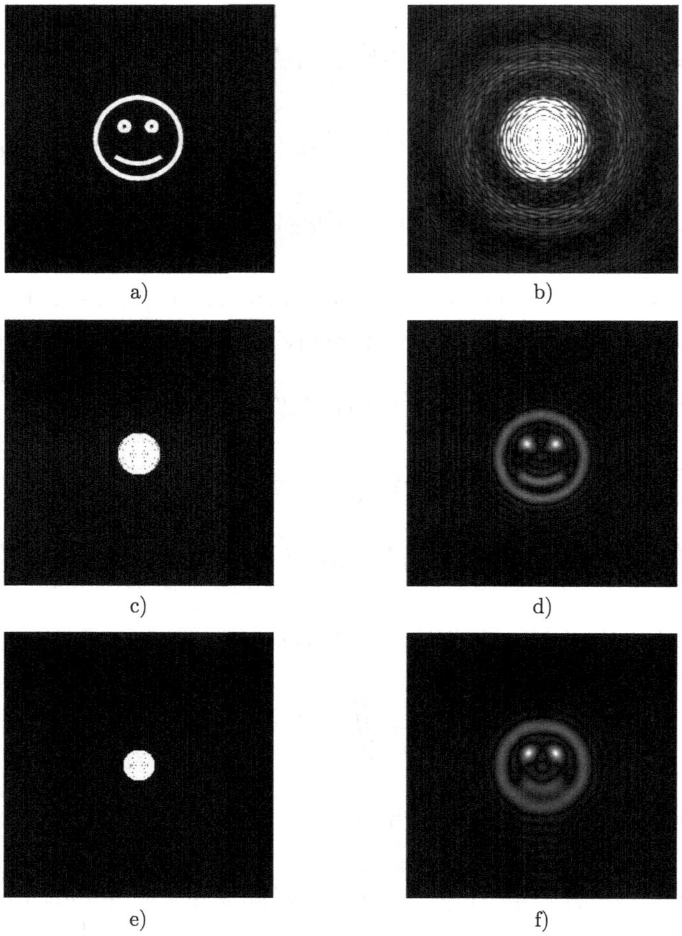

图 4.2-3 低通滤波举例: a) 原始图像; b) 图 a) 的频谱; c) 低通滤波后的频谱; d) 低通滤波频谱为图 c) 中情况时图 a) 的低通滤波图像; e) 当开口设计 r_0k_0/f 比图 c) 中的小时, 得到的低通滤波后的频谱; f) 低通滤波频谱为图 e) 中情况时图 a) 的低通滤波图像

Example: Low-Pass Filtering, $p(x, y) = circ\left(\frac{r}{r_0}\right)$

According to Eq. (4.2-5), the coherent transfer function is

$$H_c(k_x, k_y) = p(x, y)\Big|_{x=\frac{-fk_x}{k_0},\, y=\frac{-fk_y}{k_0}}$$
$$= circ\left(\frac{r}{r_0}\right)\Big|_{x=\frac{-fk_x}{k_0},\, y=\frac{-fk_y}{k_0}}$$
$$= circ\left(\frac{\sqrt{x^2+y^2}}{r_0}\right)\Bigg|_{x=\frac{-fk_x}{k_0},\, y=\frac{-fk_y}{k_0}} = circ\left(\frac{k_r}{r_0k_0/f}\right),$$

where $k_r = \sqrt{k_x^2 + k_y^2}$. This translates to *low-pass filtering* as the opening of the circle on the pupil plane only allows spatial frequencies up to r_0k_0/f to transmit. Figure 4.2-3 shows an example of low-pass filtering. In Figure 4.2-3a and b, we show the original image and its spectrum, respectively. In Figure 4.2-3c and e, we show the low-pass filtered spectra for different values of r_0, and their corresponding filtered images are shown in Figure 4.2-3d and f, respectively. The filtered spectra are obtained by multiplying the original spectrum by $circ\left(\frac{k_r}{r_0k_0/f}\right)$ [see Eq. (4.2-6)] as the radius of the "white" circle is r_0k_0/f. Note that the radius in Figure 4.2-3c is larger than that in Figure 4.2-3e, meaning we let more spatial frequencies pass through the optical system. In general, the low-pass filtered images are

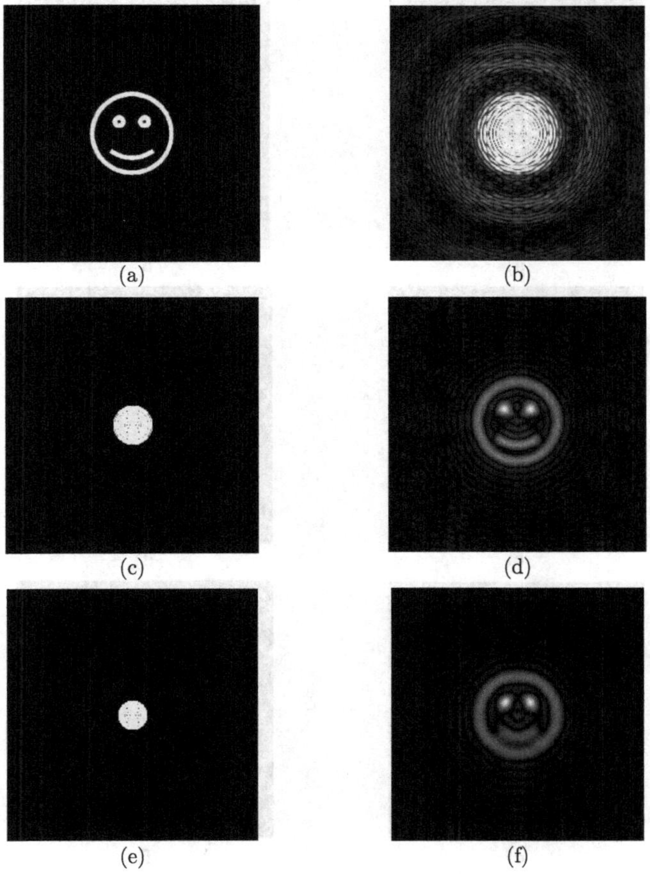

Figure 4.2-3 Low-pass filtering example: (a) original image, (b) spectrum of (a), (c) low-pass filtered spectrum, (d) low-pass filtered image of (a) for the low-pass filtered spectrum in (c), (e) low-pass filtered spectrum with r_0k_0/f designed to be smaller than that in (c), (f) low-pass filtered image of (a) for the low-pass filtered spectrum in (e)

事实证明, 一个低通滤波器也被称为 “模糊” 或 “平滑” 滤波器, 会对原始图像中的任何急剧变化进行平滑处理.

举例: 高通滤波, $p(x,y)=1-circ\left(\frac{r}{r_0}\right)$

相干传递函数为

$$\begin{aligned}H_c(k_x,k_y)&=p(x,y)\Big|_{x=\frac{-fk_x}{k_0},\,y=\frac{-fk_y}{k_0}}\\&=1-circ\left(\frac{r}{r_0}\right)\Big|_{x=\frac{-fk_x}{k_0},\,y=\frac{-fk_y}{k_0}}\\&=1-circ\left(\frac{\sqrt{x^2+y^2}}{r_0}\right)\Bigg|_{x=\frac{-fk_x}{k_0},\,y=\frac{-fk_y}{k_0}}\\&=1-circ\left(\frac{k_r}{\frac{r_0k_0}{f}}\right)=\overline{circ}\left(\frac{k_r}{\frac{r_0k_0}{f}}\right).\end{aligned}$$

当 $circ\left(\frac{k_r}{r_0k_0/f}\right)$ 为一个低通滤波器, $\overline{circ}\left(\frac{k_r}{r_0k_0/f}\right)$ 就是一个高通滤波器, 因为在光瞳平面上圆孔的遮挡限制了空间频率低于 r_0k_0/f 的光通过. 在图 4.2-4a) 和图 4.2-4b) 中, 分别给出了原始图像及其频谱. 在图 4.2-4c) 和图 4.2-4e) 中, 给出了高通滤波 (*high-pass filtering*) 后的频谱,

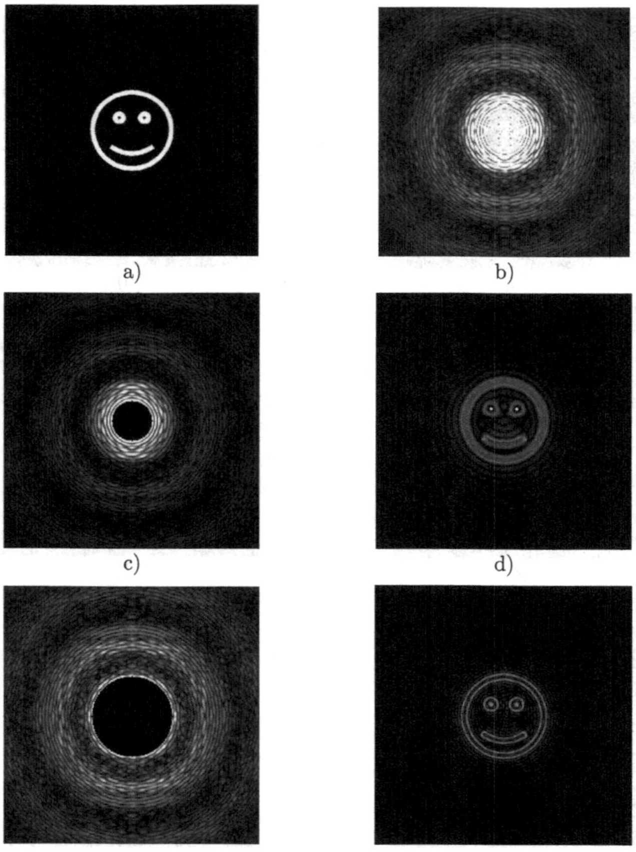

图 4.2-4 高通滤波举例: a) 原始图像; b) 图 a) 的频谱; c) 高通滤波后的频谱; d) 高通滤波频谱为图 c) 中情况时图 a) 的高通滤波图像; e) 当设计的尺寸 r_0k_0/f 比图 c) 中所示要大时, 得到的高通滤波后的频谱; f) 高通滤波的频谱为图 e) 中情况时图 a) 的高通滤波图像

blurred or have been smooth as compared to the original image. As it turns out, a low-pass filter, also called a "blurring" or "smoothing" filter, averages out any rapid changes in the original image.

Example: High-Pass Filtering, $p(x, y) = 1 - circ\left(\frac{r}{r_0}\right)$

The coherent transfer function is

$$\begin{aligned} H_c(k_x, k_y) &= p(x, y)\Big|_{x=\frac{-fk_x}{k_0},\, y=\frac{-fk_y}{k_0}} \\ &= 1 - circ\left(\frac{r}{r_0}\right)\Big|_{x=\frac{-fk_x}{k_0},\, y=\frac{-fk_y}{k_0}} \\ &= 1 - circ\left(\frac{\sqrt{x^2+y^2}}{r_0}\right)\Bigg|_{x=\frac{-fk_x}{k_0},\, y=\frac{-fk_y}{k_0}} \\ &= 1 - circ\left(\frac{k_r}{r_0k_0/f}\right) = \overline{circ}\left(\frac{k_r}{r_0k_0/f}\right). \end{aligned}$$

While $circ\left(\frac{k_r}{r_0k_0/f}\right)$ is a low-pass filter, $\overline{circ}\left(\frac{k_r}{r_0k_0/f}\right)$ is a high-pass filter as the blocking of the circle on the pupil plane forbids spatial frequencies up to r_0k_0/f to transmit. In Figure 4.2-4a and b, we show the original image and its spectrum, respectively. In Figure 4.2-4c and e, we show the high-pass filtered spectra, and their corresponding filtered images are

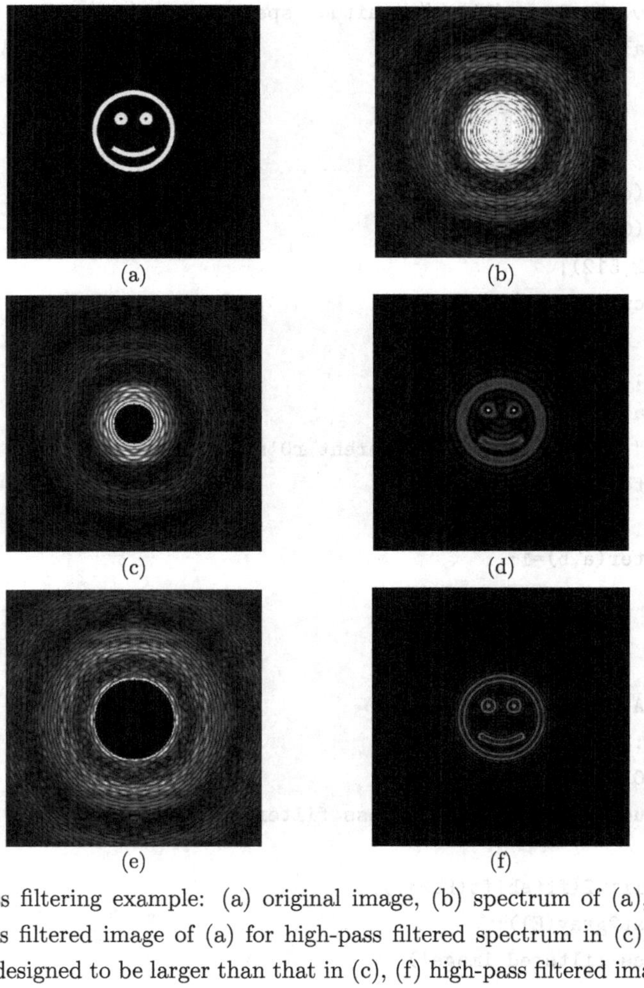

Figure 4.2-4 High-pass filtering example: (a) original image, (b) spectrum of (a), (c) high-pass filtered spectrum, (d) high-pass filtered image of (a) for high-pass filtered spectrum in (c), (e) high-pass filtered spectrum with r_0k_0/f designed to be larger than that in (c), (f) high-pass filtered image of (a) for high-pass filtered spectrum in (e)

并在图 4.2-4d) 和图 4.2-4f) 中分别给出其对应的经过滤波器后的图像. 该滤波后的频谱是从原始图像的频谱与 $\overline{circ}\left(\dfrac{k_r}{r_0k_0/f}\right)$ 的乘积得到的. 可以看出, 对于图 4.2-4e) 所给出的充分遮挡掉低空间频率的情况, 其中半径为 r_0k_0/f 的黑色圆比图 4.2-4c) 中的半径大, 高通滤波的图像就被进行边缘提取了. 事实上, 高通滤波在边缘检测方面非常有用, 如确定图像中的边界. 而边缘检测 (*edge detection*) 在模式识别的预处理 (*pre-processing*) 中亦起着至关重要的作用.

图 4.2-3 和图 4.2-4 为下面 m-文件所生成的结果.

```
========================================
    % low-pass_high-pass.m, Low-pass and high-pass filtering of an image
    % Adapted from "Introduction to Modern Digital Holography"
    % by T.-C. Poon & J.-P. Liu
    % Cambridge University Press(2014), Tables 1.5 and 1.6.
    clear all;close all;
    A=imread('front.jpg');          % read 512 X 512 8-bit image
    A=double(A);
    A=A/255;
    FA=fftshift(fft2(fftshift(A)));
    D=abs(FA);
    D=D(129:384,129:384);
    figure;imshow(A);
    % title('Original image')
    figure;imshow(30.*mat2gray(D)); %magnitude spectrum
    % title('Original spectrum')

    c=1:512;
    r=1:512;
    [C, R]=meshgrid(c, r);
    CI=((R-257).^2+(C-257).^2);
    filter=zeros(512,512);
    % produce a circular low-pass filter
    for a=1:512;
        for b=1:512;
             if CI(a,b)>=20^2;
    %Used "20" and "15" to simulate different r0's for results in text
                filter(a,b)=0;
            else
                filter(a,b)=1;
            end
        end
    end
    G=abs(filter.*FA);
    G=G(129:384,129:384);
    figure;imshow(30.*mat2gray(G));
    % title('Magnitude spectrum of low-pass filtered image')
    FLP=FA.*filter;
    E=abs(fftshift(ifft2(fftshift(FLP))));
    figure;imshow(mat2gray(E));
    % title('Low-pass filtered image')
```

shown in Figure 4.2-4d and f, respectively. The filtered spectra are obtained by multiplying the original spectrum by $\overline{circ}\left(\dfrac{k_r}{r_0k_0/f}\right)$. Note that for sufficient blocking of low spatial frequencies such as the case shown in Figure 4.2-4e, where the "black" circle of radius r_0k_0/f is larger than that in Figure 4.2-4c, the high-pass filtered image is edge-extracted. Indeed, *high-pass filtering* is useful for edge detection, that is, locating boundaries in images. *Edge detection* plays a crucial role in *pre-processing* for pattern recognition.

Figures 4.2-3 and 4.2-4 are generated using the m-file shown below.

```
==========================================
% low-pass_high-pass.m, Low-pass and high-pass filtering of an image
% Adapted from "Introduction to Modern Digital Holography"
% by T.-C. Poon & J.-P. Liu
% Cambridge University Press(2014), Tables 1.5 and 1.6.
clear all;close all;
A=imread('front.jpg');            % read 512 X 512 8-bit image
A=double(A);
A=A/255;
FA=fftshift(fft2(fftshift(A)));
D=abs(FA);
D=D(129:384,129:384);
figure;imshow(A);
% title('Original image')
figure;imshow(30.*mat2gray(D)); %magnitude spectrum
% title('Original spectrum')

c=1:512;
r=1:512;
[C, R]=meshgrid(c, r);
CI=((R-257).^2+(C-257).^2);
filter=zeros(512,512);
% produce a circular low-pass filter
for a=1:512;
    for b=1:512;
         if CI(a,b)>=20^2;
%Used "20" and "15" to simulate different r0's for results in text
            filter(a,b)=0;
        else
            filter(a,b)=1;
        end
    end
end
G=abs(filter.*FA);
G=G(129:384,129:384);
figure;imshow(30.*mat2gray(G));
% title('Magnitude spectrum of low-pass filtered image')
FLP=FA.*filter;
E=abs(fftshift(ifft2(fftshift(FLP))));
figure;imshow(mat2gray(E));
% title('Low-pass filtered image')
```

```
c=1:512;
r=1:512;
[C, R]=meshgrid(c, r);
CI=((R-257).^2+(C-257).^2);
filter=zeros(512,512);
% produce a circular high-pass filter
for a=1:512;
    for b=1:512;
         if CI(a,b)<=20^2;
%Used "20" and "40" to simulate different r0's for results in text
            filter(a,b)=0;
        else
            filter(a,b)=1;
        end
    end
end
G=abs(filter.*FA);
G=G(129:384,129:384);
figure;imshow(2.*mat2gray(G));
% title('Magnitude spectrum of high-pass filtered image')
FHP=FA.*filter;
E=abs(fftshift(ifft2(fftshift(FHP))));
figure;imshow(mat2gray(E));
% title('High-pass filtered image')
```
==

4.3　空间非相干图像处理

上节讨论了透明片物体的空间相干照明, 例如用激光产生的平面波来照明. 然而, 来自扩展光源的光, 例如荧光灯的光, 是空间非相干的. 如图 4.2-1 中所示的系统, 在非相干光的照明下变为非相干光学系统. 一个相干系统对于其复振幅来说是线性的, 因此, 当处理复振幅时, 式 (4.2-4) 和式 (4.2-7) 对于相干光学系统是成立的 [见式 (4.1-5)]. 另一方面, 非相干光学系统对于其强度来说是线性的. 换句话说, 由之前在 4.1 节中对空间相干理论的讨论可知, 对空间非相干光的处理, 可通过对其强度进行相加而实现 [见式 (4.1-6)]. 为了得到图像的强度, 对给定的强度量进行如下卷积

$$I(x,y)=|t(-x,-y)|^2*|h_c(x,y)|^2. \tag{4.3-1}$$

式 (4.3-1) 是线性空不变非相干系统 (*linear space-invariant incoherent system*) 中非相干图像处理 (*incoherent image processing*) 的基础.

4.3.1　强度点扩散函数和光学传递函数

通过观察式 (4.3-1), 可以看出, 输入强度和输出强度之间通过脉冲响应 $|h_c(x,y)|^2$ 联系了起来, 这在非相干系统中被称为强度点扩散函数 (*intensity point spread function*, IPSF)

```
c=1:512;
r=1:512;
[C, R]=meshgrid(c, r);
CI=((R-257).^2+(C-257).^2);
filter=zeros(512,512);
% produce a circular high-pass filter
for a=1:512;
    for b=1:512;
         if CI(a,b)<=20^2;
%Used "20" and "40" to simulate different r0's for results in text
              filter(a,b)=0;
         else
              filter(a,b)=1;
         end
    end
end
G=abs(filter.*FA);
G=G(129:384,129:384);
figure;imshow(2.*mat2gray(G));
% title('Magnitude spectrum of high-pass filtered image')
FHP=FA.*filter;
E=abs(fftshift(ifft2(fftshift(FHP))));
figure;imshow(mat2gray(E));
% title('High-pass filtered image')
```
==

4.3 Spatial Incoherent Image Processing

In the last section, we have discussed spatially coherent illumination of an object transparency, such as the use of a laser producing plane waves for illumination. However, light from extended sources, such as fluorescent tube lamps, is spatially incoherent. The system shown in Figure 4.2-1 becomes an incoherent optical system upon illumination from an incoherent source. A coherent system is linear with respect to the complex amplitudes and hence Eqs. (4.2-4) and (4.2-7) hold for coherent optical systems as we manipulate complex amplitudes [see Eq. (4.1-5)]. On the other hand, an incoherent optical system is linear with respect to the intensities. In other words, we manipulate intensities, which is not totally surprising as we have seen from our earlier discussion on spatial coherence in Section 4.1 that we add intensities when dealing with spatial incoherent light [see Eq. (4.1-6)]. To find the image intensity, we perform convolution with the given intensity quantities as follows:

$$I(x,y) = |t(-x,-y)|^2 * |h_c(x,y)|^2. \tag{4.3-1}$$

Equation (4.3-1) is the basis for *incoherent image processing* in a *linear space-invariant incoherent system*.

4.3.1 Intensity Point Spread Function and Optical Transfer Function

By inspecting Eq. (4.3-1), we see that the input and output intensities are related by an impulse response $|h_c(x,y)|^2$, which is known as the *intensity point spread function* (IPSF) in

$$IPSF(x,y)=|h_c(x,y)|^2. \tag{4.3-2}$$

强度点扩散函数的傅里叶变换即为传递函数, 被称为非相干系统的光学传递函数 (*optical transfer function*, OTF)

$$OTF(k_x,k_y)=\mathcal{F}\left\{|h_c(x,y)|^2\right\}. \tag{4.3-3}$$

由式 (4.2-5), 可以写出 $h_c(x,y)=\mathcal{F}^{-1}\{H_c(k_x,k_y)\}$. 再将 $h_c(x,y)$ 的表达式代入式 (4.3-3), 可以看出, OTF 与相干传递函数的关系如下

$$\begin{aligned}OTF(k_x,k_y)&=H_c(k_x,k_y)\otimes H_c(k_x,k_y)\\&=\iint\limits_{-\infty}^{\infty}H_c^*(k_x',k_y')H_c(k_x'+k_x,k_y'+k_y)\mathrm{d}k_x'\mathrm{d}k_y',\end{aligned} \tag{4.3-4}$$

其中, $\otimes$ 表示相关 [见式 (2.3-30)]. OTF 是 H_c 的自相关且通常来说为复数. OTF 的模被称为调制传递函数 (*modulation transfer function*, MTF).

OTF 一些重要的性质如下

$$OTF(-k_x,-k_y)=OTF^*(k_x,k_y), \tag{4.3-5a}$$

$$|OTF(k_x,k_y)|\leqslant|OTF(0,0)| \quad \text{或} \quad MTF(k_x,k_y)\leqslant MTF(0,0). \tag{4.3-5b}$$

式 (4.3-5b) 尤为重要. 该性质表明 MTF 总有一个中心最大值, 这表示无论在非相干系统中使用的瞳函数如何, 它始终表示低通滤波. 这就产生了问题, 例如, 如何在非相干系统中进行高通滤波, 我们将会在本章最后一节阐述这个重要的问题.

通过对式 (4.3-1) 进行傅里叶变换, 有

$$\begin{aligned}\mathcal{F}\{I(x,y)\}&=\mathcal{F}\left\{|t(-x,-y)|^2\right\}\mathcal{F}\left\{|h_c(x,y)|^2\right\}\\&=\mathcal{F}\left\{|t(-x,-y)|^2\right\}OTF(k_x,k_y),\end{aligned}$$

或

$$I(x,y)=\mathcal{F}^{-1}\left\{\mathcal{F}\left\{|t(-x,-y)|^2\right\}OTF(k_x,k_y)\right\}. \tag{4.3-6}$$

再一次, 与下面的等式进行比较 [见式 (2.2-13b)], 可以发现, 式 (4.3-6) 在滤波方面与我们在前面第二章中讨论的有着同样的意义, 即

$$y(t)=\mathcal{F}^{-1}\{Y(\omega)\}=\mathcal{F}^{-1}\{X(\omega)H(\omega)\}.$$

举例: 证明 $|OTF(k_x,k_y)|\leqslant|OTF(0,0)|$

利用施瓦茨不等式 (*Schwarz inequality*), 给定两个复函数 $P(k_x,k_y)$ 和 $Q(k_x,k_y)$, 施瓦茨不等式表明

$$\left|\iint\limits_{-\infty}^{\infty}P(k_x',k_y')Q(k_x',k_y')\mathrm{d}k_x'\mathrm{d}k_y'\right|\leqslant\left[\iint\limits_{-\infty}^{\infty}|P(k_x',k_y')|^2\mathrm{d}k_x'\mathrm{d}k_y'\right]^{\frac{1}{2}}\times\left[\iint\limits_{-\infty}^{\infty}|Q(k_x',k_y')|^2\mathrm{d}k_x'\mathrm{d}k_y'\right]^{\frac{1}{2}}.$$

incoherent systems:

$$IPSF(x,y) = |h_c(x,y)|^2. \tag{4.3-2}$$

The Fourier transform of the IPSF gives a transfer function known as the *optical transfer function* (OTF) of the incoherent system:

$$OTF(k_x,k_y) = \mathcal{F}\left\{|h_c(x,y)|^2\right\}. \tag{4.3-3}$$

According to Eq. (4.2-5), we can write $h_c(x,y) = \mathcal{F}^{-1}\{H_c(k_x,k_y)\}$. Upon substitution of this expression of $h_c(x,y)$ into Eq. (4.3-3), it can be shown that the OTF and the coherent transfer function are related as follows:

$$\begin{aligned} OTF(k_x,k_y) &= H_c(k_x,k_y) \otimes H_c(k_x,k_y) \\ &= \iint\limits_{-\infty}^{\infty} H_c^*(k_x',k_y')H_c(k_x'+k_x,k_y'+k_y)\mathrm{d}k_x'\mathrm{d}k_y', \end{aligned} \tag{4.3-4}$$

where $\otimes$ denotes correlation [see Eq. (2.3-30)]. The OTF is the autocorrelation of H_c and in general it is complex. The modulus of the OTF is called the *modulation transfer function* (MTF).

Some important properties of the OTF are as follows:

$$OTF(-k_x,-k_y) = OTF^*(k_x,k_y), \tag{4.3-5a}$$

$$|OTF(k_x,k_y)| \leqslant |OTF(0,0)| \quad \text{or} \quad MTF(k_x,k_y) \leqslant MTF(0,0). \tag{4.3-5b}$$

Equation (4.3-5b) is particularly important. The property states that the MTF always has a central maximum, which always signifies low-pass filtering regardless of the pupil function used in an incoherent system. This gives rise to a question on how to implement high-pass filtering in incoherent systems, for example. We will address this important question in the last section of this chapter.

By taking the Fourier transform of Eq. (4.3-1), we have

$$\begin{aligned} \mathcal{F}\{I(x,y)\} &= \mathcal{F}\left\{|t(-x,-y)|^2\right\}\mathcal{F}\left\{|h_c(x,y)|^2\right\} \\ &= \mathcal{F}\left\{|t(-x,-y)|^2\right\}OTF(k_x,k_y), \end{aligned}$$

or

$$I(x,y) = \mathcal{F}^{-1}\left\{\mathcal{F}\left\{|t(-x,-y)|^2\right\}OTF(k_x,k_y)\right\}. \tag{4.3-6}$$

Again, this equation has the same spirit in terms of filtering as we have discussed earlier in Chapter 2 when comparing to the equation below [see Eq. (2.2-13b)]:

$$y(t) = \mathcal{F}^{-1}\{Y(\omega)\} = \mathcal{F}^{-1}\{X(\omega)H(\omega)\}.$$

Example: Show that $|OTF(k_x,k_y)| \leqslant |OTF(0,0)|$

We use *Schwarz inequality*. Given two complex functions $P(k_x,k_y)$ and $Q(k_x,k_y)$, Schwarz inequality states that

$$\left|\iint\limits_{-\infty}^{\infty} P(k_x',k_y')Q(k_x',k_y')\mathrm{d}k_x'\mathrm{d}k_y'\right| \leqslant \left[\iint\limits_{-\infty}^{\infty}|P(k_x',k_y')|^2\mathrm{d}k_x'\mathrm{d}k_y'\right]^{\frac{1}{2}} \times \left[\iint\limits_{-\infty}^{\infty}|Q(k_x',k_y')|^2\mathrm{d}k_x'\mathrm{d}k_y'\right]^{\frac{1}{2}}.$$

通过令 $P(k_x', k_y') = H_c^*(k_x', k_y')$ 且 $Q(k_x', k_y') = H_c(k_x' + k_x, k_y' + k_y)$, 有

$$\begin{aligned}&\left|\iint_{-\infty}^{\infty} H_c^*(k_x', k_y') H_c(k_x' + k_x, k_y' + k_y) \mathrm{d}k_x' \mathrm{d}k_y'\right| \\ \leqslant & \left[\iint_{-\infty}^{\infty} |H_c^*(k_x', k_y')|^2 \mathrm{d}k_x' \mathrm{d}k_y'\right]^{\frac{1}{2}} \times \left[\iint_{-\infty}^{\infty} \left| H_c(k_x' + k_x, k_y' + k_y)\right|^2 \mathrm{d}k_x' \mathrm{d}k_y'\right]^{\frac{1}{2}}.\end{aligned} \tag{4.3-7}$$

然而, 不等式右边的第一项为

$$\iint_{-\infty}^{\infty} |H_c^*(k_x', k_y')|^2 \mathrm{d}k_x' \mathrm{d}k_y' = \iint_{-\infty}^{\infty} |H_c(k_x', k_y')|^2 \mathrm{d}k_x' \mathrm{d}k_y', \tag{4.3-8}$$

且第二项可以表示为

$$\iint_{-\infty}^{\infty} |H_c(k_x' + k_x, k_y' + k_y)|^2 \mathrm{d}k_x' \mathrm{d}k_y' = \iint_{-\infty}^{\infty} |H_c(k_x', k_y')|^2 \mathrm{d}k_x' \mathrm{d}k_y'. \tag{4.3-9}$$

对于函数的移动, 即移动了 k_x 和 k_y 的位移不会带来积分值的改变. 将式 (4.3-8) 与式 (4.3-9) 的结果代入式 (4.3-7), 得

$$\begin{aligned}&\left|\iint_{-\infty}^{\infty} H_c^*(k_x', k_y') H_c(k_x' + k_x, k_y' + k_y) \mathrm{d}k_x' \mathrm{d}k_y'\right| \\ \leqslant & \left[\iint_{-\infty}^{\infty} |H_c(k_x', k_y')|^2 \mathrm{d}k_x' \mathrm{d}k_y'\right]^{\frac{1}{2}} \times \left[\iint_{-\infty}^{\infty} \left|H_c(k_x', k_y')\right|^2 \mathrm{d}k_x' \mathrm{d}k_y'\right]^{\frac{1}{2}} \\ = & \iint_{-\infty}^{\infty} |H_c(k_x', k_y')|^2 \mathrm{d}k_x' \mathrm{d}k_y'.\end{aligned}$$

用 OTF 来表示上述结果, 有

$$|OTF(k_x, k_y)| \leqslant |OTF(0, 0)|.$$

在式 (4.3-5b) 中的这一公式, 是 OTF 的重要性质.

4.3.2 非相干图像处理实例

举例: 单缝光瞳, $p(x, y) = rect\left(\frac{x}{x_0}\right)$

在如图 4.2-1 所示的标准 4-f 光学图像处理系统中, 以沿 y 方向伸展宽度为 x_0 的狭缝作为光瞳. 根据式 (4.2-5), 其相干传递函数为

$$H_c(k_x, k_y) = p(x, y)\Big|_{x=\frac{-fk_x}{k_0},\, y=\frac{-fk_y}{k_0}} = rect\left(\frac{fk_x}{x_0 k_0}\right) = rect\left(\frac{k_x}{x_0 k_0/f}\right).$$

其 OTF 为

$$OTF(k_x, k_y) = H_c(k_x, k_y) \otimes H_c(k_x, k_y)$$

By letting $P(k_x', k_y') = H_c^*(k_x', k_y')$ and $Q(k_x', k_y') = H_c(k_x' + k_x, k_y' + k_y)$, we have

$$\left| \iint_{-\infty}^{\infty} H_c^*(k_x', k_y') H_c(k_x' + k_x, k_y' + k_y) \mathrm{d}k_x' \mathrm{d}k_y' \right|$$

$$\leqslant \left[\iint_{-\infty}^{\infty} |H_c^*(k_x', k_y')|^2 \mathrm{d}k_x' \mathrm{d}k_y' \right]^{\frac{1}{2}} \times \left[\iint_{-\infty}^{\infty} |H_c(k_x' + k_x, k_y' + k_y)|^2 \mathrm{d}k_x' \mathrm{d}k_y' \right]^{\frac{1}{2}}. \tag{4.3-7}$$

However, the first term of the right side of the inequality is

$$\iint_{-\infty}^{\infty} |H_c^*(k_x', k_y')|^2 \mathrm{d}k_x' \mathrm{d}k_y' = \iint_{-\infty}^{\infty} |H_c(k_x', k_y')|^2 \mathrm{d}k_x' \mathrm{d}k_y', \tag{4.3-8}$$

and in the second term we can express

$$\iint_{-\infty}^{\infty} |H_c(k_x' + k_x, k_y' + k_y)|^2 \mathrm{d}k_x' \mathrm{d}k_y' = \iint_{-\infty}^{\infty} |H_c(k_x', k_y')|^2 \mathrm{d}k_x' \mathrm{d}k_y'. \tag{4.3-9}$$

for the reason that a shift of the function, that is, a shift by k_x and k_y cannot change the value of the integral. By substituting the results of Eqs. (4.3-8) and (4.3-9) into Eq. (4.3-7), we obtain

$$\left| \iint_{-\infty}^{\infty} H_c^*(k_x', k_y') H_c(k_x' + k_x, k_y' + k_y) \mathrm{d}k_x' \mathrm{d}k_y' \right|$$

$$\leqslant \left[\iint_{-\infty}^{\infty} |H_c(k_x', k_y')|^2 \mathrm{d}k_x' \mathrm{d}k_y' \right]^{\frac{1}{2}} \times \left[\iint_{-\infty}^{\infty} |H_c(k_x', k_y')|^2 \mathrm{d}k_x' \mathrm{d}k_y' \right]^{\frac{1}{2}}$$

$$= \iint_{-\infty}^{\infty} |H_c(k_x', k_y')|^2 \mathrm{d}k_x' \mathrm{d}k_y'.$$

Expressing the above result in terms of the OTF, we have

$$|OTF(k_x, k_y)| \leqslant |OTF(0, 0)|.$$

which is the important property of the OTF in Eq. (4.3-5b).

4.3.2 Incoherent Image Processing Examples

Example: Single-Slit Pupil, $p(x, y) = rect\left(\dfrac{x}{x_0}\right)$

In this example, we use a slit of width x_0 along the y-direction as the pupil in the standard 4-f optical image processing system shown in Figure 4.2-1. According to Eq. (4.2-5), the coherent transfer function is

$$H_c(k_x, k_y) = p(x, y)\Big|_{x = \frac{-fk_x}{k_0},\, y = \frac{-fk_y}{k_0}} = rect\left(\frac{fk_x}{x_0 k_0}\right) = rect\left(\frac{k_x}{x_0 k_0 / f}\right).$$

The OTF is

$$OTF(k_x, k_y) = H_c(k_x, k_y) \otimes H_c(k_x, k_y)$$

$$= rect\left(\frac{k_x}{x_0k_0/f}\right)\otimes rect\left(\frac{k_x}{x_0k_0/f}\right)$$

$$= \frac{x_0k_0}{f}tri\left(\frac{k_x}{x_0k_0/f}\right), \tag{4.3-10}$$

其中, 定义三角函数 (*triangle function*) $tri(x/a)$ 为

$$tri\left(\frac{x}{a}\right) = \begin{cases} 0, & |x/a| \geqslant 1 \\ 1-|x/a|, & |x/a|<1, \end{cases} \tag{4.3-11}$$

函数的宽度为 $2a$, 三角函数如图 4.3-1 所示.

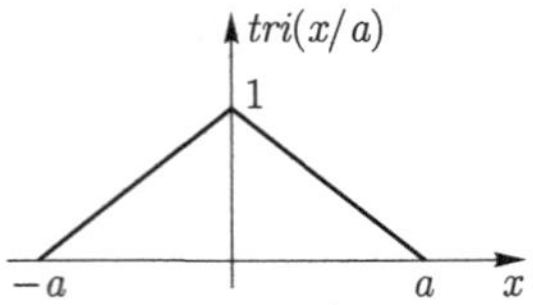

图 4.3-1 三角函数

让 k_y 为常数, 在图 4.3-2a) 中画出沿 k_x 方向的 $H_c(k_x,k_y)$, 在图 4.3-2b) 中画出相应的 OTF. 可以发现, 图 4.3-2 中的这两种情况皆对输入图像沿 x 方向进行了低通滤波. 然而, 相比于用相干光进行照明, 在非相干照明下, 可以传播超过输入图像空间频率两倍的范围.

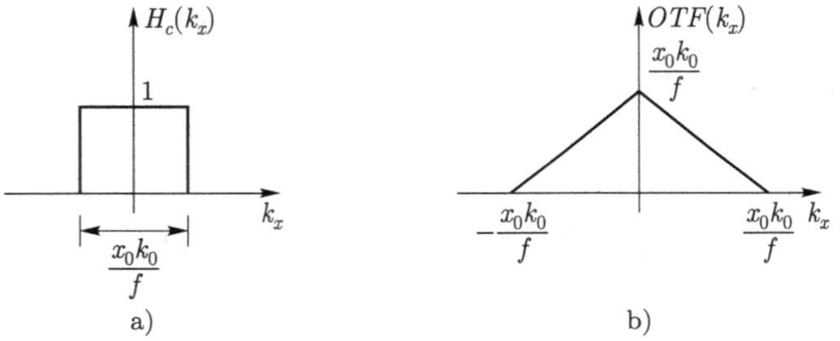

图 4.3-2 a) 相干传递函数; b) $p(x) = rect\left(\frac{x}{x_0}\right)$ 下双透镜系统的 OTF

举例: 双缝光瞳, $p(x,y) = rect\left(\frac{x-b}{x_0}\right) + rect\left(\frac{x+b}{x_0}\right), b > x_0/2$

在如图 4.2-1 所示的标准的 4-f 光学图像处理系统中, 用沿 y 方向扩展宽度为 x_0 的双缝光瞳, 其沿 x 方向的相干传递函数为

$$H_c(k_x,k_y) = p(x,y)\Big|_{x=\frac{-fk_x}{k_0},\, y=\frac{-fk_y}{k_0}}$$

$$= rect\left(\frac{\frac{-fk_x}{k_0}-b}{x_0}\right) + rect\left(\frac{\frac{-fk_x}{k_0}+b}{x_0}\right)$$

$$= rect\left(\frac{-k_x-\frac{bk_0}{f}}{\frac{x_0k_0}{f}}\right) + rect\left(\frac{-k_x+\frac{bk_0}{f}}{\frac{x_0k_0}{f}}\right),$$

如图 4.3-3a) 所示. 所给出的 $H_c(k_x)$ 的自相关函数为 $OTF(k_x)$, 如图 4.3-3b) 所示. 可以看出,

$$= rect\left(\frac{k_x}{x_0k_0/f}\right) \otimes rect\left(\frac{k_x}{x_0k_0/f}\right)$$
$$= \frac{x_0k_0}{f} tri\left(\frac{k_x}{x_0k_0/f}\right), \quad (4.3\text{-}10)$$

where we define a *triangle function* $tri(x/a)$ as

$$tri\left(\frac{x}{a}\right) = \begin{cases} 0, & |x/a| \geqslant 1 \\ 1-|x/a|, & |x/a| < 1, \end{cases} \quad (4.3\text{-}11)$$

with $2a$ being the width of the function. The triangle function is shown in Figure 4.3-1.

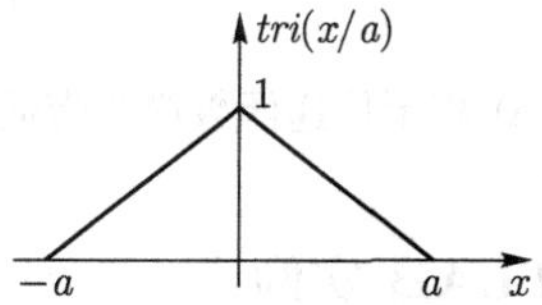

Figure 4.3-1 Triangle function

We plot $H_c(k_x, k_y)$ along k_x with k_y constant in Figure 4.3-2a along with the OTF in Figure 4.3-2b. We observe that both situation perform low-pass filtering along the x-direction on an input image. However, under incoherent illumination, it is possible to transmit twice the range of spatial frequencies of the input image as compared to the use of coherent illumination.

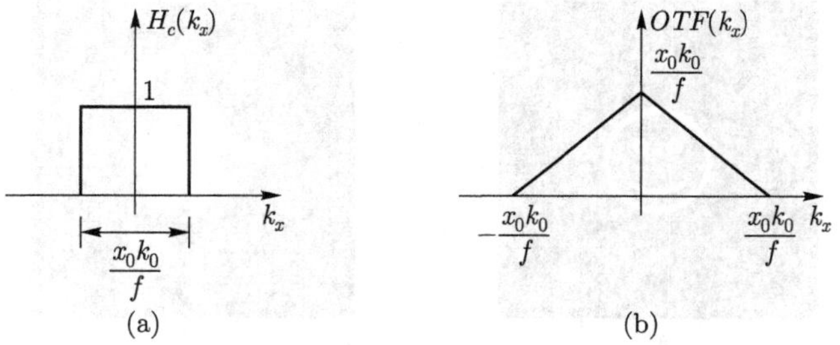

Figure 4.3-2 (a) The coherent transfer function and (b) the OTF of the two-lens system with $p(x) = rect\left(\frac{x}{x_0}\right)$

Example: Double-Slit Pupil, $p(x,y)=rect\left(\frac{x-b}{x_0}\right)+rect\left(\frac{x+b}{x_0}\right)$, $b > x_0/2$

We use a double-slit pupil of width x_0 along the y-direction in the standard 4-f optical image processing system shown in Figure 4.2-1. The coherent transfer function along the x-direction is

$$H_c(k_x, k_y) = p(x,y)\Big|_{x=\frac{-fk_x}{k_0},\, y=\frac{-fk_y}{k_0}}$$
$$= rect\left(\frac{\frac{-fk_x}{k_0} - b}{x_0}\right) + rect\left(\frac{\frac{-fk_x}{k_0} + b}{x_0}\right)$$
$$= rect\left(\frac{-k_x - \frac{bk_0}{f}}{\frac{x_0k_0}{f}}\right) + rect\left(\frac{-k_x + \frac{bk_0}{f}}{\frac{x_0k_0}{f}}\right),$$

and its plot is shown in Figure 4.3-3a. The $OTF(k_x)$ is the autocorrelation of $H_c(k_x)$ given above and its plot is shown in Figure 4.3-3b. Note that while the coherent transfer function

当相干传递函数表现为带通滤波 (*bandpass filtering*) 特性时, 其传递函数允许一定范围的空间频率通过并阻止该范围以外的空间频率通过, OTF 仍然展现出低通滤波特性, 且 $OTF(0)$ 具有最大值.

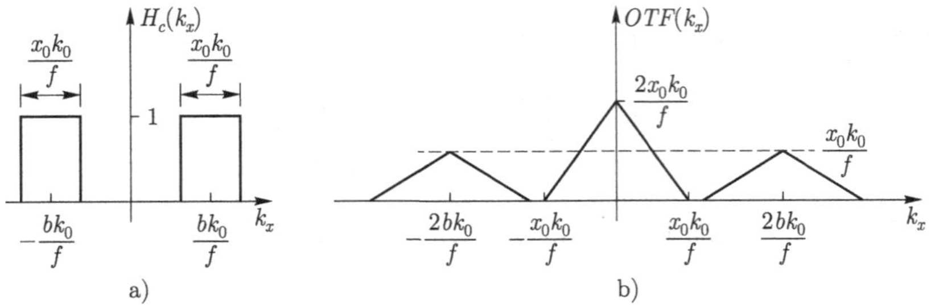

图 4.3-3 具有双缝光瞳的双透镜系统: a) 相干传递函数说明带通滤波特性; b) OTF 说明低通滤波特性

举例: 非相干图像处理的 MATLAB 实例

上两个例子阐述了非相干空间滤波总是带来低通滤波 [见图 4.3-2b) 与图 4.3-3b)], 在本例中, 我们将进行两类光瞳的仿真: $circ(r/r_0)$ 和 $\overline{circ}(r/r_0) = 1 - circ(r/r_0)$.

在图 4.3-4a) 与图 4.3-4b) 中, 分别给出了原始图像及其频谱. 在图 4.3-4c) 和图 4.3-4e) 中, 分别给出了经光瞳 $circ(r/r_0)$ 和 $\overline{circ}(r/r_0) = 1-circ(r/r_0)$ 滤波后的频谱, 并在图 4.3-4d)

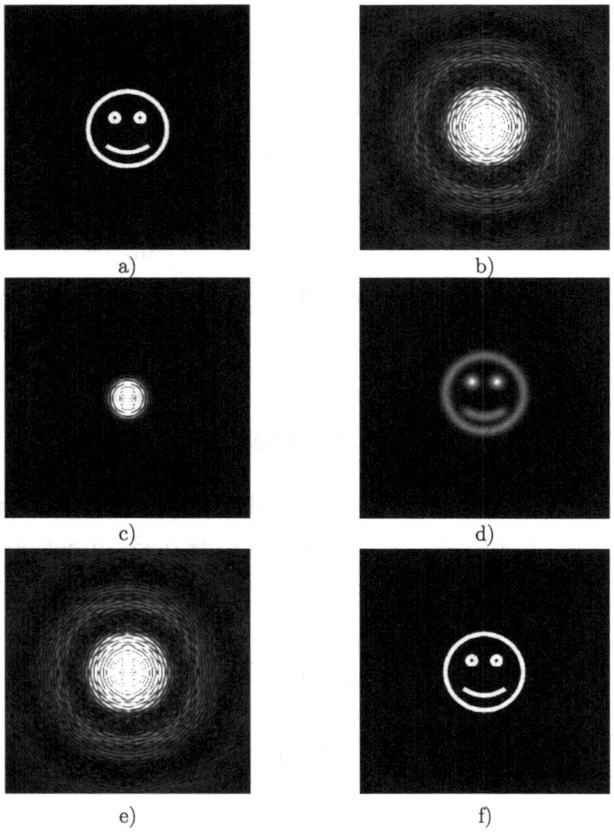

图 4.3-4 非相干空间滤波: a) 原始图像; b) 图 a) 的频谱; c) 瞳函数为 $circ(r/r_0)$ 的滤波后频谱; d) 经过瞳函数为 $circ(r/r_0)$ 的图 a) 的滤波后图像; e) 瞳函数为 $\overline{circ}(r/r_0)$ 的滤波后频谱; f) 经过瞳函数为 $\overline{circ}(r/r_0)$ 的图 a) 的滤波后图像

shows *bandpass filtering* characteristics as the transfer function passes spatial frequencies within a certain range and rejects frequencies outside that range, the OTF still exhibits low-pass filtering characteristics with $OTF(0)$ being maximum.

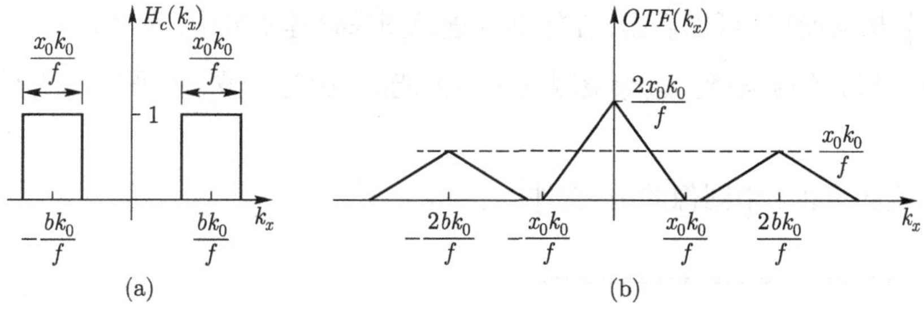

Figure 4.3-3 Two-lens system with double-slit pupil: (a) coherent transfer function illustrating bandpass filtering characteristics and (b) OTF illustrating low-pass filtering characteristics

Example: MATLAB Example of Incoherent Image Processing

While we have illustrated in the last two examples that incoherent spatial filtering always leads to low-pass filtering (see Figures 4.3-2b and 4.3-3b). In this example, we perform some simulations for two types of pupils: $circ(r/r_0)$ and $\overline{circ}(r/r_0) = 1 - circ(r/r_0)$.

In Figure 4.3-4a and b, we show the original image and its spectrum, respectively. In Figure 4.3-4c and e, we show the filtered spectra for pupils $circ(r/r_0)$ and $\overline{circ}(r/r_0)$, respec-

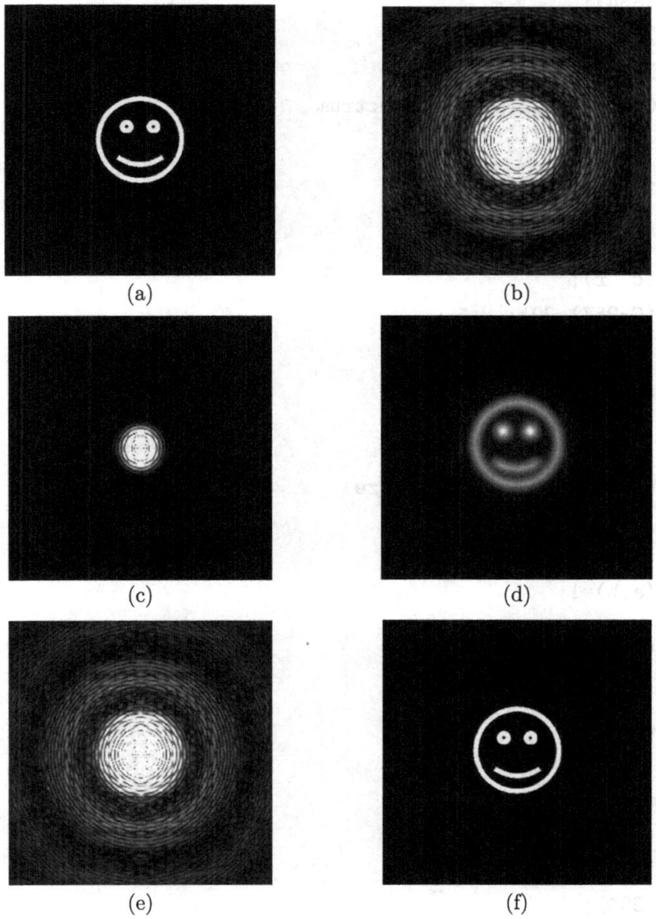

Figure 4.3-4 Incoherent spatial filtering: (a) original image, (b) spectrum of (a), (c) filtered spectrum with $circ(r/r_0)$ as the pupil, (d) filtered image of (a) with $circ(r/r_0)$ as the pupil, (e) filtered spectrum with $\overline{circ}(r/r_0)$ as the pupil, (f) filtered image of (a) with $\overline{circ}(r/r_0)$ as the pupil

和图 4.3-4f) 中分别给出了相应的滤波后的图像. 图 4.3-4c) 和图 4.3-4e) 中所滤波后的频谱, 是原始频谱分别与 $OTF = circ\left(\dfrac{k_r}{r_0k_0/f}\right) \otimes circ\left(\dfrac{k_r}{r_0k_0/f}\right)$ 和 $OTF = \overline{circ}\left(\dfrac{k_r}{r_0k_0/f}\right) \otimes \overline{circ}\left(\dfrac{k_r}{r_0k_0/f}\right)$ 相乘而得. 可以看出, 在标准双透镜图像处理系统中, 在非相干照明下, 即便是使用 $\overline{circ}(r/r_0)$ 形式的瞳函数, 边缘检测也是不可能的, 因为滤波后的图像只是输入图像的一个低通情况.

以下为生成图 4.3-4 中图像的 m-文件.

```
==========================================
    % Incoherent spatial filtering
    % circ(r/r0) and  1-circ(r/r0) as pupil functions
    % Adapted from "Introduction to Modern Digital Holography"
    % by T.-C. Poon & J.-P. Liu
    % Cambridge University Press (2014), Table 1.7 and Table 1.8.

    clear all;close all;
    A=imread('front.jpg');          % read image file 512x512 8-bit
    A=double(A);
    A=A/255;
    SP=fftshift(fft2(fftshift(A)));
    D=abs(SP);
    D=D(129:384,129:384);
    figure;imshow(A);
    %title('Original image')
    figure;imshow(30.*mat2gray(D)); % spectrum
    title('Original spectrum')

    c=1:512;
    r=1:512;
    [C, R]=meshgrid(c, r);
    CI=((R-257).^2+(C-257).^2);
    pup=zeros(512,512);
    % produce pupil circ(r/r0)
    for a=1:512;
        for b=1:512;
             if CI(a,b)>=15^2;  %pupil size
                pup(a,b)=0;
             else
                pup(a,b)=1;
             end
        end
    end
    h=ifft2(fftshift(pup));

    OTF=fftshift(fft2(h.*conj(h)));
    OTF=OTF/max(max(abs(OTF)));
    G=abs(OTF.*SP);
    G=G(129:384,129:384);
    figure;imshow(30.*mat2gray(G));
    %title('Spectrum filtered by circ(r/r0)')
```

tively, and their corresponding filtered images are shown in Figure 4.3-4d and f, respectively. The filtered spectra in Figure 4.3-4c and e are obtained by multiplying the original spectrum by $OTF = circ\left(\frac{k_r}{r_0 k_0/f}\right) \otimes circ\left(\frac{k_r}{r_0 k_0/f}\right)$ and $OTF = \overline{circ}\left(\frac{k_r}{r_0 k_0/f}\right) \otimes \overline{circ}\left(\frac{k_r}{r_0 k_0/f}\right)$, respectively. Note that even with the pupil of the form of $\overline{circ}(r/r_0)$, edge detection is not possible with incoherent illumination in the standard two-lens image processing system, as the filtered image is simply a lowpass version of the input image.

The images shown in Figure 4.3-4 are generated from the m-file below.

==

```
% Incoherent spatial filtering
% circ(r/r0) and  1-circ(r/r0) as pupil functions
% Adapted from "Introduction to Modern Digital Holography"
% by T.-C. Poon & J.-P. Liu
% Cambridge University Press (2014), Table 1.7 and Table 1.8.

clear all;close all;
A=imread('front.jpg');             % read image file 512x512 8-bit
A=double(A);
A=A/255;
SP=fftshift(fft2(fftshift(A)));
D=abs(SP);
D=D(129:384,129:384);
figure;imshow(A);
%title('Original image')
figure;imshow(30.*mat2gray(D)); % spectrum
title('Original spectrum')

c=1:512;
r=1:512;
[C, R]=meshgrid(c, r);
CI=((R-257).^2+(C-257).^2);
pup=zeros(512,512);
% produce pupil circ(r/r0)
for a=1:512;
    for b=1:512;
         if CI(a,b)>=15^2;  %pupil size
            pup(a,b)=0;
        else
            pup(a,b)=1;
        end
    end
end
h=ifft2(fftshift(pup));

OTF=fftshift(fft2(h.*conj(h)));
OTF=OTF/max(max(abs(OTF)));
G=abs(OTF.*SP);
G=G(129:384,129:384);
figure;imshow(30.*mat2gray(G));
%title('Spectrum filtered by circ(r/r0)')
```

```
I=abs(fftshift(ifft2(fftshift(OTF.*SP))));
figure;imshow(mat2gray(I));
%title('Image filtered by circ(r/r0)')

c=1:512;
r=1:512;
[C, R]=meshgrid(c, r);
CI=((R-257).^2+(C-257).^2);
pup=zeros(512,512);
% produce pupil 1-circ(r/r0)
for a=1:512;
    for b=1:512;
         if CI(a,b)>=15^2;  %pupil size
            pup(a,b)=1;
        else
            pup(a,b)=0;
        end
    end
end
h=ifft2(fftshift(pup));

OTF=fftshift(fft2(h.*conj(h)));
OTF=OTF/max(max(abs(OTF)));
G=abs(OTF.*SP);
G=G(129:384,129:384);
figure;imshow(30.*mat2gray(G));
%title('Spectrum filtered by 1-circ(r/r0)')

I=abs(fftshift(ifft2(fftshift(OTF.*SP))));
figure;imshow(mat2gray(I));
%title('Image filtered by 1-circ(r/r0)')
========================================
```

4.4 扫描图像处理

相干光学图像处理系统利用频谱面 (*frequency plane*) (光瞳平面) 架构来并行处理图像, 该技术在概念上比较简单, 因此受到广泛关注. 然而, 并行光学处理器往往缺乏准确性与灵活性. 准确性的缺失是由于相干系统对相位 (因为是对复振幅进行处理) 非常敏感, 它们非常容易受到干扰从而产生噪声. 灵活性缺乏的原因是, 在通常的傅里叶分析和线性不变处理之外, 很难扩展其应用范围. 为了克服准确性的缺点, 我们可以采用非相干图像处理, 但其主要缺点是强度点扩散函数为非负实数 [见式 (4.3-2)] 且其可达到的 OTF 是严格受限的, OTF 为瞳函数 [见式 (4.3-4)] 的自相关, 因此总有中心最大值. 这就意味着即使是最简单却重要的滤波, 如高通滤波, 也是不可能实现的. 本节讨论光学扫描处理 (*optical scanning processing*), 其在某

```
I=abs(fftshift(ifft2(fftshift(OTF.*SP))));
figure;imshow(mat2gray(I));
%title('Image filtered by circ(r/r0)')

c=1:512;
r=1:512;
[C, R]=meshgrid(c, r);
CI=((R-257).^2+(C-257).^2);
pup=zeros(512,512);
% produce pupil 1-circ(r/r0)
for a=1:512;
    for b=1:512;
         if CI(a,b)>=15^2;  %pupil size
            pup(a,b)=1;
        else
            pup(a,b)=0;
        end
    end
end
h=ifft2(fftshift(pup));

OTF=fftshift(fft2(h.*conj(h)));
OTF=OTF/max(max(abs(OTF)));
G=abs(OTF.*SP);
G=G(129:384,129:384);
figure;imshow(30.*mat2gray(G));
%title('Spectrum filtered by 1-circ(r/r0)')

I=abs(fftshift(ifft2(fftshift(OTF.*SP))));
figure;imshow(mat2gray(I));
%title('Image filtered by 1-circ(r/r0)')
========================================
```

4.4 Scanning Image Processing

Coherent optical image processing systems use a frequency plane (pupil plane) architecture and processes images in parallel. The technique is conceptually simple and for this reason has received much attention. However, parallel optical processors often lack accuracy and flexibility. The lack of accuracy is owing to the fact that coherent systems are sensitive to phase (as we are dealing with the manipulation of complex amplitudes) and they are extremely susceptible to noise corruption owing to interference. The lack of flexibility comes from the fact that it is difficulty to extend the range of applications beyond the usual Fourier and linear invariant processing. To overcome the lack of accuracy, we can employ incoherent image processing but the main drawback is that the intensity point spread function is real and nonnegative [see Eq. (4.3-2)] and the achievable OTF, which is the autocorrelation of the pupil function [see Eq. (4.3-4)] and thus always has a central maximum, is strictly constrained. This means that it is not possible to implement even the simplest but important filtering such as high-pass filtering. In this section, we discuss *optical scanning processing* which offers some

些处理任务的性能上比相应的并行系统 (*parallel system*) 更具优势.

图 4.4-1 给出了一个典型的光学扫描处理器 (*scanning optical processor*). 简单来说, 该光学处理器通过 x-y 扫描镜来移动光束, 并以光栅扫描方式 (*raster scanning fashion*) 扫描一个输入 $t(x,y)$. 光电探测器 (*photodetector*) 接收到所有光信号后再转换为一个电信号 $i(t)$ 作为输出. 当这个包含了扫描输入 (物体) 的处理后信息, 即电信号, 与 x-y 扫描镜的二维扫描信号同步、数字化地储存 (如计算机中) 时, 就获得一个二维数字记录 $i(x,y)$, 即一个处理后的图像. 至此, 该处理后的图像按顺序生成, 并以 $i(x,y)$ 的形式储存在计算机上或在显示器上实时显示.

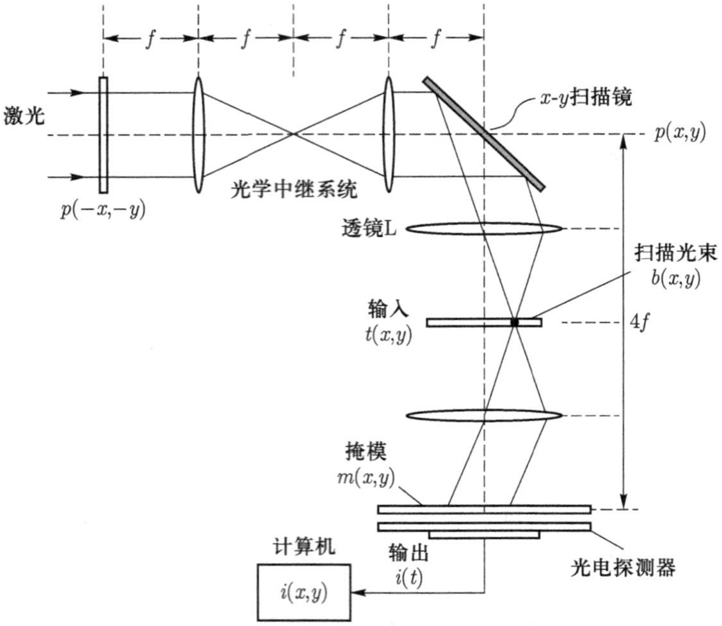

图 4.4-1　光学扫描处理

扫描的位置由下面的关系给出

$$x = x(t) = V_x t,\ \ y = y(t) = V_y t, \tag{4.4-1}$$

其中, $x(t)$ 和 $y(t)$ 是分别以均匀扫描速度 V_x 和 V_y 进行扫描的函数. 因此, 通过光学扫描, 可将图像的空间频率 $f_x = \dfrac{k_x}{2\pi}(\mathrm{mm}^{-1})$ 转换为时间频率 $f_t(\mathrm{Hz}$ 或 $\mathrm{s}^{-1})$, 根据

$$f_t = V_x f_x,$$

这里, $V_x = l/T$, 其中, l 为沿 x 方向扫描长度的线性范围, T 为沿 x 方向扫描长度所需的扫描时间. 例如, 若 $f_x = 2\ \mathrm{mm}^{-1}$, $l = 20\ \mathrm{mm}$, 且, $T = 1\ \mathrm{s}$, 则, $f_t = 40\ \mathrm{Hz}$.

现在我们来进一步详细讨论扫描系统. 瞳函数 $p(x,y)$ 首先以光学的方式传输到 x-y 扫描镜的表面, 即透镜 L 的前焦面上. 假设在 x-y 扫描镜的光瞳由 $p(x,y)$ 给出, 这只需将 $p(-x,-y)$ 放在如图 4.4-1 所示的光学中继系统 (*relay system*) 的前面即可. 因此, 在输入平面上有振幅为 $b(x,y)$ 的扫描光束, 且 $b(x,y) = \mathcal{F}\{p(x,y)\}\big|_{k_x=\frac{k_0 x}{f},\, k_y=\frac{k_0 y}{f}}$.

advantage over the corresponding parallel system in the performance of certain processing tasks.

In Figure 4.4-1, we show a typical *scanning optical processor*. Briefly, the optical processor scans out an input $t(x, y)$ in a *raster scanning fashion* with an optical beam by moving the beam through the x-y scanner. A *photodetector* accepts all the light and gives an electrical signal $i(t)$ as an output. When this electrical signal, which contains the processed information of the scanned input, is digitally stored (i.e., in a computer) in synchronization with 2-D scan signals of the x-y scanner, we have a 2-D digital record, $i(x, y)$, as a processed image. Here, the processed image is built sequentially to give $i(x, y)$ to be stored on a computer or real-time displayed on a monitor.

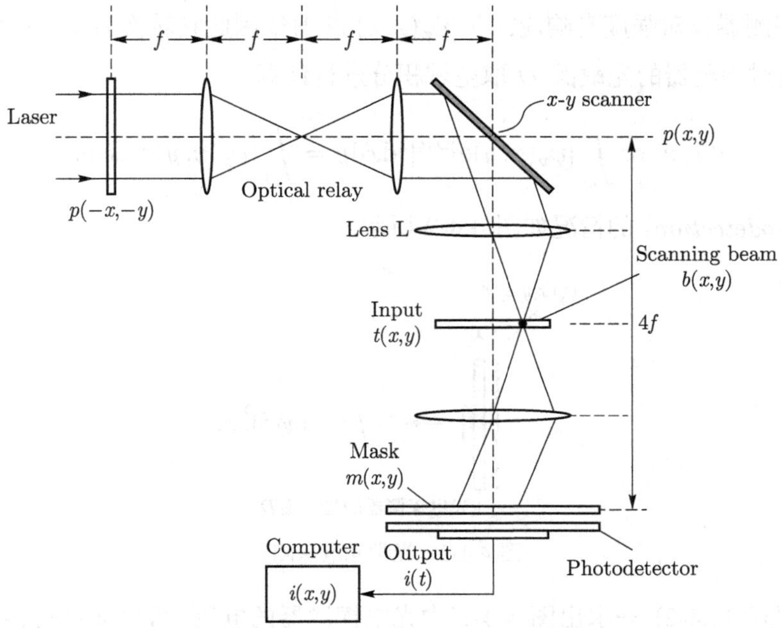

Figure 4.4-1 Optical scanning processing

The scanning position is described by the following relationships:

$$x = x(t) = V_x t, \quad y = y(t) = V_y t, \tag{4.4-1}$$

where $x(t)$ and $y(t)$ are a function of the scan with a uniform scan at speeds V_x and V_y, respectively. Hence, by optical scanning, we translate spatial frequencies, $f_x = \dfrac{k_x}{2\pi}(\text{mm}^{-1})$, of an image into temporal frequencies, $f_t(\text{Hz or s}^{-1})$, according to

$$f_t = V_x f_x,$$

where $V_x = l/T$ with l being the linear range of the scan length and T the scan time over the scan length along x. For example, if $f_x = 2\ \text{mm}^{-1}$ with $l = 20$ mm and $T = 1$ s, $f_t = 40$ Hz.

Let us now further discuss the scanning system in detail. A pupil function $p(x, y)$ is first optically relayed to the surface of the x-y scanner, which is in the front focal plane of lens L. We assume the pupil on the x-y scanner is given by $p(x, y)$ and this can be done by simply putting $p(-x, -y)$ in front of the optical *relay system* shown in Figure 4.4-1. We, therefore, have a scanning beam with amplitude $b(x, y)$ on the input plane, and $b(x, y) = \mathcal{F}\{p(x, y)\}\big|_{k_x = \frac{k_0 x}{f},\, k_y = \frac{k_0 y}{f}}$.

为简单起见，这里进行一维的分析. 输入 $t(x)$ 后的复振幅可由 $b(x'-x)t(x')$ 给出，因为 $b(x'-x)$ 这一项表示光束根据 $x=x(t)$ 进行传输 (或扫描). 此时，该复振幅经傅里叶变换到掩模 $m(x)$ 处，从掩模出来的复振幅为

$$\psi(x,x_m)\propto\left[\int b(x'-x)t(x')\mathrm{e}^{\frac{\mathrm{j}k_0}{f}x_m x'}\mathrm{d}x'\right]m(x_m),\tag{4.4-2}$$

其中，x_m 是掩模面的坐标位置. 紧挨着掩模后是光电探测器，它将光信号转换为电流或电压形式的电信号.

所有的光电探测器是对光的强度响应的平方律探测器 (*square-law detector*)，而不是对光信号的复振幅. 假设紧贴在光电探测器前表面上的复振幅为 $\psi_p(x,y)\mathrm{e}^{\mathrm{j}\omega_0 t}$，其中 ω_0 是光的频率. 由于光电探测器仅对强度有响应，即 $|\psi_p(x,y)\mathrm{e}^{\mathrm{j}\omega_0 t}|^2$，因此它转换后给出的电流 i 将作为输出，这可通过对探测器的光敏面 D 取空域积分进行计算

$$i(x,y)\propto\int_D|\psi_p(x,y)\mathrm{e}^{\mathrm{j}\omega_0 t}|^2\mathrm{d}x\mathrm{d}y=\int_D|\psi_p(x,y)|^2\mathrm{d}x\mathrm{d}y.\tag{4.4-3}$$

光电探测 (*photodetection*) 的情况如图 4.4-2 所示.

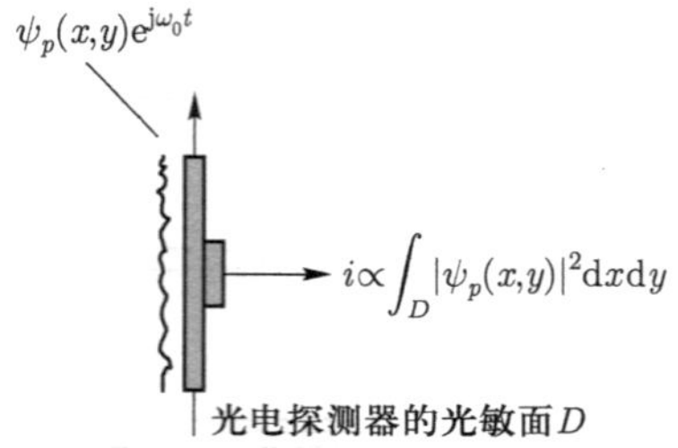

图 4.4-2　光学光电探测

现在，利用式 (4.4-2) 来求出图 4.4-1 中光电探测器的电流，可以通过对整个掩模上的坐标进行积分，得

$$\begin{aligned}i(x)&=\int_D|\psi(x,x_m)|^2\mathrm{d}x_m=\int_D\psi(x,x_m)\psi^*(x,x_m)\mathrm{d}x_m\\&=\int_D\left[\int b(x'-x)t(x')\mathrm{e}^{\frac{\mathrm{j}k_0}{f}x_m x'}\mathrm{d}x'\right]m(x_m)\\&\quad\times\left[\int b^*(x''-x)t^*(x'')\mathrm{e}^{\frac{-\mathrm{j}k_0}{f}x_m x''}\mathrm{d}x''\right]m^*(x_m)\mathrm{d}x_m\\&=\int_D\left[\iint t(x')t^*(x'')b(x'-x)b^*(x''-x)\mathrm{e}^{\frac{\mathrm{j}k_0}{f}x_m(x'-x'')}\mathrm{d}x'\mathrm{d}x''|m(x_m)|^2\right]\mathrm{d}x_m.\end{aligned}\tag{4.4-4}$$

上式中，$i(x)$ 表示在计算机中处理后的一幅图像. 扫描处理器可实现一般需要通过光学手段并行架构完成的相干和非相干图像处理运算.

4.4.1　相干成像

现在考虑点探测器 (*point detector*). 取 $m(x)=\delta(x)$，物理上，是通过在探测器前放置一个针孔来实现的. 因此，该探测器被称为点探测器，式 (4.4-4) 变为

$$i(x)=\int_D\left[\iint t(x')t^*(x'')b(x'-x)b^*(x''-x)\mathrm{e}^{\frac{\mathrm{j}k_0}{f}x_m(x'-x'')}\delta(x_m)\mathrm{d}x'\mathrm{d}x''\right]\mathrm{d}x_m.$$

For brevity, let us now perform analysis in one dimension. The complex field after the input $t(x)$ is now given by $b(x'-x)t(x')$ as the term $b(x'-x)$ simply means that the optical beam is translating (or scanning) according to $x=x(t)$. Now, this field is Fourier transform onto the mask $m(x)$, and the complex field exiting from the mask is

$$\psi(x,x_m)\propto\left[\int b(x'-x)t(x')\mathrm{e}^{\frac{\mathrm{j}k_0}{f}x_m x'}\mathrm{d}x'\right]m(x_m), \tag{4.4-2}$$

where x_m is the coordinate on the plane of the mask. Immediately behind the mask is a photodetector that converts optical signals into electrical signals, in the form of either current or voltage.

All photodetectors are *square-law detectors* that respond to the intensity, rather than the complex field, of an optical signal. Assume that the complex field just in front of the photodetector's surface is $\psi_p(x,y)\mathrm{e}^{\mathrm{j}\omega_0 t}$, where ω_0 is the frequency of the light. Since the photodetector could only respond to intensity, that is, $|\psi_p(x,y)\mathrm{e}^{\mathrm{j}\omega_0 t}|^2$, it gives out a current i as an output by spatially integrating over the active area, D, of the detector:

$$i(x,y)\propto\int_D|\psi_p(x,y)\mathrm{e}^{\mathrm{j}\omega_0 t}|^2\mathrm{d}x\mathrm{d}y=\int_D|\psi_p(x,y)|^2\mathrm{d}x\mathrm{d}y. \tag{4.4-3}$$

The situation of *photodetection* is illustrated in Figure 4.4-2.

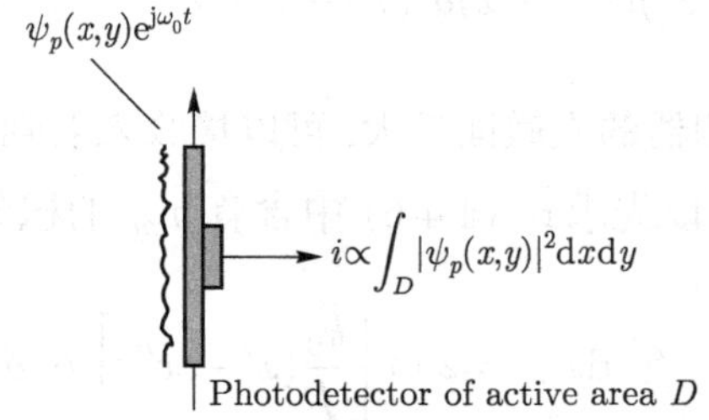

Figure 4.4-2 Optical photodetection

We can now use Eq. (4.4-2) to find the current from the photodetector in Figure 4.4-1 by integrating over the mask's coordinate:

$$\begin{aligned}
i(x)&=\int_D|\psi(x,x_m)|^2\mathrm{d}x_m=\int_D\psi(x,x_m)\psi^*(x,x_m)\mathrm{d}x_m\\
&=\int_D\left[\int b(x'-x)t(x')\mathrm{e}^{\frac{\mathrm{j}k_0}{f}x_m x'}\mathrm{d}x'\right]m(x_m)\\
&\quad\times\left[\int b^*(x''-x)t^*(x'')\mathrm{e}^{\frac{-\mathrm{j}k_0}{f}x_m x''}\mathrm{d}x''\right]m^*(x_m)\mathrm{d}x_m\\
&=\int_D\left[\iint t(x')t^*(x'')b(x'-x)b^*(x''-x)\mathrm{e}^{\frac{\mathrm{j}k_0}{f}x_m(x'-x'')}\mathrm{d}x'\mathrm{d}x''|m(x_m)|^2\right]\mathrm{d}x_m.
\end{aligned} \tag{4.4-4}$$

In Eq. (4.4-4), $i(x)$ represents a processed image on a computer. The scanning processor can perform all the usual coherent and incoherent image processing operations achievable by optical means under a parallel architecture.

4.4.1 Coherent Imaging

Let us consider a *point detector*. We take $m(x)=\delta(x)$. Physically, this is accomplished by putting a pinhole in front of the detector. Hence, the detector is termed as a point detector. Eq. (4.4-4) becomes

$$i(x)=\int_D\left[\iint t(x')t^*(x'')b(x'-x)b^*(x''-x)\mathrm{e}^{\frac{\mathrm{j}k_0}{f}x_m(x'-x'')}\delta(x_m)\mathrm{d}x'\mathrm{d}x''\right]\mathrm{d}x_m.$$

假定光电探测器的光敏面很大, 比如, $D \to \infty$, 则在 δ 函数积分后, 有

$$i(x) = \iint t(x')t^*(x'')b(x'-x)b^*(x''-x)\mathrm{d}x'\mathrm{d}x''.$$

因为 x' 和 x'' 的积分可以分别完成, 将上式重新写为

$$\begin{aligned} i(x) &= \int t(x')b(x'-x)\mathrm{d}x' \times \int t^*(x'')b^*(x''-x)\mathrm{d}x'' \\ &= \left| \int t(x')b(x'-x)\mathrm{d}x' \right|^2 \\ &= \left| \int t(x')b[-(x-x')]\mathrm{d}x' \right|^2 = |t(x) * b(-x)|^2. \end{aligned} \tag{4.4-5}$$

对比式 (4.2-7), 可以看到, 由于先处理复振幅, 然后计算强度, 因此 $b(-x)$ 是扫描系统所对应相干图像处理的相干点扩散函数.

4.4.2　非相干成像

考虑积分探测器 (*integrating detector*), 取 $m(x) = 1$, 式 (4.4-4) 变为

$$i(x) = \int_D \left[\iint t(x')t^*(x'')b(x'-x)b^*(x''-x)\mathrm{e}^{\frac{\mathrm{j}k_0}{f}x_m(x'-x'')}\mathrm{d}x'\mathrm{d}x'' \right] \mathrm{d}x_m. \tag{4.4-6}$$

当 $m(x) = 1$ 时, 假设光电探测器的光敏面较大, 可以接收入射到探测器上的所有光 (因此, 该探测器被称为积分探测器). 可以先求式 (4.4-6) 中含有 x_m 的积分 [见式 (2.3-4)和式 (2.1-6)]

$$\int_{-\infty}^{\infty} \mathrm{e}^{\frac{\mathrm{j}k_0}{f}x_m(x'-x'')}\mathrm{d}x_m = 2\pi\delta\left[\frac{k_0}{f}(x'-x'')\right] \propto \delta(x'-x'').$$

基于这一结果, 式 (4.4-6) 变为

$$i(x) \propto \iint t(x')t^*(x'')b(x'-x)b^*(x''-x)\delta(x'-x'')\mathrm{d}x'\mathrm{d}x''.$$

利用 δ 函数的筛选性质 [见式 (2.1-5)], 对 x' 进行积分得

$$\begin{aligned} i(x) &\propto \int t(x'')t^*(x'')b(x''-x)b^*(x''-x)\mathrm{d}x'' = \int |t(x'')|^2|b(x''-x)|^2\mathrm{d}x'' \\ &= |t(x)|^2 * |b(-x)|^2. \end{aligned} \tag{4.4-7}$$

对比式 (4.3-1), 可以看到, $|b(-x)|^2$ 是扫描系统中对应非相干图像处理的强度点扩散函数.

可以看到, 扫描系统可以通过仅对光电探测器前的掩模的处理来进行相干和非相干处理. 当 $m(x) = \delta(x)$ 时, 系统为相干图像处理. 当 $m(x) = 1$ 时, 系统为非相干图像处理. δ 函数和单值函数的使用对应掩模尺寸大小的两个极端情况. 不难想象, 利用一些有限大小尺寸的掩模可进行部分相干图像处理 (*partial coherent image processing*), 但该内容超出了本书的范围.

Assuming that the active area of the photodetector is large, that is, $D \to \infty$, and after the delta function integration, we have

$$i(x) = \iint t(x')t^*(x'')b(x'-x)b^*(x''-x)\mathrm{d}x'\mathrm{d}x''.$$

Since the x' and x'' integrations can be done separately, we rewrite the above equation to become

$$\begin{aligned} i(x) &= \int t(x')b(x'-x)\mathrm{d}x' \times \int t^*(x'')b^*(x''-x)\mathrm{d}x'' \\ &= \left| \int t(x')b(x'-x)\mathrm{d}x' \right|^2 \\ &= \left| \int t(x')b[-(x-x')]\mathrm{d}x' \right|^2 = |t(x) * b(-x)|^2. \end{aligned} \tag{4.4-5}$$

Comparing to Eq. (4.2-7), we can see that corresponds to coherent image processing with a coherent point spread function in the scanning system given by $b(-x)$ as we manipulate (or process) complex amplitudes first and then compute the intensity.

4.4.2 Incoherent Imaging

Let us consider an *integrating detector*. We take $m(x) = 1$. Equation (4.4-4) becomes

$$i(x) = \int_D \left[\iint t(x')t^*(x'')b(x'-x)b^*(x''-x)\mathrm{e}^{\frac{\mathrm{j}k_0}{f}x_m(x'-x'')}\mathrm{d}x'\mathrm{d}x'' \right] \mathrm{d}x_m. \tag{4.4-6}$$

Along with $m(x) = 1$ and assuming the active area of the photodetector is large, we accept all the light incident on the detector (hence, the detector is termed as an *integrating detector*). The integral in Eq. (4.4-6) involving x_m can be evaluated first to get [see Eqs. (2.3-4) and (2.1-6)]

$$\int_{-\infty}^{\infty} \mathrm{e}^{\frac{\mathrm{j}k_0}{f}x_m(x'-x'')}\mathrm{d}x_m = 2\pi\delta\left[\frac{k_0}{f}(x'-x'')\right] \propto \delta(x'-x'').$$

With this result, Eq. (4.4-6) becomes

$$i(x) \propto \iint t(x')t^*(x'')b(x'-x)b^*(x''-x)\delta(x'-x'')\mathrm{d}x'\mathrm{d}x''.$$

Using the sampling property of a delta function [see Eq. (2.1-5)], we perform the integration over x' to obtain

$$\begin{aligned} i(x) \propto \int t(x'')t^*(x'')b(x''-x)b^*(x''-x)\mathrm{d}x'' &= \int |t(x'')|^2|b(x''-x)|^2\mathrm{d}x'' \\ &= |t(x)|^2 * |b(-x)|^2. \end{aligned} \tag{4.4-7}$$

Comparing to Eq. (4.3-1), we can see that this is incoherent image processing with the intensity point spread function in the scanning system given by $|b(-x)|^2$.

We have seen that the scanning system can perform coherent or incoherent processing by simply manipulating the mask in front of the photodetector. Using $m(x) = \delta(x)$, we have coherent image processing. For $m(x) = 1$, we have incoherent image processing. The use of a delta function and a uniform function is two extreme cases on the size of the mask. It is not hard to envision *partial coherent image processing* by using a mask of some finite size but

在结束本节内容之前, 要指出一些重要的用扫描方法进行成像的潜在应用.

在扫描方法中, 相干光 $b(x,y)$ 生成并扫描输入物体, 其中光电探测器前掩模的大小决定了扫描系统的相干性. 因此, 相比于并行系统, 扫描技术的优点是能更好地利用光能, 且其输出为将要被处理或串行传输 (*transmitted serially*) 的电信号形式. 在许多情况下, 输入数据 [例如, 来自扫描仪 (*scanner*)] 已经以一种扫描的格式存在了, 在此情况下, 以这种形式来处理信号似乎很自然, 而不是将信号记录在膜上或用昂贵的空间光调制器来并行处理 (将在第七章介绍空间光调制器). 相比于非扫描技术, 光学扫描的另一个优点是能够有效地对大尺寸物体进行图像处理, 比如, 遥感方面的应用.

扫描系统比并行系统更为灵活. 式 (4.4-1) 表示的 x-y 扫描镜的函数定义了从空间到时间的转换. 对于 $x=x(t)=V_xt$, 有一个在空间频率与时间频率的线性映射关系. 如果 $x(t)$ 和 t 是非线性的关系, 那么在原理上其映射也可以是非线性的, 这表示通过非线性扫描 (*nonlinear scanning*) 实现一些特殊的操作. 此外, 因为图像处理是通过扫描进行的, 可以设想, 在扫描输入物体时只需调整其瞳函数 $p(x,y)$, 从而给出空变滤波 (*space-variant filtering*) 的可能性, 则扫描光束即可被多次修改——这是相应的并行技术很难完成的任务.

4.5 光学传递函数的双光瞳合成

在式 (4.3-2) 和式 (4.3-3) 中看到, 非相干系统中的强度点扩散函数 (IPSF) 是非负实函数, 这对 OTF 的振幅和相位都有限制, 因此, 它限制了我们之前所讨论的可处理范围. 许多诸如高通滤波、边缘检测等重要的处理运算都需要双极性 (*bipolar*) 的点扩散函数 (有正值和负值). 为了超过正实数 IPSF 的限制, 引入了所谓的光学传递函数的双光瞳合成 (*two-pupil synthesis*) [Lohmann 和 Rhodes (1978)]. 因其在实际中的重要性, 在过去的几十年间进行了大量关于双光瞳合成的相关工作. 这里将仅讨论一种被称为非光瞳交互合成 (*non-pupil interaction synthesis*) 的方法. 我们先在频域来讨论这一概念.

在非相干系统中, 由式 (4.3-3) 或相当于式 (4.3-4) 给出

$$
\begin{aligned}
OTF(k_x,k_y) &= \mathcal{F}\left\{|h_c(x,y)|^2\right\} = H_c(k_x,k_y)\otimes H_c(k_x,k_y)\\
&= \iint\limits_{-\infty}^{\infty} H_c^*(k_x',k_y')H_c(k_x'+k_x,k_y'+k_y)\mathrm{d}k_x'\mathrm{d}k_y'.
\end{aligned}
\tag{4.5-1}
$$

现在, 根据式 (4.2-5), 相干传递函数可用瞳函数 $p(x,y)$ 表示为

$$
H_c(k_x,k_y)=p\left(\frac{-fk_x}{k_0},\frac{-fk_y}{k_0}\right).
$$

the topic is beyond the scope of the present book. To end this subsection, we want to point out some important potential applications with the scanning approach to imaging.

In the scanning approach, it is the coherent optical beam $b(x, y)$ that is generated and scanned over the input. The size of the mask in front of the photodetector determines the coherency of the scanning system. Some of the advantages of the scanning technique over the parallel counterpart are, therefore, a better utilization of the light energy and the fact that the output signal is an electronic form, which is ready to be processed or *transmitted serially.* In many instances, the input already exists in a scanned format (e.g., from a *scanner*) and in such case it seems natural to process the signal in that form rather than put it on film or use an expensive spatial light modulator for parallel processing (we will discuss spatial light modulators in Chapter 7). Another advantage of optical scanning as compared to non-scanning optical techniques is the capability of efficiently performing image processing of large-scale objects such as in remote sensing applications.

The scanning system is much more flexible than the parallel counterpart. The function of the x-y scanner represented by Eq. (4.4-1) defines the spatial to temporal conversion. For $x = x(t) = V_x t$, we have a linear mapping between the spatial and temporal frequencies. The mapping can, in principle, be nonlinear if $x(t)$ and t are related nonlinearly, thus representing some unusual manipulating operation through *nonlinear scanning.* Also, since image processing is performed via scanning, we can envision that the scanning beam can be modified continuously as it scans over the input simply by updating the pupil function $p(x, y)$, giving the possibility of *space-variant filtering*—a task that is difficult with the parallel counterpart.

4.5 Two-Pupil Synthesis of Optical Transfer Functions

As we have seen from Eqs. (4.3-2) and (4.3-3), the intensity point spread function (IPSF) of incoherent systems is a real and nonnegative function which imposes restrictions on both the magnitude and the phase of the OTF. Consequently, it limits the range of processing as we have discussed before. Many important processing operations, such as high-pass filtering, edge extraction, etc., require point spread functions that are *bipolar* (having positive as well as negative values). In order to overcome the limitation of a positive and real IPSF, the so-called *two-pupil synthesis* of the optical transfer function has been introduced [Lohmann and Rhodes (1978)]. Considerable work has been done in past decades on two-pupil synthesis owing to its practical importance. We shall limit ourselves to discuss one approach called the *non-pupil interaction synthesis.* Let us discuss this concept in the frequency domain first.

In incoherent systems, the OTF is given by Eq. (4.3-3) or equivalently by Eq. (4.3-4):

$$OTF(k_x, k_y) = \mathcal{F}\left\{|h_c(x, y)|^2\right\} = H_c(k_x, k_y) \otimes H_c(k_x, k_y)$$

$$= \iint_{-\infty}^{\infty} H_c^*(k_x', k_y') H_c(k_x' + k_x, k_y' + k_y) \mathrm{d}k_x' \mathrm{d}k_y'. \tag{4.5-1}$$

Now, according to Eq. (4.2-5), the coherent transfer function is expressed in terms of the pupil $p(x, y)$ as

$$H_c(k_x, k_y) = p\left(\frac{-fk_x}{k_0}, \frac{-fk_y}{k_0}\right).$$

因此, 可以根据瞳函数重新将 OTF 写为

$$OTF(k_x,k_y)=p\left(\frac{-fk_x}{k_0},\frac{-fk_y}{k_0}\right)\otimes p\left(\frac{-fk_x}{k_0},\frac{-fk_y}{k_0}\right). \tag{4.5-2}$$

可以看出, 瞳函数的自相关即为光学系统的 OTF. 这被称为光学传递函数的单光瞳合成 (*one-pupil synthesis*). 因为任意的自相关函数总有一个中心最大值 [见式 (4.3-5b)], 因此我们不会得到一个如带通滤波的函数. 为了克服单光瞳合成的局限性, 引入了双光瞳合成.

根据式 (4.5-1), 先定义归一化的 OTF 为

$$\begin{aligned}\overline{OTF}(k_x,k_y)&=\frac{H_c(k_x,k_y)\otimes H_c(k_x,k_y)}{OTF(0,0)}\\&=\frac{\iint\limits_{-\infty}^{\infty}H_c^*(k_x',k_y')H_c(k_x'+k_x,k_y'+k_y)\mathrm{d}k_x'\mathrm{d}k_y'}{\iint\limits_{-\infty}^{\infty}|H_c(k_x',k_y')|^2\mathrm{d}k_x'\mathrm{d}k_y'}.\end{aligned} \tag{4.5-3}$$

因此, $\overline{OTF}(0,0)=1$. 同样地, 用瞳函数表示, 有

$$\overline{OTF}(k_x,k_y)=\frac{p\left(\frac{-fk_x}{k_0},\frac{-fk_y}{k_0}\right)\otimes p\left(\frac{-fk_x}{k_0},\frac{-fk_y}{k_0}\right)}{OTF(0,0)}. \tag{4.5-4}$$

在非光瞳交互合成中, 可以通过将 IPSF 的两个不同的 $\overline{OTF}$ 即 $\overline{OTF_1}$ 和 $\overline{OTF_2}$ 相减, 得到一个双极性 (*bipolar*) IPSF

$$\begin{aligned}OTF_{eff}&=\overline{OTF_1}-\overline{OTF_2}\\&=\frac{p_1\left(\frac{-fk_x}{k_0},\frac{-fk_y}{k_0}\right)\otimes p_1\left(\frac{-fk_x}{k_0},\frac{-fk_y}{k_0}\right)}{OTF_1(0,0)}\\&\quad-\frac{p_2\left(\frac{-fk_x}{k_0},\frac{-fk_y}{k_0}\right)\otimes p_2\left(\frac{-fk_x}{k_0},\frac{-fk_y}{k_0}\right)}{OTF_2(0,0)},\end{aligned} \tag{4.5-5}$$

其中, p_1 和 p_2 为不同的瞳函数. 可以看出, OTF_{eff} 在 $(0,0)$ 处不再是最大值, 事实上, $OTF_{eff}(0,0)=0$, 其给出了光学系统的非低通滤波特性. 根据式 (4.3-6), 处理后的强度为

$$I(x,y)=\mathcal{F}^{-1}\left\{\mathcal{F}\left\{|t(-x,-y)|^2\right\}OTF_{eff}(k_x,k_y)\right\}. \tag{4.5-6}$$

为了计算 $OTF_{eff}(k_x,k_y)$, 首先需要利用 p_1 获取处理后的强度, 并将其储存起来, 然后用 p_2 获取另一个处理后的强度. 最后, 将这两个储存的强度相减, 以得到双极性非相干图像处理 (*bipolar incoherent image processing*). 术语 "双极性" 用在这里表示可获得的系统的点扩散函数是双极性的. 因为不能直接将强度减掉, 该方法需要用特殊的技术, 比如用计算机进行储存. 在 OTF 的光瞳交互合成 (*pupil interaction synthesis*) 中, 来自 p_1 和 p_2 的光会产生干涉. 由于操作中干涉测量的特性, 这些方法相对比较复杂. 尽管如此, 因为在相对一般的双极性

Therefore, we can rewrite the OTF in terms of the pupil function as

$$OTF(k_x, k_y) = p\left(\frac{-fk_x}{k_0}, \frac{-fk_y}{k_0}\right) \otimes p\left(\frac{-fk_x}{k_0}, \frac{-fk_y}{k_0}\right). \tag{4.5-2}$$

We can see that the autocorrelation of the pupil function is the OTF of the optical system. This is called *one-pupil synthesis* of the optical transfer functions. As any autocorrelation function always has a central maximum [see Eq. (4.3-5b)], we could not obtain, for example, a bandpass filter function. In order to circumvent the limitations of one-pupil syntheses, *two-pupil synthesis* is introduced.

According to Eq. (4.5-1), we first define the normalized OTF by

$$\begin{aligned}\overline{OTF}(k_x, k_y) &= \frac{H_c(k_x, k_y) \otimes H_c(k_x, k_y)}{OTF(0,0)} \\ &= \frac{\iint_{-\infty}^{\infty} H_c^*(k'_x, k'_y) H_c(k'_x + k_x, k'_y + k_y) \mathrm{d}k'_x \mathrm{d}k'_y}{\iint_{-\infty}^{\infty} |H_c(k'_x, k'_y)|^2 \mathrm{d}k'_x \mathrm{d}k'_y}.\end{aligned} \tag{4.5-3}$$

Therefore, $\overline{OTF}(0,0) = 1$. By the same token, in terms of the pupil function, we have

$$\overline{OTF}(k_x, k_y) = \frac{p\left(\frac{-fk_x}{k_0}, \frac{-fk_y}{k_0}\right) \otimes p\left(\frac{-fk_x}{k_0}, \frac{-fk_y}{k_0}\right)}{OTF(0,0)}. \tag{4.5-4}$$

In the non-pupil interaction synthesis, we can obtain a *bipolar* IPSF by subtracting the IPSFs of two different $\overline{OTF}$s, $\overline{OTF_1}$ and $\overline{OTF_2}$:

$$\begin{aligned}OTF_{eff} &= \overline{OTF_1} - \overline{OTF_2} \\ &= \frac{p_1\left(\frac{-fk_x}{k_0}, \frac{-fk_y}{k_0}\right) \otimes p_1\left(\frac{-fk_x}{k_0}, \frac{-fk_y}{k_0}\right)}{OTF_1(0,0)} \\ &\quad - \frac{p_2\left(\frac{-fk_x}{k_0}, \frac{-fk_y}{k_0}\right) \otimes p_2\left(\frac{-fk_x}{k_0}, \frac{-fk_y}{k_0}\right)}{OTF_2(0,0)},\end{aligned} \tag{4.5-5}$$

where p_1 and p_2 are the different pupil functions. Note that OTF_{eff} is no longer maximum at (0,0) and in fact, $OTF_{eff}(0,0) = 0$, giving the optical system of non-low-pass filtering characteristics. The processed intensity, according to Eq. (4.3-6), is

$$I(x,y) = \mathcal{F}^{-1}\left\{\mathcal{F}\left\{|t(-x,-y)|^2\right\} OTF_{eff}(k_x, k_y)\right\}. \tag{4.5-6}$$

To implement $OTF_{eff}(k_x, k_y)$, we need to first capture the processed intensity using p_1, store it, and then capture another processed intensity using p_2. Finally we perform subtraction of the two acquired stored intensities to obtain *bipolar incoherent image processing.* The term "bipolar" is used here to represent that the achievable point spread function of the system is bipolar. Because we cannot subtract intensities directly, this method requires special techniques such as, for instance, the use of computer storage. In *pupil interaction synthesis* of the OTFs, interference of light from p_1 and p_2 is involved. These methods are relatively complicated because of the interferometric nature of their operation. They are nevertheless

点扩散函数合成中允许更大的灵活性, 因此其还是很受欢迎. 对于该研究领域的深入回顾, 感兴趣的读者可参考本章末尾引用的参考文献, 以获取该方法的更多细节 [Indebetouw 和 Poon (1992)].

举例: 双极性非相干图像处理的 MATLAB 实例

取两个不同的圆域函数作为进行双极性非相干图像处理的例子: $p_1(x,y)=circ\left(\dfrac{r}{r_1}\right)$ 和 $p_2(x,y)=circ\left(\dfrac{r}{r_2}\right)$.根据式 (4.5-6) 进行仿真. 图 4.5-1a) 显示了原始图像, 图 4.5-1b) 给出了滤波后的图像, 其清晰地说明了非低通滤波, 并且的确阐明双极性非相干滤波 (*bipolar incoherent filtering*) 实现了边缘提取.

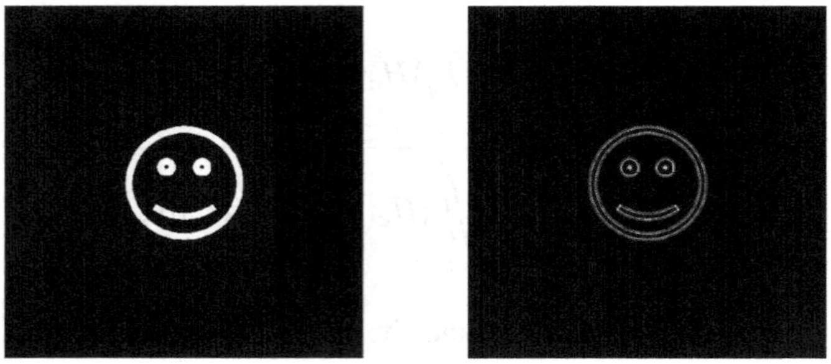

a) b)

图 4.5-1 a) 原始输入图像; b) 双极性滤波后的输出图像

图 4.5-1 中的图像是利用以下 m-文件生成的.

```
==========================================
% Bipolar_incoherent_image_processing
% circ(r/r1) and  circ(r/r2) as pupil functions
% Adapted from "Introduction to Modern Digital Holography"
% by T.-C. Poon & J.-P. Liu
% Cambridge University Press (2014), Table 1.7 .
clear all;close all;
A=imread('front.jpg');           % read image file 512x512 8-bit
A=double(A);
A=A/255;
SP=fftshift(fft2(fftshift(A)));
D=abs(SP);
D=D(129:384,129:384);
figure;imshow(A);
%title('Original image')
figure;imshow(30.*mat2gray(D)); % spectrum
title('Original spectrum')

c=1:512;
r=1:512;
[C, R]=meshgrid(c, r);
CI=((R-257).^2+(C-257).^2);
pup=zeros(512,512);
% produce pupil circ(r/r1)
for a=1:512;
    for b=1:512;
```

attractive because of the greater flexibility they allow in the synthesis of relatively general bipolar point spread functions. For an in-depth review of the research area, the interested reader is referred to the cited references at the end of this chapter for detailed description of the method [Indebetouw and Poon (1992)].

Example: MATLAB Example of Bipolar Incoherent Image Processing

We take two different circular functions as an example to perform bipolar incoherent image processing: $p_1(x, y) = circ(r/r_1)$ and $p_2(x, y) = circ(r/r_2)$. The simulation is performed according to Eq. (4.5-6). Figure 4.5-1a shows the original image, and Figure 4.5-1b shows the filtered image and it clearly illustrates non-low-pass filtering, and indeed edge extraction has been performed with bipolar incoherent filtering.

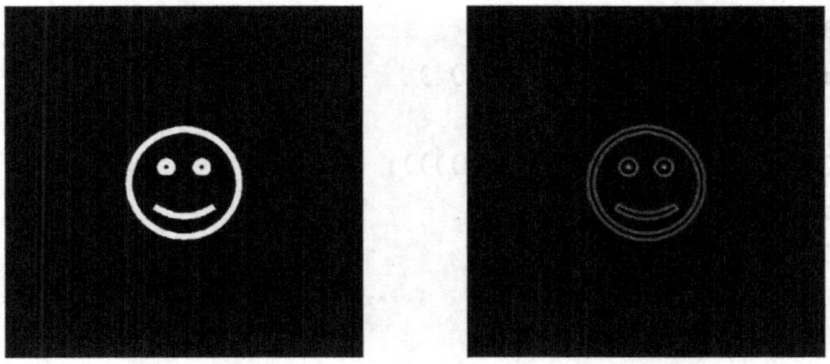

(a) (b)

Figure 4.5-1 (a) Original input image and (b) bipolar filtered output image

The images produced in Figure 4.5-1 are generated using the m-file shown below.

```
==========================================
% Bipolar_incoherent_image_processing
% circ(r/r1) and  circ(r/r2) as pupil functions
% Adapted from "Introduction to Modern Digital Holography"
% by T.-C. Poon & J.-P. Liu
% Cambridge University Press (2014), Table 1.7 .
clear all;close all;
A=imread('front.jpg');             % read image file 512x512 8-bit
A=double(A);
A=A/255;
SP=fftshift(fft2(fftshift(A)));
D=abs(SP);
D=D(129:384,129:384);
figure;imshow(A);
%title('Original image')
figure;imshow(30.*mat2gray(D)); % spectrum
title('Original spectrum')


c=1:512;
r=1:512;
[C, R]=meshgrid(c, r);
CI=((R-257).^2+(C-257).^2);
pup=zeros(512,512);
% produce pupil circ(r/r1)
for a=1:512;
    for b=1:512;
```

```
          if CI(a,b)>=100^2;  %pupil size
             pup(a,b)=0;
         else
             pup(a,b)=1;
         end
    end
end
h=ifft2(fftshift(pup));

OTF1=fftshift(fft2(h.*conj(h)));
OTF1=OTF1/max(max(abs(OTF1)));
G=abs(OTF1.*SP);
G=G(129:384,129:384);
figure;imshow(30.*mat2gray(G));
%title('Spectrum filtered by circ(r/r1)')

I=abs(fftshift(ifft2(fftshift(OTF1.*SP))));
figure;imshow(mat2gray(I));
%title('Image filtered by circ(r/r1)')

c=1:512;
r=1:512;
[C, R]=meshgrid(c, r);
CI=((R-257).^2+(C-257).^2);
pup=zeros(512,512);
% produce pupil circ(r/r2)
for a=1:512;
    for b=1:512;
          if CI(a,b)>=105^2;  %pupil size
             pup(a,b)=0;
         else
             pup(a,b)=1;
         end
    end
end
h=ifft2(fftshift(pup));

OTF2=fftshift(fft2(h.*conj(h)));
OTF2=OTF2/max(max(abs(OTF2)));
G=abs(OTF2.*SP);
G=G(129:384,129:384);
figure;imshow(30.*mat2gray(G));
%title('Spectrum filtered by circ(r/r2)')

I=abs(fftshift(ifft2(fftshift(OTF2.*SP))));
figure;imshow(mat2gray(I));
%title('Image filtered by circ(r/r2)')
%-------------------
TOTF=OTF1-OTF2;
TOTF=TOTF/max(max(abs(TOTF)));
G=abs(TOTF.*SP);
```

```
          if CI(a,b)>=100^2;  %pupil size
             pup(a,b)=0;
         else
             pup(a,b)=1;
         end
    end
end
h=ifft2(fftshift(pup));

OTF1=fftshift(fft2(h.*conj(h)));
OTF1=OTF1/max(max(abs(OTF1)));
G=abs(OTF1.*SP);
G=G(129:384,129:384);
figure;imshow(30.*mat2gray(G));
%title('Spectrum filtered by circ(r/r1)')

I=abs(fftshift(ifft2(fftshift(OTF1.*SP))));
figure;imshow(mat2gray(I));
%title('Image filtered by circ(r/r1)')

c=1:512;
r=1:512;
[C, R]=meshgrid(c, r);
CI=((R-257).^2+(C-257).^2);
pup=zeros(512,512);
% produce pupil circ(r/r2)
for a=1:512;
    for b=1:512;
          if CI(a,b)>=105^2;  %pupil size
             pup(a,b)=0;
         else
             pup(a,b)=1;
         end
    end
end
h=ifft2(fftshift(pup));

OTF2=fftshift(fft2(h.*conj(h)));
OTF2=OTF2/max(max(abs(OTF2)));
G=abs(OTF2.*SP);
G=G(129:384,129:384);
figure;imshow(30.*mat2gray(G));
%title('Spectrum filtered by circ(r/r2)')

I=abs(fftshift(ifft2(fftshift(OTF2.*SP))));
figure;imshow(mat2gray(I));
%title('Image filtered by circ(r/r2)')
%-------------------
TOTF=OTF1-OTF2;
TOTF=TOTF/max(max(abs(TOTF)));
G=abs(TOTF.*SP);
```

```
G=G(129:384,129:384);
figure;imshow(10.*mat2gray(G));
title('Spectrum filtered by TOTF')

I=abs(fftshift(ifft2(fftshift(TOTF.*SP))));
figure;imshow(mat2gray(I));
```
==

习题

4.1 在如图 4.2-1 所示的相干图像处理系统中, 相干点扩散函数为 $h_c(x,y) = \mathcal{F}\{p(x,y)\}\big|_{k_x=\frac{k_0x}{f},k_y=\frac{k_0y}{f}}$. 证明相干传递函数由 $H_c(k_x,k_y) = p\left(\dfrac{-fk_x}{k_0}, \dfrac{-fk_y}{k_0}\right)$ 给出.

4.2 参考图 4.2-1 中标准的 4-f 光学图像处理系统, 设计 $p(x,y)$ 来完成以下数学运算:

a) $\dfrac{\partial t(x,y)}{\partial x}, \dfrac{\partial t(x,y)}{\partial y}$.

b) $\dfrac{\partial t(x,y)}{\partial x} + \dfrac{\partial t(x,y)}{\partial y}$.

c) $\dfrac{\partial^2 t(x,y)}{\partial x^2} + \dfrac{\partial^2 t(x,y)}{\partial y^2}$.

d) $\dfrac{\partial^2 t(x,y)}{\partial x\partial y}$.

4.3 在一个 4-f 相干光学图像处理系统中, 对于 $t(x,y) = rect\left(\dfrac{x}{a}, \dfrac{y}{a}\right)$, 编写 MATLAB 代码来实现习题 4.2 中的数学运算, 并显示输出的强度图.

4.4 $t(x,y)$ 的二维希尔伯特变换 (*Hilbert transform*) 定义如下

$$\mathcal{H}\{t(x,y)\} = \frac{1}{\pi^2}\iint\limits_{-\infty}^{\infty} \frac{t(x',y')}{(x-x')(y-y')}\mathrm{d}x'\mathrm{d}y'.$$

a) 对于如图 4.2-1 所示的标准相干图像处理系统, 求其能够实现二维希尔伯特变换的瞳函数. 提示: $\dfrac{1}{\pi x}$ 的一维傅里叶变换为 $\mathcal{F}\left\{\dfrac{1}{\pi x}\right\} = \mathrm{j}sgn(k_x)$, 其中 $sgn(x)$ 为符号函数 (*signum function*), 定义如下

$$sgn(x) = \begin{cases} 1, & x > 0 \\ -1, & x < 0. \end{cases}$$

b) 编写可以实现 $t(x,y) = rect\left(\dfrac{x}{a}, \dfrac{y}{a}\right)$ 的希尔伯特变换的 MATLAB 代码, 并显示输出的强度图.

4.5 假设在如图 4.2-1 所示的标准 4-f 光学图像处理系统中, 有一个纯相位物体 $t(x,y) = \mathrm{e}^{\mathrm{j}\phi(x,y)}$, 其中 $\phi(x,y) \ll 1$, 如生物细胞, 那么

a) 对于 $p(x,y) = 1$, 求出其输出平面上的强度分布 $|\psi_{p,out}(x,y)|^2$.

```
G=G(129:384,129:384);
figure;imshow(10.*mat2gray(G));
title('Spectrum filtered by TOTF')

I=abs(fftshift(ifft2(fftshift(TOTF.*SP))));
figure;imshow(mat2gray(I));
```
=======================================

Problems

4.1 In the coherent image processing system shown in Figure 4.2-1, the coherent point spread function is $h_c(x,y) = \mathcal{F}\{p(x,y)\}\big|_{k_x=\frac{k_0 x}{f},k_y=\frac{k_0 y}{f}}$. Show that the coherent transfer function is given by $H_c(k_x,k_y) = p\left(\frac{-fk_x}{k_0}, \frac{-fk_y}{k_0}\right)$.

4.2 With reference to the standard 4-f optical image processing system shown in Figure 4.2-1, design $p(x,y)$ to implement the following mathematical operations:

(a) $\dfrac{\partial t(x,y)}{\partial x}, \dfrac{\partial t(x,y)}{\partial y}$.

(b) $\dfrac{\partial t(x,y)}{\partial x} + \dfrac{\partial t(x,y)}{\partial y}$.

(c) $\dfrac{\partial^2 t(x,y)}{\partial x^2} + \dfrac{\partial^2 t(x,y)}{\partial y^2}$.

(d) $\dfrac{\partial^2 t(x,y)}{\partial x \partial y}$.

4.3 For $t(x,y) = rect\left(\frac{x}{a}, \frac{y}{a}\right)$ in a 4-f coherent optical image processing system, write MATLAB codes to implement mathematical operations in P.4.2. Display the intensity of the output.

4.4 The two-dimensional *Hilbert transform* of $t(x,y)$ is defined as follows:

$$\mathcal{H}\{t(x,y)\} = \frac{1}{\pi^2}\iint_{-\infty}^{\infty} \frac{t(x',y')}{(x-x')(y-y')}\mathrm{d}x'\mathrm{d}y'.$$

(a) For a standard coherent image processing system shown in Figure 4.2-1, find the pupil function that is able to implement the 2-D Hilbert transform. Hint: The 1-D Fourier transform of $\frac{1}{\pi x}$ is $\mathcal{F}\left\{\frac{1}{\pi x}\right\} = \mathrm{j}sgn(k_x)$, where $sgn(x)$ is the *signum function* defined by

$$sgn(x) = \begin{cases} 1, & x > 0 \\ -1, & x < 0. \end{cases}$$

(b) Write a MATLAB code to implement the Hilbert transform of $t(x,y) = rect\left(\frac{x}{a}, \frac{y}{a}\right)$. Display the intensity of the output.

4.5 Suppose we have a pure phase object $t(x,y) = \mathrm{e}^{\mathrm{j}\phi(x,y)}$, where $\phi(x,y) \ll 1$, such as biological cells, in the standard 4-f optical image processing system shown in Figure 4.2-1.
(a) For $p(x,y) = 1$, find the intensity distribution $|\psi_{p,out}(x,y)|^2$ on the output plane.

b) 对于

$$p(x,y)=p(r=\sqrt{x^2+y^2})=\begin{cases}\mathrm{j}a, & r=\varepsilon\approx 0\\ 1, & \text{其他},\end{cases}$$

其中, a 为一个正的实常数, 证明输出平面上的强度分布由下式给出

$$|\psi_{p,out}(x,y)|^2\approx a^2+2\phi(-x,-y).$$

这是 1935 年由泽尼克 (Frits Zernike) 发明的相衬显微镜 (*phase contrast microscope*) 的基础, 为了实现 $\pi/2$ 的相移, 我们可以在光瞳平面的原点处放置一个矫正厚度的玻璃片, 这时, $a=1$.

4.6 参考如图 4.2-1 所示的 4-f 光学图像处理系统, 求出并画出其相干传递函数. 同时求出由以下相应的光瞳给出的相干点扩散函数:

a) $p(x,y)=rect\left(\dfrac{x-x_c}{x_0}\right)+rect\left(\dfrac{x+x_c}{x_0}\right)$, 一个双缝.

b) $p(x,y)=rect\left(\dfrac{x-x_c}{x_0}\right)$, 一个偏离了光轴的单缝.

4.7 在图题 4.7 中, 有一个被单位振幅平面波照射的输入 $t(x,y)$.

a) 证明输出平面上的复振幅由下式给出

$$\psi_{p,out}(x,y)=t\left(\frac{x}{M},\frac{y}{M}\right)*h_c(x,y),$$

其中, 相干点扩散函数 $h_c(x,y)=\mathcal{F}\{p(x,y)\}\big|_{k_x=\frac{k_0x}{f_2},\,k_y=\frac{k_0y}{f_2}}$, 且, $M=-f_2/f_1$.

b) 证明相干传递函数由下式给出

$$H_c(k_x,k_y)=p(x,y)\big|_{x=\frac{-f_2k_x}{k_0},\,y=\frac{-f_2k_y}{k_0}}.$$

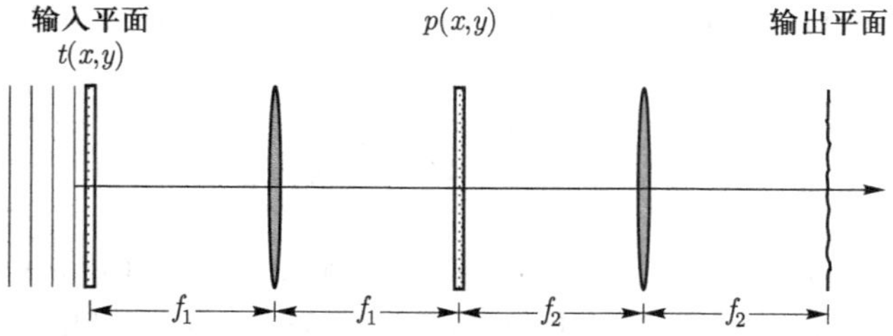

图题 4.7

4.8 参考如图 4.2-1 所示的 4-f 光学图像处理系统, 求出下述条件时输出平面上的复振幅:

a) $p(x,y)=T_2\left(\dfrac{k_0x}{f},\dfrac{k_0y}{f}\right)$.

b) $p(x,y)=T_2^*\left(-\dfrac{k_0x}{f},-\dfrac{k_0y}{f}\right)$.

其中, $\mathcal{F}\{t_2(x,y)\}=T_2(k_x,k_y)$. 此系统表示了哪种类型的数学运算?

4.9 相干传递函数由下式给出

(b) For

$$p(x, y) = p(r = \sqrt{x^2 + y^2}) = \begin{cases} ja, & r = \varepsilon \approx 0 \\ 1, & \text{otherwise}, \end{cases}$$

where a is a real and positive constant, show that the intensity distribution on the output plane is given by

$$|\psi_{p,out}(x, y)|^2 \approx a^2 + 2\phi(-x, -y).$$

This is the basis of the *phase contrast microscope*, invented by Frits Zernike in 1935. To implement the phase shift of $\pi/2$, we can put a glass plate of a correct thickness at the origin of the pupil plane, and in this case, $a = 1$.

4.6 With reference to the 4-f optical image processing system shown in Figure 4.2-1, find and plot the coherent transfer function. Also find the corresponding coherent point spread function for the pupil given below:

(a) $p(x, y) = rect\left(\dfrac{x - x_c}{x_0}\right) + rect\left(\dfrac{x + x_c}{x_0}\right)$, a double slit.

(b) $p(x, y) = rect\left(\dfrac{x - x_c}{x_0}\right)$, a single slit offset from the optical axis.

4.7 In Figure P.4.7, we have an input $t(x, y)$ illuminated by a plane wave of unit amplitude.

(a) Show that the complex amplitude on the output plane is given by

$$\psi_{p,out}(x, y) = t\left(\frac{x}{M}, \frac{y}{M}\right) * h_c(x, y),$$

where the coherent point spread function $h_c(x, y) = \mathcal{F}\{p(x, y)\}\big|_{k_x = \frac{k_0 x}{f_2},\, k_y = \frac{k_0 y}{f_2}}$, and $M = -f_2/f_1$.

(b) Show that the coherent transfer function is given by

$$H_c(k_x, k_y) = p(x, y)\big|_{x = \frac{-f_2 k_x}{k_0},\, y = \frac{-f_2 k_y}{k_0}}.$$

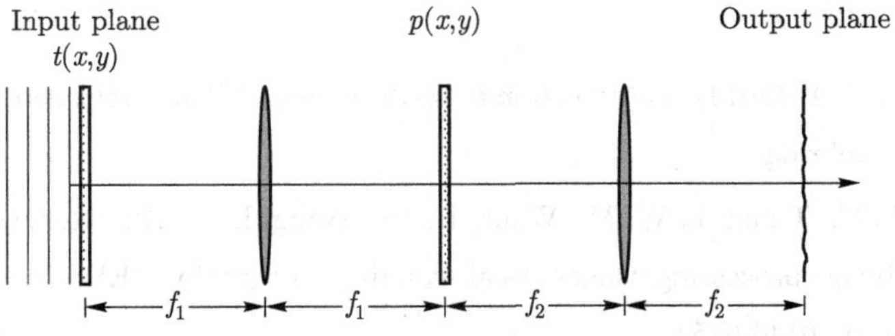

Figure P.4.7

4.8 Referring to the 4-f optical image processing system shown in Figure 4.2-1, find the complex field on the output plane for

(a) $p(x, y) = T_2\left(\dfrac{k_0 x}{f}, \dfrac{k_0 y}{f}\right)$.

(b) $p(x, y) = T_2^*\left(-\dfrac{k_0 x}{f}, -\dfrac{k_0 y}{f}\right)$.

where $\mathcal{F}\{t_2(x, y)\} = T_2(k_x, k_y)$. What type of mathematical operations does this system represent?

4.9 The coherent transfer function is given by

$$H_c(k_x, k_y) = rect\left(\frac{k_x}{K}\right),$$

证明其 OTF 为

$$OTF(k_x, k_y) = Ktri\left(\frac{k_x}{K}\right).$$

4.10 非相干光学系统的光学传递函数由下式给出

$$OTF(k_x, k_y) = \mathcal{F}\left\{|h_c(x, y)|^2\right\},$$

其中, $h_c(x, y) = \mathcal{F}^{-1}\{H_c(k_x, k_y)\}$, $H_c(k_x, k_y)$ 为相干传递函数, 证明

$$OTF(k_x, k_y) = \iint_{-\infty}^{\infty} H_c^*(k_x', k_y') H_c(k_x' + k_x, k_y' + k_y) \mathrm{d}k_x' \mathrm{d}k_y'.$$

参考文献

[1] Francon, M. (1974). *Holography.* Acadmic Press, New York and London.

[2] Indebetouw, G. and Poon, T.-C. (1992). "Novel approaches of incoherent image processing with emphasis on scanning methods," Optcial Engineering 31, pp.2159-2167.

[3] Indebetouw, G. (1981). "Scanning optical data processor," Optics and Laser Technology 13, pp.197-202.

[4] Lauterborn, W., Hurz, T. and Wiesenfeldt, M. (1995). *Coherent Optics Fundamentals and Applications.* Springer-Verlag, Berlin Heidelberg.

[5] Lohmann, A. W. and Rhodes, W. T. (1978). "Two-pupil synthesis of optical transfer functions," Applied Optics 17, pp.1141-1150.

[6] Poon, T.-C. (2007). *Optical Scanning Holography with MATLAB®.* Springer, New York.

[7] Poon, T.-C. and Kim, T. (2018). *Engineering Optics with MATLAB®*, 2nd ed. World Scientific, New Jersey.

[8] Poon, T.-C. and Liu, J.-P. (2014). *Introduction to Modern Digital Holography with MATLAB.* Cambridge University Press, Cambridge.

[9] Zhang, Y., Poon, T.-C., Tsang, P. W. M., Wang, R. and Wang, L. (2019). "Review on feature extraction for 3-D incoherent image processing using optical scanning holography," IEEE Transactions on Industrial Informatics 15 (11), pp.6146-6154.

$$H_c(k_x, k_y) = rect\left(\frac{k_x}{K}\right),$$

show that the OTF is

$$OTF(k_x, k_y) = Ktri\left(\frac{k_x}{K}\right).$$

4.10 The optical transfer function of incoherent optical systems is given by

$$OTF(k_x, k_y) = \mathcal{F}\left\{|h_c(x, y)|^2\right\},$$

whereh $h_c(x, y) = \mathcal{F}^{-1}\left\{H_c(k_x, k_y)\right\}$ with $H_c(k_x, k_y)$ being the coherent transfer function. Show that

$$OTF(k_x, k_y) = \iint\limits_{-\infty}^{\infty} H_c^*(k_x', k_y')H_c(k_x' + k_x, k_y' + k_y)\mathrm{d}k_x'\mathrm{d}k_y'.$$

Bibliography

[1] Francon, M. (1974). *Holography.* Acadmic Press, New York and London.

[2] Indebetouw, G. and Poon, T.-C. (1992). "Novel approaches of incoherent image processing with emphasis on scanning methods," Optcial Engineering 31, pp. 2159-2167.

[3] Indebetouw, G. (1981). "Scanning optical data processor," Optics and Laser Technology 13, pp. 197-202.

[4] Lauterborn, W., Hurz, T. and Wiesenfeldt, M. (1995). *Coherent Optics Fundamentals and Applications.* Springer-Verlag, Berlin Heidelberg.

[5] Lohmann, A. W. and Rhodes, W. T. (1978). "Two-pupil synthesis of optical transfer functions," Applied Optics 17, pp. 1141-1150.

[6] Poon, T.-C. (2007). *Optical Scanning Holography with MATLAB®.* Springer, New York.

[7] Poon, T.-C. and Kim, T. (2018). *Engineering Optics with MATLAB®*, 2nd ed. World Scientific, New Jersey.

[8] Poon, T.-C. and Liu, J.-P. (2014). *Introduction to Modern Digital Holography with MATLAB.* Cambridge University Press, Cambridge.

[9] Zhang, Y., Poon, T.-C., Tsang, P. W. M., Wang, R. and Wang, L. (2019). "Review on feature extraction for 3-D incoherent image processing using optical scanning holography," IEEE Transactions on Industrial Informatics 15 (11), pp. 6146-6154.

第五章　相干全息理论

在照相技术中, 三维物体的强度被成像并记录在一个二维记录介质中, 如照相底片或电荷耦合器件 (*charge-coupled device*, CCD) 相机, 它们只对光的强度有反应. 由于在记录过程中没有干涉发生, 因此波场的相位信息未能保存. 物体光场相位信息的丢失破坏了所记录场景的三维特征, 因此三维物体的视差 (*parallax*) 和深度信息是不可能通过照片看到的. 视差是指从不同角度观察物体时所看到的物体视觉位置的差异. 全息术 (*holography*) 是丹尼斯·伽博 (Dennis Gabor) 在 1948 年发明的一项技术, 该技术通过干涉记录了物体光场的振幅和相位信息. 其相位被编码在干涉图样中. 所记录的干涉图样就是全息图 (*hologram*). 回顾杨氏 (Young's) 干涉实验, 在此实验中, 干涉条纹的位置取决于两个光源之间的相位差 [(见式 (4.1-3) 和式 (4.1-4)], 其中, 干涉图样的第一个峰值 x_p 是两个点源相位差 $(\theta_1 - \theta_2)$ 的函数. 一旦一个三维物体的全息图被记录下来, 我们就可以通过简单地照亮该全息图 (如果干涉图样被记录在全息膜上) 或通过数字重建 (如果干涉条纹被 CCD 记录) 来重建该物体的三维像. 在相干全息术 (*coherent holography*) 中, 记录三维物体的复振幅信息, 而在非相干全息术 (*incoherent holography*) 中, 记录三维物体的强度分布. 本章主要讨论相干全息理论, 非相干全息术将在第六章讨论.

5.1　菲涅耳波带板作为点源全息图

本节讨论点源的全息记录. 理解了点源在光学系统中的行为, 就会深刻理解如何处理一个三维物体, 因为任何物体都可以看作许多点源的集合.

5.1.1　同轴记录

全息术含有两个过程: 记录 (*recording*) 和重建 (*reconstruction*). 我们将分开进行每一个过程的讨论.

记录

图 5.1-1a) 所示为一个点源的全息记录, 利用两个反射镜 (M) 和两个分束器 (BS), 准直后的激光先被分成两束平面波, 然后再合成. 其中一束平面波被用来照亮作为点物的针孔. 点源向记录介质射出发散的球面波, 该球面波在全息术中被称为物波 (*object wave*). 直接照射在记录介质的平面波被称为参考波 (*reference wave*). 图 5.1-1b) 所示为朝向记录介质传播的球

5 Principles of Coherent Holography

In photography, the intensity of a 3-D object is imaged and recorded in a 2-D recording medium such as a photographic film or a charge-coupled device (CCD) camera, which responds only to light intensity. Since there is no interference during recording, the phase information of the wave field is not preserved. The loss of the phase information of the light field from the object destroys the 3-D characteristics of the recorded scene, and therefore *parallax* and depth information of the 3-D object cannot be observed by viewing a photograph. Parallax is a difference in the apparent position of an object viewed from different angles. *Holography*, invented by Dennis Gabor in 1948, is a technique in which the amplitude and phase information of the light field of the object are recorded through interference. The phase is coded in the interference pattern. The recorded interference pattern is a *hologram*. It is reminiscent of Young's interference experiment in which the position of the interference fringes depends on the phase difference between the two sources [see Eqs. (4.1-3) and (4.1-4), where the first peak of the interference pattern x_p is a function of the phase difference of the two point sources, $\theta_1 - \theta_2$]. Once the hologram of a 3-D object has been recorded, we can reconstruct the 3-D image of the object by simply illuminating the hologram (if the interference pattern is recorded on a film) or through digital reconstruction (if the interference pattern is captured by a CCD). We record the complex amplitude of the 3-D object in *coherent holography*, whereas in *incoherent holography*, we record the intensity distribution of the 3-D object. In this chapter, we discuss the principles of coherent holography. Incoherent holography will be discussed in Chapter 6.

5.1 Fresnel Zone Plate as a Point-Source Hologram

In this section, we discuss the holographic recording of a point source. Once we understand how a point source behaves in an optical system, we will get much insight into how to deal with a 3-D object as any object is considered as a collection of many point sources.

5.1.1 On-axis Recording

Holography is a two-step process: *recording* and *reconstruction*. We will discuss each process separately.

Recording

Figure 5.1-1a illustrates holographic recording of a point source. A collimated laser is split into two plane waves and then recombined by using two mirrors (M) and two beam splitters (BS). One of the plane waves is used to illuminate a pinhole, which acts as our point source object. The point source gives a diverging spherical wave toward the recording medium. This spherical wave is known as the *object wave* in holography. The plane wave that directly illuminates the recording medium is known as the *reference wave*. Figure 5.1-1b shows the spherical object wave fronts and the plane reference wave fronts toward the recording medium.

面物波波前和参考平面波波前, 由于两个波前沿同一方向传播, 即沿 z 方向传播, 这种记录条件被称为同轴记录 (*on-axis recording*), 即所谓的同轴全息术 (*on-axis holography*).

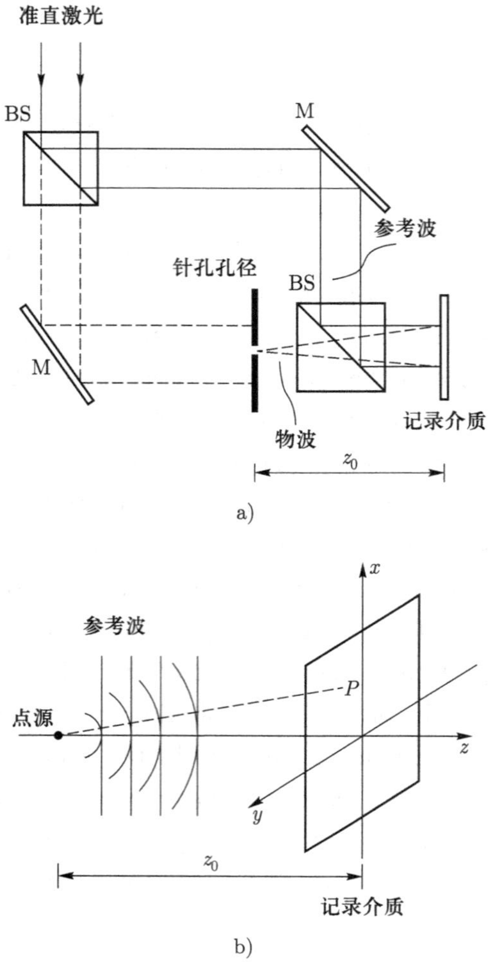

图 5.1-1　点物的全息记录: a) 实验装置; b) 参考平面波波前和球面物波波前的说明

分别用 ψ_0 和 ψ_r 表示记录介质平面上物波和参考波的复振幅, 因此记录平面上两个光波的干涉为 $\psi_0+\psi_r$. 假定参考波和物波在整个记录介质上是相干的, 由于记录介质只记录强度, 因此所记录的强度为 $I=|\psi_0+\psi_r|^2$. 实验中, 物参光两路的光程差小于激光器相干长度, 使用这样的激光是为了保证光波的相干性. 这种记录被称为全息记录 (*holographic recording*). 在摄影照相记录中, 参考波不存在, 因此只有物波被记录下来.

如果强度被照相底片所记录, 那么冲洗底片后的全息图可由透过率函数给出

$$t(x,y)\propto I=|\psi_0+\psi_r|^2. \tag{5.1-1}$$

然而, 如果使用 CCD 进行记录, 则记录的数字文件为 $t(x,y)$. 将针孔模型化为点源 δ 函数, 并假设点源位于与记录介质之间的距离为 z_0 的原点 ($z=0$) 处. 根据菲涅耳衍射公式 [见

Since both of the wave fronts propagate along the same direction, that is, along the z-direction, this recording condition is called *on-axis recording*, giving what is known as *on-axis holography*.

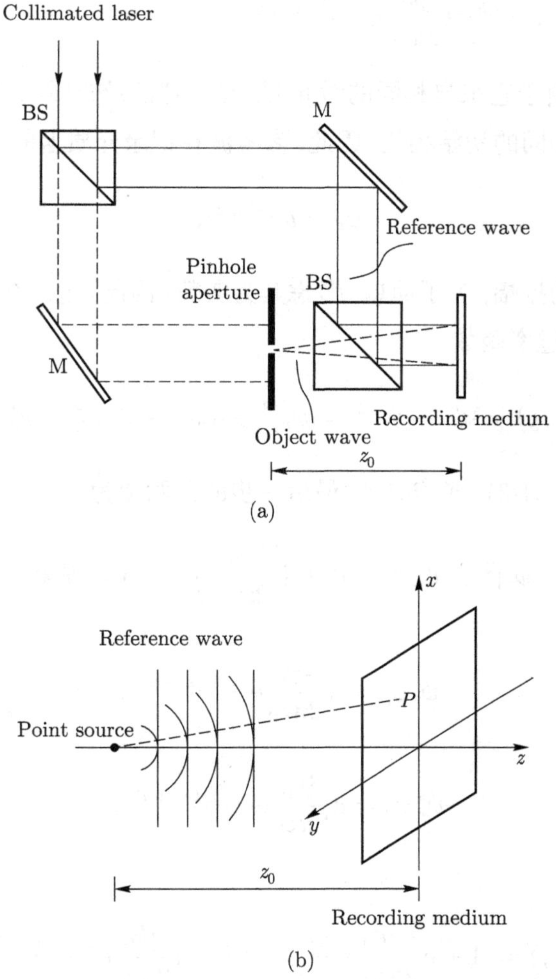

Figure 5.1-1 Holographic recording of a point source object: (a) experimental setup and (b) illustration of wave fronts of plane reference wave and spherical object wave

Let ψ_0 and ψ_r represent the complex fields of the object wave and the reference wave on the plane of the recording medium, respectively. The interference of the two waves on the recording plane is given by $\psi_0 + \psi_r$. Since the recording medium only records intensity, what is recorded is given by the intensity $I = |\psi_0 + \psi_r|^2$, provided the reference wave and the object wave are mutually coherent over the recording medium. The coherence of the light waves is guaranteed by the use of a laser with the difference between the two light paths being less than the coherence length of the laser. This kind of recording is known as *holographic recording*. In photographic recording, the reference wave does not exist and hence only the object wave is recorded.

If the intensity is recorded by a photographic film, what is developed is a hologram given by a transparency function

$$t(x, y) \propto I = |\psi_0 + \psi_r|^2. \tag{5.1-1}$$

However, if a CCD is used for recording, the recorded digital file is $t(x, y)$. Let us model the pinhole as a delta function point source and assume that the point source is located at the

式 (3.4-17)], 记录介质上的物波为

$$\psi_0(x,y;z_0)=\delta(x,y)*h(x,y;z_0)=\delta(x,y)*\mathrm{e}^{-\mathrm{j}k_0z_0}\frac{\mathrm{j}k_0}{2\pi z_0}\mathrm{e}^{-\mathrm{j}\frac{k_0}{2z_0}(x^2+y^2)}$$
$$=\mathrm{e}^{-\mathrm{j}k_0z_0}\frac{\mathrm{j}k_0}{2\pi z_0}\mathrm{e}^{-\mathrm{j}\frac{k_0}{2z_0}(x^2+y^2)}. \tag{5.1-2}$$

对于参考平面波而言, 由于它来自相同的激光源, 并且到记录介质的光程相同, 它和距离记录平面 z_0 处的点物有着相同的初始相位. 因此, 参考波在记录介质上的复振幅为

$$\psi_r=a\mathrm{e}^{-\mathrm{j}k_0z_0}, \tag{5.1-3}$$

其中, a 为参考平面波的振幅, 为了简单, 考虑其为实数. 因此, 记录介质上所记录的强度分布有一个如下式所给的透过率函数

$$t(x,y)=|\psi_r+\psi_0|^2=|\psi_r|^2+|\psi_0|^2+\psi_r\psi_0^*+\psi_r^*\psi_0. \tag{5.1-4}$$

再根据式 (5.1-2) 和式 (5.1-3), 式 (5.1-4) 最后一步的前两项为

$$|\psi_r|^2+|\psi_0|^2=a^2+\left(\frac{k_0}{2\pi z_0}\right)^2=A=\text{常数}. \tag{5.1-5a}$$

第三项为

$$\psi_r\psi_0^*=a\frac{-\mathrm{j}k_0}{2\pi z_0}\mathrm{e}^{\mathrm{j}\frac{k_0}{2z_0}(x^2+y^2)}, \tag{5.1-5b}$$

最后一项为

$$\psi_r^*\psi_0=a\frac{\mathrm{j}k_0}{2\pi z_0}\mathrm{e}^{-\mathrm{j}\frac{k_0}{2z_0}(x^2+y^2)}. \tag{5.1-5c}$$

最终, 式 (5.1-4) 为

$$t(x,y)=A+a\frac{-\mathrm{j}k_0}{2\pi z_0}\mathrm{e}^{\mathrm{j}\frac{k_0}{2z_0}(x^2+y^2)}+a\frac{\mathrm{j}k_0}{2\pi z_0}\mathrm{e}^{-\mathrm{j}\frac{k_0}{2z_0}(x^2+y^2)}, \tag{5.1-6}$$

上式可以简化成一个实函数, 因此就有一个由下式给出的实数全息图 (*real hologram*)

$$t(x,y)=A+B\sin\left[\frac{k_0}{2z_0}(x^2+y^2)\right]=FZP(x,y;z_0), \tag{5.1-7}$$

其中, $B=\dfrac{ak_0}{\pi z_0}$. 式 (5.1-7) 中的表达式通常被称为正弦菲涅耳波带板 (*Fresnel zone plate*, FZP), 这是点物在距离记录介质 $z=z_0$ 处的全息图. 如图 5.1-2 所示为 FZP 的图示, 图中只将 k_0 设为一个常量, 并取 $z=z_0$ 和 $z=2z_0$ 的情况.

接下来介绍 FZP 的二次空间相关性. 一个正弦信号的瞬时角频率 (*instantaneous angular frequency*) 是由其相位的时间导数 (即时间变化率) 定义的. 因此, 对于一个给定的正弦信号 $x(t)=\sin(2\pi f_0t)$, 其相位为 $\theta(t)=2\pi f_0t$. 正弦波的瞬时频率 f_i 为

$$f_i=\frac{1}{2\pi}\frac{\mathrm{d}\theta(t)}{\mathrm{d}t}=\frac{1}{2\pi}\frac{\mathrm{d}(2\pi f_0t)}{\mathrm{d}t}=f_0. \tag{5.1-8}$$

origin ($z = 0$) with a distance of z_0 away from the recording medium. According to Fresnel diffraction [see Eq. (3.4-17)], on the recording medium the object wave is

$$\begin{aligned}\psi_0(x, y; z_0) &= \delta(x, y) * h(x, y; z_0) = \delta(x, y) * \mathrm{e}^{-\mathrm{j}k_0 z_0} \frac{\mathrm{j}k_0}{2\pi z_0} \mathrm{e}^{-\mathrm{j}\frac{k_0}{2z_0}(x^2+y^2)} \\ &= \mathrm{e}^{-\mathrm{j}k_0 z_0} \frac{\mathrm{j}k_0}{2\pi z_0} \mathrm{e}^{-\mathrm{j}\frac{k_0}{2z_0}(x^2+y^2)}. \end{aligned} \quad (5.1\text{-}2)$$

For the reference plane wave, since it comes from the same laser source and has the same optical path length to the recording medium, it has the same initial phase as the point object at a distance of z_0 away from the recording medium. Therefore, its complex field on the recording medium is

$$\psi_r = a\mathrm{e}^{-\mathrm{j}k_0 z_0}, \quad (5.1\text{-}3)$$

where a, considered real for simplicity here, is the amplitude of the plane wave. Hence, the recorded intensity distribution on the recording medium has a transparency function given by

$$t(x, y) = |\psi_r + \psi_0|^2 = |\psi_r|^2 + |\psi_0|^2 + \psi_r \psi_0^* + \psi_r^* \psi_0. \quad (5.1\text{-}4)$$

Upon using Eqs. (5.1-2) and (5.1-3), the first two terms are

$$|\psi_r|^2 + |\psi_0|^2 = a^2 + \left(\frac{k_0}{2\pi z_0}\right)^2 = A = \text{constant}. \quad (5.1\text{-}5\text{a})$$

The third term is

$$\psi_r \psi_0^* = a \frac{-\mathrm{j}k_0}{2\pi z_0} \mathrm{e}^{\mathrm{j}\frac{k_0}{2z_0}(x^2+y^2)}, \quad (5.1\text{-}5\text{b})$$

and the last term is

$$\psi_r^* \psi_0 = a \frac{\mathrm{j}k_0}{2\pi z_0} \mathrm{e}^{-\mathrm{j}\frac{k_0}{2z_0}(x^2+y^2)}. \quad (5.1\text{-}5\text{c})$$

Finally, Eq. (5.1-4) becomes

$$t(x, y) = A + a \frac{-\mathrm{j}k_0}{2\pi z_0} \mathrm{e}^{\mathrm{j}\frac{k_0}{2z_0}(x^2+y^2)} + a \frac{\mathrm{j}k_0}{2\pi z_0} \mathrm{e}^{-\mathrm{j}\frac{k_0}{2z_0}(x^2+y^2)}, \quad (5.1\text{-}6)$$

which can be simplified to become a real function and we have a *real hologram* given by

$$t(x, y) = A + B \sin\left[\frac{k_0}{2z_0}(x^2 + y^2)\right] = FZP(x, y; z_0), \quad (5.1\text{-}7)$$

where $B = ak_0/\pi z_0$. The expression in Eq. (5.1-7) is often called the sinusoidal *Fresnel zone plate* (FZP), which is the hologram of the point source object at a distance $z = z_0$ away from the recording medium. Plots of the FZP are shown in Figure 5.1-2, where we have set k_0 to be some constant but for $z = z_0$ and $z = 2z_0$.

Let us investigate the quadratic spatial dependence of the FZP. The *instantaneous angular frequency* of a sinusoidal signal is defined by the time derivative (i.e., time rate of change) of its phase. Hence, for a given sinusoidal signal $x(t) = \sin(2\pi f_0 t)$, its phase is $\theta(t) = 2\pi f_0 t$. The instantaneous frequency, f_i, of the sine wave is, therefore, given by

$$f_i = \frac{1}{2\pi}\frac{\mathrm{d}\theta(t)}{\mathrm{d}t} = \frac{1}{2\pi}\frac{\mathrm{d}(2\pi f_0 t)}{\mathrm{d}t} = f_0. \quad (5.1\text{-}8)$$

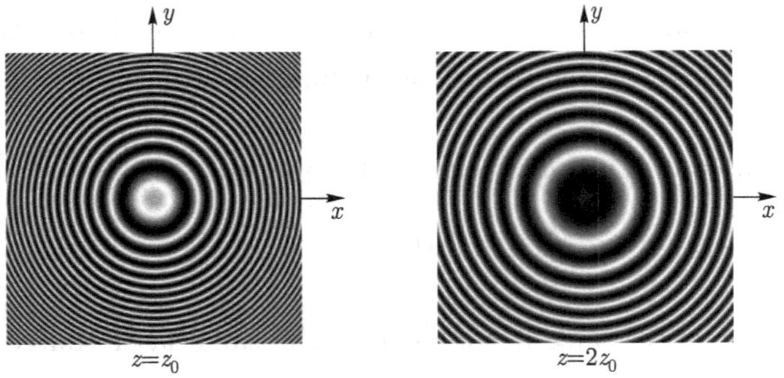

图 5.1-2 非涅耳波带板图示

对于 FZP, 相位的空间变化率, 比如沿 x 方向的, 被称为局部频率 (*local frequency*), 如下表示

$$f_{local}=\frac{1}{2\pi}\frac{\mathrm{d}}{\mathrm{d}x}\left(\frac{k_0}{2z_0}x^2\right)=\frac{k_0x}{2\pi z_0}. \tag{5.1-9}$$

可以看出, 局部条纹的频率随空间坐标 x 呈线性增加. 也可以发现, 当 $x<0$ 时, 取这个量的绝对值. 换句话说, 从图 5.1-2 中可以明显看出, 距离中心区域越远, 局部空间频率会越高. 从图中还可以看出, 当 z 值翻倍时, 比如从 $z=z_0$ 变为 $z=2z_0$, 局部频率变低了, 这点从式 (5.1-9) 也可以看出. 因此, 局部频率携带深度信息, 即 z 信息, 因此, 从局部频率中, 可以推断点物距离记录介质有多远. 换句话说, 深度信息以 "条纹密度" 的方式被编码进 FZP 中.

对于距记录平面 z_0 处的点源 $\delta(x,y)$, 可以得到由 $FZP(x,y;z_0)=A+B\sin\left[\frac{k_0}{2z_0}(x^2+y^2)\right]$ 给出的全息图. 不难看出, 对于距记录平面 z_0 处的离轴点源 $\delta(x-x_0,y-y_0)$, 离轴 FZP 全息图可以表示为

$$FZP(x-x_0,y-y_0;z_0)=A+B\sin\left\{\frac{k_0}{2z_0}\left[(x-x_0)^2+(y-y_0)^2\right]\right\}. \tag{5.1-10}$$

离轴 FZP 如图 5.1-3 所示. 可以看出, 区域的中心指定了点源物体的位置 x_0 和 y_0, 以及其深度信息 z_0, z_0 已被编码进 "条纹密度" 中. 因此, 可以看出, 离轴 FZP 完全含有 δ 函数中三维坐标的全部信息, 即 (x_0,y_0,z_0).

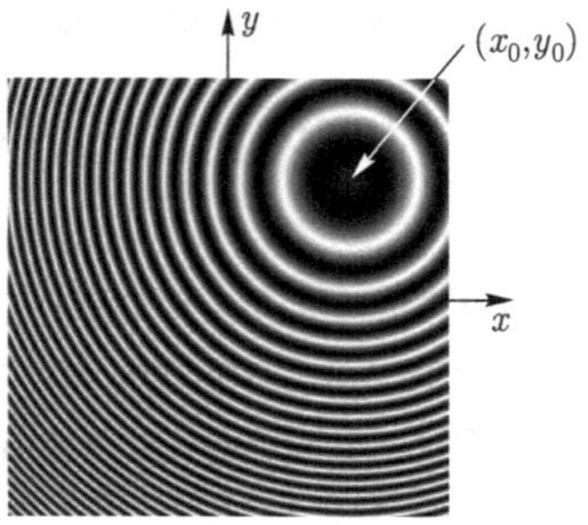

图 5.1-3 离轴点源物体的非涅耳波带板图示

图 5.1-2 和图 5.1-3 中所示图像由以下 m-文件生成.

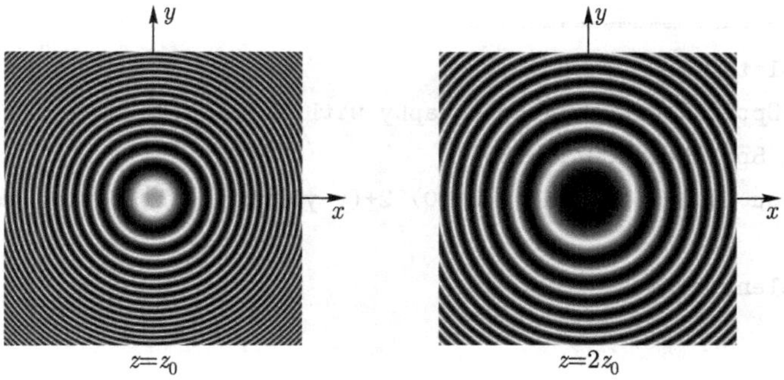

Figure 5.1-2 Plots of Fresnel zone plates

For the FZP, the spatial rate of change of the phase, say, along the x-direction is called the *local frequency* and is given by

$$f_{local} = \frac{1}{2\pi}\frac{\mathrm{d}}{\mathrm{d}x}\left(\frac{k_0}{2z_0}x^2\right) = \frac{k_0 x}{2\pi z_0}. \tag{5.1-9}$$

Note that the local fringe frequency increases linearly with the spatial coordinate, x. Note that, for $x < 0$, we take the absolute value of the quantity. In other words, the further we are away from the zone's center, the higher the local spatial frequency, which is obvious from Figure 5.1-2. Note also from the figure, when we double the z value, say from $z = z_0$ to $z = 2z_0$, the local frequency has become less as evident from Eq. (5.1-9) as well. Hence, the local frequency carries the information on depth, that is, z, and from the local frequency, therefore, we can deduce how far the object point source is away from the recording medium. In other words, the depth information is coded by the "fringe density" in the FZP.

For a point source $\delta(x, y)$ which is z_0 away from the recording medium, we have a hologram given by $FZP(x, y; z_0) = A + B\sin\left[\frac{k_0}{2z_0}(x^2 + y^2)\right]$. It is not hard to see that for an off-axis point source $\delta(x - x_0, y - y_0)$ at a distance of z_0 from the recording medium, its hologram is given by the *off-axis FZP*:

$$FZP(x - x_0, y - y_0; z_0) = A + B\sin\left\{\frac{k_0}{2z_0}\left[(x - x_0)^2 + (y - y_0)^2\right]\right\}. \tag{5.1-10}$$

Figure 5.1-3 shows the off-axis FZP. Note that the center of the zone specifies the location, x_0 and y_0, of the point source object, and the depth information, that is, z_0 is coded by the "fringe density." Hence, we see that the off-axis FZP completely contains the complete information of the 3-D coordinates of the delta function, that is, (x_0, y_0, z_0).

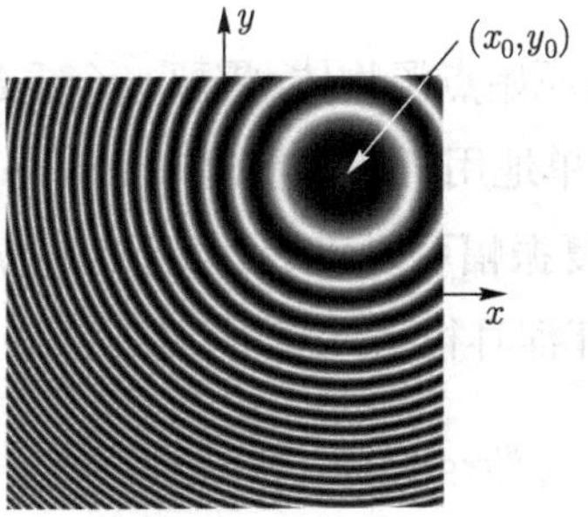

Figure 5.1-3 Fresnel zone plate due to off-axis point source object

The images produced in Figures 5.1-2 and 5.1-3 are generated using the m-file shown below.

```
========================================
    %Fresnel_zone_plate
    %Adapted from "Optical Scanning Holography with MATLAB®,"
    %by T.-C. Poon, Springer 2007, p.55.
    %display function is 1+sin(sigma*((x-x0)^2+(y-y0)^2)). All scales are
arbitrary.
    %sigma=pi/(wavelength*z)
    %
    clear;

    z0=4 %z0 proportional to the distance from the point object to
recording medium
    x0=7 % x0=y0, center of the FZP

    ROWS=256;
    COLS=256;
    colormap(gray(256))
    sigma=1/z0;
    y0=-x0;
    y=-12.8;
    for r=1:COLS,
     x=-12.8;
       for c=1:ROWS,      %compute Fresnel zone plate
           fFZP(r,c)=exp(j*sigma*(x-x0)*(x-x0)+j*sigma*(y-y0)*(y-y0));
          x=x+.1;
          end
      y=y+.1;
    end
    %normalization
    max1=max(fFZP);
    max2=max(max1);
    scale=1.0/max2;
    fFZP=fFZP.*scale;
    image(127*(1+imag(fFZP)));
    axis square on
    axis off
========================================
```

重建

为了从全息图 $t(x,y)$ 重建出原始点源物体, 需要一个重建 (*reconstruction*) 或是解码 (*decoding*) 的过程, 这在全息术中简单地用所谓的重建光 (*reconstruction wave*) ψ_{rec} 照射全息图即可实现. 因此紧贴全息图后的复振幅可以表示为 $\psi_{rec}t(x,y)$, 为了求出全息图后方距离为 z 处的光场, 仅需用下式对其进行菲涅耳衍射计算 [见式 (3.4-17)]

$$\psi_{rec}t(x,y) * h(x,y;z). \tag{5.1-11}$$

以同轴 FZP 的全息图为例, 重建光为垂直入射 (*normally incident*) 的平面波, 因此为简单起见, 取 $\psi_{rec}=1$. 这里不用式 (5.1-11) 中已有的计算方法 (当然, 该方法也可以进行计算),

```
========================================
    %Fresnel_zone_plate
    %Adapted from "Optical Scanning Holography with MATLAB®,"
    %by T.-C. Poon, Springer 2007, p.55.
    %display function is 1+sin(sigma*((x-x0)^2+(y-y0)^2)). All scales are
arbitrary.
    %sigma=pi/(wavelength*z)
    %
    clear;

    z0=4 %z0 proportional to the distance from the point object to
recording medium
    x0=7 % x0=y0, center of the FZP

    ROWS=256;
    COLS=256;
    colormap(gray(256))
    sigma=1/z0;
    y0=-x0;
    y=-12.8;
    for r=1:COLS,
     x=-12.8;
       for c=1:ROWS,        %compute Fresnel zone plate
           fFZP(r,c)=exp(j*sigma*(x-x0)*(x-x0)+j*sigma*(y-y0)*(y-y0));
          x=x+.1;
          end
      y=y+.1;
    end
    %normalization
    max1=max(fFZP);
    max2=max(max1);
    scale=1.0/max2;
    fFZP=fFZP.*scale;
    image(127*(1+imag(fFZP)));
    axis square on
    axis off
========================================
```

Reconstruction

To reconstruct the original point source object from the hologram $t(x, y)$, we need a *reconstruction* or *decoding* process. This is done simply by illuminating the hologram with the so-called *reconstruction wave*, ψ_{rec}, in holography. Therefore, the complex field immediately after the hologram is $\psi_{rec}t(x, y)$. To find the field after the hologram at some distance z, we just perform Fresnel diffraction as follows [see Eq. (3.4-17)]:

$$\psi_{rec}t(x, y) * h(x, y; z). \tag{5.1-11}$$

Let us take our hologram to be an on-axis FZP as an example with the reconstruction wave being a plane wave *normally incident* on it. Therefore, we take $\psi_{rec} = 1$ for simplicity. Instead of explicitly calculating Eq. (5.1-11) (which can be done, of course), let us take

我们用另一种方法来求, 当全息图被平面波照亮时会发生什么? 同轴点源全息图表示为

$$t(x,y) = FZP(x,y;z_0) = |\psi_r + \psi_0|^2 = |\psi_r|^2 + |\psi_0|^2 + \psi_r\psi_0^* + \psi_r^*\psi_0$$
$$= A + B\sin\left[\frac{k_0}{2z_0}(x^2+y^2)\right].$$

这 [见式 (5.1-5)] 在前面已经计算过, 即

$$|\psi_r|^2 + |\psi_0|^2 = a^2 + \left(\frac{k_0}{2\pi z_0}\right)^2 = A = \text{常数},$$
$$\psi_r\psi_0^* = a\frac{-\mathrm{j}k_0}{2\pi z_0}\mathrm{e}^{\mathrm{j}\frac{k_0}{2z_0}(x^2+y^2)},\quad \psi_r^*\psi_0 = a\frac{\mathrm{j}k_0}{2\pi z_0}\mathrm{e}^{-\mathrm{j}\frac{k_0}{2z_0}(x^2+y^2)}.$$

这三项分别对应于可忽略厚度的平板玻璃、正透镜和负透镜的效应. 第一个常数项仅表示: 如果一平面波入射到它上面, 该平面波会没有任何偏离地直线传播, 这就是所谓的零级光 (*zeroth-order beam*).

回顾一下, 焦距为 f 的理想透镜的透过率函数为 $\mathrm{e}^{\mathrm{j}\frac{k_0}{2f}(x^2+y^2)}$ [见式 (3.4-24)]. 因此, 第二项 $\psi_r\psi_0^*$ 正比于焦距为 $f = z_0$ 的理想透镜的透过率函数. 同样, 第三项表示焦距为 $f = -z_0$ 的负透镜. 总的来说, 可以把 FZP 看作如图 5.1-4 所示三个光学元件的组合. 如果一个平面波照射到 FZP 上, 将有三束光从全息图射出. 其中零级光依旧是直线传播没有偏离, 会聚的光束在 $z = z_0$ 处生成一个实像 (*real image*), 如果一个观察者看向全息图, 会在全息图后 z_0 处看到一个虚像 (*virtual image*), 因为正好在位置 z_0 处重建该物点, 该情况如图 5.1-5 所示. 重建

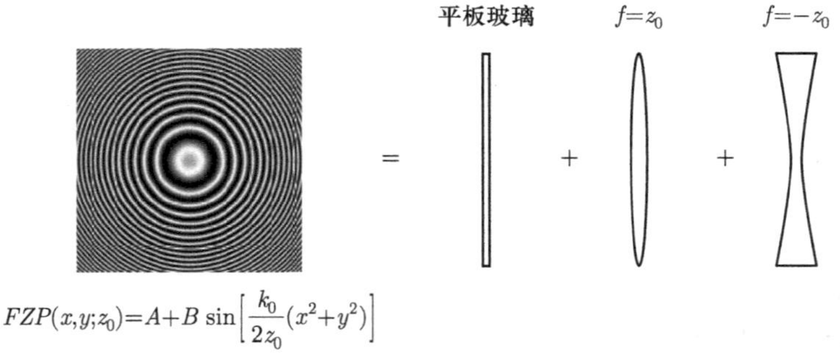

图 5.1-4 三个光学元件的组合作为菲涅耳波带板

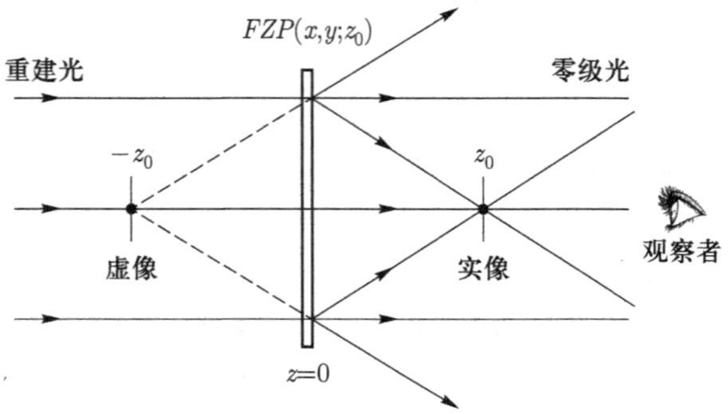

图 5.1-5 重建平面波照明下 FZP 的重建光路图

another approach to find what happens when the hologram is illuminated by a plane wave. The on-axis point-source hologram is

$$t(x,y) = FZP(x,y;z_0) = |\psi_r + \psi_0|^2 = |\psi_r|^2 + |\psi_0|^2 + \psi_r\psi_0^* + \psi_r^*\psi_0$$
$$= A + B\sin\left[\frac{k_0}{2z_0}(x^2+y^2)\right].$$

where we have calculated that [see Eq. (5.1-5)]

$$|\psi_r|^2 + |\psi_0|^2 = a^2 + \left(\frac{k_0}{2\pi z_0}\right)^2 = A = \text{constant},$$
$$\psi_r\psi_0^* = a\frac{-\mathrm{j}k_0}{2\pi z_0}\mathrm{e}^{\mathrm{j}\frac{k_0}{2z_0}(x^2+y^2)} \text{ and } \psi_r^*\psi_0 = a\frac{\mathrm{j}k_0}{2\pi z_0}\mathrm{e}^{-\mathrm{j}\frac{k_0}{2z_0}(x^2+y^2)}.$$

These three terms correspond to the effect of a flat glass of negligible thickness, a positive lens and a negative lens, respectively. The first constant term simply means that if a plane wave is incident on it, the plane wave travels straight without any deviation and that gives the so-called *zeroth-order beam.*

Recall that the transparency function of an idea lens of focal length f is $\mathrm{e}^{\mathrm{j}\frac{k_0}{2f}(x^2+y^2)}$ [see Eq. (3.4-24)]. Hence, the second term $\psi_r\psi_0^*$ is proportional to a transparency function of an idea lens of focal length $f = z_0$. Similarly, the third term represents a negative lens of focal length $f = -z_0$. As a whole, we can consider the FZP as a combination of these three pieces of optical elements as illustrated in Figure 5.1-4. If a plane wave is illuminated onto the FZP, we will have three beams exiting from the hologram. The zeroth-order beam just goes straight without deviation. The focused beam at $z = z_0$ generates a *real image*, and the *virtual image* is observed z_0 behind the hologram if an observer looks into the hologram as we reconstruct the original point source object at the exact location z_0. The situation is shown in Figure 5.1-5. The reconstructed real image is known as the *twin image.* Besides the

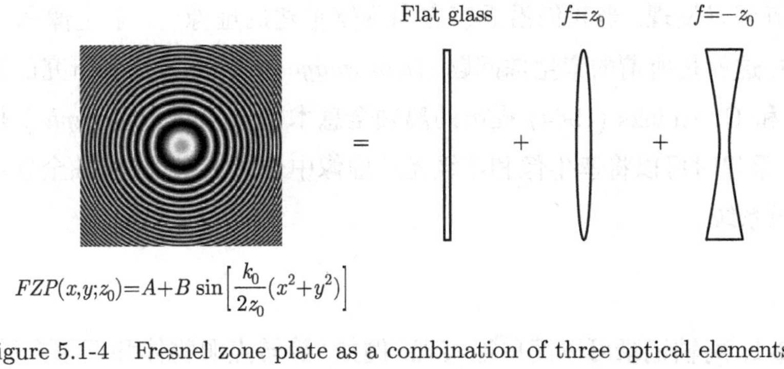

Figure 5.1-4 Fresnel zone plate as a combination of three optical elements

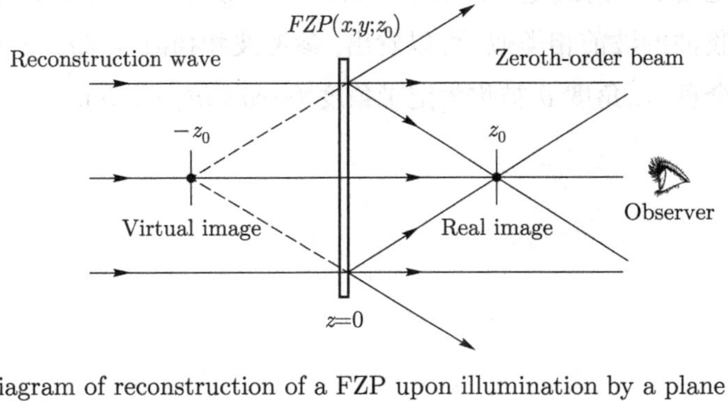

Figure 5.1-5 Ray diagram of reconstruction of a FZP upon illumination by a plane reconstruction wave

的实像被称为孪生像 (*twin image*). 除了零级光以外, 孪生像也会产生我们不希望的且会影响观察虚像的光线. 在下一小节中, 我们将讨论一种避免零级光和孪生像干扰的方法.

一旦我们理解了点物是如何全息记录和重建的, 就容易处理一个三维物体了. 图 5.1-6 表示了一个三点表示的物体的全息记录和重建. 其虚像出现在和原始图像相同的三维位置处, 观测者将在与原物相同的视角处看到原物的重建像, 该像被称为无畸变像 (*orthoscopic image*). 实像 (孪生像) 是原始物体的镜像, 其反射轴在全息图的平面上, 这样的像被称为赝像 (*pseudoscopic image*). 由于赝像不能给观察者提供正确的视差, 因此不适合用于三维显示. 值得注意的是, 全息图中的这一项 ψ_0 [见式 (5.1-4)] 会产生虚像重建, 而它在全息图中的复共轭 ψ_0^* 对应于实像重建.

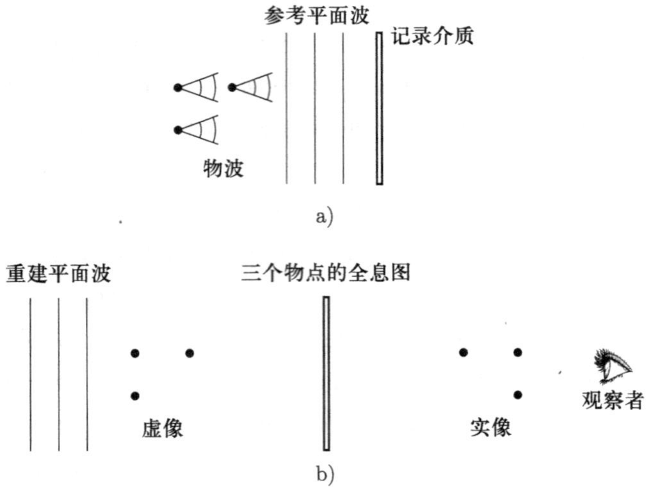

图 5.1-6 三点物体: a) 记录; b) 重建

5.1.2 离轴记录

从上一小节可以发现, 当我们沿观察方向观察重建的虚像时, 孪生像和零级光产生了干扰. 在全息术中, 这就是所谓的孪生像问题 (*twin image problem*), 很多研究已经开展以解决这个问题. Leith 和 Upatnieks (1964) 提出的离轴全息术 (*off-axis holography*) 是一种开创性的方法, 该方法在重建时可以将孪生像和零级光从虚像中分离出来. 故离轴全息术的思想是建立一个离轴的记录系统.

记录

参考图 5.1-1, 离轴记录可以简单地实现, 例如, 旋转点源物体和记录介质之间的分束器, 使参考平面波以一定角度入射到记录介质上. 如图 5.1-7 所示的这种情况与图 5.1-1b) 给出的参考平面波和球面物波的波前相类似. 可以看出, 参考波和物波并未沿 z 轴相同的方向传播, 因此它被称为离轴全息术, 角度 θ 被称为记录角度 (*recording angle*).

zeroth-order beam, the twin image will also create unwanted light affecting the viewing of the virtual image. In the next subsection, we discuss a way to avoid the zeroth-order beam and the twin image disturbances.

Once we understand how a point object is holographically recorded and reconstructed, it is easy to deal with a 3-D object. Figure 5.1-6 shows the holographic recording and reconstruction of a three-point object. The virtual image appears at the correct 3-D location as the original image, and the observer will see a reconstructed image with the same perspective as the original object and the image is called the *orthoscopic image*. The real image (the twin image) is the mirror image of the original object, with the axis of reflection on the plane of the hologram. Such an image is called the *pseudoscopic image*. Since the pseudoscopic image cannot provide the correct parallax to the observer, it is not suitable for 3-D display. It is worthwhile to note that the term ψ_0 in the hologram [see Eq. (5.1-4)] gives rise to a virtual image reconstruction, while its complex conjugate ψ_0^* in the hologram corresponds to a real image reconstruction.

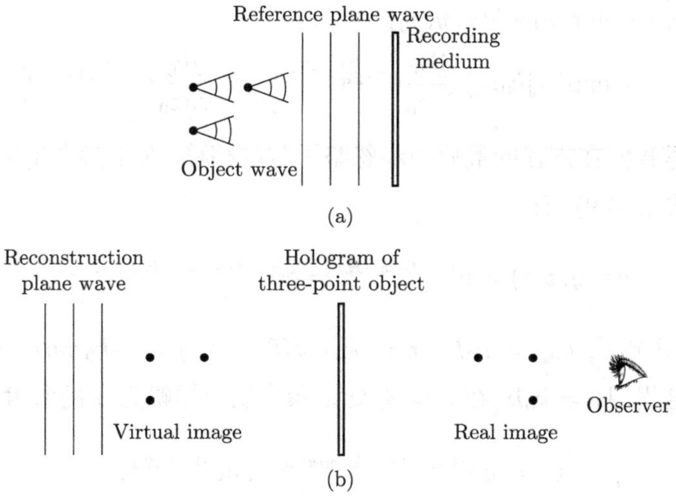

Figure 5.1-6 Three-point object: (a) recording and (b) reconstruction

5.1.2 Off-axis Recording

We have seen from the last section that the twin image and the zeroth-order beam create an annoying effect when we view the reconstructed virtual image along the viewing direction. In holography, this is infamously known as the *twin image problem*. Much research has been done to solve this problem. *Off-axis holography* is an original method devised by Leith and Upatnieks (1964) to separate the twin image and the zeroth-order beam from the virtual image upon reconstruction. The idea of *off-axis holography* is to set up an off-axis recording geometry.

Recording

With reference to Figure 5.1-1, off-axis recording can be done by simply, for example, rotating the beam splitter between the point source object and the recording medium so that the reference plane wave is incident on the recording medium at an angle. The situation, similar to Figure 5.1-1b in that we show the wave fronts of the reference plane wave and the spherical wave, is shown in Figure 5.1-7. Note that the reference wave and the object wave do not propagate in the same direction along z, and hence it is called off-axis holography. The angle

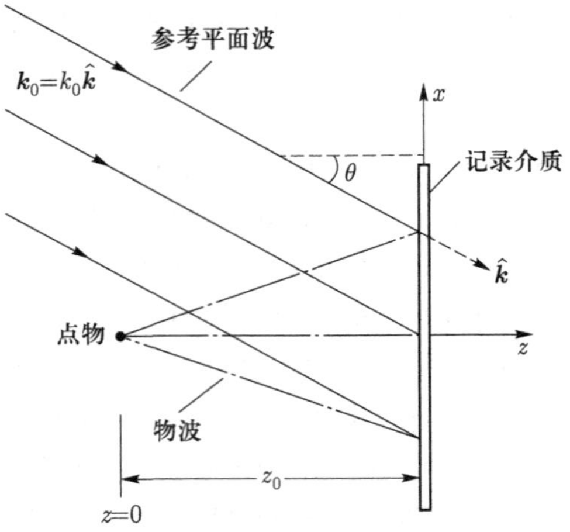

图 5.1-7 离轴记录

同样, 记录介质上的物波与式 (5.1-2) 相同

$$\psi_0(x,y;z_0)=\delta(x,y)*h(x,y;z_0)$$
$$=\exp(-\mathrm{j}k_0z_0)\frac{\mathrm{j}k_0}{2\pi z_0}\mathrm{e}^{-\mathrm{j}\frac{k_0}{2z_0}(x^2+y^2)}\propto\frac{\mathrm{j}k_0}{2\pi z_0}\mathrm{e}^{-\mathrm{j}\frac{k_0}{2z_0}(x^2+y^2)},\tag{5.1-12}$$

为了简单起见, 这里我们在方程的最后一步忽略了常数相移. 对于参考平面波来说, 且为一般传播方向时, 根据式 (3.3-9), 有

$$\psi(x,y,z;t)=\mathrm{e}^{\mathrm{j}(\omega_0t-\boldsymbol{k}_0\cdot\boldsymbol{R})}=\mathrm{e}^{\mathrm{j}(\omega_0t-k_{0x}x-k_{0y}y-k_{0z}z)}.$$

在图 5.1-7 所示的情况下, $k_{0x}=k_0\hat{\boldsymbol{k}}\cdot\hat{\boldsymbol{x}}=k_0\cos(\theta+90°)=-k_0\sin\theta$, $k_{0y}=0$, 且 $k_{0z}=k_0\hat{\boldsymbol{k}}\cdot\hat{\boldsymbol{z}}=k_0\cos\theta$, 这里, $\boldsymbol{k}_0=k_0\hat{\boldsymbol{k}}$. 在 $z=z_0$ 处记录介质上振幅为 a 的参考平面波的复振幅为

$$\psi_r=a\mathrm{e}^{\mathrm{j}k_0\sin\theta\,x-\mathrm{j}k_0\cos\theta\,z_0}\propto a\mathrm{e}^{\mathrm{j}k_0\sin\theta\,x}.\tag{5.1-13}$$

可以看出, 离轴参考平面波在全息面上沿 x 方向有一个线性相移量. 记录介质上所记录的强度分布可用一个透过率函数表示为

$$t(x,y)=|\psi_r+\psi_0|^2=|\psi_r|^2+|\psi_0|^2+\psi_r\psi_0^*+\psi_r^*\psi_0.\tag{5.1-14}$$

根据式 (5.1-12) 和式 (5.1-13), 前两项为

$$|\psi_r|^2+|\psi_0|^2=a^2+\left(\frac{k_0}{2\pi z_0}\right)^2=A=\text{常数}.\tag{5.1-15a}$$

第三项为

$$\psi_r\psi_0^*=a\mathrm{e}^{\mathrm{j}k_0\sin\theta\,x}\frac{-\mathrm{j}k_0}{2\pi z_0}\mathrm{e}^{\mathrm{j}\frac{k_0}{2z_0}(x^2+y^2)},\tag{5.1-15b}$$

最后一项为

$$\psi_r^*\psi_0=a\mathrm{e}^{-\mathrm{j}k_0\sin\theta\,x}\frac{\mathrm{j}k_0}{2\pi z_0}\mathrm{e}^{-\mathrm{j}\frac{k_0}{2z_0}(x^2+y^2)}.\tag{5.1-15c}$$

最后, 式 (5.1-14) 变为

$$t(x,y)=A+B\sin\left[\frac{k_0}{2z_0}(x^2+y^2)+k_0\sin\theta\,x\right],\tag{5.1-16}$$

θ is called the *recording angle.*

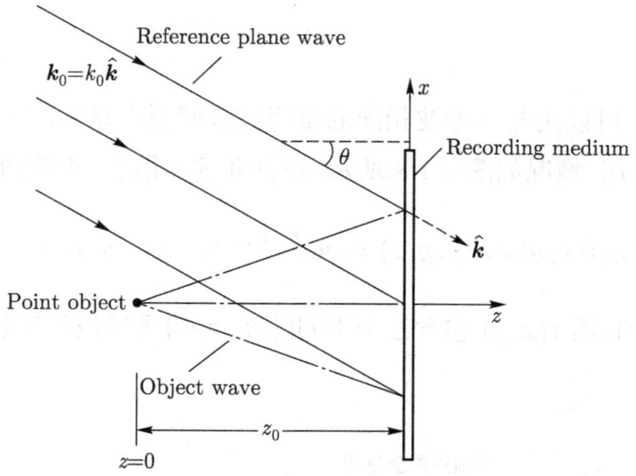

Figure 5.1-7 Off-axis recording

Again, the object wave on the recording medium is the same as in Eq. (5.1-2):

$$\begin{aligned}\psi_0(x,y;z_0) &= \delta(x,y) * h(x,y;z_0)\\ &= \exp(-\mathrm{j}k_0 z_0)\frac{\mathrm{j}k_0}{2\pi z_0}\mathrm{e}^{-\mathrm{j}\frac{k_0}{2z_0}(x^2+y^2)} \propto \frac{\mathrm{j}k_0}{2\pi z_0}\mathrm{e}^{-\mathrm{j}\frac{k_0}{2z_0}(x^2+y^2)},\end{aligned} \tag{5.1-12}$$

where we have neglected the constant phase shift in the last step of the equation for simplicity. As for the reference plane wave, according to Eq. (3.3-9) for a general propagating direction, we have

$$\psi(x,y,z;t) = \mathrm{e}^{\mathrm{j}(\omega_0 t - \boldsymbol{k}_0\cdot\boldsymbol{R})} = \mathrm{e}^{\mathrm{j}(\omega_0 t - k_{0x}x - k_{0y}y - k_{0z}z)}.$$

Under our current situation shown in Figure 5.1-7, $k_{0x} = k_0\hat{\boldsymbol{k}}\cdot\hat{\boldsymbol{x}} = k_0\cos(\theta+90°) = -k_0\sin\theta$, $k_{0y} = 0$, and $k_{0z} = k_0\hat{\boldsymbol{k}}\cdot\hat{\boldsymbol{z}} = k_0\cos\theta$, where $\boldsymbol{k}_0 = k_0\hat{\boldsymbol{k}}$. The complex amplitude of the reference plane wave with amplitude a on the recording medium at $z = z_0$ is

$$\psi_r = a\mathrm{e}^{\mathrm{j}k_0\sin\theta\, x - \mathrm{j}k_0\cos\theta\, z_0} \propto a\mathrm{e}^{\mathrm{j}k_0\sin\theta\, x}. \tag{5.1-13}$$

Note that the off-axis reference plane wave has a linear phase shift variation along x on the hologram plane. The recorded intensity distribution on the recording medium gives a transparency function given by

$$t(x,y) = |\psi_r + \psi_0|^2 = |\psi_r|^2 + |\psi_0|^2 + \psi_r\psi_0^* + \psi_r^*\psi_0. \tag{5.1-14}$$

Upon using Eqs. (5.1-12) and (5.1-13), the first two terms are

$$|\psi_r|^2 + |\psi_0|^2 = a^2 + \left(\frac{k_0}{2\pi z_0}\right)^2 = A = \text{constant}. \tag{5.1-15a}$$

The third term is

$$\psi_r\psi_0^* = a\mathrm{e}^{\mathrm{j}k_0\sin\theta\, x}\frac{-\mathrm{j}k_0}{2\pi z_0}\mathrm{e}^{\mathrm{j}\frac{k_0}{2z_0}(x^2+y^2)}, \tag{5.1-15b}$$

and the last term is

$$\psi_r^*\psi_0 = a\mathrm{e}^{-\mathrm{j}k_0\sin\theta\, x}\frac{\mathrm{j}k_0}{2\pi z_0}\mathrm{e}^{-\mathrm{j}\frac{k_0}{2z_0}(x^2+y^2)}. \tag{5.1-15c}$$

Finally, Eq. (5.1-14) becomes

$$t(x,y) = A + B\sin\left[\frac{k_0}{2z_0}(x^2+y^2) + k_0\sin\theta\, x\right], \tag{5.1-16}$$

其中, $B=\dfrac{ak_0}{\pi z_0}$. 式 (5.1-16) 被称为离轴点源全息图 (*off-axis point-source hologram*).

重建

为了重建全息图, 可以用与参考波相同的重建波来照射全息图, 令重建波 $\psi_{rec}=\psi_r=a\mathrm{e}^{\mathrm{j}k_0\sin\theta x}$ [见式 (5.1-13)]. 情况如图 5.1-8 所示. 根据菲涅耳衍射, 重建图由下式给出

$$\psi_{rec}t(x,y)*h(x,y;z)=a\mathrm{e}^{\mathrm{j}k_0\sin\theta x}t(x,y)*h(x,y;z). \tag{5.1-17}$$

由式 (5.1-14) 可知, 全息图 $t(x,y)$ 包含三个不同的项, 经过重建波的照射, 这三个项会产生如下三个不同的波.

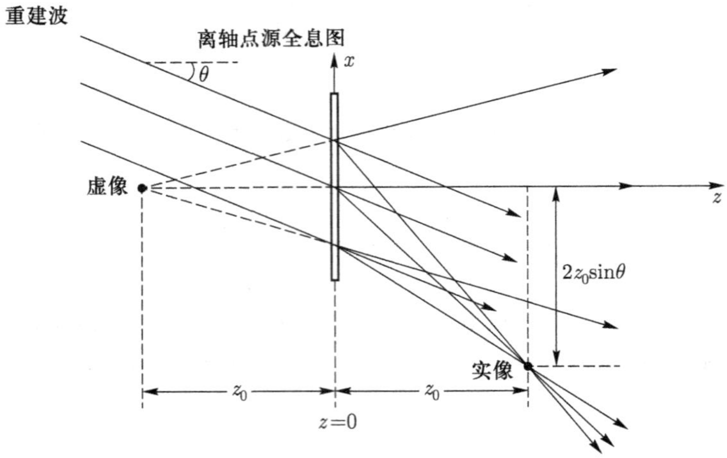

图 5.1-8 离轴点源全息图的全息重建

a) 零级光

全息图的前两项由式 (5.1-15a) 给出, 它是一个常数 A. 该常数 A 表示如图 5.1-4 所示同轴点源全息图中的一个平板玻璃. 因此, 对于入射到它上面的平面波, 我们期望平面波传播时没有任何偏离. 式 (5.1-14) 里全息图的前两项所产生的零级光, 正是与零级光 $\mathrm{e}^{\mathrm{j}k_0\sin\theta x-\mathrm{j}k_0\cos\theta z}$ 成正比的一束平面波.

b) 实像 (或孪生像)

全息图的第三项由式 (5.1-15b) 给出, 即

$$\psi_r\psi_0^*=a\mathrm{e}^{\mathrm{j}k_0\sin\theta x}\frac{-\mathrm{j}k_0}{2\pi z_0}\mathrm{e}^{\mathrm{j}\frac{k_0}{2z_0}(x^2+y^2)}.$$

经 ψ_{rec} 照射后, 全息图后的复振幅为

$$[\psi_{rec}\times\psi_r\psi_0^*]*h(x,y;z)=[\psi_r\times\psi_r\psi_0^*]*h(x,y;z). \tag{5.1-18}$$

在图 5.1-8 所示的情况下, ψ_{rec} 与 $\psi_r=a\mathrm{e}^{\mathrm{j}k_0\sin\theta x}$ 相同. 同样, 不需要再计算式 (5.1-17), 而是认识到

$$\psi_r\times\psi_r\psi_0^*\propto\mathrm{e}^{\mathrm{j}k_0\sin\theta x}\times\mathrm{e}^{\mathrm{j}k_0\sin\theta x}\mathrm{e}^{\mathrm{j}\frac{k_0}{2z_0}(x^2+y^2)}=\mathrm{e}^{2\mathrm{j}k_0\sin\theta x}\mathrm{e}^{\mathrm{j}\frac{k_0}{2z_0}(x^2+y^2)},$$

where $B = \dfrac{ak_0}{\pi z_0} \cdot t(x, y)$ given by the above equation is called an off-axis point-source hologram.

Reconstruction

To reconstruct the hologram, we can illuminate the hologram with a reconstruction wave identical to the reference wave. We then have the reconstruction wave $\psi_{rec} = \psi_r = ae^{jk_0 \sin\theta\, x}$ [see Eq. (5.1-13)]. The situation is shown in Figure 5.1-8. The reconstruction, according to Fresnel diffraction, is given by

$$\psi_{rec}t(x, y) * h(x, y; z) = ae^{jk_0 \sin\theta\, x}t(x, y) * h(x, y; z). \tag{5.1-17}$$

From Eq. (5.1-14), the hologram $t(x, y)$ contains three distinct terms and upon illumination by the reconstruction wave, these three terms create three different waves as follows.

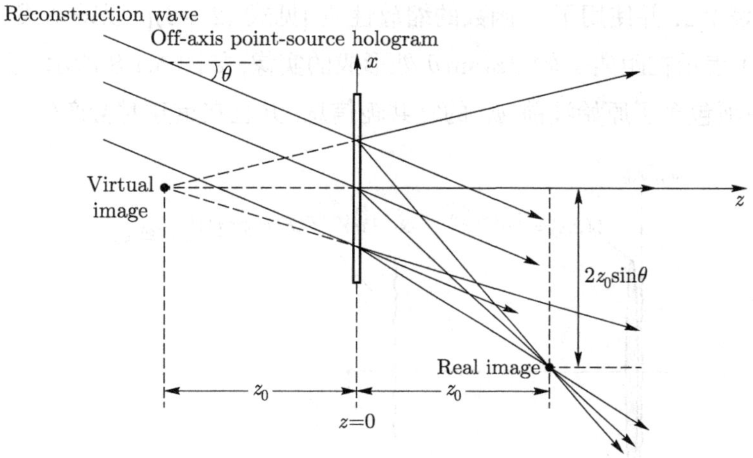

Figure 5.1-8 Holographic reconstruction of off-axis point-source hologram

a) Zeroth-Order Beam

The first two terms of the hologram are given by Eq. (5.1-15a), which is a constant A. The constant A represents a flat glass as illustrated in Figure 5.1-4 for the on-axis point-source hologram. Therefore, for a plane wave incident on it, we expect the plane wave propagates without any deviation. The zeroth-order beam due to the first two terms of the hologram in Eq. (5.1-14) is a plane wave proportional to $e^{jk_0 \sin\theta\, x - jk_0 \cos\theta\, z}$, and this is the zeroth-order beam.

b) Real Image (or the Twin Image):

The third term of the hologram is given by Eq. (5.1-15b), which is

$$\psi_r\psi_0^* = ae^{jk_0 \sin\theta\, x}\frac{-jk_0}{2\pi z_0}e^{j\frac{k_0}{2z_0}(x^2+y^2)}.$$

Upon illumination by ψ_{rec}, the complex field after the hologram is

$$[\psi_{rec} \times \psi_r\psi_0^*] * h(x, y; z) = [\psi_r \times \psi_r\psi_0^*] * h(x, y; z) \tag{5.1-18}$$

as ψ_{rec} is identical to $\psi_r = ae^{jk_0 \sin\theta\, x}$ in our situation shown in Figure 5.1-8. Again, instead of explicitly calculating Eq. (5.1-17), we recognize that

$$\psi_r \times \psi_r\psi_0^* \propto e^{jk_0 \sin\theta\, x} \times e^{jk_0 \sin\theta\, x}e^{j\frac{k_0}{2z_0}(x^2+y^2)} = e^{2jk_0 \sin\theta\, x}e^{j\frac{k_0}{2z_0}(x^2+y^2)},$$

它对应于一个焦距为 z_0 的理想透镜的透过率函数，被 $\mathrm{e}^{2\mathrm{j}k_0\sin\theta\, x}$ 的复光波照射.

这里回顾一下图 3.4-5, 并在图 5.1-9 中总结它所示的含义, 其中 $\psi_{fl}(x,y)$ 表示紧贴透镜前方的光场分布. 利用该结果去处理 $\psi_r\times\psi_r\psi_0^*$, $\psi_{fl}=\mathrm{e}^{2\mathrm{j}k_0\sin\theta\, x}$, 因为 $\mathrm{e}^{\mathrm{j}\frac{k_0}{2z_0}(x^2+y^2)}$ 起透镜中透过率函数的作用. 在距离全息图 z_0 处, 有

$$
\begin{aligned}
\psi_p(x,y;z_0) &\propto \mathrm{e}^{-\mathrm{j}\frac{k_0}{2z_0}(x^2+y^2)}\mathcal{F}\left\{\psi_{fl}(x,y)\right\}\Big|_{k_x=\frac{k_0x}{z_0},\,k_y=\frac{k_0y}{z_0}}\\
&= \mathrm{e}^{-\mathrm{j}\frac{k_0}{2z_0}(x^2+y^2)}\mathcal{F}\left\{\mathrm{e}^{2\mathrm{j}k_0\sin\theta\, x}\right\}\Big|_{k_x=\frac{k_0x}{z_0},\,k_y=\frac{k_0y}{z_0}}\\
&= \mathrm{e}^{-\mathrm{j}\frac{k_0}{2z_0}(x^2+y^2)}4\pi^2\delta(k_x+2k_0\sin\theta,k_y)\Big|_{k_x=\frac{k_0x}{z_0},\,k_y=\frac{k_0y}{z_0}}\\
&\propto \mathrm{e}^{-\mathrm{j}\frac{k_0}{2f}(x^2+y^2)}\delta\left(\frac{k_0x}{z_0}+2k_0\sin\theta,\frac{k_0y}{z_0}\right)\propto\delta(x+2z_0\sin\theta,y),
\end{aligned}
\tag{5.1-19}
$$

这里, 我们用了表 2.2, 并使用了 δ 函数的缩放性质 [见式 (2.1-6)], 以得到式 (5.1-19) 的最终结果. 式 (5.1-19) 表示在距离 z 轴 $2z_0\sin\theta$ 处形成的实像, 如图 5.1-8 所示. 总之, 在全息图公式里, $\psi_r\psi_0^*$ 这一项包含了原始波前 ψ_0 的复共轭信息, 并且在重建时会产生一个实像.

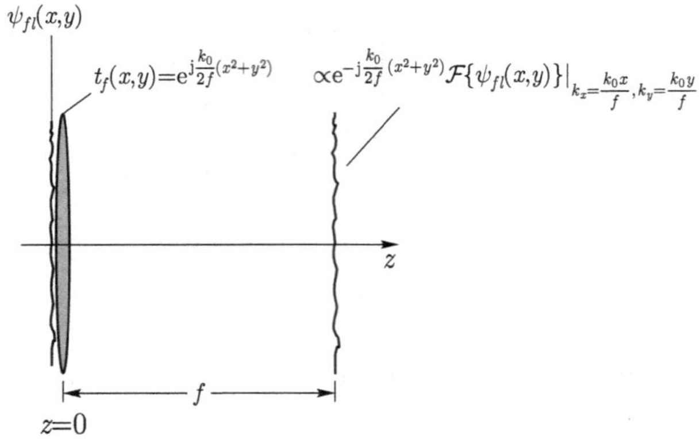

图 5.1-9　紧贴理想透镜前复振幅的傅里叶变换

c) 虚像

由 (5.1-15c) 给出的全息图的第四项为

$$\psi_r^*\psi_0=a\mathrm{e}^{-\mathrm{j}k_0\sin\theta\, x}\frac{\mathrm{j}k_0}{2\pi z_0}\mathrm{e}^{-\mathrm{j}\frac{k_0}{2z_0}(x^2+y^2)}.$$

经 ψ_{rec} 照明后, 全息图后的复振幅为

$$\psi_{rec}\times\psi_r^*\psi_0 * h(x,y;z)=\psi_r\times\psi_r^*\psi_0 * h(x,y;z). \tag{5.1-20}$$

同样, ψ_{rec} 等同于 $\psi_r=a\mathrm{e}^{\mathrm{j}k_0\sin\theta\, x}$, 和前面一样, 不需要再次计算上式, 而是认识到

$$\psi_r\times\psi_r^*\psi_0\propto\mathrm{e}^{\mathrm{j}k_0\sin\theta\, x}\times\mathrm{e}^{-\mathrm{j}k_0\sin\theta\, x}\mathrm{e}^{-\mathrm{j}\frac{k_0}{2z_0}(x^2+y^2)}=\mathrm{e}^{-\mathrm{j}\frac{k_0}{2z_0}(x^2+y^2)},$$

这对应于一个焦距为 $-z_0$ 的理想透镜的透过率函数. 因此, 在全息图中, 这一项代表一个负透镜, 它在 z 轴 (初始点源位置处) 上会成一个虚像, 如图 5.1-8 所示. 总而言之, 在全息图公式中, $\psi_r^*\psi_0$ 这一项包含原始波前 ψ_0, 且它在重建时会产生一个虚像.

which corresponds to a transparency function of an idea lens of focal length z_0, illuminated by complex wave $e^{2jk_0 \sin\theta\, x}$.

Let us recall Figure 3.4-5 and summarize its result in Figure 5.1-9, where $\psi_{fl}(x, y)$ represents the field distribution just in front of the lens. Using this result for $\psi_r \times \psi_r\psi_0^*$, $\psi_{fl} = e^{2jk_0 \sin\theta\, x}$ because $e^{j\frac{k_0}{2z_0}(x^2+y^2)}$ serves as a transparency function of a lens. At a distance z_0 away from the hologram, we have

$$\begin{aligned}\psi_p(x, y; z_0) &\propto e^{-j\frac{k_0}{2z_0}(x^2+y^2)}\mathcal{F}\{\psi_{fl}(x, y)\}\Big|_{k_x=\frac{k_0x}{z_0},\, k_y=\frac{k_0y}{z_0}} \\ &= e^{-j\frac{k_0}{2z_0}(x^2+y^2)}\mathcal{F}\left\{e^{2jk_0 \sin\theta\, x}\right\}\Big|_{k_x=\frac{k_0x}{z_0},\, k_y=\frac{k_0y}{z_0}} \\ &= e^{-j\frac{k_0}{2z_0}(x^2+y^2)}4\pi^2\delta(k_x + 2k_0 \sin\theta, k_y)\Big|_{k_x=\frac{k_0x}{z_0},\, k_y=\frac{k_0y}{z_0}} \\ &\propto e^{-j\frac{k_0}{2f}(x^2+y^2)}\delta\left(\frac{k_0x}{z_0} + 2k_0 \sin\theta, \frac{k_0y}{z_0}\right) \propto \delta(x + 2z_0 \sin\theta, y),\end{aligned} \tag{5.1-19}$$

where we have used Table 2.2 and employed the scaling property of the delta function [see Eq. (2.1-6)] to arrive the final result in Eq. (5.1-19). Equation (5.1-19) means that we have a real image formed $2z_0 \sin\theta$ away from the z-axis (see Figure 5.1-8). In summary, the term $\psi_r\psi_0^*$ in the hologram equation contains the complex conjugate of the original wave front ψ_0, and it gives rise to a real image upon reconstruction.

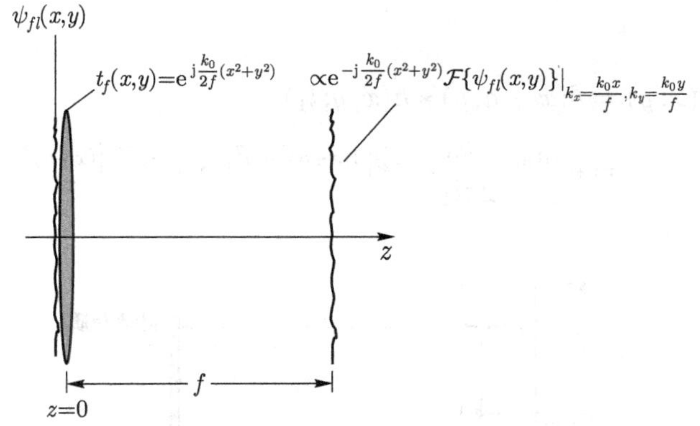

Figure 5.1-9 Fourier transformation of the complex field just in front of an ideal lens

c) Virtual Image

The fourth term of the hologram given by Eq. (5.1-15c) is

$$\psi_r^*\psi_0 = ae^{-jk_0 \sin\theta\, x}\frac{jk_0}{2\pi z_0}e^{-j\frac{k_0}{2z_0}(x^2+y^2)}.$$

Upon illumination by ψ_{rec}, the complex field after the hologram is

$$\psi_{rec} \times \psi_r^*\psi_0 * h(x, y; z) = \psi_r \times \psi_r^*\psi_0 * h(x, y; z). \tag{5.1-20}$$

Again, ψ_{rec} is identical to $\psi_r = ae^{jk_0 \sin\theta\, x}$. As before, instead of explicitly evaluating the above equation, we recognize that

$$\psi_r \times \psi_r^*\psi_0 \propto e^{jk_0 \sin\theta\, x} \times e^{-jk_0 \sin\theta\, x}e^{-j\frac{k_0}{2z_0}(x^2+y^2)} = e^{-j\frac{k_0}{2z_0}(x^2+y^2)},$$

which corresponds to a transparency function of an idea lens of focal length $-z_0$. So this term in the hologram represents a negative lens, which gives a virtual image on the z-axis (the location of the original point source. See Figure 5.1-8). In summary, the term $\psi_r^*\psi_0$ in the hologram equation contains the original wave front ψ_0, and it gives rise to a virtual image upon reconstruction.

5.2 三维全息成像

本节通过两个点源的记录和重建来研究全息成像中的横向和纵向放大率. 这里利用球面波来进行全息记录和重建.

5.2.1 记录与重建

记录

如图 5.2-1 所示为记录的几何示意图. 标为 1 和 2 的两个点物, 及标为 R 的参考点源在记录介质的平面上产生球面波, 它们所形成的复振幅分别为 ψ_{p1}, ψ_{p2} 和 $\psi_{p\mathrm{R}}$, 分别表示为

$$\begin{aligned}\psi_{p1}(x,y)&=\delta\left(x-\frac{h}{2},y\right)*h(x,y;R)\\&=\mathrm{e}^{-\mathrm{j}k_0R}\frac{\mathrm{j}k_0}{2\pi R}\mathrm{e}^{-\mathrm{j}\frac{k_0}{2R}\left[(x-h/2)^2+y^2\right]}\propto\mathrm{e}^{-\mathrm{j}\frac{k_0}{2R}\left[(x-h/2)^2+y^2\right]},\end{aligned}\tag{5.2-1}$$

$$\begin{aligned}\psi_{p2}(x,y)&=\delta\left(x+\frac{h}{2},y\right)*h(x,y;R+d)\\&=\mathrm{e}^{-\mathrm{j}k_0(R+d)}\frac{\mathrm{j}k_0}{2\pi(R+d)}\mathrm{e}^{-\mathrm{j}\frac{k_0}{2(R+d)}\left[(x+h/2)^2+y^2\right]}\propto\mathrm{e}^{-\mathrm{j}\frac{k_0}{2(R+d)}\left[(x+h/2)^2+y^2\right]},\end{aligned}\tag{5.2-2}$$

和

$$\begin{aligned}\psi_{p\mathrm{R}}(x,y)&=\delta(x+a,y)*h(x,y;l_1)\\&=\mathrm{e}^{-\mathrm{j}k_0l_1}\frac{\mathrm{j}k_0}{2\pi l_1}\mathrm{e}^{-\mathrm{j}\frac{k_0}{2l_1}\left[(x+a)^2+y^2\right]}\propto\mathrm{e}^{-\mathrm{j}\frac{k_0}{2l_1}\left[(x+a)^2+y^2\right]}.\end{aligned}\tag{5.2-3}$$

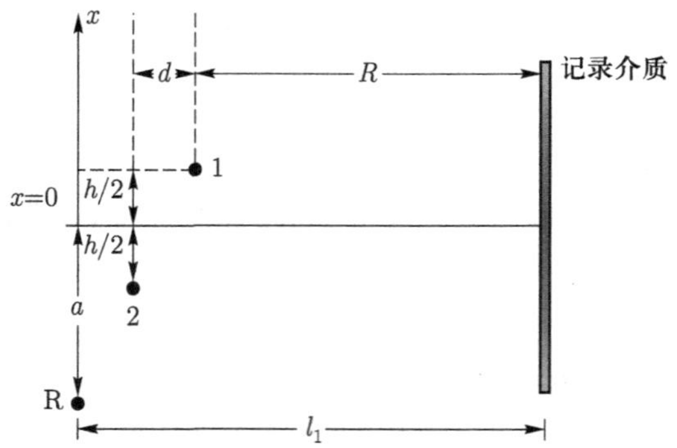

图 5.2-1 记录的几何示意图: 两个点物被标为 1 和 2, 参考点源被标为 R

这些球面波在记录介质上干涉而产生全息图

$$\begin{aligned}t(x,y)&=|\psi_{p1}(x,y)+\psi_{p2}(x,y)+\psi_{p\mathrm{R}}(x,y)|^2\\&=\left[\psi_{p1}(x,y)+\psi_{p2}(x,y)+\psi_{p\mathrm{R}}(x,y)\right]\left[\psi_{p1}^*(x,y)+\psi_{p2}^*(x,y)+\psi_{p\mathrm{R}}^*(x,y)\right].\end{aligned}\tag{5.2-4}$$

重建

全息图中共有九项, 这里要在前期关于点源物体记录和重建知识的基础上, 找出一些

5.2 Three-Dimensional Holographic Imaging

In this section, we study the lateral and longitudinal magnifications in holographic imaging through the recording and reconstruction of two point sources. We consider holographic recording and reconstruction using spherical waves.

5.2.1 Recording and Reconstruction

Recording

The geometry for recording is shown in Figure 5.2-1. The two point objects, labeled 1 and 2, and the reference wave, labeled R, generate spherical waves that, on the plane of the recording medium, contribute to complex fields, ψ_{p1}, ψ_{p2}, and $\psi_{p\mathrm{R}}$, respectively, given by

$$\begin{aligned}\psi_{p1}(x,y) &= \delta\left(x-\frac{h}{2},y\right)*h(x,y;R)\\ &= \mathrm{e}^{-\mathrm{j}k_0R}\frac{\mathrm{j}k_0}{2\pi R}\mathrm{e}^{-\mathrm{j}\frac{k_0}{2R}\left[(x-h/2)^2+y^2\right]}\propto \mathrm{e}^{-\mathrm{j}\frac{k_0}{2R}\left[(x-h/2)^2+y^2\right]},\end{aligned} \tag{5.2-1}$$

$$\begin{aligned}\psi_{p2}(x,y) &= \delta\left(x+\frac{h}{2},y\right)*h(x,y;R+d)\\ &= \mathrm{e}^{-\mathrm{j}k_0(R+d)}\frac{\mathrm{j}k_0}{2\pi(R+d)}\mathrm{e}^{-\mathrm{j}\frac{k_0}{2(R+d)}\left[(x+h/2)^2+y^2\right]}\propto \mathrm{e}^{-\mathrm{j}\frac{k_0}{2(R+d)}\left[(x+h/2)^2+y^2\right]},\end{aligned} \tag{5.2-2}$$

and

$$\begin{aligned}\psi_{p\mathrm{R}}(x,y) &= \delta(x+a,y)*h(x,y;l_1)\\ &= \mathrm{e}^{-\mathrm{j}k_0l_1}\frac{\mathrm{j}k_0}{2\pi l_1}\mathrm{e}^{-\mathrm{j}\frac{k_0}{2l_1}\left[(x+a)^2+y^2\right]}\propto \mathrm{e}^{-\mathrm{j}\frac{k_0}{2l_1}\left[(x+a)^2+y^2\right]}.\end{aligned} \tag{5.2-3}$$

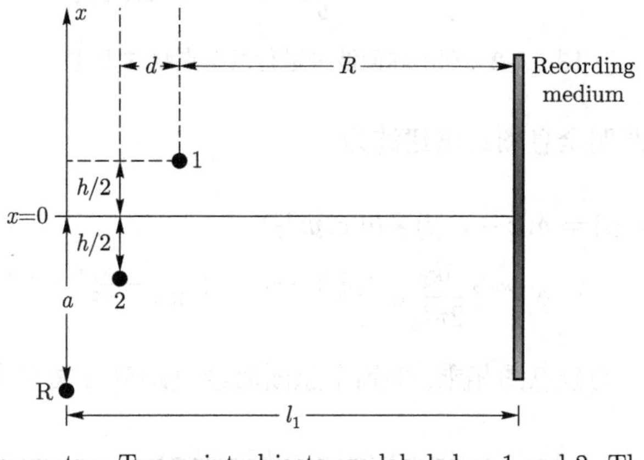

Figure 5.2-1 Recording geometry: Two point objects are labeled as 1 and 2. The reference point source is labeled R

These spherical waves interfere on the recording medium to yield a hologram given by

$$\begin{aligned}t(x,y) &= |\psi_{p1}(x,y)+\psi_{p2}(x,y)+\psi_{p\mathrm{R}}(x,y)|^2\\ &= \left[\psi_{p1}(x,y)+\psi_{p2}(x,y)+\psi_{p\mathrm{R}}(x,y)\right]\left[\psi_{p1}^*(x,y)+\psi_{p2}^*(x,y)+\psi_{p\mathrm{R}}^*(x,y)\right].\end{aligned} \tag{5.2-4}$$

Reconstruction

There are nine terms in the holograms and we shall, on the basis of our previous knowledge regarding point source object recording and reconstruction, pick out some relevant terms

与图像重建相关的项. 同样, 这些项包含全息图中物波的复共轭, 即, $\psi_{p\mathrm{R}}(x,y)\psi^*_{p1}(x,y)$ 和 $\psi_{p\mathrm{R}}(x,y)\psi^*_{p2}(x,y)$ 会产生实像的重建, 它们是

$$\begin{aligned}t_{rel_1}(x,y) &= \psi_{p\mathrm{R}}(x,y)\psi^*_{p1}(x,y)\\ &= \mathrm{e}^{-\mathrm{j}\frac{k_0}{2l_1}\left[(x+a)^2+y^2\right]}\mathrm{e}^{\mathrm{j}\frac{k_0}{2R}\left[(x-h/2)^2+y^2\right]},\end{aligned} \tag{5.2-5}$$

和

$$\begin{aligned}t_{rel_2}(x,y) &= \psi_{p\mathrm{R}}(x,y)\psi^*_{p2}(x,y)\\ &= \mathrm{e}^{-\mathrm{j}\frac{k_0}{2l_1}\left[(x+a)^2+y^2\right]}\mathrm{e}^{\mathrm{j}\frac{k_0}{2(R+d)}\left[(x+h/2)^2+y^2\right]}.\end{aligned} \tag{5.2-6}$$

包含原始物波的项, 即 $\psi^*_{p\mathrm{R}}(x,y)\psi_{p1}(x,y)$ 和 $\psi^*_{p\mathrm{R}}(x,y)\psi_{p2}(x,y)$ 会产生虚像, 它们是

$$t_{rel_3}(x,y) = \psi^*_{p\mathrm{R}}(x,y)\psi_{p1}(x,y) = \left[t_{rel_1}(x,y)\right]^*, \tag{5.2-7}$$

和

$$t_{rel_4}(x,y) = \psi^*_{p\mathrm{R}}(x,y)\psi_{p2}(x,y) = \left[t_{rel_2}(x,y)\right]^*. \tag{5.2-8}$$

在重建时, 我们使用来自点源 r 的球面波作为重建波来照射全息图, 如图 5.2-2 所示. 接下来举例说明如何找到实像位置.

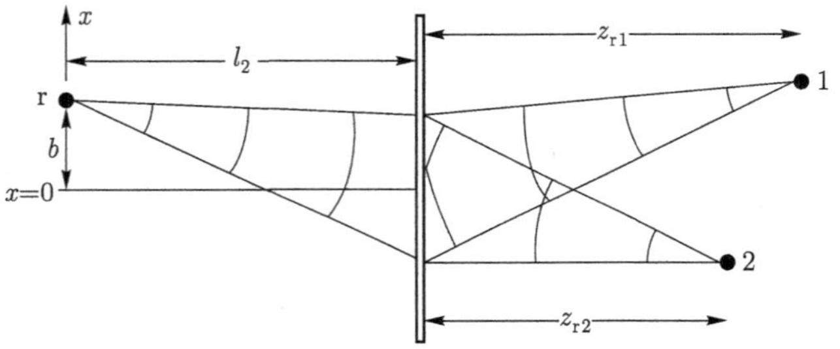

图 5.2-2　利用球面波来进行两个点物的重建

根据非涅耳衍射, 照明全息图的重建波为

$$\begin{aligned}\psi_{p\mathrm{r}}(x,y) &= \delta(x-b,y) * h(x,y;l_2)\\ &= \mathrm{e}^{-\mathrm{j}k_0l_2}\frac{\mathrm{j}k_0}{2\pi l_2}\mathrm{e}^{-\mathrm{j}\frac{k_0}{2l_2}\left[(x-b)^2+y^2\right]} \propto \mathrm{e}^{-\mathrm{j}\frac{k_0}{2l_2}\left[(x-b)^2+y^2\right]}.\end{aligned} \tag{5.2-9}$$

将该场与式 (5.2-4) 给出的全息图相乘, 得到全息图后经过距离 z 处的复振幅

$$\psi_{p\mathrm{r}}(x,y)t(x,y) * h(x,y;z),$$

但是在全息图中负责点物 1 的实像重建的项是

$$\begin{aligned}\psi_{p\mathrm{r}}(x,y)t_{rel_1}(x,y) &= \psi_{p\mathrm{r}}(x,y)\psi_{p\mathrm{R}}(x,y)\psi^*_{p1}(x,y)\\ &= \mathrm{e}^{-\mathrm{j}\frac{k_0}{2l_2}\left[(x-b)^2+y^2\right]}\mathrm{e}^{-\mathrm{j}\frac{k_0}{2l_1}\left[(x+a)^2+y^2\right]}\mathrm{e}^{\mathrm{j}\frac{k_0}{2R}\left[\left(x-\frac{h}{2}\right)^2+y^2\right]}.\end{aligned}$$

因此需要找到

$$\psi_{p,rel_1}(x,y;z) = \psi_{p\mathrm{r}}(x,y)t_{rel_1}(x,y) * h(x,y;z). \tag{5.2-10}$$

responsible for image reconstruction. Again, the terms containing the complex conjugate of the object waves in the hologram, that is, $\psi_{pR}(x,y)\psi^*_{p1}(x,y)$ and $\psi_{pR}(x,y)\psi^*_{p2}(x,y)$ will give rise to real image reconstructions, and they are

$$t_{rel_1}(x,y) = \psi_{pR}(x,y)\psi^*_{p1}(x,y) = e^{-j\frac{k_0}{2l_1}\left[(x+a)^2+y^2\right]}e^{j\frac{k_0}{2R}\left[(x-h/2)^2+y^2\right]}, \tag{5.2-5}$$

and

$$t_{rel_2}(x,y) = \psi_{pR}(x,y)\psi^*_{p2}(x,y) = e^{-j\frac{k_0}{2l_1}\left[(x+a)^2+y^2\right]}e^{j\frac{k_0}{2(R+d)}\left[(x+h/2)^2+y^2\right]}. \tag{5.2-6}$$

The term containing the original object waves, that is, $\psi^*_{pR}(x,y)\psi_{p1}(x,y)$ and $\psi^*_{pR}(x,y)\psi_{p2}(x,y)$ will give rise to virtual images, and they are

$$t_{rel_3}(x,y) = \psi^*_{pR}(x,y)\psi_{p1}(x,y) = \left[t_{rel_1}(x,y)\right]^*, \tag{5.2-7}$$

and

$$t_{rel_4}(x,y) = \psi^*_{pR}(x,y)\psi_{p2}(x,y) = \left[t_{rel_2}(x,y)\right]^*. \tag{5.2-8}$$

On reconstruction, we use a spherical wave from a point source labeled r as our reconstruction wave to illuminate the hologram, shown in Figure 5.2-2. We shall demonstrate how to find the real image locations as an example.

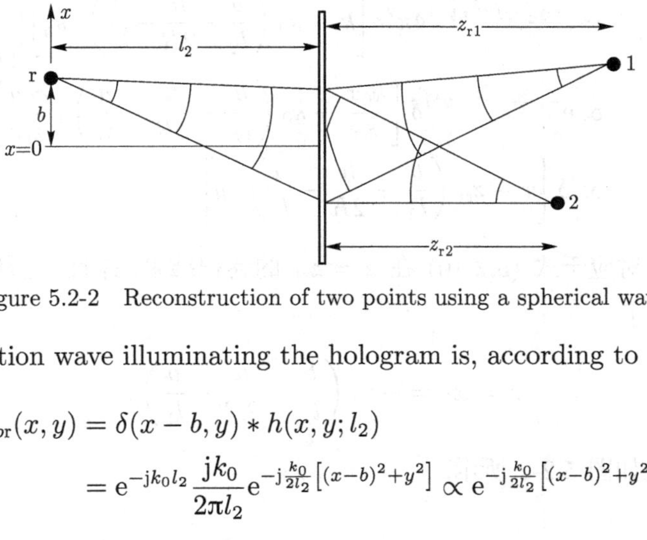

Figure 5.2-2 Reconstruction of two points using a spherical wave

The reconstruction wave illuminating the hologram is, according to Fresnel diffraction,

$$\psi_{pr}(x,y) = \delta(x-b,y) * h(x,y;l_2) = e^{-jk_0l_2}\frac{jk_0}{2\pi l_2}e^{-j\frac{k_0}{2l_2}\left[(x-b)^2+y^2\right]} \propto e^{-j\frac{k_0}{2l_2}\left[(x-b)^2+y^2\right]}. \tag{5.2-9}$$

The complex field after propagation through a distance z behind the hologram is given by multiplying this field with the hologram given by Eq. (5.2-4):

$$\psi_{pr}(x,y)t(x,y) * h(x,y;z),$$

but the term in the hologram responsible for real image reconstruction of point 1 is

$$\psi_{pr}(x,y)t_{rel_1}(x,y) = \psi_{pr}(x,y)\psi_{pR}(x,y)\psi^*_{p1}(x,y) = e^{-j\frac{k_0}{2l_2}\left[(x-b)^2+y^2\right]}e^{-j\frac{k_0}{2l_1}\left[(x+a)^2+y^2\right]}e^{j\frac{k_0}{2R}\left[\left(x-\frac{h}{2}\right)^2+y^2\right]}.$$

So we need to find

$$\psi_{p,rel_1}(x,y;z) = \psi_{pr}(x,y)t_{rel_1}(x,y) * h(x,y;z). \tag{5.2-10}$$

同样, 不再直接计算上式, 而是研究 $\psi_{\mathrm{pr}}(x,y)t_{rel_1}(x,y)$ 项, 并试着根据一个理想透镜来求其透过率函数.

通过展开二次项并把系数 x^2+y^2 和 x 分组, $\psi_{\mathrm{pr}}(x,y)t_{rel_1}(x,y)$ 变成

$$\begin{aligned}\psi_{\mathrm{pr}}(x,y)t_{rel_1}(x,y) &\propto \mathrm{e}^{\mathrm{j}k_0\left(-\frac{1}{2l_2}+\frac{1}{2R}-\frac{1}{2l_1}\right)(x^2+y^2)} \times \mathrm{e}^{\mathrm{j}k_0\left(\frac{b}{l_2}-\frac{h}{2R}-\frac{a}{l_1}\right)x}\\ &= \mathrm{e}^{\mathrm{j}\frac{k_0}{2z_{\mathrm{r1}}}(x^2+y^2)}\mathrm{e}^{\mathrm{j}k_0\left(\frac{b}{l_2}-\frac{h}{2R}-\frac{a}{l_1}\right)x},\end{aligned}$$

其中

$$z_{\mathrm{r1}} = \left(\frac{1}{R}-\frac{1}{l_1}-\frac{1}{l_2}\right)^{-1} = \frac{Rl_1l_2}{l_1l_2-(l_1+l_2)R}, \tag{5.2-11}$$

上式可以被认为是一个理想透镜的焦距. 因此, 利用图 5.1-9 所示的结果, 对于 $\psi_{\mathrm{pr}}(x,y)t_{rel_1}(x,y)$, $\psi_{fl}(x,y) = \mathrm{e}^{\mathrm{j}k_0\left(\frac{b}{l_2}-\frac{h}{2R}-\frac{a}{l_1}\right)x}$, 因为 $\mathrm{e}^{\mathrm{j}\frac{k_0}{2z_{\mathrm{r1}}}(x^2+y^2)}$ 等同于一个透镜的透过率函数. 在距离全息图 z_{r1} 处, 与式 (5.2-10) 类似, 有

$$\begin{aligned}\psi_{p,rel_1}(x,y;z_{\mathrm{r1}}) &= \psi_{\mathrm{pr}}(x,y)t_{rel_1}(x,y) * h(x,y;z_{\mathrm{r1}})\\ &\propto \mathrm{e}^{-\mathrm{j}\frac{k_0}{2z_{\mathrm{r1}}}(x^2+y^2)}\mathcal{F}\left\{\psi_{fl}(x,y)\right\}\Big|_{k_x=\frac{k_0x}{z_{\mathrm{r1}}},\,k_y=\frac{k_0y}{z_{\mathrm{r1}}}}\\ &= \mathrm{e}^{-\mathrm{j}\frac{k_0}{2z_{\mathrm{r1}}}(x^2+y^2)}\mathcal{F}\left\{\mathrm{e}^{\mathrm{j}k_0\left(\frac{b}{l_2}-\frac{h}{2R}-\frac{a}{l_1}\right)x}\right\}\Bigg|_{k_x=\frac{k_0x}{z_{\mathrm{r1}}},\,k_y=\frac{k_0y}{z_{\mathrm{r1}}}}\\ &= \mathrm{e}^{-\mathrm{j}\frac{k_0}{2z_{\mathrm{r1}}}(x^2+y^2)}4\pi^2\delta\left[k_x+k_0\left(\frac{b}{l_2}-\frac{h}{2R}-\frac{a}{l_1}\right),k_y\right]\Bigg|_{k_x=\frac{k_0x}{z_{\mathrm{r1}}},\,k_y=\frac{k_0y}{z_{\mathrm{r1}}}}\\ &\propto \mathrm{e}^{-\mathrm{j}\frac{k_0}{2z_{\mathrm{r1}}}(x^2+y^2)}\delta\left[\frac{k_0x}{z_{\mathrm{r1}}}+k_0\left(\frac{b}{l_2}-\frac{h}{2R}-\frac{a}{l_1}\right),\frac{k_0y}{z_{\mathrm{r1}}}\right]\\ &\propto \delta\left[x+z_{\mathrm{r1}}\left(\frac{b}{l_2}-\frac{h}{2R}-\frac{a}{l_1}\right),y\right].\end{aligned}$$

可以看出, 这个结果对应于式 (5.2-10) 在 $z=z_{\mathrm{r1}}$ 的卷积结果, 并且, 这是重建的点物 1 的实像, 沿横向位置移动了

$$x = x_1 = -z_{\mathrm{r1}}\left(\frac{b}{l_2}-\frac{h}{2R}-\frac{a}{l_1}\right), \tag{5.2-12}$$

在距离全息图 z_{r1} 处如图 5.2-2 所示.

刚刚已经求得

$$\psi_{\mathrm{pr}}(x,y)t_{rel_1}(x,y) = \psi_{\mathrm{pr}}(x,y)\psi_{\mathrm{pR}}(x,y)\psi_{p1}^*(x,y)$$

会产生点物 1 的实像重建. 同样, 这一项

$$\psi_{\mathrm{pr}}(x,y)t_{rel_2}(x,y) = \psi_{\mathrm{pr}}(x,y)\psi_{\mathrm{pR}}(x,y)\psi_{p2}^*(x,y)$$

在全息图中会产生点物 2 的实像重建. 对点物 2 重复相同的过程, 得出如下结果

$$\begin{aligned}\psi_{p,rel_2}(x,y;z_{\mathrm{r2}}) &= \psi_{\mathrm{pr}}(x,y)t_{rel_2}(x,y) * h(x,y;z_{\mathrm{r2}})\\ &\propto \delta\left[x+z_{\mathrm{r2}}\left(\frac{b}{l_2}+\frac{h}{2(R+d)}-\frac{a}{l_1}\right),y\right],\end{aligned} \tag{5.2-13}$$

Again, instead of straightforward calculation of the above, we investigate the term $\psi_{\text{pr}}(x,y)t_{rel_1}(x,y)$ and try to find its transparency function in terms of an ideal lens.

By expanding the quadratic terms and grouping the coefficients of x^2+y^2 and x, $\psi_{\text{pr}}(x,y)t_{rel_1}(x,y)$ becomes

$$\begin{aligned}\psi_{\text{pr}}(x,y)t_{rel_1}(x,y) &\propto \mathrm{e}^{\mathrm{j}k_0\left(-\frac{1}{2l_2}+\frac{1}{2R}-\frac{1}{2l_1}\right)(x^2+y^2)}\times \mathrm{e}^{\mathrm{j}k_0\left(\frac{b}{l_2}-\frac{h}{2R}-\frac{a}{l_1}\right)x}\\ &=\mathrm{e}^{\mathrm{j}\frac{k_0}{2z_{\text{r}1}}(x^2+y^2)}\mathrm{e}^{\mathrm{j}k_0\left(\frac{b}{l_2}-\frac{h}{2R}-\frac{a}{l_1}\right)x},\end{aligned}$$

where

$$z_{\text{r}1}=\left(\frac{1}{R}-\frac{1}{l_1}-\frac{1}{l_2}\right)^{-1}=\frac{Rl_1l_2}{l_1l_2-(l_1+l_2)R}, \tag{5.2-11}$$

which can be identified as a focal length of an ideal lens. So employing the result shown in Figure 5.1-9, for $\psi_{\text{pr}}(x,y)t_{rel_1}(x,y)$, $\psi_{fl}(x,y)=\mathrm{e}^{\mathrm{j}k_0\left(\frac{b}{l_2}-\frac{h}{2R}-\frac{a}{l_1}\right)x}$ because $\mathrm{e}^{\mathrm{j}\frac{k_0}{2z_{\text{r}1}}(x^2+y^2)}$ serves as a transparency function of a lens. At a distance $z_{\text{r}1}$ away from the hologram, that is, equivalently from Eq. (5.2-10), we have

$$\begin{aligned}\psi_{p,rel_1}(x,y;z_{\text{r}1}) &= \psi_{\text{pr}}(x,y)t_{rel_1}(x,y)*h(x,y;z_{\text{r}1})\\ &\propto \mathrm{e}^{-\mathrm{j}\frac{k_0}{2z_{\text{r}1}}(x^2+y^2)}\mathcal{F}\{\psi_{fl}(x,y)\}\Big|_{k_x=\frac{k_0x}{z_{\text{r}1}},\,k_y=\frac{k_0y}{z_{\text{r}1}}}\\ &= \mathrm{e}^{-\mathrm{j}\frac{k_0}{2z_{\text{r}1}}(x^2+y^2)}\mathcal{F}\left\{\mathrm{e}^{\mathrm{j}k_0\left(\frac{b}{l_2}-\frac{h}{2R}-\frac{a}{l_1}\right)x}\right\}\Bigg|_{k_x=\frac{k_0x}{z_{\text{r}1}},\,k_y=\frac{k_0y}{z_{\text{r}1}}}\\ &= \mathrm{e}^{-\mathrm{j}\frac{k_0}{2z_{\text{r}1}}(x^2+y^2)}4\pi^2\delta\left[k_x+k_0\left(\frac{b}{l_2}-\frac{h}{2R}-\frac{a}{l_1}\right),k_y\right]\Bigg|_{k_x=\frac{k_0x}{z_{\text{r}1}},\,k_y=\frac{k_0y}{z_{\text{r}1}}}\\ &\propto \mathrm{e}^{-\mathrm{j}\frac{k_0}{2z_{\text{r}1}}(x^2+y^2)}\delta\left[\frac{k_0x}{z_{\text{r}1}}+k_0\left(\frac{b}{l_2}-\frac{h}{2R}-\frac{a}{l_1}\right),\frac{k_0y}{z_{\text{r}1}}\right]\\ &\propto \delta\left[x+z_{\text{r}1}\left(\frac{b}{l_2}-\frac{h}{2R}-\frac{a}{l_1}\right),y\right].\end{aligned}$$

Note that this result corresponds to the convolution result of Eq. (5.2-10) for $z=z_{\text{r}1}$, and this is the real image of the reconstructed point object 1, shifted along the lateral location at

$$x=x_1=-z_{\text{r}1}\left(\frac{b}{l_2}-\frac{h}{2R}-\frac{a}{l_1}\right), \tag{5.2-12}$$

and located $z_{\text{r}1}$ away from the hologram (see Figure 5.2-2).

While we have just found out that

$$\psi_{\text{pr}}(x,y)t_{rel_1}(x,y)=\psi_{\text{pr}}(x,y)\psi_{p\text{R}}(x,y)\psi_{p1}^*(x,y)$$

will give rise to the real image reconstruction of point 1 . Similarly, the term

$$\psi_{\text{pr}}(x,y)t_{rel_2}(x,y)=\psi_{\text{pr}}(x,y)\psi_{p\text{R}}(x,y)\psi_{p2}^*(x,y)$$

in the hologram will give rise to the real image reconstruction of point 2. By carrying out the same procedures done for point 1, we state the following result:

$$\begin{aligned}\psi_{p,rel_2}(x,y;z_{\text{r}2}) &= \psi_{\text{pr}}(x,y)t_{rel_2}(x,y)*h(x,y;z_{\text{r}2})\\ &\propto \delta\left[x+z_{\text{r}2}\left(\frac{b}{l_2}+\frac{h}{2(R+d)}-\frac{a}{l_1}\right),y\right],\end{aligned} \tag{5.2-13}$$

这里

$$z_{r2}=\left(\frac{1}{R+d}-\frac{1}{l_1}-\frac{1}{l_2}\right)^{-1}=\frac{(R+d)l_1l_2}{l_1l_2-(l_1+l_2)(R+d)}.$$

z_{r2} 是点物 2 在全息图后重建实像的距离, 横向方向上像点位于

$$x=x_2=-z_{r2}\left[\frac{b}{l_2}+\frac{h}{2(R+d)}-\frac{a}{l_1}\right]. \tag{5.2-14}$$

点物 2 的实像重建如图 5.2-2 所示.

5.2.2 全息像的横向和纵向放大率

横向重建

由上节的结论, 现在可以分析全息成像的横向全息放大率 (*lateral holographic magnification*). 两个实像点 1 和 2 之间的横向距离 (沿 x 方向) 为 x_1-x_2, 这两点物之间的初始距离为 h, 因此, 该实像的横向 (全息) 放大率为

$$\begin{aligned}M_{lat}^{r}&=\frac{x_1-x_2}{h}\\&=\frac{-z_{r1}\left(\dfrac{b}{l_2}-\dfrac{h}{2R}-\dfrac{a}{l_1}\right)+z_{r2}\left(\dfrac{b}{l_2}+\dfrac{h}{2(R+d)}-\dfrac{a}{l_1}\right)}{h}\\&\cong\frac{(z_{r2}-z_{r1})\left(\dfrac{b}{l_2}-\dfrac{a}{l_1}\right)+(z_{r2}+z_{r1})\dfrac{h}{2R}}{h},\end{aligned} \tag{5.2-15}$$

当 $R\gg d$ 时. 放大率是一个关于 h 的函数, 这并不是我们想要的. 为了使放大率不依赖于横向距离 h, 设

$$\frac{b}{l_2}-\frac{a}{l_1}=0,$$

或

$$\frac{b}{l_2}=\frac{a}{l_1}. \tag{5.2-16}$$

则, 当 $R\gg d$ 时, 式 (5.2-15) 变为

$$M_{lat}^{r}=\frac{z_{r2}+z_{r1}}{2R}\simeq\frac{l_1l_2}{l_1l_2-(l_1+l_2)R}. \tag{5.2-17}$$

纵向重建

两个实像点之间的纵向距离 (沿 z 方向) 为 $z_{r2}-z_{r1}$, 因此, 其全息像的纵向全息放大率 (*longitudinal holographic magnification*) 被定义为

$$M_{long}^{r}=\frac{z_{r2}-z_{r1}}{d}. \tag{5.2-18}$$

根据式 (5.2-11) 和式 (5.2-13), 且假定 $R\gg d$, 纵向 (全息) 放大率变为

$$M_{long}^{r}\cong\frac{(l_1l_2)^2}{(l_1l_2-Rl_1-Rl_2)^2}. \tag{5.2-19}$$

where

$$z_{r2} = \left(\frac{1}{R+d} - \frac{1}{l_1} - \frac{1}{l_2}\right)^{-1} = \frac{(R+d)l_1l_2}{l_1l_2 - (l_1+l_2)(R+d)}.$$

z_{r2} is the distance of the real image reconstruction of point 2 behind the hologram and the image point is located laterally at

$$x = x_2 = -z_{r2}\left[\frac{b}{l_2} + \frac{h}{2(R+d)} - \frac{a}{l_1}\right]. \tag{5.2-14}$$

Figure 5.2-2 shows the real image reconstruction of point 2.

5.2.2 Lateral and Longitudinal Holographic Magnifications

Lateral Reconstruction

From the results of last subsection, we can now evaluate *lateral holographic magnification* of holographic imaging. The lateral distance (along x) between the two real image points 1 and 2 is $x_1 - x_2$ with the original separation between the two points being h. Therefore, the lateral holographic magnification for the real image is

$$\begin{aligned} M_{lat}^{r} &= \frac{x_1 - x_2}{h} \\ &= \frac{-z_{r1}\left(\frac{b}{l_2} - \frac{h}{2R} - \frac{a}{l_1}\right) + z_{r2}\left(\frac{b}{l_2} + \frac{h}{2(R+d)} - \frac{a}{l_1}\right)}{h} \\ &\cong \frac{(z_{r2} - z_{r1})\left(\frac{b}{l_2} - \frac{a}{l_1}\right) + (z_{r2} + z_{r1})\frac{h}{2R}}{h}, \end{aligned} \tag{5.2-15}$$

for $R \gg d$. The magnification is a function of h, which is undesirable. To make this magnification independent of the lateral separation, h, we set

$$\frac{b}{l_2} - \frac{a}{l_1} = 0,$$

or

$$\frac{b}{l_2} = \frac{a}{l_1}. \tag{5.2-16}$$

Then, Eq. (5.2-15) becomes

$$M_{lat}^{r} = \frac{z_{r2} + z_{r1}}{2R} \simeq \frac{l_1l_2}{l_1l_2 - (l_1+l_2)R}, \tag{5.2-17}$$

for $R \gg d$.

Longitudinal Reconstruction

The longitudinal distance (along z) between the two real point images is $z_{r2} - z_{r1}$, so that the *longitudinal holographic magnification* is defined as

$$M_{long}^{r} = \frac{z_{r2} - z_{r1}}{d}. \tag{5.2-18}$$

Using Eqs. (5.2-11) and (5.2-13) and assuming $R \gg d$, the longitudinal magnification becomes

$$M_{long}^{r} \cong \frac{(l_1l_2)^2}{(l_1l_2 - Rl_1 - Rl_2)^2}. \tag{5.2-19}$$

通过比较式 (5.2-17) 和式 (5.2-19), 可以得到三维成像中放大率之间的关系

$$M_{long}^{\mathrm{r}} = (M_{lat}^{\mathrm{r}})^2. \tag{5.2-20}$$

这一结果与单透镜三维成像的结果相对应 (见第一章中单透镜的三维成像), 这在体成像 (*volume imaging*) 中会引起失真.

举例: 平面波作为参考光和重建光的情况

当参考光和重建光都为平面波时, 令式 (5.2-17) 和式 (5.2-18) 中的 l_1 和 l_2 趋于无穷, 则, 当 $R \gg d$ 时, 有

$$\begin{aligned} M_{lat}^{\mathrm{r}} &= \lim_{l_1, l_2 \to \infty} \frac{z_{\mathrm{r}2} + z_{\mathrm{r}1}}{2R} \\ &= \lim_{l_1, l_2 \to \infty} \frac{\left(\dfrac{1}{R+d} - \dfrac{1}{l_1} - \dfrac{1}{l_2}\right)^{-1} + \left(\dfrac{1}{R} - \dfrac{1}{l_1} - \dfrac{1}{l_2}\right)^{-1}}{2R} = \frac{2R+d}{2R} \approx 1, \end{aligned}$$

且

$$\begin{aligned} M_{long}^{\mathrm{r}} &= \lim_{l_1, l_2 \to \infty} \frac{z_{\mathrm{r}2} - z_{\mathrm{r}1}}{d} \\ &= \lim_{l_1, l_2 \to \infty} \frac{\left(\dfrac{1}{R+d} - \dfrac{1}{l_1} - \dfrac{1}{l_2}\right)^{-1} - \left(\dfrac{1}{R} - \dfrac{1}{l_1} - \dfrac{1}{l_2}\right)^{-1}}{d} = 1. \end{aligned}$$

这样, $M_{lat}^{\mathrm{r}} = M_{long}^{\mathrm{r}}$, 且三维成像就没有体失真了. 可以发现, 当 $l_1 = l_2 \to \infty$ 时, 式 (5.2-16) 也是满足的.

举例: 利用球面波做参考光而用平面波进行重建时的横向放大率

让我们从式 (5.2-15) 中横向 (全息) 放大率的定义开始, 即

$$M_{lat}^{\mathrm{r}} = \frac{x_1 - x_2}{h} = \frac{-z_{\mathrm{r}1}\left(\dfrac{b}{l_2} - \dfrac{h}{2R} - \dfrac{a}{l_1}\right) + z_{\mathrm{r}2}\left[\dfrac{b}{l_2} + \dfrac{h}{2(R+d)} - \dfrac{a}{l_1}\right]}{h}.$$

考虑点源的记录和重建都在 z 轴上的情况, 即 $a = b = 0$. 同时, 取 $d = 0$, 因为考虑的是一个平面图像, 此时, M_{lat}^{r} 变为

$$M_{lat}^{\mathrm{r}} = \frac{z_{\mathrm{r}2} + z_{\mathrm{r}1}}{2R}.$$

这里, $z_{\mathrm{r}2} = z_{\mathrm{r}1} = \left(\dfrac{1}{R} - \dfrac{1}{l_1} - \dfrac{1}{l_2}\right)^{-1}$. 当用平面波重建时, $l_2 \to \infty$, 上式可以化简为

$$M_{lat}^{\mathrm{r}} = \left(1 - \frac{R}{l_1}\right)^{-1}.$$

例如, 当 $l_1 = 2R$ 时, $M_{lat}^{\mathrm{r}} = 2$, 则放大率因子为 2; 当 $l_1 = R/4 < R$ 时, $M_{lat}^{\mathrm{r}} = -1/3$, 则图像会缩小. 可以发现, 如果记录的参考光也是平面波, 即 $l_1 \to \infty$ 时, 那么如我们前面例子所讨论的, 利用平面波来记录和重建时是没有放大的.

By comparing Eqs. (5.2-17) and (5.2-19), we have the following relationship between the magnifications in 3-D imaging:

$$M_{long}^{\mathrm{r}} = (M_{lat}^{\mathrm{r}})^2. \tag{5.2-20}$$

This result corresponds to the result obtained in 3-D imaging of a single lens (see 3-D imaging using a single lens in Chapter 1), which introduces distortion in *volume imaging.*

Example: Plane Wave Reference and Plane Wave Reconstruction

When both the reference and the reconstruction waves are plane waves, we let l_1 and l_2 approach infinity in Eqs. (5.2-17) and (5.2-18):

$$M_{lat}^{\mathrm{r}} = \lim_{l_1,l_2\to\infty} \frac{z_{\mathrm{r}2}+z_{\mathrm{r}1}}{2R}$$

$$= \lim_{l_1,l_2\to\infty} \frac{\left(\dfrac{1}{R+d}-\dfrac{1}{l_1}-\dfrac{1}{l_2}\right)^{-1}+\left(\dfrac{1}{R}-\dfrac{1}{l_1}-\dfrac{1}{l_2}\right)^{-1}}{2R} = \frac{2R+d}{2R} \approx 1,$$

for $R \gg d$, and

$$M_{long}^{\mathrm{r}} = \lim_{l_1,l_2\to\infty} \frac{z_{\mathrm{r}2}-z_{\mathrm{r}1}}{d}$$

$$= \lim_{l_1,l_2\to\infty} \frac{\left(\dfrac{1}{R+d}-\dfrac{1}{l_1}-\dfrac{1}{l_2}\right)^{-1}-\left(\dfrac{1}{R}-\dfrac{1}{l_1}-\dfrac{1}{l_2}\right)^{-1}}{d} = 1.$$

Thus, we see that $M_{lat}^{\mathrm{r}} = M_{long}^{\mathrm{r}}$ and there is no volume distortion for 3-D imaging. Note that Eq. (5.2-16) is satisfied when $l_1 = l_2 \to \infty$.

Example: Lateral Magnification Using Spherical Wave Reference and Plane Wave Reconstruction

Let us start from the definition of the lateral holographic magnification in Eq. (5.2-15), which is

$$M_{lat}^{\mathrm{r}} = \frac{x_1-x_2}{h} = \frac{-z_{\mathrm{r}1}\left(\dfrac{b}{l_2}-\dfrac{h}{2R}-\dfrac{a}{l_1}\right)+z_{\mathrm{r}2}\left[\dfrac{b}{l_2}+\dfrac{h}{2(R+d)}-\dfrac{a}{l_1}\right]}{h}.$$

We consider the recording and reconstruction point sources are on the z-axis, that is, $a = b = 0$. Also we take $d = 0$ as we are considering a planar image, M_{lat}^{r} then becomes

$$M_{lat}^{\mathrm{r}} = \frac{z_{\mathrm{r}2}+z_{\mathrm{r}1}}{2R}$$

with $z_{\mathrm{r}2} = z_{\mathrm{r}1} = \left(\dfrac{1}{R}-\dfrac{1}{l_1}-\dfrac{1}{l_2}\right)^{-1}$. For plane wave reconstruction, $l_2 \to \infty$, we have a simple expression given by

$$M_{lat}^{\mathrm{r}} = \left(1-\frac{R}{l_1}\right)^{-1}.$$

For example, taking $l_1 = 2R$, $M_{lat}^{\mathrm{r}} = 2$, we have a magnification of a factor of 2. For $l_1 = \left(\dfrac{1}{4}\right)R < R$, $M_{lat}^{\mathrm{r}} = -\dfrac{1}{3}$, we have demagnification. Note that if the recording reference wave is also a plane wave, that is, $l_1 \to \infty$, there is no magnification for using plane wave for recording and reconstruction as discussed in our earlier example.

5.3 典型的全息图

本节介绍一些常见的全息图类型及其基本理论.

5.3.1 伽博全息图和同轴全息图

记录

先来看伽博最初的方法, 如图 5.3-1a) 所示, 假定物体有些透明但有些小的偏差 $\Delta(x,y)$, 因此, 物体的总透过率 $\sigma(x,y)$ 可以表示为

$$\sigma(x,y)=\sigma_0+\Delta(x,y). \tag{5.3-1}$$

当条件为 $\Delta(x,y)\ll\sigma_0$, 这里 σ_0 为某常数, 表示该透明片的均匀背景, 由于 $\Delta(x,y)$ 非常小, 其散射光不会干扰到均匀的参考光.

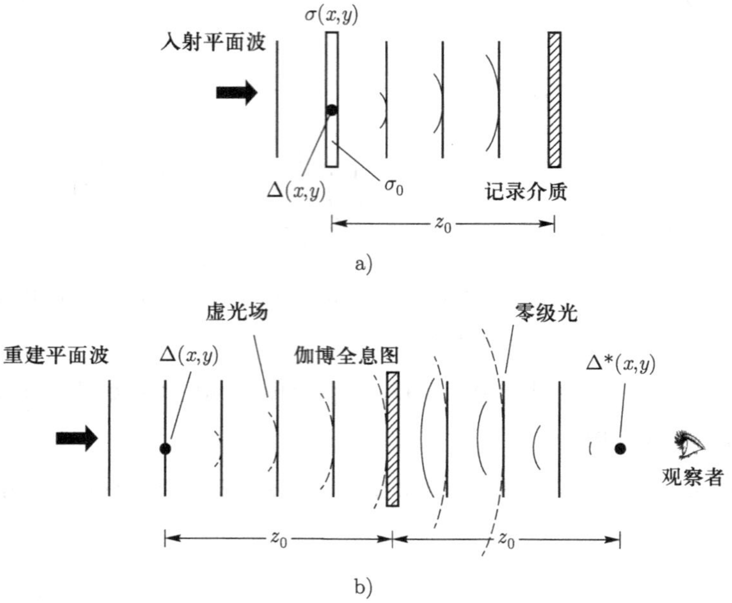

图 5.3-1 伽博全息图: a) 记录; b) 重建

根据菲涅耳衍射, 全息面上距离物体 z_0 处的复振幅为

$$A[\sigma_0+\Delta(x,y)]*h(x,y;z_0)=\psi_c+\psi_0(x,y), \tag{5.3-2}$$

其中, A 为入射平面波的振幅, 且 $h(x,y;z_0)$ 为傅里叶光学中的空间脉冲响应 [见式 (3.4-16)]. 在式 (5.3-2) 中, 有

$$\psi_c=A\sigma_0*h(x,y;z_0), \tag{5.3-3}$$

这基本上是一个复振幅为 $A\sigma_0$ 的平面波, 传播一段距离 z_0 后到达记录介质. 为了证明它, 这里用傅里叶变换的方法. 对式 (5.3-3) 进行傅里叶变换

$$\begin{aligned}\mathcal{F}\{\psi_c\}&=\mathcal{F}\{A\sigma_0\}\mathcal{F}\{h(x,y;z_0)\}=A\sigma_0 4\pi^2\delta(k_x,k_y)H(k_x,k_y;z_0)\\&=A\sigma_0 4\pi^2\delta(k_x,k_y)\mathrm{e}^{-\mathrm{j}k_0z_0}\mathrm{e}^{\mathrm{j}\frac{(k_x^2+k_y^2)z_0}{2k_0}}=A\sigma_0 4\pi^2\delta(k_x,k_y)\mathrm{e}^{-\mathrm{j}k_0z_0},\end{aligned}$$

5.3 Types of Holograms

In this section, we will introduce some common types of holograms and their basic principles.

5.3.1 Gabor Hologram and On-axis Hologram

Recording

Let us consider Gabor's original method, which is shown in Figure 5.3-1a. The object is assumed to be rather transparent but with a small variation of $\Delta(x,y)$. Hence, the total transparency of the object $\sigma(x,y)$ can be expressed as

$$\sigma(x,y)=\sigma_0+\Delta(x,y). \tag{5.3-1}$$

with the condition that $\Delta(x,y)\ll\sigma_0$, where σ_0 is some constant, representing a uniform background of the transparency. Since $\Delta(x,y)$ is small, its scattered wave does not disturb the uniform reference wave.

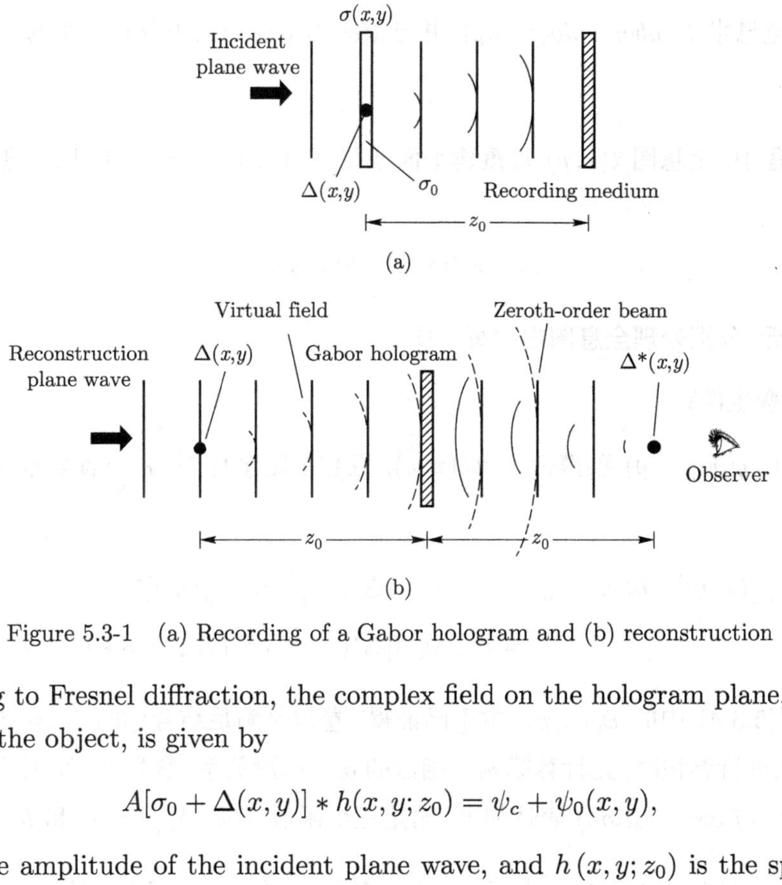

Figure 5.3-1 (a) Recording of a Gabor hologram and (b) reconstruction

According to Fresnel diffraction, the complex field on the hologram plane, at a distance z_0 away from the object, is given by

$$A[\sigma_0+\Delta(x,y)]*h(x,y;z_0)=\psi_c+\psi_0(x,y), \tag{5.3-2}$$

where A is the amplitude of the incident plane wave, and $h(x,y;z_0)$ is the spatial impulse response in Fourier optics [see Eq. (3.4-16)]. In Eq. (5.3-2),

$$\psi_c=A\sigma_0*h(x,y;z_0), \tag{5.3-3}$$

which is basically a plane wave of complex amplitude of $A\sigma_0$, propagating a distance z_0 to reach the recording medium. To show it, let us use the Fourier transform method. Taking the Fourier transform of the equation, we have

$$\begin{aligned}\mathcal{F}\{\psi_c\}&=\mathcal{F}\{A\sigma_0\}\mathcal{F}\{h(x,y;z_0)\}=A\sigma_0 4\pi^2\delta(k_x,k_y)H(k_x,k_y;z_0)\\&=A\sigma_0 4\pi^2\delta(k_x,k_y)\mathrm{e}^{-\mathrm{j}k_0z_0}\mathrm{e}^{\mathrm{j}\frac{(k_x^2+k_y^2)z_0}{2k_0}}=A\sigma_0 4\pi^2\delta(k_x,k_y)\mathrm{e}^{-\mathrm{j}k_0z_0},\end{aligned}$$

其中, $H(k_x,k_y;z)$ 为傅里叶光学中的空间频率传递函数 [见式 (3.4-18)], 我们使用了 δ 函数的乘法性质 [见式 (2.1-4)] 来得到上式的最后一步. 现在, 求其逆变换, 有

$$\mathcal{F}^{-1}\left\{\mathcal{F}\left\{\psi_c\right\}\right\}=\psi_c=A\sigma_0\mathrm{e}^{-\mathrm{j}k_0z_0},$$

这是平面波传播一段距离 z_0 后的复振幅. 这种平面波在全息术中充当参考光. 现在, 根据式 (5.3-2), 有

$$\psi_0(x,y)=A\Delta(x,y)*h(x,y;z_0),\tag{5.3-4}$$

这是由于物体的 $\Delta(x,y)$ 所产生的散射场, 被视为物光波.

因此, 全息面上的强度或全息图的透过率函数为

$$t(x,y)=|\psi_c+\psi_0(x,y)|^2=|\psi_c|^2+|\psi_0(x,y)|^2+\psi_c\psi_0^*(x,y)+\psi_c^*\psi_0(x,y).\tag{5.3-5}$$

如今, 这种类型的全息图被众所周知为伽博全息图 (*Gabor hologram*). 这种类型的全息术即被称为伽博全息术 (*Gabor holography*), 并已被应用于粒子大小及其分布的分析.

重建

在重建过程中, 全息图 $t(x,y)$ 被重建平面波照射, 即, $\psi_{rec}=$ 常数. 从全息图中出来的复振幅为

$$\psi_{rec}\times t(x,y)*h(x,y;z).\tag{5.3-6}$$

根据前面的分析, 分别处理全息图中的每一项.

a) 实像 (孪生像)

全息图中的 $\psi_c\psi_0^*(x,y)$ 项携带了 $\psi_0^*(x,y)$, 且它在距全息图 z_0 位置处重建一个实像. 数学上, 有

$$\begin{aligned}[\psi_{rec}\times\psi_c\psi_0^*(x,y)]*h(x,y;z_0)&=\psi_{rec}\psi_c[A\Delta(x,y)*h(x,y;z_0)]^**h(x,y;z_0)\\&=\psi_{rec}\psi_cA[\Delta^*(x,y)*h^*(x,y;z_0)]*h(x,y;z_0).\end{aligned}\tag{5.3-7}$$

上式利用了式 (5.3-4) 中的 $\psi_0(x,y)$. 由于已证得, 卷积是满足结合律的 (*associative*), 也就是说, 将三个函数进行卷积时, 先计算哪两个函数的卷积都没关系. 数学上, 卷积的结合律 (*associative property of convolution*) 表述如下. 给定三个函数 $f(x,y)$, $g(x,y)$ 和 $h(x,y)$

$$[f(x,y)*g(x,y)]*h(x,y)=f(x,y)*[g(x,y)*h(x,y)].\tag{5.3-8}$$

利用结合律, 式 (5.3-7) 可以被重新写为

$$\psi_{rec}\psi_c\psi_0^*(x,y)*h(x,y;z_0)=\psi_{rec}\psi_cA\Delta^*(x,y)*[h^*(x,y;z_0)*h(x,y;z_0)].\tag{5.3-9}$$

先来分析 $K(x,y)=h^*(x,y;z_0)*h(x,y;z_0)$. 进行傅里叶变换得

$$\begin{aligned}\mathcal{F}\left\{K(x,y)\right\}&=\mathcal{F}\left\{h^*(x,y;z_0)*h(x,y;z_0)\right\}\\&=\mathcal{F}\left\{h^*(x,y;z_0)\right\}\mathcal{F}\left\{h(x,y;z_0)\right\}\end{aligned}$$

where we recall that $H(k_x, k_y; z)$ is the spatial frequency transfer function in Fourier optics [see Eq. (3.4-18)], and we have used the Product Property of the delta function [see Eq. (2.1-4)] to obtain the last step of the above equation. Now, taking the inverse transform, we have

$$\mathcal{F}^{-1}\{\mathcal{F}\{\psi_c\}\} = \psi_c = A\sigma_0 \mathrm{e}^{-\mathrm{j}k_0 z_0},$$

which is the complex amplitude of a plane wave after propagating a distance z_0. This plane wave serves as the reference light in holography. Now from Eq. (5.3-2),

$$\psi_0(x, y) = A\Delta(x, y) * h(x, y; z_0) \tag{5.3-4}$$

is the scattered field due to the object $\Delta(x, y)$ and is regarded as the object wave.

As a result, the intensity on the hologram plane or the transparency function $t(x, y)$ of the hologram is given by

$$t(x, y) = |\psi_c + \psi_0(x, y)|^2 = |\psi_c|^2 + |\psi_0(x, y)|^2 + \psi_c\psi_0^*(x, y) + \psi_c^*\psi_0(x, y). \tag{5.3-5}$$

Nowadays, this type of hologram is known as a *Gabor hologram.* This type of holography is called *Gabor holography* and has found use in particle sizing as well as distribution analysis.

Reconstruction

In the reconstruction process, the hologram $t(x, y)$ is illuminated by a reconstruction plane wave, that is, $\psi_{rec} =$ constant. The complex field emerging from the hologram is given by

$$\psi_{rec} \times t(x, y) * h(x, y; z). \tag{5.3-6}$$

According to our previous analysis, let us deal with each term in the hologram separately.

a) Real Image (Twin Image)

The term $\psi_c\psi_0^*(x, y)$ in the hologram carries $\psi_0^*(x, y)$ and it reconstructs a real image at a distance z_0 from the hologram. Mathematically, we have

$$\begin{aligned}[\psi_{rec} \times \psi_c\psi_0^*(x, y)] * h(x, y; z_0) &= \psi_{rec}\psi_c[A\Delta(x, y) * h(x, y; z_0)]^* * h(x, y; z_0) \\ &= \psi_{rec}\psi_c A[\Delta^*(x, y) * h^*(x, y; z_0)] * h(x, y; z_0).\end{aligned} \tag{5.3-7}$$

where we have used $\psi_0(x, y)$ from Eq. (5.3-4). As it turns out convolution is *associative*, that is, when convolving three functions together, it does not matter which two functions we perform convolution first. Mathematically, the *Associative Property of Convolution* is stated as follows. Given three functions $f(x, y)$, $g(x, y)$, and $h(x, y)$,

$$[f(x, y) * g(x, y)] * h(x, y) = f(x, y) * [g(x, y) * h(x, y)]. \tag{5.3-8}$$

Using the Associative Property, Eq. (5.3-7) can be rewritten as

$$\psi_{rec}\psi_c\psi_0^*(x, y) * h(x, y; z_0) = \psi_{rec}\psi_c A\Delta^*(x, y) * [h^*(x, y; z_0) * h(x, y; z_0)]. \tag{5.3-9}$$

Let us first evaluate $K(x, y) = h^*(x, y; z_0) * h(x, y; z_0)$. We take the Fourier transform of the equation to get

$$\begin{aligned}\mathcal{F}\{K(x, y)\} &= \mathcal{F}\{h^*(x, y; z_0) * h(x, y; z_0)\} \\ &= \mathcal{F}\{h^*(x, y; z_0)\}\,\mathcal{F}\{h(x, y; z_0)\}\end{aligned}$$

$$= H^*(-k_x, -k_y; z_0)H(k_x, k_y; z_0),$$

这里利用表 2.2 求得 $\mathcal{F}\{h^*(x, y; z_0)\}$.

现在, 由于 $H(k_x, k_y; z_0) = \mathrm{e}^{-\mathrm{j}k_0 z_0}\mathrm{e}^{\mathrm{j}\frac{(k_x^2+k_y^2)z_0}{2k_0}}$, 有

$$\mathcal{F}\{K(x, y)\} = H^*(-k_x, -k_y; z_0)H(k_x, k_y; z_0) = 1.$$

因此, 得

$$K(x, y) = h^*(x, y; z_0) * h(x, y; z_0) = \delta(x, y). \tag{5.3-10}$$

基于这个结果, 式 (5.3-9) 变为

$$\psi_{rec}\psi_c\psi_0^*(x, y) * h(x, y; z_0) = \psi_{rec}\psi_c A\Delta^*(x, y) * \delta(x, y) \propto \Delta^*(x, y). \tag{5.3-11}$$

如图 5.3-1b) 所示, 在全息图右侧距离 z_0 处重建其实像.

b) 虚像

类似地, 全息图公式 (5.3-5) 中的 $\psi_c^*\psi_0(x, y)$ 项, 在初始物体的位置处重建振幅波动 $\Delta(x, y)$ 的一个虚像. 为了说明这一点, 首先来讨论在波动光学中如何求虚光场 (*virtual field*).

这里用一个负透镜来说明虚光场的计算, 该负透镜的透过率函数为

$$t_f(x, y) = \mathrm{e}^{-\mathrm{j}\frac{k_0}{2|f|}(x^2+y^2)}, \tag{5.3-12}$$

因为 $f < 0$ [见式 (3.4-24)]. 对于入射到该透镜上的单位振幅平面波, 根据菲涅耳衍射, 在距离透镜 z_0 处的实光场 (*real field*) 为

$$t_f(x, y) * h(x, y; z_0) = \mathrm{e}^{-\mathrm{j}\frac{k_0}{2|f|}(x^2+y^2)} * h(x, y; z_0).$$

为求该透镜前方 z_0 处的虚光场, 即被观察者看到的在其初始位置的光场, 进行反向传播计算

$$\mathrm{e}^{-\mathrm{j}\frac{k_0}{2|f|}(x^2+y^2)} * h(x, y; -z_0). \tag{5.3-13}$$

如图 5.3-2 所示。根据 $h(x, y; z)$ 的定义 [见式 (3.4-16)], 可推出 $h^*(x, y; z_0) = h(x, y; -z_0)$. 基于这个等式, 计算虚光场可根据下式

$$\mathrm{e}^{-\mathrm{j}\frac{k_0}{2|f|}(x^2+y^2)} * h(x, y; -z_0) = \mathrm{e}^{-\mathrm{j}\frac{k_0}{2|f|}(x^2+y^2)} * h^*(x, y; z_0).$$

情况如图 5.3-2 所示.

当全息图被平面波 ψ_{rec} 照射时, 我们重新来看全息图的 $\psi_c^*\psi_0(x, y)$ 这一项, 当从全息图反向传播 (*back-propagating*) 一段距离 z_0 后, 虚光场变为

$$\begin{aligned}\psi_{rec}\psi_c^*\psi_0(x, y) * h^*(x, y; z_0) &= \psi_{rec}\psi_c^*[A\Delta(x, y) * h(x, y; z_0)] * h^*(x, y; z_0)\\ &= \psi_{rec}\psi_c^*[A\Delta(x, y) * \delta(x, y)]\\ &\propto \Delta(x, y),\end{aligned} \tag{5.3-14}$$

其中, 用式 (5.3-10) 对上式进行了简化, 并将式 (5.3-4) 给出的物光场 $\psi_0(x, y)$ 代入上式. 结果如图 5.3-1b) 所示, 得到在 $z = -z_0$ 处的虚光场 $\Delta(x, y)$.

$$= H^*(-k_x, -k_y; z_0)H(k_x, k_y; z_0),$$

where we have used Table 2.2 to find $\mathcal{F}\{h^*(x, y; z_0)\}$.

Now, since $H(k_x, k_y; z_0) = \mathrm{e}^{-\mathrm{j}k_0 z_0}\mathrm{e}^{\mathrm{j}\frac{(k_x^2+k_y^2)z_0}{2k_0}}$,

$$\mathcal{F}\{K(x, y)\} = H^*(-k_x, -k_y; z_0)H(k_x, k_y; z_0) = 1.$$

Therefore,

$$K(x, y) = h^*(x, y; z_0) * h(x, y; z_0) = \delta(x, y). \quad (5.3\text{-}10)$$

With this result, Eq. (5.3-9) becomes

$$\psi_{rec}\psi_c\psi_0^*(x, y) * h(x, y; z_0) = \psi_{rec}\psi_c A\Delta^*(x, y) * \delta(x, y) \propto \Delta^*(x, y). \quad (5.3\text{-}11)$$

This is the real image reconstructed at a distance z_0 on the right side of the hologram as shown in Figure 5.3-1b.

b) Virtual Image

Similarly, $\psi_c^*\psi_0(x, y)$ from the hologram in Eq. (5.3-5) reconstructs a virtual image of the amplitude fluctuation $\Delta(x, y)$ at the location of the original object. To show this is the case, let us first discuss how to find a *virtual field* in wave optics.

Let us take a negative lens to illustrate the calculation of a virtual field. The transparency function of the negative lens is

$$t_f(x, y) = \mathrm{e}^{-\mathrm{j}\frac{k_0}{2|f|}(x^2+y^2)} \quad (5.3\text{-}12)$$

as $f < 0$ [see Eq. (3.4-24)]. For a plane wave of unit amplitude incident on the lens, the *real field* z_0 away from the lens, according to Fresnel diffraction, is given by

$$t_f(x, y) * h(x, y; z_0) = \mathrm{e}^{-\mathrm{j}\frac{k_0}{2|f|}(x^2+y^2)} * h(x, y; z_0).$$

To find the virtual field z_0 in front of the lens, that is, the field from where it appears to be originating as seen by an observer, we perform the back-propagation calculation as follows:

$$\mathrm{e}^{-\mathrm{j}\frac{k_0}{2|f|}(x^2+y^2)} * h(x, y; -z_0). \quad (5.3\text{-}13)$$

The situation is shown in Figure 5.3-2. From the definition of $h(x, y; z_0)$ [see Eq. (3.4-16)], we deduce that $h^*(x, y; z_0) = h(x, y; -z_0)$. With this identity, the virtual field can be calculated according to

$$\mathrm{e}^{-\mathrm{j}\frac{k_0}{2|f|}(x^2+y^2)} * h(x, y; -z_0) = \mathrm{e}^{-\mathrm{j}\frac{k_0}{2|f|}(x^2+y^2)} * h^*(x, y; z_0).$$

The situation is summarized in Figure 5.3-2.

We now return to the term $\psi_c^*\psi_0(x, y)$ in the hologram when the hologram is illuminated by plane wave ψ_{rec}. After *back propagating* a distance of z_0 from the hologram, the virtual field is

$$\begin{aligned}\psi_{rec}\psi_c^*\psi_0(x, y) * h^*(x, y; z_0) &= \psi_{rec}\psi_c^*[A\Delta(x, y) * h(x, y; z_0)] * h^*(x, y; z_0)\\ &= \psi_{rec}\psi_c^*[A\Delta(x, y) * \delta(x, y)]\\ &\propto \Delta(x, y),\end{aligned} \quad (5.3\text{-}14)$$

where we have used Eq. (5.3-10) to simplify the equation and substituted the object field $\psi_0(x, y)$ given by Eq. (5.3-4) into Eq. (5.3-14). The result indicates that we have the virtual field $\Delta(x, y)$ at $z = -z_0$ as shown in Figure 5.3-1b.

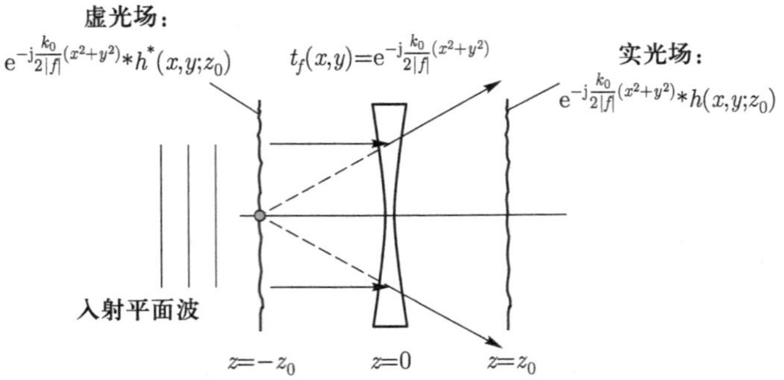

图 5.3-2 单位振幅平面波入射下实光场和虚光场的计算

c) 零级光

从式 (5.3-5) 得到的全息图公式的前两项对应于传输的零级光, 由下式给出

$$\begin{aligned}\psi_{rec}[|\psi_c|^2+|\psi_0(x,y)|^2]*h(x,y;z)&\approx\psi_{rec}|\psi_c|^2*h(x,y;z)\\&=\psi_{rec}(A\sigma_0)^2*h(x,y;z).\end{aligned}\qquad(5.3\text{-}15)$$

因为 $\Delta(x,y)\ll\sigma_0$, 所以 $|\psi_c|^2=(A\sigma_0)^2\gg|\psi_0(x,y)|^2=|A\Delta(x,y)*h(x,y;z_0)|^2$. 上式表示一束平面波沿 z 方向传播, 如图 5.3-1b) 所示.

伽博 (Gabor) 全息术的优点是实验装置非常简单. 然而, 其中一个缺点就是所传输的零级光总是使重建的虚像变得模糊, 另一个缺点是孪生像的问题, 这是因为伽博全息图是同轴全息图. 在全息术中, 这是众所周知的 "孪生像问题", 该问题引起对孪生像消除的研究. 伽博全息图还有一个问题, 为了使该技术有用, 物体的振幅变化必须足够小, 即 $\Delta(x,y)\ll\sigma_0$, 为了克服这一缺点, 独立光源可被用来作为参考波, 如图 5.1-1a) 所示, 且对所用的物体的类型没有限制. 然而, 零级光和孪生像的问题仍然存在. 由于参考波和物波沿同一方向传播, 这种光学装置被称为同轴的 (*on-axis*) 或轴向的 (*in-line*) 光路, 因此这种全息术被称为同轴全息术 (*on-axis holography*). 由于在全息面上的物波是用物体的菲涅耳衍射来描述的, 因此我们称这种全息图为菲涅耳全息图 (*Fresnel hologram*).

5.3.2 傅里叶全息图

如上节所述, 当全息面上的物波由物体的菲涅耳衍射来描述时, 就得到一幅菲涅耳全息图 (*Fresnel hologram*). 如果在全息面上的物波用物体的傅里叶变换来描述, 就得到一幅傅里叶全息图 (*Fourier hologram*). 傅里叶变换透镜可以简单地实现物体的傅里叶变换.

记录

使用如图 5.3-3a) 所示的装置来记录傅里叶全息图. 在该装置中, 输入或者物体 $s(x,y)$ 位于透镜 2 的前焦面. 因此, 透镜 2 后焦面的复振幅即为物光的傅里叶变换. 同时, 通过透镜 1 的使用, 物体旁边的会聚光点在记录介质上形成倾斜的参考平面波.

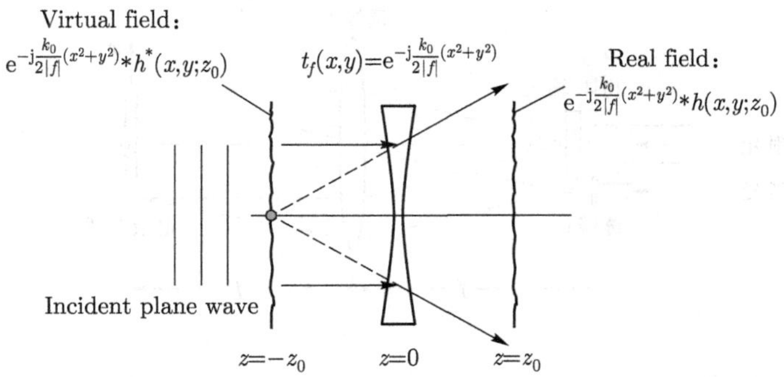

Figure 5.3-2 Calculation of real field and virtual field for an incident plane wave of unit amplitude

c) Zeroth-Order Beam

The first two terms of the hologram equation from Eq. (5.3-5) correspond to the transmitted zeroth-order beam given by

$$\psi_{rec}[|\psi_c|^2 + |\psi_0(x,y)|^2] * h(x,y;z) \approx \psi_{rec}|\psi_c|^2 * h(x,y;z)$$
$$= \psi_{rec}(A\sigma_0)^2 * h(x,y;z) \qquad (5.3\text{-}15)$$

as $|\psi_c|^2 = (A\sigma_0)^2 \gg |\psi_0(x,y)|^2 = |A\Delta(x,y) * h(x,y;z_0)|^2$ because $\Delta(x,y) \ll \sigma_0$. The above equation represents a plane wave propagation along the z-direction (see Figure 5.3-1b).

The merit of Gabor holography is that the setup is very simple. However, one of the drawbacks is that the transmitted zeroth-order beam always blurs the reconstructed virtual image. Another drawback is the problem of twin image as the Gabor hologram is an on-axis hologram. In holography, this is known as the "twin-image problem," leading to research known as twin-image elimination. Another problem of the Gabor hologram is that the amplitude variation of the object must be small enough, that is, $\Delta(x,y) \ll \sigma_0$, to make the technique useful. To overcome this shortcoming, an independent light is used as a reference wave, as shown in Figure 5.1-1a, and there is no limitation on the types of the object being used. However, the problem of the zeroth-order beam and the twin image remains. Since the reference light and the object light propagate along the same direction, this optical setup is called the *on-axis* or *in-line* geometry and we have *on-axis holography.* Since the object wave on the hologram plane is described by Fresnel diffraction of the object, we call this type of holograms a *Fresnel hologram.*

5.3.2 Fourier Hologram

While the object wave on the hologram plane is described by Fresnel diffraction of the object, we have a *Fresnel hologram*, as discussed in the last section. If the object wave on the hologram plane is described by the Fourier transform of the object, we have a *Fourier hologram.* A Fourier transform lens can conveniently provide Fourier transformation of the object.

Recording

We use the setup shown in Figure 5.3-3a to record a Fourier hologram. In the setup the input or the object $s(x,y)$ is located at the front focal plane of lens 2. Hence, the complex field at the back focal plane of lens 2 is the Fourier transform of the object light. Meanwhile, a focused light spot next to the object, through the use of lens 1, makes a tilted reference plane wave on the recording medium.

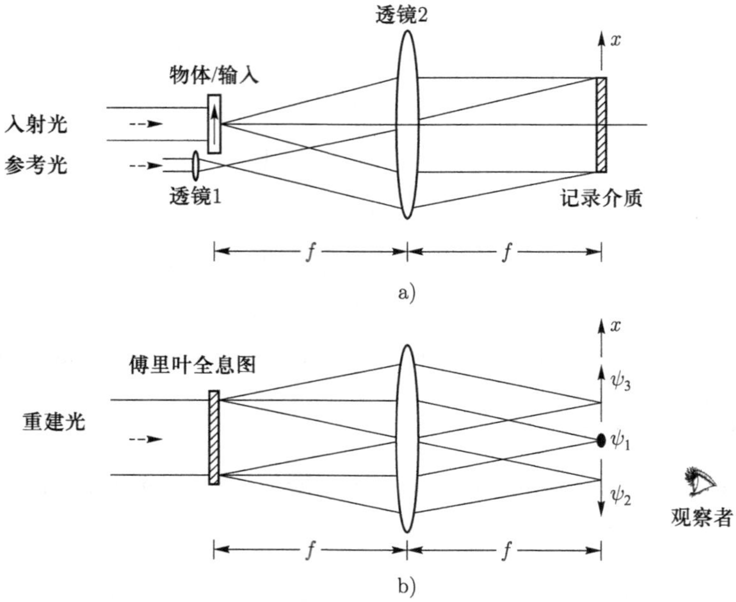

图 5.3-3 a) 全息图记录的几何示意图; b) 全息图重建的几何示意图

根据图 3.4-7 所示的结果, 可以将透镜 2 的后焦面处的总复振幅写为

$$\begin{aligned}\psi_t(x,y) &= \mathcal{F}\left\{s(x,y)+A\delta(x+x_0,y)\right\}\Big|_{k_x=k_0x/f,k_y=k_0y/f}\\ &= S\left(\frac{k_0x}{f},\frac{k_0y}{f}\right)+A\mathrm{e}^{-\mathrm{j}\frac{k_0x_0x}{f}},\end{aligned}\tag{5.3-16}$$

其中, $s(x,y)$ 为物体透过率函数, 且其傅里叶变换为 $S(k_x,k_y)$; $\delta(x+x_0,y)$ 为位于 $x=-x_0$, $y=0$ 处的焦点; f 为透镜 2 的焦距; A 为倾斜的参考平面波的振幅.

最终, 傅里叶全息图可以表示为

$$\begin{aligned}t_{FH}(x,y) &= |\psi_t(x,y)|^2\\ &= \left|S\left(\frac{k_0x}{f},\frac{k_0y}{f}\right)\right|^2+|A|^2+S\left(\frac{k_0x}{f},\frac{k_0y}{f}\right)\times A^*\mathrm{e}^{\mathrm{j}k_0x_0x/f}\\ &\quad+S^*\left(\frac{k_0x}{f},\frac{k_0y}{f}\right)\times A\mathrm{e}^{-\mathrm{j}k_0x_0x/f}.\end{aligned}\tag{5.3-17}$$

重建

在重建过程中, 如图 5.3-3b) 所示, 用单位振幅的垂直入射重建平面波照射放置在透镜前焦面处的全息图. 经过透镜的傅里叶变换后, 透镜后焦面的总复振幅包含如下三项.

a) 零级光

$$\begin{aligned}\psi_1(x,y) &= \mathcal{F}\left\{\left|S\left(\frac{k_0x}{f},\frac{k_0y}{f}\right)\right|^2+|A|^2\right\}\Bigg|_{k_x=k_0x/f,k_y=k_0y/f}\\ &= \frac{4\pi^2f^2}{k_0^2}s(-x,-y)\otimes s(-x,-y)+\frac{4\pi^2f^2}{k_0^2}|A|^2\delta(x,y),\end{aligned}\tag{5.3-18a}$$

其中, $\otimes$ 表示相关. 这表示零级光位于后焦面上 $x=0$, $y=0$ 处的位置.

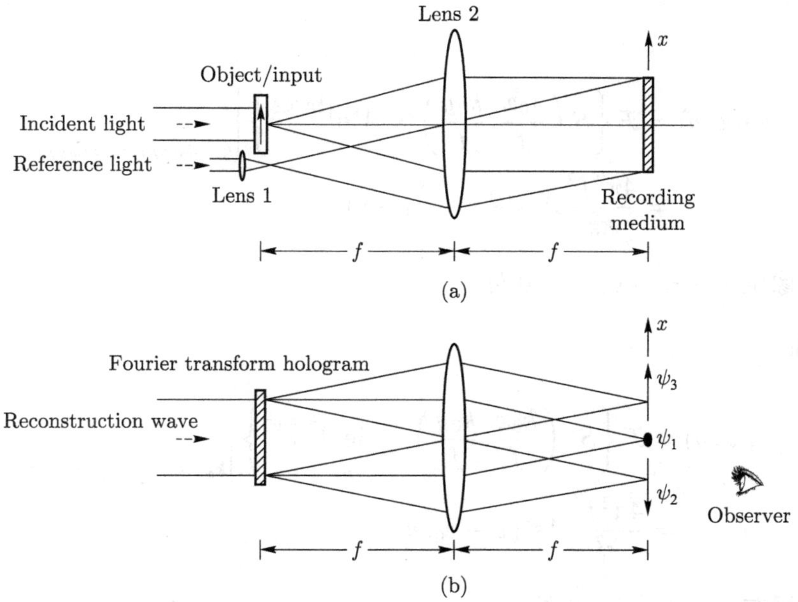

Figure 5.3-3 (a) Recording geometry and (b) reconstruction geometry for Fourier hologram

According to the result shown in Figure 3.4-7, we can write the total complex field at the back focal plane of lens 2 as

$$\begin{aligned}\psi_t(x,y) &= \mathcal{F}\{s(x,y)+A\delta(x+x_0,y)\}\big|_{k_x=k_0x/f,k_y=k_0y/f}\\ &= S\left(\frac{k_0x}{f},\frac{k_0y}{f}\right)+A\mathrm{e}^{-\mathrm{j}\frac{k_0x_0x}{f}},\end{aligned} \tag{5.3-16}$$

where $s(x,y)$ is the transparency function of the object and its Fourier transform is $S(k_x,k_y)$; $\delta(x+x_0,y)$ is the focused spot at $x=-x_0$, $y=0$; f is the focal length of lens 2, and A is the amplitude of the reference tilted plane wave.

Consequently, the Fourier hologram can be expressed as

$$\begin{aligned}t_{FH}(x,y) &= |\psi_t(x,y)|^2\\ &= \left|S\left(\frac{k_0x}{f},\frac{k_0y}{f}\right)\right|^2+|A|^2+S\left(\frac{k_0x}{f},\frac{k_0y}{f}\right)\times A^*\mathrm{e}^{\mathrm{j}k_0x_0x/f}\\ &\quad+S^*\left(\frac{k_0x}{f},\frac{k_0y}{f}\right)\times A\mathrm{e}^{-\mathrm{j}k_0x_0x/f}.\end{aligned} \tag{5.3-17}$$

Reconstruction

In the reconstruction process, the hologram placed at the front focal plane of a lens is illuminated using a normal-incident reconstruction plane wave with unit amplitude, as shown in Figure 5.3-3b. After Fourier transformation by the lens, the total complex field at the back focal plane of the lens contains three terms:

a) Zeroth-Order Beam

$$\begin{aligned}\psi_1(x,y) &= \mathcal{F}\left\{\left|S\left(\frac{k_0x}{f},\frac{k_0y}{f}\right)\right|^2+|A|^2\right\}\Bigg|_{k_x=k_0x/f,k_y=k_0y/f}\\ &= \frac{4\pi^2f^2}{k_0^2}s(-x,-y)\otimes s(-x,-y)+\frac{4\pi^2f^2}{k_0^2}|A|^2\delta(x,y),\end{aligned} \tag{5.3-18a}$$

where $\otimes$ denotes correlation. This is the zeroth-order beam located at $x=0, y=0$ on the back focal plane.

b) 倒像

$$\psi_2(x,y)=\mathcal{F}\left\{S\left(\frac{k_0x}{f},\frac{k_0y}{f}\right)\times A^*\mathrm{e}^{\mathrm{j}\frac{k_0x_0x}{f}}\right\}\bigg|_{k_x=k_0x/f,\,k_y=k_0y/f}$$
$$=\frac{4\pi^2f^2}{k_0^2}A^*s(-x-x_0,-y). \tag{5.3-18b}$$

该倒立的重建像位于 $x=-x_0,\ y=0$ 处.

c) 共轭像

$$\psi_3(x,y)=\mathcal{F}\left\{S^*\left(\frac{k_0x}{f},\frac{k_0y}{f}\right)\times A\mathrm{e}^{-\mathrm{j}\frac{k_0x_0x}{f}}\right\}\bigg|_{k_x=k_0x/f,\,k_y=k_0y/f}$$
$$=\frac{4\pi^2f^2}{k_0^2}As^*(x-x_0,y). \tag{5.3-18c}$$

该共轭像的位置在 $x=x_0,\ y=0$ 处.

图 5.3-4 给出了输入图像 $s(x,y)$、其傅里叶变换后的全息图 $t_{FH}(x,y)$ 以及傅里叶全息图的重建像.

a)

b)

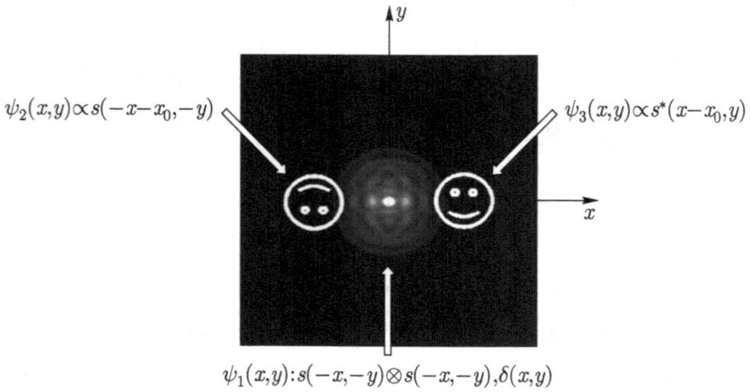

c)

图 5.3-4 a) 输入图像; b) 图 a) 的傅里叶变换后的全息图; c) 图 b) 中傅里叶全息图的重建像 [为了显示出条纹对比度, 从 MATLAB 输出的图 b) 中的结果被放大了]

5.3.3 像全息图

记录

一幅像全息图 (*image hologram*) 可以用如图 5.3-5a) 所示的光路来记录, 其中物体的实像通过成像透镜被记录在介质上.

b) Inverted Image

$$\psi_2(x,y) = \mathcal{F}\left\{ S\left(\frac{k_0 x}{f}, \frac{k_0 y}{f}\right) \times A^* \mathrm{e}^{\mathrm{j}\frac{k_0 x_0 x}{f}} \right\}\Bigg|_{k_x = k_0 x/f,\, k_y = k_0 y/f}$$

$$= \frac{4\pi^2 f^2}{k_0^2} A^* s(-x - x_0, -y). \tag{5.3-18b}$$

This is an inverted reconstructed image at $x = -x_0, y = 0$.

c) Conjugate Image

$$\psi_3(x,y) = \mathcal{F}\left\{ S^*\left(\frac{k_0 x}{f}, \frac{k_0 y}{f}\right) \times A \mathrm{e}^{-\mathrm{j}\frac{k_0 x_0 x}{f}} \right\}\Bigg|_{k_x = k_0 x/f,\, k_y = k_0 y/f}$$

$$= \frac{4\pi^2 f^2}{k_0^2} A s^*(x - x_0, y). \tag{5.3-18c}$$

This is the conjugate image at $x = x_0,\ y = 0$.

Figure 5.3-4 illustrates the input image $s(x, y)$, its Fourier transform hologram $t_{FH}(x, y)$, and the reconstruction of the transform hologram.

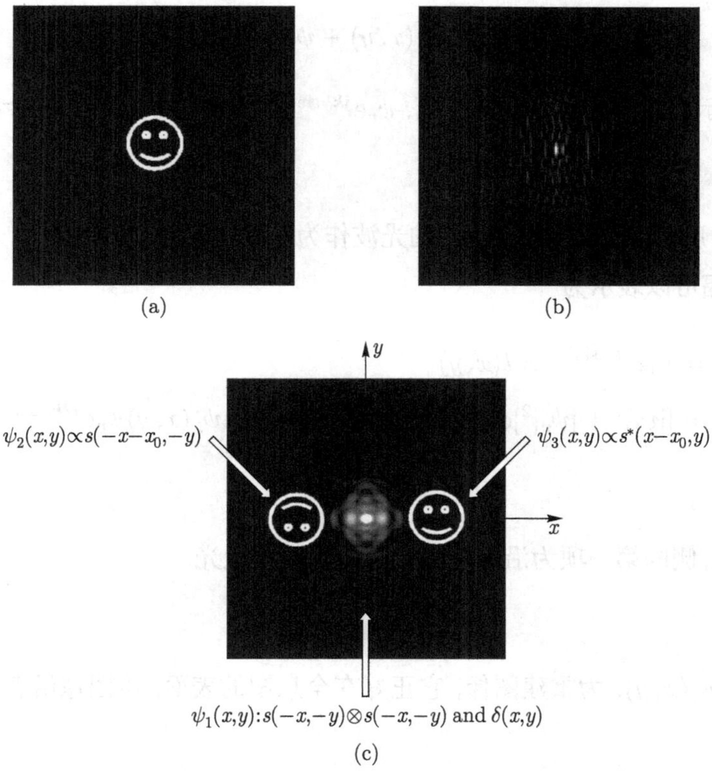

Figure 5.3-4 (a) Input image, (b) Fourier transform hologram of (a), and (c) reconstruction of the Fourier transform hologram in (b). Note that the image in (b) has been zoomed in from the MATLAB's output to show fringe contrast

5.3.3 Image Hologram

Recording

An *image hologram* can be recorded as shown in Figure 5.3-5a, where a real image of the object is formed on the recording medium by an imaging lens.

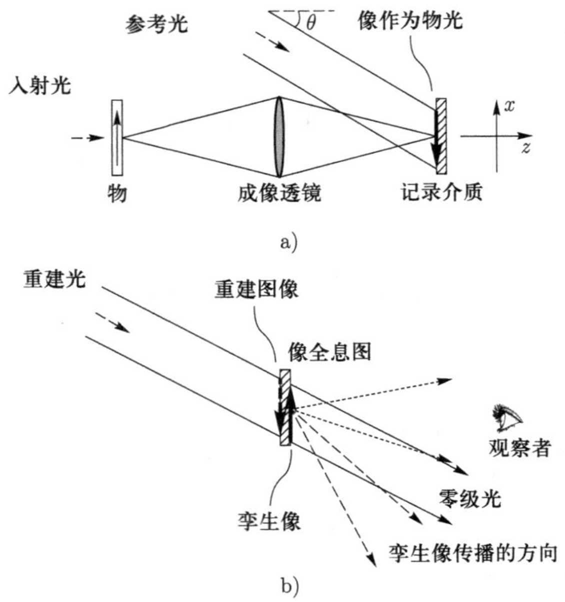

图 5.3-5　a) 像全息图记录的示意图; b) 像全息图重建的示意图

实像被作为物光波记录时, 像全息图由下式给出

$$t(x,y)=|\psi_i(x,y)+\psi_r \mathrm{e}^{\mathrm{j}k_0\sin\theta\, x}|^2, \tag{5.3-19}$$

其中, $\psi_i(x,y)$ 表示全息图上实像的复振幅, $\psi_r \mathrm{e}^{\mathrm{j}k_0\sin\theta\, x}$ 为全息图上的离轴参考平面波.

重建

如图 5.3-5b) 所示, 取与参考光相同的光波作为重建光 ψ_{rec}, 因此, $\psi_{rec}=\psi_r \mathrm{e}^{\mathrm{j}k_0\sin\theta\, x}$, 全息图之后的复振幅可以表示为

$$\begin{aligned}\psi_{rec}t(x,y)&=\psi_r \mathrm{e}^{\mathrm{j}k_0\sin\theta\, x}\times t(x,y)\\&=[|\psi_i|^2+|\psi_r|^2]\mathrm{e}^{\mathrm{j}k_0\sin\theta\, x}+\psi_i(x,y)\psi_r^*+\psi_i^*(x,y)\psi_r \mathrm{e}^{\mathrm{j}2k_0\sin\theta\, x}.\end{aligned} \tag{5.3-20}$$

a) 零级光

式 (5.3-20) 右侧的第一项为沿重建光方向传播的零级光.

b) 重建图像

第二项包含 $\psi_i(x,y)$, 为重建图像, 它正好在全息图的表面, 该图像沿着 z 方向远离全息图传播.

c) 共轭像 (孪生像)

第三项为孪生像, 它也重建在全息面上, 但其传播方向为沿着离开 z 轴 2θ 的方向. 如图 5.3-5b) 所示, 由于孪生像也在全息面上重建, 因此需要离轴记录才能获得像全息图, 则重建像可以通过垂直于全息图的方向来观察.

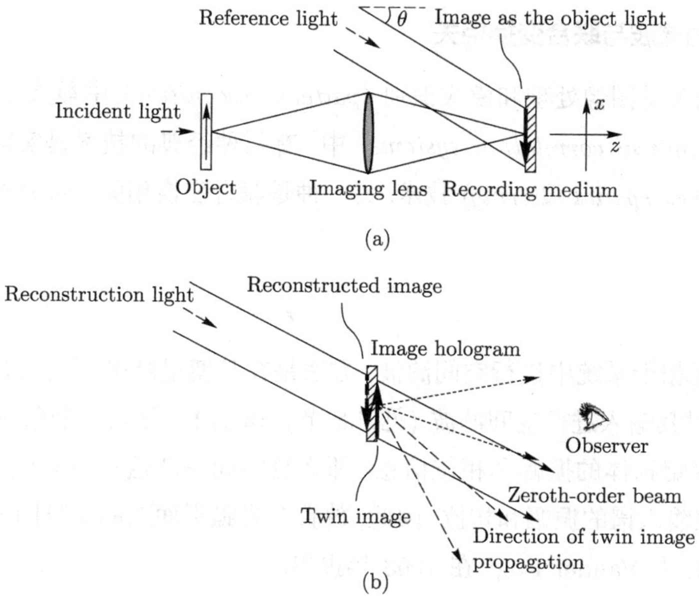

(a)

(b)

Figure 5.3-5 (a) Recording geometry and (b) reconstruction geometry for image hologram

The real image serves as the object wave. Upon recording, the image hologram is given by

$$t(x,y) = |\psi_i(x,y) + \psi_r \mathrm{e}^{\mathrm{j}k_0 \sin\theta\, x}|^2, \tag{5.3-19}$$

where $\psi_i(x,y)$ represents the complex field of the real image on the hologram and $\psi_r \mathrm{e}^{\mathrm{j}k_0 \sin\theta\, x}$ is the off-axis reference plane wave on the hologram.

Reconstruction

Let us take the reconstruction wave ψ_{rec} the same as the reference wave as shown in Figure 5.3-5b. Hence, $\psi_{rec} = \psi_r \mathrm{e}^{\mathrm{j}k_0 \sin\theta\, x}$, and the complex field just behind the hologram can be expressed as

$$\begin{aligned}\psi_{rec} t(x,y) &= \psi_r \mathrm{e}^{\mathrm{j}k_0 \sin\theta\, x} \times t(x,y) \\ &= [|\psi_i|^2 + |\psi_r|^2]\mathrm{e}^{\mathrm{j}k_0 \sin\theta\, x} + \psi_i(x,y)\psi_r^* + \psi_i^*(x,y)\psi_r \mathrm{e}^{\mathrm{j}2k_0 \sin\theta\, x}.\end{aligned} \tag{5.3-20}$$

a) Zeroth-Order Beam

The first term of the right hand side of Eq. (5.3-20) is the zeroth-order beam propagating along the direction of the reconstruction wave.

b) Reconstructed Image

The second term contains $\psi_i(x,y)$ and is the reconstructed image. It is right on the surface of the hologram. The image propagates away from the hologram along the z-direction.

c) Conjugate Image (Twin Image)

The third term is the twin image, which is reconstructed also on the hologram plane but propagating along the direction that is 2θ away from the z-axis. Since the twin image is also reconstructed on the hologram plane as shown in Figure 5.3-5b, off-axis recording is necessary in obtaining the image hologram, and the reconstructed image can be observed normal to the hologram.

5.3.4 复空间滤波与联合变换相关

两幅图像的相关是图像处理和模式识别 (*pattern recognition*) 中最重要的数学运算之一. 在光学相关系统 (*optical correlation systems*) 中, 有两种经典的技术是实际可用的: 一种是复空间滤波 (*complex spatial filtering*) 技术, 另一种是联合变换相关 (*joint-transform correlation*) 技术.

复空间滤波

在 4-f 双透镜相干系统中进行空间滤波, 方法是在其傅里叶平面上放置一个空间滤波器 (*spatial filter*) 来对其输入进行空间滤波 (见 4.2 节). 事实上, 可以用全息图作为空间滤波器, 由于全息图包含给定物体的振幅和相位信息, 那么复空间滤波这一概念就变得清晰起来, 因为全息图可以修改输入谱的振幅和相位. 因此, 放置在光瞳平面上的全息图就被称为复数滤波器 (*complex filter*), 由 Vander Lugt 在 1963 年提出.

将式 (5.3-17) 所给出的 $s(x,y)$ 的傅里叶全息图插入到图 5.3-6 所示的双透镜系统的傅里叶平面上.

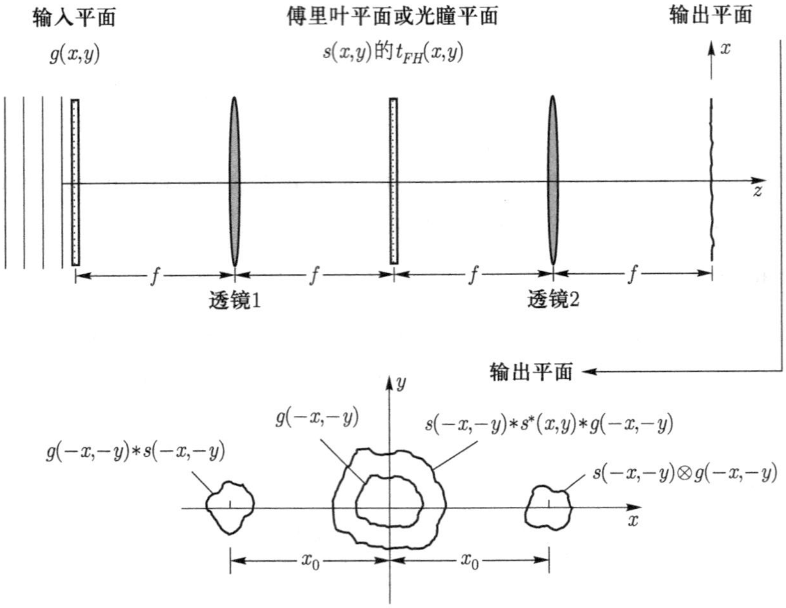

图 5.3-6 复空间滤波

现在, 将一个透射系数为 $g(x,y)$ 的透明片放置在透镜 1 的前焦面处, 并用一个单位振幅的平面波照射. 在复数滤波器之后的复振幅, 即全息图立刻变为 $G\left(\frac{k_0x}{f},\frac{k_0y}{f}\right)\times t_{FH}(x,y)$, 其中, $G\left(\frac{k_0x}{f},\frac{k_0y}{f}\right)$ 为 $g(x,y)$ 在 4-f 系统中傅里叶平面上的频谱. 因此, 在复数滤波后的输出平面, 即透镜 2 的后焦面上, 其复振幅为

$$\begin{aligned}\psi_p(x,y)&\propto\mathcal{F}\left\{G\left(\frac{k_0x}{f},\frac{k_0y}{f}\right)\times t_{FH}(x,y)\right\}\bigg|_{k_x=k_0x/f,\,k_y=k_0y/f}\\&=\mathcal{F}\bigg\{G\left(\frac{k_0x}{f},\frac{k_0y}{f}\right)\bigg[\left|S\left(\frac{k_0x}{f},\frac{k_0y}{f}\right)\right|^2+|A|^2+S\left(\frac{k_0x}{f},\frac{k_0y}{f}\right)\times A^*\mathrm{e}^{\mathrm{j}k_0x_0x/f}\\&\quad+S^*\left(\frac{k_0x}{f},\frac{k_0y}{f}\right)\times A\mathrm{e}^{-\mathrm{j}k_0x_0x/f}\bigg]\bigg\}\bigg|_{k_x=k_0x/f,\,k_y=k_0y/f}.\end{aligned}\tag{5.3-21}$$

5.3.4 Complex Spatial Filtering and Joint-Transform Correlation

The correlation of two images is one of the most important mathematical operations in image processing and *pattern recognition.* In *optical correlation systems,* there are two classical techniques available in practice: one is to utilize *complex spatial filtering* and the other is to utilize *joint-transform correlation.*

Complex Spatial Filtering

We perform spatial filtering in the 4-f two-lens coherent system by putting a *spatial filter* in the Fourier plane to perform spatial filtering of the input (see Chapter 4, Section 4.2). Indeed, we could use holograms as spatial filters. Since a hologram contains the amplitude and phase information of a given object, the notion of *complex spatial filtering* becomes clear as the hologram can modify the amplitude and phase of the input spectrum. For this reason, the hologram placed on the pupil plane is called a *complex filter*, introduced by Vander Lugt in 1963.

Let us insert the Fourier hologram of $s(x, y)$, given by Eq. (5.3-17), on the Fourier plane of the two-lens system shown in Figure 5.3-6.

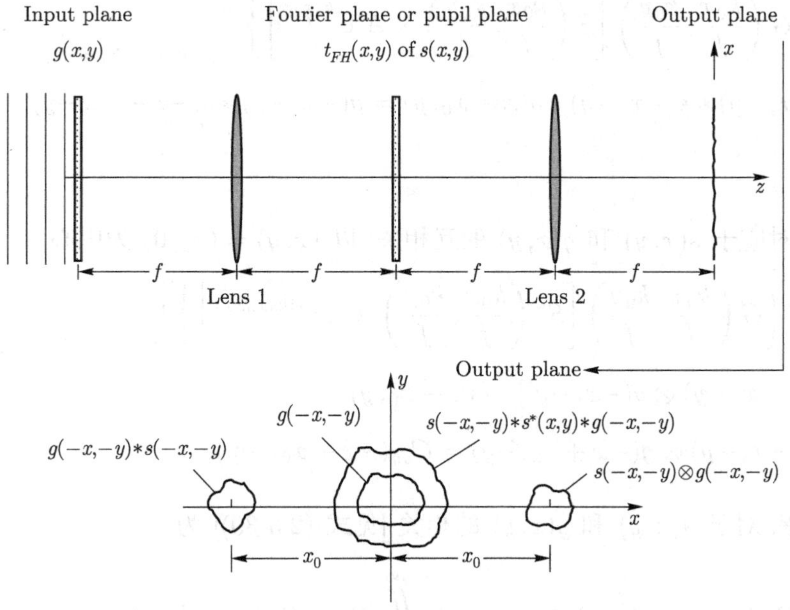

Figure 5.3-6 Complex spatial filtering

Now, we let a transparency $g(x, y)$ be placed in the front focal plane of lens 1 and illuminated by a plane wave of unit amplitude. Then the complex field immediately after the complex filter, that is, the hologram, is $G\left(\frac{k_0 x}{f}, \frac{k_0 y}{f}\right) \times t_{FH}(x, y)$, where $G\left(\frac{k_0 x}{f}, \frac{k_0 y}{f}\right)$ is the spectrum of $g(x, y)$ on the Fourier plane of the 4-f system. Hence, on the complex filtered output plane, that is, the back focal plane of lens 2, the complex field is

$$\psi_p(x, y) \propto \mathcal{F}\left\{G\left(\frac{k_0 x}{f}, \frac{k_0 y}{f}\right) \times t_{FH}(x, y)\right\}\bigg|_{k_x = k_0 x/f,\, k_y = k_0 y/f}$$

$$= \mathcal{F}\left\{G\left(\frac{k_0 x}{f}, \frac{k_0 y}{f}\right)\left[\left|S\left(\frac{k_0 x}{f}, \frac{k_0 y}{f}\right)\right|^2 + |A|^2 + S\left(\frac{k_0 x}{f}, \frac{k_0 y}{f}\right) \times A^* \mathrm{e}^{\mathrm{j} k_0 x_0 x/f}\right.\right.$$

$$\left.\left. + S^*\left(\frac{k_0 x}{f}, \frac{k_0 y}{f}\right) \times A \mathrm{e}^{-\mathrm{j} k_0 x_0 x/f}\right]\right\}\bigg|_{k_x = k_0 x/f,\, k_y = k_0 y/f} . \qquad (5.3\text{-}21)$$

接下来分别看一下全息图中每一项的作用.

a) 零级光

前两项在复振幅中的作用为

$$\mathcal{F}\left\{G\left(\frac{k_0x}{f},\frac{k_0y}{f}\right)\left|S\left(\frac{k_0x}{f},\frac{k_0y}{f}\right)\right|^2\right\}\Bigg|_{k_x=k_0x/f,\,k_y=k_0y/f}$$
$$\propto s(-x,-y)*s^*(x,y)*g(-x,-y), \tag{5.3-22a}$$

及

$$\mathcal{F}\left\{G\left(\frac{k_0x}{f},\frac{k_0y}{f}\right)|A|^2\right\}\Bigg|_{k_x=k_0x/f,\,k_y=k_0y/f}\propto g(-x,-y), \tag{5.3-22b}$$

两个场都以输出平面的原点为中心.

b) 卷积

第三项对应于 $g(x,y)$ 和 $s(x,y)$ 的卷积, 以 $(x,y)=(-x_0,0)$ 为中心

$$\mathcal{F}\left\{G\left(\frac{k_0x}{f},\frac{k_0y}{f}\right)\left[S\left(\frac{k_0x}{f},\frac{k_0y}{f}\right)\times A^*\mathrm{e}^{\frac{\mathrm{j}k_0x_0x}{f}}\right]\right\}\Bigg|_{k_x=k_0x/f,\,k_y=k_0y/f}$$
$$\propto g(-x,-y)*s(-x,-y)*\delta(x+x_0,y)=g(-x,-y)*s(-x-x_0,-y). \tag{5.3-23}$$

c) 相关

最后一项对应于 $s(x,y)$ 和 $g(x,y)$ 的互相关, 以 $(x,y)=(x_0,0)$ 为中心

$$\mathcal{F}\left\{G\left(\frac{k_0x}{f},\frac{k_0y}{f}\right)\left[S^*\left(\frac{k_0x}{f},\frac{k_0y}{f}\right)\times A\mathrm{e}^{-\mathrm{j}k_0x_0x/f}\right]\right\}\Bigg|_{k_x=k_0x/f,\,k_y=k_0y/f}$$
$$\propto [s(-x,-y)\otimes g(-x,-y)]*\delta(x-x_0,y)$$
$$=s(-x,-y)\otimes g(-x+x_0,-y)=C_{sg}(-x+x_0,\,-y), \tag{5.3-24}$$

这里, 作为提醒, 对于 $s(x,y)$ 和 $g(x,y)$ 的相关 [见式 (2.3-30)] 为

$$C_{sg}(x,y)=s(x,y)\otimes g(x,y)=\iint_{-\infty}^{\infty}s^*(x',y')g(x+x',y+y')\mathrm{d}x'\mathrm{d}y'.$$

在图 5.3-6 的下半个图中, 我们总结了这些项在输出平面上的位置. 可以看出, 当两个函数进行相关运算时, 复空间滤波 (*complex spatial filtering*) 也被称为第二章介绍过的匹配滤波 (*matched filtering*).

MATLAB 实例: 复空间滤波

本例给出了一些关于复空间滤波的 MATLAB 结果, 如图 5.3-7 所示. 首先, 得到由式 (5.3-17) 给出的 $s(x,y)$ 的傅里叶全息图, 接着为了复数滤波, 根据式 (5.3-21), 对输入 $g(x,y)$ 进行傅里叶变换, 以提供 4-f 相干系统的输出. 在复数滤波的输出上, 我们期望零级光在输出坐标的原点处. 零级光的两侧是 $s(x,y)$ 和 $g(x,y)$ 的卷积和相关. 图 5.3-7a) 给出了 $s(x,y)=g(x,y)$,

Let us look at the contributions due to each term in the hologram separately.

a) Zeroth-Order Beam

The first two terms contribute to complex fields of

$$\mathcal{F}\left\{G\left(\frac{k_0x}{f},\frac{k_0y}{f}\right)\left|S\left(\frac{k_0x}{f},\frac{k_0y}{f}\right)\right|^2\right\}\Bigg|_{k_x=k_0x/f,\,k_y=k_0y/f}$$
$$\propto s(-x,-y)*s^*(x,y)*g(-x,-y), \tag{5.3-22a}$$

and

$$\mathcal{F}\left\{G\left(\frac{k_0x}{f},\frac{k_0y}{f}\right)|A|^2\right\}\Bigg|_{k_x=k_0x/f,\,k_y=k_0y/f} \propto g(-x,-y), \tag{5.3-22b}$$

Both of the fields are centered on the origin of the output plane.

b) Convolution

The third term corresponds to the convolution of $g(x,y)$ and $s(x,y)$, centered at $(x,y) = (-x_0,0)$:

$$\mathcal{F}\left\{G\left(\frac{k_0x}{f},\frac{k_0y}{f}\right)\left[S\left(\frac{k_0x}{f},\frac{k_0y}{f}\right)\times A^*\mathrm{e}^{\frac{\mathrm{j}k_0x_0x}{f}}\right]\right\}\Bigg|_{k_x=k_0x/f,\,k_y=k_0y/f}$$
$$\propto g(-x,-y)*s(-x,-y)*\delta(x+x_0,y) = g(-x,-y)*s(-x-x_0,-y). \tag{5.3-23}$$

c) Correlation

The last term corresponds to the cross-correlation of $s(x,y)$ and $g(x,y)$, centered at $(x,y) = (x_0,0)$.

$$\mathcal{F}\left\{G\left(\frac{k_0x}{f},\frac{k_0y}{f}\right)\left[S^*\left(\frac{k_0x}{f},\frac{k_0y}{f}\right)\times A\mathrm{e}^{-\mathrm{j}k_0x_0x/f}\right]\right\}\Bigg|_{k_x=k_0x/f,\,k_y=k_0y/f}$$
$$\propto [s(-x,-y)\otimes g(-x,-y)]*\delta(x-x_0,y)$$
$$= s(-x,-y)\otimes g(-x+x_0,-y) = C_{sg}(-x+x_0,\,-y), \tag{5.3-24}$$

where, as a reminder, the correlation of $s(x,y)$ and $g(x,y)$ is [see Eq. (2.3-30)]

$$C_{sg}(x,y) = s(x,y)\otimes g(x,y) = \iint_{-\infty}^{\infty} s^*(x',y')g(x+x',y+y')\mathrm{d}x'\mathrm{d}y'.$$

On the bottom of Figure 5.3-6, we summarize the locations of these terms on the output plane. Note that *complex spatial filtering* is also called *matched filtering* introduced in Chapter 2, as correlation of two functions are performed.

MATLAB Example: Complex Filtering

In this example, we illustrate some MATLAB results on complex spatial filtering. The results are shown in Figure 5.3-7. We first obtain the Fourier transform hologram of $s(x,y)$, given by Eq. (5.3-17). Subsequently an input $g(x,y)$ is Fourier transformed for complex filtering, according to Eq. (5.3-21), to provide the output of the 4-f coherent system. On the output of complex filtering, we expect to have the zeroth-order beam on the origin of the output coordinates. On both side of the zeroth-order beam, there are convolution and correlation of $s(x,y)$ and $g(x,y)$. In Figure 5.3-7a, we show $s(x,y) = g(x,y)$, and we clearly see the

在图 5.3-7d) 的复数滤波输出中, 可以清楚地看到零级光左侧的卷积结果和右侧的自相关结果. 记住, 自相关应该给出一个类似峰值样的输出, 这也是我们在结果中所看到的. 在这种情况下, 卷积也显示了一些类似峰值样的输出, 这是因为 $s(x,y)$ 和 $g(x,y)$ 非常对称, 故两个图像的外环应该给出一个相当强的卷积结果. 另一方面, 在图 5.3-7e) 中, 图 5.3-7b) 中的 $s(x,y)$ 和 $g(x,y)$ 的卷积和互相关结果皆没有清晰的亮点. 图 5.3-7c) 给出了 $s(x,y)$ 和 $g(x,y)$ 完全相同的随机图案 (*random pattern*). 显然, 在图 5.3-7f) 所示的复数滤波后的结果中, 其卷积的结果显示输出为一个散射的光斑, 而自相关的结果显示为一个清晰的点. 其实, 在比较两幅图像的相似性时, 相关是有用的, 且已利用复空间滤波实现, 并在光学模式识别 (*optical pattern recognition*) 中得以应用.

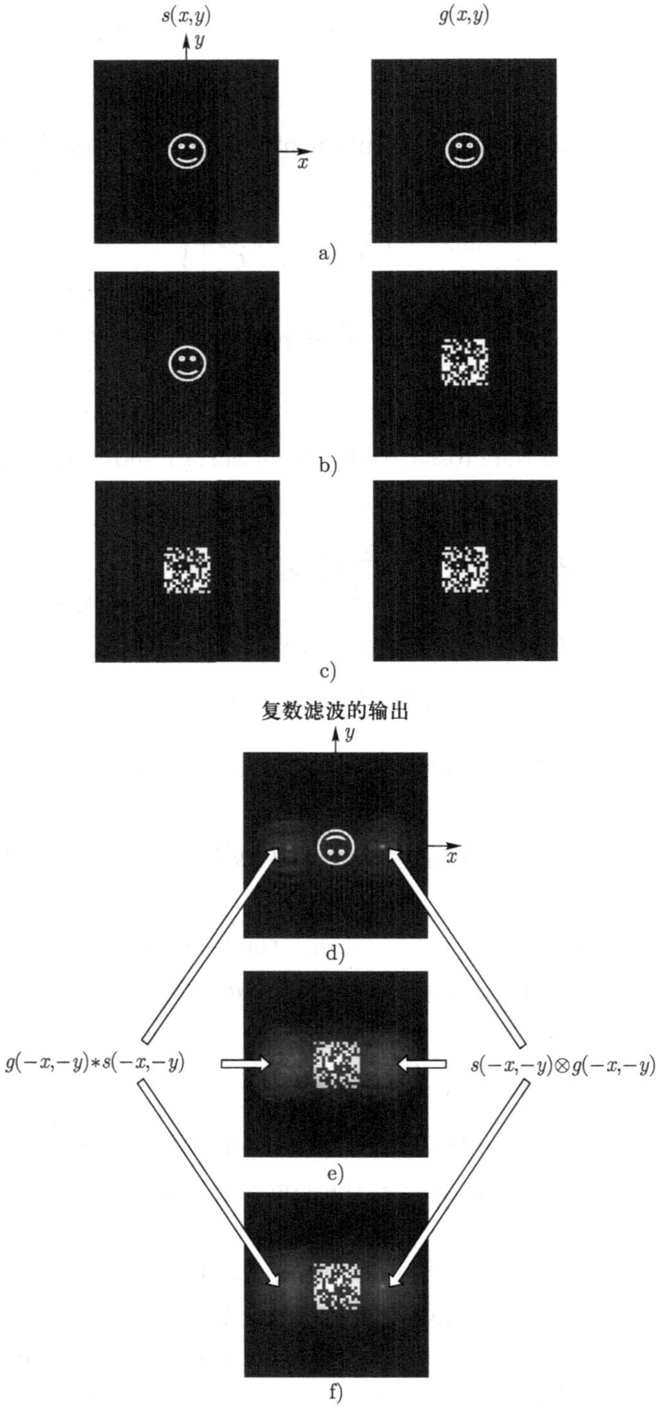

图 5.3-7 复空间滤波例子

convolution result on the left side of the zeroth-order beam with the auto-correlation result on the right side, as shown in the output of complex filtering shown in Figure 5.3-7d. Remember that auto-correlation should give a peak-like output and that is what we see in the result. In this case, convolution also shows some peak-like output, that is because $s(x, y)$ and $g(x, y)$ are very symmetrical, as the outer ring of the two images should give a fairly strong convolution response. On the other hand, Figure 5.3-7e shows no distinct spots for convolution and cross-correlation of $s(x, y)$ and $g(x, y)$ given by Figure 5.3-7b. Finally, Figure 5.3-7c shows identical *random patterns* for $s(x, y)$ and $g(x, y)$. Clearly, in the output of complex filtering shown in Figure 5.3-7f, the convolution result shows a diffused patch of light as output, while the autocorrelation result shows a distinct spot. Indeed, correlation is useful when comparing the similarity of two images, and it has been implemented through complex spatial filtering and used for *optical pattern recognition* applications.

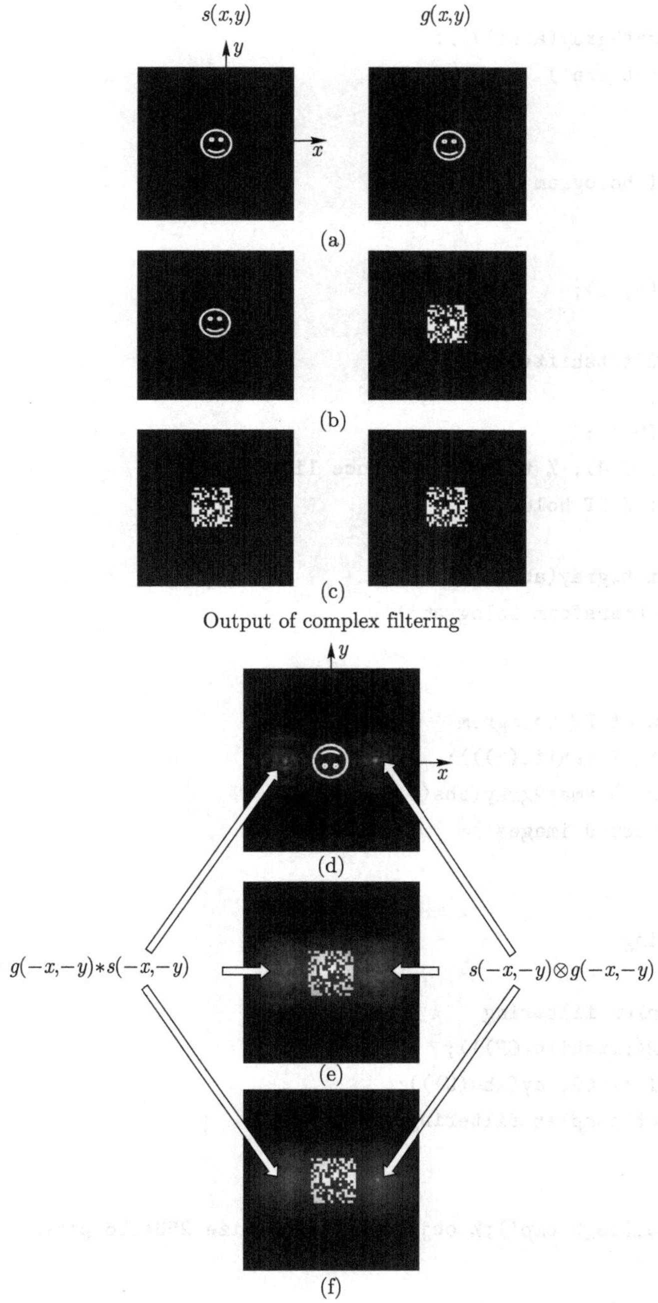

Figure 5.3-7 Complex spatial filtering examples

图 5.3-4 和图 5.3-7 中的结果利用以下 m-文件生成.

==

```
% FT_hologram_complex_filtering2
% Adapted from "Introduction to Modern Digital Holography"
% by T.-C. Poon & J.-P. Liu
% Cambridge University Press (2014), Table 3.3.

clear all, close all;
Ii=imread('smiley3.bmp');% object pattern, size 256*256 pixels
Ii=double(Ii);
M=512;
I=zeros(512);
I(128:383,128:383)=Ii;  % zero-padding

figure; imshow(mat2gray(abs(I)));
%title('Object pattern')
axis off

% Produce the FT hologram
r=1:M;
c=1:M;
[C, R]=meshgrid(c, r);

O=fftshift(ifft2(fftshift(I)));
R=ones(512,512);
R=R*max(max(abs(O)));
R=R.*exp(-2i*pi.*C/4); % tilted reference light
H=(abs(O+R)).^2; % FT hologram

figure; imshow(mat2gray(abs(H)));
%title('Fourier transform hologram')
axis off

% Reconstruction of FT hologram
 U=fftshift(ifft2(fftshift(H)));
figure; imshow(9000.*mat2gray(abs(U)));
%title('Reconstructed images')
axis off

%complex filtering
CF=O.*H;
% Output of complex filtering
U=fftshift(ifft2(fftshift(CF)));
figure; imshow(1.*mat2gray(abs(U)));
%title('Output of complex filtering')
axis off

Ii=imread('random256_2.bmp');% object pattern, size 256*256 pixels
Ii=double(Ii);

M=512;
```

The results obtained in Figures 5.3-4 and 5.3-7 use the m-file below.

==

```
% FT_hologram_complex_filtering2
% Adapted from "Introduction to Modern Digital Holography"
% by T.-C. Poon & J.-P. Liu
% Cambridge University Press (2014), Table 3.3.

clear all, close all;
Ii=imread('smiley3.bmp');% object pattern, size 256*256 pixels
Ii=double(Ii);
M=512;
I=zeros(512);
I(128:383,128:383)=Ii;  % zero-padding

figure; imshow(mat2gray(abs(I)));
%title('Object pattern')
axis off

% Produce the FT hologram
r=1:M;
c=1:M;
[C, R]=meshgrid(c, r);

O=fftshift(ifft2(fftshift(I)));
R=ones(512,512);
R=R*max(max(abs(O)));
R=R.*exp(-2i*pi.*C/4); % tilted reference light
H=(abs(O+R)).^2; % FT hologram

figure; imshow(mat2gray(abs(H)));
%title('Fourier transform hologram')
axis off

% Reconstruction of FT hologram
 U=fftshift(ifft2(fftshift(H)));
figure; imshow(9000.*mat2gray(abs(U)));
%title('Reconstructed images')
axis off

%complex filtering
CF=O.*H;
% Output of complex filtering
U=fftshift(ifft2(fftshift(CF)));
figure; imshow(1.*mat2gray(abs(U)));
%title('Output of complex filtering')
axis off

Ii=imread('random256_2.bmp');% object pattern, size 256*256 pixels
Ii=double(Ii);

M=512;
```

```
I=zeros(512);
I(128:383,128:383)=Ii;  % zero-padding

figure; imshow(mat2gray(abs(I)));
%title('Object pattern2')
axis off

% Produce FT of another image, O2
r=1:M;
c=1:M;
[C, R]=meshgrid(c, r);

O2=fftshift(ifft2(fftshift(I)));
%complex filtering 2
CF=H.*O2;
% Output of complex filtering
U=fftshift(ifft2(fftshift(CF)));
figure; imshow(1.*mat2gray(abs(U)));
%title('Output of complex filtering')
axis off
========================================
```

联合变换相关 (JTC)

联合变换相关 (*joint-transform correlation*, JTC) 是由 Weaver 和 Goodman 在 20 世纪 60 年代首次提出的. 图 5.3-8 所示为一个典型的 JTC 系统.

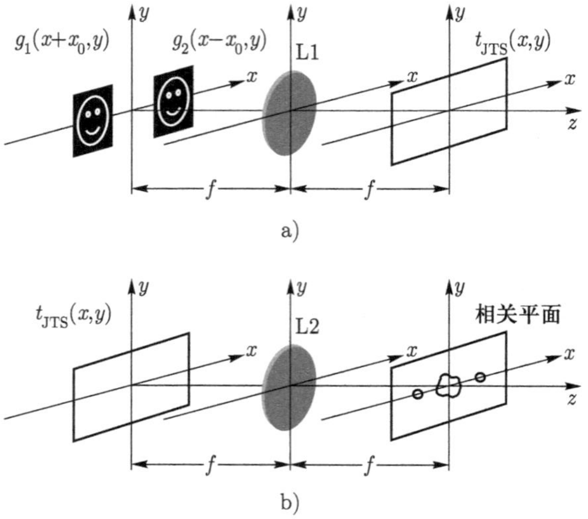

图 5.3-8 联合变换相关系统: a) 联合变换频谱记录; b) 各种相关输出显示

$g_1(x,y)$ 和 $g_2(x,y)$ 为要进行相关操作的两幅图, 且它们以透明片的形式存在, 由平面波照射. 两幅图在焦距为 f 的傅里叶变换透镜 L1 的前焦面上且间距为 $2x_0$, 如图 5.3-8a) 所示. 两幅图的频谱由记录介质共同记录下来, 由此记录下来的强度被称为联合变换频谱 (*joint-transform spectrum*, JTS), 所产生的透射系数为

$$t_{JTS}(x,y) = |\mathcal{F}\{g_1(x+x_0,y)\} + \mathcal{F}\{g_2(x-x_0,y)\}|^2$$

```
I=zeros(512);
I(128:383,128:383)=Ii;  % zero-padding

figure; imshow(mat2gray(abs(I)));
%title('Object pattern2')
axis off

% Produce FT of another image, O2
r=1:M;
c=1:M;
[C, R]=meshgrid(c, r);

O2=fftshift(ifft2(fftshift(I)));
%complex filtering 2
CF=H.*O2;
% Output of complex filtering
U=fftshift(ifft2(fftshift(CF)));
figure; imshow(1.*mat2gray(abs(U)));
%title('Output of complex filtering')
axis off
==========================================
```

Joint-Transform Correlation (JTC)

The *joint-transform correlation* (JTC) was first proposed by Weaver and Goodman in the 1960s. A typical JTC system is shown in Figure 5.3-8.

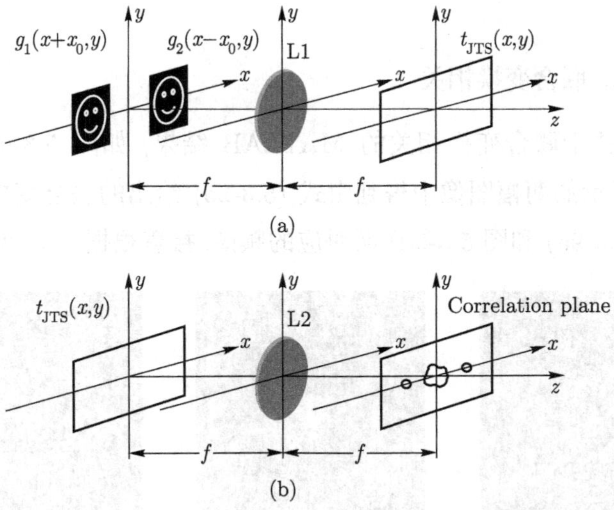

Figure 5.3-8 Joint-transform correlation: (a) recording the joint-transform spectrum and (b) display the various correlation outputs

$g_1(x, y)$ and $g_2(x, y)$ are the two images to be correlated, and they are in the form of transparencies, which are illuminated by plane waves. The two images are separated by a distance of $2x_0$ in the front focal plane of Fourier transform lens L1 of focal length f, as shown in Figure 5.3-8a. The spectra of the two images are jointly recorded by a recording medium. The resulting recorded intensity, which is called the *joint-transform spectrum* (JTS), yields a transparency given by

$$t_{\mathrm{JTS}}(x, y) = |\mathcal{F}\{g_1(x + x_0, y)\} + \mathcal{F}\{g_2(x - x_0, y)\}|^2$$

$$= |G_1(k_x,k_y)|^2 + |G_2(k_x,k_y)|^2 + G_1^*(k_x,k_y)G_2(k_x,k_y)\mathrm{e}^{\mathrm{j}2k_x x_0} + G_1(k_x,k_y)G_2^*(k_x,k_y)\mathrm{e}^{-\mathrm{j}2k_x x_0}. \tag{5.3-25}$$

其中，$G_1(k_x,k_y)$ 和 $G_2(k_x,k_y)$ 分别为 $g_1(x,y)$ 和 $g_2(x,y)$ 的傅里叶变换，$k_x = k_0x/f$ 及 $k_y = k_0y/f$，k_0 为照明平面波的波数. 一旦 JTS 被记录，即可将 $t_{\mathrm{JTS}}(x,y)$ 放置在傅里叶变换透镜 L2 的前焦面处，如图 5.3-8b) 所示，则透镜的后焦面即为其相关平面 (*correlation plane*)，因为它包含了所有相关的输出. 至此，就得到了傅里叶变换透镜后焦面上的光场分布 (见习题 5.8)

$$\begin{aligned}
&\mathcal{F}\{t_{\mathrm{JTS}}(x,y)\}\big|_{k_x=k_0x/f,\,k_y=k_0y/f}\\
&=\mathcal{F}\left\{\left|G_1\left(\frac{k_0x}{f},\frac{k_0y}{f}\right)\right|^2+\left|G_2\left(\frac{k_0x}{f},\frac{k_0y}{f}\right)\right|^2+G_1^*\left(\frac{k_0x}{f},\frac{k_0y}{f}\right)G_2\left(\frac{k_0x}{f},\frac{k_0y}{f}\right)\mathrm{e}^{\mathrm{j}2\frac{k_0x}{f}x_0}\right.\\
&\quad\left.+G_1\left(\frac{k_0x}{f},\frac{k_0y}{f}\right)G_2^*\left(\frac{k_0x}{f},\frac{k_0y}{f}\right)\mathrm{e}^{-\mathrm{j}2\frac{k_0x}{f}x_0}\right\}\Bigg|_{k_x=k_0x/f,\,k_y=k_0y/f}\\
&=\frac{4\pi^2f^2}{k_0^2}[C_{11}(-x,-y)+C_{22}(-x,-y)+C_{12}(-x-2x_0,-y)+C_{21}(-x+2x_0,-y)],
\end{aligned} \tag{5.3-26}$$

其中

$$C_{mn}(x,y) = g_m(x,y)\otimes g_n(x,y),$$

当 $m=1$ 或 2，$n=1$ 或 2，且当 $m=n$ 时，$C_{mn}(x,y)$ 为自相关. 当 $m\neq n$ 时，$C_{mn}(x,y)$ 为互相关. 因此，如果两幅图像 (或图案) 相同，除了有一个由式 (5.3-26) 的前两项所引起的相关平面原点处的强峰值以外，式 (5.3-26) 的后两项还会有两个中心点分别位于 $x=2x_0$ 和 $x=-2x_0$ 的强峰值.

MATLAB 实例：联合变换相关

本例给出了一些关于联合变换相关的 MATLAB 结果，如图 5.3-9所示. 首先从图 5.3-9a) 和图 5.3-9d) 中所示的两幅图像中得到由式 (5.3-25) 给出的联合变换频谱，图 5.3-9b) 和图 5.3-9e) 分别为图 5.3-9a) 和图 5.3-9d) 所对应的频谱. 接着根据式 (5.3-26) 对频谱进行傅里

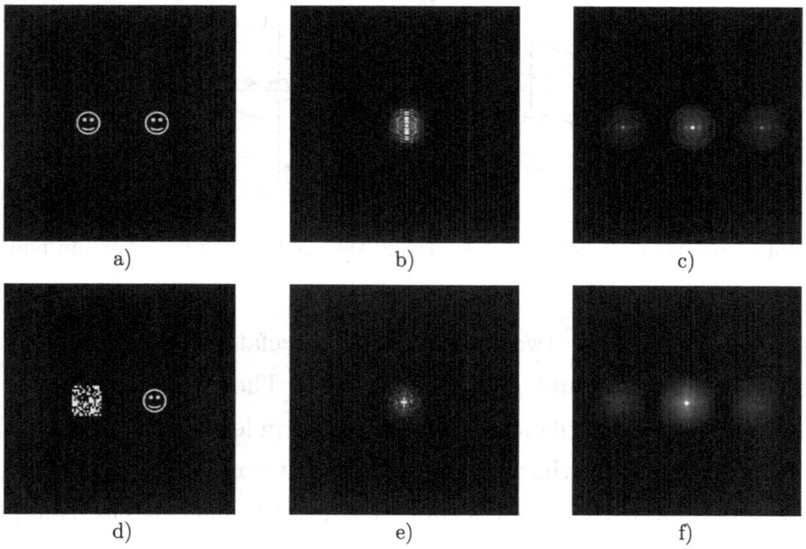

图 5.3-9 联合变换相关结果：a) 和 d) 为图像对；b) 和 e) 分别为图 a) 和图 d) 所对应的联合变换频谱；c) 和 f) 分别为在相关平面上相应于图 a) 和图 d) 中图像对的相关输出结果

$$= |G_1(k_x,k_y)|^2 + |G_2(k_x,k_y)|^2 + G_1^*(k_x,k_y)G_2(k_x,k_y)\mathrm{e}^{\mathrm{j}2k_x x_0} + G_1(k_x,k_y)G_2^*(k_x,k_y)\mathrm{e}^{-\mathrm{j}2k_x x_0}. \quad (5.3\text{-}25)$$

$G_1(k_x,k_y)$ and $G_2(k_x,k_y)$ are the Fourier transforms of $g_1(x,y)$ and $g_2(x,y)$, respectively, with $k_x = k_0x/f$ and $k_y = k_0y/f$, where k_0 is the wavenumber of the illumination plane wave. Once the JTS has been recorded, we place $t_{\mathrm{JTS}}(x,y)$ in the front focal plane of Fourier transform lens L2, as shown in Figure 5.3-8b. The back focal plane of the lens is the *correlation plane* as it contains all the correlation outputs. To see that, we find the field distribution in the back focal plane of the Fourier transform lens (see P.5.8):

$$\mathcal{F}\{t_{\mathrm{JTS}}(x,y)\}\big|_{k_x=k_0x/f,\,k_y=k_0y/f}$$

$$= \mathcal{F}\left\{\left|G_1\left(\frac{k_0x}{f},\frac{k_0y}{f}\right)\right|^2 + \left|G_2\left(\frac{k_0x}{f},\frac{k_0y}{f}\right)\right|^2 + G_1^*\left(\frac{k_0x}{f},\frac{k_0y}{f}\right)G_2\left(\frac{k_0x}{f},\frac{k_0y}{f}\right)\mathrm{e}^{\mathrm{j}2\frac{k_0x}{f}x_0} + G_1\left(\frac{k_0x}{f},\frac{k_0y}{f}\right)G_2^*\left(\frac{k_0x}{f},\frac{k_0y}{f}\right)\mathrm{e}^{-\mathrm{j}2\frac{k_0x}{f}x_0}\right\}\Bigg|_{k_x=k_0x/f,\,k_y=k_0y/f}$$

$$= \frac{4\pi^2f^2}{k_0^2}[C_{11}(-x,-y) + C_{22}(-x,-y) + C_{12}(-x-2x_0,-y) + C_{21}(-x+2x_0,-y)], \quad (5.3\text{-}26)$$

where

$$C_{mn}(x,y) = g_m(x,y) \otimes g_n(x,y),$$

with $m = 1$ or 2, and $n = 1$ or 2. $C_{mn}(x,y)$ is the autocorrelation when $m = n$ and is cross-correlation when $m \neq n$. Hence if the two images (or patterns) are the same, besides a strong peak at the origin of the correlation plane due to the first two terms of Eq. (5.3-26), we have two strong peaks centered at $x = \pm 2x_0$ due to the last two terms of Eq. (5.3-26).

MATLAB Example: Joint-Transform Correlation

In this example, we illustrate some MATLAB results on the joint-transform correlation. The results are shown in Figure 5.3-9. We first obtain the joint-transform spectrum given by Eq. (5.3-25) from the two images shown in Figure 5.3-9a and d. Figure 5.3-9b and e shows

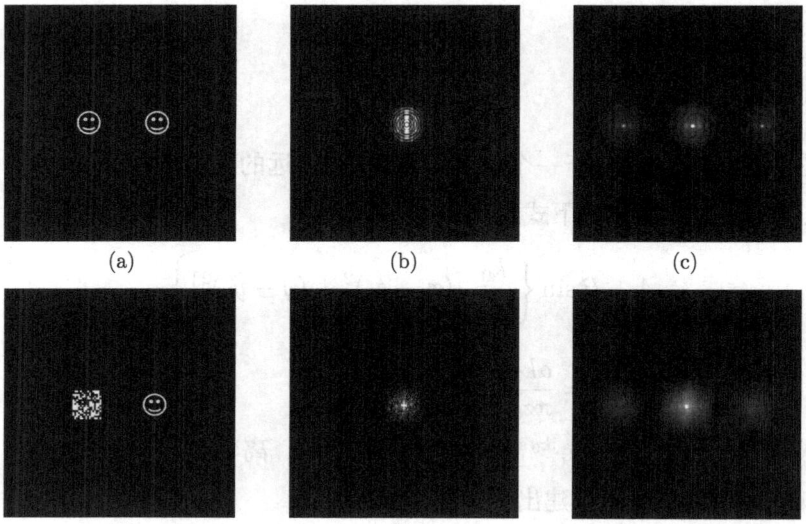

(a) (b) (c) (d) (e) (f)

Figure 5.3-9 Joint-transform correlation results: (a) and (d) are the image pairs; (b) and (e) are the joint-transform spectrum corresponding to (a) and (d), respectively; (c) and (f) show the correlation outputs on the correlation plane corresponding to the image pairs in (a) and (d), respectively

叶变换得到相关后的输出. 图 5.3-9c) 和图 5.3-9f) 分别对应了图 5.3-9a) 和图 5.3-9d) 中图像对的相关输出. 对于匹配的一对, 可以清楚地观察到, 在相关平面原点处的光斑附近有两个强的自相关峰.

上述结果利用以下 m-文件生成.

```
=========================================
% JTC
clear, close all;
Ii=zeros(1024,1024);
A=imread('smiley3.bmp');
A=imresize(A, [1024, 1024]);
A=double(A(:,:,1));
A=imresize(A, 0.25);
A=A./(max(max(A)));
B=imread('random256_2.bmp');
B=imresize(B, [1024, 1024]);
B=double(B);
B=imresize(B, 0.25);
B=B./(max(max(B)));
Ii(385:640,237:492)=A;
Ii(385:640,533:788)=B;
FT=fftshift(fft2(Ii));
CO=abs(FT).*abs(FT);
figure; imshow(mat2gray(Ii)); % two images on front focal plane of lens
figure; imshow(100.*mat2gray(CO)); % Joint-transform spectrum (JTS)
IFT=fftshift(fft2(CO));
figure; imshow(2.*mat2gray(3.*abs(IFT))); % FT of JTC
=========================================
```

习题

5.1 如图 5.1-1a) 所示, 对于一个距离记录介质 z_0 远的离轴点源 $\delta(x-x_0, y-y_0)$, 证明所记录的全息图的透过率函数由下式给出

$$A+B\sin\left\{\frac{k_0}{2z_0}\left[(x-x_0)^2+(y-y_0)^2\right]\right\},$$

其中, $A=a^2+\left(\dfrac{k_0}{2\pi z_0}\right)^2$ 且 $B=\dfrac{ak_0}{\pi z_0}$.

5.2 当被重建光 $\psi_{rec}=a\mathrm{e}^{\mathrm{j}k_0\sin\theta\, x}=\psi_r$ 照明时, 离轴点源全息图中的 $\psi_r\psi_0^*$ 项 [见式 (5.1-14) 和式 (5.1-15b)], 会重建出一个实像. 证明

$$[\psi_{rec}\times\psi_r\psi_0^*]*h(x,y;z=z_0)\propto\delta(x+2z_0\sin\theta, y)$$

可以直接通过卷积进行计算.

the spectrum corresponding to Figure 5.3-9a and d, respectively. Subsequently, the spectrum is Fourier transformed for correlation outputs according to Eq. (5.3-26). Figure 5.3-9c and f shows the correlation outputs corresponding to the image pairs shown in Figure 5.3-9a and d, respectively. For a matched pair, we clearly observe the two strong autocorrelation peaks next to the patch of light at the origin of the correlation plane.

The above results are obtained using the m-file below.

```
===========================================
% JTC
clear, close all;
Ii=zeros(1024,1024);
A=imread('smiley3.bmp');
A=imresize(A, [1024, 1024]);
A=double(A(:,:,1));
A=imresize(A, 0.25);
A=A./(max(max(A)));
B=imread('random256_2.bmp');
B=imresize(B, [1024, 1024]);
B=double(B);
B=imresize(B, 0.25);
B=B./(max(max(B)));
Ii(385:640,237:492)=A;
Ii(385:640,533:788)=B;
FT=fftshift(fft2(Ii));
CO=abs(FT).*abs(FT);
figure; imshow(mat2gray(Ii)); % two images on front focal plane of lens
figure; imshow(100.*mat2gray(CO)); % Joint-transform spectrum (JTS)
IFT=fftshift(fft2(CO));
figure; imshow(2.*mat2gray(3.*abs(IFT))); % FT of JTC
===========================================
```

Problems

5.1 Show that for an off-axis point source $\delta(x - x_0, y - y_0)$ that is z_0 away from the recording medium as shown in Figure 5.1-1a, the recorded hologram has a transparency function given by

$$A + B\sin\left\{\frac{k_0}{2z_0}\left[(x-x_0)^2 + (y-y_0)^2\right]\right\},$$

where $A = a^2 + \left(\dfrac{k_0}{2\pi z_0}\right)^2$ and $B = \dfrac{ak_0}{\pi z_0}$.

5.2 The term $\psi_r\psi_0^*$ in the off-axis point-source hologram [see Eqs. (5.1-14) and (5.1-15b)] is responsible for a real image reconstruction upon illumination by the reconstruction wave $\psi_{rec} = ae^{jk_0\sin\theta\, x} = \psi_r$. Show that

$$[\psi_{rec} \times \psi_r\psi_0^*] * h(x, y; z = z_0) \propto \delta(x + 2z_0\sin\theta, y)$$

by carrying out directly the convolution involved.

5.3　当被重建光 $\psi_{rec} = a\mathrm{e}^{\mathrm{j}k_0 \sin\theta\, x} = \psi_r$ 照明时, 离轴点源全息图中的 $\psi_r^*\psi_0$ 项 [见式 (5.1-14) 和式 (5.1-15c)], 会重建出一个虚像. 证明

$$\psi_{rec} \times \psi_r^*\psi_0 * h(x, y; z = -z_0) \propto \delta(x, y)$$

可以直接通过卷积进行计算.

5.4　一个离轴点源全息图如图 5.1-7 所示记录, 该记录的全息图的透过率函数由式 (5.1-16) 给出, 即

$$t(x, y) = A + B\sin\left[\frac{k_0}{2z_0}(x^2 + y^2) + k_0 \sin\theta\, x\right].$$

该全息图被如图题 5.4 所示的一束平面波来重建.

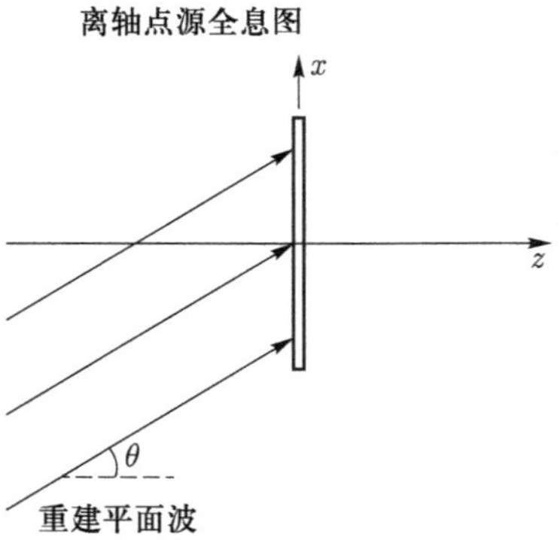

图题 5.4

a) 求出其实像和虚像的位置.

b) 画出类似图 5.1-8 所示的实像和虚像的重建光路图.

5.5　在 5.2 节中, 我们已经利用球面波对两个点源的记录与重建讨论了全息成像.

a) 对于点 1 的实像重建, 有 [见式 (5.2-10)]

$$\psi_{p,\,rel_1}(x, y; z) = \psi_{\mathrm{pr}}(x, y)t_{rel_1}(x, y) * h(x, y; z).$$

通过对上式直接卷积计算, 证明其实像被重建在位置

$$z = z_{\mathrm{r1}} = \left(\frac{1}{R} - \frac{1}{l_1} - \frac{1}{l_2}\right)^{-1}$$

处, 且

$$\psi_{p,\,rel_1}(x, y; z = z_{\mathrm{r1}}) \propto \delta\left[x + z_{\mathrm{r1}}\left(\frac{b}{l_2} - \frac{h}{2R} - \frac{a}{l_1}\right), y\right].$$

b) 对于点 2 的实像重建, 有 [见式 (5.2-13)]

$$\psi_{p,\,rel_2}(x, y; z) = \psi_{\mathrm{pr}}(x, y)t_{rel_2}(x, y) * h(x, y; z).$$

通过对上式直接卷积计算, 证明其实像被重建在位置

$$z = z_{\mathrm{r2}} = \left(\frac{1}{R + d} - \frac{1}{l_1} - \frac{1}{l_2}\right)^{-1}$$

5.3 The term $\psi_r^*\psi_0$ in the off-axis point-source hologram [see Eqs. (5.1-14) and (5.1-15c)] is responsible for a virtual image reconstruction upon illumination by a reconstruction wave $\psi_{rec} = ae^{jk_0 \sin\theta\, x} = \psi_r$. Show that

$$\psi_{rec} \times \psi_r^*\psi_0 * h(x, y; z = -z_0) \propto \delta(x, y)$$

by carrying out directly the convolution involved.

5.4 We have an off-axis point source hologram recorded as shown in Figure 5.1-7 and the recorded hologram of a transparency function is given by Eq. (5.1-16):

$$t(x, y) = A + B \sin\left[\frac{k_0}{2z_0}(x^2 + y^2) + k_0 \sin\theta\, x\right].$$

The hologram is now reconstructed by a plane wave shown in Figure P.5.4.

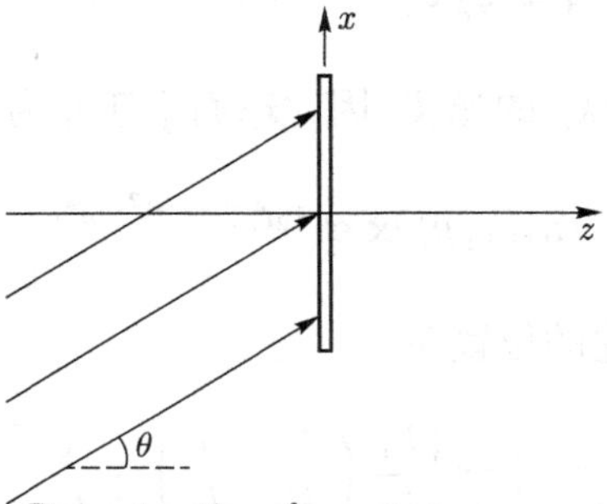

Figure P.5.4

(a) Find the location of the real and virtual images.

(b) Draw a reconstruction ray diagram similar to that in Figure 5.1-8 for both the real and virtual image.

5.5 We have discussed holographic imaging through the recording and reconstruction of two point sources using spherical waves in Section 5.2.

(a) For real image reconstruction of point 1, we calculate [see Eq. (5.2-10)]

$$\psi_{p,\,rel_1}(x, y; z) = \psi_{\rm pr}(x, y) t_{rel_1}(x, y) * h(x, y; z).$$

By carrying out directly the above convolution, show that the real image is reconstructed at

$$z = z_{\rm r1} = \left(\frac{1}{R} - \frac{1}{l_1} - \frac{1}{l_2}\right)^{-1},$$

and at this location,

$$\psi_{p,\,rel_1}(x, y; z = z_{\rm r1}) \propto \delta\left[x + z_{\rm r1}\left(\frac{b}{l_2} - \frac{h}{2R} - \frac{a}{l_1}\right), y\right].$$

(b) For real image reconstruction of point 2, we calculate [see Eq. (5.2-13)]

$$\psi_{p,\,rel_2}(x, y; z) = \psi_{\rm pr}(x, y) t_{rel_2}(x, y) * h(x, y; z).$$

By carrying out directly the above convolution, show that the real image is reconstructed at

$$z = z_{\rm r2} = \left(\frac{1}{R+d} - \frac{1}{l_1} - \frac{1}{l_2}\right)^{-1},$$

处, 且

$$\psi_{p,\,rel_2}(x,y;z=z_{r2})\propto\delta\left\{x+z_{r2}\left[\frac{b}{l_2}+\frac{h}{2(R+d)}-\frac{a}{l_1}\right],y\right\}.$$

5.6 理论上, 全息图可以用一种特定颜色的光波进行记录, 然后用另一种颜色的光波来重建, 本习题来探讨这一方面. 图 5.2-1 和图 5.2-2 所示分别为两个点源的记录与重建, 在记录时, 使用的记录波长为 $\lambda_1=2\pi/k_1$, 对于点 1 和点 2 在记录介质上的复振幅, 以及参考球面波分别由下列式子给出

$$\psi_{p1}(x,y)\propto \mathrm{e}^{-\mathrm{j}\frac{k_1}{2R}[(x-h/2)^2+y^2]},$$
$$\psi_{p2}(x,y)\propto \mathrm{e}^{-\mathrm{j}\frac{k_1}{2(R+d)}[(x+h/2)^2+y^2]},$$

以及

$$\psi_{pR}(x,y)\propto \mathrm{e}^{-\mathrm{j}\frac{k_1}{2l_1}\left[(x+a)^2+y^2\right]},$$

在重建时, 利用波长为 $\lambda_2=2\pi/k_2$ 的光波, 因此球面重建光为

$$\psi_{pr}(x,y)\propto \mathrm{e}^{-\mathrm{j}\frac{k_2}{2l_2}[(x-b)^2+y^2]}.$$

a) 证明: 点 1 的实像被重建的位置在

$$z=z_{r1}=\left[\frac{k_1}{k_2}\left(\frac{1}{R}-\frac{1}{l_1}\right)-\frac{1}{l_2}\right]^{-1}$$

处, 且

$$\psi_{p,rel_1}(x,y;z=z_{r1})\propto\delta\left[x+z_{r1}\left(\frac{b}{l_2}-\frac{k_1}{k_2}\frac{h}{2R}-\frac{k_1}{k_2}\frac{a}{l_1}\right),y\right].$$

b) 证明: 点 2 的实像被重建的位置在

$$z_{r2}=\left[\frac{k_1}{k_2}\left(\frac{1}{R+d}-\frac{1}{l_1}\right)-\frac{1}{l_2}\right]^{-1}$$

处, 且

$$\psi_{p,rel_2}(x,y;z=z_{r2})\propto\delta\left\{x+z_{r2}\left[\frac{b}{l_2}+\frac{k_1}{k_2}\frac{h}{2(R+d)}-\frac{k_1}{k_2}\frac{a}{l_1}\right],y\right\}.$$

c) 证明: 对于放大率, 当其与横向间距 h 无关时, 须有以下条件

$$\frac{b}{a}=\frac{\lambda_2}{\lambda_1}\frac{l_2}{l_1},$$

且在此条件下, 有

$$M_{lat}^{\mathrm{r}}=\frac{\lambda_2}{\lambda_1}\frac{(z_{r2}+z_{r1})}{2R}.$$

d) 证明: 当 $R\gg d$ 时, 纵向放大率为

$$M_{long}^{\mathrm{r}}\cong\frac{\lambda_1\lambda_2(l_1l_2)^2}{(\lambda_2l_1l_2-\lambda_1Rl_1-\lambda_2Rl_2)^2}.$$

e) 证明: 当 $R\gg d$ 时, 有

$$M_{long}^{\mathrm{r}}=\frac{\lambda_1}{\lambda_2}(M_{lat}^{\mathrm{r}})^2.$$

and at this location,

$$\psi_{p,\,rel_2}(x, y; z = z_{r2}) \propto \delta\left\{x + z_{r2}\left[\frac{b}{l_2} + \frac{h}{2(R+d)} - \frac{a}{l_1}\right], y\right\}.$$

5.6 In principle, a hologram can be recorded using a particular color and reconstructed by some other color. This problem explores this aspect. The recording and reconstruction of two point sources are shown in Figures 5.2-1 and 5.2-2, respectively. For recording, let us now record with wavelength $\lambda_1 = 2\pi/k_1$. The complex fields on the recording medium for point 1, point 2, and the spherical reference wave are given by

$$\psi_{p1}(x, y) \propto \mathrm{e}^{-\mathrm{j}\frac{k_1}{2R}[(x-h/2)^2+y^2]},$$

$$\psi_{p2}(x, y) \propto \mathrm{e}^{-\mathrm{j}\frac{k_1}{2(R+d)}[(x+h/2)^2+y^2]},$$

and

$$\psi_{pR}(x, y) \propto \mathrm{e}^{-\mathrm{j}\frac{k_1}{2l_1}\left[(x+a)^2+y^2\right]},$$

respectively. On reconstruction, we use wavelength $\lambda_2 = 2\pi/k_2$, and hence the spherical reconstruction wave is

$$\psi_{pr}(x, y) \propto \mathrm{e}^{-\mathrm{j}\frac{k_2}{2l_2}[(x-b)^2+y^2]}.$$

(a) Show that the real image of point 1 is reconstructed at

$$z = z_{r1} = \left[\frac{k_1}{k_2}\left(\frac{1}{R} - \frac{1}{l_1}\right) - \frac{1}{l_2}\right]^{-1},$$

and at this location,

$$\psi_{p,rel_1}(x, y; z = z_{r1}) \propto \delta\left[x + z_{r1}\left(\frac{b}{l_2} - \frac{k_1}{k_2}\frac{h}{2R} - \frac{k_1}{k_2}\frac{a}{l_1}\right), y\right].$$

(b) Show that the real image of point 2 is reconstructed at

$$z_{r2} = \left[\frac{k_1}{k_2}\left(\frac{1}{R+d} - \frac{1}{l_1}\right) - \frac{1}{l_2}\right]^{-1},$$

and at this location,

$$\psi_{p,rel_2}(x, y; z = z_{r2}) \propto \delta\left\{x + z_{r2}\left[\frac{b}{l_2} + \frac{k_1}{k_2}\frac{h}{2(R+d)} - \frac{k_1}{k_2}\frac{a}{l_1}\right], y\right\}.$$

(c) Show that for magnification that is independent of the lateral separation, h, we must have the following condition:

$$\frac{b}{a} = \frac{\lambda_2}{\lambda_1}\frac{l_2}{l_1},$$

and under this condition,

$$M^{\mathrm{r}}_{lat} = \frac{\lambda_2}{\lambda_1}\frac{(z_{r2} + z_{r1})}{2R}.$$

(d) Show that the longitudinal magnification is

$$M^{\mathrm{r}}_{long} \cong \frac{\lambda_1\lambda_2(l_1 l_2)^2}{(\lambda_2 l_1 l_2 - \lambda_1 R l_1 - \lambda_2 R l_2)^2},$$

for $R \gg d$.

(e) Verify that $M^{\mathrm{r}}_{long} = \frac{\lambda_1}{\lambda_2}(M^{\mathrm{r}}_{lat})^2$, for $R \gg d$.

f) 设置 $M_{lat}^{\mathrm{r}}=\dfrac{\lambda_2}{\lambda_1}$ 的优点是什么?

5.7 $s(x,y)$ 的傅里叶全息图由式 (5.3-17) 给出, 在重建时,

a) 证明其零级光为

$$\mathcal{F}\left\{\left|S\left(\frac{k_0x}{f},\frac{k_0y}{f}\right)\right|^2+|A|^2\right\}\bigg|_{k_x=k_0x/f,\,k_y=k_0y/f}$$
$$=\frac{4\pi^2f^2}{k_0^2}s(-x,-y)\otimes s(-x,-y)+\frac{4\pi^2f^2}{k_0^2}|A|^2\delta(x,y).$$

b) 证明其倒像为

$$\psi_2(x,y)=\mathcal{F}\left\{S\left(\frac{k_0x}{f},\frac{k_0y}{f}\right)\times A^*\exp\left(\frac{\mathrm{j}k_0x_0x}{f}\right)\right\}\bigg|_{k_x=k_0x/f,\,k_y=k_0y/f}$$
$$=\frac{4\pi^2f^2}{k_0^2}A^*s(-x-x_0,-y).$$

c) 证明其共轭像为

$$\psi_3(x,y)=\mathcal{F}\left\{S^*\left(\frac{k_0x}{f},\frac{k_0y}{f}\right)\times A\exp\left(\frac{-\mathrm{j}k_0x_0x}{f}\right)\right\}\bigg|_{k_x=k_0x/f,\,k_y=k_0y/f}$$
$$=\frac{4\pi^2f^2}{k_0^2}As^*(x-x_0,y).$$

5.8 参考图 5.3-8, 联合变换频谱由下式给出

$$\begin{aligned}t_{\mathrm{JTS}}(x,y)=&\left|\mathcal{F}\{g_1(x+x_0,y)\}\big|_{k_x=k_0x/f,\,k_y=k_0y/f}+\mathcal{F}\{g_2(x+x_0,y)\}\big|_{k_x=k_0x/f,\,k_y=k_0y/f}\right|^2\\=&\left|G_1\left(\frac{k_0x}{f},\frac{k_0y}{f}\right)\right|^2+\left|G_2\left(\frac{k_0x}{f},\frac{k_0y}{f}\right)\right|^2\\&+G_1^*\left(\frac{k_0x}{f},\frac{k_0y}{f}\right)G_2\left(\frac{k_0x}{f},\frac{k_0y}{f}\right)\mathrm{e}^{\mathrm{j}2\frac{k_0x}{f}x_0}\\&+G_1\left(\frac{k_0x}{f},\frac{k_0y}{f}\right)G_2^*\left(\frac{k_0x}{f},\frac{k_0y}{f}\right)\mathrm{e}^{-\mathrm{j}2\frac{k_0x}{f}x_0}.\end{aligned}$$

证明在其相关平面上, 相关输出由下式给出

$$\mathcal{F}\{t_{\mathrm{JTS}}(x,y)\}\big|_{k_x=k_0x/f,\,k_y=k_0y/f}$$
$$=\frac{4\pi^2f^2}{k_0^2}[C_{11}(-x,-y)+C_{22}(-x,-y)+C_{12}(-x-2x_0,-y)+C_{21}(-x+2x_0,-y)],$$

这里

$$C_{mn}(x,y)=g_m(x,y)\otimes g_n(x,y).$$

参考文献

[1] Gabor, D. (1948). "A new microscopic principle," Nature 161, pp.777-778.

[2] Leith. E. N. and Upatnieks, J. (1964). "Wavefront reconstruction with diffused illumination and three-dimensional objects," Joural of the Optical Society of America 54, pp.1295-1301.

(f) What is the advantage of setting $M_{lat}^{r} = \frac{\lambda_2}{\lambda_1}$?

5.7 The Fourier hologram of $s(x, y)$ is given by Eq. (5.3-17). Upon reconstruction,
(a) verify that the zeroth-order beam

$$\mathcal{F}\left\{\left|S\left(\frac{k_0 x}{f}, \frac{k_0 y}{f}\right)\right|^2 + |A|^2\right\}\Big|_{k_x=k_0x/f,\, k_y=k_0y/f}$$
$$= \frac{4\pi^2 f^2}{k_0^2} s(-x, -y) \otimes s(-x, -y) + \frac{4\pi^2 f^2}{k_0^2} |A|^2 \delta(x, y),$$

(b) verify that the inverted image

$$\psi_2(x, y) = \mathcal{F}\left\{S\left(\frac{k_0 x}{f}, \frac{k_0 y}{f}\right) \times A^* \exp\left(\frac{\mathrm{j}k_0 x_0 x}{f}\right)\right\}\Big|_{k_x=k_0x/f,\, k_y=k_0y/f}$$
$$= \frac{4\pi^2 f^2}{k_0^2} A^* s(-x - x_0, -y),$$

(c) verify that the conjugate image

$$\psi_3(x, y) = \mathcal{F}\left\{S^*\left(\frac{k_0 x}{f}, \frac{k_0 y}{f}\right) \times A \exp\left(\frac{-\mathrm{j}k_0 x_0 x}{f}\right)\right\}\Big|_{k_x=k_0x/f,\, k_y=k_0y/f}$$
$$= \frac{4\pi^2 f^2}{k_0^2} A s^*(x - x_0, y).$$

5.8 With reference to Figure 5.3-8, the joint-transform spectrum is given by

$$t_{\mathrm{JTS}}(x, y) = \left|\mathcal{F}\{g_1(x + x_0, y)\}\big|_{k_x=k_0x/f,\, k_y=k_0y/f} + \mathcal{F}\{g_2(x + x_0, y)\}\big|_{k_x=k_0x/f,\, k_y=k_0y/f}\right|^2$$
$$= \left|G_1\left(\frac{k_0 x}{f}, \frac{k_0 y}{f}\right)\right|^2 + \left|G_2\left(\frac{k_0 x}{f}, \frac{k_0 y}{f}\right)\right|^2$$
$$+ G_1^*\left(\frac{k_0 x}{f}, \frac{k_0 y}{f}\right) G_2\left(\frac{k_0 x}{f}, \frac{k_0 y}{f}\right) \mathrm{e}^{\mathrm{j}2\frac{k_0 x}{f} x_0}$$
$$+ G_1\left(\frac{k_0 x}{f}, \frac{k_0 y}{f}\right) G_2^*\left(\frac{k_0 x}{f}, \frac{k_0 y}{f}\right) \mathrm{e}^{-\mathrm{j}2\frac{k_0 x}{f} x_0}.$$

Show that on the correlation plane, the correlation outputs are given by

$$\mathcal{F}\{t_{\mathrm{JTS}}(x, y)\}\big|_{k_x=k_0x/f,\, k_y=k_0y/f}$$
$$= \frac{4\pi^2 f^2}{k_0^2}[C_{11}(-x, -y) + C_{22}(-x, -y) + C_{12}(-x - 2x_0, -y) + C_{21}(-x + 2x_0, -y)],$$

where

$$C_{mn}(x, y) = g_m(x, y) \otimes g_n(x, y).$$

Bibliography

[1] Gabor, D. (1948). "A new microscopic principle," Nature 161, pp. 777-778.

[2] Leith, E. N. and Upatnieks, J. (1964). "Wavefront reconstruction with diffused illumination and three-dimensional objects," Journal of the Optical Society of America 54, pp. 1295-1301.

[3] Poon, T.-C. and Banerjee, P. P. (2001). *Contemporary Optical Image Processing with MATLAB®*. Elsevier, Oxford.

[4] Poon, T.-C.and Liu, J.-P. (2014). *Introduction to Modern Digital Holography with MATLAB*. Cambridge University Press, Cambridge.

[5] Poon, T.-C. (2007). *Optical Scanning Holography with MATLAB®*. Springer, New York.

[6] Poon, T.-C. and Kim, T.(2018). *Engineering Optics with MATLAB®*, 2nd ed. World Scientific, New Jersey.

[7] Vander Lugt, A. (1963). "Signal detection by complex spatial filtering," IEEE Transactions on Information Theory. IT-10, pp.139-145.

[8] Weaver, C. J. and Goodman, J. W. (1966). "A technique for optically convolving two functions," Applied Optics 5, pp.1248-1249.

[9] Yu, F. T. S (1982). *Optical Information Processing*. John Wiley & Sons, New York.

[3] Poon, T.-C. and Banerjee, P. P. (2001). *Contemporary Optical Image Processing with MATLAB®*. Elsevier, Oxford.

[4] Poon, T.-C. and Liu, J.-P. (2014). *Introduction to Modern Digital Holography with MATLAB.* Cambridge University Press, Cambridge.

[5] Poon, T.-C. (2007). *Optical Scanning Holography with MATLAB®*. Springer, New York.

[6] Poon, T.-C. and Kim, T. (2018). *Engineering Optics with MATLAB®*, 2nd ed. World Scientific, New Jersey.

[7] Vander Lugt, A. (1963). "Signal detection by complex spatial filtering," IEEE Transactions on Information Theory, IT-10, pp. 139-145.

[8] Weaver, C. J. and Goodman, J. W. (1966). "A technique for optically convolving two functions," Applied Optics 5, pp. 1248-1249.

[9] Yu, F. T. S (1982). *Optical Information Processing.* John Wiley & Sons, New York.

第六章　数字全息术

基于全息膜的全息术主要利用如光聚合物 (*photopolymer*) 或光折变材料 (*photorefractive material*) 等高分辨率的膜来记录, 每厘米有上千个线对的分辨率是很常见的. 这些材料尽管有高的分辨率, 但有几个缺点. 基于全息膜的技术基本上很难实时应用, 并且对于所记录的全息图很难允许直接访问操作及后续处理. 随着近年来高分辨率二维固体传感器 [如电荷耦合器件 (*charge-coupled device*, CCD) 或互补金属氧化物半导体 (*complementary metal-oxide semiconductor*, CMOS) 传感器] 的发展, 以及日益提高的计算机能力与数据储存能力, 全息术结合电子/数字器件已经在如计量学 (*metrology*)、无损检测 (*nondestructive testing*)、三维成像等领域中成为一项应用不断增长的新兴技术.

1966 年, Enloe 等首次通过电视摄像管对全息图进行电子获取, 而全息图的数字重建是由 Goodman 和 Lawrence (1967) 提出的. 自 2005 年以来, 随着高质量 CCD 的出现及计算能力的提高, 数字全息术被光学度量界及全息界广泛接受 [Schnars 和 Jueptner (2005)]. 数字全息术 (*digital holography*), 它意味着三维物体的全息信息被 CCD 捕获, 随后全息图的重建被数字化地计算出来. 当前, 数字全息术也旨在用于如下情况: 通过一个电子设备 [如 CCD、CMOS 传感器或光电探测器 (*photodetector*)] 完成全息的记录, 且该记录的全息图可以被数字化地重建或被发送到一个显示设备 [如空间光调制器 (*spatial light modulator*, SLM), 将在第七章介绍] 以光学重建; 或全息图的整个生成过程都是数字模拟的, 全息图生成结果被顺序送往显示设备再进行光学重建, 这类数字全息术通常被称为计算全息或计算机生成全息 (*computer-generated holography*).

6.1　相干数字全息术

6.1.1　CCD 的局限

传统的同轴或者离轴相干全息术需要一个电子设备来获得全息条纹, 例如 CCD 或 CMOS 传感器. 本节将会讨论这些传感器阵列的特性和限制.

CCD 和 CMOS 传感器是由大量的光敏单元 [称为像素 (*pixel*)] 组成的, 其按二维阵列如图 6.1-1 所示排列. 每个像素的面积 $x_0 \times x_0$ (假设为一个方形像素) 即为一个对光照敏感的有效面积. 在像素的有效面积周围有电磁屏蔽. 像素中心点与中心点的距离被称为像素间距 (*pixel pitch*), 在图 6.1-1 中表示为 Δx. 其总的 (单个的) 光敏面积与整个 (单个的) 阵列面积的比值被称为填充因子 (*fill factor*) F, 比如, 如图 6.11 所示的, $F = 25x_0^2/(5\Delta x)^2 =$

6 Digital Holography

Film-based holography employs the use of high-resolution films such as the use of *photopolymers* or *photorefractive materials* for recording. Resolution of several thousand-line pairs per millimeter is common. These materials, while having high resolution, have a couple of drawbacks. The film-based techniques are typically slow for real-time applications and difficult to allow direct access to the recorded hologram for manipulation and subsequent processing. With recent advances in high-resolution solid-state 2-D sensors [e.g., the charge-coupled device (CCD) or *complementary metal-oxide semiconductor* (CMOS) sensor] and the availability of ever-increasing power of computers and digital data storage capabilities, holography coupled with electronic/digital devices has become an emerging technology with an increasing number of applications such as in *metrology*, *nondestructive testing*, and three-dimensional (3-D) imaging.

While electronic detection of holograms by a TV camera was first performed by Enloe et al. in 1966, hologram numerical reconstruction was initiated by Goodman and Lawrence (1967). With the advent of high-quality CCDs and faster computing power, digital holography has been widely accepted by the optical metrology community as well as holography community since 2005 [Schnars and Jueptner (2005)]. In *digital holography*, it has meant that holographic information of 3-D objects is captured by a CCD, and reconstruction of holograms is subsequently calculated numerically. Nowadays, digital holography means the following situations as well. Holographic recording is done by an electronic device (such as a CCD, CMOS sensor or a *photodetector*), and the recorded hologram can be numerically reconstructed or sent to a display device [called a *spatial light modulator* (SLM), to be discussed in Chapter 7] for optical reconstruction. Or, hologram construction is completely numerically simulated. The resulting hologram is sent subsequently to a display device for optical reconstruction. This aspect of digital holography is often known as *computer-generated holography.*

6.1 Coherent Digital Holography

6.1.1 CCD Limitations

Conventional on-axis and off-axis coherent holography employ an electronic device, such as a CCD or a CMOS sensor, to capture holographic fringes. In this section, we discuss the properties and limitations of these sensor arrays.

A CCD sensor and a CMOS sensor are composed of numerous light sensing units, namely pixels, arranged as a two-dimensional array, shown in Figure 6.1-1. Each pixel of area $x_0 \times x_0$ (assuming a square pixel) is an active area sensitive to light illumination. There is a shielding around the active area of the pixel. The centerto-center distance of the pixels is called the *pixel pitch*, denoted as Δx in Figure 6.1-1. The ratio of the light sensitive area to the whole array area is called the *fill factor*, F. For example, as shown in Figure 6.11, $F = 25x_0^2/(5\Delta x)^2 =$

$x_0^2/\Delta x^2$. 对于实际的二维阵列传感器, 像素间距 Δx 为 $4\sim 8$ μm, 填充因子 F 为 $0.8\sim 0.9$. 一个普通的 CCD 传感器有 1024×768 个像素. 若忽略电磁屏蔽区域, 即 $F=1$, 则其像素间距的线性尺寸与像素尺寸 x_0 相同. 那么, 像素尺寸就决定了其空间分辨率. 若像素大小为 6 μm, 芯片尺寸大约为 6.0 mm × 4.5 mm. 一个线对 (*line pair*, lp) 是由一条亮线和一条黑线组成的, 如果一条线是 0.5 μm, 那么一个线对的宽度为 1 μm, 因此, 1 μm 即为一个线对, 即 1 lp/μm, 也就是 1000 lp/mm. 再次忽略电磁屏蔽区域, 对于一个给定的像素尺寸, 式 (6.1-1) 可用来估算其空间分辨率 (*spatial resolution*)

$$\text{分辨率}\left[\frac{\text{lp}}{\text{mm}}\right]=\frac{1[\text{lp}]}{2\times\text{像素尺寸}[\text{mm}]}. \tag{6.1-1}$$

对于像素尺寸 5 μm, 其分辨率为 $1/(2\times 5\times 10^{-3})$ lp/mm = 100 lp/mm. 可以看出, 该分辨率相比一般可用的全息膜分辨率小了一个数量级. 换句话说, 当利用 CCD 进行数字全息时, 必须在实验中慎重考虑这一不利因素.

另一个需要强调的重要问题是, CCD 芯片由离散的单元 (像素) 所构成, 如图 6.1-1 所示. 当一个连续的条纹强度图案 $g(x,y)$ 落在 CCD 的二维阵列 [单元格 (*grid*)] 上时, 这些像素点就采集到一个该条纹的空间采样图案.

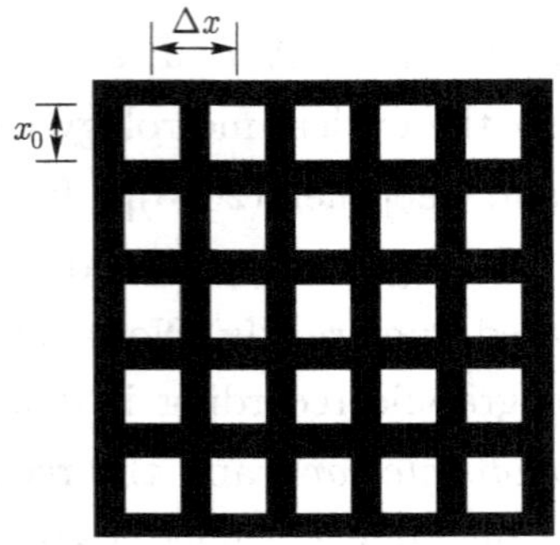

图 6.1-1 CCD 或 CMOS 二维阵列传感结构

因为像素足够小, 假设采样像素点由 $\delta_{\Delta x}(x)=\sum\limits_{n=-\infty}^{\infty}\delta(x-n\Delta x)$ 给出. 因此, 采样图案为

$$g_s(x)=g(x)\times\sum_{n=-\infty}^{\infty}\delta(x-n\Delta x), \tag{6.1-2}$$

为简单起见, 这里假设进行一维分析. 考虑一个带宽限制为 B (周期/mm) 的信号 $g(x)$, 即信号没有高于 B 的空间频率; 因此, 就说该信号有 B 的带宽 (*bandwidth*). 在图 6.1-2a) 中, 给出 $g(x)$ 的频谱 $G(k_x)$. 求出采样图案的频谱 $G_s(k_x)$ 为

$$G_s(k_x)=\mathcal{F}\{g_s(x)\}=\mathcal{F}\left\{g(x)\times\sum_{n=-\infty}^{\infty}\delta(x-n\Delta x)\right\}. \tag{6.1-3}$$

由于 $\delta_{\Delta x}(x)$ 是采样周期 (*sampling period*) 为 $T_s=\Delta x$ 的周期性函数, 因此可将该函数表示为傅里叶级数的形式 [见式 (2.2-4a) 傅里叶级数的定义]

$$\delta_{\Delta x}(x)=\sum_{n=-\infty}^{\infty}d_n\mathrm{e}^{\mathrm{j}n\frac{2\pi}{\Delta x}x}, \tag{6.1-4a}$$

$x_0^2/\Delta x^2$. For practical 2-D array sensors, the pixel pitch Δx is $4 \sim 8$ μm, and the fill factor F is $0.8 \sim 0.9$. A common CCD sensor has 1024×768 pixels. Neglecting the shielding area, that is, $F = 1$, the pixel pitch has the same linear dimension of the pixel size x_0. The pixel size then determines spatial resolution. With the pixel size of 6 μm, the chip size is about 6.0 mm$\times$4.5 mm. A line pair consists of a dark line and a bright line. If one line is 0.5 micron (μm), then a line pair is 1 micron width. Hence, there is one line pair (lp) per micron, that is, 1 lp/μm, which is 1000 lp/mm. Again neglecting the shielding area, Eq. (6.1-1) allows us to approximate the *spatial resolution* in lp/mm for a given pixel size

$$\text{resolution}\left[\frac{\text{lp}}{\text{mm}}\right] = \frac{1[\text{lp}]}{2 \times \text{pixel size}[\text{mm}]}. \tag{6.1-1}$$

For a pixel size of 5 μm, resolution $= 1/(2 \times 5 \times 10^{-3})$ lp/mm $= 100$ lp/mm. Note that this resolution is an order of magnitude smaller than that typically available with films. In other words, when we work with digital holography using CCDs, we have to deal with this detrimental issue and care must be taken in performing experiments.

Another important issue that needs to be addressed is the fact that CCD chip is made of discrete elements (pixels) as shown in Figure 6.1-1. When a continuous fringe intensity pattern $g(x, y)$ falls on the 2-D array (grid) formed by a CCD, the pixels capture a spatially sampled version of the pattern.

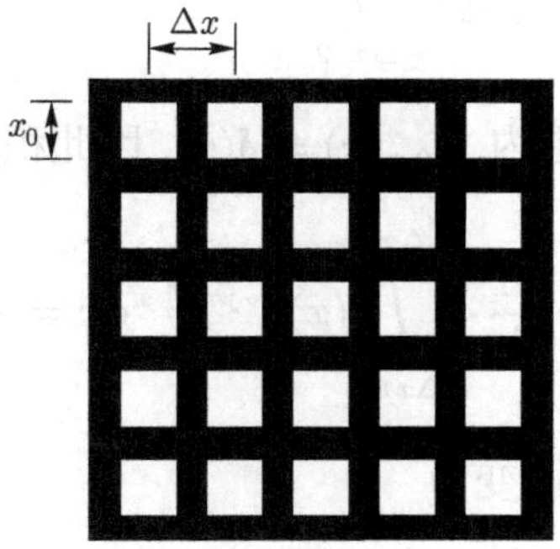

Figure 6.1-1 Structure of a CCD/CMOS 2-D array sensor

Since the pixel is small enough, we assume the sampling pixels are given by $\delta_{\Delta x}(x) = \sum_{n=-\infty}^{\infty} \delta(x - n\Delta x)$. Therefore, the sampled pattern is

$$g_s(x) = g(x) \times \sum_{n=-\infty}^{\infty} \delta(x - n\Delta x), \tag{6.1-2}$$

where we have assumed a 1-D analysis for simplicity. Let us consider a signal $g(x)$ that is bandlimited to B (cycle/mm), that is, the signal has no spatial frequencies higher than B; therefore the signal is said to have a *bandwidth* of B. In Figure 6.1-2a, we show the spectrum $G(k_x)$ of $g(x)$. Let us find the spectrum of the sampled pattern $G_s(k_x)$:

$$G_s(k_x) = \mathcal{F}\{g_s(x)\} = \mathcal{F}\left\{g(x) \times \sum_{n=-\infty}^{\infty} \delta(x - n\Delta x)\right\}. \tag{6.1-3}$$

Since $\delta_{\Delta x}(x)$ is a periodic function with *sampling period* $T_s = \Delta x$, we can express the function in terms of Fourier series as [see the definition of Fourier series in Eq. (2.2-4a)]

$$\delta_{\Delta x}(x) = \sum_{n=-\infty}^{\infty} d_n e^{\mathrm{j}n\frac{2\pi}{\Delta x}x}, \tag{6.1-4a}$$

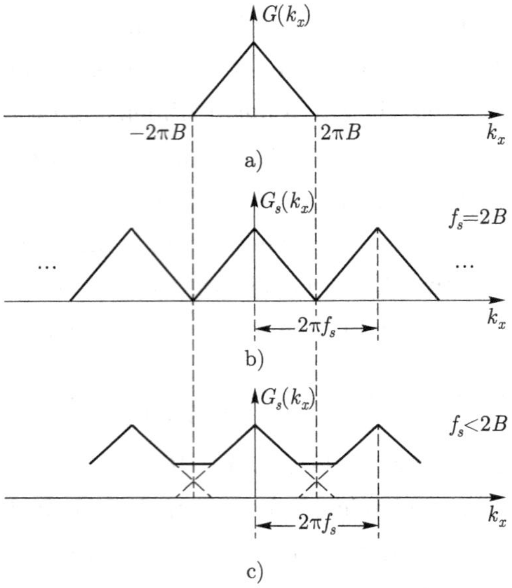

图 6.1-2 a) $g(x)$ 的频谱; b) $f_s = 2B$ 时的采样频谱; c) $f_s < 2B$ 时的欠采样频谱

其中

$$d_n = \frac{1}{\Delta x}\int_{-\Delta x/2}^{\Delta x/2} \delta_{\Delta x}(x)\mathrm{e}^{-\mathrm{j}n\frac{2\pi}{\Delta x}x}\mathrm{d}x. \tag{6.1-4b}$$

在从 $-\Delta x/2$ 到 $\Delta x/2$ 的积分范围内, $\delta_{\Delta x}(x) = \delta(x)$. 因此积分变为

$$d_n = \frac{1}{\Delta x}\int_{-\Delta x/2}^{\Delta x/2} \delta(x)\mathrm{e}^{-\mathrm{j}n\frac{2\pi}{\Delta x}x}\mathrm{d}x = \frac{1}{\Delta x}. \tag{6.1-5}$$

利用式 (6.1-5) 的结果, 式 (6.1-4a) 变为

$$\delta_{\Delta x}(x) = \sum_{n=-\infty}^{\infty} \delta(x - n\Delta x) = \frac{1}{\Delta x}\sum_{n=-\infty}^{\infty} \mathrm{e}^{\mathrm{j}n\frac{2\pi}{\Delta x}x}. \tag{6.1-6}$$

将式 (6.1-6) 代入式 (6.1-3), 有

$$\begin{aligned} G_s(k_x) &= \frac{1}{\Delta x}\mathcal{F}\left\{g(x) \times \sum_{n=-\infty}^{\infty} \mathrm{e}^{\mathrm{j}n\frac{2\pi}{\Delta x}x}\right\} = \frac{1}{\Delta x}\mathcal{F}\left\{\sum_{n=-\infty}^{\infty} g(x)\mathrm{e}^{\mathrm{j}n\frac{2\pi}{\Delta x}x}\right\} \\ &= \frac{1}{\Delta x}\sum_{n=-\infty}^{\infty} \mathcal{F}\left\{g(x)\mathrm{e}^{\mathrm{j}n\frac{2\pi}{\Delta x}x}\right\} = \frac{1}{\Delta x}\sum_{n=-\infty}^{\infty} G\left(k_x + \frac{2\pi n}{\Delta x}\right). \end{aligned} \tag{6.1-7}$$

当 $T_s = \Delta x$ 是采样周期时, 其倒数被称为采样频率 (*sampling frequency*), 即 $f_s = 1/T_s = 1/\Delta x$. 根据这一定义, 采样函数的频谱可以写为

$$G_s(k_x) = f_s \sum_{n=-\infty}^{\infty} G(k_x + 2\pi n f_s). \tag{6.1-8}$$

在图 6.1-2b) 与图 6.1-2c) 中, 分别给出了 $f_s = 2B$ 和 $f_s < 2B$ 时的采样频谱. 图 6.1-2b)

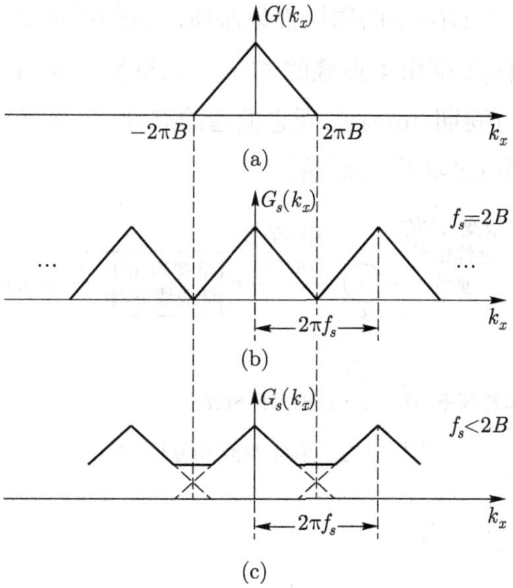

Figure 6.1-2 (a) Spectrum of $g(x)$, (b) sampled spectrum for $f_s = 2B$, and (c) under-sampled spectrum when $f_s < 2B$

where

$$d_n = \frac{1}{\Delta x}\int_{-\Delta x/2}^{\Delta x/2} \delta_{\Delta x}(x)\mathrm{e}^{-\mathrm{j}n\frac{2\pi}{\Delta x}x}\mathrm{d}x. \tag{6.1-4b}$$

Over the range of integration from $-\Delta x/2$ to $\Delta x/2$, $\delta_{\Delta x}(x) = \delta(x)$. Hence, the integration becomes

$$d_n = \frac{1}{\Delta x}\int_{-\Delta x/2}^{\Delta x/2} \delta(x)\mathrm{e}^{-\mathrm{j}n\frac{2\pi}{\Delta x}x}\mathrm{d}x = \frac{1}{\Delta x}. \tag{6.1-5}$$

Using the result from Eq. (6.1-5), Eq. (6.1-4a) becomes

$$\delta_{\Delta x}(x) = \sum_{n=-\infty}^{\infty} \delta(x - n\Delta x) = \frac{1}{\Delta x}\sum_{n=-\infty}^{\infty} \mathrm{e}^{\mathrm{j}n\frac{2\pi}{\Delta x}x}. \tag{6.1-6}$$

Substituting Eq. (6.1-6) into Eq. (6.1-3), we have

$$\begin{aligned} G_s(k_x) &= \frac{1}{\Delta x}\mathcal{F}\left\{g(x) \times \sum_{n=-\infty}^{\infty} \mathrm{e}^{\mathrm{j}n\frac{2\pi}{\Delta x}x}\right\} = \frac{1}{\Delta x}\mathcal{F}\left\{\sum_{n=-\infty}^{\infty} g(x)\mathrm{e}^{\mathrm{j}n\frac{2\pi}{\Delta x}x}\right\} \\ &= \frac{1}{\Delta x}\sum_{n=-\infty}^{\infty} \mathcal{F}\left\{g(x)\mathrm{e}^{\mathrm{j}n\frac{2\pi}{\Delta x}x}\right\} \\ &= \frac{1}{\Delta x}\sum_{n=-\infty}^{\infty} G\left(k_x + \frac{2\pi n}{\Delta x}\right). \end{aligned} \tag{6.1-7}$$

While $T_s = \Delta x$ is the sampling period, its reciprocal is called the *sampling frequency* $f_s = 1/T_s = 1/\Delta x$. With this definition, the spectrum of the sampled function can be written as

$$G_s(k_x) = f_s \sum_{n=-\infty}^{\infty} G(k_x + 2\pi n f_s). \tag{6.1-8}$$

In Figure 6.1-2b and 6.1-2c, we show the sampled spectra for $f_s = 2B$ and $f_s < 2B$, respectively. Figure 6.1-2b displays the spectrum $G_s(k_x)$ that consists of a periodic replica of

所示为原始信号 $g(x)$ 的频谱 $G(k_x)$ 的周期性复制所组成的频谱 $G_s(k_x)$. 只要采样频率 f_s 大于信号带宽 B 的两倍, $G_s(k_x)$ 即由不重叠的 $G(k_x)$ 的重复组成. 让采样信号经过一个截止频率 (*cutoff frequency*) 为 B (周期/mm) 的理想低通滤波器, 可以准确无误地恢复原始信号, 其中采样滤波和复原过程如图 6.1-3 中所概括.

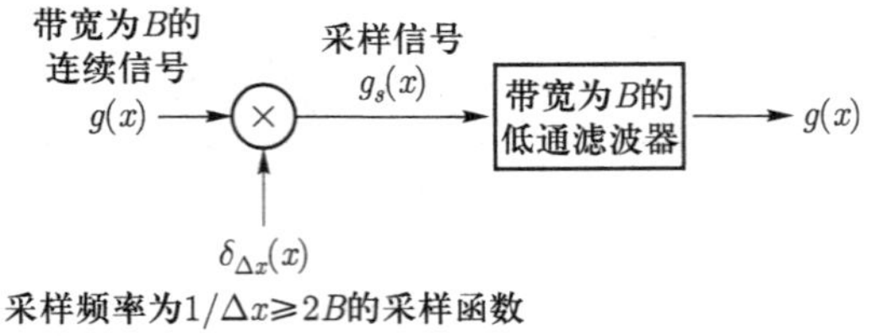

图 6.1-3 采样滤波和复原过程

其采样频率为

$$f_s = 2B, \tag{6.1-9}$$

被称为奈奎斯特采样频率 (*Nyquist sampling frequency*) 或者奈奎斯特率 (*Nyquist rate*), 它表示为了准确地恢复原始信号而应该对 $g(x)$ 进行采样的一个下限频率. 即, 为了准确地恢复原始信号, 必须以一个比奈奎斯特率更高的频率来采样.

当信号被欠采样时, 即采样频率低于奈奎斯特采样频率时, 恢复的信号是不准确的, 所导致的错误被称为混叠 (*aliasing*) 或谱折叠 (*spectral folding*), 如图 6.1-2c) 所示.

6.1.2 数字全息图的光学记录

通过上一节可以发现, 因为有限的像素尺寸, 一般 CCD 阵列的有限分辨率约为 100 lp/mm. 为了避免混叠, 需要以高于奈奎斯特率的频率来对全息图的精细条纹图案进行采样, 这由像素尺寸 Δx 来决定 [见式 (6.1-9)].

在物体上的最远位置处取一个点源. 对于该点源, 会有一个离轴点源全息图 [见式 (5.1-10)]

$$FZP(x-x_0, y-y_0; z_0) = A + B\sin\left\{\frac{k_0}{2z_0}[(x-x_0)^2+(y-y_0)^2]\right\}.$$

其局部频率为 [见式 (5.1-9)]

$$f_{local} = \frac{1}{2\pi}\frac{\mathrm{d}}{\mathrm{d}x}\left[\frac{k_0}{2z_0}(x-x_0)^2\right] = \frac{1}{\lambda_0 z_0}(x-x_0). \tag{6.1-10}$$

在全息面上的 $x<0$ 时 (见图 5.1-3), 可以看到, 其局部条纹频率 $|f_{local}|$ 会越来越大, 但我们可以通过限制物体的尺寸来减小该局部条纹频率, 即减小 x_0. 因此, 可以看到物体的横向尺寸越大, 全息记录时的局部条纹频率就越精细. 因此, CCD 只能记录横向尺寸小的物体, 否则记录就会出现混叠现象. 或者, 由于 z_0 变大时, 局部频率会减小, 因此可通过将物体放至离 CCD 更远的距离来避免频谱混叠.

$G(k_x)$, the spectrum of the original signal $g(x)$. As long as the sampling frequency f_s is greater than twice the signal bandwidth B, $G_s(k_x)$ consists of nonoverlapping repetitions of $G(k_x)$. The signal can be recovered faithfully without error by passing the sampled signal through an ideal low-pass filter with *cutoff frequency* B (cycle/mm). The filtering and recovering processes are summarized in Figure 6.1-3.

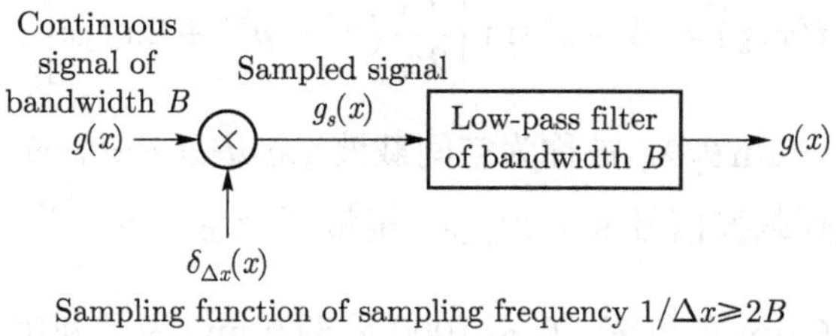

Figure 6.1-3 Sampling process and recovery process

The sampling frequency

$$f_s = 2B \tag{6.1-9}$$

is called the *Nyquist sampling frequency* or *Nyquist rate* and it represents a lower bound to the sampling frequency at which $g(x)$ should be sampled in order to faithfully recover the original signal. Sampling must be at a rate higher than the Nyquist rate in order to faithfully recover the original signal.

When the signal is under-sampled, that is, the sampling frequency is less than the Nyquist sampling frequency, the retrieved signal is incorrect, resulting in errors known as *aliasing* or *spectral folding*, illustrated in Figure 6.1-2c.

6.1.2 Optical Recording of Digital Holograms

We have found out from the last section that typically CCD arrays have limited resolution of the order of about 100 lp/mm due to the finite size of the pixel. In order to avoid aliasing, we need to sample the fine fringe pattern of the hologram higher than the Nyquist rate, which is dictated by the pixel size Δx [see Eq. (6.1-9)].

Let us take a point source at the furthest extent of the object. For that point source, we have an off-axis point-source hologram [see Eq. (5.1-10)]:

$$FZP(x - x_0, y - y_0; z_0) = A + B\sin\left\{\frac{k_0}{2z_0}[(x - x_0)^2 + (y - y_0)^2]\right\},$$

and its local frequency is [see Eq. (5.1-9)]

$$f_{local} = \frac{1}{2\pi}\frac{\mathrm{d}}{\mathrm{d}x}\left[\frac{k_0}{2z_0}(x - x_0)^2\right] = \frac{1}{\lambda_0 z_0}(x - x_0). \tag{6.1-10}$$

For $x < 0$ on the hologram plane (see Figure 5.1-3), we see that the local fringe frequency $|f_{local}|$ is getting larger. However, we can lower the local fringe frequency by limiting the extent of the object, that is, letting x_0 smaller. Therefore, we see that the larger the object in lateral dimension, the finer the local fringe frequency is for holographic recording. Hence, the CCD could only capture objects with small lateral dimension if we do not want to record with aliasing. Alternatively, we can avoid aliasing by placing the object further away from the CCD as z_0 is getting larger and hence, the local frequency gets smaller.

当利用离轴全息记录时, 会发生什么情况? 对于一个离轴点源全息图, 有 [见式 (5.1-16)]

$$t(x,y)=A+B\sin\left[\frac{k_0}{2z_0}(x^2+y^2)+k_0\sin\theta\, x\right].$$

将上式重写, 有

$$t(x,y)=A+B\sin\left[\frac{k_0}{2z_0}(x^2+y^2)+2\pi f_c x\right],\tag{6.1-11}$$

其中, $f_c=k_0\sin\theta/(2\pi)=\sin\theta/\lambda_0$ 被称为空间载波 (*spatial carrier*). 载波这个术语实际来源于通信理论. 在这里, 高频载波信号 $\sin(2\pi f_c x)$ 携带了全息信息 $\frac{k_0}{2z_0}(x^2+y^2)$. 对于一个记录角为 $\theta=45°$ 且 $\lambda_0=0.6\ \mu\text{m}$ 的红光, $f_c\sim 1000$ 周期/mm. 为了利用这一技术进行全息记录, 这么高的空间频率即是要求记录介质的分辨率至少达到 1000 lp/mm. 根据式 (5.1-9), 全息图的局部频率为

$$f_{local}=\frac{1}{2\pi}\frac{\mathrm{d}}{\mathrm{d}x}\left(\frac{k_0}{2z_0}x^2+2\pi f_c x\right)=\frac{k_0 x}{2\pi z_0}+f_c=\frac{x}{\lambda_0 z_0}+f_c.\tag{6.1-12}$$

从上述讨论可以总结出, 当用 CCD 阵列来获取全息信息时, 一定要注意避免混叠误差. 为此, 可以限制物体的横向尺寸, 或将物体置于离 CCD 较远的位置 (使 z_0 更大), 或者, 若使用离轴记录光路的话, 使用更小的记录角 (使 f_c 更小).

同轴菲涅耳全息图

要记录同轴菲涅耳全息图, 光路如图 6.1-4 所示. 在接下来的分析中, 只讨论 x-z 平面的情况. 假定 CCD 含有 $M\times M$ 个像素点, 因此其尺寸为 $M\Delta x\times M\Delta x$. 物体的尺寸为 D 且其中心在光轴上. 物体与 CCD 传感器之间的距离为 z, 平面参考光与物光经过分束器 (*beam splitter*, BS) 合成. 在 CCD 平面上, 每一个物点的光与参考光形成一个菲涅耳波带板 (FZP).

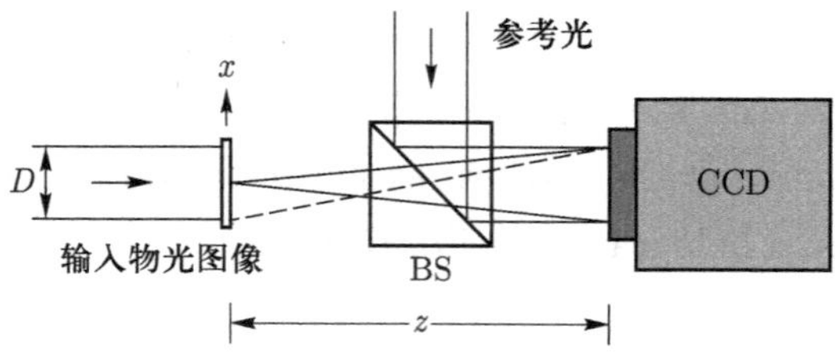

图 6.1-4 同轴菲涅耳全息的典型光路

从物体下方边缘处的一个物点发出至 CCD 传感器上方边缘的光线 (由图 6.1-4 中的虚线表示) 表示在 CCD 传感器上该物点的菲涅耳波带板的最大横向尺寸. 因此, 同轴菲涅耳波带板的最大横向尺寸为 $D/2+(M\Delta x)/2$, 且相应的局部条纹频率最大值为 [见式 (6.1-10), 当 $x_0=0$ 时]

$$f_{local}=\left.\frac{x}{\lambda_0 z}\right|_{x=D/2+(M\Delta x)/2}=\frac{1}{2z\lambda_0}(D+M\Delta x).$$

What happens when we want to employ off-axis holographic recording? For an off-axis point-source hologram, we have [see Eq. (5.1-16)]

$$t(x,y) = A + B\sin\left[\frac{k_0}{2z_0}(x^2+y^2) + k_0\sin\theta\, x\right].$$

Let use rewrite the above equation, and we have

$$t(x,y) = A + B\sin\left[\frac{k_0}{2z_0}(x^2+y^2) + 2\pi f_c x\right], \qquad (6.1\text{-}11)$$

where $f_c = k_0\sin\theta/(2\pi) = \sin\theta/\lambda_0$ is called the *spatial carrier*. The terminology of carrier is actually borrowed from communication theory. In the context of our situation, the high-frequency carrier signal $\sin(2\pi f_c x)$ carries the holographic information $\frac{k_0}{2z_0}(x^2+y^2)$. For a recording angle $\theta = 45°$ and $\lambda_0 = 0.6$ μm for a red laser, $f_c \sim 1000$ cycle/mm. This high spatial frequency translates into a recording medium of resolution at least 1000 lp/mm in order to employ this technique for holographic recording. Indeed, the local frequency of the hologram is, according to Eq. (5.1-9),

$$f_{local} = \frac{1}{2\pi}\frac{\mathrm{d}}{\mathrm{d}x}\left(\frac{k_0}{2z_0}x^2 + 2\pi f_c x\right) = \frac{k_0 x}{2\pi z_0} + f_c = \frac{x}{\lambda_0 z_0} + f_c. \qquad (6.1\text{-}12)$$

From the above discussion, we can conclude that when capturing holographic information with CCD arrays, we must guard against aliasing errors. To do this, we can limit the lateral dimension of the object, place the object farther away from the CCD (making z_0 larger), or make the recording angle small (making f_c smaller) if off-axis recording is employed.

On-axis Fresnel Hologram

For the recording of on-axis Fresnel holograms, the setup is shown in Figure 6.1-4. In the analysis presented below, we involve the x-z plane only. Assume that the CCD contains $M \times M$ pixels, so its size is $M\Delta x \times M\Delta x$. The size of the object is D and centered on the optical axis. The distance between the object and the CCD sensor is z, and the plane-wave reference wave is coupled to the object light via a *beam splitter* (BS). At the CCD plane, each object point together with the reference wave forms a FZP.

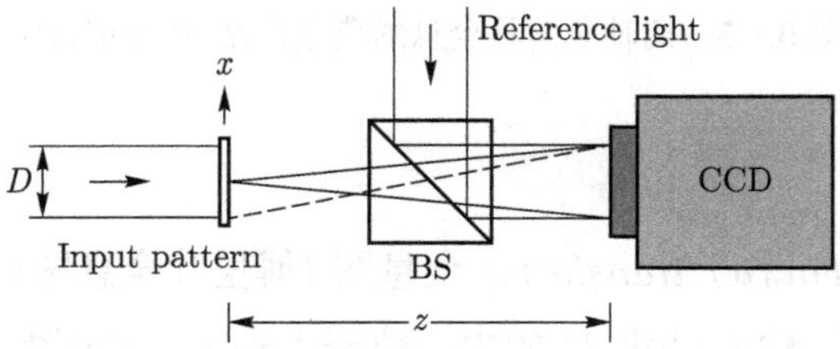

Figure 6.1-4 Typical setup of on-axis Fresnel holography

The ray of an object point emitting from the object's lower edge to the upper edge of the CCD sensor (indicating by a dashed line in Figure 6.1-4) indicates a maximum lateral extent of the Fresnel zone plate due to the object point on the CCD sensor. Thus, the maximum lateral dimension of the on-axis Fresnel zone plate is $D/2 + (M\Delta x)/2$, and the corresponding maximum local fringe frequency [see Eq. (6.1-10) for $x_0 = 0$] is

$$f_{local} = \left.\frac{x}{\lambda_0 z}\right|_{x=D/2+(M\Delta x)/2} = \frac{1}{2z\lambda_0}(D + M\Delta x).$$

为了避免全息条纹的欠采样, CCD 的采样频率必须至少是接收到的条纹最大局部频率的两倍, 即为 $2B$ [见式 (6.1-9)]

$$f_s = \frac{1}{\Delta x} \geqslant 2B = 2\frac{1}{2z\lambda_0}(D + M\Delta x), \tag{6.1-13}$$

给定

$$z \geqslant \frac{\Delta x}{\lambda_0}(D + M\Delta x). \tag{6.1-14}$$

因此, 为了避免混叠, 物体与 CCD 传感器之间的距离要有一个最小值.

离轴菲涅耳全息图

要进行离轴全息图的记录, 光路如图 6.1-5 所示. 此分析类似上文的同轴记录. 再次, 对于一个从物体下方边缘处物点发出的光线, 菲涅耳波带板的最大横向尺寸为 $D/2 + (M\Delta x)/2$, 相应的局部条纹频率的最大值为 [见式 (6.1-12)]

$$f_{local} = \left.\frac{x}{\lambda_0 z}\right|_{x=D/2+(M\Delta x)/2} + f_c = \frac{D/2 + (M\Delta x)/2}{\lambda_0 z} + \frac{\sin\theta}{\lambda_0}. \tag{6.1-15}$$

为避免混叠, 需

$$f_s = \frac{1}{\Delta x} \geqslant 2B, \tag{6.1-16}$$

其中, $B = f_{local}$ 由式 (6.1-15) 给出. 结合式 (6.1-15) 与式 (6.1-16), 得

$$z \geqslant \frac{\Delta x(D + M\Delta x)}{\lambda_0 - 2\sin\theta\Delta x},$$

这是物体必须放置的距 CCD 传感器位置的最小距离.

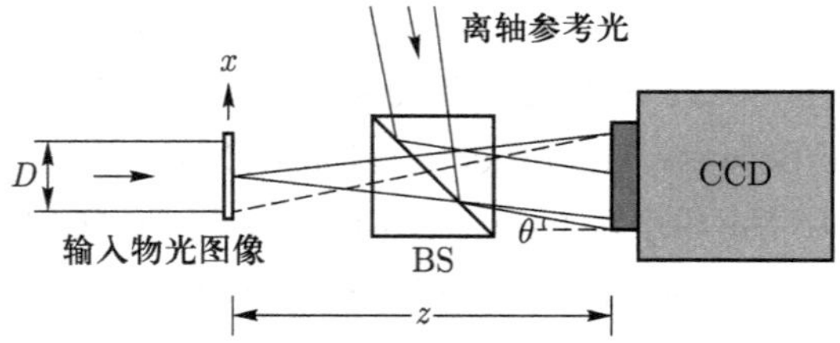

图 6.1-5　离轴菲涅耳全息的典型光路. θ 为记录角

傅里叶全息图

要进行傅里叶变换 (*Fourier transform*) 全息图 (傅里叶全息图) 的记录, 其光路如图 6.1-6 所示 [同时见图 5.3-3a)]. 输入尺寸 D 的中心位于光轴上, 同时置于傅里叶变换透镜的前焦面处. 一个 CCD 置于傅里叶透镜的后焦面处. 一个单点的入射将会在 CCD 上产生一束平面波, 同时参考光的一个参考点光源 $\delta(x + x_0)$ 在 CCD 平面上产生一个倾斜平面波 $\mathrm{e}^{-\mathrm{j}k_0 x_0 x/f}$ (见 5.3.2 节). 两个平面波在 CCD 上产生一个均匀的正弦条纹. 然而, 最细的条纹来自位于 $\delta(x - D/2)$ 处的点源, 其由在 $x = D/2$ 处入射的边缘光线产生. 该点光源在 CCD 上产生一束平面波 $\mathrm{e}^{\mathrm{j}k_0 Dx/(2f)}$. 因此, CCD 上最细的条纹由下式给出

$$\left|\mathrm{e}^{\mathrm{j}k_0 Dx/(2f)} + \mathrm{e}^{-\mathrm{j}k_0 x_0 x/f}\right|^2 = 2 + 2\cos\left[\pi\frac{(D + 2x_0)x}{\lambda_0 f}\right]. \tag{6.1-17}$$

To avoid under-sampling of the holographic fringes, the sampling frequency of the CCD must be at least twice the highest local frequency of the incoming fringe pattern, that is, $2B$ [see Eq. (6.1-9)]:

$$f_s = \frac{1}{\Delta x} \geqslant 2B = 2\frac{1}{2z\lambda_0}(D + M\Delta x), \tag{6.1-13}$$

giving

$$z \geqslant \frac{\Delta x}{\lambda_0}(D + M\Delta x). \tag{6.1-14}$$

Therefore, there is a minimum distance between the object and the CCD senor if aliasing is to be avoided.

Off-axis Fresnel Hologram

For recording of off-axis Fresnel holograms, the setup is shown in Figure 6.1-5. The analysis is similar to that in the last section for on-axis recording. Again, for an object point emitting from the object's lower edge, the maximum lateral dimension of the Fresnel zone plate is $D/2 + (M\Delta x)/2$. The corresponding maximum local fringe frequency [see Eq. (6.1-12)] is

$$f_{local} = \left.\frac{x}{\lambda_0 z}\right|_{x=D/2+(M\Delta x)/2} + f_c = \frac{D/2 + (M\Delta x)/2}{\lambda_0 z} + \frac{\sin\theta}{\lambda_0}. \tag{6.1-15}$$

To avoid aliasing, we need

$$f_s = \frac{1}{\Delta x} \geqslant 2B, \tag{6.1-16}$$

where $B = f_{local}$ is given by Eq. (6.1-15). Combining Eqs. (6.1-15) and (6.1-16), we derive

$$z \geqslant \frac{\Delta x(D + M\Delta x)}{\lambda_0 - 2\sin\theta\Delta x},$$

which is the minimum distance the object must be positioned away from the CCD sensor.

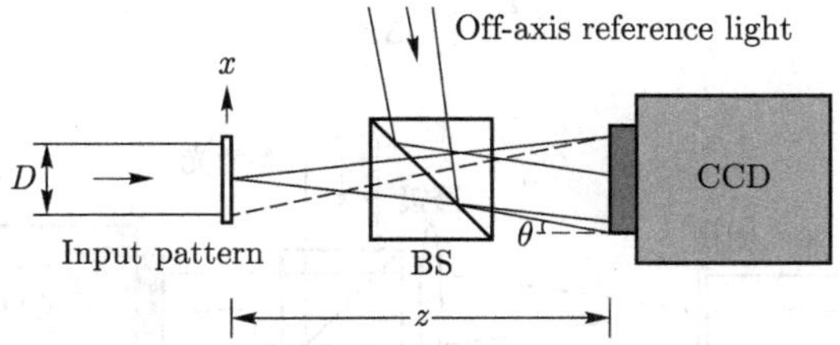

Figure 6.1-5 Typical setup of off-axis Fresnel holography. θ is the recording angle

Fourier Hologram

For recording of *Fourier transforms*, the setup is shown in Figure 6.1-6 (also see Figure 5.3-3a). The input of size D is centered on the optical axis and is on the front focal plane of the Fourier transform lens. A CCD is at the back focal plane of the Fourier lens. A single point of the input will produce a plane wave on the CCD, and at the same time, a reference point source $\delta(x + x_0)$ produced by the reference light provides a tilted plane wave $e^{-jk_0 x_0 x/f}$ on the CCD plane (see Section 5.3.2). The two plane waves result in a uniform sinusoidal fringe on the CCD. However, the finest fringe is contributed from the point source at $\delta(x - D/2)$, which emits at the edge of the input at $x = D/2$. This point source produces a plane wave $e^{jk_0 Dx/(2f)}$ on the CCD. Hence, the finest fringe on the CCD plane is given by

$$\left|e^{jk_0 Dx/(2f)} + e^{-jk_0 x_0 x/f}\right|^2 = 2 + 2\cos\left[\pi\frac{(D + 2x_0)x}{\lambda_0 f}\right]. \tag{6.1-17}$$

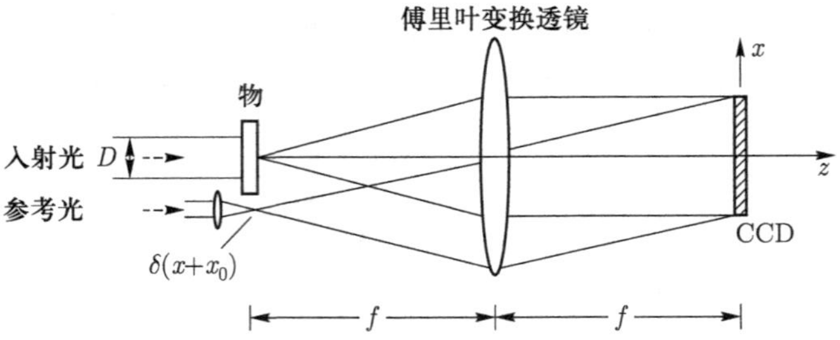

图 6.1-6 傅里叶全息的典型光路

为了求解最精细的条纹, 与 CCD 相关的奈奎斯特率必须至少是局部条纹频率的两倍, 即

$$f_s = \frac{1}{\Delta x} \geqslant 2B, \tag{6.1-18}$$

其中, 由式 (6.1-17) 可得, $B = f_{local} = \frac{1}{2\pi}\frac{\mathrm{d}}{\mathrm{d}x}\left[\pi\frac{(D+2x_0)x}{\lambda_0 f}\right] = \frac{D+2x_0}{2\lambda_0 f}$. 因此, 有

$$\frac{1}{\Delta x} \geqslant \frac{D+2x_0}{\lambda_0 f},$$

或

$$D \leqslant \frac{\lambda_0 f}{\Delta x} - 2x_0. \tag{6.1-19}$$

为了避免在记录过程中出现混叠, 输入的尺寸, 即 D, 必须小于 $\frac{\lambda_0 f}{\Delta x} - 2x_0$. 为了扩大记录尺寸, x_0 可以被设置为零. 然而, 利用如图 6.1-6 所示的记录光路在实际中是不现实的. 因此, 一个实际可行的记录傅里叶全息图的方法如图 6.1-7 所示, 在该情况下, x_0 为有效零, 有下式

$$D \leqslant \frac{\lambda_0 f}{\Delta x}. \tag{6.1-20}$$

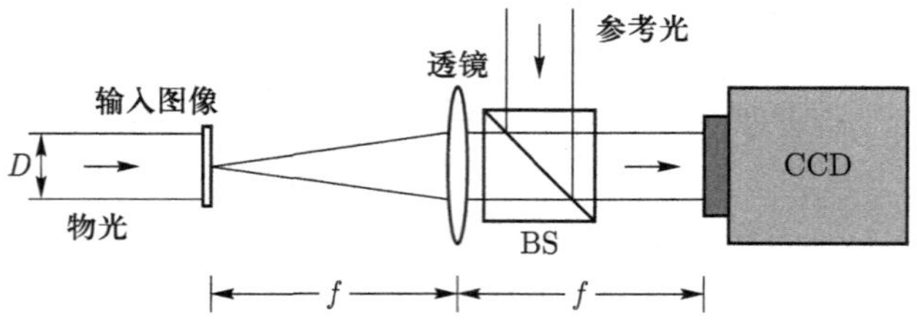

图 6.1-7 可供选择的扩大物体的记录尺寸的傅里叶全息光路

6.2 现代数字全息术

从前文已经看到, 离轴全息术是一种有效分离零级光和孪生像的技术. 然而, 由于 CCD 有限的分辨率, 很难记录高质量的离轴数字菲涅耳全息图. 数字相移全息术 (*phase-shifting holography*, PSH) 是一种充分利用 CCD 有限分辨率的现代技术. 本节还将介绍光学扫描全息术, 这是一种完全绕过 CCD 而进行记录的单像素全息记录技术.

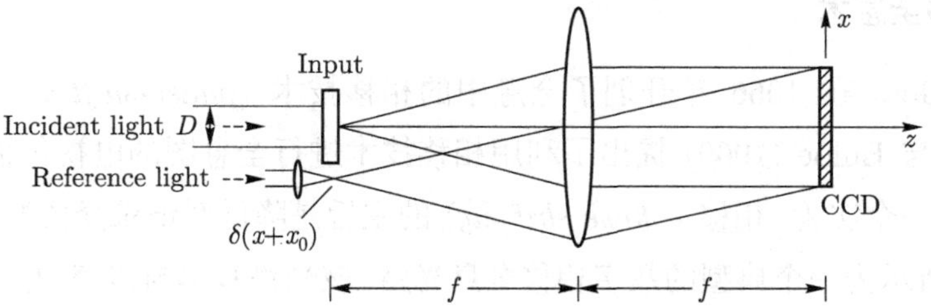

Figure 6.1-6 Typical setup of Fourier holography

In order to resolve the finest fringe, the Nyquist frequency associated with the CCD must be at least twice the local frequency, which is

$$f_s = \frac{1}{\Delta x} \geqslant 2B, \tag{6.1-18}$$

where $B = f_{local} = \frac{1}{2\pi}\frac{\mathrm{d}}{\mathrm{d}x}\left[\pi\frac{(D+2x_0)x}{\lambda_0 f}\right] = \frac{D+2x_0}{2\lambda_0 f}$ is obtained from Eq. (6.1-17). Therefore, we have

$$\frac{1}{\Delta x} \geqslant \frac{D+2x_0}{\lambda_0 f},$$

or

$$D \leqslant \frac{\lambda_0 f}{\Delta x} - 2x_0. \tag{6.1-19}$$

To avoid aliasing during recording, the size of the input, that is, D must be less than $\frac{\lambda_0 f}{\Delta x} - 2x_0$. To maximize the recording size, x_0 can be set to zero. However, this is not possible in practice using the recoding geometry shown in Figure 6.1-6. An alternative and practical way to record a Fourier hologram is shown in Figure 6.1-7, and in this case, x_0 is effective zero, giving

$$D \leqslant \frac{\lambda_0 f}{\Delta x}. \tag{6.1-20}$$

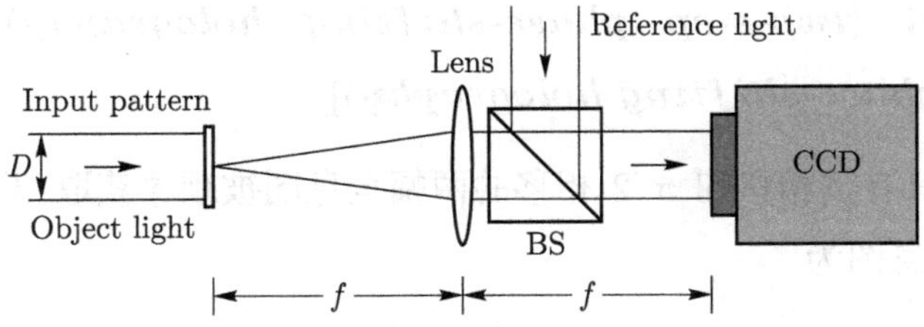

Figure 6.1-7 Alternative setup of Fourier holography to maximize the recording size of the object

6.2 Modern Digital Holographic Techniques

We have seen that off-axis holography is an effective technique to separate the zeroth-order light and the twin image. However, because of the limited resolution of the CCD, it is difficult to record a high-quality off-axis digital Fresnel hologram. Digital *phase-shifting holography* is a modern-day technique to fully utilize the limited resolution of the CCD. In this section, we also cover optical scanning holography that is a single-pixel holographic recording technique, completely bypassing the use of a CCD for recording.

6.2.1 相移全息术

Gabor 与 Goss 在 1966 年开创了全息中的相移技术 (*phase-shifting technique*). 几年后, Burckhardt 与 Enloe (1969) 提出了利用相移技术进行全息图的电视传输, 最终 Berrang (1970) 演示了第一个实验. 相移 (*phase shifting*) 的主旨是降低对记录介质的分辨率要求.

如图 6.2-1 所示为一个典型的数字相移全息光路, 该光路基本与传统的同轴菲涅耳数字全息光路相同, 除了光学系统的参考光路, 在压电式换能器 (*piezoelectric transducer*, PZT) 上安装了一个反射镜作为移相器来实现相移. 当 PZT 移动时, 它会在物光与参考光之间产生光程差.

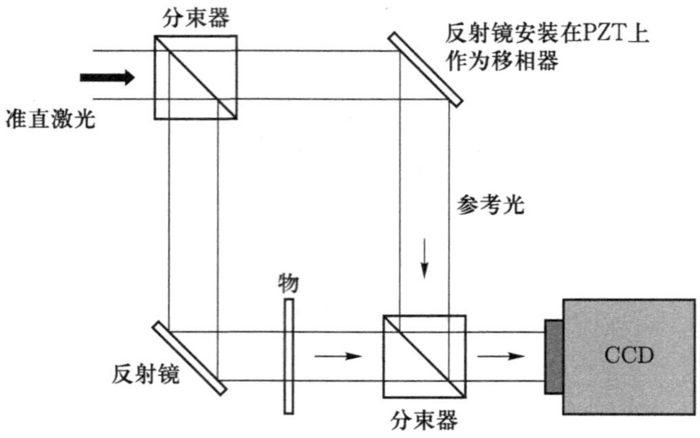

图 6.2-1 典型的数字相移全息光路

在 PSH 中, 我们需要根据物光与参考光之间的不同相位差, 得到多幅全息图, 可以将相移全息图 (*phase-shifted hologram*) 表示为

$$I_\delta = |\psi_0 + \psi_r \mathrm{e}^{-\mathrm{j}\delta}|^2 = |\psi_0|^2 + |\psi_r|^2 + \psi_0\psi_r^*\mathrm{e}^{\mathrm{j}\delta} + \psi_0^*\psi_r\mathrm{e}^{-\mathrm{j}\delta}, \tag{6.2-1}$$

其中, ψ_0 和 ψ_r 分别为物光波与参考光波在 CCD 上的复振幅, δ 表示移相器 (*phase shifter*) 产生的相位.

两步相移全息术 (*two-step phase-shifting holography*) [正交相移全息术 (*quadrature phase-shifting holography*)]

在这种情况下, 只有零相移和 π/2 相移的两幅全息图被顺序获取. 根据式 (6.2-1), $\delta = 0$ 和 $\delta = \pi/2$ 的两幅全息图为

$$I_0 = |\psi_0|^2 + |\psi_r|^2 + \psi_0\psi_r^* + \psi_0^*\psi_r, \tag{6.2-2a}$$

和

$$I_{\pi/2} = |\psi_0 + \psi_r \mathrm{e}^{-\mathrm{j}\pi/2}|^2 = |\psi_0|^2 + |\psi_r|^2 + \mathrm{j}\psi_0\psi_r^* - \mathrm{j}\psi_0^*\psi_r. \tag{6.2-2b}$$

式 (6.2-2a) 和式 (6.2-2b) 可以被写为

$$I_0 - |\psi_0|^2 - |\psi_r|^2 = \psi_0\psi_r^* + \psi_0^*\psi_r,$$

和

$$I_{\pi/2} - |\psi_0|^2 - |\psi_r|^2 = \mathrm{j}\psi_0\psi_r^* - \mathrm{j}\psi_0^*\psi_r.$$

6.2.1 Phase-Shifting Holography

Gabor and Goss pioneered the *phase-shifting technique* in holography in 1966. A few years later, Burckhardt and Enloe (1969) proposed the television transmission of a hologram using phase shifting and finally the first experiments were demonstrated by Berrang (1970). The central theme of *phase shifting* was to reduce the resolution requirements on the recording medium.

A typical setup for digital phase-shifting holography (PSH) is shown in Figure 6.2-1. The setup is basically identical to that of conventional on-axis Fresnel digital holography with the exception that on the reference arm of the optical system, a mirror is mounted on a *piezoelectric transducer* (PZT) as a phase shifter to effectuate phase shifting. As the PZT is moved, it creates an optical path difference between the object wave and the reference wave.

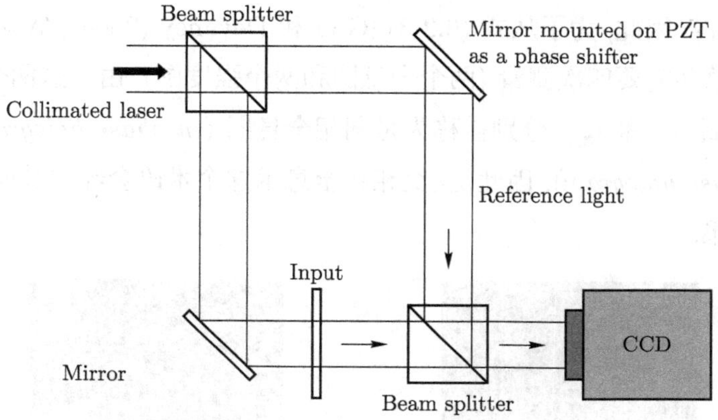

Figure 6.2-1 Typical setup of phase-shifting digital holography

In PSH, we need to take multiple holograms corresponding to the various phase differences between the object wave and the reference wave. We can express the *phase-shifted hologram* as

$$I_\delta = |\psi_0 + \psi_r \mathrm{e}^{-\mathrm{j}\delta}|^2 = |\psi_0|^2 + |\psi_r|^2 + \psi_0\psi_r^*\mathrm{e}^{\mathrm{j}\delta} + \psi_0^*\psi_r\mathrm{e}^{-\mathrm{j}\delta}, \tag{6.2-1}$$

where ψ_0 and ψ_r are the complex amplitude of the object wave and the reference wave, respectively on the CCD; δ stands for the phase induced by the *phase shifter.*

Two-Step Phase-Shifting Holography (Quadrature Phase-Shifting Holography)

In this case, only two holograms with a zero phase shift and a $\pi/2$ phase shift are acquired sequentially. According to Eq. (6.2-1), the two holograms for $\delta = 0$ and $\delta = \pi/2$ are

$$I_0 = |\psi_0|^2 + |\psi_r|^2 + \psi_0\psi_r^* + \psi_0^*\psi_r, \tag{6.2-2a}$$

and

$$I_{\pi/2} = |\psi_0 + \psi_r \mathrm{e}^{-\mathrm{j}\pi/2}|^2 = |\psi_0|^2 + |\psi_r|^2 + \mathrm{j}\psi_0\psi_r^* - \mathrm{j}\psi_0^*\psi_r. \tag{6.2-2b}$$

Equations (6.2-2a) and (6.2-2b) can be written as

$$I_0 - |\psi_0|^2 - |\psi_r|^2 = \psi_0\psi_r^* + \psi_0^*\psi_r,$$

and

$$I_{\pi/2} - |\psi_0|^2 - |\psi_r|^2 = \mathrm{j}\psi_0\psi_r^* - \mathrm{j}\psi_0^*\psi_r.$$

因此, 从上面两个式子, 有

$$I_0 - |\psi_0|^2 - |\psi_r|^2 - \mathrm{j}(I_{\pi/2} - |\psi_0|^2 + |\psi_r|^2) = 2\psi_0\psi_r^*,$$

或

$$\psi_0 = \frac{(I_0 - |\psi_0|^2 - |\psi_r|^2) - \mathrm{j}(I_{\pi/2} - |\psi_0|^2 - |\psi_r|^2)}{2\psi_r^*}. \tag{6.2-3}$$

这是物体在全息面上的复振幅, 即全息面上的物光波. 在数字全息中, 这通常被称为物体的复全息图 (*complex hologram*), 因为可以通过反向传播 (*back-propagation*) 在物面上重新恢复物体的振幅分布. 在式 (6.2-3) 中, ψ_r 通常被认为是参考光波, 实际中它可以是平面波或球面波. 然而, 为了完全求解出 ψ_0, 仍需要两个量, 即物光波的强度 $|\psi_0|^2$ 与参考光波的强度 $|\psi_r|^2$. 这两个强度很容易测得. 因此, 为了从式 (6.2-3) [Guo 和 Devaney (2004), Wang 等 (2004)] 求得物体的复全息图, 总共需要四次测量 (两个全息图和两个强度图). 由于这两个全息图都有零相移和 $\pi/2$ 相移, 因此 I_0 和 $I_{\pi/2}$ 分别被称为是同相全息图 (*in-phase hologram*) 和正交相全息图 (*quadrature-phase hologram*). 因此, 正交相移全息术这个术语会经常被使用, 图 6.2-2 为两步 PSH 的仿真结果.

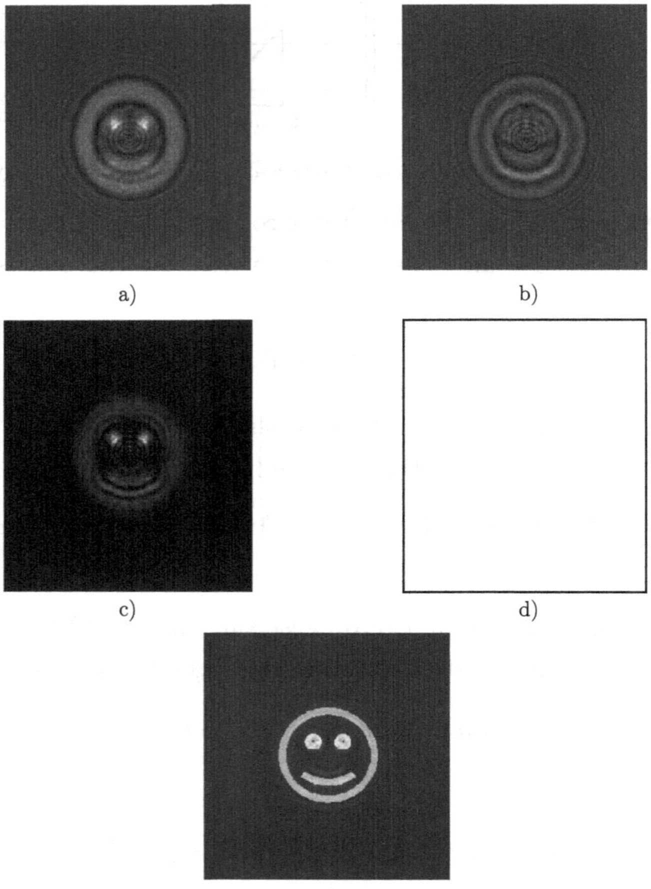

图 6.2-2 两步 PSH 的仿真 a) 同相全息图; b) 正交相全息图; c) 物光波的强度图; d) 参考光波的强度图 (全白图案表示均匀照明); e) 复全息图 [即式 (6.2-3)] 的重建图像

Hence, from the above two equations, we have

$$I_0 - |\psi_0|^2 - |\psi_r|^2 - \mathrm{j}(I_{\pi/2} - |\psi_0|^2 + |\psi_r|^2) = 2\psi_0\psi_r^*,$$

or

$$\psi_0 = \frac{(I_0 - |\psi_0|^2 - |\psi_r|^2) - \mathrm{j}(I_{\pi/2} - |\psi_0|^2 - |\psi_r|^2)}{2\psi_r^*}. \tag{6.2-3}$$

This is the complex amplitude of the object at the hologram plane, that is, the object wave on the hologram plane. In digital holography, it is often referred to as the *complex hologram* of the object as we can retrieve the amplitude distribution of the object in the object plane by performing *back propagation.* In Eq. (6.2-3), ψ_r is usually known as the reference wave, which is either a plane wave or a spherical wave in practice. However, we still need two quantities, that is, the intensities of the object wave, $|\psi_0|^2$, and the reference wave, $|\psi_r|^2$, in order to completely solve for ψ_0. These two intensities can be measured easily. We, therefore, need a total of four measurements (two holograms and two intensity patterns) in order to find the complex hologram of the object from Eq. (6.2-3) [Guo and Devaney (2004), Wang et al. (2004)]. Since the two holograms are with a zero phase shift and a $\pi/2$ phase shift, I_0 and $I_{\pi/2}$ are called the *in-phase hologram* and the *quadrature-phase hologram*, respectively. Hence, the term quadrature phase shifting holography is often used. Figure 6.2-2 shows the simulation results for two-step PSH.

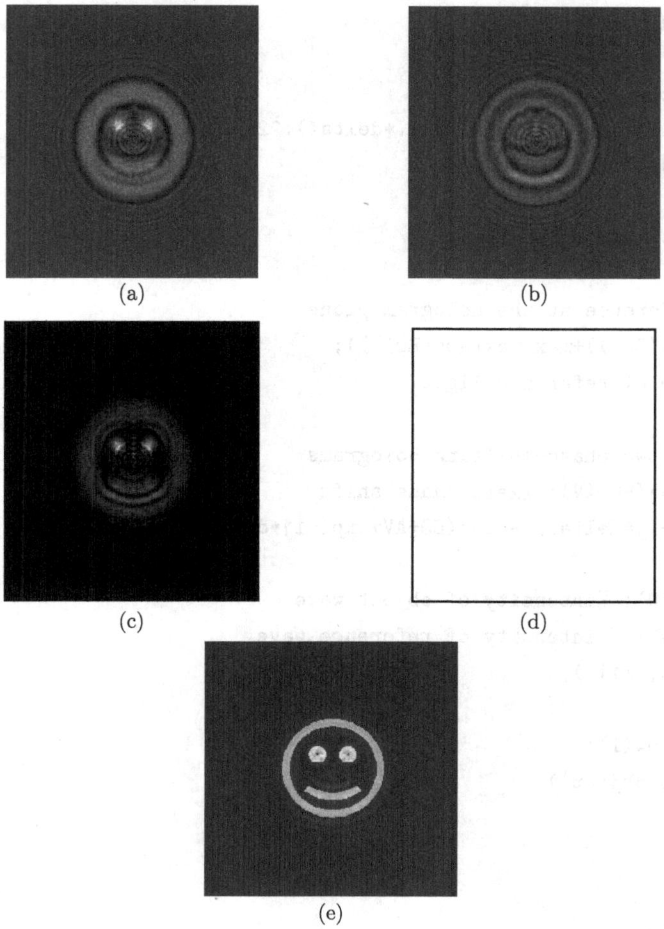

Figure 6.2-2 Simulation of two-step PSH. (a) In-phase hologram, (b) quadrature-phase hologram, (c) intensity pattern of object wave, (d) intensity pattern of reference wave (all-white pattern denoting uniform illumination), and (e) reconstructed image from the complex hologram, Eq. (6.2-3)

图 6.2-2 中的图像是使用如下所示的 m-文件生成的.

```
==========================================
    % simulation_of_two_step_PSH.m
    % Adapted from "Introduction to Modern Digital Holography"
    % by T.-C. Poon & J.-P. Liu
    % Cambridge University Press(2014), Tables 5.2
    clear all; close all;

    %%Reading input bitmap file
    I=imread('smiley3.bmp','bmp'); % original object, 256x256
    I=double(I);

    % parameter setup
    M=256;
    deltax=0.001; % pixel pitch 0.001 cm (10 um)
    w=633*10^-8; % wavelength 633 nm
    z=25; % 25 cm, propagation distance
    delta=pi/2; % phase step

    %Step 1: simulation of propagation
    r=1:M;
    c=1:M;
    [C, R]=meshgrid(c, r);
    A0=fftshift(ifft2(fftshift(I)));
    deltaf=1/M/deltax;
    p=exp(-2i*pi*z.*((1/w)^2-((R-M/2-1).*deltaf).^2-...
      ((C-M/2-1).*deltaf).^2).^0.5);
    Az=A0.*p;
    EO=fftshift(fft2(fftshift(Az)));

    %Step 2: Interference at the hologram plane
    AV=(min(min(abs(EO)))+max(max(abs(EO))));
    % the amplitude of reference light

    % Recording of two phase-shifting holograms
    I0=(EO+AV).*conj(EO+AV); %zero phase shift
    I1=(EO+AV*exp(-1j*delta)).*conj(EO+AV*exp(-1j*delta)); %90 degrees
phase shift
    I2=(EO).*conj(EO); %intensity of object wave
    I3=(AV).*conj(AV);% intensity of reference wave
    MAX=max(max([I0, I1]));

    figure(1); imshow(I);
    title('Original object')
    axis off

    figure(2)
    imshow(I0/MAX);
    title('In-phase hologram ')

    figure(3)
```

The images in Figure 6.2-2 are generated using the m-file shown below.

```
========================================
    % simulation_of_two_step_PSH.m
    % Adapted from "Introduction to Modern Digital Holography"
    % by T.-C. Poon & J.-P. Liu
    % Cambridge University Press(2014), Tables 5.2
    clear all; close all;

    %%Reading input bitmap file
    I=imread('smiley3.bmp','bmp'); % original object, 256x256
    I=double(I);

    % parameter setup
    M=256;
    deltax=0.001; % pixel pitch 0.001 cm (10 um)
    w=633*10^-8; % wavelength 633 nm
    z=25; % 25 cm, propagation distance
    delta=pi/2; % phase step

    %Step 1: simulation of propagation
    r=1:M;
    c=1:M;
    [C, R]=meshgrid(c, r);
    A0=fftshift(ifft2(fftshift(I)));
    deltaf=1/M/deltax;
    p=exp(-2i*pi*z.*((1/w)^2-((R-M/2-1).*deltaf).^2-...
      ((C-M/2-1).*deltaf).^2).^0.5);
    Az=A0.*p;
    EO=fftshift(fft2(fftshift(Az)));

    %Step 2: Interference at the hologram plane
    AV=(min(min(abs(EO)))+max(max(abs(EO))));
    % the amplitude of reference light

    % Recording of two phase-shifting holograms
    I0=(EO+AV).*conj(EO+AV); %zero phase shift
    I1=(EO+AV*exp(-1j*delta)).*conj(EO+AV*exp(-1j*delta)); %90 degrees
phase shift
    I2=(EO).*conj(EO); %intensity of object wave
    I3=(AV).*conj(AV);% intensity of reference wave
    MAX=max(max([I0, I1]));

    figure(1); imshow(I);
    title('Original object')
    axis off

    figure(2)
    imshow(I0/MAX);
    title('In-phase hologram ')

    figure(3)
```

```
imshow(I1/MAX);
axis off
title('Q-phase hologram ')

I2=I2./max(max(I2));
figure (4)
imshow(I2);
axis off
title('intensity pattern of object wave')

I3=I3./max(max(I3));
figure (5)
imshow(I3);
axis off
title('intensity pattern of reference wave')

%Step 3: Reconstruction
CH=(I0-I2-I3)-1j*(I1-I2-I3); % the complex hologram (2-step PSH)
A1=fftshift(ifft2(fftshift(CH)));
Az1=A1.*conj(p);
EI=fftshift(fft2(fftshift(Az1)));
EI=(EI.*conj(EI));
EI=EI/max(max(EI));

figure(6);
imshow(EI);
title('Reconstructed image')
axis off
========================================
```

三步相移全息术 (*three-step phase-shifting holography*)

在三步 PSH 中, 依次获取三幅全息图. 三幅全息图的相位差分别为 $\delta=0$, $\delta=\pi/2$, $\delta=\pi$, 则三幅全息图分别为

$$I_0 = |\psi_0|^2 + |\psi_r|^2 + \psi_0\psi_r^* + \psi_0^*\psi_r, \tag{6.2-4a}$$

$$I_{\pi/2} = |\psi_0 + \psi_r \mathrm{e}^{-\mathrm{j}\pi/2}|^2 = |\psi_0|^2 + |\psi_r|^2 + \mathrm{j}\psi_0\psi_r^* - \mathrm{j}\psi_0^*\psi_r, \tag{6.2-4b}$$

及

$$I_{\pi} = |\psi_0 + \psi_r \mathrm{e}^{-\mathrm{j}\pi}|^2 = |\psi_0|^2 + |\psi_r|^2 - \psi_0\psi_r^* - \psi_0^*\psi_r. \tag{6.2-4c}$$

与上一节的一些数学运算类似, 可求出物体的复全息图为

$$\psi_0 = \frac{(1+\mathrm{j})(I_0 - I_{\pi/2}) + (\mathrm{j}-1)(I_\pi - I_{\pi/2})}{4\psi_r^*}. \tag{6.2-5}$$

四步相移全息术 (*four-step phase-shifting holography*)

在四步 PSH 中, 依次获取四幅全息图. 四幅全息图的相位差分别为 $\delta=0$, $\delta=\pi/2$, $\delta=\pi$, $\delta=3\pi/2$. 因此, 根据式 (6.2-1), 这四幅全息图分别为

$$I_0 = |\psi_0|^2 + |\psi_r|^2 + \psi_0\psi_r^* + \psi_0^*\psi_r, \tag{6.2-6a}$$

```
imshow(I1/MAX);
axis off
title('Q-phase hologram ')

I2=I2./max(max(I2));
figure (4)
imshow(I2);
axis off
title('intensity pattern of object wave')

I3=I3./max(max(I3));
figure (5)
imshow(I3);
axis off
title('intensity pattern of reference wave')

%Step 3: Reconstruction
CH=(I0-I2-I3)-1j*(I1-I2-I3); % the complex hologram (2-step PSH)
A1=fftshift(ifft2(fftshift(CH)));
Az1=A1.*conj(p);
EI=fftshift(fft2(fftshift(Az1)));
EI=(EI.*conj(EI));
EI=EI/max(max(EI));

figure(6);
imshow(EI);
title('Reconstructed image')
axis off
==========================================
```

Three-Step Phase-Shifting Holography

In three-step PSH, three holograms are acquired sequentially. The phase differences for the three holograms are $\delta = 0$, $\delta = \pi/2$, and $\delta = \pi$, and the three holograms are

$$I_0 = |\psi_0|^2 + |\psi_r|^2 + \psi_0\psi_r^* + \psi_0^*\psi_r, \tag{6.2-4a}$$

$$I_{\pi/2} = |\psi_0 + \psi_r e^{-j\pi/2}|^2 = |\psi_0|^2 + |\psi_r|^2 + j\psi_0\psi_r^* - j\psi_0^*\psi_r, \tag{6.2-4b}$$

and

$$I_\pi = |\psi_0 + \psi_r e^{-j\pi}|^2 = |\psi_0|^2 + |\psi_r|^2 - \psi_0\psi_r^* - \psi_0^*\psi_r. \tag{6.2-4c}$$

Similar to some mathematical manipulations in the last section, we can find that the complex hologram of the object is given by

$$\psi_0 = \frac{(1+j)(I_0 - I_{\pi/2}) + (j-1)(I_\pi - I_{\pi/2})}{4\psi_r^*}. \tag{6.2-5}$$

Four-Step Phase-Shifting Holography

In four-step PSH, four holograms are acquired sequentially. The phase differences for the four holograms are $\delta = 0$, $\delta = \pi/2$, $\delta = \pi$, and $\delta = 3\pi/2$. Therefore, according to Eq. (6.2-1), the four holograms are

$$I_0 = |\psi_0|^2 + |\psi_r|^2 + \psi_0\psi_r^* + \psi_0^*\psi_r, \tag{6.2-6a}$$

$$I_{\pi/2} = |\psi_0|^2 + |\psi_r|^2 + \mathrm{j}\psi_0\psi_r^* - \mathrm{j}\psi_0^*\psi_r, \tag{6.2-6b}$$

$$I_{\pi} = |\psi_0|^2 + |\psi_r|^2 - \psi_0\psi_r^* - \psi_0^*\psi_r, \tag{6.2-6c}$$

$$I_{3\pi/2} = |\psi_0|^2 + |\psi_r|^2 - \mathrm{j}\psi_0\psi_r^* + \mathrm{j}\psi_0^*\psi_r. \tag{6.2-6d}$$

从这四幅全息图中, 可以推导出物体的复全息图

$$\psi_0 = \frac{(I_0 - I_\pi) - \mathrm{j}(I_{\pi/2} - I_{3\pi/2})}{4\psi_r^*}. \tag{6.2-7}$$

在标准的正交 PSH 中, 需要两幅全息图和两幅强度图才能完全计算出无零级光和无孪生像的复全息图的重建. 在四步 PSH 中, 也需要四次测量. 在三步 PSH 中, 可以省去对物光的强度图的测量, 从而减少整个全息记录过程. 近年来 PSH 的研究进展表明, 只需要两幅全息图及一幅参考光强度图即可获取复全息图 [Meng 等 (2006)]. 尽管这也需要三次测量, 但相比三步 PSH, 它更具优势, 因为强度图的记录对引起物光和参考光之间相位变化的任何振动都不敏感. 在两步 PSH 方面的最新进展表明所谓的仅两步正交 (*two-step-only quadrature*) PSH 仅需要两幅全息图来恢复物体的复全息图 [Liu 和 Poon (2009)]. 之后, Liu 等 (2011) 进行了二次、三次和四次曝光的正交 PSH 的比较, 感兴趣的读者请参阅本章参考文献.

6.2.2 相干模式的光学扫描全息术

双光瞳外差图像处理器

光学扫描全息术 (*optical scanning holography*, OSH) 是一种截然不同的全息记录方法, 因为它避开了使用 CCD 进行记录. 实际上, 它是一种单像素全息记录技术, 使用光电二极管 (*photodiode*) 等光电探测器进行记录. OSH 的最初想法可以追溯到 20 世纪 70 年代末, 当时 Poon 和 Korpel 用他们的双光瞳外差图像处理器 (*two-pupil heterodyning image processor*) 研究了双极性图像处理 (*bipolar image processing*). 本书 4.5 节中简要地介绍过双光瞳系统 (*two-pupil system*). 本节将先介绍一个广义的双光瞳系统, 如图 6.2-3 所示, 然后将展示该系统是如何构架成为一个全息记录系统的, 及这种技术为何被称为光学扫描全息术. 其整个系统是光电混合的. 在解释系统的电学处理部分之前, 需先描述其光学系统.

如图 6.2-3 所示, 在光学系统中, 当 z_0 等于零时, 本质上是一个其透镜为 L1 和 L2 的 4-f 系统, z_0 为从透镜 L1 的后焦面到物体测量面的距离. 将三维物体建模为一堆薄且弱散射的横向切片, 其中每个切片由一个透过率函数 $T(x, y; z)$ 表示. 位于透镜 L1 的前焦面的两个光瞳 $p_1(x, y)$ 和 $p_2(x, y)$, 分别被频率为 ω_0 和 $\omega_0 + \Omega$ 的平面波照射, 其中, ω_0 为激光的频率. 其中一个平面波的频率从 ω_0 移动到 $\omega_0 + \Omega$, 这可以很方便地利用移频器来实现, 如声光或电光调制器 (将在第七章介绍). 分束器用于合成从两个光瞳发出的两束光, 合成后的光通过 x-y 扫描

$$I_{\pi/2} = |\psi_0|^2 + |\psi_r|^2 + \mathrm{j}\psi_0\psi_r^* - \mathrm{j}\psi_0^*\psi_r, \tag{6.2-6b}$$

$$I_{\pi} = |\psi_0|^2 + |\psi_r|^2 - \psi_0\psi_r^* - \psi_0^*\psi_r, \tag{6.2-6c}$$

$$I_{3\pi/2} = |\psi_0|^2 + |\psi_r|^2 - \mathrm{j}\psi_0\psi_r^* + \mathrm{j}\psi_0^*\psi_r. \tag{6.2-6d}$$

From the four holograms, we can derive the complex hologram of the object and it is given by

$$\psi_0 = \frac{(I_0 - I_\pi) - \mathrm{j}(I_{\pi/2} - I_{3\pi/2})}{4\psi_r^*}. \tag{6.2-7}$$

In standard quadrature PSH, two holograms and two intensity patterns are needed to completely calculate the complex hologram for zeroth-order-free and twin image-free reconstruction. In four-step PSH, also four measurements are needed. The measurement of the intensity pattern of the object wave can be omitted in three-step PSH, thus reducing the overall holographic recording process. Recent advances in PSH research lead to the use of only two holograms and a reference intensity to extract the complex hologram [Meng et al. (2006)]. While it also takes three measurements, it is advantageous as compared to those in three-step PSH because the recording of the intensity pattern is not sensitive to any vibration that causes phase variation between the object wave and the reference wave. Most recent development in two-step PSH has led to what is known as *two-step-only quadrature PSH* in which only two holograms are needed to retrieve the complex hologram of the object [Liu and Poon (2009)]. Subsequently, Liu et al. (2011) have provided a report on the comparison of two-, three- and four-exposure quadrature PSH. Interested readers are referred to the references in the biography list at the end of this chapter.

6.2.2 Optical Scanning Holography in Coherent Mode

Two-Pupil Heterodyning Image Processor

Optical scanning holography (OSH) is a radically different approach to holographic recording as it bypasses the use of a CCD for recording. Indeed, it is a single-pixel holographic recording technique where a photodetector such as a *photodiode* is used for recording. The original idea of OSH was dated back to the late 1970s when Poon and Korpel investigated bipolar image processing with their *two-pupil heterodyning image processor* (1979). We have briefly encountered a *two-pupil system* in Section 4.5. In this section, we will first describe a generalized version of the *two-pupil system*, shown in Figure 6.2-3. We will then show how the system is configured to become a holographic recording system and such a technique is known as optical scanning holography. The overall system is hybrid (optical/electrical). We shall describe the optical system first before explaining the electrical processing part of the system.

We essentially have a 4-f system with lenses L1 and L2 when z_0 is zero in the optical system. z_0 is a distance measured away from the back focal plane of lens L1. We model the 3-D object as a stack of transverse slices where each slide is represented by a transparency function $T(x, y; z)$, which is thin and weakly scattering. $p_1(x, y)$ and $p_2(x, y)$ are the two pupils located at the front focal plane of lens L1, and they are illuminated by plane waves at frequencies ω_0 and $\omega_0 + \Omega$, respectively, where ω_0 is the frequency of a laser. The frequency shift of one of the plane waves from ω_0 to $\omega_0 + \Omega$ can be conveniently done by the use of a frequency shifter, such as an acousto-optic or electro-optic modulator (to be covered in

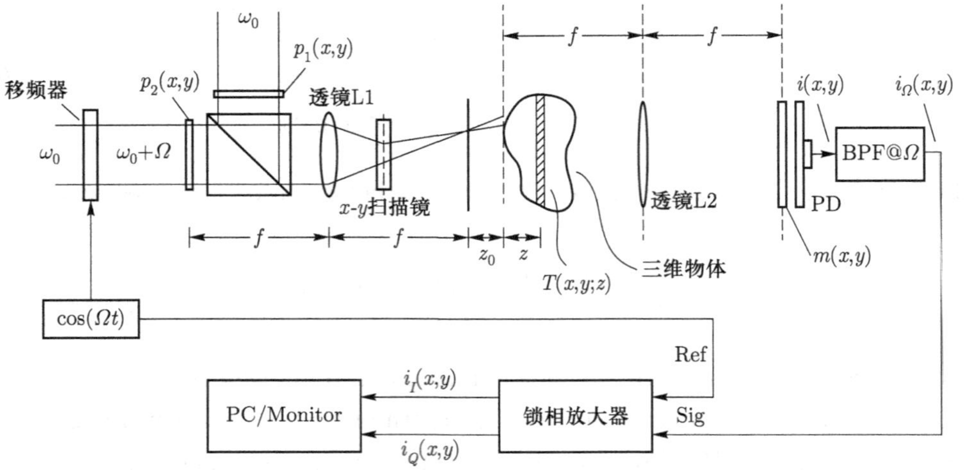

图 6.2-3 典型的双光瞳外差图像处理装置. BPF@Ω 为调频在 Ω 处的带通滤波器, PD 为光电探测器 [Zhang 等 (2019)]

镜投射到三维物体上从而进行光栅扫描 (*raster scan*).

从 $p_1(x,y)$ 发出的振荡频率为 ω_0 的复光场, 在通过透镜 L1 的后焦面, 再到达物体切片位置处即 z_0+z 面上时为

$$s_1(x,y;z+z_0)\mathrm{e}^{\mathrm{j}\omega_0 t}=\left[\mathcal{F}\{p_1(x,y)\}\Big|_{k_x=\frac{k_0x}{f},k_y=\frac{k_0y}{f}} * h(x,y;z+z_0)\right]\mathrm{e}^{\mathrm{j}\omega_0 t}, \tag{6.2-8}$$

因为光瞳已经被透镜 L1 进行了傅里叶变换, 然后传播了 $z+z_0$ 的距离. 类似地, 从 $p_2(x,y)$ 出射的振荡频率为 $\omega_0+\Omega$ 的复光场到达物体切片时为

$$s_2(x,y;z+z_0)\mathrm{e}^{\mathrm{j}(\omega_0+\Omega)t}=\left[\mathcal{F}\{p_2(x,y)\}\Big|_{k_x=\frac{k_0x}{f},k_y=\frac{k_0y}{f}} * h(x,y;z+z_0)\right]\mathrm{e}^{\mathrm{j}(\omega_0+\Omega)t}. \tag{6.2-9}$$

因此, 在物体切片处总的复光场为

$$s(x,y;z)=s_1(x,y;z+z_0)\mathrm{e}^{\mathrm{j}\omega_0 t}+s_2(x,y;z+z_0)\mathrm{e}^{\mathrm{j}(\omega_0+\Omega)t}. \tag{6.2-10}$$

这是总的扫描光场, 它由两个扫描光束 s_1 和 s_2 组成. 根据 4.4 节中提出的光学扫描原理, 紧贴物体切片 $T(x,y;z)$ 后方的光场由 $s(x''-x(t),y''-y(t);z)T(x'',y'';z)$ 给出, 或由下式给出

$$s(x',y';z)T(x'+x(t),y'+y(t);z),$$

如果使用了 $x''-x=x'$ 和 $y''-y=y'$. 同样, $x(t)$ 和 $y(t)$ 表示物体相对于扫描场的瞬时二维位置. 该场现在通过傅里叶变换透镜 L2 传播并到达光电探测器 (PD) 前方的掩模 (*mask*) $m(x,y)$ 处.

根据图 3.4-6, 复光场 $t(x,y)$, 位于焦距为 f 的傅里叶变换透镜前方 d_0 处时, 所给出的在透镜后焦面上的光场分布 [见式 (3.4-29)] 正比于下式

$$\mathrm{e}^{-\mathrm{j}\frac{k_0}{2f}\left(1-\frac{d_0}{f}\right)(x^2+y^2)}\mathcal{F}\{t(x,y)\}\Big|_{k_x=\frac{k_0x}{f},k_y=\frac{k_0y}{f}}$$

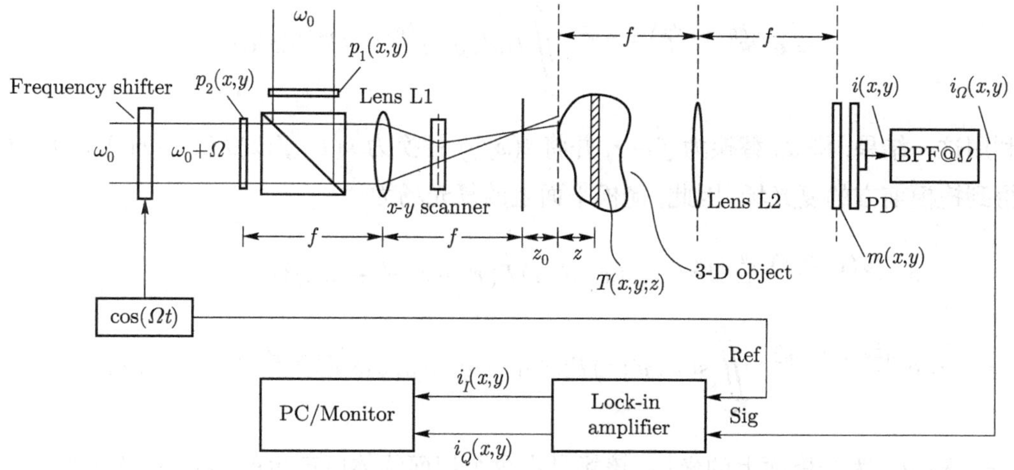

Figure 6.2-3 Typical setup of two-pupil heterodyning image processing. BPF@Ω: bandpass filter tuned at frequency Ω, PD: photodetector. Adapted from Zhang et al. (2019)

Chapter 7). The beam splitter is used to combine the two beams exiting from the two pupils and the combined beam is projected onto the 3-D object through the x-y scanner for a raster scan as shown in Figure 6.2-3.

The complex field oscillating at frequency ω_0 from $p_1(x, y)$ just before the object slice at position $z_0 + z$ from the focal plane of lens L1 is given by

$$s_1(x, y; z + z_0)e^{j\omega_0 t} = \left[\mathcal{F}\{p_1(x, y)\}\Big|_{k_x=\frac{k_0 x}{f}, k_y=\frac{k_0 y}{f}} * h(x, y; z + z_0)\right] e^{j\omega_0 t}, \tag{6.2-8}$$

as the pupil has been Fourier transformed by lens L1 and then propagated for a distance $z + z_0$. Similarly, the complex field exiting from $p_2(x, y)$ with a time oscillation $\omega_0 + \Omega$ at the object slice is

$$s_2(x, y; z + z_0)e^{j(\omega_0+\Omega)t} = \left[\mathcal{F}\{p_2(x, y)\}\Big|_{k_x=\frac{k_0 x}{f}, k_y=\frac{k_0 y}{f}} * h(x, y; z + z_0)\right] e^{j(\omega_0+\Omega)t}. \tag{6.2-9}$$

Hence, the total complex field at the object slide is

$$s(x, y; z) = s_1(x, y; z + z_0)e^{j\omega_0 t} + s_2(x, y; z + z_0)e^{j(\omega_0+\Omega)t}. \tag{6.2-10}$$

This is the total scanning field, which consists of two scanning beams s_1 and s_2. According to the principle of optical scanning developed in Section 4.4, the field just after the object slide, $T(x, y; z)$, is given by $s(x'' - x(t), y'' - y(t); z)T(x'', y''; z)$ or alternatively given by

$$s(x', y'; z)T(x' + x(t), y' + y(t); z),$$

if $x'' - x = x'$ and $y'' - y = y'$ have been used. Again, $x(t)$ and $y(t)$ represent the instantaneous 2-D position of the object with respect to the scanning field. This field now propagates through the Fourier transform lens L2 and reaches a *mask* $m(x, y)$ placed in front of the photodetector, PD.

According to Figure 3.4-6, the complex field, $t(x, y)$, d_0 in front of the Fourier transform lens of focal length f, gives the field distribution on the focal plane proportional to [see Eq. (3.4-29)]

$$e^{-j\frac{k_0}{2f}\left(1-\frac{d_0}{f}\right)(x^2+y^2)}\mathcal{F}\{t(x, y)\}\Big|_{k_x=\frac{k_0 x}{f}, k_y=\frac{k_0 y}{f}}$$

$$= \mathrm{e}^{-\mathrm{j}\frac{k_0}{2f}\left(1-\frac{d_0}{f}\right)(x^2+y^2)} \iint\limits_{-\infty}^{\infty} t(x',y')\mathrm{e}^{\frac{\mathrm{j}k_0}{f}(xx'+yy')}\mathrm{d}x'\mathrm{d}y'.$$

此处利用这一结果, 将 d_0 替换为 $f-z$, 并将 $t(x,y)$ 替换为 $s(x',y';z)T(x'+x(t),y'+y(t);z)$, 从而得到掩模前方的复光场. 因此, 掩模平面上的复光场为

$$\begin{aligned}&\mathrm{e}^{-\mathrm{j}\frac{k_0}{2f}\left(1-\frac{f-z}{f}\right)(x_m^2+y_m^2)}\mathcal{F}\{s(x',y';z)T(x'+x,y'+y;z)\}\big|_{k_x=\frac{k_0x_m}{f},k_y=\frac{k_0y_m}{f}}\\ &= \mathrm{e}^{-\mathrm{j}\frac{k_0z}{2f^2}(x_m^2+y_m^2)}\iint\limits_{-\infty}^{\infty} s(x',y';z)T(x'+x,y'+y;z)\mathrm{e}^{\frac{\mathrm{j}k_0}{f}(x_mx'+y_my')}\mathrm{d}x'\mathrm{d}y',\end{aligned}$$

其中, x_m 和 y_m 为掩模面上的坐标. 该场是由单个切面物体所产生的. 对于一个三维物体, 只需对场在三维物体的厚度 z 上积分, 紧贴掩模前方的总光场为

$$\int \mathrm{e}^{-\mathrm{j}\frac{k_0z}{2f^2}(x_m^2+y_m^2)}\iint\limits_{-\infty}^{\infty} s(x',y';z)T(x'+x,y'+y;z)\mathrm{e}^{\frac{\mathrm{j}k_0}{f}(x_mx'+y_my')}\mathrm{d}x'\mathrm{d}y'\mathrm{d}z.$$

最终, 掩模后方的复光场为

$$\begin{aligned}\psi(x,y;x_m,y_m) = \Bigg\{\int \mathrm{e}^{-\mathrm{j}\frac{k_0z}{2f^2}(x_m^2+y_m^2)}\iint\limits_{-\infty}^{\infty} s(x',y';z)T(x'+x,y'+y;z)\\ \times\mathrm{e}^{\frac{\mathrm{j}k_0}{f}(x_mx'+y_my')}\mathrm{d}x'\mathrm{d}y'\mathrm{d}z\Bigg\}m(x_m,y_m).\end{aligned} \tag{6.2-11}$$

响应强度的光电探测器 (PD) 通过对探测器的光敏面进行空间积分输出电流 $i(x,y)$

$$i(x,y) \propto \iint |\psi(x,y;x_m,y_m)|^2\mathrm{d}x_m\mathrm{d}y_m. \tag{6.2-12}$$

利用点探测器进行相干处理 (*coherent processing using point detector*)

来看一个简单的情况. 当掩模为一个针孔时, 即 $m(x_m,y_m)=\delta(x_m,y_m)$, 在这种情况下, 有一个点探测器 (*point detector*). 换句话说, 该探测器仅在它的光敏面内一个单点位置处接收光. 当出现这种情况时, 根据式 (6.2-11), 掩模后的复光场为 [也见 δ 函数的乘法性质, 即式 (2.1-4)]

$$\psi(x,y;x_m,y_m) = \left\{\iiint\limits_{-\infty}^{\infty} s(x',y';z)T(x'+x,y'+y;z)\mathrm{d}x'\mathrm{d}y'\mathrm{d}z\right\}\delta(x_m,y_m).$$

将该光场代入式 (6.4-12), 并对 x_m 和 y_m 积分, 式 (6.2-12) 变为

$$i(x,y) \propto \left\{\iiint\limits_{-\infty}^{\infty} s(x',y';z')T(x'+x,y'+y;z')\mathrm{d}x'\mathrm{d}y'\mathrm{d}z'\right\}$$

$$= \mathrm{e}^{-\mathrm{j}\frac{k_0}{2f}\left(1-\frac{d_0}{f}\right)(x^2+y^2)} \iint_{-\infty}^{\infty} t(x',y')\mathrm{e}^{\frac{\mathrm{j}k_0}{f}(xx'+yy')}\mathrm{d}x'\mathrm{d}y'.$$

Using this result in our situation, we replace d_0 with $f-z$, and $t(x,y)$ with the field given by $s(x',y';z)T(x'+x(t),y'+y(t);z)$ to obtain the complex field just in front of the mask. Hence, the complex field on the plane of the mask is

$$\mathrm{e}^{-\mathrm{j}\frac{k_0}{2f}\left(1-\frac{f-z}{f}\right)(x_m^2+y_m^2)} \mathcal{F}\{s(x',y';z)T(x'+x,y'+y;z)\}\big|_{k_x=\frac{k_0x_m}{f},k_y=\frac{k_0y_m}{f}}$$

$$= \mathrm{e}^{-\mathrm{j}\frac{k_0z}{2f^2}(x_m^2+y_m^2)} \iint_{-\infty}^{\infty} s(x',y';z)T(x'+x,y'+y;z)\mathrm{e}^{\frac{\mathrm{j}k_0}{f}(x_mx'+y_my')}\mathrm{d}x'\mathrm{d}y',$$

where we denote x_m and y_m as the coordinates in the plane of the mask. This field is caused by a single object slide. For a 3-D object, we simply need to integrate the field over the thickness, z, of the 3-D object, giving the total field just before the mask as

$$\int \mathrm{e}^{-\mathrm{j}\frac{k_0z}{2f^2}(x_m^2+y_m^2)} \iint_{-\infty}^{\infty} s(x',y';z)T(x'+x,y'+y;z)\mathrm{e}^{\frac{\mathrm{j}k_0}{f}(x_mx'+y_my')}\mathrm{d}x'\mathrm{d}y'\mathrm{d}z.$$

Finally, the complex field after the mask is

$$\psi(x,y;x_m,y_m) = \left\{\int \mathrm{e}^{-\mathrm{j}\frac{k_0z}{2f^2}(x_m^2+y_m^2)} \iint_{-\infty}^{\infty} s(x',y';z)T(x'+x,y'+y;z)\right.$$
$$\left.\times\, \mathrm{e}^{\frac{\mathrm{j}k_0}{f}(x_mx'+y_my')}\mathrm{d}x'\mathrm{d}y'\mathrm{d}z\right\} m(x_m,y_m). \tag{6.2-11}$$

The photodetector, PD, responding to intensity gives out a current $i(x,y)$ as an output by spatially integrating over the active area of the detector [see Eq. (4.4-3)]:

$$i(x,y) \propto \iint |\psi(x,y;x_m,y_m)|^2\mathrm{d}x_m\mathrm{d}y_m. \tag{6.2-12}$$

Coherent Processing Using Point Detector

Let us examine a simple case that when the mask is a pinhole, that is, $m(x_m,y_m)=\delta(x_m,y_m)$. In this case, we have a *point detector*. In other words, the detector only receives light at a single point location within its active area. When this happens, the complex field after the mask is, according to Eq. (6.2-11) [also see the product property of the delta function, Eq. (2.1-4)],

$$\psi(x,y;x_m,y_m) = \left\{\iiint_{-\infty}^{\infty} s(x',y';z)T(x'+x,y'+y;z)\mathrm{d}x'\mathrm{d}y'\mathrm{d}z\right\}\delta(x_m,y_m).$$

Putting this field into Eq. (6.4-12) and after integrating over x_m and y_m, Eq. (6.2-12) becomes

$$i(x,y) \propto \left\{\iiint_{-\infty}^{\infty} s(x',y';z')T(x'+x,y'+y;z')\mathrm{d}x'\mathrm{d}y'\mathrm{d}z'\right\}$$

$$\times\left\{\iiint_{-\infty}^{\infty} s^*(x'',y'';z'')T^*(x''+x,y''+y;z'')\mathrm{d}x''\mathrm{d}y''\mathrm{d}z''\right\}.$$

将式 (6.2-10) 中的扫描场 $s(x,y;z)$ 代入上式, 有

$$\begin{aligned}i(x,y)\propto&\left\{\iiint_{-\infty}^{\infty}[s_1(x',y';z'+z_0)\mathrm{e}^{\mathrm{j}\omega_0 t}+s_2(x',y';z'+z_0)\mathrm{e}^{\mathrm{j}(\omega_0+\Omega)t}]\right.\\&\left.\times T(x'+x,y'+y;z')\mathrm{d}x'\mathrm{d}y'\mathrm{d}z'\right\}\\&\times\left\{\iiint_{-\infty}^{\infty}[s_1^*(x'',y'';z''+z_0)\mathrm{e}^{-\mathrm{j}\omega_0 t}+s_2^*(x'',y'';z''+z_0)\mathrm{e}^{-\mathrm{j}(\omega_0+\Omega)t}]\right.\\&\left.\times T^*(x''+x,y''+y;z'')\mathrm{d}x''\mathrm{d}y''\mathrm{d}z''\right\}.\end{aligned}$$

重新组合上式, 有

$$\begin{aligned}&\int[s_1(x',y';z'+z_0)\mathrm{e}^{\mathrm{j}\omega_0 t}+s_2(x',y';z'+z_0)\mathrm{e}^{\mathrm{j}(\omega_0+\Omega)t}]\\&\times[s_1^*(x'',y'';z''+z_0)\mathrm{e}^{-\mathrm{j}\omega_0 t}+s_2^*(x'',y'';z''+z_0)\mathrm{e}^{-\mathrm{j}(\omega_0+\Omega)t}]\\&\times T(x'+x,y'+y;z')T^*(x''+x,y''+y;z'')\mathrm{d}x'\mathrm{d}y'\mathrm{d}x''\mathrm{d}y''\mathrm{d}z'\mathrm{d}z''.\end{aligned}$$

包含 $s_1s_1^*$ 和 $s_2s_2^*$ 的项给出的基带电流 (*baseband current*) $i_{base}(t)$ 为

$$\begin{aligned}i_{base}(x,y)=&\int(s_1s_1^*+s_2s_2^*)T(x'+x,y'+y;z')\\&\times T^*(x''+x,y''+y;z'')\mathrm{d}x'\mathrm{d}y'\mathrm{d}x''\mathrm{d}y''\mathrm{d}z'\mathrm{d}z'',\end{aligned}$$

包含频率 Ω 的项给出了外差电流 (*heterodyne current*) $i_\Omega(t)$ 为

$$\begin{aligned}i_\Omega(x,y)=&\int(s_2s_1^*\mathrm{e}^{\mathrm{j}\Omega t}+s_1s_2^*\mathrm{e}^{-\mathrm{j}\Omega t})T(x'+x,y'+y;z')\\&\times T^*(x''+x,y''+y;z'')\mathrm{d}x'\mathrm{d}y'\mathrm{d}x''\mathrm{d}y''\mathrm{d}z'\mathrm{d}z''.\end{aligned}\tag{6.2-13a}$$

把带有单撇与双撇的变量分组并重组, 得

$$\begin{aligned}i_\Omega(x,y)=&\left[\int s_2(x',y';z'+z_0)T(x'+x,y'+y;z')\mathrm{d}x'\mathrm{d}y'\mathrm{d}z'\right.\\&\left.\times\int s_1^*(x'',y'';z''+z_0)T^*(x''+x,y''+y;z'')\mathrm{d}x''\mathrm{d}y''\mathrm{d}z''\right]\mathrm{e}^{\mathrm{j}\Omega t}\\&+\left[\int s_2^*(x'',y'';z''+z_0)T^*(x''+x,y''+y;z'')\mathrm{d}x''\mathrm{d}y''\mathrm{d}z''\right.\\&\left.\times\int s_1(x',y';z'+z_0)T(x'+x,y'+y;z')\mathrm{d}x'\mathrm{d}y'\mathrm{d}z'\right]\mathrm{e}^{-\mathrm{j}\Omega t}.\end{aligned}\tag{6.2-13b}$$

可以发现, 其中的第二项仅仅是第一项的复共轭. 由于 $2\mathrm{Re}[\mathrm{e}^{\mathrm{j}\theta}]=\mathrm{e}^{\mathrm{j}\theta}+\mathrm{e}^{-\mathrm{j}\theta}$, 其中 $\mathrm{Re}[\cdot]$ 表示取括号内复值的实部, 忽略常数因子, 可以写出 $i_\Omega(t)$ 为

$$i_\Omega(x,y)=\mathrm{Re}\left[\int s_2(x',y';z'+z_0)T(x'+x,y'+y;z')\mathrm{d}x'\mathrm{d}y'\mathrm{d}z'\right.$$

$$\times \left\{ \iiint_{-\infty}^{\infty} s^*(x'', y''; z'') T^*(x'' + x, y'' + y; z'') \mathrm{d}x''\mathrm{d}y''\mathrm{d}z'' \right\}.$$

Substituting the scanning field, $s(x, y; z)$, from Eq. (6.2-10) into the above equation, we have

$$i(x,y) \propto \left\{ \iiint_{-\infty}^{\infty} [s_1(x', y'; z' + z_0)\mathrm{e}^{\mathrm{j}\omega_0 t} + s_2(x', y'; z' + z_0)\mathrm{e}^{\mathrm{j}(\omega_0+\Omega)t}] \right.$$

$$\left. \times T(x' + x, y' + y; z')\mathrm{d}x'\mathrm{d}y'\mathrm{d}z' \right\}$$

$$\times \left\{ \iiint_{-\infty}^{\infty} [s_1^*(x'', y''; z'' + z_0)\mathrm{e}^{-\mathrm{j}\omega_0 t} + s_2^*(x'', y''; z'' + z_0)\mathrm{e}^{-\mathrm{j}(\omega_0+\Omega)t}] \right.$$

$$\left. \times T^*(x'' + x, y'' + y; z'')\mathrm{d}x''\mathrm{d}y''\mathrm{d}z'' \right\}.$$

Rearranging the above equation, we have

$$\int [s_1(x', y'; z' + z_0)\mathrm{e}^{\mathrm{j}\omega_0 t} + s_2(x', y'; z' + z_0)\mathrm{e}^{\mathrm{j}(\omega_0+\Omega)t}]$$

$$\times [s_1^*(x'', y''; z'' + z_0)\mathrm{e}^{-\mathrm{j}\omega_0 t} + s_2^*(x'', y''; z'' + z_0)\mathrm{e}^{-\mathrm{j}(\omega_0+\Omega)t}]$$

$$\times T(x' + x, y' + y; z')T^*(x'' + x, y'' + y; z'')\mathrm{d}x'\mathrm{d}y'\mathrm{d}x''\mathrm{d}y''\mathrm{d}z'\mathrm{d}z''.$$

The terms involving $s_1 s_1^*$ and $s_2 s_2^*$ give a *baseband current* $i_{base}(t)$ as

$$i_{base}(x,y) = \int (s_1 s_1^* + s_2 s_2^*) T(x' + x, y' + y; z')$$

$$\times T^*(x'' + x, y'' + y; z'')\mathrm{d}x'\mathrm{d}y'\mathrm{d}x''\mathrm{d}y''\mathrm{d}z'\mathrm{d}z'',$$

whereas the terms involving frequency Ω give a *heterodyne current* $i_\Omega(t)$ as

$$i_\Omega(x,y) = \int (s_2 s_1^* \mathrm{e}^{\mathrm{j}\Omega t} + s_1 s_2^* \mathrm{e}^{-\mathrm{j}\Omega t}) T(x' + x, y' + y; z')$$

$$\times T^*(x'' + x, y'' + y; z'')\mathrm{d}x'\mathrm{d}y'\mathrm{d}x''\mathrm{d}y''\mathrm{d}z'\mathrm{d}z''. \tag{6.2-13a}$$

Grouping the primed and double primed coordinates, and rearranging, we have

$$i_\Omega(x,y) = \left[\int s_2(x', y'; z' + z_0) T(x' + x, y' + y; z')\mathrm{d}x'\mathrm{d}y'\mathrm{d}z' \right.$$

$$\left. \times \int s_1^*(x'', y''; z'' + z_0) T^*(x'' + x, y'' + y; z'')\mathrm{d}x''\mathrm{d}y''\mathrm{d}z'' \right] \mathrm{e}^{\mathrm{j}\Omega t}$$

$$+ \left[\int s_2^*(x'', y''; z'' + z_0) T^*(x'' + x, y'' + y; z'')\mathrm{d}x''\mathrm{d}y''\mathrm{d}z'' \right.$$

$$\left. \times \int s_1(x', y'; z' + z_0) T(x' + x, y' + y; z')\mathrm{d}x'\mathrm{d}y'\mathrm{d}z' \right] \mathrm{e}^{-\mathrm{j}\Omega t}. \tag{6.2-13b}$$

We now recognize that the second term is simply the complex conjugate of the first term. We know $2\mathrm{Re}[\mathrm{e}^{\mathrm{j}\theta}] = \mathrm{e}^{\mathrm{j}\theta} + \mathrm{e}^{-\mathrm{j}\theta}$, where $\mathrm{Re}[\cdot]$ denotes taking the real part of the quantity being bracketed. By neglecting a constant factor, we can write $i_\Omega(t)$ as

$$i_\Omega(x,y) = \mathrm{Re} \left[\int s_2(x', y'; z' + z_0) T(x' + x, y' + y; z')\mathrm{d}x'\mathrm{d}y'\mathrm{d}z' \right.$$

$$\times \int s_1^*(x'',y'';z''+z_0)T^*(x''+x,y''+y;z'')\mathrm{d}x''\mathrm{d}y''\mathrm{d}z''\mathrm{e}^{\mathrm{j}\Omega t}\Bigg]. \tag{6.2-14}$$

尽管式 (6.2-14) 很普遍适用, 但其求解相当复杂. 可以观察到, 物体切片的透过率函数 $T(x,y;z)$ 和 $T^*(x,y;z)$ 一般为复数, 分别被 s_2 与 s_1^* 处理. 因此, 系统进行了相干处理 (*coherent processing*), 即对物体的复振幅进行操作和处理.

A) 其中一个瞳函数为 δ 函数的特殊情况

a) 记录

对于一个特定的情况, 这里指定一些瞳函数. 假定 $p_2(x,y)$ 保持原样, 但是令 $p_1(x,y)=\delta(x,y)$, 即一个点源, 通过观察图 6.2-3 可知, 这会得到一个照明物体的均匀平面波. 数学上, 根据式 (6.2-8), 点源孔径会给出一个均匀照射的平面波

$$\begin{aligned} s_1(x,y;z+z_0) &= \mathcal{F}\{p_1(x,y)\}\big|_{k_x=\frac{k_0x}{f},k_y=\frac{k_0y}{f}} * h(x,y;z+z_0) \\ &= \mathcal{F}\{\delta(x,y)\}\big|_{k_x=\frac{k_0x}{f},k_y=\frac{k_0y}{f}} * h(x,y;z+z_0) \propto \text{常数}, \end{aligned}$$

从而使式 (6.2-14) 中的下列项变为一个常数, 即

$$\int s_1^*(x'',y'';z''+z_0)T^*(x''+x,y''+y;z'')\mathrm{d}x''\mathrm{d}y''\mathrm{d}z'' \propto \text{常数},$$

因为该积分是一个相关积分, 即一个函数与一个常数的相关将会得到一个常数. 由上述结果, 式 (6.2-14) 变为

$$i_\Omega(x,y) \propto \mathrm{Re}[i_p^\delta(x,y)\mathrm{e}^{\mathrm{j}\Omega t}], \tag{6.2-15}$$

其中

$$i_p^\delta(x,y) = \int s_2(x',y';z+z_0)T(x'+x,y'+y;z)\mathrm{d}x'\mathrm{d}y'\mathrm{d}z,$$

这是一个复数 [电子工程中的相量 (*phasor*)]. 事实上, i_Ω 的振幅和相位信息包含了被 $p_2(x,y)$ 扫描和处理了的三维物体信息, 该三维物体信息由复函数 $T(x,y;z)$ 给出.

b) 电子解调 (*electronic demodulation*)

$i(x,y)$ 含有一个基带信号 $i_{base}(x,y)$ 和一个外差信号 $i_\Omega(x,y)$. 可让 $i(x,y)$ 通过一个如图 6.2-3 所示调谐外差频率为 Ω 的电子带通滤波器来滤掉该基带信号, 而外差信号 $i_\Omega(x,y)$ 可根据如图 6.2-4 所示的锁相解调 (*lock-in demodulation*) 方式来处理, 其中 "Sig" 与 "Ref" 分别是放大器的输入信号终端和参考信号终端. 参考信号 $\cos(\Omega t)$ 一般由外部信号源给定. 令

$$i_p^\delta(x,y) = |i_p^\delta(x,y)|\mathrm{e}^{\mathrm{j}\theta(x,y)},$$

则

$$i_\Omega(x,y) = |i_p^\delta(x,y)|\cos[\Omega t+\theta(x,y)]. \tag{6.2-16}$$

该输入信号分为两个通道, 从而获得两个输出 i_I 与 i_Q, 如图 6.2-4 所示. 基本上每一个通道都

$$\times \int s_1^*(x'', y''; z'' + z_0)T^*(x'' + x, y'' + y; z'')\mathrm{d}x''\mathrm{d}y''\mathrm{d}z''\mathrm{e}^{\mathrm{j}\Omega t}\Big] . \tag{6.2-14}$$

Equation (6.2-14) is fairly complicated, albeit quite general. We observe that the transparency function of the object slide $T(x, y; z)$ and $T^*(x, y; z)$, which are complex in general, are processed by s_2 and s_1^*, respectively. The system therefore performs *coherent processing*, that is, the manipulation and processing of the complex amplitude of the object.

A) Specific Case with One of the Pupils Being a Delta Function

a) Recording For a specific case, let us specify some pupil functions. We leave $p_2(x, y)$ as is but let $p_1(x, y) = \delta(x, y)$, that is, a point source, and this will give a uniform plane wave illuminating the object by inspecting Figure 6.2-3. Mathematically, the point source pupil, according to Eq. (6.2-8), gives a uniform plane wave illumination:

$$\begin{aligned} s_1(x, y; z + z_0) &= \mathcal{F}\{p_1(x, y)\}\big|_{k_x = \frac{k_0 x}{f}, k_y = \frac{k_0 y}{f}} * h(x, y; z + z_0) \\ &= \mathcal{F}\{\delta(x, y)\}\big|_{k_x = \frac{k_0 x}{f}, k_y = \frac{k_0 y}{f}} * h(x, y; z + z_0) \propto \text{constant}, \end{aligned}$$

and this subsequently gives the following term in Eq. (6.2-14) to become a constant, that is,

$$\int s_1^*(x'', y''; z'' + z_0)T^*(x'' + x, y'' + y; z'')\mathrm{d}x''\mathrm{d}y''\mathrm{d}z'' \propto \text{constant},$$

as the integral is a correlation integral where the correlation of a function with a constant will give a constant. With the above result, Eq. (6.2-14) becomes

$$i_\Omega(x, y) \propto \mathrm{Re}[i_p^\delta(x, y)\mathrm{e}^{\mathrm{j}\Omega t}], \tag{6.2-15}$$

where

$$i_p^\delta(x, y) = \int s_2(x', y'; z + z_0)T(x' + x, y' + y; z)\mathrm{d}x'\mathrm{d}y'\mathrm{d}z,$$

is a complex quantity (*phasor* in electrical engineering). Indeed, the amplitude and phase information of i_Ω contains the scanned and processed [by $p_2(x, y)$] information of the 3-D object given by the complex function $T(x, y; z)$.

b) Electronic Demodulation Remember that $i(x, y)$ contains a baseband signal $i_{base}(x, y)$ and a heterodyned signal $i_\Omega(x, y)$. We can reject the baseband signal by passing $i(x, y)$ through an electronic bandpass filter tuned at the heterodyne frequency, Ω, as shown in Figure 6.2-3. The heterodyne signal, $i_\Omega(x, y)$, can now be processed according to the *lock-in demodulation* scheme shown in Figure 6.2-4, where "Sig" and "Ref" are the input signal and reference signal terminals of the amplifier, respectively. The reference signal $\cos(\Omega t)$ is typically provided from an external source. Let

$$i_p^\delta(x, y) = |i_p^\delta(x, y)|\mathrm{e}^{\mathrm{j}\theta(x,y)},$$

then

$$i_\Omega(x, y) = |i_p^\delta(x, y)|\cos[\Omega t + \theta(x, y)]. \tag{6.2-16}$$

This input signal splits into two channels to obtain two outputs, i_I and i_Q, as shown in Figure 6.2-4. Basically, each channel performs *lock-in detection*, which consists of electronically

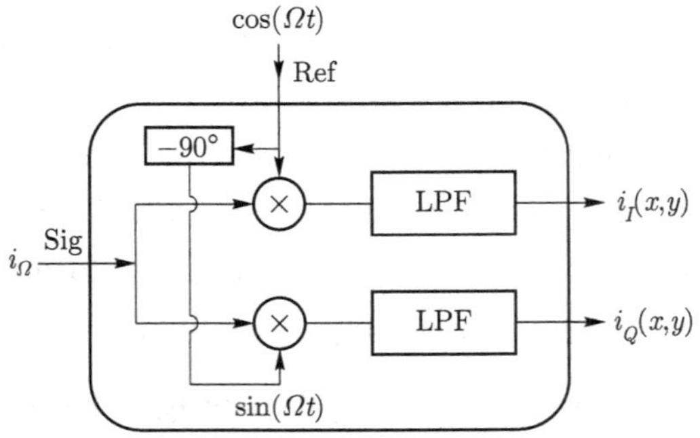

图 6.2-4 锁相放大器, 解调方式图示说明. ⊗ 为电子乘法器, LPF 为低通滤波器, 模块 $\boxed{-90^\circ}$ 表示此处可以对输入的正弦信号进行 -90° 的相移

进行锁相检测 (*lock-in detection*), 其中包括以电子的方式乘以带有外差频率 Ω 的正弦或者余弦输入信号, 并在此之后进行低通滤波, 以提取输入电子信号的振幅与相位.

考虑电子乘法器给出的上通道, 有

$$\begin{aligned} i_\Omega(x,y)\times\cos(\Omega t) &= |i_p^\delta(x,y)|\cos[\Omega t+\theta(x,y)]\times\cos(\Omega t) \\ &= |i_p^\delta(x,y)|\frac{1}{2}\left\{\cos\theta(x,y)+\cos[2\Omega t+\theta(x,y)]\right\}. \end{aligned}$$

在低通滤波后 (意味着滤掉了频率 2Ω), 除了某常数外, 所谓的输入外差信号 $i_\Omega(x,y)$ 的同相分量 (*in-phase component*) 为

$$i_I(x,y)=|i_p^\delta(x,y)|\cos\theta(x,y). \tag{6.2-17}$$

同样, 下通道给出外差输入信号的正交分量 (*quadrature component*) 为

$$i_Q(x,y)=|i_p^\delta(x,y)|\sin\theta(x,y). \tag{6.2-18}$$

一旦 $i_I(x,y)$ 与 $i_Q(x,y)$ 被提取出来, 就以两个数字记录的形式被存储在计算机中, 计算机可以对其进行以下的复数相加以获得复记录 (*complex record*) $i_C(x,y)$

$$i_C(x,y)=i_I(x,y)+\mathrm{j}i_Q(x,y)=|i_p^\delta(x,y)|\mathrm{e}^{\mathrm{j}\theta(x,y)}=i_p^\delta(x,y). \tag{6.2-19}$$

可以看到, 这里已经从外差电流中提取出了全部的复数信息, 即 $i_p^\delta(x,y)$ 可以由式 (6.2-15) 给出

$$\begin{aligned} i_C(x,y) &= i_p^\delta(x,y)=i_I(x,y)+\mathrm{j}i_Q(x,y) \\ &= \int s_2(x',y';z+z_0)T(x'+x,y'+y;z)\mathrm{d}x'\mathrm{d}y'\mathrm{d}z, \end{aligned} \tag{6.2-20}$$

其中, 三维复数信息 $T(x,y;z)$ 已被扫描光束 s_2 处理.

B) 相干模式的全息记录 (*holographic recording in coherent mode*)

这里来看整个系统是如何进行全息记录的. 根据式 (6.2-9), 可以知道, $p_2(x,y)$ 与扫描光束 s_2 之间的关系为

$$s_2(x,y;z+z_0)=\mathcal{F}\{p_2(x,y)\}\big|_{k_x=\frac{k_0x}{f},k_y=\frac{k_0y}{f}}*h(x,y;z+z_0).$$

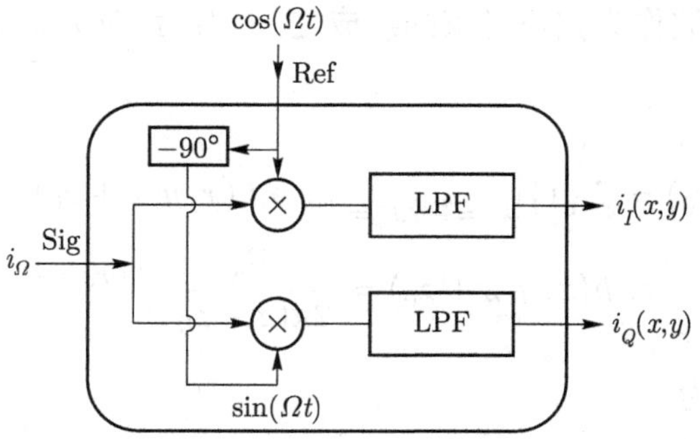

Figure 6.2-4 Lock-in amplifier, illustrating demodulation scheme. ⊗: electronic multiplier, LPF: low-pass filter. The block showing −90° represents a phase shifter of −90° of the incoming sinusoidal signal

multiplying the incoming signal with the sine or cosine of the heterodyne frequency, Ω, and then performing low-pass filtering to extract the amplitude and phase of the incoming electrical signal.

Consider the upper channel where the electronic multiplier gives

$$\begin{aligned} i_\Omega(x,y) \times \cos(\Omega t) &= |i_p^\delta(x,y)| \cos[\Omega t + \theta(x,y)] \times \cos(\Omega t) \\ &= |i_p^\delta(x,y)| \frac{1}{2} \{\cos\theta(x,y) + \cos[2\Omega t + \theta(x,y)]\}. \end{aligned}$$

After low-pass filtering (which means we are rejecting the frequency of 2Ω), we have, apart from some constant, the so-called *in-phase component* of the incoming heterodyne signal $i_\Omega(x,y)$:

$$i_I(x,y) = |i_p^\delta(x,y)| \cos\theta(x,y). \tag{6.2-17}$$

Similarly, the lower channel gives the *quadrature component* of the heterodyne incoming signal, which is given by

$$i_Q(x,y) = |i_p^\delta(x,y)| \sin\theta(x,y). \tag{6.2-18}$$

Once $i_I(x,y)$ and $i_Q(x,y)$ have been extracted, they can be stored in a computer as two digital records, and the computer can perform the following complex addition to get a *complex record* $i_C(x,y)$:

$$i_C(x,y) = i_I(x,y) + \mathrm{j} i_Q(x,y) = |i_p^\delta(x,y)| \mathrm{e}^{\mathrm{j}\theta(x,y)} = i_p^\delta(x,y). \tag{6.2-19}$$

Note that we have extracted the full complex information from the heterodyne current, that is, $i_p^\delta(x,y)$ given by Eq. (6.2-15) as follows:

$$\begin{aligned} i_C(x,y) &= i_p^\delta(x,y) = i_I(x,y) + \mathrm{j} i_Q(x,y) \\ &= \int s_2(x',y';z+z_0) T(x'+x,y'+y;z) \mathrm{d}x'\mathrm{d}y'\mathrm{d}z, \end{aligned} \tag{6.2-20}$$

where the 3-D complex information $T(x,y;z)$ has been processed by the scanning beam s_2.

B) Holographic Recording in Coherent Mode

Let us see how the overall system can perform holographic recording. Remember that $p_2(x,y)$, according to Eq. (6.2-9), relates to the scanning beam, s_2, as

$$s_2(x,y;z+z_0) = \mathcal{F}\{p_2(x,y)\}\Big|_{k_x=\frac{k_0 x}{f}, k_y=\frac{k_0 y}{f}} * h(x,y;z+z_0).$$

取 $p_2(x,y)=1$, 即让平面波作为扫描光束的组成之一. 然后, 对 $h(x,y;z+z_0)$ 应用式 (3.4-16), 有

$$s_2(x,y;z+z_0)=\mathcal{F}\{1\}\big|_{k_x=\frac{k_0x}{f},k_y=\frac{k_0y}{f}}*h(x,y;z+z_0)$$
$$\propto h(x,y;z+z_0)=\frac{\mathrm{j}k_0}{2\pi(z+z_0)}\mathrm{e}^{-\mathrm{j}k_0(z+z_0)}\mathrm{e}^{\frac{-\mathrm{j}k_0(x^2+y^2)}{2(z+z_0)}}.$$

由此结果, 式 (6.2-20) 变为

$$i_C(x,y)=\int\frac{\mathrm{j}k_0}{2\pi(z+z_0)}\mathrm{e}^{-\mathrm{j}k_0(z+z_0)}\mathrm{e}^{\frac{-\mathrm{j}k_0}{2(z+z_0)}(x'^2+y'^2)}T(x'+x,y'+y;z)\mathrm{d}x'\mathrm{d}y'\mathrm{d}z.$$
$$=\int\frac{\mathrm{j}k_0}{2\pi(z+z_0)}\mathrm{e}^{-\mathrm{j}k_0(z+z_0)}\left[\mathrm{e}^{\frac{-\mathrm{j}k_0(x^2+y^2)}{2(z+z_0)}}*T(x,y;z)\right]\mathrm{d}z, \tag{6.2-21}$$

其中用卷积运算方便地表示出了上面的结果. 这是三维物体 $T(x,y;z)$ 的复菲涅耳全息图 (*complex Fresnel hologram*). 这种单像素数字全息记录技术被称为相干模式的光学扫描全息术 (*optical scanning holography in a coherent mode*) [Poon (2007)].

举例: 点源的复菲涅耳全息图

考虑 $T(x,y;z)$ 为一点物, 在如图 6.2-3 所示的双光瞳系统中, 位于距透镜 L1 后焦面 z_1 处. 对于该点物, 建模 $T(x,y;z)=\delta(x,y)\delta(z-z_1)$, 现在求其复全息图, 根据式 (6.2-21), 有

$$i_C(x,y)=\int\frac{\mathrm{j}k_0}{2\pi(z+z_0)}\mathrm{e}^{-\mathrm{j}k_0(z+z_0)}\mathrm{e}^{\frac{-\mathrm{j}k_0}{2(z+z_0)}(x'^2+y'^2)}$$
$$\times\delta(x'+x,y'+y)\delta(z-z_1)\mathrm{d}x'\mathrm{d}y'\mathrm{d}z.$$

在对物体深度进行积分后, 得

$$i_C(x,y)=\frac{\mathrm{j}k_0}{2\pi(z_1+z_0)}\mathrm{e}^{-\mathrm{j}k_0(z_1+z_0)}\int\mathrm{e}^{\frac{-\mathrm{j}k_0}{2(z_1+z_0)}(x'^2+y'^2)}\delta(x'+x,y'+y)\mathrm{d}x'\mathrm{d}y'$$
$$\propto\mathrm{e}^{\frac{-\mathrm{j}k_0}{2(z_1+z_0)}(x^2+y^2)}=H^c(x,y). \tag{6.2-22}$$

这是一个点源的复菲涅耳全息图. 该复全息图不包含零级光和孪生像信息 (见图 5.1-5, FZP 重建出零级光、虚像以及孪生像). 通过 $H^c(x,y)*h^*(x,y;z_1+z_0)\propto\delta(x,y)$ 在 $z=-z_1-z_0$ 处获得虚像重建 (见图 5.3-2). 可以看出, 如果想要重建一个实像, 可以取 $H^c(x,y)$ 的复共轭得到 $H^{c*}(x,y)$, 从而通过 $H^{c*}(x,y)*h(x,y;z_1+z_0)\propto\delta(x,y)$ 在 $z=z_1+z_0$ 处得到实像重建.

图 6.2-5 是光学扫描全息术在相干模式下实现定量相位成像 (*quantitative phase imaging*, QPI) 的一个例子. 相比于传统相位成像方法, 如泽尼克相衬成像 (*Zernike phase contrast imaging*) (见第四章的习题 4.5) 可定性地提供生物结构相位的可视化, 而 QPI 是一组显微方法的集合名, 其涉及对弱散射和吸收标本 [Popescu (2011)] 的研究, 并提供定量测量. 图 6.2-5a) 为三叉海绵刺针 (*three-pronged spongilla spicule*) 重建振幅的绝对值, 图 6.2-5b) 为物体的三维相位轮廓. 该标本几乎是一个纯相位物体, 且其振幅图像由于标本边缘的衍射而被渲染后是可见的. 为了计算出图 6.2-5b) 中的三维相位轮廓, 利用菲涅耳衍射公式 [见式 (3.4-17)]

We take $p_2(x, y) = 1$, that is, it is a plane wave as one of the scanning beams. Then, using Eq. (3.4-16) for $h(x, y; z + z_0)$,

$$s_2(x, y; z + z_0) = \mathcal{F}\{1\}\big|_{k_x=\frac{k_0 x}{f}, k_y=\frac{k_0 y}{f}} * h(x, y; z + z_0)$$
$$\propto h(x, y; z + z_0) = \frac{\mathrm{j}k_0}{2\pi(z + z_0)} \mathrm{e}^{-\mathrm{j}k_0(z+z_0)} \mathrm{e}^{\frac{-\mathrm{j}k_0(x^2+y^2)}{2(z+z_0)}}.$$

With this result, Eq. (6.2-20) becomes

$$i_C(x, y) = \int \frac{\mathrm{j}k_0}{2\pi(z + z_0)} \mathrm{e}^{-\mathrm{j}k_0(z+z_0)} \mathrm{e}^{\frac{-\mathrm{j}k_0}{2(z+z_0)}(x'^2+y'^2)} T(x' + x, y' + y; z)\mathrm{d}x'\mathrm{d}y'\mathrm{d}z.$$
$$= \int \frac{\mathrm{j}k_0}{2\pi(z + z_0)} \mathrm{e}^{-\mathrm{j}k_0(z+z_0)} \left[\mathrm{e}^{\frac{-\mathrm{j}k_0(x^2+y^2)}{2(z+z_0)}} * T(x, y; z)\right] \mathrm{d}z, \quad (6.2\text{-}21)$$

where we have expressed the above result conveniently using the convolution operation. This is a *complex Fresnel hologram* of the 3-D object $T(x, y; z)$. This single-pixel digital holographic recording technique is known as *optical scanning holography* in a coherent mode [Poon (2007)].

Example: Complex Fresnel Hologram of a Point Source

Consider $T(x, y; z)$ as a point object, located z_1 away from the back focal plane of lens L1 in the two-pupil system shown in Figure 6.2-3. For the point object, we model $T(x, y; z) = \delta(x, y)\delta(z - z_1)$. Let us find its complex hologram. According to Eq. (6.2-21), we have

$$i_C(x, y) = \int \frac{\mathrm{j}k_0}{2\pi(z + z_0)} \mathrm{e}^{-\mathrm{j}k_0(z+z_0)} \mathrm{e}^{\frac{-\mathrm{j}k_0}{2(z+z_0)}(x'^2+y'^2)} \times \delta(x' + x, y' + y)\delta(z - z_1)\mathrm{d}x'\mathrm{d}y'\mathrm{d}z.$$

After integrating over the depth of the object, we obtain

$$i_C(x, y) = \frac{\mathrm{j}k_0}{2\pi(z_1 + z_0)} \mathrm{e}^{-\mathrm{j}k_0(z_1+z_0)} \int \mathrm{e}^{\frac{-\mathrm{j}k_0}{2(z_1+z_0)}(x'^2+y'^2)} \delta(x' + x, y' + y)\mathrm{d}x'\mathrm{d}y'$$
$$\propto \mathrm{e}^{\frac{-\mathrm{j}k_0}{2(z_1+z_0)}(x^2+y^2)} = H^c(x, y). \quad (6.2\text{-}22)$$

This is a complex Fresnel hologram of the point source. The complex hologram does not contain the information of the zeroth-order beam and the twin image (see Figure 5.1-5 where the FZP reconstructs the zeroth-order beam, the virtual image as well as the twin image). We obtain the virtual image reconstruction by $H^c(x, y) * h^*(x, y; z_1 + z_0) \propto \delta(x, y)$ at $z = -z_1 - z_0$ (see Figure 5.3-2). Note that if we want to reconstruct a real image, we could take a complex conjugate of $H^c(x, y)$ to obtain $H^{c*}(x, y)$, and that leads to real image reconstruction as $H^{c*}(x, y) * h(x, y; z_1 + z_0) \propto \delta(x, y)$ at $z = z_1 + z_0$.

Figure 6.2-5 is an example of *quantitative phase imaging* (QPI) achievable with optical scanning holography operating in a coherent mode. In contrast to conventional phase imaging methods, such as *Zernike phase contrast imaging* (see P.4.5), which provide visualization of the phase of biological structures in a qualitative way, QPI is a collective name for a group of microscopy methods that involves the study of weakly scattering and absorbing specimens [Popescu (2011)] and provides quantitative measurements. Figure 6.2-5a shows the absolute value of the reconstruction amplitude of a *three-pronged spongilla spicule*, and Figure 6.2-5b is the 3-D phase profile of the object. The specimen is almost a pure phase object and its amplitude image is rendered visible due to diffraction from the specimen's edges. To calculate the 3-D phase profile in Figure 6.2-5b, we simply first reconstruct the complex field $\psi_p(x, y)$

或角谱法 [见式 (3.4-8)], 先由复全息图 [见式 (6.2-21)] 重建其复振幅 $\psi_p(x,y)$, 然后计算其相位 $\theta(x,y)$, 得下式

$$\theta(x,y)=\arctan\left\{\frac{\mathrm{Im}[\psi_p(x,y)]}{\mathrm{Re}[\psi_p(x,y)]}\right\}. \tag{6.2-23}$$

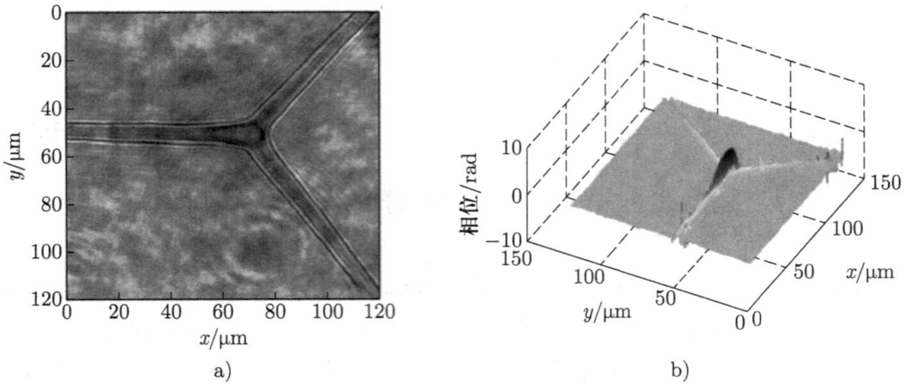

图 6.2-5 相干模式下光学扫描全息术获得的复菲涅耳全息图的重建, 物体为一个三叉海绵刺针: a) 重建振幅的绝对值; b) 三维相位轮廓. 转载自 T.-C. Poon, Optical scanning holography- a review of recent progress, Journal of the Optical Society of Korea 13, pp. 406-415 (2009), 经许可. 原创工作来自 G. Indebetouw et al., Quantitative phase imaging with scanning holographic microscopy: an experimental assessment, BioMedical Engineering OnLine, 5:60 (2006)

6.3 非相干数字全息术

6.1 节和 6.2 节讨论了相干全息术. "相干" 一词意味着记录物体的复振幅. 在非相干全息术 (*incoherent holography*) 中, 目的是获取三维物体全息信息的强度分布, 相对于相干全息图, 允许更高的信噪比. 目前, 有两种众所周知的非相干数字全息术: 光学扫描全息术 (*optical scanning holography*, OSH) 和菲涅耳非相干相关全息术 (*Fresnel incoherent correlation holography*, FINCH). 本节将进一步展开上节所讨论的 OSH 的理论, 并集中讨论其非相干记录能力的研究. 之后将对 FINCH 的原理进行解释. FINCH 的一般情况, 被称为编码孔径相关全息术 (*coded aperture correlation holography*, COACH), 将在本节的最后讨论.

6.3.1 非相干模式光学扫描全息术

上文已经讨论过其中一个光瞳为针孔时的情况, 即 $p_1(x,y)=\delta(x,y)$, 另一个光瞳为全通, 即 $p_2(x,y)=1$, 在一个用点探测器的双光瞳系统中, 即 $m(x,y)=\delta(x,y)$ (见图 6.2-3), 因为物体的复振幅被记录, 所以有相干全息记录. 本节中, 先让掩模保持不变, 并在某种程度上发展一个更普遍的情况, 然后令掩模为 $m(x,y)=1$, 来研究非相干处理.

根据式 (6.2-11), 在掩模后的复振幅为

$$\psi(x,y;x_m,y_m)=\left\{\int \mathrm{e}^{-\mathrm{j}\frac{k_0 z}{2f^2}(x_m^2+y_m^2)}\iint_{-\infty}^{\infty}s(x',y';z)T(x'+x,y'+y;z)\right.$$

from the complex hologram [see Eq. (6.2-21)] by the Fresnel diffraction formula [see Eq. (3.4-17)] or the angular spectrum method [see Eq. (3.4-8)] and then calculate the phase $\theta(x,y)$ according to

$$\theta(x,y)=\arctan\left\{\frac{\mathrm{Im}[\psi_p(x,y)]}{\mathrm{Re}[\psi_p(x,y)]}\right\}. \tag{6.2-23}$$

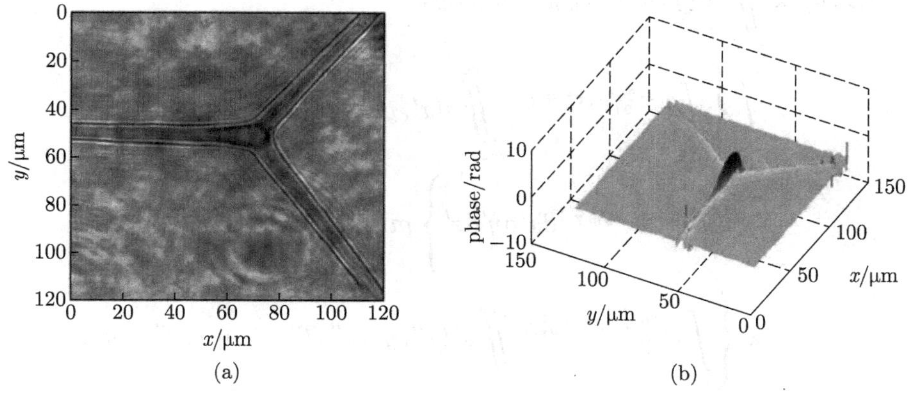

Figure 6.2-5 Reconstruction of a complex Fresnel hologram obtained by optical scanning holography operating in a coherent mode. The object is a three-pronged spongilla spicule. (a) Absolute value of the reconstruction amplitude and (b) 3-D phase profile. Reprinted from T.-C. Poon, Optical scanning holography-a review of recent progress, Journal of the Optical Society of Korea 13, pp. 406-415 (2009), with permission. Original work is from G. Indebetouw et al., Quantitative phase imaging with scanning holographic microscopy: an experimental assessment, BioMedical Engineering OnLine, 5:60 (2006)

6.3 Incoherent Holography

In Sections 6.1 and 6.2, we have discussed coherent holography. The word "coherent" means that we record the complex amplitude of the object. In *incoherent holography*, we capture the intensity distribution of a 3-D object holographically, allowing a higher signal-to-noise ratio as compared to its coherent counterpart. Nowadays, there are two major and well-known incoherent digital holographic techniques: *optical scanning holography* (OSH) and *Fresnel incoherent correlation holography* (FINCH). In this section, we will further develop the theory of OSH discussed in the last section, and concentrate on the development of its incoherent recording capability. We will then explain the principle of FINCH. A generalization of FINCH, called *coded aperture correlation holography* (COACH), will then be discussed at the end of this subsection.

6.3.1 Optical Scanning Holography in Incoherent Mode

We have discussed that with one of the pupils being a pinhole, that is, $p_1(x,y)=\delta(x,y)$ and the other pupil being open, that is, $p_2(x,y)=1$, in the two-pupil system with a point detector, that is, $m(x,y)=\delta(x,y)$ (see Figure 6.2-3), we have coherent holographic recording in that the complex amplitude of the object is being recorded. In this section, we first leave the mask as is and develop a more general situation to some extent. We then let the mask $m(x,y)=1$ to investigate incoherent processing.

According to Eq. (6.2-11), the complex field after the mask is

$$\psi(x,y;x_m,y_m)=\left\{\int \mathrm{e}^{-\mathrm{j}\frac{k_0 z}{2f^2}(x_m^2+y_m^2)}\iint_{-\infty}^{\infty} s(x',y';z)T(x'+x,y'+y;z)\right.$$

$$\left.\times \mathrm{e}^{\frac{\mathrm{j}k_0}{f}(x_m x'+y_m y')}\mathrm{d}x'\mathrm{d}y'\mathrm{d}z\right\} m(x_m, y_m),$$

则光电探测器中的电流为

$$\begin{aligned}
i(x,y) &\propto \iint |\psi(x,y;x_m,y_m)|^2 \mathrm{d}x_m \mathrm{d}y_m \\
&= \int \left\{ \int \mathrm{e}^{-\mathrm{j}\frac{k_0 z'}{2f^2}(x_m^2+y_m^2)} \iint_{-\infty}^{\infty} s(x',y';z')T(x'+x,y'+y;z') \right. \\
&\quad \left. \times \mathrm{e}^{\frac{\mathrm{j}k_0}{f}(x_m x'+y_m y')}\mathrm{d}x'\mathrm{d}y'\mathrm{d}z' \right\} m(x_m,y_m) \\
&\quad \times \left\{ \int \mathrm{e}^{\mathrm{j}\frac{k_0 z''}{2f^2}(x_m^2+y_m^2)} \iint_{-\infty}^{\infty} s^*(x'',y'';z'')T^*(x''+x,y''+y;z'') \right. \\
&\quad \left. \times \mathrm{e}^{\frac{-\mathrm{j}k_0}{f}(x_m x''+y_m y'')}\mathrm{d}x''\mathrm{d}y''\mathrm{d}z'' \right\} m^*(x_m,y_m)\mathrm{d}x_m\mathrm{d}y_m. \qquad (6.3\text{-}1)
\end{aligned}$$

根据式 (6.2-10), 扫描光束为

$$s(x,y;z) = s_1(x,y;z+z_0)\mathrm{e}^{\mathrm{j}\omega_0 t} + s_2(x,y;z+z_0)\mathrm{e}^{\mathrm{j}(\omega_0+\Omega)t},$$

将这个量代入式 (6.3-1), 将如上节说明的那样, 获得一个基带电流 $i_{base}(x,y)$ 和一个外差电流 $i_\Omega(x,y)$. 由于基带电流将会被放置在光电探测器之后的带通滤波器滤掉, 因此我们只需关注式 (6.3-1) 中出来的外差项. 将外差项合并, 类似上节所示 [见式 (6.2-13)], 得到其外差电流为

$$\begin{aligned}
i_\Omega(x,y) = &\int (s_2 s_1^* \mathrm{e}^{\mathrm{j}\Omega t} + s_1 s_2^* \mathrm{e}^{-\mathrm{j}\Omega t})\mathrm{e}^{\mathrm{j}\frac{k_0}{f}[x_m(x'-x'')+y_m(y'-y'')]}\mathrm{e}^{-\mathrm{j}\frac{k_0(z'-z'')}{2f^2}(x_m^2+y_m^2)} \\
&\times T(x'+x,y'+y;z') \\
&\times T^*(x''+x,y''+y;z'')\mathrm{d}x'\mathrm{d}y'\mathrm{d}x''\mathrm{d}y''\mathrm{d}z'\mathrm{d}z''|m(x_m,y_m)|^2\mathrm{d}x_m\mathrm{d}y_m. \qquad (6.3\text{-}2)
\end{aligned}$$

将所有的 x_m 和 y_m 变量合并, 其积分结果被称为扫描系统的相干函数 (*coherence function of the scanning system*)

$$\begin{aligned}
&\Gamma(x'-x'',y'-y'';z'-z'') \\
&= \int |m(x_m,y_m)|^2 \mathrm{e}^{\mathrm{j}\frac{k_0}{f}[x_m(x'-x'')+y_m(y'-y'')]}\mathrm{e}^{-\mathrm{j}\frac{k_0(z'-z'')}{2f^2}(x_m^2+y_m^2)}\mathrm{d}x_m\mathrm{d}y_m. \qquad (6.3\text{-}3)
\end{aligned}$$

结合相干函数的定义, 式 (6.3-2) 变为

$$\begin{aligned}
i_\Omega(x,y) = &\int (s_2 s_1^* \mathrm{e}^{\mathrm{j}\Omega t} + s_1 s_2^* \mathrm{e}^{-\mathrm{j}\Omega t})\Gamma(x'-x'',y'-y'';z'-z'') \\
&\times T(x'+x,y'+y;z')T^*(x''+x,y''+y;z'')\mathrm{d}x'\mathrm{d}y'\mathrm{d}x''\mathrm{d}y''\mathrm{d}z'\mathrm{d}z''. \qquad (6.3\text{-}4)
\end{aligned}$$

实际上, 相干函数控制着扫描系统的相干性. 例如, 当 $|m(x,y)|^2=\delta(x,y)$ 时

$$\left.\times\, \mathrm{e}^{\frac{\mathrm{j}k_0}{f}(x_m x'+y_m y')}\mathrm{d}x'\mathrm{d}y'\mathrm{d}z\right\} m(x_m, y_m),$$

and the current from the photodetector is

$$\begin{aligned}i(x,y) &\propto \iint |\psi(x,y;x_m,y_m)|^2 \mathrm{d}x_m \mathrm{d}y_m \\ &= \int\left\{\int \mathrm{e}^{-\mathrm{j}\frac{k_0 z'}{2f^2}(x_m^2+y_m^2)} \iint_{-\infty}^{\infty} s(x',y';z')T(x'+x,y'+y;z')\right. \\ &\quad \left.\times\, \mathrm{e}^{\frac{\mathrm{j}k_0}{f}(x_m x'+y_m y')}\mathrm{d}x'\mathrm{d}y'\mathrm{d}z'\right\} m(x_m,y_m) \\ &\quad \times\left\{\int \mathrm{e}^{\mathrm{j}\frac{k_0 z''}{2f^2}(x_m^2+y_m^2)} \iint_{-\infty}^{\infty} s^*(x'',y'';z'')T^*(x''+x,y''+y;z'')\right. \\ &\quad \left.\times\, \mathrm{e}^{\frac{-\mathrm{j}k_0}{f}(x_m x''+y_m y'')}\mathrm{d}x''\mathrm{d}y''\mathrm{d}z''\right\} m^*(x_m,y_m)\mathrm{d}x_m\mathrm{d}y_m. \qquad (6.3\text{-}1)\end{aligned}$$

According to Eq. (6.2-10), the scanning beam is

$$s(x,y;z) = s_1(x,y;z+z_0)\mathrm{e}^{\mathrm{j}\omega_0 t} + s_2(x,y;z+z_0)\mathrm{e}^{\mathrm{j}(\omega_0+\Omega)t},$$

and with this quantity substituted into Eq. (6.3-1), we will have a baseband current $i_{base}(x,y)$ as well as a heterodyne current, $i_\Omega(x,y)$, as demonstrated earlier in the last section. Since the baseband current will be rejected by the bandpass filter placed after the photodetector, we only need to concentrate on the heterodyne term coming out from Eq. (6.3-1). By collecting the heterodyne terms, we have, similar to what has been shown earlier in the last section [see Eq. (6.2-13)], the heterodyne current is

$$\begin{aligned}i_\Omega(x,y) &= \int (s_2 s_1^* \mathrm{e}^{\mathrm{j}\Omega t} + s_1 s_2^* \mathrm{e}^{-\mathrm{j}\Omega t})\mathrm{e}^{\mathrm{j}\frac{k_0}{f}[x_m(x'-x'')+y_m(y'-y'')]}\mathrm{e}^{-\mathrm{j}\frac{k_0(z'-z'')}{2f^2}(x_m^2+y_m^2)} \\ &\quad \times T(x'+x,y'+y;z') \\ &\quad \times T^*(x''+x,y''+y;z'')\mathrm{d}x'\mathrm{d}y'\mathrm{d}x''\mathrm{d}y''\mathrm{d}z'\mathrm{d}z''|m(x_m,y_m)|^2\mathrm{d}x_m\mathrm{d}y_m. \qquad (6.3\text{-}2)\end{aligned}$$

Let us group all the x_m and y_m variables together and the resulting integral is called the *coherence function of the scanning system*:

$$\begin{aligned}&\Gamma(x'-x'',y'-y'';z'-z'') \\ &= \int |m(x_m,y_m)|^2 \mathrm{e}^{\mathrm{j}\frac{k_0}{f}[x_m(x'-x'')+y_m(y'-y'')]}\mathrm{e}^{-\mathrm{j}\frac{k_0(z'-z'')}{2f^2}(x_m^2+y_m^2)}\mathrm{d}x_m\mathrm{d}y_m. \qquad (6.3\text{-}3)\end{aligned}$$

With definition of the coherence function, Eq. (6.3-2) becomes

$$\begin{aligned}i_\Omega(x,y) &= \int (s_2 s_1^* \mathrm{e}^{\mathrm{j}\Omega t} + s_1 s_2^* \mathrm{e}^{-\mathrm{j}\Omega t})\Gamma(x'-x'',y'-y'';z'-z'') \\ &\quad \times T(x'+x,y'+y;z')T^*(x''+x,y''+y;z'')\mathrm{d}x'\mathrm{d}y'\mathrm{d}x''\mathrm{d}y''\mathrm{d}z'\mathrm{d}z''. \qquad (6.3\text{-}4)\end{aligned}$$

Indeed, the coherence function controls the coherency of the optical scanning system. For example, when $|m(x,y)|^2 = \delta(x,y)$,

$$\begin{aligned}&\Gamma(x'-x'',y'-y'';z'-z'')\\&=\int\delta(x_m,y_m)\mathrm{e}^{\mathrm{j}\frac{k_0}{f}[x_m(x'-x'')+y_m(y'-y'')]}\mathrm{e}^{-\mathrm{j}\frac{k_0(z'-z'')}{2f^2}(x_m^2+y_m^2)}\mathrm{d}x_m\mathrm{d}y_m=1,\end{aligned}$$

式 (6.3-4) 变为

$$\begin{aligned}i_\Omega(x,y)=&\int(s_2s_1^*\mathrm{e}^{\mathrm{j}\Omega t}+s_1s_2^*\mathrm{e}^{-\mathrm{j}\Omega t})\\&\times T(x'+x,y'+y;z')T^*(x''+x,y''+y;z'')\mathrm{d}x'\mathrm{d}y'\mathrm{d}x''\mathrm{d}y''\mathrm{d}z'\mathrm{d}z'',\end{aligned}$$

这与式 (6.2-13a) 相同, 这让我们想到了相干处理中点探测器的使用. 因此, 对于点探测器, 即 $|m(x,y)|^2=\delta(x,y)$, 有一个相干处理系统. 现在来看掩模的另一种极端情况, 即 $|m(x,y)|^2=1$ 时, 根据式 (6.3-3), 有

$$\begin{aligned}\Gamma(x'-x'',y'-y'';z'-z'')&=\int\mathrm{e}^{\mathrm{j}\frac{k_0}{f}[x_m(x'-x'')+y_m(y'-y'')]}\mathrm{e}^{-\mathrm{j}\frac{k_0(z'-z'')}{2f^2}(x_m^2+y_m^2)}\mathrm{d}x_m\mathrm{d}y_m\\&=\mathcal{F}\left\{\mathrm{e}^{-\mathrm{j}\frac{k_0(x_m^2+y_m^2)}{2f^2/(z'-z'')}}\right\}\Bigg|_{k_x=k_0(x'-x'')/f,k_y=k_0(y'-y'')/f}\\&\sim\frac{1}{z'-z''}\mathrm{e}^{\mathrm{j}\frac{k_0[(x'-x'')^2+(y'-y'')^2]}{2(z'-z'')}}\\&\sim\delta(x'-x'',y'-y'';z'-z''),\end{aligned}\qquad(6.3\text{-}5)$$

其中, 可以看出, 在上述积分中的二次项代表一个曲率半径为 $R=f^2/(z'-z'')$ 的球面波, 该曲率半径的值可以为任意大 (见 3.5 节中的曲率半径). 因此, 最终会在 $z'\to z''$ 的极限下取 1 的傅里叶变换, 得到一个三维 δ 函数. 有了这一结果, 式 (6.3-2) 被简化为

$$\begin{aligned}i_\Omega(x,y)=&\int(s_2s_1^*\mathrm{e}^{\mathrm{j}\Omega t}+s_1s_2^*\mathrm{e}^{-\mathrm{j}\Omega t})\delta(x'-x'',y'-y'';z'-z'')\\&\times T(x'+x,y'+y;z')T^*(x''+x,y''+y;z'')\mathrm{d}x'\mathrm{d}y'\mathrm{d}x''\mathrm{d}y''\mathrm{d}z'\mathrm{d}z''.\end{aligned}$$

现在可以写出 $s_2s_1^*$ 项和 $s_1s_2^*$ 项, 其分别由式 (6.2-8) 和式 (6.2-9) 给出. 因此, 上式变为

$$\begin{aligned}i_\Omega(x,y)=&\int[s_2(x',y';z'+z_0)s_1^*(x'',y'';z''+z_0)\mathrm{e}^{\mathrm{j}\Omega t}\\&+s_1(x',y';z'+z_0)s_2^*(x'',y'';z''+z_0)\mathrm{e}^{-\mathrm{j}\Omega t}]\\&\times\delta(x'-x'',y'-y'';z'-z'')T(x'+x,y'+y;z')\\&\times T^*(x''+x,y''+y;z'')\mathrm{d}x'\mathrm{d}y'\mathrm{d}x''\mathrm{d}y''\mathrm{d}z'\mathrm{d}z''.\end{aligned}$$

在求出包含 δ 函数的积分后, 将所有带两撇的变量替换为带单撇的变量

$$\begin{aligned}i_\Omega(x,y)=&\int[s_2(x',y';z'+z_0)s_1^*(x',y';z'+z_0)\mathrm{e}^{\mathrm{j}\Omega t}\\&+s_1(x',y';z'+z_0)s_2^*(x',y';z'+z_0)\mathrm{e}^{-\mathrm{j}\Omega t}]\\&\times T(x'+x,y'+y;z')T^*(x'+x,y'+y;z')\mathrm{d}x'\mathrm{d}y'\mathrm{d}z'\\=&\int[s_2(x',y';z'+z_0)s_1^*(x',y';z'+z_0)\mathrm{e}^{\mathrm{j}\Omega t}\end{aligned}$$

$$
\begin{aligned}
&\Gamma(x'-x'',y'-y'';z'-z'')\\
&=\int \delta(x_m,y_m)e^{j\frac{k_0}{f}[x_m(x'-x'')+y_m(y'-y'')]}e^{-j\frac{k_0(z'-z'')}{2f^2}(x_m^2+y_m^2)}dx_m dy_m=1,
\end{aligned}
$$

and Eq. (6.3-4) becomes

$$
\begin{aligned}
i_\Omega(x,y)=\int&(s_2s_1^*e^{j\Omega t}+s_1s_2^*e^{-j\Omega t})T(x'+x,y'+y;z')\\
&\times T^*(x''+x,y''+y;z'')dx'dy'dx''dy''dz'dz'',
\end{aligned}
$$

which is identical to Eq. (6.2-13a), bringing us to the use of a point detector for coherent processing. So, for a point detector, that is, $|m(x,y)|^2=\delta(x,y)$, we have a coherent processing system. Now, let us look at another extreme case for the mask, which is when $|m(x,y)|^2=1$. According to Eq. (6.3-3),

$$
\begin{aligned}
\Gamma(x'-x'',y'-y'';z'-z'')&=\int e^{j\frac{k_0}{f}[x_m(x'-x'')+y_m(y'-y'')]}e^{-j\frac{k_0(z'-z'')}{2f^2}(x_m^2+y_m^2)}dx_m dy_m\\
&=\mathcal{F}\left\{e^{-j\frac{k_0(x_m^2+y_m^2)}{2f^2/(z'-z'')}}\right\}\Bigg|_{k_x=k_0(x'-x'')/f,k_y=k_0(y'-y'')/f}\\
&\sim\frac{1}{z'-z''}e^{j\frac{k_0[(x'-x'')^2+(y'-y'')^2]}{2(z'-z'')}}\\
&\sim\delta(x'-x'',y'-y'';z'-z''),
\end{aligned}
\tag{6.3-5}
$$

where we recognize that the quadratic term in the above integral represents a spherical wave with a radius of curvature $R=f^2/(z'-z'')$, which can be made arbitrarily large (see Section 3.5 on radius of curvature). Therefore, we have ended up taking the Fourier transform of 1 in the limit that $z'\to z''$, giving us a 3-D delta function. With this result, Eq. (6.3-2) is simplified to

$$
\begin{aligned}
i_\Omega(x,y)=\int&(s_2s_1^*e^{j\Omega t}+s_1s_2^*e^{-j\Omega t})\delta(x'-x'',y'-y'';z'-z'')\\
&\times T(x'+x,y'+y;z')T^*(x''+x,y''+y;z'')dx'dy'dx''dy''dz'dz''.
\end{aligned}
$$

We can now write out the terms $s_2s_1^*$ and $s_1s_2^*$, where they are given by Eqs. (6.2-8) and (6.2-9), respectively. Hence, the above equation becomes

$$
\begin{aligned}
i_\Omega(x,y)=\int&[s_2(x',y';z'+z_0)s_1^*(x'',y'';z''+z_0)e^{j\Omega t}\\
&+s_1(x',y';z'+z_0)s_2^*(x'',y'';z''+z_0)e^{-j\Omega t}]\\
&\times\delta(x'-x'',y'-y'';z'-z'')T(x'+x,y'+y;z')\\
&\times T^*(x''+x,y''+y;z'')dx'dy'dx''dy''dz'dz''.
\end{aligned}
$$

After evaluating the integral involving the delta function, we replace all the double primed coordinates by primed coordinates, giving

$$
\begin{aligned}
i_\Omega(x,y)&=\int[s_2(x',y';z'+z_0)s_1^*(x',y';z'+z_0)e^{j\Omega t}\\
&\quad+s_1(x',y';z'+z_0)s_2^*(x',y';z'+z_0)e^{-j\Omega t}]\\
&\quad\times T(x'+x,y'+y;z')T^*(x'+x,y'+y;z')dx'dy'dz'\\
&=\int[s_2(x',y';z'+z_0)s_1^*(x',y';z'+z_0)e^{j\Omega t}
\end{aligned}
$$

$$
+ s_1(x',y';z'+z_0)s_2^*(x',y';z'+z_0)\mathrm{e}^{-\mathrm{j}\Omega t}]
\times |T(x'+x,y'+y;z')|^2\mathrm{d}x'\mathrm{d}y'\mathrm{d}z'.
$$

最后, 写出其外差电流为

$$
i_\Omega(x,y) \propto \mathrm{Re}[i_p(x,y)\mathrm{e}^{\mathrm{j}\Omega t}],
$$

其中

$$
i_p(x,y) = \int s_2(x',y';z'+z_0)s_1^*(x',y';z'+z_0)|T(x'+x,y'+y;z')|^2\mathrm{d}x'\mathrm{d}y'\mathrm{d}z'. \tag{6.3-6}
$$

该等式应用比较广泛, 它代表了被两个扫描光束 s_1 和 s_2 处理的三维物体的强度分布.

与相干情况相似 [见式 (6.2-20)], 外差电流被电子解调, 且最后给出了一个复记录

$$
\begin{aligned}
i_C(x,y) &= i_p(x,y) = i_I(x,y) + \mathrm{j}i_Q(x,y) \\
&= \int s_2(x',y';z+z_0)s_1^*(x',y';z+z_0)|T(x'+x,y'+y;z)|^2\mathrm{d}x'\mathrm{d}y'\mathrm{d}z,
\end{aligned} \tag{6.3-7}
$$

其中, $i_I(x,y)$ 与 $i_Q(x,y)$ 是分别从锁相放大器输出的同相和正交相信号, 三维强度信息 $|T(x,y;z)|^2$ 被扫描光束 s_1 和 s_2 处理. 由于其操作和处理三维物体的强度分布, 故该系统进行非相干处理 (*incoherent processing*).

全息记录

这里特意选择 $p_1(x,y)=\delta(x,y)$ 且 $p_2(x,y)=1$, 根据式 (6.2-8) 与式 (6.2-9), s_1 和 s_2 变为

$$
s_1(x,y;z+z_0) = \mathcal{F}\{\delta(x,y)\}\big|_{k_x=\frac{k_0x}{f},k_y=\frac{k_0y}{f}} * h(x,y;z+z_0) \propto \mathrm{e}^{-\mathrm{j}k_0(z+z_0)}, \tag{6.3-8}
$$

和

$$
\begin{aligned}
s_2(x,y;z+z_0) &= \mathcal{F}\{1\}\big|_{k_x=\frac{k_0x}{f},k_y=\frac{k_0y}{f}} * h(x,y;z+z_0) \\
&\propto \mathrm{e}^{-\mathrm{j}k_0(z+z_0)}\frac{\mathrm{j}k_0}{2\pi(z+z_0)}\mathrm{e}^{-\mathrm{j}\frac{k_0(x^2+y^2)}{2(z+z_0)}},
\end{aligned} \tag{6.3-9}
$$

式 (6.3-7) 的复记录变为

$$
\begin{aligned}
i_C(x,y) &= i_I(x,y) + \mathrm{j}i_Q(x,y) \\
&= \int \frac{\mathrm{j}k_0}{2\pi(z+z_0)}\mathrm{e}^{\frac{-\mathrm{j}k_0}{2(z+z_0)}(x'^2+y'^2)}|T(x'+x,y'+y;z)|^2\mathrm{d}x'\mathrm{d}y'\mathrm{d}z \\
&= \int \frac{\mathrm{j}k_0}{2\pi(z+z_0)}\left[\mathrm{e}^{\frac{-\mathrm{j}k_0(x^2+y^2)}{2(z+z_0)}} * |T(x,y;z)|^2\right]\mathrm{d}z.
\end{aligned} \tag{6.3-10}
$$

这是三维物体 $|T(x,y;z)|^2$ 的复全息记录. 这种技术就是非相干模式的光学扫描全息术.

举例: 余弦与正弦全息图

从式 (6.3-10), 有

$$
i_C(x,y) = \int \frac{\mathrm{j}k_0}{2\pi(z+z_0)}\mathrm{e}^{\frac{-\mathrm{j}k_0(x^2+y^2)}{2(z+z_0)}} * |T(x,y;z)|^2\mathrm{d}z = i_I(x,y) + \mathrm{j}i_Q(x,y). \tag{6.3-11a}
$$

$$+ s_1(x', y'; z' + z_0) s_2^*(x', y'; z' + z_0) \mathrm{e}^{-\mathrm{j}\Omega t}]$$
$$\times |T(x' + x, y' + y; z')|^2 \mathrm{d}x'\mathrm{d}y'\mathrm{d}z'.$$

Finally, we write the heterodyne current as

$$i_\Omega(x, y) \propto \mathrm{Re}[i_p(x, y)\mathrm{e}^{\mathrm{j}\Omega t}],$$

where

$$i_p(x, y) = \int s_2(x', y'; z' + z_0) s_1^*(x', y'; z' + z_0) |T(x' + x, y' + y; z')|^2 \mathrm{d}x'\mathrm{d}y'\mathrm{d}z'. \tag{6.3-6}$$

This equation is fairly general and it represents the intensity distribution of the 3-D object being processing by the two scanning beams, s_1 and s_2.

Similar to the coherent case [see Eq. (6.2-20)], the heterodyne current is demodulated electronically and finally gives a complex record as follows:

$$\begin{aligned} i_C(x, y) &= i_p(x, y) = i_I(x, y) + \mathrm{j} i_Q(x, y) \\ &= \int s_2(x', y'; z + z_0) s_1^*(x', y'; z + z_0) |T(x' + x, y' + y; z)|^2 \mathrm{d}x'\mathrm{d}y'\mathrm{d}z, \end{aligned} \tag{6.3-7}$$

where $i_I(x, y)$ and $i_Q(x, y)$ are the in-phase and the quadrature-phase outputs from the lock-in amplifier, and the 3-D intensity information, $|T(x, y; z)|^2$, has been processed by the scanning beams s_1 and s_2. The system performs *incoherent processing* as it manipulates and processes the intensity distribution of the 3-D object.

Holographic Recording

Let us specifically choose $p_1(x, y) = \delta(x, y)$ and $p_2(x, y) = 1$. According to Eqs. (6.2-8) and (6.2-9), s_1 and s_2 become

$$s_1(x, y; z + z_0) = \mathcal{F}\{\delta(x, y)\}\big|_{k_x = \frac{k_0 x}{f}, k_y = \frac{k_0 y}{f}} * h(x, y; z + z_0) \propto \mathrm{e}^{-\mathrm{j}k_0(z+z_0)}, \tag{6.3-8}$$

and

$$\begin{aligned} s_2(x, y; z + z_0) &= \mathcal{F}\{1\}\big|_{k_x = \frac{k_0 x}{f}, k_y = \frac{k_0 y}{f}} * h(x, y; z + z_0) \\ &\propto \mathrm{e}^{-\mathrm{j}k_0(z+z_0)} \frac{\mathrm{j}k_0}{2\pi(z + z_0)} \mathrm{e}^{-\mathrm{j}\frac{k_0(x^2+y^2)}{2(z+z_0)}}, \end{aligned} \tag{6.3-9}$$

and the complex record from Eq. (6.3-7) becomes

$$\begin{aligned} i_C(x, y) &= i_I(x, y) + \mathrm{j} i_Q(x, y) \\ &= \int \frac{\mathrm{j}k_0}{2\pi(z + z_0)} \mathrm{e}^{\frac{-\mathrm{j}k_0}{2(z+z_0)}(x'^2+y'^2)} |T(x' + x, y' + y; z)|^2 \mathrm{d}x'\mathrm{d}y'\mathrm{d}z \\ &= \int \frac{\mathrm{j}k_0}{2\pi(z + z_0)} \left[\mathrm{e}^{\frac{-\mathrm{j}k_0(x^2+y^2)}{2(z+z_0)}} * |T(x, y; z)|^2 \right] \mathrm{d}z. \end{aligned} \tag{6.3-10}$$

This is a complex holographic record of the 3-D object $|T(x, y; z)|^2$. This is optical scanning holography operating in the incoherent mode.

Example: Cosine and Sine Holograms

From Eq.(6.3-10), we have

$$i_C(x, y) = \int \frac{\mathrm{j}k_0}{2\pi(z + z_0)} \mathrm{e}^{\frac{-\mathrm{j}k_0(x^2+y^2)}{2(z+z_0)}} * |T(x, y; z)|^2 \mathrm{d}z = i_I(x, y) + \mathrm{j} i_Q(x, y). \tag{6.3-11a}$$

从式 (6.3-11a), 有

$$i_I(x,y)=\mathrm{Re}[i_C(x,y)]=\int\frac{k_0}{2\pi(z+z_0)}\sin\left(\frac{k_0(x^2+y^2)}{2(z+z_0)}\right)*|T(x,y;z)|^2\mathrm{d}z$$
$$=H_{\sin}(x,y),\tag{6.3-11b}$$

和

$$i_Q(x,y)=\mathrm{Im}[i_C(x,y)]=\int\frac{k_0}{2\pi(z+z_0)}\cos\left(\frac{k_0(x^2+y^2)}{2(z+z_0)}\right)*|T(x,y;z)|^2\mathrm{d}z$$
$$=H_{\cos}(x,y).\tag{6.3-11c}$$

$H_{\sin}(x,y)$ 和 $H_{\cos}(x,y)$ 分别被称为正弦全息图 (*sine hologram*) 和余弦全息图 (*cosine hologram*), 因为这些全息图是通过正弦和余弦波带板与物体的强度分布卷积得到的. 可以看到, 正弦和余弦全息图是由锁相放大器的两个输出得到的. 一个复菲涅耳波带板全息图 (*complex Fresnel zone plate hologram*) $H^c(x,y)$ 可以根据下式来重建

$$H^c_{\pm}(x,y)=H_{\cos}(x,y)\pm\mathrm{j}H_{\sin}(x,y)=\int\frac{k_0}{2\pi(z+z_0)}\mathrm{e}^{\pm\mathrm{j}\frac{k_0(x^2+y^2)}{2(z+z_0)}}*|T(x,y;z)|^2\mathrm{d}z.\tag{6.3-11d}$$

与相干情况下的式 (6.2-22) 相似, 因为 $H^c_+(x,y)=[H^c_-(x,y)]^*$, 我们可以通过选择式 (6.3-11d) 中指数的符号来重建虚像或实像.

图 6.3-1 给出了一个典型的非相干模式的光学扫描全息装置, 其中可以看到, 在光电探测器前的掩模为全通状态, 即 $|m(x,y)|^2=1$.

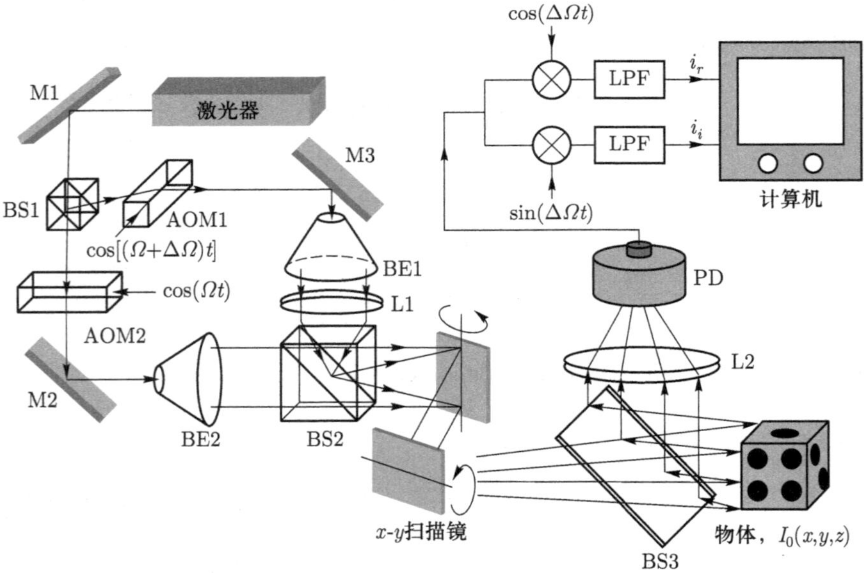

图 6.3-1　典型的非相干模式的光学扫描全息装置. BE 为扩束器, BS 为分束器, AOM 为声光调制器, PD 为光电探测器. 转载自 Kim et al., Speckle-free digital holographic recording of a diffusely reflecting object, Optics Express 21, pp. 8183-8189 (2013), 经 OSA© 许可

从图 6.2-3 中建立的双光瞳系统的基本理论可以看出, 在 $p_1(x,y)=\delta(x,y)$ 处, 光波以频率 ω_0 的一束平面波照射物体, 在 $p_2(x,y)=1$ 处, 光波以振荡频率 $\omega_0+\Omega$ 的球面波照射物体,

From Eq. (6.3-11a), we see that

$$i_I(x,y) = \mathrm{Re}[i_C(x,y)] = \int \frac{k_0}{2\pi(z+z_0)} \sin\left(\frac{k_0(x^2+y^2)}{2(z+z_0)}\right) * |T(x,y;z)|^2 \mathrm{d}z$$
$$= H_{\sin}(x,y), \tag{6.3-11b}$$

and

$$i_Q(x,y) = \mathrm{Im}[i_C(x,y)] = \int \frac{k_0}{2\pi(z+z_0)} \cos\left(\frac{k_0(x^2+y^2)}{2(z+z_0)}\right) * |T(x,y;z)|^2 \mathrm{d}z$$
$$= H_{\cos}(x,y). \tag{6.3-11c}$$

$H_{\sin}(x,y)$ and $H_{\cos}(x,y)$ are called the *sine and cosine holograms*, respectively as these holograms are obtained through the convolution of the sine and cosine zone plates with the intensity distribution of the object. Note that the sine and cosine holograms are obtained from the two outputs of the lock-in amplifier. A *complex Fresnel zone plate hologram* $H^c(x,y)$ can be constructed according to

$$H^c_\pm(x,y) = H_{\cos}(x,y) \pm \mathrm{j}H_{\sin}(x,y)$$
$$= \int \frac{k_0}{2\pi(z+z_0)} \mathrm{e}^{\pm \mathrm{j}\frac{k_0(x^2+y^2)}{2(z+z_0)}} * |T(x,y;z)|^2 \mathrm{d}z. \tag{6.3-11d}$$

Similar to Eq. (6.2-22) in the coherent case, we can reconstruct a virtual image or real image by selecting the sign of the exponent in Eq. (6.3-11d) as $H^c_+(x,y) = [H^c_-(x,y)]^*$.

Figure 6.3-1 shows a typical setup of optical scanning holography (OSH) operating in the incoherent mode, where we see that we have an open mask, that is, $|m(x,y)|^2 = 1$ in front of the photodetector PD.

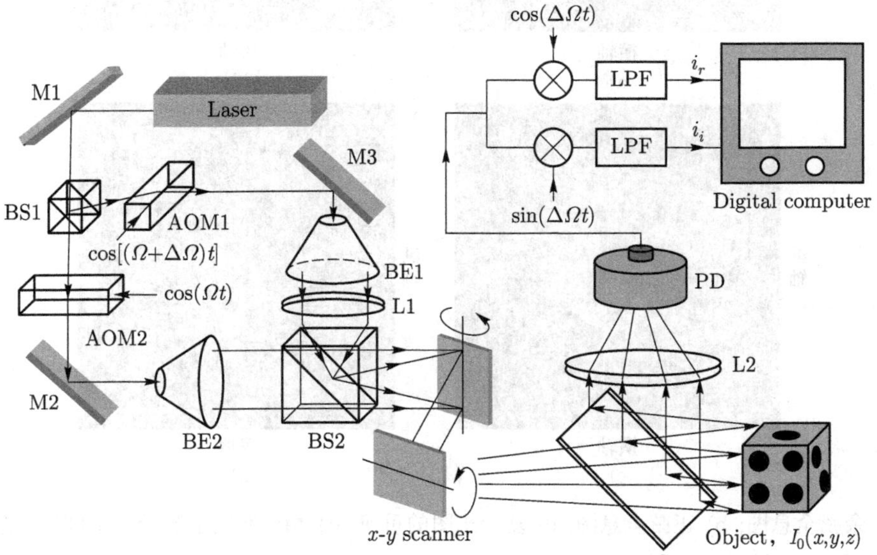

Figure 6.3-1 Typical setup of optical scanning holography operating in incoherent mode. BE: beam expander, BS: beam splitter, AOM: acousto-optic modulator, PD: photodetector. Reprinted from Kim et al., Speckle-free digital holographic recording of a diffusely reflecting object, Optics Express 21, pp. 8183-8189 (2013), with permission. © OSA

From the general theory developed for the two-pupil system in Figure 6.2-3, we see that optical scanning holography (OSH) is achieved when $p_1(x,y) = \delta(x,y)$, at light frequency ω_0, sends a plane wave illuminating the object, while $p_2(x,y) = 1$ oscillating at $\omega_0 + \Omega$ illuminates

从而使光学扫描全息术 (OSH) 得以实现. 因此, 在实际应用中, 对于 OSH, 我们采用在不同时间频率 (*temporal frequencies*) 的平面波和球面波的干涉来扫描物体, 以获取三维物体的全息信息. OSH 的相干性取决于光电探测器前方的掩模, 外差 (*heterodyning*) 是在光电探测器上进行的, 从而发出外差频率为 Ω (即两个扫描光束的频率差) 的外差信号. 图 6.3-1 所示的装置中, 激光器的频率为 ω_0. 一个从扩束器 2(BE2) 发出频率为 $\omega_0+\Omega$ 的平面波照射物体, 通过声光调制器 2 (*acousto-optic modulator* 2, AOM2) 来调节频率 Ω, 光波频率从 ω_0 上移到 $\omega_0+\Omega$. 我们将会在第七章介绍 AOM 的频移功能. 在干涉仪的另一臂上, 通过扩束器 1(BE1) 和透镜 L1 的使用, 会产生一束频率为 $\omega_0+\Omega+\Delta\Omega$ 的球面波以照射物体, 这里 $\Omega \gg \Delta\Omega$, 通过 AOM1 调节频率 $\Omega+\Delta\Omega$ 以实现光频率的上移. 因此, 实验中会产生外差频率 $\Delta\Omega$ 并进入锁相检测系统. 在该实验中, 两个分别运行在 $\Omega/(2\pi)=40$ MHz 和 $(\Omega+\Delta\Omega)/(2\pi)=40.01$ MHz 的 AOM 共同实现一个较低的外差频率, 即 0.01 MHz, 以便后续的电信号处理过程.

锁相系统输出时, 同相和正交相位输出 i_r 和 i_i, 分别给出正弦和余弦全息图. 图 6.3-2a) 与图 6.3-2b) 给出了骰子的余弦全息图和正弦全息图. 从这两个全息图, 再由式 $H_{\cos}(x,y) \pm \mathrm{j}H_{\sin}(x,y)$, 即可得到一个复全息图, 从而得到实像或虚像的重建. 图 6.3-2c) 所示为没有任何

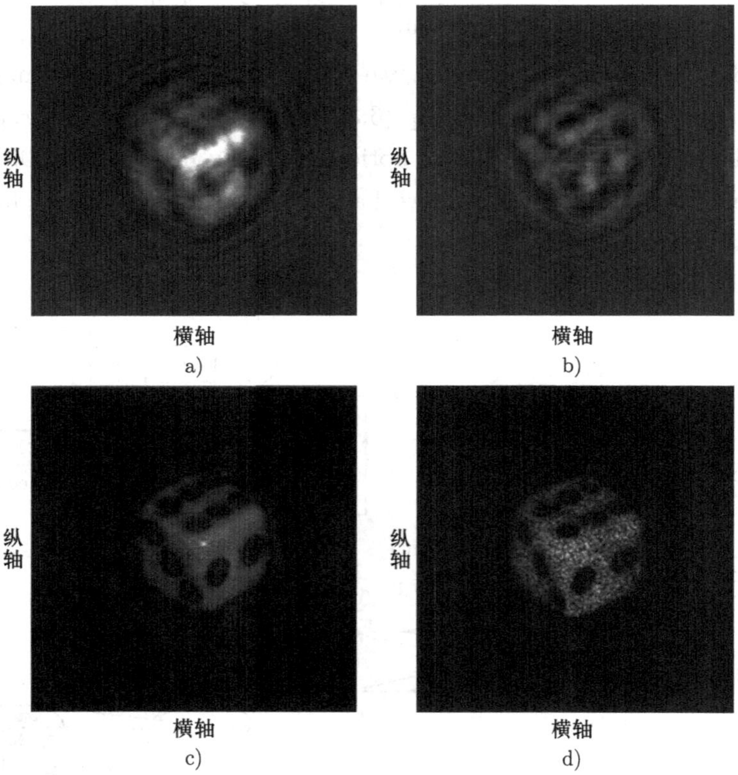

图 6.3-2 a) 余弦全息图; b) 正弦全息图; c) 复全息图的重建; d) 相干照明下骰子的 CCD 成像, 说明散斑噪声由相干照明产生. 转载自 Kim et al., Speckle-free digital holographic recording of a diffusely reflecting object, Optics Express 21, pp. 8183-8189 (2013), 经 OSA© 许可

the object with a spherical wave. Therefore, in practice for OSH, we have the interference of a plane wave and a spherical wave of different temporal frequencies to scan the object to acquire the holographic information of a 3-D object. Coherency of OSH depends on the mask in front of the photodetector. *Heterodyning* is performed at the photodetector to give out a heterodyne signal at heterodyne frequency Ω (being the frequency difference between the two scanning light beams). In Figure 6.3-1, a typical experimental setup of OSH is shown. The laser is at frequency ω_0. A plane wave at frequency $\omega_0 + \Omega$ exiting from beam expander 2 (BE2) illuminates the object. The upshifting of the light frequency from ω_0 to $\omega_0 + \Omega$ is with *acousto-optic modulator* 2 (AOM2) operating at frequency Ω. We will cover frequency-shifting capabilities of AOM in Chapter 7. From the other arm of the interferometer, through the use of beam expander 1 (BE1) and lens L1 we generate a spherical wave at frequency $\omega_0 + \Omega + \Delta\Omega$ toward the object, where $\Omega \gg \Delta\Omega$. The upshifting of the light frequency is with AOM1 operating at frequency $\Omega+\Delta\Omega$. So in the experiment, a heterodyne frequency of $\Delta\Omega$ is generated and inputted to the lock-in detection system. In the experiment, two AOMs operating at $\Omega/(2\pi) = 40$ MHz and $(\Omega + \Delta\Omega)/(2\pi) = 40.01$ MHz are used so as to achieve a lower heterodyne frequency at 0.01 MHz for ease of subsequent electronic processing.

At the outputs of the lock-in system, i_r and i_i are the in-phase and quadrature-phase outputs, giving the sine and cosine holograms, respectively. Figure 6.3-2a and b shows the cosine and sine holograms of the dice. From the two holograms, a complex hologram is constructed according to $H_{\cos}(x, y) \pm \mathrm{j}H_{\sin}(x, y)$ for either a real image reconstruction or virtual image reconstruction. Figure 6.3-2c shows a reconstruction of a complex hologram

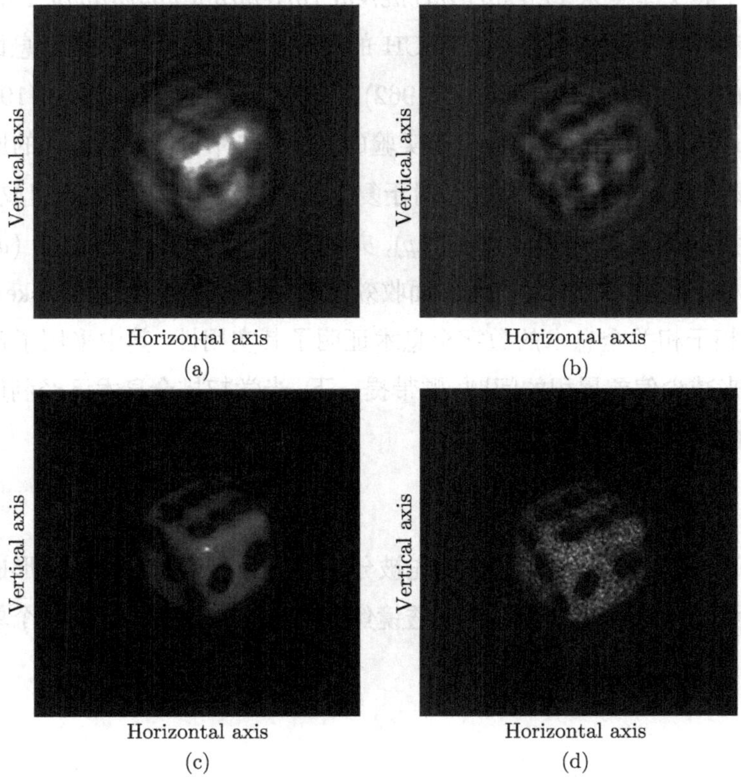

Figure 6.3-2 (a) Cosine hologram, (b) sine hologram, (c) reconstruction of a complex hologram, (d) CCD imaging of a coherently illuminated dice, illustrating speckle noise due to coherent illumination. Reprinted from Kim et al., Speckle-free digital holographic recording of a diffusely reflecting object, Optics Express 21, pp. 8183-8189 (2013), with permission. © OSA

相干散斑噪声的复全息图的重建. 最后, 为了比较, 图 6.3-2d) 给出一个在相干照明下骰子的 CCD 成像, 以说明散斑噪声是由相干照明引起的.

基于非相干模式下的光学扫描全息原理, 首次通过数字全息术记录荧光标本的全息图是在 1997 年 [Schilling 等 (1997)]. 在混入浑浊 (*turbid*) 介质的荧光不均匀体 (*inhomogeneities*) 的全息成像得到了证明 [Indebetouw 等 (1998)]. 分辨率大于 1 μm 的荧光颤藻丝 (*fluorescent Oscillatoria strands*) 的复全息图的三维重建也被报道 [Indebetouw 和 Zhong (2006)].

与 4.4 节中研究的扫描图像处理一样, 光学扫描全息术获得的数字全息图的相干性也取决于光电探测器光敏面前方的掩模的大小. 对于一个点探测器, 即 $|m(x,y)|^2 = \delta(x,y)$, 有相干数字全息记录. 对于另一种极端的情况, 当使用的掩模为全通时, 即 $|m(x,y)|^2 = 1$, 有非相干数字全息处理. 显然, 可以推测, 当使用的掩模尺寸为有限大小时, 就得到部分相干数字全息记录 (*partial coherent digital holographic recording*). 近年来, 部分相干数字全息成像已经得到证实 [Liu 等 (2015)].

最后, 作为本节结尾, 这里想要指出, 近些年在实现光学扫描全息的方法中, 还有采用 SLM 技术的静态光学扫描全息 [Yoneda 等 (2020); Yoneda, Saita 和 Nomura (2020)] 及利用几何相位透镜 (*geometric phase lens*) 的同轴扫描全息 [Kim 和 Kim (2020); Tsai 等 (2021)], 同时, 一种用于记录曲面数字全息图 (*curved digital hologram*) 的全息系统也被提出并被验证 [Liu 等 (2020)].

6.3.2 菲涅耳非相干相关全息术

菲涅耳非相干相关全息术 (*Fresnel incoherent correlation holography*, FINCH) 是目前被广泛研究的第二种非相干数字全息术. FINCH 的原理是基于每个物点在全息面上产生其自身的菲涅耳波带板的概念 [Mertz 和 Young (1962)]. 这一想法通过 Lohmann (1965) 的干涉概念得以延展, 而 Cochran (1966) 首次给出了实验的结果. 这种非相干全息术的原理现在被称为自干涉 (*self-interference*) [Poon (2008)]. 对于复杂的物体, 如由许多点所组成的物体, 自干涉会自动地产生大量的偏差累积 (*bias buildup*), 并限制记录介质的动态范围 (*dynamic range*). 基于自干涉的非相干全息术在实际应用方面收效甚微, 直到 Rosen 和 Brooker (2007) 通过他们称为菲涅耳非相干相关全息术的数字全息术证明了其实用性. 其中采用了现代空间光调制技术与相移技术来减少偏差累积的问题. 顺带提一下, 光学扫描全息术已经利用光学外差避免了此类偏差累积的问题 [见式 (6.3-4)].

自干涉

自干涉的原理是基于每个点物的波前先被分为两部分再重新合成从而形成干涉条纹的概念. 图 6.3-3 所示为一个自干涉装置. 位于透镜焦面上三维物体表面上的一个单物点向该透镜

without any coherent speckle noise. Finally, for comparison, in Figure 6.3-2d, an image taken with CCD imaging of a coherently illuminated dice is shown, illustrating speckle noise due to coherent illumination.

Based on the principle of optical scanning holography operating in the incoherent mode, holograms of fluorescent specimens were recorded by a digital holographic technique for the first time since 1997 [Schilling et al. (1997)]. Holographic imaging of fluorescent *inhomogeneities* embedded in a *turbid* medium was subsequently demonstrated [Indebetouw et al. (1998)]. Three-dimensional reconstructions of a complex hologram of *fluorescent Oscillatoria strands* of resolution better than 1 μm were also reported [Indebetouw and Zhong (2006)].

As with scanning image processing studied in Section 4.4, the coherence properties of the acquired digital holograms in optical scanning holography also depend on the size of the mask in front of the active area of the photodetector. For a point detector, that is, $|m(x,y)|^2 = \delta(x,y)$, we have coherent digital holographic recording. For another extreme case when we use an open mask, that is, $|m(x,y)|^2 = 1$, we have incoherent digital holographic processing. Clearly, we can envision *partial coherent digital holographic recording* when the mask employed is finite in size. Recently partial coherent digital holographic imaging has been demonstrated [Liu et al. (2015)].

To conclude this subsection, we want to point out that recent approaches to the implementation of optical scanning holography include motionless optical scanning holography using SLM technology [Yoneda et al. (2020), Yoneda, Saita and Nomura (2020)] and coaxial scanning holography using a *geometric phase lens* [Kim and Kim (2020), Tsai et al. (2021)]. A holographic system for recording a *curved digital hologram* has also been proposed and demonstrated [Liu et al. (2020)].

6.3.2 Fresnel Incoherent Correlation Holography

Fresnel incoherent correlation holography (FINCH) is the second incoherent digital holographic technique widely researched nowadays. The principle of FINCH is based on the concept that each object point generates its own Fresnel zone plate on the hologram plane [Mertz and Young (1962)]. The idea was extended through the concept of interference by Lohmann (1965) and Cochran reported the first experimental result (1966). This principle of incoherent holography is now known as *self-interference* [Poon (2008)]. Self-interference automatically creates enormous *bias buildup* for complicated objects, that is, object being made up of many points, and limits the *dynamic range* of the recording medium. Incoherent holography based on self-interference had little success in terms of practical applications until Rosen and Brooker (2007) demonstrated its usefulness through their digital holographic technique called *Fresnel incoherent correlation holography* (FINCH). They have employed modern spatial-light-modulator technology as well as phase-shifting technique to alleviate the issue of bias buildup. In passing, we mention that optical scanning holography has avoided the bias buildup using optical heterodyning [see Eq. (6.3-4)].

Self-Interference

The principle of self-interference is based on the idea that each point object's wave front is split into two parts and then re-combined to form an inference pattern. Figure 6.3-3 shows a version of the setup for self-interference. A single point object in the 3-D surface on the focal plane of the lens generates a spherical wave toward the lens, which then sends a plane

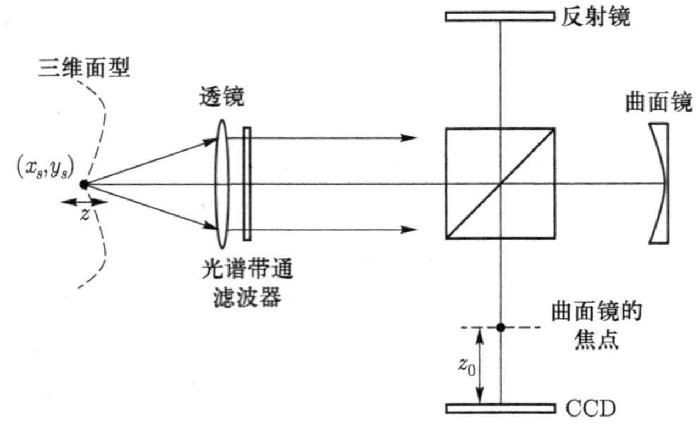

图 6.3-3 自干涉装置

发出一束球面波, 该球面波通过透镜后发出平面波到其中一个光路最终为曲面镜的改进迈克耳孙干涉仪结构上. 在透镜后, 放置一个光谱带通滤波器 (*spectral bandpass filter*) 以增加光的相干长度, 从而使系统更加有效地产生高对比度的干涉条纹. 回想一下, 当光源的谱宽变窄, 光源的相干长度就会增加 [见式 (4.1-1)]. 反射镜和曲面镜分别反射平面波和球面波 (假定该曲面镜的焦点在 CCD 前方 z_0 处) 朝向记录干涉条纹的 CCD 平面, 该干涉条纹即为菲涅耳波带板, 由下式给出 [见图 5.1-1a) 和式 (5.1-7)]

$$FZP(x,y;z_0)=A+B\sin\left[\frac{k_0}{2z_0}(x^2+y^2)\right].$$

如图 6.3-3 所示, 在该干涉仪 "展开" 的情况下, 利用图 6.3-4a) 会更容易分析. "展开" 一词的意思是, 在反射镜和曲面镜的表面处利用光学元件 (*optical element*) 取代反射, 这里仅考虑光波继续向前传播. 因此, 可以看到, 取代反射镜和曲面镜效果的光学元件具有以下形式的透过率函数

$$t(x,y)=1+\mathrm{e}^{\mathrm{j}\frac{k_0}{2f}(x^2+y^2)}, \tag{6.3-12}$$

其中, f 为曲面镜的焦距.

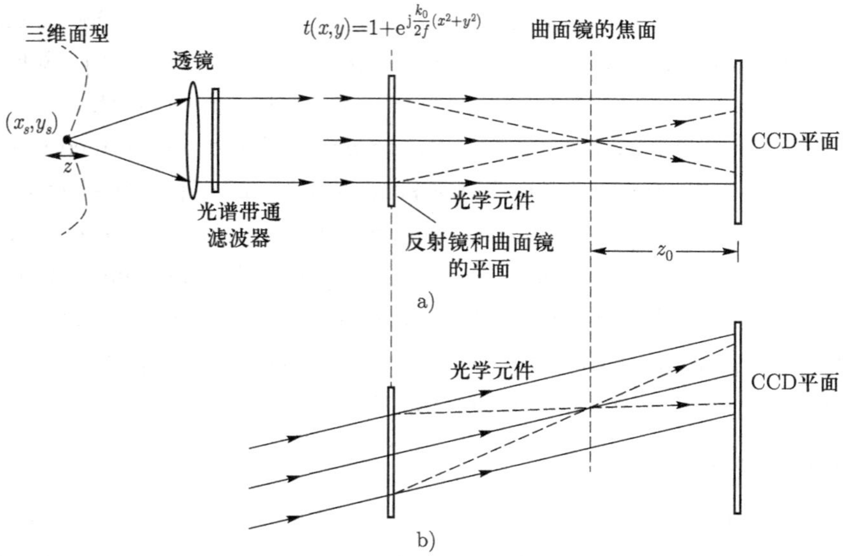

图 6.3-4 "展开" 干涉仪来解释自干涉条件下菲涅耳波带板的产生: a) 当物点位于光轴上; b) 当物点在横向方向上偏离时

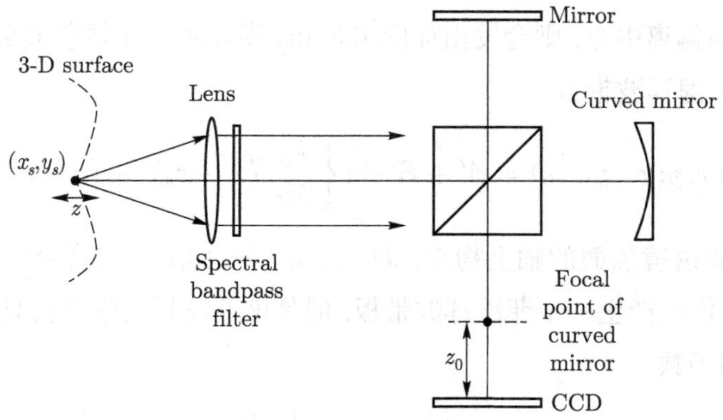

Figure 6.3-3 Self-interference

wave toward a modified Michelson interferometer configuration with one of the paths ending up with a curved mirror. After the lens, a *spectral bandpass filter* is placed to increase the coherence length of the light so as to make the system more efficient to produce high-contrast interference patterns. Recall that as we make the spectral width of the source narrower, the coherence length of the source increases [see Eq. (4.1-1)]. The mirror and the curved mirror reflect a plane wave and a spherical wave (assuming the focal point of the curved mirror is at a distance z_0 in front of the CCD), respectively toward the CCD plane for recording the interference pattern, which is a Fresnel zone plate given by [see Figure 5.1-1a and Eq. (5.1-7)]

$$FZP(x, y; z_0) = A + B\sin\left[\frac{k_0}{2z_0}(x^2 + y^2)\right].$$

The situation, shown in Figure 6.3-3, is easier to analyze using Figure 6.3-4a if the interferometer is "unfolded". The word "unfolded" means that on the plane of the mirror and the curved mirror, instead of allowing reflections, we simply consider that the light waves continue propagating forward. Hence, we see that the *optical element* serving to replace the effect of the plane and curved mirrors has a transparency function of the form

$$t(x, y) = 1 + e^{j\frac{k_0}{2f}(x^2+y^2)}, \tag{6.3-12}$$

where f is the focal length of the curved mirror.

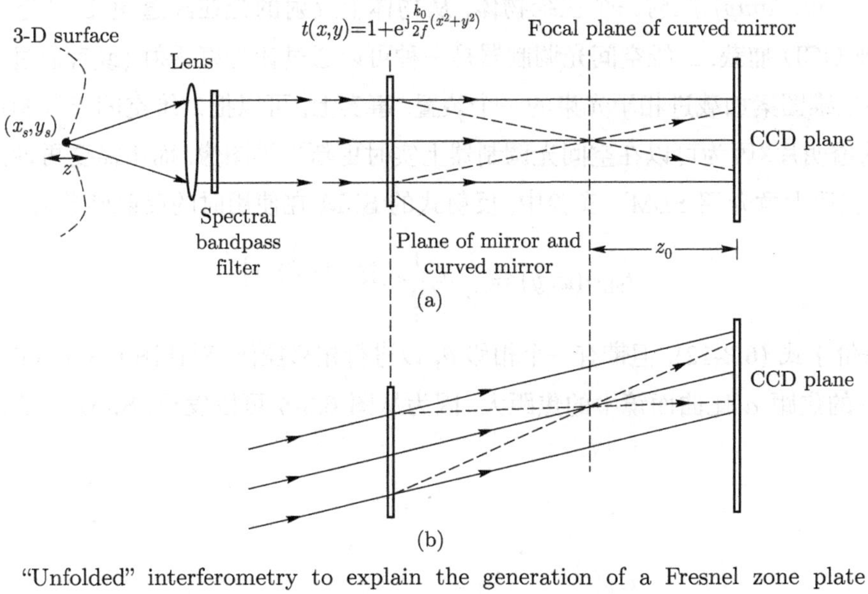

Figure 6.3-4 "Unfolded" interferometry to explain the generation of a Fresnel zone plate under self-interference (a) when the object point is on the optical axis and (b) when the object point is off in the transverse direction

如果该物点横向偏离中心, 则会发出如图 6.3-4b) 所示的一个倾斜光波, 在全息记录平面上形成一个离轴的菲涅耳波带板

$$FZP(x,y;x_s,y_s,z_0)=A'+B'\sin\left\{\frac{k_0}{2z_0}[(x-x_s)^2+(y-y_s)^2]\right\}.$$

对于另一个位于偏离透镜焦面的轴上物点, 即 $(x_s,y_s)=(0,0)$, 它将会记录两个不同曲率半径的球面波的干涉, 仍会产生一个菲涅耳波带板, 但是该菲涅耳波带板会具有不同的深度参数 $\gamma(z)$, 该参数是 z 的函数

$$FZP(x,y;z)=A''+B''\sin\left[\frac{k_0}{2\gamma(z)}(x^2+y^2)\right].$$

$\gamma(z)$ 的实际函数形式取决于所考虑的几何光路情况. 由于光源是非相干的, 这意味着这些物点之间没有干涉, 因为所有的物点来自三维物体, 因此, 仅需将全息面上 FZP 的强度条纹进行相加. 因此, 对于由上述三个点所组成的物体, 我们只是对全息面上的所有 FZP 进行相加. 显然, 可以看到恒定偏差项 A, A' 和 A'' 被加起来. 简言之, 有用的信号, 即正弦变化, 被隐藏在一个巨大的常数中, 而该巨大的常数会降低记录设备的动态范围. 一般来说, 对于一个强度分布为 $I(x_s,y_s,z)$ 的三维物体, 自干涉全息图会有这样的形式

$$H(x,y)\cong C+\iiint I(x_s,y_s,z)\sin\left\{\frac{k_0}{2\gamma(z)}\left[(x-x_s)^2+(y-y_s)^2\right]\right\}\mathrm{d}x_s\mathrm{d}y_s\mathrm{d}z,$$

其中, C 为某恒定偏差.

FINCH 描述

Rosen 和 Brooker 使用了现代空间光调制器技术与相移技术以缓解偏置累积的问题, 这一技术被称为菲涅耳非相干相关全息术 (FINCH), 并验证了基于自干涉的非相干数字全息术 (*incoherent digital holography*) 的实际应用. 其系统如图 6.3-5 所示. 可以看出, 其光学结构基本上与图 6.3-4a) 相同.

弧光灯 (*arc lamp*) 照射一个三维物体, 从物体上反射的光在经透镜 L 和空间光调制器 (SLM) 后被 CCD 捕获. 二维空间光调制器是一种可以通过让光束透射 (或者通过它反射该光束) 从而将二维图案加载进相干光束的一个装置. 事实上, 可以把二维空间光调制器看作一个实时更新的透明片, 因为可以在空间光调制器上实时更新二维图像, 而不需要再冲洗胶片制作透明片 (将在第七章介绍 SLM). 实验中, 反射式的 SLM 在使用时的反射函数为

$$t_{\mathrm{SLM}}(x,y)=\frac{1}{2}+\frac{1}{2}\mathrm{e}^{\mathrm{j}\left[\frac{k_0}{2a}(x^2+y^2)-\theta\right]},$$

这基本上等价于式 (6.3-12), 但带有一个相移 θ, 以进行相移操作. 对比图 6.3-5 和图 6.3-4a) 发现, SLM 中的焦距 a 比曲面镜中的焦距大, 因为从图 6.3-5 可以发现, SLM 上的入射平面波

If the object point is off-center transversely, it will send inclined waves as shown in Figure 6.3-4b, giving an off-axis Fresnel zone plate on the hologram recording plane:

$$FZP(x, y; x_s, y_s, z_0) = A' + B' \sin\left\{\frac{k_0}{2z_0}[(x - x_s)^2 + (y - y_s)^2]\right\}.$$

For an object point that is located away from the focal plane of the lens but on the optical axis, that is, $(x_s, y_s) = (0, 0)$, it will record interference of two spherical waves with different radii of curvature, still giving a Fresnel zone plate but with a different depth parameter $\gamma(z)$ that is a function of z:

$$FZP(x, y; z) = A'' + B'' \sin\left[\frac{k_0}{2\gamma(z)}(x^2 + y^2)\right].$$

The actual functional form of $\gamma(z)$ depends on the optical geometry under consideration. Since the source is incoherent, which means there is no interference between object points, we simply add up all the intensity patterns [see Eq. (4.1-6)] from the FZPs on the hologram plane due to all the points from the 3-D object. Hence, for an object that consists of three points given above, we simply add all the FZPs on the hologram plane. Clearly, we can see the constant bias terms, A, A', and A'' being added up. In short, the useful signals, that is, the sine variations, are being buried within a huge constant and that huge constant will lower the dynamic range of the recording device. In general, for a 3-D object with intensity distribution $I(x_s, y_s, z)$, a self-interference hologram will have a form

$$H(x, y) \cong C + \iiint I(x_s, y_s, z) \sin\left\{\frac{k_0}{2\gamma(z)}\left[(x - x_s)^2 + (y - y_s)^2\right]\right\} dx_s dy_s dz,$$

where C is some constant bias.

FINCH Description

Rosen and Brooker employed modern spatial-light-modulator technology as well as phase-shifting technique to alleviate the issue of bias buildup with the technique called FINCH and demonstrated the practical use of *incoherent digital holography* based on self-reference. Their system is shown in Figure 6.3-5. Note that in principle the optical configuration is identical to that in Figure 6.3-4a.

An *arc lamp* illuminates a 3-D object, and a CCD camera captures the reflected light from the object after passing through a lens L and a spatial light modulator SLM. A 2-D spatial light modulator is a device with which one can imprint a 2-D pattern on a coherent light beam by passing the beam through it (or by reflecting the beam off the device). In fact, we can think of a 2-D spatial light modulator as a real-time transparency because one can update 2-D images upon the spatial light modulator in real time without developing films into transparencies (We will cover SLMs in Chapter 7). In the experiment, the SLM operates in reflection mode with reflection function of the form given by

$$t_{\mathrm{SLM}}(x, y) = \frac{1}{2} + \frac{1}{2} e^{j\left[\frac{k_0}{2a}(x^2+y^2)-\theta\right]},$$

which is basically equivalent to Eq. (6.3-12) but with a phase shift θ, allowing phase-shifting operation. Note that when comparing Figures 6.3-5 and 6.3-4a, the focal length, a, of the

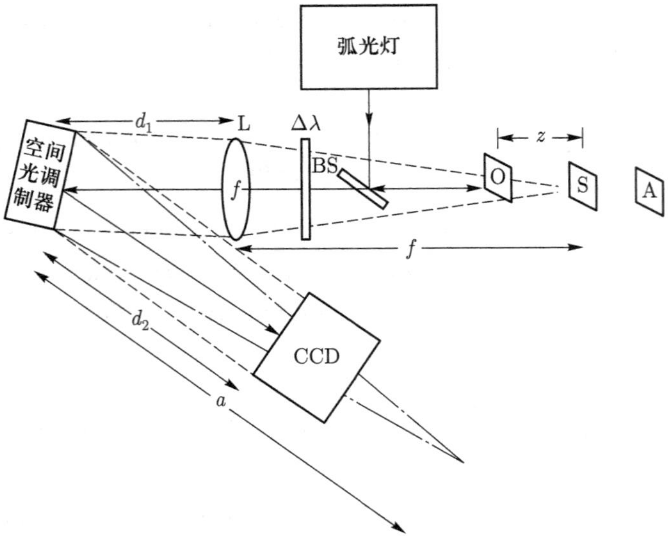

图 6.3-5 FINCH 系统示意图. 转载自 Rosen and Brooker, Digital spatially incoherent Fresnel holography, Optics Letters 32, pp. 912-914 (2007), 经 OSA© 许可

被聚焦在 CCD 的后面, 如图 6.3-5 所示.

为了完全地分析图 6.3-5 所示的光学系统. 作者计算了系统的点扩散函数 (*point spread function*, PSF), 并将其推广到三维物体. 现在概述其数学过程. 考虑焦距为 f 的透镜 L 距离 $f-z$ 处一个离轴点源 $\delta(x-x_s, y-y_s)$, 来求其紧贴 CCD 相机前方的复光场. 根据菲涅耳衍射, 紧贴透镜前方的光场分布为 $\delta(x-x_s, y-y_s)*h(x,y;f-z)$, 在透镜之后, 有 $[\delta(x-x_s, y-y_s)*h(x,y;f-z)]\mathrm{e}^{\mathrm{j}\frac{k_0}{2f}(x^2+y^2)}$, 在传播距离 d_1 后到达 SLM 平面, 该复光场为

$$\psi_{\text{front,SLM}}(x,y)=\left\{[\delta(x-x_s,y-y_s)*h(x,y;f-z)]\mathrm{e}^{\mathrm{j}\frac{k_0}{2f}(x^2+y^2)}\right\}*h(x,y;d_1).$$

在 SLM 后, 有 $\psi_{\text{front,SLM}}(x,y)t_{\text{SLM}}(x,y)$. 最后, 在 SLM 后距离 d_2 处的 CCD 平面上, 系统的点扩散函数为

$$I_p(x,y)=|\psi_{\text{front,SLM}}(x,y)t_{\text{SLM}}(x,y)*h(x,y;d_2)|^2,$$

上式已被作者计算过, 当 $f\gg z$ 时, 有

$$I_p(x,y)\propto 2+\mathrm{e}^{\mathrm{j}\frac{k_0}{2\gamma(z)}\left[\left(x-\frac{ax_s}{f}\right)^2+\left(y-\frac{ay_s}{f}\right)^2\right]+\mathrm{j}\theta}+c.c.,$$

其中, $c.c.$ 表示复共轭, 且 $\gamma(z)=\dfrac{d_2-a-z(d_1a+d_2f-af+d_2a-d_2d_1)f^{-2}}{1-z(a+f-d_1)f^{-2}}$.

对于一般的三维物体, 记录的全息图可以简单地通过对物体强度分布 $g(x_s,y_s,z)$ 的积分给出

$$\begin{aligned}H_\theta(x,y)\propto D&+\iiint g(x_s,y_s,z)\mathrm{e}^{\mathrm{j}\left\{\frac{k_0}{2\gamma(z)}\left[\left(x-\frac{ax_s}{f}\right)^2+\left(y-\frac{ay_s}{f}\right)^2+\theta\right]\right\}}\mathrm{d}x_s\mathrm{d}y_s\mathrm{d}z\\&+\iiint g(x_s,y_s,z)\mathrm{e}^{-\mathrm{j}\left\{\frac{k_0}{2\gamma(z)}\left[\left(x-\frac{ax_s}{f}\right)^2+\left(y-\frac{ay_s}{f}\right)^2+\theta\right]\right\}}\mathrm{d}x_s\mathrm{d}y_s\mathrm{d}z,\end{aligned}$$

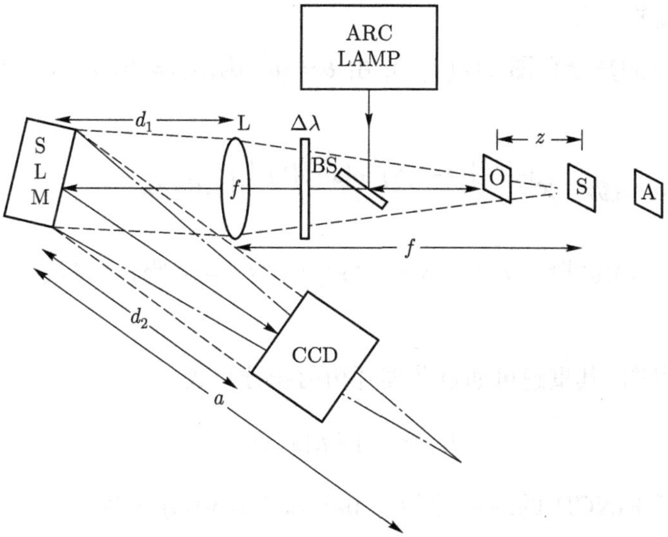

Figure 6.3-5 Schematic of FINCH. Reprinted from Rosen and Brooker, Digital spatially incoherent Fresnel holography, Optics Letters 32, pp. 912-914 (2007), with permission. © OSA

SLM is larger than that in the curved mirror as we notice that an incident plane wave on the SLM is focused behind the CCD shown in Figure 6.3-5.

To analyze the optical system shown in Figure 6.3-5 completely, the authors have calculated the *point spread function* (PSF) of the system and generalized it for a 3-D object. We shall outline the mathematical description of their procedures. Consider an off-axis point source $\delta(x-x_s, y-y_s)$ located at a distance $f-z$ away from lens L of focal length f. We shall find the complex field just in front of the CCD camera. Now, the field distribution just before the lens is $\delta(x-x_s, y-y_s)*h(x,y;f-z)$ according to Fresnel diffraction. After the lens, we have $[\delta(x-x_s, y-y_s)*h(x,y;f-z)]\mathrm{e}^{\mathrm{j}\frac{k_0}{2f}(x^2+y^2)}$. After propagating a distance of d_1 to the SLM plane, the complex field is

$$\psi_{\text{front,SLM}}(x,y)=\left\{[\delta(x-x_s, y-y_s)*h(x,y;f-z)]\mathrm{e}^{\mathrm{j}\frac{k_0}{2f}(x^2+y^2)}\right\}*h(x,y;d_1).$$

After the SLM, we have $\psi_{\text{front,SLM}}(x,y)t_{\text{SLM}}(x,y)$. Finally, in the CCD plane, a distance d_2 from the SLM, the PSF of the system is

$$I_p(x,y)=|\psi_{\text{front,SLM}}(x,y)t_{\text{SLM}}(x,y)*h(x,y;d_2)|^2,$$

which has been evaluated by the authors to be, for $f \gg z$,

$$I_p(x,y)\propto 2+\mathrm{e}^{\mathrm{j}\frac{k_0}{2\gamma(z)}\left[\left(x-\frac{ax_s}{f}\right)^2+\left(y-\frac{ay_s}{f}\right)^2\right]+\mathrm{j}\theta}+c.c.,$$

where *c.c.* denotes the complex conjugate, and

$$\gamma(z)=\frac{d_2-a-z(d_1a+d_2f-af+d_2a-d_2d_1)f^{-2}}{1-z(a+f-d_1)f^{-2}}.$$

For a general 3-D object, the record hologram is simply given by integrating over the object intensity distribution $g(x_s, y_s, z)$:

$$H_\theta(x,y)\propto D+\iiint g(x_s,y_s,z)\mathrm{e}^{\mathrm{j}\left\{\frac{k_0}{2\gamma(z)}\left[\left(x-\frac{ax_s}{f}\right)^2+\left(y-\frac{ay_s}{f}\right)^2+\theta\right]\right\}}\mathrm{d}x_s\mathrm{d}y_s\mathrm{d}z$$
$$+\iiint g(x_s,y_s,z)\mathrm{e}^{-\mathrm{j}\left\{\frac{k_0}{2\gamma(z)}\left[\left(x-\frac{ax_s}{f}\right)^2+\left(y-\frac{ay_s}{f}\right)^2+\theta\right]\right\}}\mathrm{d}x_s\mathrm{d}y_s\mathrm{d}z,$$

其中, D 为某恒定偏差.

为了提取出物体的复全息图 $H_F(x,y)$, 由 $\theta=(\theta_1,\theta_2,\theta_3)=(0,2\pi/3,4\pi/3)$, 得到三步相移全息图

$$H_F(x,y)=\iiint g(x_s,y_s,z)\mathrm{e}^{\mathrm{j}\frac{k_0}{2\gamma(z)}\left[\left(x-\frac{ax_s}{f}\right)^2+\left(y-\frac{ay_s}{f}\right)^2\right]}\mathrm{d}x_s\mathrm{d}y_s\mathrm{d}z$$

$$=H_{\theta_1}(x,y)(\mathrm{e}^{-\mathrm{j}\theta_3}-\mathrm{e}^{-\mathrm{j}\theta_2})+H_{\theta_2}(x,y)(\mathrm{e}^{-\mathrm{j}\theta_1}-\mathrm{e}^{-\mathrm{j}\theta_3})+H_{\theta_3}(x,y)(\mathrm{e}^{-\mathrm{j}\theta_2}-\mathrm{e}^{-\mathrm{j}\theta_1}). \tag{6.3-13}$$

这是一幅非涅耳全息图, 其重建可通过非涅耳衍射积分完成

$$H_F(x,y)*h(x,y;z).$$

图 6.3-6 给出了 FINCH 的结果, 图 6.3-6a) 和图 6.3-6b) 分别表示了 $H_F(x,y)$ 的强度和相位. 图 6.3-6c)、图 6.3-6d) 和图 6.3-6f) 是在最佳聚焦位置处的重建结果. 自 FINCH 问世以来, 利用自干涉非相干数字全息术的报道层出不穷. 一篇比较 OSH 与 FINCH 的综述文章已经发表 [Liu 等 (2018)].

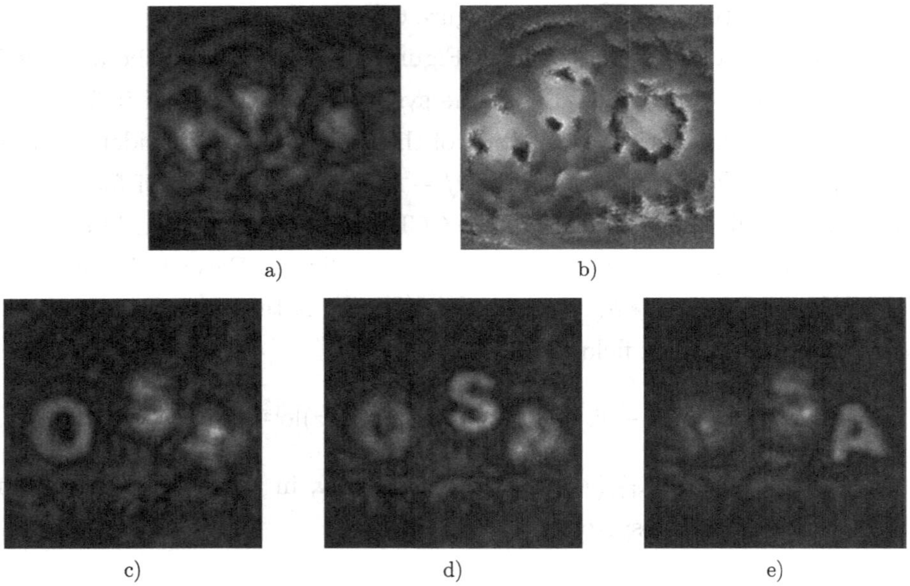

图 6.3-6 FINCH 结果: a) $H_F(x,y)$ 的强度, 即 $|H_F(x,y)|$; b) $H_F(x,y)$ 的相位; c) 对 "O" 在最佳聚焦位置处 $H_F(x,y)$ 的重建; d) 对 "S" 在最佳聚焦位置处的重建; e) 对 "A" 在最佳聚焦位置处的重建. 转载自 Rosen and Brooker, Digital spatially incoherent Fresnel holography, Optics Letters 32, pp. 912-914 (2007), 经 OSA© 许可

6.3.3 编码孔径成像与编码孔径相关全息术

编码孔径成像

FINCH 和编码孔径相关全息术 (*coded aperture correlation holography*, COACH) 的起源可以追溯到 20 世纪 60 年代的编码孔径成像 (*coded aperture imaging*) 技术 [Mertz 和 Young (1962)]. 编码孔径相机 (*coded aperture camera*) 类似于针孔相机, 在编码孔径相机中, 针孔相机的单一开口被带有针孔的图案 [被称为编码孔径 (*coded aperture*, CA)] 所取代. 编码图像是多个针孔图像的叠加, 多针孔孔径第一次是在编码孔径成像中被提出的 [Dicke (1968)].

where D is some constant bias.

To extract the complex hologram $H_F(x,y)$ of the object, three phase-shifted holograms are obtained with $\theta = (\theta_1, \theta_2, \theta_3) = (0, 2\pi/3, 4\pi/3)$:

$$H_F(x,y) = \iiint g(x_s, y_s, z) \mathrm{e}^{\mathrm{j}\frac{k_0}{2\gamma(z)}\left[\left(x-\frac{ax_s}{f}\right)^2+\left(y-\frac{ay_s}{f}\right)^2\right]} \mathrm{d}x_s \mathrm{d}y_s \mathrm{d}z$$
$$= H_{\theta_1}(x,y)(\mathrm{e}^{-\mathrm{j}\theta_3} - \mathrm{e}^{-\mathrm{j}\theta_2}) + H_{\theta_2}(x,y)(\mathrm{e}^{-\mathrm{j}\theta_1} - \mathrm{e}^{-\mathrm{j}\theta_3}) + H_{\theta_3}(x,y)(\mathrm{e}^{-\mathrm{j}\theta_2} - \mathrm{e}^{-\mathrm{j}\theta_1}). \tag{6.3-13}$$

This is a Fresnel hologram and reconstruction of it is done by performing the Fresnel diffraction integral:

$$H_F(x,y) * h(x,y;z).$$

Figure 6.3-6 shows the FINCH results. Figure 6.3-6a and b shows the magnitude and the phase of $H_F(x,y)$, respectively. Figure 6.3-6c, d, and f are the reconstructions at the best focus distances. Since the introduction of FINCH, there has been a plethora of reports in *incoherent digital holography* using self-interference. Most recently, a review comparing the differences between OSH and FINCH has been reported [Liu et al. (2018)].

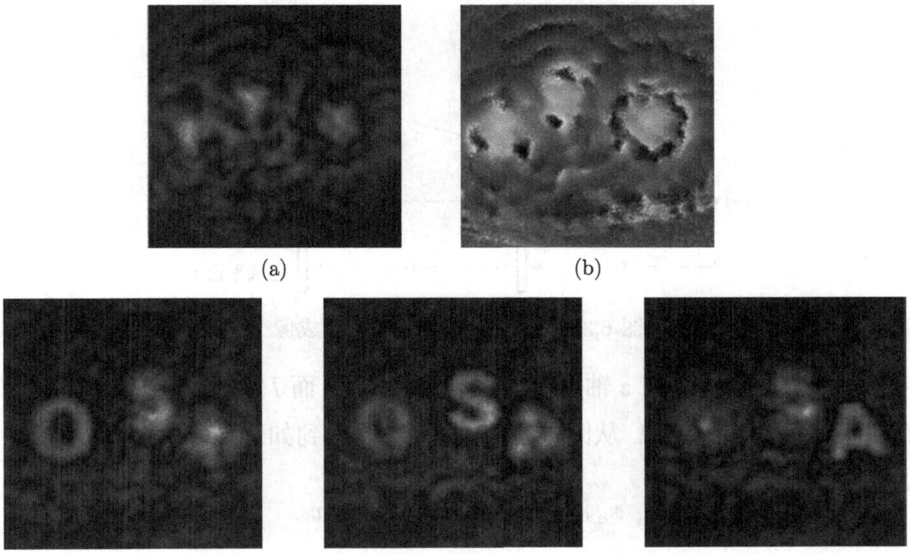

(a) (b) (c) (d) (e)

Figure 6.3-6 FINCH results: (a) magnitude of $H_F(x,y)$, that is, $|H_F(x,y)|$, (b) phase of $H_F(x,y)$, (c) reconstruction of $H_F(x,y)$ at the best focus distance for "O", (d) reconstruction at the best focus distance for "S" and (e) reconstruction at the best focus distance for "A". Reprinted from Rosen and Brooker, Digital spatially incoherent Fresnel holography, Optics Letters 32, pp. 912-914 (2007), with permission. © OSA

6.3.3 Coded Aperture Imaging and Coded Aperture Correlation Holography

Coded Aperture Imaging

FINCH and *coded aperture correlation holography* (COACH) have their roots in *coded aperture imaging* dated back to the 1960s [Mertz and Young (1962)]. A coded aperture camera is similar to a pinhole camera. In *coded aperture camera*, the single opening of the pinhole camera is replaced with a pattern of pinholes [called the *coded aperture* (CA)]. The coded image is then the superposition of many pinhole images. A multiple-pinhole aperture was

图 6.3-7a) 和图 6.3-7b) 分别表示了针孔成像和编码孔径成像. 图 6.3-7b) 仅给出了考虑两个针孔情况下的两幅图像. 我们来阐述图 6.3-8 中 x-z 平面上的编码孔径成像, 其中 $C(x)$ 为编码孔径, 从单点 $I(x)$ 发出的两束光, 穿过两个针孔并在记录平面上给出两幅图像, 因此, 可以将编码成像写为这两幅图像的求和

$$I(-x)+I(-x+x_{c,i}),$$

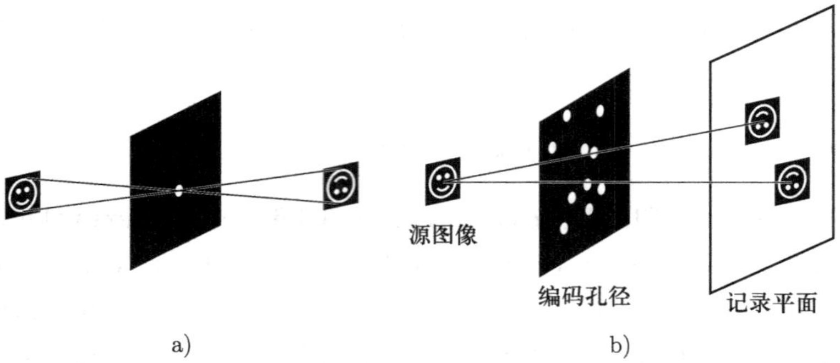

图 6.3-7 a) 针孔成像; b) 编码孔径成像

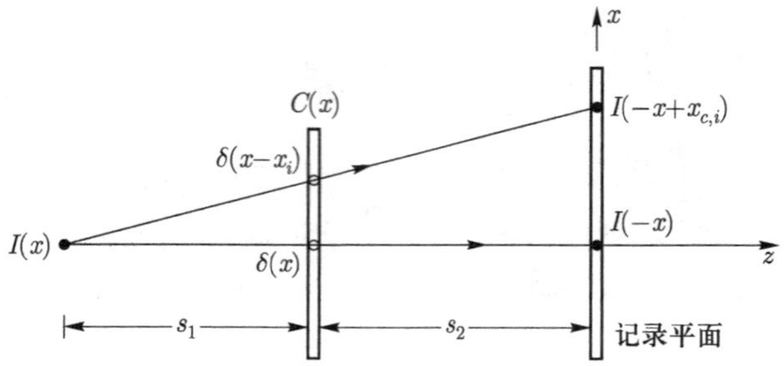

图 6.3-8 两个针孔的编码孔径成像

其中, $I(-x)$ 是由于针孔 $\delta(x)$ 在 z 轴上而产生的强度图像, 而 $I(-x+x_{c,i})$ 是由于针孔 $\delta(x-x_i)$ 离开 z 轴 x_i 距离而产生的图像. 从图中几何关系, 可以得到如下 $x_{c,i}$ 和 x_i 的关系

$$x_{c,i}=\frac{s_1+s_2}{s_1}x_i=M(z)x_i. \tag{6.3-14}$$

其中, $M(z)=(s_1+s_2)/s_1$ 为放大系数, 这取决于光源到孔径的轴向距离, 因为 s_1 可以写为 s_1+z, 其中 z 是三维物体的深度.

对于一个 n−孔的编码孔径 $C(x)=\sum_{i=1}^{n}\delta(x-x_i)$, 其编码图像 (*coded image*) $I_{coded}(x)$ 为

$$\begin{aligned}I_{coded}(x)&=\sum_i I(-x+x_{c,i})=\sum_i I(-x+Mx_i),\\&=I(-x)*\sum_i\delta(x-Mx_i),\end{aligned} \tag{6.3-15a}$$

上式可写成相关的形式

$$I_{coded}(x)=I(x)\otimes\sum_i\delta(x-Mx_i). \tag{6.3-15b}$$

first proposed in coded aperture imaging [Dicke (1968)]. Figure 6.3-7a and b shows pinhole imaging and coded aperture imaging, respectively. In Figure 6.3-7b, we just show two images due to the two pinholes under consideration. Let us formulate coded aperture imaging in the x-z plane shown in Figure 6.3-8. $C(x)$ is the coded aperture. We show two light rays from a single point of $I(x)$ going through the two pinholes to give two images on the recording plane. Hence, we can write the coded image as the sum of the two images given by

$$I(-x) + I(-x + x_{c,i}),$$

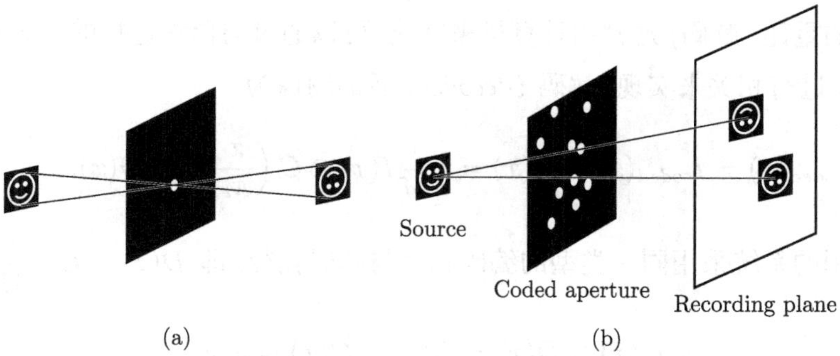

Figure 6.3-7 (a) Pinhole imaging and (b) coded aperture imaging

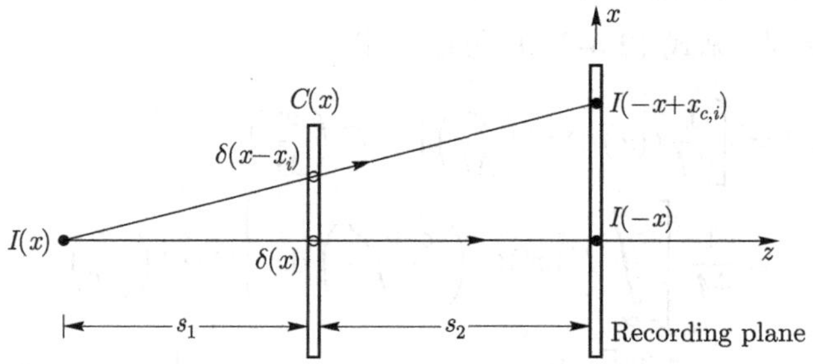

Figure 6.3-8 Coded aperture imaging (illustrated with two pinholes)

where $I(-x)$ is the intensity image due to the pinhole $\delta(x)$ on the z-axis, and $I(-x + x_{c,i})$ is the image due to the pinhole $\delta(x - x_i)$ that is x_i away from the z-axis. From the geometry of the figure, we can relate $x_{c,i}$ to x_i as follows:

$$x_{c,i} = \frac{s_1 + s_2}{s_1} x_i = M(z)x_i. \tag{6.3-14}$$

$M(z) = (s_1 + s_2)/s_1$ is the magnification factor, which depends on the axial distance of the source from the aperture as s_1 can be written as $s_1 + z$, where z is the depth of the 3-D object.

For an n -pinhole coded aperture $C(x) = \sum_{i=1}^{n} \delta(x - x_i)$, the *coded image* $I_{coded}(x)$ is

$$\begin{aligned} I_{coded}(x) &= \sum_i I(-x + x_{c,i}) = \sum_i I(-x + Mx_i), \\ &= I(-x) * \sum_i \delta(x - Mx_i), \end{aligned} \tag{6.3-15a}$$

which can be written in terms of correlation as

$$I_{coded}(x) = I(x) \otimes \sum_i \delta(x - Mx_i). \tag{6.3-15b}$$

注意到 $C(x)$ 的放大倍数 M 为

$$C\left(\frac{x}{M}\right)=\sum_{i=1}^{n}\delta\left(\frac{x}{M}-x_i\right)=\sum_{i=1}^{n}\delta\left(\frac{1}{M}(x-Mx_i)\right)=M\sum_{i=1}^{n}\delta(x-Mx_i).$$

利用这个结果, 式 (6.3-15b) 变为

$$I_{coded}(x)=\frac{1}{M}I(x)\otimes C\left(\frac{x}{M}\right). \tag{6.3-16}$$

因此, 可以看到, 编码图像被表示为强度 $I(x)$ 与放大了 M 倍的编码孔径 $C(x)$ 的相关.

编码图像的重建 (解码) 通常由计算机来完成, 可以通过与慎重选择的解码孔径 (*decoding aperture*) $D(x)$ 进行相关来实现. 解码 (*decoded*) 后的图像为

$$I_{decoded}(x)=I_{coded}(x)\otimes D(x)=\left[\frac{1}{M}I(x)\otimes C\left(\frac{x}{M}\right)\right]\otimes D(x). \tag{6.3-17}$$

最简单与最常用的系统采用同一类型的编码孔径与解码孔径, 即 $D(x)=C\left(\frac{x}{M}\right)$, 条件是

$$C(x)\otimes D(x)=C(x)\otimes C(x)\sim\delta(x). \tag{6.3-18}$$

换句话说, 编码孔径的自相关近似等于 δ 函数. 利用式 (6.3-18), 现在从式 (6.3-17) 中求出其解码图像. 从相关的定义 [见式 (2.3-30)], 可以写出

$$\begin{aligned}I_{decoded}(x)&=\left[\frac{1}{M}I(x)\otimes C\left(\frac{x}{M}\right)\right]\otimes C\left(\frac{x}{M}\right)\\&=\frac{1}{M}\left[\int_{-\infty}^{\infty}I^*(x')C\left(\frac{x+x'}{M}\right)\mathrm{d}x'\right]\otimes C\left(\frac{x}{M}\right)\\&=\frac{1}{M}\int_{-\infty}^{\infty}\left[\int_{-\infty}^{\infty}I^*(x')C\left(\frac{x''+x'}{M}\right)\mathrm{d}x'\right]^*C\left(\frac{x+x''}{M}\right)\mathrm{d}x''\\&=\frac{1}{M}\iint_{-\infty}^{\infty}I(x')C^*\left(\frac{x''+x'}{M}\right)C\left(\frac{x+x''}{M}\right)\mathrm{d}x'\mathrm{d}x''.\end{aligned} \tag{6.3-19}$$

从式 (6.3-18), 可以建立如下关系

$$C(x)\otimes C(x)=\int_{-\infty}^{\infty}C^*(x')C(x+x')\mathrm{d}x'=\delta(x).$$

利用上述关系, 通过将式 (6.3-19) 中带两撇的变量进行重组, 得 (见习题 6.6)

$$\int_{-\infty}^{\infty}C^*\left(\frac{x''+x'}{M}\right)C\left(\frac{x+x''}{M}\right)\mathrm{d}x''=\delta\left(\frac{x-x'}{M}\right). \tag{6.3-20}$$

利用上述结果, 求式 (6.1-19), 得

$$I_{decoded}(x)=\frac{1}{M}\int_{-\infty}^{\infty}I(x')\delta\left(\frac{x-x'}{M}\right)\mathrm{d}x'=I(x),$$

Note that the magnification M of $C(x)$ is

$$C\left(\frac{x}{M}\right)=\sum_{i=1}^{n}\delta\left(\frac{x}{M}-x_i\right)=\sum_{i=1}^{n}\delta\left(\frac{1}{M}(x-Mx_i)\right)=M\sum_{i=1}^{n}\delta(x-Mx_i).$$

Using this result, Eq. (6.3-15b) becomes

$$I_{coded}(x)=\frac{1}{M}I(x)\otimes C\left(\frac{x}{M}\right). \tag{6.3-16}$$

We see that the coded image, therefore, is expressed as the correlation of the source intensity $I(x)$ with the coded aperture $C(x)$ magnified by a factor M.

The reconstruction (or decoding) of the coded image is typically done by a computer and may be achieved by correlation with a carefully chosen *decoding aperture* $D(x)$. The *decoded* image is given by

$$I_{decoded}(x)=I_{coded}(x)\otimes D(x)=\left[\frac{1}{M}I(x)\otimes C\left(\frac{x}{M}\right)\right]\otimes D(x). \tag{6.3-17}$$

The simplest and most commonly used systems employ the same type of coded aperture and decoding aperture, that is, $D(x)=C\left(\frac{x}{M}\right)$, with the condition that

$$C(x)\otimes D(x)=C(x)\otimes C(x)\sim\delta(x). \tag{6.3-18}$$

In other words, the autocorrelation of the coded aperture is approximately equal to the delta function. With Eq. (6.3-18), let us evaluate the decoded image from Eq. (6.3-17). From the definition of correlation [see Eq. (2.3-30)], we write

$$\begin{aligned}
I_{decoded}(x)&=\left[\frac{1}{M}I(x)\otimes C\left(\frac{x}{M}\right)\right]\otimes C\left(\frac{x}{M}\right)\\
&=\frac{1}{M}\left[\int_{-\infty}^{\infty}I^*(x')C\left(\frac{x+x'}{M}\right)\mathrm{d}x'\right]\otimes C\left(\frac{x}{M}\right)\\
&=\frac{1}{M}\int_{-\infty}^{\infty}\left[\int_{-\infty}^{\infty}I^*(x')C\left(\frac{x''+x'}{M}\right)\mathrm{d}x'\right]^*C\left(\frac{x+x''}{M}\right)\mathrm{d}x''\\
&=\frac{1}{M}\int_{-\infty}^{\infty}\int_{-\infty}^{\infty}I(x')C^*\left(\frac{x''+x'}{M}\right)C\left(\frac{x+x''}{M}\right)\mathrm{d}x'\mathrm{d}x''.
\end{aligned} \tag{6.3-19}$$

From Eq. (6.3-18), we establish the following relationship:

$$C(x)\otimes C(x)=\int_{-\infty}^{\infty}C^*(x')C(x+x')\mathrm{d}x'=\delta(x).$$

Using the above relationship and by grouping the double-primed coordinates in Eq. (6.3-19), we obtain (see P. 6.6)

$$\int_{-\infty}^{\infty}C^*\left(\frac{x''+x'}{M}\right)C\left(\frac{x+x''}{M}\right)\mathrm{d}x''=\delta\left(\frac{x-x'}{M}\right). \tag{6.3-20}$$

With the above result, we evaluate Eq. (6.1-19) to obtain

$$I_{decoded}(x)=\frac{1}{M}\int_{-\infty}^{\infty}I(x')\delta\left(\frac{x-x'}{M}\right)\mathrm{d}x'=I(x),$$

现在已经从编码图像 $I_{coded}(x)$ 中恢复了原始图像 $I(x)$. 因此, 需要设计满足式 (6.3-18) 的 $C\left(\frac{x}{M}\right)$. 无限随机针孔阵列 (*infinite random pinhole array*) 原则上是最好的选择, 因为其自相关是一个 δ 函数 [Cannon 和 Fenimore (1980)]. 一些常用的编码孔径有菲涅耳波带板 [Mertz 和 Young (1962)]、随机针孔阵列 [Dicke (1968)], 以及环形孔径 [Simpson 等 (1975)]. 在近年来的编码孔径成像技术中, 吴等 (2020) 使用菲涅耳波带孔径在非相干照明下进行无透镜成像 (*lensless imaging*).

编码孔径相关全息术

近年来, 一种基于编码孔径的自干涉非相干数字全息术得到了发展. 该技术被称为编码孔径相关全息术 (*coded aperture correlation holography*, COACH). 它是 FINCH 的自干涉非相干数字全息术的更一般化的形式, 而不是像图 6.3-5 所示的 FINCH 那样使用一个二次相位掩模, 即 $t_{\mathrm{SLM}}(x,y)=\frac{1}{2}+\frac{1}{2}\mathrm{e}^{\mathrm{j}\left[\frac{k_0}{2a}(x^2+y^2)-\theta\right]}$, 原理上只要 COACH 的相位掩模具有自相关近似一个 δ 函数的特性, 则它可以采用任意随机相位掩模进行操作. COACH 技术需要记录两幅复全息图, 一幅来自点源作为点扩散函数 (*point spread function*, PSF) 全息图, 另一幅作为物全息图, 将这两个复全息图进行相关即可重建该物体. 全息图是用 COACH 技术记录的, 因此, 不可以被归类为傅里叶全息图或菲涅耳全息图, 因为无论是傅里叶变换或是菲涅耳后向传输都不能重建该图像. 参考图 6.3-9, 我们来讨论 COACH 的基本原理.

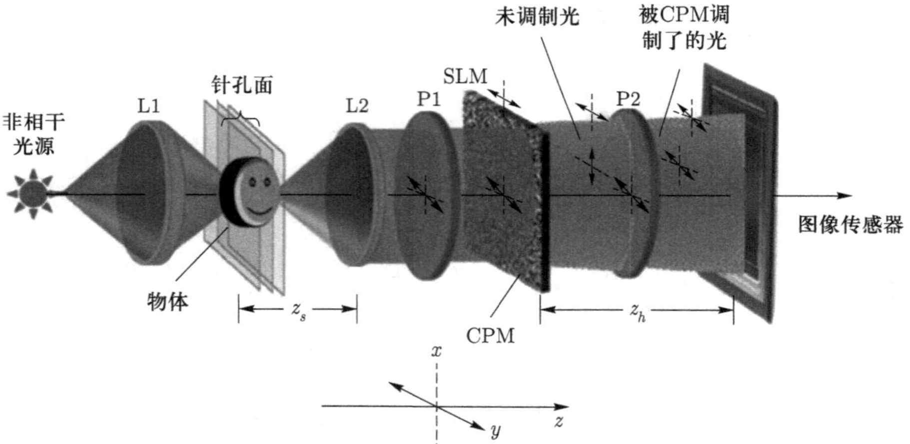

图 6.3-9　COACH 的光路图. CPM 为编码相位掩模, L1 和 L2 为透射透镜, P1 和 P2 为偏振片, SLM 为空间光调制器; 短箭头表示偏振方向. 该光学系统转载自 Rosen 等 (2019)

除了两个额外的偏振片外, 该光学系统与图 6.3-5 所示的 FINCH 基本相同. 照明光通过透镜 L1 直接聚焦在标本平面上. 这种类型的照明在显微镜中被称为光源聚焦 (*source-focused*) 或临界照明 (*critical illumination*). 编码相位掩模 (*coded phase mask*, CPM) 在纯相位的 SLM 上显示. 相位调制 (*phase modulation*) 沿着 y 轴进行 (通过空间光调制器进行相位调制, 见图 7.3-6). 如图 6.3-9 所示, 偏振片 P1 和 P2 的偏振轴方向相对于 y 轴方向为 45°. 从每个物点上发出的光沿着 P1 的偏振轴方向偏振, 其偏振波被分解为两个正交的方向, 一个方向沿 x 轴,

and we have recovered the original image $I(x)$ from the coded image $I_{coded}(x)$. Therefore, one needs to design $C\left(\frac{x}{M}\right)$ such that Eq. (6.3-18) is satisfied. An *infinite random pinhole array* is in principle a perfect choice as its autocorrelation is a delta function [Cannon and Fenimore (1980)]. Some of the popular employed coded apertures are Fresnel zone plates [Mertz and Young (1962)], random pinhole arrays [Dicke (1968)], and annular apertures [Simpson et al. (1975)]. In one of the most recent coded aperture imaging techniques, Wu et al. (2020) have employed a Fresnel zone aperture for *lensless imaging* under incoherent illumination.

Coded Aperture Correlation Holography (COACH)

Recently, a self-interference incoherent digital holography technique using a coded aperture has been developed. The technique is known as COACH and it is a generalized version of the self-interference incoherent digital holography technique of FINCH. Instead of using a quadratic phase mask, that is, $t_{\mathrm{SLM}}(x,y) = \frac{1}{2} + \frac{1}{2}e^{j\left[\frac{k_0}{2a}(x^2+y^2)-\theta\right]}$, in FINCH as shown in Figure 6.3-5, COACH is operational with any random phase mask in principle as long as the phase mask has the property that its auto-correlation approximates a delta function. In COACH, it requires to record two complex holograms, one from a point source as the PSF hologram and the other as the object hologram. Reconstruction of the object is to perform correlation of the two complex holograms. The holograms recorded by COACH, therefore, cannot be classified as Fourier or Fresnel holograms as neither Fourier transform nor Fresnel back propagation can reconstruct the image. With reference to Figure 6.3-9, we discuss the basic principle of COACH.

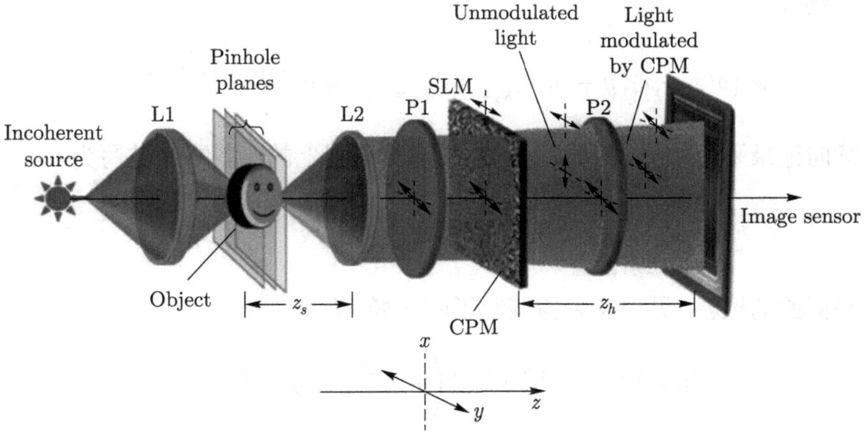

Figure 6.3-9 Optical configuration of COACH. CPM – Coded phase mask; L1, L2 – Refractive lenses; P1, P2 – Polarizers; SLM – Spatial light modulator; blue arrows indicate polarization orientations. The optical system is reprinted from Rosen et al. (2019)

Apart from the two additional polarizers, the optical system is basically identical to that of FINCH shown in Figure 6.3-5. The illumination is focused by lens L1 directly into the plane of the specimen. This type of illumination is called *source-focused or critical illumination* in microscopy. A *coded phase mask* (CPM) is displayed on the phase-only SLM. *Phase modulation* occurs along the y-axis (see Figure 7.3-6 on phase modulation by a spatial light modulator). The polarization axes of polarizers P1 and P1 are oriented 45° with respect to the y-axis, as shown in the figure. The light emitted from each object point is polarized along the polarization axis of P1. The polarized wave is decomposed into the two

另一个沿 y 轴. y 偏振波是被相位调制的, 但 x 偏振波是没有被调制的. 偏振片 P2 让两波以相对于 y 轴 45° 的相同偏振通过, 然后在图像传感器上干涉, 从而实现自干涉.

现在我们来推导其数学公式. 之前提到, COACH 需要记录两幅复全息图, 点源全息图和物全息图. 对位于焦距为 f 的透镜 L2 前焦面处的一个点物, 即 $z_s = f$, 其强度分布 $I_\delta(x,y)$ 是通过振幅为 A 的未调制的平面波与 CPM 的调制衍射图案 $G(x,y)$ 两者所形成的, 最终该强度分布到达图像传感器上, 为

$$I_\delta(x,y) = |A + G(x,y)\mathrm{e}^{\mathrm{j}\delta}|^2, \tag{6.3-21}$$

其中, δ 为相移. 一种可能的三步算法是, 当 $\delta = 0, \pi/2$ 和 π 时 [见式 (6.2-4) 和式 (6.2-5)], 根据式 (6.2-5), 得到点源全息图 $H_{\mathrm{PSH}}(x,y)$, 这也被称为复点扩散全息图 (*complex point spread hologram*)

$$H_{\mathrm{PSH}}(x,y) = A^* G(x,y). \tag{6.3-22}$$

接下来, 一个强度为 $I(x,y)$ 的平面物体距离透镜 L2 位置与点源的轴向位置相同. 由于可以将该强度物体分解为点的集合, 则

$$I(x,y) = \int I(x_i,y_i)\delta(x-x_i, y-y_i)\mathrm{d}x_i\mathrm{d}y_i,$$

每个位于物面上 (x_i,y_i) 处的第 i 个物点在图像传感器上产生两个彼此相干的光束. 一束为非调制的倾斜平面波, 可以看到, 离轴物点 $I(x_i,y_i)\delta(x-x_i,y-y_i)$, 向透镜焦面上的记录平面发出一束平面波, 根据

$$\mathcal{F}\{\delta(x-x_i,y-y_i)\}\big|_{k_x=\frac{k_0 x}{f},k_y=\frac{k_0 y}{f}} = \mathrm{e}^{\mathrm{j}2\pi(xx_i+yy_i)/(\lambda_0 f)},$$

该平面波继续向记录平面传播. 因此, 可以将图像传感器上的倾斜平面波写为

$$A_i\mathrm{e}^{\mathrm{j}2\pi(xx_i+yy_i)/(\lambda_0 f)}, \tag{6.3-23}$$

其中, A_i 是平面波的振幅. 另一光束为调制了的光场, 为

$$B_i\mathrm{e}^{\mathrm{j}2\pi(xx_i+yy_i)/(\lambda_0 f)}G\left(x+\frac{x_i z_h}{f}, y+\frac{y_i z_h}{f}\right), \tag{6.3-24}$$

其中, 用一个平移了的 $G(x,y)$ 乘以倾斜的平面波, 这是因为调制了的光束同样沿着未调制平面波的方向进行传播. 因为点源 $I(x_i,y_i)\delta(x-x_i,y-y_i)$ 通过 z 轴上的针孔投射了一个平移了的图像, 平移了的 $G(x,y)$ 可以通过图 6.3-10 看出, 这与图 6.3-7b 中编码孔径成像的概念类似.

点源 $I(x_i,y_i)\delta(x-x_i,y-y_i)$ 在记录传感器上的总强度, 由式 (6.3-23) 和式 (6.3-24) 所表示的两个光场的和给出

$$\left|A_i\mathrm{e}^{\mathrm{j}2\pi(xx_i+yy_i)/(\lambda_0 f)} + B_i\mathrm{e}^{\mathrm{j}2\pi(xx_i+yy_i)/(\lambda_0 f)}G\left(x+\frac{x_i z_h}{f}, y+\frac{y_i z_h}{f}\right)\mathrm{e}^{\mathrm{j}\delta}\right|^2.$$

orthogonal directions, one along the x-axis and the other along the y-axis. The y-polarized wave is phase-modulated but the x-polarized wave is unmodulated. Polarized P2 allows the two waves with the same polarization at 45° away from the y-axis to interfere on the image sensor, thereby accomplishing self-interference.

Let us now develop the mathematical formulation. As mentioned before, COACH needs to record two complex holograms, the point-source hologram and the object hologram. For a point object located at the front focal plane of lens L2 of focal length f, that is, $z_s = f$, the intensity distribution $I_\delta(x, y)$ is formed between the unmodulated plane wave with amplitude A and the modulated diffracted pattern of CPM, $G(x, y)$, on the image sensor:

$$I_\delta(x, y) = |A + G(x, y)\mathrm{e}^{\mathrm{j}\delta}|^2, \tag{6.3-21}$$

where δ is the phase shift. A possible three-step algorithm has $\delta = 0$, $\pi/2$, and π [see Eqs. (6.2-4) and (6.2-5)] and, according to Eq. (6.2-5), we obtain the point source hologram $H_{\mathrm{PSH}}(x, y)$, also called the *complex point spread hologram*:

$$H_{\mathrm{PSH}}(x, y) = A^* G(x, y). \tag{6.3-22}$$

Next, a planar intensity object $I(x, y)$ at the same axial location from lens L2 of the point source is placed. Since we can decompose the intensity object as a collection of points as

$$I(x, y) = \int I(x_i, y_i)\delta(x - x_i, y - y_i)\mathrm{d}x_i\mathrm{d}y_i,$$

each ith object point, at (x_i, y_i), on the object plane, generates two mutually coherent beams on the image sensor. One is an unmodulated tilted plane wave as we can see that the off-axis object point $I(x_i, y_i)\delta(x - x_i, y - y_i)$ sends a plane wave towards the recording plane on the focal plane of the lens according to

$$\mathcal{F}\{\delta(x - x_i, y - y_i)\}\Big|_{k_x = \frac{k_0 x}{f}, k_y = \frac{k_0 y}{f}} = \mathrm{e}^{\mathrm{j}2\pi(xx_i + yy_i)/(\lambda_0 f)},$$

and this plane wave continues travelling toward the recording plane. Therefore, we can write the tilted plane wave on the image sensor as

$$A_i \mathrm{e}^{\mathrm{j}2\pi(xx_i + yy_i)/(\lambda_0 f)}, \tag{6.3-23}$$

with A_i being the amplitude of the plane wave. Now, the other beam is the modulated field given by

$$B_i \mathrm{e}^{\mathrm{j}2\pi(xx_i + yy_i)/(\lambda_0 f)} G\left(x + \frac{x_i z_h}{f}, y + \frac{y_i z_h}{f}\right), \tag{6.3-24}$$

where we have a shifted version of $G(x, y)$ multiplied by a tilted plane wave as the modulated beam is also propagating along the same direction of the unmodulated plane wave. The shifted version of $G(x, y)$ can be recognized by inspecting Figure 6.3-10 as we can envision that the point source $I(x_i, y_i)\delta(x - x_i, y - y_i)$ projects a shifted image through the pinhole on the z-axis, reminiscent of the notion in coded aperture imaging in Figure 6.3-7b.

The total intensity on the recording sensor from the point source $I(x_i, y_i)\delta(x - x_i, y - y_i)$ is then given by the sum of the two fields from Eqs. (6.3-23) and (6.3-24)

$$\left| A_i \mathrm{e}^{\mathrm{j}2\pi(xx_i + yy_i)/(\lambda_0 f)} + B_i \mathrm{e}^{\mathrm{j}2\pi(xx_i + yy_i)/(\lambda_0 f)} G\left(x + \frac{x_i z_h}{f}, y + \frac{y_i z_h}{f}\right) \mathrm{e}^{\mathrm{j}\delta} \right|^2.$$

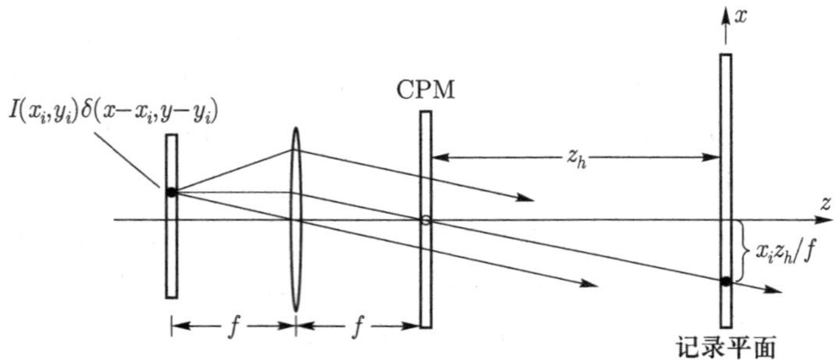

图 6.3-10　由于物体上的离轴点源而产生平移的 $G(x,y)$

对于一个包含 N 个物点集合的二维物体, 有

$$\sum_i^N \left| A_i \mathrm{e}^{\mathrm{j}2\pi(xx_i+yy_i)/(\lambda_0 f)} + B_i \mathrm{e}^{\mathrm{j}2\pi(xx_i+yy_i)/(\lambda_0 f)} G\left(x+\frac{x_i z_h}{f}, y+\frac{y_i z_h}{f}\right)\mathrm{e}^{\mathrm{j}\delta}\right|^2.$$

再次通过相移过程, 类似于式 (6.3-22) 中获得点源全息图 $H_{\mathrm{PSH}}(x,y)$ 的方法, 给出一个复物全息图 (*complex object hologram*) $H_{\mathrm{OBJ}}(x,y)$, 为

$$\begin{aligned} H_{\mathrm{OBJ}}(x,y) &= \sum_i A_i^* B_i G\left(x+\frac{x_i z_h}{f}, y+\frac{y_i z_h}{f}\right) \\ &\propto \sum_i I(x_i,y_i) G\left(x+\frac{x_i z_h}{f}, y+\frac{y_i z_h}{f}\right), \end{aligned} \tag{6.3-25}$$

其中, 振幅 A_i 与 B_i 基本上是由点源 $I(x_i,y_i)\delta(x-x_i,y-y_i)$ 导出的.

现在, 复点扩散全息图 $H_{\mathrm{PSH}}(x,y)$ 与复物全息图 $H_{\mathrm{OBJ}}(x,y)$ 已经被记录, 只需将这两幅全息图进行相关运算, 即可重建物体

$$H_{\mathrm{OBJ}}(x,y) \otimes H_{\mathrm{PSH}}(x,y) \propto \sum_i I(x_i,y_i) G\left(x+\frac{x_i z_h}{f}, y+\frac{y_i z_h}{f}\right) \otimes G(x,y). \tag{6.3-26}$$

在 COACH 中, CPM 是这样设计的

$$G(x,y) \otimes G(x,y) \sim \delta(x,y). \tag{6.3-27}$$

CPM 函数是一个纯相位函数, 有以下的约束条件: ① 其强度谱是均匀的, 因此在空域给出一个 δ 函数; ② 其在 SLM 平面上的衍射图案 $G(x,y)$ 是一个纯相位函数 [Vijayakumar 等 (2016)]. 利用式 (6.3-27) 可知 (见习题 6.7)

$$G\left(x+\frac{x_i z_h}{f}, y+\frac{y_i z_h}{f}\right) \otimes G(x,y) = \delta\left(x-\frac{x_i z_h}{f}, y-\frac{y_i z_h}{f}\right).$$

该结果可以使式 (6.3-26) 变为

$$\begin{aligned} H_{\mathrm{OBJ}}(x,y) \otimes H_{\mathrm{PSH}}(x,y) &\propto \sum_i I(x_i,y_i)\delta\left(x-\frac{x_i z_h}{f}, y-\frac{y_i z_h}{f}\right) \\ &\sim \int I(x_i,y_i)\delta\left(x-\frac{x_i z_h}{f}, y-\frac{y_i z_h}{f}\right)\mathrm{d}x_i \mathrm{d}y_i \\ &\propto I\left(\frac{x}{M}, \frac{y}{M}\right), \end{aligned}$$

其中, $M = z_h/f$ 是光学系统的放大率.

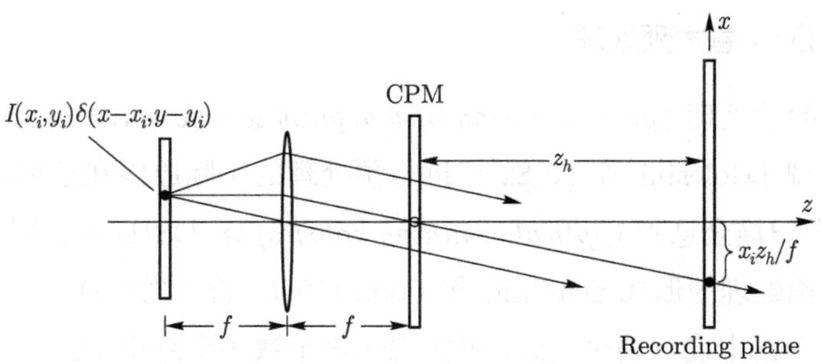

Figure 6.3-10 Shifted $G(x, y)$ due to an off-axis point source from the object

For a 2-D object with a collection of N object points, we have

$$\sum_i^N \left| A_i e^{j2\pi(xx_i+yy_i)/(\lambda_0 f)} + B_i e^{j2\pi(xx_i+yy_i)/(\lambda_0 f)} G\left(x+\frac{x_i z_h}{f}, y+\frac{y_i z_h}{f}\right) e^{j\delta}\right|^2.$$

By the phase shifting procedure again, similar to the method we obtain the point source hologram $H_{\mathrm{PSH}}(x,y)$ in Eq. (6.3-22), a complex object hologram $H_{\mathrm{OBJ}}(x,y)$ is given by

$$\begin{aligned} H_{\mathrm{OBJ}}(x,y) &= \sum_i A_i^* B_i G\left(x+\frac{x_i z_h}{f}, y+\frac{y_i z_h}{f}\right) \\ &\propto \sum_i I(x_i,y_i) G\left(x+\frac{x_i z_h}{f}, y+\frac{y_i z_h}{f}\right), \end{aligned} \qquad (6.3\text{-}25)$$

where we recognize that the amplitudes A_i and B_i basically are derived from the point source $I(x_i, y_i)\delta(x - x_i, y - y_i)$.

Now that the complex point spread hologram $H_{\mathrm{PSH}}(x,y)$ and the *complex object hologram* $H_{\mathrm{OBJ}}(x,y)$ have been recorded, we simply perform the correlation of the two holograms to reconstruct the object:

$$\begin{aligned} & H_{\mathrm{OBJ}}(x,y) \otimes H_{\mathrm{PSH}}(x,y) \\ & \propto \sum_i I(x_i,y_i) G\left(x+\frac{x_i z_h}{f}, y+\frac{y_i z_h}{f}\right) \otimes G(x,y). \end{aligned} \qquad (6.3\text{-}26)$$

In COACH, CPM is designed such that

$$G(x,y) \otimes G(x,y) \sim \delta(x,y). \qquad (6.3\text{-}27)$$

The CPM function is a pure phase function that is synthesized with the following constraints: (1) its magnitude spectrum is uniform, thereby giving a delta function in the spatial domain and (2) its diffracted pattern on the SLM plane, that is, $G(x,y)$, is to have a pure phase function [Vijayakumar et al. (2016)]. With Eq. (6.3-27), it can be shown that (see P.6.7)

$$G\left(x+\frac{x_i z_h}{f}, y+\frac{y_i z_h}{f}\right) \otimes G(x,y) = \delta\left(x-\frac{x_i z_h}{f}, y-\frac{y_i z_h}{f}\right).$$

This result allows us to write Eq. (6.3-26) as

$$\begin{aligned} H_{\mathrm{OBJ}}(x,y) \otimes H_{\mathrm{PSH}}(x,y) &\propto \sum_i I(x_i,y_i) \delta\left(x-\frac{x_i z_h}{f}, y-\frac{y_i z_h}{f}\right) \\ &\sim \int I(x_i,y_i) \delta\left(x-\frac{x_i z_h}{f}, y-\frac{y_i z_h}{f}\right) \mathrm{d}x_i \mathrm{d}y_i \propto I\left(\frac{x}{M}, \frac{y}{M}\right), \end{aligned}$$

where $M = z_h/f$ is the magnification of the optical system.

6.3.4 光学扫描全息术预处理

尽管全息信息的预处理 (*pre-processing of holographic information*) 很重要, 但在基于胶片的系统中很难实现 [Molesini 等 (1982)]. 进行预处理的实际原因在于可以在记录之前提高条纹对比度. 在光学扫描全息术 (*optical scanning holography*, OSH) 中, 物体被结构光扫描时, 即可被认为是一种预处理的形式.Schilling 和 Poon (1995) 首先发现通过在 6.2.2 节所讨论的双光瞳外差图像处理器中设计两个光瞳, 可实现非相干数字全息中全息重建时物体边缘的增强 (*edge enhancement*). 我们已经讨论过, 在双光瞳系统中, 使其中一个光瞳为 δ 函数, 另一个光瞳为单值函数来实现 OSH. 本节将在非相干模式操作下, 推导 OSH 预处理的基本原理. 这是现代非相干数字全息术的一个重要发展, 正如第四章中指出的, 传统非相干光学成像系统总会表现出低通滤波特性, 且许多重要的处理操作, 如边缘提取, 直到引入双光瞳系统才可能实现.

从简化的图 6.3-11 开始, 取 $z_0 = 0$, 且详细的电子子系统被一个称为 "电子处理" 的框图给简单替换掉. 同样, 设 $m(x, y) = 1$, 反映出此处进行非相干光学处理, 即仅对物体的强度分布 $|T(x, y; z)|^2$ 进行处理. 计算机上最终的复记录由下式 [见式 (6.3-7), $z_0 = 0$] 给出

$$\begin{aligned} i_C(x,y) &= i_I(x,y) + \mathrm{j}i_Q(x,y) \\ &= \int s_2(x',y';z)s_1^*(x',y';z)|T(x'+x,y'+y;z)|^2\mathrm{d}x'\mathrm{d}y'\mathrm{d}z, \end{aligned} \tag{6.3-28a}$$

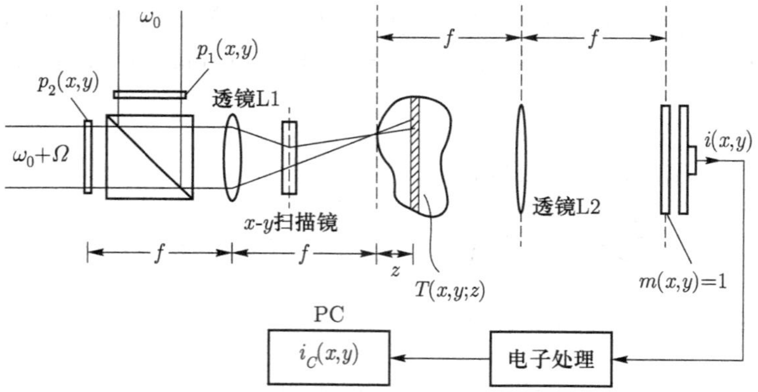

图 6.3-11 非相干模式光学扫描全息术

其中, 根据式 (6.2-8) 和式 (6.2-9), 两个扫描光束为

$$s_1(x,y;z) = \mathcal{F}\{p_1(x,y)\}\Big|_{k_x=\frac{k_0x}{f},k_y=\frac{k_0y}{f}} * h(x,y;z), \tag{6.3-28b}$$

和

$$s_2(x,y;z) = \mathcal{F}\{p_2(x,y)\}\Big|_{k_x=\frac{k_0x}{f},k_y=\frac{k_0y}{f}} * h(x,y;z). \tag{6.3-28c}$$

可以将式 (6.3-28a) 重写为含有 x 和 y 坐标变量的相关运算关系

$$i_C(x,y) = \int s_1(x,y;z)s_2^*(x,y;z) \otimes |T(x,y;z)|^2\mathrm{d}z. \tag{6.3-29}$$

6.3.4 Pre-processing in Optical Scanning Holography

Pre-processing of holographic information in film-based systems is difficult to implement, albeit important [Molesini et al. (1982)]. The practical reason of pre-processing is that we can improve the fringe contrast before recording. When the object is scanned by a structured beam in *optical scanning holography* (OSH), this is considered a form of pre-processing. Schilling and Poon (1995) first considered *edge enhancement* of an object upon holographic reconstruction in incoherent digital holography by designing the two pupils in the two-pupil heterodyning image processor discussed in Section 6.2.2. We have discussed the use of one of the pupils being a delta function and the other a uniform function to achieve OSH in the two-pupil system. In this section, we will develop a general formalism of pre-processing in OSH under the incoherent mode of operation. This is an important modern development for incoherent digital holography as we have pointed out in Chapter 4 that conventional incoherent optical imaging systems always exhibit low-pass filtering characteristics and many important processing operations, such as edge extraction, are not possible until the introduction of two-pupil systems.

Let us start with a concise version of Figure 6.2-3 in that we take $z_0 = 0$ (see Figure 6.3-11) and the detailed electronics subsystem is replaced simply by a block diagram called "electronic processing". We also set $m(x, y) = 1$ to reflect on the fact that we perform incoherent optical processing, that is, only the intensity distribution of the object $|T(x, y; z)|^2$ is being processed. The final complex record on the PC is given by [see Eq. (6.3-7) with $z_0 = 0$]

$$\begin{aligned} i_C(x, y) &= i_I(x, y) + \mathrm{j} i_Q(x, y) \\ &= \int s_2(x', y'; z) s_1^*(x', y'; z) |T(x' + x, y' + y; z)|^2 \mathrm{d}x' \mathrm{d}y' \mathrm{d}z, \end{aligned} \tag{6.3-28a}$$

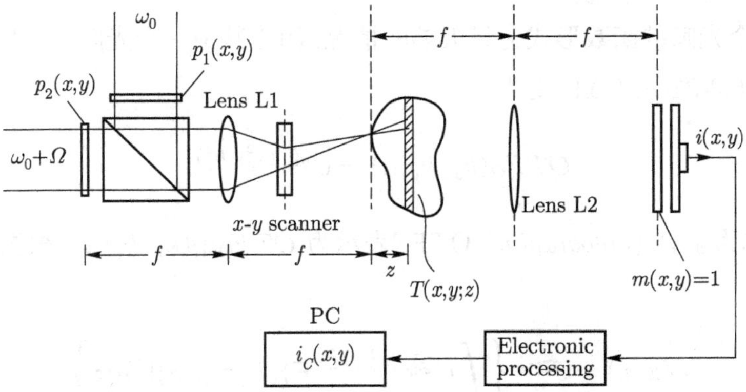

Figure 6.3-11 Optical scanning holography in incoherent mode

where, according to Eqs. (6.2-8) and (6.2-9), the two scanning beams are

$$s_1(x, y; z) = \mathcal{F}\{p_1(x, y)\}\big|_{k_x = \frac{k_0 x}{f}, k_y = \frac{k_0 y}{f}} * h(x, y; z), \tag{6.3-28b}$$

and

$$s_2(x, y; z) = \mathcal{F}\{p_2(x, y)\}\big|_{k_x = \frac{k_0 x}{f}, k_y = \frac{k_0 y}{f}} * h(x, y; z). \tag{6.3-28c}$$

We can rewrite Eq. (6.3-28a) in terms of correlation involving x and y coordinates:

$$i_C(x, y) = \int s_1(x, y; z) s_2^*(x, y; z) \otimes |T(x, y; z)|^2 \mathrm{d}z. \tag{6.3-29}$$

通过取式 (6.3-29) 的傅里叶变换, 有

$$\begin{aligned}\mathcal{F}\{i_C(x,y)\} &= \int \mathcal{F}^*\{s_1(x,y;z)s_2^*(x,y;z)\}\mathcal{F}\{|T(x,y;z)|^2\}\mathrm{d}z \\ &= \int OTF_\Omega(k_x,k_y;z)\mathcal{F}\{|T(x,y;z)|^2\}\mathrm{d}z,\end{aligned} \tag{6.3-30a}$$

其中, $OTF_\Omega(k_y,k_y;z)$ 为双光瞳外差系统的光学传递函数 (*optical transfer function of the two-pupil heterodyning system*)

$$OTF_\Omega(k_x,k_y;z) = \mathcal{F}^*\{s_1(x,y;z)s_2^*(x,y;z)\}. \tag{6.3-30b}$$

该复记录可以写为光学传递函数的形式

$$i_C(x,y) = \mathcal{F}^{-1}\left\{\int OTF_\Omega(k_x,k_y;z)\mathcal{F}\{|T(x,y;z)|^2\}\mathrm{d}z\right\}. \tag{6.3-31a}$$

将式 (6.3-28b) 和式 (6.3-28c) 代入式 (6.3-30b), Poon (1985) 用两个光瞳 p_1 和 p_2 来表示 OTF (见习题 6.8)

$$OTF_\Omega(k_x,k_y;z) = \mathrm{e}^{\mathrm{j}\frac{z}{2k_0}(k_x^2+k_y^2)}\int p_1^*(x',y')p_2\left(x'+\frac{f}{k_0}k_x, y'+\frac{f}{k_0}k_y\right)\mathrm{e}^{\mathrm{j}\frac{z}{f}(x'k_x+y'k_y)}\mathrm{d}x'\mathrm{d}y'. \tag{6.3-31b}$$

通过对两个光瞳进行操作, 可以设计出不同的 OTF, 如, 已研究的希尔伯特变换 (*Hilbert transform*) 和强度物体的带通滤波 (*bandpass filtering of intensity objects*) [Zhang 等 (2019)].

举例: 非相干模式光学扫描全息术

情况 A) $p_1(x,y)=\delta(x,y)$ 且 $p_2(x,y)=1$

这对应于两个光瞳的函数形式为最极端的情况, 因为其中一个光瞳为无限小, 而另一个为无限大. 根据式 (6.3-31b), OTF 变为

$$OTF_\Omega(k_x,k_y;z) = \mathrm{e}^{\mathrm{j}\frac{z}{2k_0}(k_x^2+k_y^2)}. \tag{6.3-32}$$

OTF 的结果被称为全息 (*holographic*) OTF, 表示为 $OTF_{\mathrm{OSH}}(k_x,k_y;z)$, 根据式 (6.3-31a), 复记录为

$$i_C(x,y) = \mathcal{F}^{-1}\left\{\int \mathrm{e}^{\mathrm{j}\frac{z}{2k_0}(k_x^2+k_y^2)}\mathcal{F}\{|T(x,y;z)|^2\}\mathrm{d}z\right\}. \tag{6.3-33}$$

对于一个在透镜 L1 的焦面距离 z_1 处强度分布为 $I(x,y)$ 的平面物体, 可以写出 $|T(x,y;z)|^2 = I(x,y)\delta(z-z_1)$, 在对 z 进行积分后, 式 (6.3-33) 变为

$$\begin{aligned}i_C(x,y) &= \mathcal{F}^{-1}\left\{\mathrm{e}^{\mathrm{j}\frac{z_1}{2k_0}(k_x^2+k_y^2)}\mathcal{F}\{I(x,y)\}\right\} \\ &= \mathcal{F}^{-1}\left\{\mathrm{e}^{\mathrm{j}\frac{z_1}{2k_0}(k_x^2+k_y^2)}\right\} * I(x,y),\end{aligned} \tag{6.3-34}$$

其中, 利用式 (2.3-27) 的卷积定理来获得最后一步. 使用表 2.1, 可以求出

$$\mathcal{F}^{-1}\left\{\mathrm{e}^{\mathrm{j}\frac{z_1}{2k_0}(k_x^2+k_y^2)}\right\} = \frac{\mathrm{j}k_0}{2\pi z_1}\mathrm{e}^{\frac{-\mathrm{j}k_0(x^2+y^2)}{2z_1}}.$$

By taking the Fourier transform of Eq. (6.3-29), we have

$$\begin{aligned}\mathcal{F}\{i_C(x,y)\} &= \int \mathcal{F}^*\{s_1(x,y;z)s_2^*(x,y;z)\}\mathcal{F}\{|T(x,y;z)|^2\}\mathrm{d}z \\ &= \int OTF_\Omega(k_x,k_y;z)\mathcal{F}\{|T(x,y;z)|^2\}\mathrm{d}z, \end{aligned} \tag{6.3-30a}$$

where $OTF_\Omega(k_y,k_y;z)$ is the *optical transfer function of the two-pupil heterodyning system:*

$$OTF_\Omega(k_x,k_y;z) = \mathcal{F}^*\{s_1(x,y;z)s_2^*(x,y;z)\}. \tag{6.3-30b}$$

The complex record can be written in terms of the optical transfer function as

$$i_C(x,y) = \mathcal{F}^{-1}\left\{\int OTF_\Omega(k_x,k_y;z)\mathcal{F}\{|T(x,y;z)|^2\}\mathrm{d}z\right\}. \tag{6.3-31a}$$

By substituting Eqs. (6.3-28b) and (6.3-28c) into Eq. (6.3-30b), Poon (1985) has expressed the OTF in terms of the two pupils p_1 and p_2 (see P.6.8):

$$OTF_\Omega(k_x,k_y;z) = \mathrm{e}^{\mathrm{j}\frac{z}{2k_0}(k_x^2+k_y^2)}\int p_1^*(x',y')p_2\left(x'+\frac{f}{k_0}k_x, y'+\frac{f}{k_0}k_y\right)\mathrm{e}^{\mathrm{j}\frac{z}{f}(x'k_x+y'k_y)}\mathrm{d}x'\mathrm{d}y'. \tag{6.3-31b}$$

By manipulating the two pupils, we can design various OTFs. For example, Hilbert transformation and *bandpass filtering of intensity objects* have been investigated [Zhang et al. (2019)].

Example: Optical Scanning Holography in Incoherent Mode

Case A $p_1(x,y) = \delta(x,y)$ and $p_2(x,y) = 1$.

This corresponds to the most extreme case in terms of the functional form of the two pupils as one of the pupils is infinitely small and the other is infinitely large. The OTF, according to Eq. (6.3-31b), becomes

$$OTF_\Omega(k_x,k_y;z) = \mathrm{e}^{\mathrm{j}\frac{z}{2k_0}(k_x^2+k_y^2)}. \tag{6.3-32}$$

The resulting OTF is called the *holographic OTF*, denoted by $OTF_{\mathrm{OSH}}(k_x,k_y;z)$, and the complex record, according to Eq. (6.3-31a), is

$$i_C(x,y) = \mathcal{F}^{-1}\left\{\int \mathrm{e}^{\mathrm{j}\frac{z}{2k_0}(k_x^2+k_y^2)}\mathcal{F}\{|T(x,y;z)|^2\}\mathrm{d}z\right\}. \tag{6.3-33}$$

For a planar object with intensity distribution of $I(x,y)$, located at a distance of z_1 away from the focal plane of lens L1, we can write $|T(x,y;z)|^2 = I(x,y)\delta(z-z_1)$ and after integrate over z, Eq. (6.3-33) becomes

$$\begin{aligned}i_C(x,y) &= \mathcal{F}^{-1}\left\{\mathrm{e}^{\mathrm{j}\frac{z_1}{2k_0}(k_x^2+k_y^2)}\mathcal{F}\{I(x,y)\}\right\} \\ &= \mathcal{F}^{-1}\left\{\mathrm{e}^{\mathrm{j}\frac{z_1}{2k_0}(k_x^2+k_y^2)}\right\} * I(x,y), \end{aligned} \tag{6.3-34}$$

where we have used the convolution theorem from Eq. (2.3-27) to obtain the last step. Using Table 2.1, we can find

$$\mathcal{F}^{-1}\left\{\mathrm{e}^{\mathrm{j}\frac{z_1}{2k_0}(k_x^2+k_y^2)}\right\} = \frac{\mathrm{j}k_0}{2\pi z_1}\mathrm{e}^{\frac{-\mathrm{j}k_0(x^2+y^2)}{2z_1}}.$$

式 (6.3-34) 可以写为

$$i_C(x,y)=\frac{\mathrm{j}k_0}{2\pi z_1}\mathrm{e}^{\frac{-\mathrm{j}k_0(x^2+y^2)}{2z_1}}*I(x,y),$$

这是一个位于透镜 L1 焦面距离 z_1 处的二维物体的复全息图. 这基本上是前面式 (6.3-10) 的结果. 给出其复全息图的虚像重建为

$$i_C(x,y)*h^*(x,y;z_1)=\frac{\mathrm{j}k_0}{2\pi z_1}\mathrm{e}^{\frac{-\mathrm{j}k_0(x^2+y^2)}{2z_1}}*I(x,y)*h^*(x,y;z_1)\propto I(x,y). \tag{6.3-35}$$

情况 B) $p_1(x,y)=1$ 且 $p_2(x,y)=\delta(x,y)$

这对应于另一种极端情况, 只需交换情况 A) 中 $p_1(x,y)$、$p_2(x,y)$ 的函数形式. 根据式 (6.3-31b), OTF 变为

$$OTF_\Omega(k_x,k_y;z)=\mathrm{e}^{-\mathrm{j}\frac{z}{2k_0}(k_x^2+k_y^2)}. \tag{6.3-36}$$

在这种情况下, $I(x,y)$ 的复记录变为

$$\begin{aligned}i_C(x,y)&=\mathcal{F}^{-1}\left\{\mathrm{e}^{-\mathrm{j}\frac{z_1}{2k_0}(k_x^2+k_y^2)}\mathcal{F}\{I(x,y)\}\right\}\\&=\mathcal{F}^{-1}\left\{\mathrm{e}^{-\mathrm{j}\frac{z_1}{2k_0}(k_x^2+k_y^2)}\right\}*I(x,y)=\frac{-\mathrm{j}k_0}{2\pi z_1}\mathrm{e}^{\frac{\mathrm{j}k_0(x^2+y^2)}{2z_1}}*I(x,y).\end{aligned}$$

该复全息图给出了一个实像的重建为

$$i_C(x,y)*h(x,y;z_1)=\frac{-\mathrm{j}k_0}{2\pi z_1}\mathrm{e}^{\frac{\mathrm{j}k_0(x^2+y^2)}{2z_1}}*I(x,y)*h(x,y;z_1)\propto I(x,y). \tag{6.3-37}$$

很明显, 在这一点上, 无论式 (6.3-32) 或式 (6.3-36) 形式的 OTF 都会产生全息记录. 因此, 可以定义光学扫描全息术中的全息 (*holographic*) OTF, 表示为 $OTF_{\mathrm{OSH}}(k_x,k_y;z)$, 则

$$OTF_{\mathrm{OSH}\pm}(k_x,k_y;z)=\mathrm{e}^{\pm\mathrm{j}\frac{z}{2k_0}(k_x^2+k_y^2)}. \tag{6.3-38}$$

随着全息 OTF 的引入, 现在可以来解释全息信息的预处理了. 从式 (6.3-31) 出发计算其一般形式. 然而, 为了简单起见, 假定一个平面物体 $|T(x,y;z)|^2=I(x,y)\delta(z-z_1)$. 由式 (6.3-31) 得到的复记录变为

$$\begin{aligned}i_C(x,y)&=\mathcal{F}^{-1}\left\{\mathrm{e}^{\mathrm{j}\frac{z_1}{2k_0}(k_x^2+k_y^2)}\left[\int p_1^*(x',y')p_2\left(x'+\frac{f}{k_0}k_x,y'+\frac{f}{k_0}k_y\right)\right.\right.\\&\qquad\left.\left.\times\mathrm{e}^{\mathrm{j}\frac{z_1}{f}(x'k_x+y'k_y)}\mathrm{d}x'\mathrm{d}y'\right]\mathcal{F}\{I(x,y)\}\right\}\\&=\mathcal{F}^{-1}\left\{OTF_{\mathrm{OSH}+}(k_x,k_y;z_1)\left[\int p_1^*(x',y')p_2\left(x'+\frac{f}{k_0}k_x,y'+\frac{f}{k_0}k_y\right)\right.\right.\\&\qquad\left.\left.\times\mathrm{e}^{\mathrm{j}\frac{z_1}{f}(x'k_x+y'k_y)}\mathrm{d}x'\mathrm{d}y'\right]\mathcal{F}\{I(x,y)\}\right\}\\&=\mathcal{F}^{-1}\left\{OTF_{\mathrm{HFE}}(k_x,k_y;z_1)\mathcal{F}\{I(x,y)\}\right\},\end{aligned} \tag{6.3-39a}$$

其中, $OTF_{\mathrm{HFE}}(k_x,k_y;z)$ 为全息特征提取 (*holographic feature extraction*) OTF, 有

Equation (6.3-34) can then be written as

$$i_C(x,y) = \frac{\mathrm{j}k_0}{2\pi z_1}\mathrm{e}^{\frac{-\mathrm{j}k_0(x^2+y^2)}{2z_1}} * I(x,y),$$

which is a complex hologram of the 2-D object located z_1 away from the focal plane of lens L1. This is basically the previous result in Eq. (6.3-10). The virtual image reconstruction of the complex hologram is given by

$$i_C(x,y) * h^*(x,y;z_1) = \frac{\mathrm{j}k_0}{2\pi z_1}\mathrm{e}^{\frac{-\mathrm{j}k_0(x^2+y^2)}{2z_1}} * I(x,y) * h^*(x,y;z_1) \propto I(x,y). \qquad (6.3\text{-}35)$$

Case B $p_1(x,y) = 1$ and $p_2(x,y) = \delta(x,y)$.

This case corresponds to another extreme case as we simply interchange the functional form of $p_1(x,y)$ and $p_2(x,y)$ from Case A. The OTF, according to Eq. (6.3-31b), becomes

$$OTF_\Omega(k_x,k_y;z) = \mathrm{e}^{-\mathrm{j}\frac{z}{2k_0}(k_x^2+k_y^2)}. \qquad (6.3\text{-}36)$$

In this case, the complex record of $I(x,y)$ becomes

$$\begin{aligned} i_C(x,y) &= \mathcal{F}^{-1}\left\{\mathrm{e}^{-\mathrm{j}\frac{z_1}{2k_0}(k_x^2+k_y^2)}\mathcal{F}\{I(x,y)\}\right\} \\ &= \mathcal{F}^{-1}\left\{\mathrm{e}^{-\mathrm{j}\frac{z_1}{2k_0}(k_x^2+k_y^2)}\right\} * I(x,y) = \frac{-\mathrm{j}k_0}{2\pi z_1}\mathrm{e}^{\frac{\mathrm{j}k_0(x^2+y^2)}{2z_1}} * I(x,y). \end{aligned}$$

This complex hologram gives a real image reconstruction as

$$i_C(x,y) * h(x,y;z_1) = \frac{-\mathrm{j}k_0}{2\pi z_1}\mathrm{e}^{\frac{\mathrm{j}k_0(x^2+y^2)}{2z_1}} * I(x,y) * h(x,y;z_1) \propto I(x,y). \qquad (6.3\text{-}37)$$

It is clear at this point that the OTF either of the form of Eq. (6.3-32) or Eq. (6.3-36) gives rise to holographic recording. Therefore, we can define the *holographic OTF* in optical scanning holography, denoted by $OTF_{\mathrm{OSH}}(k_x,k_y;z)$, as

$$OTF_{\mathrm{OSH}\pm}(k_x,k_y;z) = \mathrm{e}^{\pm\mathrm{j}\frac{z}{2k_0}(k_x^2+k_y^2)}. \qquad (6.3\text{-}38)$$

With the introduction of the holographic OTF, we can now in a position to explain pre-processing of holographic information. A general formalism starts from Eq. (6.3-31). However, for the sake of simplicity, we will assume a planar object $|T(x,y;z)|^2 = I(x,y)\delta(z - z_1)$. The complex record, from Eq. (6.3-31), then becomes

$$\begin{aligned} i_C(x,y) &= \mathcal{F}^{-1}\left\{\mathrm{e}^{\mathrm{j}\frac{z_1}{2k_0}(k_x^2+k_y^2)}\left[\int p_1^*(x',y')p_2\left(x'+\frac{f}{k_0}k_x, y'+\frac{f}{k_0}k_y\right)\right.\right. \\ &\quad \left.\left.\times\, \mathrm{e}^{\mathrm{j}\frac{z_1}{f}(x'k_x+y'k_y)}\mathrm{d}x'\mathrm{d}y'\right]\mathcal{F}\{I(x,y)\}\right\} \\ &= \mathcal{F}^{-1}\left\{OTF_{\mathrm{OSH}+}(k_x,k_y;z_1)\left[\int p_1^*(x',y')p_2\left(x'+\frac{f}{k_0}k_x, y'+\frac{f}{k_0}k_y\right)\right.\right. \\ &\quad \left.\left.\times\, \mathrm{e}^{\mathrm{j}\frac{z_1}{f}(x'k_x+y'k_y)}\mathrm{d}x'\mathrm{d}y'\right]\mathcal{F}\{I(x,y)\}\right\} \\ &= \mathcal{F}^{-1}\left\{OTF_{\mathrm{HFE}}(k_x,k_y;z_1)\mathcal{F}\{I(x,y)\}\right\}, \qquad (6.3\text{-}39\text{a}) \end{aligned}$$

where $OTF_{\mathrm{HFE}}(k_x,k_y;z)$ is the *holographic feature extraction* OTF:

$$
\begin{aligned}
&OTF_{\mathrm{HFE}}(k_x,k_y;z)\\
&=OTF_{\mathrm{OSH+}}(k_x,k_y;z)\int p_1^*(x',y')p_2\left(x'+\frac{f}{k_0}k_x,y'+\frac{f}{k_0}k_y\right)\mathrm{e}^{\mathrm{j}\frac{z}{f}(x'k_x+y'k_y)}\mathrm{d}x'\mathrm{d}y'.
\end{aligned}\tag{6.3-39b}
$$

全息特征提取 OTF 可以通过 $OTF_{\mathrm{OSH+}}(k_x,k_y;z_1)$ 进行全息记录, 与此同时, 它可以对物体的频谱进行处理, 该处理过程通过 $p_1(x,y)$ 和 $p_2(x,y)$ 的选取来控制. 为了更加清晰, 先求复记录的重建, 可通过对式 (6.3-39a) 取傅里叶变换, 乘以 $OTF_{\mathrm{OSH+}}(k_x,k_y;z_1)$ 的复共轭, 即 $\mathrm{e}^{-\mathrm{j}\frac{z_1}{2k_0}(k_x^2+k_y^2)}$, 最后再进行逆变换求得

$$
\begin{aligned}
&\mathcal{F}^{-1}\{\mathcal{F}\{i_C(x,y)\}OTF^*_{\mathrm{OSH+}}\}\\
&=\mathcal{F}^{-1}\{OTF_{\mathrm{HFE}}(k_x,k_y;z_1)\mathcal{F}\{I(x,y)\}OTF^*_{\mathrm{OSH+}}(k_x,k_y;z_1)\}\\
&=\mathcal{F}^{-1}\left\{\left[\int p_1^*(x',y')p_2\left(x'+\frac{f}{k_0}k_x,y'+\frac{f}{k_0}k_y\right)\mathrm{e}^{\mathrm{j}\frac{z_1}{f}(x'k_x+y'k_y)}\mathrm{d}x'\mathrm{d}y'\right]\mathcal{F}\{I(x,y)\}\right\}.
\end{aligned}\tag{6.3-40}
$$

上述结果表示, 经过全息重建后, 最终得到原始强度分布处理后的结果.

举例: 利用高斯环光瞳的边缘提取

Schilling 和 Poon (1995) 最早考虑用高斯环光瞳在全息重建时进行非相干物体的边缘增强. 这里以此光瞳为例. 双光瞳系统中使用的另一个光瞳是一个 δ 函数. 因此, 有 $p_1(x,y)=\mathrm{e}^{-b(\sqrt{x^2+y^2}-r_0)^2}$ 且 $p_2(x,y)=\delta(x,y)$, 其中 b 为高斯衰减 (*Gaussian falloff*), r_0 为环的半径. 在 $p_2(x,y)=\delta(x,y)$ 时, 全息特征提取 OTF 变为

$$
\begin{aligned}
OTF_{\mathrm{HFE}}(k_x,k_y;z)&=\mathrm{e}^{-\mathrm{j}\frac{z}{2k_0}(k_x^2+k_y^2)}p_1^*\left(-\frac{f}{k_0}k_x,-\frac{f}{k_0}k_y\right)\\
&=\mathrm{e}^{-\mathrm{j}\frac{z}{2k_0}(k_x^2+k_y^2)}\mathrm{e}^{-b\left[\sqrt{\left(\frac{f}{k_0}k_x\right)^2+\left(\frac{f}{k_0}k_y\right)^2}-r_0\right]^2}.
\end{aligned}\tag{6.3-41}
$$

利用该 OTF, 由式 (6.3-39a), 边缘保持的复全息图为

$$
\begin{aligned}
i_C(x,y)&=\mathcal{F}^{-1}\left\{OTF_{\mathrm{HFE}}(k_x,k_y;z_1)\mathcal{F}\{I(x,y)\}\right\}\\
&=\mathcal{F}^{-1}\left\{\mathrm{e}^{-\mathrm{j}\frac{z_1}{2k_0}(k_x^2+k_y^2)}\mathrm{e}^{-b\left[\sqrt{\left(\frac{f}{k_0}k_x\right)^2+\left(\frac{f}{k_0}k_y\right)^2}-r_0\right]^2}\mathcal{F}\{I(x,y)\}\right\}.
\end{aligned}\tag{6.3-42}
$$

显然, 强度分布的频谱被带通滤波. 图 6.3-12a) 和图 6.3-12d) 分别为原始二值图像及其谱图. 图 6.3-12b) 显示了 $p_1^*(-fk_x/k_0,-fk_y/k_0)$ 的强度大小, 其中光瞳为一个高斯环 $p_1(x,y)=\mathrm{e}^{-b(\sqrt{x^2+y^2}-r_0)^2}$. 在图 6.3-12e) 中, 我们给出了物体的原始频谱与图 6.3-12b) 所示的高斯环的乘积. 图 6.3-12c) 显示了由式 (6.3-42) 给出的复全息图的强度, 图 6.3-12f) 为复全息图的实像重建

$$
\begin{aligned}
&i_C(x,y)*h(x,y;z_1)\\
&=\mathcal{F}^{-1}\left\{\mathrm{e}^{-\mathrm{j}\frac{z_1}{2k_0}(k_x^2+k_y^2)}\mathrm{e}^{-b\left[\sqrt{\left(\frac{f}{k_0}k_x\right)^2+\left(\frac{f}{k_0}k_y\right)^2}-r_0\right]^2}\mathcal{F}\{I(x,y)\}\right\}*h(x,y;z_1)
\end{aligned}
$$

$$OTF_{\mathrm{HFE}}(k_x, k_y; z)$$
$$= OTF_{\mathrm{OSH+}}(k_x, k_y; z) \int p_1^*(x', y') p_2\left(x' + \frac{f}{k_0}k_x, y' + \frac{f}{k_0}k_y\right) \mathrm{e}^{\mathrm{j}\frac{z}{f}(x'k_x + y'k_y)} \mathrm{d}x'\mathrm{d}y'. \tag{6.3-39b}$$

The holographic feature extraction OTF can perform holographic recording through $OTF_{\mathrm{OSH+}}(k_x, k_y; z_1)$ and at the same time it can process the spectrum of the object and the processing is controlled by the selection of $p_1(x, y)$ and $p_2(x, y)$. To see it clearly, let us reconstruct the complex record by taking the Fourier transform of Eq. (6.3-39a), multiplying by the complex conjugate of $OTF_{\mathrm{OSH+}}(k_x, k_y; z_1)$, that is, $\mathrm{e}^{-\mathrm{j}\frac{z_1}{2k_0}(k_x^2+k_y^2)}$, and finally taking the inverse transform:

$$\mathcal{F}^{-1}\{\mathcal{F}\{i_C(x, y)\} OTF^*_{\mathrm{OSH+}}\}$$
$$= \mathcal{F}^{-1}\{OTF_{\mathrm{HFE}}(k_x, k_y; z_1) \mathcal{F}\{I(x, y)\} OTF^*_{\mathrm{OSH+}}(k_x, k_y; z_1)\}$$
$$= \mathcal{F}^{-1}\left\{\left[\int p_1^*(x', y') p_2\left(x' + \frac{f}{k_0}k_x, y' + \frac{f}{k_0}k_y\right) \mathrm{e}^{\mathrm{j}\frac{z_1}{f}(x'k_x + y'k_y)} \mathrm{d}x'\mathrm{d}y'\right] \mathcal{F}\{I(x, y)\}\right\}. \tag{6.3-40}$$

The above result represents that we have the processed version of the original intensity distribution upon holographic reconstruction.

Example: Edge Extraction Using a Gaussian-Ring Pupil

A Gaussian-ring pupil was first considered by Schilling and Poon (1995) to perform edge enhancement of an incoherent object upon holographic reconstruction. We shall use this pupil as an example. The other pupil used is a delta function in the two-pupil system. Hence, we have $p_1(x, y) = \mathrm{e}^{-b(\sqrt{x^2+y^2}-r_0)^2}$ and $p_2(x, y) = \delta(x, y)$, where b specifies the *Gaussian falloff*, and r_0 is the radius of the ring. With $p_2(x, y) = \delta(x, y)$, the holographic feature extraction OTF becomes

$$OTF_{\mathrm{HFE}}(k_x, k_y; z) = \mathrm{e}^{-\mathrm{j}\frac{z}{2k_0}(k_x^2+k_y^2)} p_1^*\left(-\frac{f}{k_0}k_x, -\frac{f}{k_0}k_y\right)$$
$$= \mathrm{e}^{-\mathrm{j}\frac{z}{2k_0}(k_x^2+k_y^2)} \mathrm{e}^{-b\left[\sqrt{\left(\frac{f}{k_0}k_x\right)^2 + \left(\frac{f}{k_0}k_y\right)^2} - r_0\right]^2}. \tag{6.3-41}$$

Using this OTF and from Eq. (6.3-39a), the edge-preserved complex hologram is

$$i_C(x, y) = \mathcal{F}^{-1}\{OTF_{\mathrm{HFE}}(k_x, k_y; z_1) \mathcal{F}\{I(x, y)\}\}$$
$$= \mathcal{F}^{-1}\left\{\mathrm{e}^{-\mathrm{j}\frac{z_1}{2k_0}(k_x^2+k_y^2)} \mathrm{e}^{-b\left[\sqrt{\left(\frac{f}{k_0}k_x\right)^2 + \left(\frac{f}{k_0}k_y\right)^2} - r_0\right]^2} \mathcal{F}\{I(x, y)\}\right\}. \tag{6.3-42}$$

Clearly, the spectrum of the intensity distribution is being bandpass filtered. Figure 6.3-12a and d shows the original binary image and its spectrum, respectively. Figure 6.3-12b shows the magnitude of $p_1^*(-fk_x/k_0, -fk_y/k_0)$, where the pupil is a Gaussian ring $p_1(x, y) = \mathrm{e}^{-b(\sqrt{x^2+y^2}-r_0)^2}$. In Figure 6.3-12e, we show the original spectrum of the object multiplied by the Gaussian ring shown in Figure 6.3-12b. Figure 6.3-12c shows the magnitude of the complex hologram given by Eq. (6.3-42) and Figure 6.3-12f shows the real image reconstruction of the complex hologram:

$$i_C(x, y) * h(x, y; z_1)$$
$$= \mathcal{F}^{-1}\left\{\mathrm{e}^{-\mathrm{j}\frac{z_1}{2k_0}(k_x^2+k_y^2)} \mathrm{e}^{-b\left[\sqrt{\left(\frac{f}{k_0}k_x\right)^2 + \left(\frac{f}{k_0}k_y\right)^2} - r_0\right]^2} \mathcal{F}\{I(x, y)\}\right\} * h(x, y; z_1)$$

$$
\begin{aligned}
&=\mathcal{F}^{-1}\left\{\mathrm{e}^{-\mathrm{j}\frac{z_1}{2k_0}(k_x^2+k_y^2)}\right\}*\mathcal{F}^{-1}\left\{\mathrm{e}^{-b\left[\sqrt{\left(\frac{f}{k_0}k_x\right)^2+\left(\frac{f}{k_0}k_y\right)^2}-r_0\right]^2}\mathcal{F}\{I(x,y)\}\right\}*h(x,y;z_1)\\
&=\frac{-\mathrm{j}k_0}{2\pi z_1}\mathrm{e}^{\frac{\mathrm{j}k_0(x^2+y^2)}{2z_1}}*\mathcal{F}^{-1}\left\{\mathrm{e}^{-b\left[\sqrt{\left(\frac{f}{k_0}k_x\right)^2+\left(\frac{f}{k_0}k_y\right)^2}-r_0\right]^2}\mathcal{F}\{I(x,y)\}\right\}*h(x,y;z_1)\\
&\propto\mathcal{F}^{-1}\left\{\mathrm{e}^{-b\left[\sqrt{\left(\frac{f}{k_0}k_x\right)^2+\left(\frac{f}{k_0}k_y\right)^2}-r_0\right]^2}\mathcal{F}\{I(x,y)\}\right\},
\end{aligned}
\tag{6.3-43}
$$

因为 $\mathrm{e}^{\frac{\mathrm{j}k_0(x^2+y^2)}{2z_1}}*h(x,y;z_1)\propto\mathrm{e}^{\frac{\mathrm{j}k_0(x^2+y^2)}{2z_1}}*\mathrm{e}^{-\frac{\mathrm{j}k_0(x^2+y^2)}{2z_1}}\propto\delta(x,y)$. 该结果清楚地表明了对强度物体 $I(x,y)$ 的带通滤波.

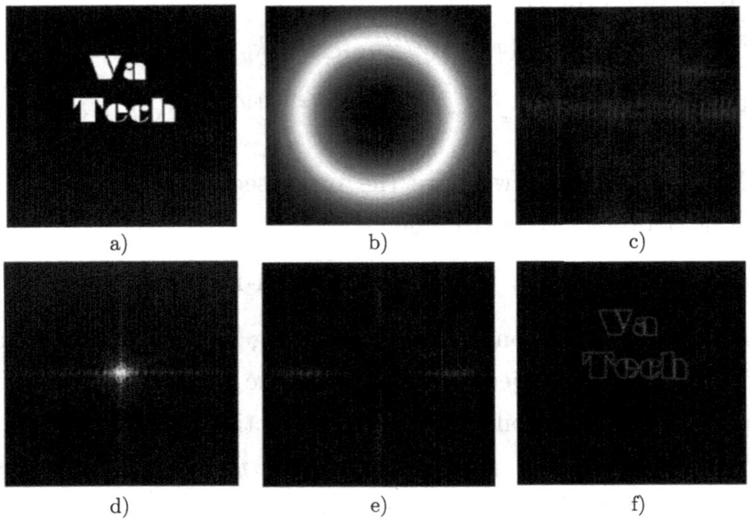

图 6.3-12 a) 原始二值图像; b) 频域里的高斯环, 即, $p_1^*(-fk_x/k_0,\ -fk_y/k_0)$, 其中 $p_1(x,y)=\mathrm{e}^{-b(\sqrt{x^2+y^2}-r_0)^2}$; c) 由式 (6.3-42) 所得的边缘保持复全息图的强度图; d) 图 a) 的强度谱; e) 原始频谱与图 b) 中高斯环的乘积; f) 图 c) 的复全息图的全息重建. 转载自 Zhang et al., Review on feature extraction for 3-D incoherent image processing using optical scanning holography, IEEE Transactions on Industrial Informatics 15, pp. 6146-6154 (2019), 经 IEEE© 许可

6.4 计算全息

计算全息 (*computer-generated holography*) 研究数字产生全息干涉条纹的方法, 由此产生的干涉条纹被称为计算全息图 (*computer-generated hologram*, CGH). CGH 随后可以被刻 (印) 制在胶片上或输入进空间光调制器 (*spatial light modulator*, SLM) 中进行光学重建. 现有两种常用的生成 CGH 的方法: 基于点源和基于面元的方法. 在点源法中, 三维物体被看作许多点源的组合, 每个点源在全息面上会产生一个球面波, 那么, 全息面上总的复振幅将是所有点源发出的球面波叠加后所得到的物光波. 在面元法中, 三维物体被表示为平面多边形的组合. 全息面上的物光波是所有多边形所发出的衍射场的叠加. 利用多边形来表示三维物体的核

$$=\mathcal{F}^{-1}\left\{\mathrm{e}^{-\mathrm{j}\frac{z_1}{2k_0}(k_x^2+k_y^2)}\right\}*\mathcal{F}^{-1}\left\{\mathrm{e}^{-b\left[\sqrt{\left(\frac{f}{k_0}k_x\right)^2+\left(\frac{f}{k_0}k_y\right)^2}-r_0\right]^2}\mathcal{F}\{I(x,y)\}\right\}*h(x,y;z_1)$$

$$=\frac{-\mathrm{j}k_0}{2\pi z_1}\mathrm{e}^{\frac{\mathrm{j}k_0(x^2+y^2)}{2z_1}}*\mathcal{F}^{-1}\left\{\mathrm{e}^{-b\left[\sqrt{\left(\frac{f}{k_0}k_x\right)^2+\left(\frac{f}{k_0}k_y\right)^2}-r_0\right]^2}\mathcal{F}\{I(x,y)\}\right\}*h(x,y;z_1)$$

$$\propto\mathcal{F}^{-1}\left\{\mathrm{e}^{-b\left[\sqrt{\left(\frac{f}{k_0}k_x\right)^2+\left(\frac{f}{k_0}k_y\right)^2}-r_0\right]^2}\mathcal{F}\{I(x,y)\}\right\}, \qquad (6.3\text{-}43)$$

as $\mathrm{e}^{\frac{\mathrm{j}k_0(x^2+y^2)}{2z_1}}*h(x,y;z_1)\propto\mathrm{e}^{\frac{\mathrm{j}k_0(x^2+y^2)}{2z_1}}*\mathrm{e}^{-\frac{\mathrm{j}k_0(x^2+y^2)}{2z_1}}\propto\delta(x,y)$. The result clearly indicates bandpass filtering of the intensity object $I(x,y)$.

Figure 6.3-12 (a) Original binary image, (b) plot of Gaussian ring in the frequency domain, that is, $p_1^*(-fk_x/k_0,-fk_y/k_0)$, where $p_1(x,y)=\mathrm{e}^{-b(\sqrt{x^2+y^2}-r_0)^2}$, (c) magnitude plot of the edge-preserved complex hologram from Eq.(6.3-42), (d) magnitude spectrum of (a), (e) original spectrum multiplied by the Gaussian ring shown in (b), (f) holographic reconstruction of the complex hologram from Figure (c). Reprinted from Zhang et al., Review on feature extraction for 3-D incoherent image processing using optical scanning holography, IEEE Transactions on Industrial Informatics 15, pp. 6146-6154 (2019), with permission. © IEEE

6.4 Computer-Generated Holography

Computer-generated holography deals with the methods for digitally generating holographic interference patterns. The resulting interference patterns are called *computer-generated holograms* (CGHs). The CGH can then be subsequently printed on a film or inputted to a *spatial light modulator* (SLM) for optical reconstruction. There are two popular approaches to generate CGHs: point-based approach and polygon-based approach. In the point-based approach, the 3-D object is considered to be made up of a collection of point sources. Each point source generates a spherical wave on the hologram plane and the total complex amplitude on the hologram plane is the object wave calculated by adding up the spherical waves emitted by all the point sources. In the polygon approach, the 3-D object is expressed as a collection of planar polygons. The object wave on the hologram plane is the summation of all the diffracted fields from each polygon. The central idea of using polygons to represent a 3-D object is

心思想是, 表示物体的多边形的数量会比点源法中要表示同一物体的点的数量少得多, 因此可以极大地加快生成全息图的计算时间. 本节将介绍点源法和面元法的基本原理, 然后将会介绍这两种计算全息图生成方法中当前的一些快速计算算法.

6.4.1 点源法

原理上, 点源法 (*point-based approach*) 非常简单. 图 6.4-1 给出了物点与全息面之间的空间关系. 三维物体被表示为一组自发光的点的集合, 且每个点发出一个指向全息面的球面波.

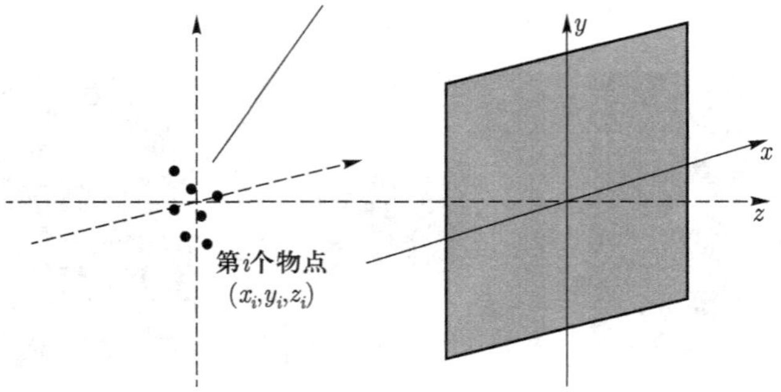

图 6.4-1 物点与全息面之间的空间关系

先前对式 (3.4-13) 讨论过, 振幅为 a_i 的球面波记为 $\dfrac{a_i}{r_i}\mathrm{e}^{-\mathrm{j}\frac{2\pi}{\lambda_0}r_i}$, 因此, 可以通过将所有点源发出的球面波在全息面上叠加得到物光波, 从而写出全息面上总的复振幅. 故, 计算复全息图由下式给出

$$H(x,y)|_{0\leqslant x<X,0\leqslant y<Y}=\sum_{i=0}^{N-1}\frac{a_i}{r_i}\mathrm{e}^{-\mathrm{j}\frac{2\pi}{\lambda_0}r_i}, \tag{6.4-1}$$

其中, X 和 Y 分别为复全息图的水平和垂直范围, λ_0 是生成全息图光波的波长, a_i 为第 i 个点源的光强, $r_i=\sqrt{(x-x_i)^2+(y-y_i)^2+z_i^2}$ 为第 i 个物点从位置 (x_i,y_i) 到全息面上 (x,y) 点之间的距离, z_i 是第 i 个物点到全息面的 (轴向) 距离.

因为传统的 SLM 不能显示复全息图, 如果使用的是振幅型 SLM, 则复全息图的实部可以被显示为

$$b_0+\mathrm{Re}[H(x,y)]=b_0+\sum_{i=0}^{N-1}\frac{a_i}{r_i}\cos\left(\frac{2\pi}{\lambda_0}r_i\right), \tag{6.4-2}$$

其中, b_0 为一直流偏置 (*DC bias*), 它使实全息图变为一个正的实值. 显示上式时, 物体的孪生像也被重建.

如果使用的是相位型 SLM, 则可以显示为

$$\theta(x,y)=\arctan\left\{\frac{\mathrm{Im}[H(x,y)]}{\mathrm{Re}[H(x,y)]}\right\}. \tag{6.4-3}$$

that the number of polygons representing an object is much smaller than that of points in the point-based approach, thus drastically speeding up the calculation time required for the generation of a hologram. In this section, we cover the basic principles of the point-based approach and the polygon approach. We then describe some of the modern fast calculation algorithms for computer-generated holograms in the two approaches.

6.4.1 Point-Based Approach

In principle, the *point-based approach* is very simple. Figure 6.4-1 shows the spatial relation between the object points and the hologram plane. The 3-D object is represented by a collection of self-illuminating points and each point is emitting a spherical wave toward the hologram plane.

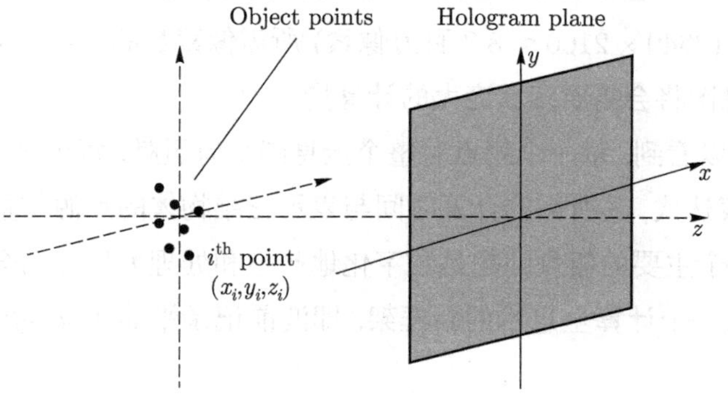

Figure 6.4-1 Spatial relation between object points and the hologram plane

Remembering that $\dfrac{a_i}{r_i}\mathrm{e}^{-\mathrm{j}\frac{2\pi}{\lambda_0}r_i}$ is a spherical wave of amplitude a_i, as discussed earlier in Eq. (3.4-13), we, therefore, can write the total complex amplitude on the hologram plane as the object wave by adding up the spherical waves emitted by all the point sources. Hence, the complex computer-generated hologram is given by

$$H(x,y)|_{0\leqslant x<X,0\leqslant y<Y} = \sum_{i=0}^{N-1} \frac{a_i}{r_i}\mathrm{e}^{-\mathrm{j}\frac{2\pi}{\lambda_0}r_i}, \tag{6.4-1}$$

where X and Y are the horizontal and vertical extents of the complex hologram, respectively. λ_0 is the wavelength of the optical beam that is used to generate the hologram. a_i is the light intensity of the ith point source. The term $r_i = \sqrt{(x-x_i)^2+(y-y_i)^2+z_i^2}$ is the distance of an ith object point at position (x_i, y_i) to a point at (x, y) on the hologram, and z_i is the distance from the ith object point to the hologram plane.

Since conventional SLMs cannot display a complex hologram, we can display the real part of the complex hologram if an amplitude-type SLM is used:

$$b_0 + \mathrm{Re}[H(x,y)] = b_0 + \sum_{i=0}^{N-1} \frac{a_i}{r_i}\cos\left(\frac{2\pi}{\lambda_0}r_i\right), \tag{6.4-2}$$

where b_0 is a DC bias added to make the real hologram become a positive real value. For this display, a twin image of the object is also constructed.

If a phase-type SLM is used, we can display

$$\theta(x,y) = \arctan\left\{\frac{\mathrm{Im}[H(x,y)]}{\mathrm{Re}[H(x,y)]}\right\}. \tag{6.4-3}$$

如果将复全息图与一个空间正弦载波相乘, 并取结果积的实部, 则也可以将复全息图转变为一个离轴全息图 (*off-axis hologram*)

$$H_{off\text{-}axis}(x,y)=b_0+\mathrm{Re}\left[H(x,y)\mathrm{e}^{\mathrm{j}\frac{2\pi}{\lambda_0}\sin\theta\, x}\right],\tag{6.4-4}$$

其中, θ 为离轴角度, 在全息重建时将所需的三维图像分离开其孪生像的角度 [见式 (5.1-16) 的离轴点源全息图与其在图 5.1-8 中所示的重建图].

计算全息很可能成为下一代三维显示技术中有希望的解决方案之一. 但与此同时, 该远景的实现因一些在短时间内难以克服的实际问题而受阻. 这些问题中的许多都可以追溯到全息图的精细像素尺寸问题, 比如, 由像素尺寸 5 μm × 5 μm 大小的正方形所形成的一个 10 mm × 10 mm 的全息图, 包含了超过 $2000\times2000=4\times10^6$ 的点, 这约为一个 4K 的超高清电视 (*4K Ultra HD TV*) 图像 ($3840\times2160\approx8.3$ 百万像素) 所需像素数的一半. 不难想象, 要数字产生和处理一个大的全息图将会需要多么庞大的计算量.

从式 (6.4-1) 可以看到, 每一个物点对整个全息图都有贡献, 每个全息图像素的计算都涉及复杂的求和运算表达式, 其所需的计算时间与表示三维物体的点源数量成正比. 实际上, 在点源计算全息中, 一个主要的瓶颈问题是数字化地产生和处理大尺寸的全息图时的密集计算. 在本节中, 我们讨论一个计算全息术的新框架, 即波前记录平面 (*wavefront recording plane*, WRP).

波前记录平面的概念, 受像全息图的方法 [Yoshikawa 等 (2009)] 启发, 被 Shimobaba 等 (2009) 提出. 传统的方法旨在提高直接从三维物体生成全息图的速度. 在 WRP 方法中, 一个二维的 WRP 是一个与全息面平行的假想平面, 且很靠近三维物体放置, 具体如图 6.4-2 所示. 自发光的物点将会产生一个球面波前到整个全息面, 并在其路径上截断该 WRP. 如果该 WRP 靠近物点, 则该 WRP 上物波前的覆盖范围会被限制在一个小的虚拟窗内, 该虚拟窗被一个小的支集约束. 这样就只需计算虚拟窗内的像素的波前, 而不需要计算全息图上每个像素的波前.

用 $w_p(x,y)$ 和 S_p 分别表示 WRP 上的球面波前和第 p 个物点的支集, 因此有

$$w_p(x,y)\big|_{(x,y)\in S_p}=\frac{a_p}{r_p}\mathrm{e}^{-\mathrm{j}\frac{2\pi}{\lambda_0}r_p},\tag{6.4-5}$$

其中, $r_p=\sqrt{(x-x_p)^2+(y-y_p)^2+z_p^2}$ 是第 p 个物点从位置 (x_p,y_p) 到 WRP 上的 (x,y) 处的距离, z_p 是从第 p 个物点到 WRP 的轴向距离, a_p 是第 p 个物点的强度. WRP 上的全息图是所有从物点发出的球面波前的求和, 为

$$W(x,y)=\sum_{p=0}^{N-1}w_p(x,y)\big|_{(x,y)\in S_p}=\sum_{p=0}^{N-1}\frac{a_p}{r_p}\mathrm{e}^{-\mathrm{j}\frac{2\pi}{\lambda_0}r_p}.\tag{6.4-6}$$

We can also convert the complex hologram to an off-axis hologram if we multiply a spatial sinusoidal carrier to the complex hologram and take the real part of the resulting product:

$$H_{\mathit{off\text{-}axis}}(x,y) = b_0 + \mathrm{Re}\left[H(x,y)\mathrm{e}^{\mathrm{j}\frac{2\pi}{\lambda_0}\sin\theta\, x}\right], \tag{6.4-4}$$

where θ is the off-axis angle that separates the desired 3-D image from the twin image upon holographic reconstruction [see Eq. (5.1-16) of an off-axis point-source hologram and its reconstruction in Figure 5.1-18].

Computer-generated holography is likely to become one of the promising solutions in the next generation of 3-D display. At the same time, such optimism is hampered by a number of practical problems which are difficult to solve in the foreseeable future. Many of these problems could be traced back to the fine pixel size of a hologram. For example, a small 10 mm × 10 mm hologram with a square pixel size of 5 μm × 5 μm, comprises over 2000 × 2000 = 4 × 10^6 points which is about half of the number of pixels in a 4K Ultra HD TV, which has 3840 × 2160 ≈ 8.3 million pixels. One can easily imagine the formidable amount of computation associated with the generation and processing of a large hologram digitally.

From Eq. (6.4-1), we can see that each object point is contributing to the entire hologram, and the evaluation of each hologram pixel involves the complicated expression enclosed in the summation operation. The calculation time required is proportional to the number of point sources representing the 3-D object. Indeed, one of the major bottlenecks is the intensive computation that is involved in the generation and processing of a large hologram digitally in point-based computer-generated holography. In this section, we discuss a modern framework of computer-generated holography known as the *wavefront recording plane* (WRP).

The concept of the wavefront recording plane (WRP), inspired by the image hologram method [Yoshikawa et al. (2009)], is suggested by Shimobaba et al. (2009). Traditional approaches aimed at enhancing the speed of generating a hologram directly from a 3-D object. In the WRP approach, a 2-D WRP is a hypothetical plane that is parallel to the hologram plane, and placed at close proximity to the 3-D object. The situation is shown in Figure 6.4-2. The self-illuminating object point will generate a spherical waveform to cover the entire hologram plane, and intercept the WRP in its path. If the WRP is near to the object point, the coverage of the object wave front on the WRP plane is limited to a small virtual window that is bounded by a small support. Instead of calculating the wave front on every pixel of the hologram, only wave front on the pixels within the virtual window is calculated.

Denoting the spherical wave front and the support of the pth object point on the WRP by $w_p(x,y)$ and S_p, respectively, we therefore have

$$w_p(x,y)\big|_{(x,y)\in S_p} = \frac{a_p}{r_p}\mathrm{e}^{-\mathrm{j}\frac{2\pi}{\lambda_0}r_p}, \tag{6.4-5}$$

where $r_p = \sqrt{(x-x_p)^2+(y-y_p)^2+z_p^2}$ is the distance of a pth object point at position (x_p, y_p) to a point at (x,y) on the WRP, and z_p is the distance from the pth object point to the WRP. a_p is the intensity of the pth point. The hologram on the WRP is the sum of the spherical wave fronts from all the object points and is given by

$$W(x,y) = \sum_{p=0}^{N-1} w_p(x,y)\big|_{(x,y)\in S_p} = \sum_{p=0}^{N-1}\frac{a_p}{r_p}\mathrm{e}^{-\mathrm{j}\frac{2\pi}{\lambda_0}r_p}. \tag{6.4-6}$$

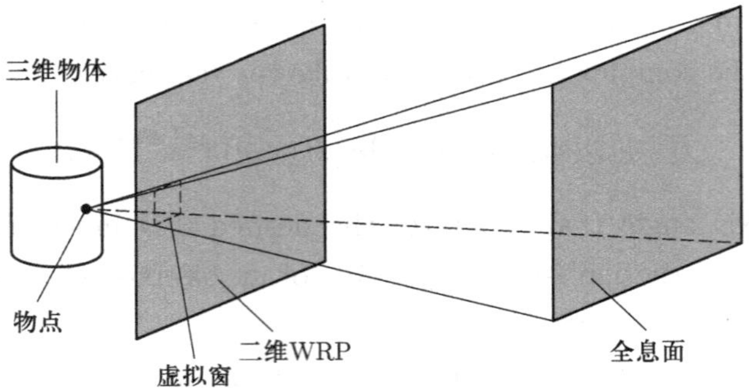

图 6.4-2 物点、二维 WRP 和全息面之间的空间关系. 改自 Tsang and Poon, Review on theory and applications of wavefront recording plane framework in generation and processing of digital holograms, Chinese Optics Letters 11, 010902 (2013)

在 WRP 图案生成后, 将光场分布扩展到全息面为

$$u(x,y) = W(x,y) * h(x,y;z_w), \tag{6.4-7}$$

其中, z_w 为 WRP 和全息面之间的距离. 因为该支集面积, 即 S_p, 远小于由式 (6.4-1) 得到的 $X \times Y$, 相比式 (6.4-1) 来说, 计算负荷显著降低. 为了从复全息图 $u(x,y)$ 中重建物体, 进行以下计算

$$u(x,y) * h^*(x,y;z+z_w), \tag{6.4-8}$$

其中, z 为从 WRP 到物体之间的距离, 这一步对应于图 5.3-2 所示的虚拟场的计算. 图 6.4-3 给出了 WRP 方法的仿真结果.

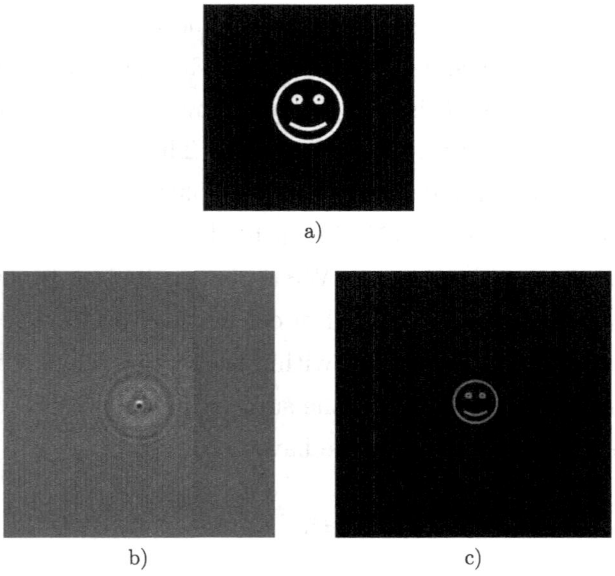

图 6.4-3 WRP 的仿真结果: a) 原物; b) WRP 上的全息图; c) 复全息图 $u(x,y)$ 的重建

图 6.4-3 中的图是使用如下所示的 m-文件生成的, 其中式 (6.4-7) 和式 (6.4-8) 是用第三章的式 (3.4-8) 中讨论的角谱法 (*angular spectrum method*, ASM) 实现的.

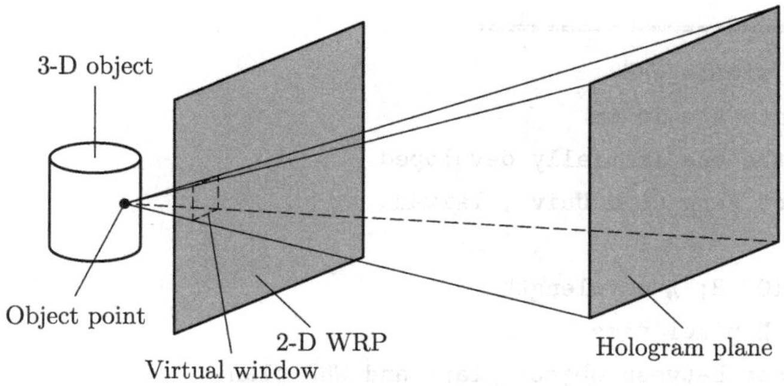

Figure 6.4-2 Spatial relation between object points, the 2-D WRP, and the hologram plane. Adapted from Tsang and Poon, Chinese Optics Letters 11 (1), 2013

After the WRP pattern is generated, the field distribution is expanded to the hologram plane given by

$$u(x,y) = W(x,y) * h(x,y;z_w), \tag{6.4-7}$$

where z_w is the separation between the WRP and the hologram plane. As the support area, that is, S_p, is much smaller than $X \times Y$ from Eq. (6.4-1), the computation load is significantly reduced as compared with that in Eq. (6.4-1). To reconstruct the object from the complex hologram $u(x,y)$, we perform

$$u(x,y) * h^*(x,y;z+z_w), \tag{6.4-8}$$

where z is the distance from the WRP to the object. This step corresponds to the virtual field calculation illustrated in Figure 5.3-2. Figure 6.4-3 shows the simulation results of the WRP method.

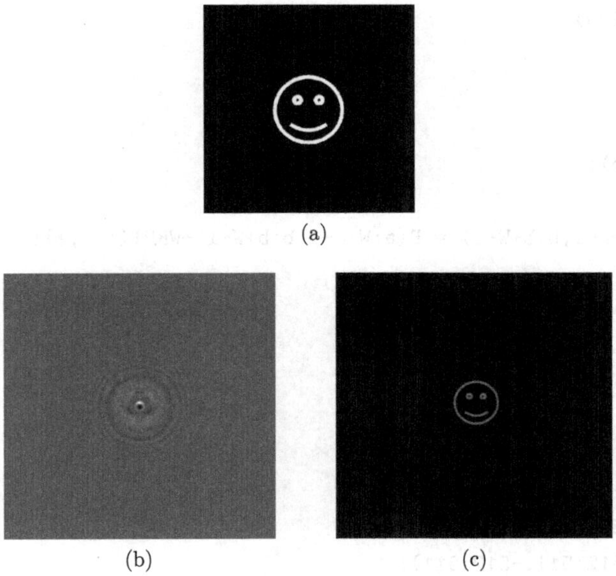

Figure 6.4-3 WRP simulation results. (a) Original object, (b) hologram on WRP, (c) reconstruction of the complex hologram $u(x,y)$

The images in Figure 6.4-3 are generated using the m-file shown below, where Eqs. (6.4-7) and (6.4-8) have been implemented using the *angular spectrum method* (ASM) discussed in Eq. (3.4-8) in Chapter 3.

```
========================================
% Simulation_of_simple_WRP
% All length units are in mm
% Adapted from the one initially developed
% by J.-P. Liu of Feng Chia Univ., Taiwan
close all;clear
lambda = 0.532*10^-3; % wavelength
dx = 6.4*10^-3; % pixel size
dz = 1; % distance between object plane and WRP plane
z = 150; %distance between WRP to hologram plane
k = 2*pi/lambda; % wavenumber
W = 5; % size of support on WRP, #of pixel=odd integer
w=floor(W/2);

% generation of the tables
[x, y]=meshgrid(-w:w,-w:w);
r=sqrt(x.^2+y.^2+(dz)^2);
WRPT = zeros(W,W,255); % the WRP tables
for g =1:255
  WRPT(:,:,g)=sqrt(g).*exp(-1j*k.*r)./r;
end

O = imread('front.jpg'); % input object, 512 by 512
figure;imshow(O);

% calculate the field on WRP
[M, N] = size(O);

P=zeros(M+2*w,N+2*w);

for a = 1:M
    for b =1:N
        s = O(a,b);
        if s>0
           P(a:W+a-1,b:b+W-1) = P(a:W+a-1,b:b+W-1)+WRPT(:,:,s);
        end

    end
end

% Propagation to the hologram plane
P2 = zeros (1024);
P2(257-w:768+w,257-w:768+w)=P;
[k,l]=meshgrid(-512:511,-512:511);
TF = exp(-1i*(z)*2*pi/lambda.*sqrt(1-(lambda*k/1024/dx).^2-(lambda*l/1024/dx).^2));
TF=fftshift(TF);
hologram=fftshift(ifft2(fft2(fftshift(P2)).*TF));
figure; imshow (real(hologram),[]); %title('Hologram')

% Reconstruction using ASM TFr = exp(1i*(z+dz)*2*pi/lambda.
*sqrt(1-(lambda*k/1024/dx).^2-(lambda*l/1024/dx).^2));
```

```
=========================================
% Simulation_of_simple_WRP
% All length units are in mm
% Adapted from the one initially developed
% by J.-P. Liu of Feng Chia Univ., Taiwan
close all;clear
lambda = 0.532*10^-3; % wavelength
dx = 6.4*10^-3; % pixel size
dz = 1; % distance between object plane and WRP plane
z = 150; %distance between WRP to hologram plane
k = 2*pi/lambda; % wavenumber
W = 5; % size of support on WRP, #of pixel=odd integer
w=floor(W/2);

% generation of the tables
[x, y]=meshgrid(-w:w,-w:w);
r=sqrt(x.^2+y.^2+(dz)^2);
WRPT = zeros(W,W,255); % the WRP tables
for g =1:255
  WRPT(:,:,g)=sqrt(g).*exp(-1j*k.*r)./r;
end

O = imread('front.jpg'); % input object, 512 by 512
figure;imshow(O);

% calculate the field on WRP
[M, N] = size(O);

P=zeros(M+2*w,N+2*w);

for a = 1:M
    for b =1:N
        s = O(a,b);
        if s>0
           P(a:W+a-1,b:b+W-1) = P(a:W+a-1,b:b+W-1)+WRPT(:,:,s);
        end


    end
end

% Propagation to the hologram plane
P2 = zeros (1024);
P2(257-w:768+w,257-w:768+w)=P;
[k,l]=meshgrid(-512:511,-512:511);
TF = exp(-1i*(z)*2*pi/lambda.*sqrt(1-(lambda*k/1024/dx).^2-(lambda*l/1024/dx).^2));
TF=fftshift(TF);
hologram=fftshift(ifft2(fft2(fftshift(P2)).*TF));
figure; imshow (real(hologram),[]); %title('Hologram')

% Reconstruction using ASM TFr = exp(1i*(z+dz)*2*pi/lambda.
*sqrt(1-(lambda*k/1024/dx).^2-(lambda*l/1024/dx).^2));
```

```
    TFr=fftshift(TFr);
    Er =fftshift(ifft2(fft2(fftshift(hologram)).*TFr));

    AFD=(abs(Er)).^2;
    AFD=AFD/max(max(AFD));
    figure; imshow (AFD); %title('Reconstructed image')
==========================================
```

如果使用查表法 [*look-up-table* (LUT) *method*], 则利用 WRP 概念可以进一步减少计算时间. 在查表法中, 需要先用式 (6.4-5) 对所有 (x_p, y_p, z_p) 的组合进行计算. 最后, 在从式 (6.4-6) 生成 $W(x,y)$ 时, 每一个组成其虚拟窗的 $w_p(x,y)\big|_{(x,y)\in S_p}$ 都可以通过相应的 LUT 中的条目得以恢复, 从而使这个过程无需计算. 一个数字全息图的视频序列 (*video sequence*), 每幅包含 2048×2048 个像素, 同时表示 3×10^4 个物点, 以每秒超过 10 帧的速率生成全息图 [Shimobaba 等 (2010)]. 尽管计算时间减少了, 但储存 LUT 的内存需求随着物点数量的增加而增加.

WRP 方法有一个大的缺点, 即, WRP 必须靠近物空间以确保对所有物点皆有一个小支集, 且该方法适用于深度范围较小的三维物体, 这种情况如图 6.4-4 所示. 可以看出, 当物点远离 WRP 时, 该支集 $S_p(p=1,2,3)$ 会变大. 为了保持 WRP 方法的计算效率, 物体的深度范围不能太大. 为了解决这一问题, 双 WRP 和多 WRP 方法得到研究 [Phan 等 (2014a), Phan 等 (2014b)]. 图 6.4-5 所示为多 WRP 的方法. 长深度的物体沿深度方向被划分为不同的横截面, 每个截面生成一个唯一的 WRP, 每个点 (图中的点) 的衍射球面波前被投射到最近的 WRP, 假设有 M 个截面, 且第 m 个截面的 WRP 上的全息图及其与全息图的距离分别被表示为 $W^m(x,y)$ 和 z^m, 则所有 WRP 的共同贡献产生了复全息图, 且对于多 WRP, 式 (6.4-7) 可重新推广为

$$u(x,y)=\sum_{m=0}^{M-1}\mathcal{F}^{-1}\{\mathcal{F}\{W^m(x,y)\}\mathcal{F}\{h(x,y;z^m)\}\}. \tag{6.4-9}$$

点源法中所固有的全息图生成的计算时间受物点数量的影响问题, 在使用 WRP 方法计

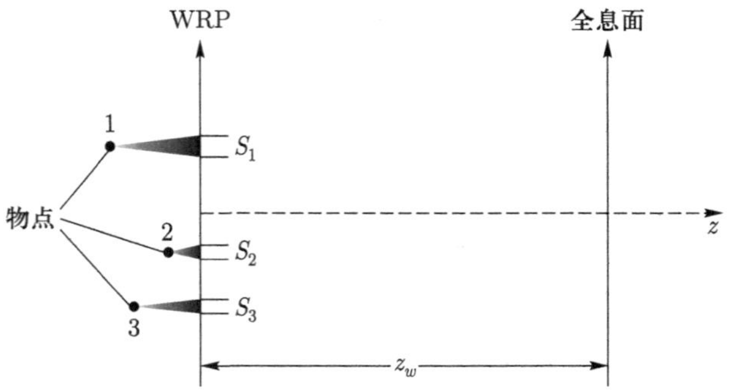

图 6.4-4 三个物点投射它们的波前在支集为 $S_p(p=1,2,3)$ 的 WRP 上的示例, 可以看出, 物点离 WRP 越近, 支集越小

```
    TFr=fftshift(TFr);
    Er =fftshift(ifft2(fft2(fftshift(hologram)).*TFr));

    AFD=(abs(Er)).^2;
    AFD=AFD/max(max(AFD));
    figure; imshow (AFD); %title('Reconstructed image')
==========================================
```

The calculation time using the concept of WRP can be further reduced if we employ a *look-up-table (LUT) method.* In a look-up table, we pre-compute Eq. (6.4-5) for all combinations of (x_p, y_p, z_p). As a result, in the generation of $W(x, y)$ from Eq. (6.4-6), each of its constituting virtual window $w_p(x, y)\big|_{(x,y)\in S_p}$ can be retrieved from the corresponding entries in the LUT, giving the process being computational free. A video sequence of digital holograms, each comprising 2048×2048 pixels and representing 3×10^4 object points, have been generated at a rate of over 10 frames per second [Shimobaba et al. (2010)]. Although the computation time is decreased, the memory requirement for storing the LUT is increased as the number of object points get larger.

The WRP approach has one major drawback. The WRP must be close to the object space to ensure a small support for all the object points and, therefore, the method is suitable for 3-D objects with a narrow depth range. The situation is illustrated in Figure 6.4-4. Note that the support $S_p (p = 1, 2, 3)$ is larger when the object point is further away from the WRP. In order to maintain the computation efficiency of the WRP method, the depth range of the object must not be too large. To address this issue, double WRPs and multiple WRPs have been investigated [Phan et al. (2014a, 2014b)]. Figure 6.4-5 illustrates the multiple WRP approach. The long-depth object is divided into different cross sections along the depth direction, and a unique WRP is generated for each section. The diffracted spherical wave front of each point (dot in the figure) is projected onto the nearest WRP. Suppose there are M sections, and the hologram on the WRP of the mth section and its distance from the hologram are denoted by $W^m(x, y)$ and z^m, respectively. The complex hologram is, therefore, generated for the collective contributions of all the WRPs, and Eq. (6.4-7) can be re-generalized for multiple WRPs as

$$u(x, y) = \sum_{m=0}^{M-1} \mathcal{F}^{-1}\{\mathcal{F}\{W^m(x, y)\}\mathcal{F}\{h(x, y; z^m)\}\}. \tag{6.4-9}$$

Inherent in the point-based method, the computation time of hologram generation in the WRP approach also suffers from the number of object points used in the calculation. In

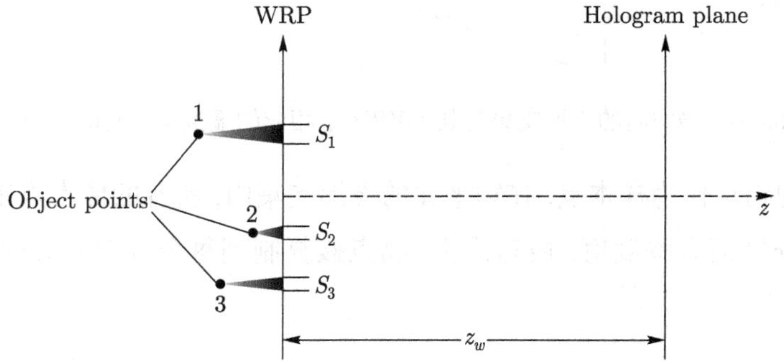

Figure 6.4-4 Example of three object points casting three wavefronts on a WRP with support S_p, where $p = 1, 2, 3$. Note that the nearer the object point to the WRP, the smaller is the support

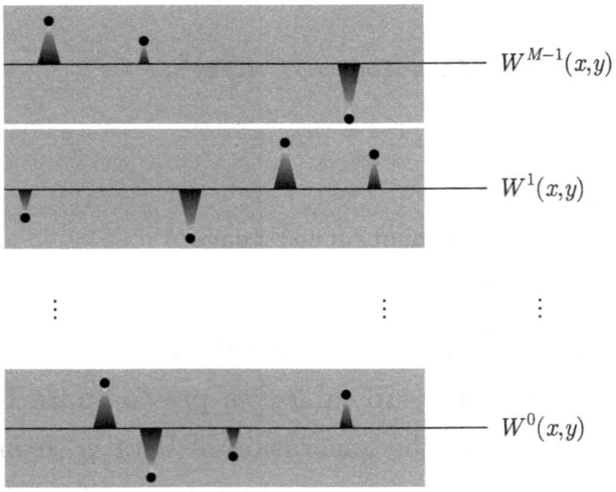

图 6.4-5 多 WRP 结构: 图中的点为物点, $W^m(x, y)$ 为第 m 个截面 WRP 的全息图. 转载自 Tsang et al., Review on the state-of-the-art technologies for acquisition and display of digital holograms, IEEE Transactions on Industrial Informatics 12, pp. 886-901 (2016), 经 IEEE© 许可

算时同样会遇到. 鉴于此, Tsang 等 (2011) 提出了插值 WRP 方法 [*interpolated WRP* (IWRP) *method*] 来缓解这一问题. 在此方法中, 假定场景图像的分辨率比全息图的分辨率小得多, 在实际中一般也是这样的. 因此, 聚集在一个小邻域范围内的物点将具有相似的光学和空间特性, 在此基础上, 物体强度分布均匀地划分为不重叠的方形支集, 每个支集被其组成像素的平均强度和平均深度值均匀地填充, 这种情况如图 6.4-6 所示.

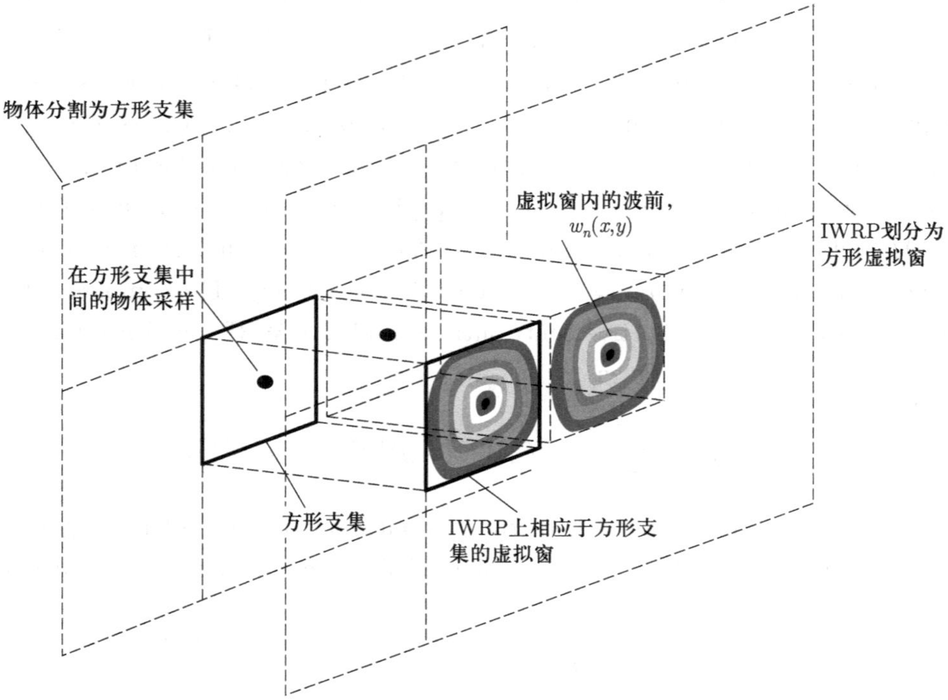

图 6.4-6 物空间的方形支集与其 IWRP 上相应的虚拟窗之间的空间关系

将支集中心的物点作为样本点, 在物平面的方形支集内, 所有的像素点都与方形中心的样本点具有相等的强度值和深度值. 换句话说, 物点被复制到每个方形支集内的所有像素点处,

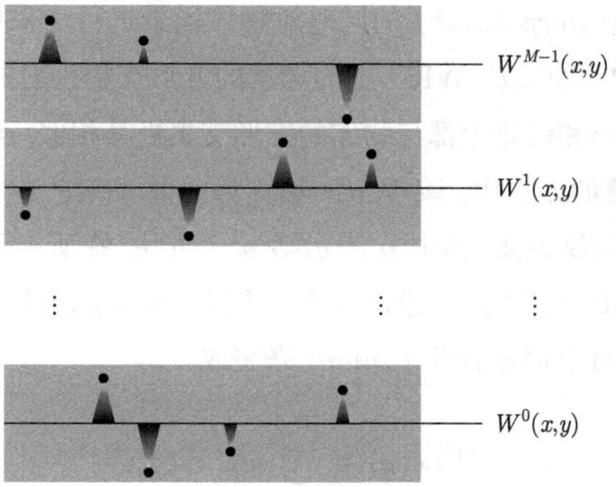

Figure 6.4-5 Multiple WRP configuration: Dots in the figure are object points. $W^m(x, y)$ is the hologram on the mth section WRP. Reprinted from Tsang et al., Review on the state-of-the-art technologies for acquisition and display of digital holograms, IEEE Transactions on Industrial Informatics, 12, pp. 886-901 (2016), with permission. © IEEE

light of this, Tsang et al. (2011) have developed the *interpolated WRP (IWRP) method* to alleviate this problem. In the approach, it is assumed that the scene image is considerably smaller in resolution than that of the hologram, which is in general true in practice. As such, object points that are clustered around a small neighborhood will share similar optical and spatial properties. On this basis, the object intensity profile is evenly partitioned into nonoverlapping square supports with each support uniformly filled with the mean intensity and mean depth values of its constituting pixels. The situation is illustrated in Figure 6.4-6.

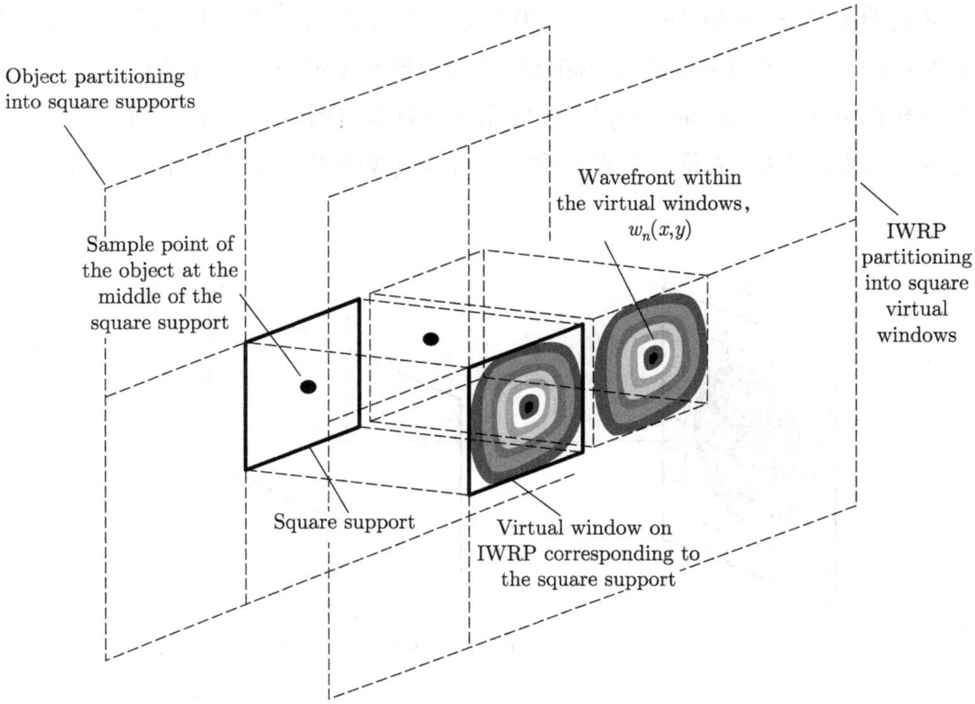

Figure 6.4-6 Spatial relation between the square support on the object space and its corresponding virtual window on the IWRP

The object point at the center of the support is taken as a sample point and within the square support on the object plane, all the pixels have identical intensity and depth values as the sample point in the center of the square. In other words, the object point is duplicated to

整个过程相当于在插值 (*interpolation*) 的同时对物体场景进行子采样 (*sub-sampling*). 现在, 由于物体场景距离 WRP 很近, 从 WRP 上的像素的方形支集投射的波前将只会覆盖一个小的虚拟窗, 该虚拟窗在大小和位置上都与物体的方形支集极其相似, 换句话说, 物空间和 WRP 都被均匀地分配到不重叠的方块中, 该 WRP 被称为插值 WRP 或 IWRP. 我们记 $w_n(x,y)$ 为 IWRP 中虚拟窗上的条纹图案, 其中 n 表示第 n 个方块. 在所有不重叠的虚拟窗的全息条纹图案被算出来后, IWRP 上总的全息图, 仅为不重叠的单独的虚拟窗的条纹图案的求和. 由于是不重叠的窗口, 求和可以通过合并 (*union*) 来完成

$$W(x,y)=\bigcup_{n=0}^{N-1} w_n(x,y). \tag{6.4-10}$$

重要的是要指出, 求和可以用合并运算来代替. 使用合并的计算量可以忽略不计, 因为它只是从 LUT 进行的一个内存复制操作 (Peter Tsang, 香港城市大学, 私人通信). 这样大全息图视频的帧速率就有可能实现, 通过 IWRP 方法, 实验结果显示了以每秒 40 帧的速率生成了约 400 万个物点的 2048×2048 像素的全息图 [Tsang 等 (2011)]. 在结束本小节之前, 这里需要指出, 近期一些关于点源法的综述文章可供学习参考 [Tsang 和 Poon (2013), Tsang 等 (2018)].

6.4.2 面元法

与点源法相比, 面元法 (*polygon-based method*) 减少了大量的采样单元. 该方法将三维物体分为二维多边形 (*polygon*) (该方法通常用三角形) 的集合. 面元法出现的部分原因也是受可视化工具或如 3ds Max 类渲染软件的推动. 3ds Max 是一个专业的用于制作三维模型和图像的三维计算机图形程序, 网格 (*mesh*) 是三维物体的几何模型, 其基本形状由边和顶点的连接组成. 在 3ds Max 中, 可以通过添加或删除各种多边形来编辑一个网格. 因此, 此方法的特点使得将计算机图形学 (*computer graphics*) 应用于计算全息图成为可能. 在本小节中, 所使用的多边形为三角形. 图 6.4-7 是一个典型的由三角形组成的半球体三维网格, 其全息图在 x-y 平面上.

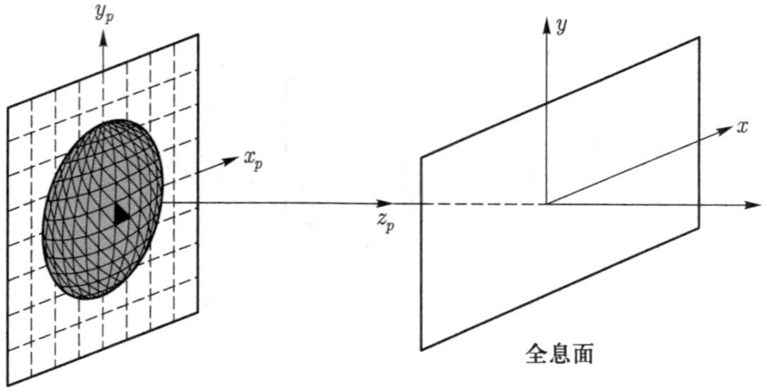

图 6.4-7 半球体三维网格

all the pixels within each square support. The overall process is equivalent to *sub-sampling* the object scene along with *interpolation.* Now, due to the close proximity of the object scene and the WRP, the wave front projected from a square support of pixels on the WRP will only cover a small virtual window, which is practically similar in size and location as the object's square support. In other words, both the object space and the WRP are evenly portioned into nonoverlapping square blocks. The WRP is now called the interpolated WRP or IWRP. We denote $w_n(x, y)$ as the fringe pattern on the virtual window on the IWRP with n denoting the nth square block. After the hologram fringe patterns of all the nonoverlapping virtual window have been calculated, the overall hologram on the IWRP is simply given by the sum of the fringe pattern of the nonoverlapping individual virtual window. Since we have nonoverlapping windows, summation can be done by *union* as

$$W(x, y) = \bigcup_{n=0}^{N-1} w_n(x, y). \tag{6.4-10}$$

It is important to point out that summation can be replaced by the union operation. The amount of computation is negligible with union as it is only a memory copying operation from the LUT (Peter Tsang, City University of Hong Kong, private communications). Video frame rate is thus possible for large holograms. With the IWRP approach, experimental result demonstrating the generation of a 2048 × 2048 hologram of around 4 million object points at 40 frames per second has been demonstrated [Tsang et al. (2011)]. To end this subsection, we want to point out that there are a number of recent review papers on the point-based approach [Tsang and Poon (2013), Tsang et al. (2018)].

6.4.2 Polygon-Based Approach

The *polygon-based method* reduces a vast amount of sampling units as compared with the point-based method. The method divides a 3-D object into a collection of 2-D *polygons* (a triangle is typically used for the method). The polygon method is also partly motivated by the availability of visualization tools or rendering software such as *3ds Max.* 3ds Max is a professional 3-D computer graphics program for making 3-D models and images. A *mesh* is a geometric model of a 3-D object in which the basic shape is made up of vertices connected by edges. In 3ds Max, we can edit a mesh by adding or deleting the various polygons. The feature of the method, therefore, makes it possible to apply *computer graphics* to computer-generated holograms. In the development of this subsection, the polygon employed is a triangle. Figure 6.4-7 illustrates a typical 3-D mesh of a half sphere consisting of triangles and the hologram is in the x-y plane.

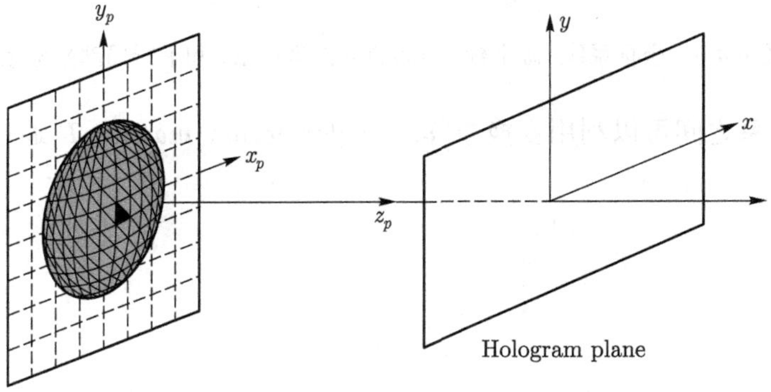

Figure 6.4-7 3-D mesh of a half sphere

每个面元发出的复光场即为面元场 (*polygon field*). 因此, 总的复光场可以写为所有面元的面元场的求和所得到的全息图上的物光波, 即

$$u(x,y)=\sum_{i=1}^{N}u_i(x,y), \tag{6.4-11}$$

其中, N 为面元数量, $u_i(x,y)$ 为第 i 个面元在全息面上的光场. 式 (6.4-11) 会让人想到点源法中式 (6.4-1) 所给出的复全息图的表示. 在本小节中, 我们将讨论面元法计算全息 [Zhang et al (2022)] 中的两种基本方法: 传统法 [Matsushima (2020), Shimobaba 和 Ito (2019)] 和解析法 [Ahrenberg 等 (2008)].

传统法

涉及两个平行平面之间的衍射计算已经很确定了. 例如, 可以通过传输的空间传递函数将一个平面的频谱与另一个平面的频谱联系起来 [见式 (3.4-7)], 或者可以通过在小角度近似下的菲涅耳衍射公式将两个相互平行的平面间的复光场联系起来 [见式 (3.4-17)]. 然而, 在一般组成物体的网格中, 大多数的面元与全息面并不平行.

传统法的主要目标首先是求出面元场 $u_i(x,y)$. 图 6.4-8 所示为一个任意倾斜面元. 将在倾斜平面上的任意面元的表面函数 (*surface function of an arbitrary polygon*) 表示为 $u_s(x_s,y_s)=A(x_s,y_s)\mathrm{e}^{\mathrm{j}\phi(x_s,x_s)}$, 其中 $A(x_s,y_s)$ 和 $\phi(x_s,x_s)$ 分别为振幅与相位分布. 振幅分布给出了面元的形状、纹理 (*texture*) 以及亮度等信息, 而相位分布被随机化来模拟光扩散. 假设倾斜量足够小, 以便全息图能从所有包含的面元中获得其全部的衍射光. 定义坐标系统 (x_s,y_s,z_s) 是倾斜局部坐标系 (*tilted local coordinate system*) 或源坐标系 (*source coordinate system*). 同时定义一个平行局部坐标系 (*parallel local coordinate system*) (x_p,y_p,z_p), 它与倾斜局部坐标系共用同一个原点. 平行局部坐标系平行于全息面 (x,y). 给定倾斜局部坐标系中的复光场 $u_s(x_s,y_s)$, 求出其在全息面的面元场 $u_i(x,y)$. 方法是首先通过旋转变换将倾斜局部坐标系上的光场分布与平行局部坐标系下的光场分布联系起来, 然后, 通过两个平行平面间的衍射获得全息面上的面元场.

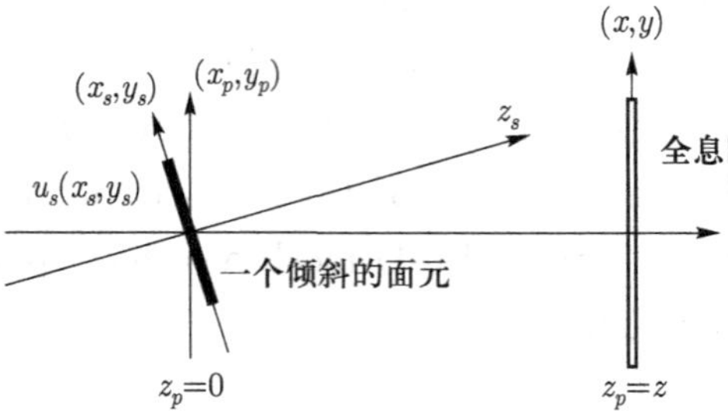

图 6.4-8 坐标系统: 源坐标系 (倾斜局部坐标系) 和平行局部坐标系

两个局部坐标系之间可以利用变换矩阵(*transformation matrix*) T 通过坐标旋转进行相互转换, 如下所示

The optical complex field emitted by each polygon is referred to as the *polygon field.* Hence, we can write the total complex field as the object wave on the hologram by summing all the polygon fields of each polygon:

$$u(x,y)=\sum_{i=1}^{N}u_i(x,y), \tag{6.4-11}$$

where N is the number of polygons and $u_i(x,y)$ is the polygon field on the hologram plane from the ith polygon. Equation (6.4-11) is reminiscent of the complex hologram given by Eq. (6.4-1) in the point-based method. In this subsection, we discuss two classical methods in polygon-based computer-generated holography [Zhang et al. (2022)]: the traditional method [Matsushima (2020), Shimobaba and Ito (2019)] and the analytical method [Ahrenberg et al. (2008)].

Traditional Method

Diffraction involving two parallel planes is well established. For example, we can relate the spectrum on one plane to that on the other plane by the spatial transfer function of propagation [see Eq. (3.4-7)], or we can relate the complex fields between the two parallel planes through the Fresnel diffraction formula under the small-angle approximations [see Eq. (3.4-17)]. However, in a general mesh, most polygons do not lie parallel to the hologram plane.

The main objective first is to find the polygon field $u_i(x,y)$. Figure 6.4-8 shows a general single tilted polygon. We represent the *surface function of an arbitrary polygon* on a tilted plane as $u_s(x_s,y_s) = A(x_s,y_s)\mathrm{e}^{\mathrm{j}\phi(x_s,x_s)}$, where $A(x_s,y_s)$ and $\phi(x_s,x_s)$ are amplitude and phase distributions. The amplitude distribution gives the shape, texture, and brightness of the polygon, and the phase distribution is taken to be randomized to simulate a light diffuser. We also assume that the amount of tilt is small enough so that the hologram captures all the diffracted light from all the polygons involved. Coordinate system (x_s,y_s,z_s) is referred to as a *tilted local coordinate system* or a *source coordinate system.* We also define a *parallel local coordinate system* (x_p,y_p,z_p), which shares the origin of the tilted local coordinates. The parallel local coordinate system is parallel to the hologram plane (x,y). Given the complex field $u_s(x_s,y_s)$ in the tilted local coordinate system, we want to find the polygon field $u_i(x,y)$ on the hologram plane. The approach is to first relate the field distribution in the tilted local coordinate system to that in the parallel local coordinate system through rotation transformation. Subsequent diffraction between the two parallel planes is then performed to obtain the polygon field on the hologram.

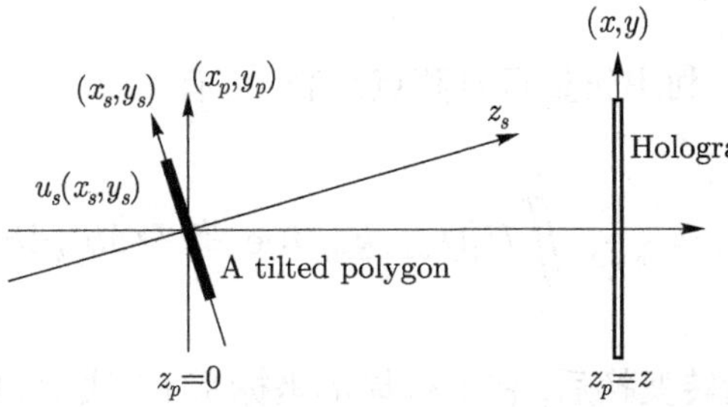

Figure 6.4-8 Coordinate systems: source coordinates (tilted local coordinates) and parallel local coordinates

The two local coordinates can be mutually transformed by coordinate rotation using *transformation matrix* T as follows:

$$\boldsymbol{r}_{\boldsymbol{p}}^{\mathrm{t}}=\begin{pmatrix}x_p\\y_p\\z_p\end{pmatrix}=\begin{pmatrix}a_1&a_4&a_7\\a_2&a_5&a_8\\a_3&a_6&a_9\end{pmatrix}\begin{pmatrix}x_s\\y_s\\z_s\end{pmatrix}=T\boldsymbol{r}_{\boldsymbol{s}}^{\mathrm{t}},\tag{6.4-12a}$$

和

$$\boldsymbol{r}_{\boldsymbol{s}}^{\mathrm{t}}=T^{-1}\boldsymbol{r}_{\boldsymbol{p}}^{\mathrm{t}}\tag{6.4-12b}$$

其中, $\boldsymbol{r_p}=(x_p,y_p,z_p)$ 和 $\boldsymbol{r_s}=(x_s,y_s,z_s)$ 是由行矩阵 (*row matrix*) 定义的位置矢量 (*position vector*). 矩阵 $\mathcal{M}$ 的转置 (*transpose*) 和逆 (*inverse*) 分别用 $\mathcal{M}^{\mathrm{t}}$ 和 $\mathcal{M}^{-1}$ 表示. 一般来说, 矩阵 T 为一个旋转矩阵 $R(\theta)$ 或是旋转矩阵如 $R_x(\theta)$、$R_y(\theta)$ 和 $R_z(\theta)$ 的乘积, 其中下标 x,y 和 z 表示旋转轴, θ 是绕这些轴旋转的角度. 图 6.4-9 给出了三个旋转矩阵. 角度 θ 的判定按右手定则. 当为正的旋转, 即 $\theta>0$ 时, 意味着如果右手的大拇指指向旋转轴的正向, 比如 x 轴, 那么, 其余手指弯曲方向为正. 因此, 可以得出此处绕轴逆时针旋转. 容易证明, 任何变换矩阵 T 都是一个正交矩阵 (*orthogonal matrix*), 即 $T^{\mathrm{t}}=T^{-1}$, 且 T 的行列式为 1, 即 $|T|=1$.

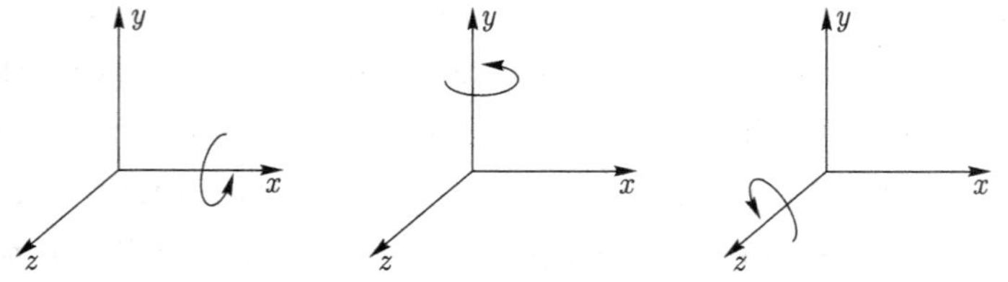

$$R_x(\theta)=\begin{pmatrix}1&0&0\\0&\cos\theta&\sin\theta\\0&-\sin\theta&\cos\theta\end{pmatrix}\quad R_y(\theta)=\begin{pmatrix}\cos\theta&0&-\sin\theta\\0&1&0\\\sin\theta&0&\cos\theta\end{pmatrix}\quad R_z(\theta)=\begin{pmatrix}\cos\theta&\sin\theta&0\\-\sin\theta&\cos\theta&0\\0&0&1\end{pmatrix}$$

图 6.4-9 旋转矩阵

现在来建立单个面元在全息面上的面元场. 参考图 6.4-8, 面元的平面波角谱 (*angular plane wave spectrum*) $U_s(k_{sx},k_{sy})=U_s(k_{sx},k_{sy};z_s=0)$ 沿 z_s 轴传播的复光场为 [见式 (3.4-8)]

$$\begin{aligned}u_s(x_s,y_s,z_s)&=\frac{1}{4\pi^2}\iint_{-\infty}^{\infty}U_s(k_{sx},k_{sy};0)\mathrm{e}^{-\mathrm{j}(k_{sx}x_s+k_{sy}y_s+k_{sz}z_s)}\mathrm{d}k_{sx}\mathrm{d}k_{sy}\\&=\frac{1}{4\pi^2}\iint_{-\infty}^{\infty}U_s(k_{sx},k_{sy};0)\mathrm{e}^{-\mathrm{j}z_s\sqrt{k_0^2-k_{sx}^2-k_{sy}^2}}\mathrm{e}^{-\mathrm{j}(k_{sx}x_s+k_{sy}y_s)}\mathrm{d}k_{sx}\mathrm{d}k_{sy}\\&=\mathcal{F}^{-1}\left\{U_s(k_{sx},k_{sy};0)\mathrm{e}^{-\mathrm{j}z_s\sqrt{k_0^2-k_{sx}^2-k_{sy}^2}}\right\},\end{aligned}\tag{6.4-13}$$

因为 $k_{sz}=\sqrt{k_0^2-k_{sx}^2-k_{sy}^2}$. 利用矢量符号和点乘表示, 有

$$u_s(\boldsymbol{r_s})=\frac{1}{4\pi^2}\iint_{-\infty}^{\infty}U_s(k_{sx},k_{sy};0)\mathrm{e}^{-\mathrm{j}\boldsymbol{k_s}\cdot\boldsymbol{r_s}}\mathrm{d}k_{sx}\mathrm{d}k_{sy},\tag{6.4-14}$$

其中, $\boldsymbol{k_s}=(k_{sx},k_{sy},k_{sz})$. 旋转变换后, 在平行局部坐标系下, 式 (6.4-14) 中的复光场变为

$$\boldsymbol{r}_p^{\mathrm{t}} = \begin{pmatrix} x_p \\ y_p \\ z_p \end{pmatrix} = \begin{pmatrix} a_1 & a_4 & a_7 \\ a_2 & a_5 & a_8 \\ a_3 & a_6 & a_9 \end{pmatrix} \begin{pmatrix} x_s \\ y_s \\ z_s \end{pmatrix} = T\boldsymbol{r}_s^{\mathrm{t}}, \tag{6.4-12a}$$

and

$$\boldsymbol{r}_s^{\mathrm{t}} = T^{-1}\boldsymbol{r}_p^{\mathrm{t}} \tag{6.4-12b}$$

where $\boldsymbol{r_p} = (x_p, y_p, z_p)$ and $\boldsymbol{r_s} = (x_s, y_s, z_s)$ are *position vectors* defined by *row matrices.* The *transpose* and *inverse* of a matrix $\mathcal{M}$ are denoted by $\mathcal{M}^{\mathrm{t}}$ and $\mathcal{M}^{-1}$, respectively. In general, the matrix T is a rotation matrix $R(\theta)$ or the product of rotation matrices such as $R_x(\theta)$, $R_y(\theta)$, and $R_z(\theta)$, where the subscripts x, y, and z denote the axes of rotation and θ is the angle of rotation around the axes. Figure 6.4-9 gives the three rotation matrices. The sense of the angle θ is defined by the right-hand rule. A positive rotation, that is, $\theta > 0$, means that if the thumb of the right hand is pointed along the positive direction of the rotation axis, say, the x-axis, then the fingers curl in the positive direction. Therefore, we have a counter-clockwise rotation about the axis. It can be verified readily that any transformation matrix T is an *orthogonal matrix*, that is, $T^{\mathrm{t}} = T^{-1}$, and the determinant of T is unity, that is, $|T| = 1$.

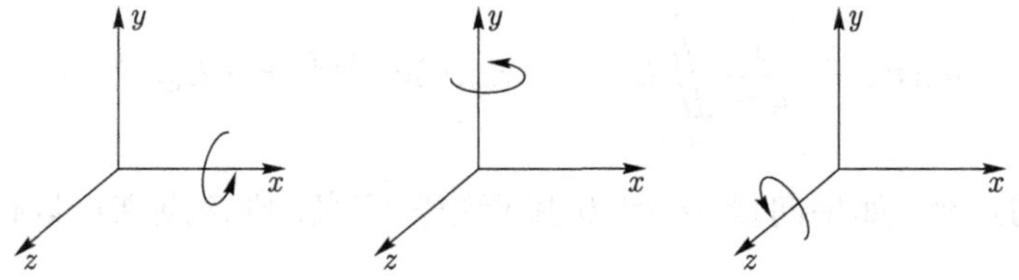

$$R_x(\theta)=\begin{pmatrix} 1 & 0 & 0 \\ 0 & \cos\theta & \sin\theta \\ 0 & -\sin\theta & \cos\theta \end{pmatrix} \quad R_y(\theta)=\begin{pmatrix} \cos\theta & 0 & -\sin\theta \\ 0 & 1 & 0 \\ \sin\theta & 0 & \cos\theta \end{pmatrix} \quad R_z(\theta)=\begin{pmatrix} \cos\theta & \sin\theta & 0 \\ -\sin\theta & \cos\theta & 0 \\ 0 & 0 & 1 \end{pmatrix}$$

Figure 6.4-9 Rotation matrices

Let us now establish the polygon field due to a single polygon on the hologram plane. With reference to Figure 6.4-8, the complex polygon field with *angular plane wave spectrum* $U_s(k_{sx}, k_{sy}) = U_s(k_{sx}, k_{sy}; z_s = 0)$ propagating along z_s is [see Eq. (3.4-8)]

$$\begin{aligned} u_s(x_s, y_s, z_s) &= \frac{1}{4\pi^2} \iint_{-\infty}^{\infty} U_s(k_{sx}, k_{sy}; 0)\mathrm{e}^{-\mathrm{j}(k_{sx}x_s + k_{sy}y_s + k_{sz}z_s)}\mathrm{d}k_{sx}\mathrm{d}k_{sy} \\ &= \frac{1}{4\pi^2} \iint_{-\infty}^{\infty} U_s(k_{sx}, k_{sy}; 0)\mathrm{e}^{-\mathrm{j}z_s\sqrt{k_0^2 - k_{sx}^2 - k_{sy}^2}}\mathrm{e}^{-\mathrm{j}(k_{sx}x_s + k_{sy}y_s)}\mathrm{d}k_{sx}\mathrm{d}k_{sy} \\ &= \mathcal{F}^{-1}\left\{U_s(k_{sx}, k_{sy}; 0)\mathrm{e}^{-\mathrm{j}z_s\sqrt{k_0^2 - k_{sx}^2 - k_{sy}^2}}\right\}, \end{aligned} \tag{6.4-13}$$

as $k_{sz} = \sqrt{k_0^2 - k_{sx}^2 - k_{sy}^2}$. In vector notations and using the dot product, we have

$$u_s(\boldsymbol{r_s}) = \frac{1}{4\pi^2} \iint_{-\infty}^{\infty} U_s(k_{sx}, k_{sy}; 0)\mathrm{e}^{-\mathrm{j}\boldsymbol{k_s}\cdot\boldsymbol{r_s}}\mathrm{d}k_{sx}\mathrm{d}k_{sy}, \tag{6.4-14}$$

where $\boldsymbol{k_s} = (k_{sx}, k_{sy}, k_{sz})$. Upon rotation transformation, the complex field in Eq. (6.4-14)

$$u_p(\boldsymbol{r_p}) = u_s(\boldsymbol{r_s})|_{\boldsymbol{r_s}=\boldsymbol{r_p}T} = \frac{1}{4\pi^2}\iint\limits_{-\infty}^{\infty} U_s(k_{sx}, k_{sy}; 0)\mathrm{e}^{-\mathrm{j}\boldsymbol{k_s}\cdot\boldsymbol{r_p}T}\mathrm{d}k_{sx}\mathrm{d}k_{sy}, \tag{6.4-15a}$$

这里利用了式 (6.4-12b) 中的 $\boldsymbol{r_s}^{\mathrm{t}}$ 的转置来建立

$$\boldsymbol{r_s} = \boldsymbol{r_p}T. \tag{6.4-15b}$$

类似地, 源坐标系和平行局部坐标系中的传播矢量 (*propagation vector*) 同样也可以像位置矢量一样进行如下转换

$$\boldsymbol{k_p}^{\mathrm{t}} = T\boldsymbol{k_s}^{\mathrm{t}}, \tag{6.4-16a}$$

和

$$\boldsymbol{k_s}^{\mathrm{t}} = T^{-1}\boldsymbol{k_p}^{\mathrm{t}}, \tag{6.4-16b}$$

其中, $\boldsymbol{k_p} = (k_{px}, k_{py}, k_{pz})$. 式 (6.4-16b) 经转置得

$$\boldsymbol{k_s} = \boldsymbol{k_p}(T^{-1})^{\mathrm{t}} = \boldsymbol{k_p}T. \tag{6.4-16c}$$

基于此式, 式 (6.4-15a) 变为

$$u_p(\boldsymbol{r_p}) = \frac{1}{4\pi^2}\iint\limits_{-\infty}^{\infty} U_s(k_{sx}, k_{sy}; 0)\mathrm{e}^{-\mathrm{j}\boldsymbol{k_p}T\cdot\boldsymbol{r_p}T}\mathrm{d}k_{sx}\mathrm{d}k_{sy}. \tag{6.4-17}$$

计算出指数项中的点乘. 如果向量 $\boldsymbol{a}$ 和 $\boldsymbol{b}$ 用行矩阵定义, 则该点乘可以被写为如下矩阵积的形式

$$\boldsymbol{a}\cdot\boldsymbol{b} = \boldsymbol{a}\boldsymbol{b}^{\mathrm{t}}.$$

因此

$$\boldsymbol{k_s}\cdot\boldsymbol{r_s} = \boldsymbol{k_p}T\cdot\boldsymbol{r_p}T = \boldsymbol{k_p}T(\boldsymbol{r_p}T)^{\mathrm{t}} = \boldsymbol{k_p}T(T^{\mathrm{t}}\boldsymbol{r_p}^{\mathrm{t}}) = \boldsymbol{k_p}\boldsymbol{r_p}^{\mathrm{t}} = \boldsymbol{k_p}\cdot\boldsymbol{r_p}, \tag{6.4-18}$$

因为 $TT^{\mathrm{t}} = TT^{-1} = I$, 其中 I 为单位矩阵 (*identity matrix*). 式 (6.4-18) 的结果实际上就说得通了, 因为坐标旋转并不改变向量间的长度和角度. 基于这个结果, 式 (6.4-17) 变为

$$\begin{aligned} u_p(\boldsymbol{r_p}) &= \frac{1}{4\pi^2}\iint\limits_{-\infty}^{\infty} U_s(k_{sx}, k_{sy}; 0)\mathrm{e}^{-\mathrm{j}\boldsymbol{k_p}\cdot\boldsymbol{r_p}}\mathrm{d}k_{sx}\mathrm{d}k_{sy} \\ &= \frac{1}{4\pi^2}\iint\limits_{-\infty}^{\infty} U_s(k_{sx}, k_{sy}; 0)\mathrm{e}^{-\mathrm{j}(k_{px}x_p + k_{py}y_p + k_{pz}z_p)}\mathrm{d}k_{sx}\mathrm{d}k_{sy}, \end{aligned} \tag{6.4-19a}$$

其中

$$k_{pz}(k_{px}, k_{py}) = \sqrt{k_0^2 - k_{px}^2 - k_{py}^2}. \tag{6.4-19b}$$

可以看出, 式 (6.4-19a) 中的指数函数表示, 在平行局部坐标系下, 平面波沿着 z_p 方向 (或者 z 方向) 进行传播. 为了在平行局部坐标系中完整地描述 $u_p(\boldsymbol{r_p})$, 需要将式 (6.4-19a) 完全变换

in the parallel local coordinates becomes

$$u_p(\boldsymbol{r_p}) = u_s(\boldsymbol{r_s})|_{\boldsymbol{r_s}=\boldsymbol{r_p}T} = \frac{1}{4\pi^2}\iint_{-\infty}^{\infty} U_s(k_{sx}, k_{sy}; 0)\mathrm{e}^{-\mathrm{j}\boldsymbol{k_s}\cdot\boldsymbol{r_p}T}\mathrm{d}k_{sx}\mathrm{d}k_{sy}, \tag{6.4-15a}$$

where we have used the transpose of $\boldsymbol{r_s^{\mathrm{t}}}$ in Eq. (6.4-12b) to establish

$$\boldsymbol{r_s} = \boldsymbol{r_p}T. \tag{6.4-15b}$$

Similarly, a *propagation vector* in the source and parallel local coordinates can also be transformed like position vectors as follows:

$$\boldsymbol{k_p^{\mathrm{t}}} = T\boldsymbol{k_s^{\mathrm{t}}}, \tag{6.4-16a}$$

and

$$\boldsymbol{k_s^{\mathrm{t}}} = T^{-1}\boldsymbol{k_p^{\mathrm{t}}}, \tag{6.4-16b}$$

where $\boldsymbol{k_p} = (k_{px}, k_{py}, k_{pz})$. From Eq. (6.4-16b), we take the transpose to have

$$\boldsymbol{k_s} = \boldsymbol{k_p}(T^{-1})^{\mathrm{t}} = \boldsymbol{k_p}T. \tag{6.4-16c}$$

With this expression, Eq. (6.4-15a) becomes

$$u_p(\boldsymbol{r_p}) = \frac{1}{4\pi^2}\iint_{-\infty}^{\infty} U_s(k_{sx}, k_{sy}; 0)\mathrm{e}^{-\mathrm{j}\boldsymbol{k_p}T\cdot\boldsymbol{r_p}T}\mathrm{d}k_{sx}\mathrm{d}k_{sy}. \tag{6.4-17}$$

Let us try to work out the dot product in the exponent. If vectors $\boldsymbol{a}$ and $\boldsymbol{b}$ are defined with row matrices, the dot product can be written as a matrix product as follows:

$$\boldsymbol{a}\cdot\boldsymbol{b} = \boldsymbol{a}\boldsymbol{b}^{\mathrm{t}}.$$

Hence,

$$\boldsymbol{k_s}\cdot\boldsymbol{r_s} = \boldsymbol{k_p}T\cdot\boldsymbol{r_p}T = \boldsymbol{k_p}T(\boldsymbol{r_p}T)^{\mathrm{t}} = \boldsymbol{k_p}T(T^{\mathrm{t}}\boldsymbol{r_p^{\mathrm{t}}}) = \boldsymbol{k_p}\boldsymbol{r_p^{\mathrm{t}}} = \boldsymbol{k_p}\cdot\boldsymbol{r_p}, \tag{6.4-18}$$

as $TT^{\mathrm{t}} = TT^{-1} = I$ with I being an *identity matrix.* The result in Eq. (6.4-18) actually makes sense as coordinates rotation does not change the length and angles between vectors. With the result, Eq. (6.4-17) becomes

$$\begin{aligned} u_p(\boldsymbol{r_p}) &= \frac{1}{4\pi^2}\iint_{-\infty}^{\infty} U_s(k_{sx}, k_{sy}; 0)\mathrm{e}^{-\mathrm{j}\boldsymbol{k_p}\cdot\boldsymbol{r_p}}\mathrm{d}k_{sx}\mathrm{d}k_{sy} \\ &= \frac{1}{4\pi^2}\iint_{-\infty}^{\infty} U_s(k_{sx}, k_{sy}; 0)\mathrm{e}^{-\mathrm{j}(k_{px}x_p + k_{py}y_p + k_{pz}z_p)}\mathrm{d}k_{sx}\mathrm{d}k_{sy}, \end{aligned} \tag{6.4-19a}$$

where

$$k_{pz}(k_{px}, k_{py}) = \sqrt{k_0^2 - k_{px}^2 - k_{py}^2}. \tag{6.4-19b}$$

Note that the exponential function in Eq. (6.4-19a) represents plane wave propagating along the z_p-direction (or the z-direction) on the parallel local coordinates. To completely describe

成变量为 (k_{px},k_{py},k_{pz}) 的式子. 从式 (6.4-16b) 开始, 先来计算 $U_s(k_{sx},k_{sy};0)$, 有

$$\boldsymbol{k}_{\boldsymbol{s}}^{\mathrm{t}}=T^{-1}\boldsymbol{k}_{\boldsymbol{p}}^{\mathrm{t}},$$

这相当于

$$\begin{pmatrix}k_{sx}\\k_{sy}\\k_{sz}\end{pmatrix}=\begin{pmatrix}a_1&a_2&a_3\\a_4&a_5&a_6\\a_7&a_8&a_9\end{pmatrix}\begin{pmatrix}k_{px}\\k_{py}\\k_{pz}\end{pmatrix}.$$

因此, 有

$$k_{sx}=k_{sx}(k_{px},k_{py})=a_1k_{px}+a_2k_{py}+a_3k_{pz}(k_{px},k_{py}),\tag{6.4-20a}$$

和

$$k_{sy}=k_{sy}(k_{px},k_{py})=a_4k_{px}+a_5k_{py}+a_6k_{pz}(k_{px},k_{py}),\tag{6.4-20b}$$

这给出

$$\begin{aligned}U_s(k_{sx},k_{sy};0)&=U_s(k_{sx}(k_{px},k_{py}),k_{sy}(k_{px},k_{py});0)\\&=U_s(a_1k_{px}+a_2k_{py}+a_3k_{pz},a_4k_{px}+a_5k_{py}+a_6k_{pz};0),\end{aligned}\tag{6.4-21}$$

其中, 再一次, $U_s(k_{sx},k_{sy};0)$ 为给定面元的源函数 $u_s(x_s,y_s)=u_s(x_s,y_s;z_s=0)$ 的频谱, 明确地, 有

$$\begin{aligned}U_s(k_{sx},k_{sy};z_s=0)&=\mathcal{F}\{u_s(x_s,y_s;z_s=0)\}\\&=\iint\limits_{-\infty}^{\infty}u_s(x_s,y_s)\mathrm{e}^{\mathrm{j}(k_{sx}x_s+k_{sy}y_s)}\mathrm{d}x_s\mathrm{d}y_s.\end{aligned}\tag{6.4-22}$$

现在, 式 (6.4-19a) 中的微元 (*differential element*) 可通过下式实现从 k_{sx} 和 k_{sy} 到 k_{px} 和 k_{py} 的变换

$$\mathrm{d}k_{sx}\mathrm{d}k_{sy}=|J(k_{px},k_{py})|\mathrm{d}k_{px}\mathrm{d}k_{py},\tag{6.4-23}$$

其中, $J(k_{px},k_{py})$ 是表示 k_{sx} 和 k_{sy} 关于 k_{px} 和 k_{py} 的坐标变换的雅可比行列式 (*Jacobian*)

$$J(k_{px},k_{py})=\frac{\partial(k_{sx},k_{sy})}{\partial(k_{px},k_{py})}=\begin{vmatrix}\dfrac{\partial k_{sx}}{\partial k_{px}}&\dfrac{\partial k_{sx}}{\partial k_{py}}\\\dfrac{\partial k_{sy}}{\partial k_{px}}&\dfrac{\partial k_{sy}}{\partial k_{py}}\end{vmatrix}.\tag{6.4-24}$$

利用式 (6.4-20a)、式 (6.4-20b) 和式 (6.4-19b), 该雅可比行列式可以被准确地计算出

$$J(k_{px},k_{py})=\frac{(a_2a_6-a_3a_5)k_{px}}{k_{pz}(k_{px},k_{py})}+\frac{(a_3a_4-a_1a_6)k_{py}}{k_{pz}(k_{px},k_{py})}+(a_1a_5-a_2a_4).\tag{6.4-25}$$

在傍轴近似下, k_{px} 和 k_{py} 要比 k_{pz} 小得多, 因此, 雅可比行列式变为一个常数

$$J(k_{px},k_{py})\approx a_1a_5-a_2a_4.\tag{6.4-26}$$

$u_p(\boldsymbol{r_p})$ by the parallel local coordinates, we need to fully convert Eq. (6.4-19a) in terms of (k_{px}, k_{py}, k_{pz}). Let us first work on $U_s(k_{sx}, k_{sy}; 0)$. Starting from Eq. (6.4-16b), we have

$$\boldsymbol{k}_{\boldsymbol{s}}^{\mathrm{t}} = T^{-1}\boldsymbol{k}_{\boldsymbol{p}}^{\mathrm{t}},$$

which is equivalently equal to

$$\begin{pmatrix} k_{sx} \\ k_{sy} \\ k_{sz} \end{pmatrix} = \begin{pmatrix} a_1 & a_2 & a_3 \\ a_4 & a_5 & a_6 \\ a_7 & a_8 & a_9 \end{pmatrix} \begin{pmatrix} k_{px} \\ k_{py} \\ k_{pz} \end{pmatrix}.$$

Therefore, we have

$$k_{sx} = k_{sx}(k_{px}, k_{py}) = a_1 k_{px} + a_2 k_{py} + a_3 k_{pz}(k_{px}, k_{py}), \tag{6.4-20a}$$

and

$$k_{sy} = k_{sy}(k_{px}, k_{py}) = a_4 k_{px} + a_5 k_{py} + a_6 k_{pz}(k_{px}, k_{py}), \tag{6.4-20b}$$

which give

$$\begin{aligned} U_s(k_{sx}, k_{sy}; 0) &= U_s(k_{sx}(k_{px}, k_{py}), k_{sy}(k_{px}, k_{py}); 0) \\ &= U_s(a_1 k_{px} + a_2 k_{py} + a_3 k_{pz}, a_4 k_{px} + a_5 k_{py} + a_6 k_{pz}; 0), \end{aligned} \tag{6.4-21}$$

where again $U_s(k_{sx}, k_{sy}; 0)$ is the spectrum of the given source function $u_s(x_s, y_s) = u_s(x_s, y_s; z_s = 0)$ of the polygon. Explicitly,

$$\begin{aligned} U_s(k_{sx}, k_{sy}; z_s = 0) &= \mathcal{F}\{u_s(x_s, y_s; z_s = 0)\} \\ &= \iint\limits_{-\infty}^{\infty} u_s(x_s, y_s) \mathrm{e}^{\mathrm{j}(k_{sx}x_s + k_{sy}y_s)} \mathrm{d}x_s \mathrm{d}y_s. \end{aligned} \tag{6.4-22}$$

Now, changing of variables from k_{sx} and k_{sy} to k_{px} and k_{py} in the *differential element* of Eq. (6.4-19a) is achieved by

$$\mathrm{d}k_{sx}\mathrm{d}k_{sy} = |J(k_{px}, k_{py})| \mathrm{d}k_{px}\mathrm{d}k_{py}, \tag{6.4-23}$$

where $J(k_{px}, k_{py})$ is the *Jacobian* of the coordinate transformation of k_{sx} and k_{sy} with respect to k_{px} and k_{py}, given by

$$J(k_{px}, k_{py}) = \frac{\partial(k_{sx}, k_{sy})}{\partial(k_{px}, k_{py})} = \begin{vmatrix} \dfrac{\partial k_{sx}}{\partial k_{px}} & \dfrac{\partial k_{sx}}{\partial k_{py}} \\ \dfrac{\partial k_{sy}}{\partial k_{px}} & \dfrac{\partial k_{sy}}{\partial k_{py}} \end{vmatrix}. \tag{6.4-24}$$

Using Eqs. (6.4-20a), (6.4-20b) and (6.4-19b), the Jacobian can be evaluated explicitly to be

$$J(k_{px}, k_{py}) = \frac{(a_2 a_6 - a_3 a_5)k_{px}}{k_{pz}(k_{px}, k_{py})} + \frac{(a_3 a_4 - a_1 a_6)k_{py}}{k_{pz}(k_{px}, k_{py})} + (a_1 a_5 - a_2 a_4). \tag{6.4-25}$$

For paraxial approximations, k_{px} and k_{py} are much smaller than k_{pz}. As a result, the Jacobian becomes a constant as

$$J(k_{px}, k_{py}) \approx a_1 a_5 - a_2 a_4. \tag{6.4-26}$$

将式 (6.4-21) 和式 (6.4-23) 合并为式 (6.4-19a), 由已知面元的表面函数 $u_s(x_s, y_s; z_s = 0) = u_s(x_s, y_s)$ 可知, 沿 z_p 方向的复光场为

$$\begin{aligned} u_p(x_p, y_p, z_p) &= \frac{1}{4\pi^2}\iint\limits_{-\infty}^{\infty} U_s(k_{sx}(k_{px}, k_{py}), k_{sy}(k_{px}, k_{py}); 0)\times \\ &\quad \mathrm{e}^{-\mathrm{j}(k_{px}x_p + k_{py}y_p + z_p\sqrt{k_0^2 - k_{px}^2 - k_{py}^2})}|J(k_{px}, k_{py})|\mathrm{d}k_{px}\mathrm{d}k_{py} \\ &= \mathcal{F}^{-1}\left\{U_p(k_{px}, k_{py}; 0)\mathrm{e}^{-\mathrm{j}z_p\sqrt{k_0^2 - k_{px}^2 - k_{py}^2}}\right\}, \end{aligned} \tag{6.4-27}$$

其中, $U_p(k_{px}, k_{py}; 0)$ 定义为 $u_p(x_p, y_p, z_p)$ 在 $z_p = 0$ 的平面波角谱, 从式 (6.4-27), 可以给出

$$U_p(k_{px}, k_{py}; 0) = U_s(k_{sx}(k_{px}, k_{py}), k_{sy}(k_{px}, k_{py}); 0)|J(k_{px}, k_{py})|. \tag{6.4-28}$$

显然, 有

$$\begin{aligned} &U_s(k_{sx}(k_{px}, k_{py}), k_{sy}(k_{px}, k_{py}); 0) \\ =&\mathcal{F}\{u_s(x_s, y_s; z_s = 0)\}|_{k_{sx}=a_1k_{px}+a_2k_{py}+a_3k_{pz}, k_{sy}=a_4k_{px}+a_5k_{py}+a_6k_{pz}} \\ =&U_s(a_1k_{px} + a_2k_{py} + a_3k_{pz}, a_4k_{px} + a_5k_{py} + a_6k_{pz}; 0). \end{aligned}$$

式 (6.4-28) 是一个重要的结果, 因为它将面元在局部坐标系下的频谱 $U_s(k_{sx}, k_{sy}; 0)$ 与平行局部坐标系下的频谱 $U_p(k_{px}, k_{py}; 0)$ 通过坐标变换联系了起来. 在解析法中, 将依然利用这一重要结果.

因为我们刚刚建立了任意面元在平行局部坐标系下的面元场, 即式 (6.4-27) 中的 $u_p(x_p, y_p, z_p)$, 当 $z_p = z_i$ 时, 可以写出全息面上的面元场 $u_i(x, y)$ (第 i 个面元产生的), 用 (x, y) 来替代式 (6.4-27) 中的 (x_p, y_p), 有

$$\begin{aligned} u_i(x, y) = \mathcal{F}^{-1}\Big\{&U_{s,i}(a_{1,i}k_{px} + a_{2,i}k_{py} + a_{3,i}k_{pz}, a_{4,i}k_{px} + \\ &a_{5,i}k_{py} + a_{6,i}k_{pz}; 0)|J_i(k_{px}, k_{py})|\mathrm{e}^{-\mathrm{j}z_i\sqrt{k_0^2 - k_{px}^2 - k_{py}^2}}\Big\}, \end{aligned} \tag{6.4-29}$$

其中, $U_{s,i}(k_{sx}, k_{sy}) = \mathcal{F}\{u_{s,i}(x_s, y_s; z_s = 0)\}$, 且 $u_{s,i}(x_s, y_s; z_s = 0)$ 表示离全息面 z_i 处的第 i 个面元. 可以看出, 变换矩阵 T 和雅可比行列式中的元素 a 也用下标 i 表示, 因为不同的面元将进行不同的旋转, 直至其平行于全息面. 从式 (6.4-29) 可观察到, 全息图上每个面元场的数值计算都需要两次 FFT, 且全息图上总的面元场由式 (6.4-11) 计算得出.

解析法

在传统法中, 式 (6.4-29) 利用了两次 FFT 进行数值求解. 第一次 FFT 只是在局部坐标系中对规则采样网格进行 FFT 以求得每个表面函数的频谱, 即求 $\mathcal{F}\{u_{s,i}(x_s, y_s; z_s = 0)\}$. 然而, 第二次 FFT 因为包含了平行局部坐标系中的采样, 造成了失真, 其原因是坐标系在由倾斜局部坐标系旋转到平行局部坐标系时, 两个频谱之间有了非线性映射关系 (*nonlinear mapping*)

Incorporating Eqs. (6.4-21) and (6.4-23) into Eq. (6.4-19a), the complex field propagating along the z_p direction, caused by the known polygon surface function $u_s(x_s, y_s; z_s = 0) = u_s(x_s, y_s)$, is

$$u_p(x_p, y_p, z_p) = \frac{1}{4\pi^2} \iint_{-\infty}^{\infty} U_s(k_{sx}(k_{px}, k_{py}), k_{sy}(k_{px}, k_{py}); 0) \times$$

$$e^{-j(k_{px}x_p + k_{py}y_p + z_p\sqrt{k_0^2 - k_{px}^2 - k_{py}^2})} |J(k_{px}, k_{py})| dk_{px} dk_{py}$$

$$= \mathcal{F}^{-1}\left\{U_p(k_{px}, k_{py}; 0) e^{-jz_p\sqrt{k_0^2 - k_{px}^2 - k_{py}^2}}\right\}, \tag{6.4-27}$$

where we recognize that $U_p(k_{px}, k_{py}; 0)$ is the definition of the angular plane wave spectrum of $u_p(x_p, y_p, z_p)$ at $z_p = 0$, and, from Eq. (6.4-27), it is given by

$$U_p(k_{px}, k_{py}; 0) = U_s(k_{sx}(k_{px}, k_{py}), k_{sy}(k_{px}, k_{py}); 0) |J(k_{px}, k_{py})|. \tag{6.4-28}$$

Explicitly,

$$U_s(k_{sx}(k_{px}, k_{py}), k_{sy}(k_{px}, k_{py}); 0)$$

$$= \mathcal{F}\{u_s(x_s, y_s; z_s = 0)\}|_{k_{sx} = a_1 k_{px} + a_2 k_{py} + a_3 k_{pz}, k_{sy} = a_4 k_{px} + a_5 k_{py} + a_6 k_{pz}}$$

$$= U_s(a_1 k_{px} + a_2 k_{py} + a_3 k_{pz}, a_4 k_{px} + a_5 k_{py} + a_6 k_{pz}; 0).$$

Equation (6.4-28) is an important result as it relates the spectrum of the polygon on the local coordinates $U_s(k_{sx}, k_{sy}; 0)$ to that on the parallel local coordinates $U_p(k_{px}, k_{py}; 0)$ upon coordinate transformation. In the analytical method, we will also employ this important result.

Since we have just established the polygon field due to an arbitrary polygon in the parallel local coordinates, that is, $u_p(x_p, y_p, z_p)$ in Eq. (6.4-27), we can now write the polygon field $u_i(x, y)$ (due to the ith polygon) on the hologram plane at $z_p = z_i$ with (x_p, y_p) replaced by (x, y) in Eq. (6.4-27):

$$u_i(x, y) = \mathcal{F}^{-1}\Big\{U_{s,i}(a_{1,i}k_{px} + a_{2,i}k_{py} + a_{3,i}k_{pz}, a_{4,i}k_{px} +$$

$$a_{5,i}k_{py} + a_{6,i}k_{pz}; 0) |J_i(k_{px}, k_{py})| e^{-jz_i\sqrt{k_0^2 - k_{px}^2 - k_{py}^2}}\Big\}, \tag{6.4-29}$$

where $U_{s,i}(k_{sx}, k_{sy}) = \mathcal{F}\{u_{s,i}(x_s, y_s; z_s = 0)\}$ and $u_{s,i}(x_s, y_s; z_s = 0)$ denotes the ith polygon that is z_i away from the hologram. Note that elements as from the transformation matrix T and the Jacobian are also indicated by subscript i as different polygons undergo different rotations till parallel to the hologram plane. From Eq. (6.4-29), we see that numerical calculations of each of the polygon field on the hologram require two *FFTs* and the total polygon field on the hologram is computed by Eq. (6.4-11).

Analytical Method

In the traditional method, Eq. (6.4-29) is numerically calculated using two FFTs. The first FFT is simply executed on a regular sampling grid in the local coordinate system to find the spectrum of each surface function, that is, the calculation of $\mathcal{F}\{u_{s,i}(x_s, y_s; z_s = 0)\}$. However, the second FFT involving sampling in the parallel local coordinate system gets distorted. The reason is that coordinate system rotation from the tilted local coordinates to parallel local

[见式 (6.4-20)]. 因此, 通常使用插值来确保由于旋转导致的两个采样网格之间的不匹配问题. 此外, 由于每个面元关于全息面的几何关系, 因此插值是唯一的. 简言之, 插值降低了全息图的整体计算速度.

Ahrenberg 等 (2008) 开创了一种利用仿射变换 (*affine transformation*) 来解析计算任意面元频谱即 $\mathcal{F}\{u_{s,i}\}$ 的方法. 因此, 不再需要像传统法中一样对每一个面元进行 FFT, 从而也无需进行后续的插值. 此外, 每个面元的频谱可以被预先计算出来. 该解析法的核心思想是, 将任意三角形的傅里叶变换 (或是频谱) 与一个已知的单位直角三角形 (*unit right triangle*) (后面会给出其定义) 的解析频谱关联起来, 而这种关联是利用二维仿射变换建立的. 一旦获得 $u_{s,i}$ 的频谱解析解, 即可通过式 (6.4-28) 将该频谱旋转到平行局部坐标系中, 最终, 通过处理两个平行平面之间的衍射而得到全息面上的面元场.

A) 单位直角三角形的频谱

图 6.4-10 所示为一个单位直角三角形 (*unit right triangle*) △, 三个顶点的坐标为 $(0,0)$, $(1,0)$ 和 $(1,1)$. 那么, 二维单位三角形函数 (*2-D unit triangle function*) 被定义为

$$f_{\triangle}(x,y)=\begin{cases}1, & \text{若 } (x,y) \text{ 在 } \triangle \text{ 内}\\ 0, & \text{其他}.\end{cases} \tag{6.4-30}$$

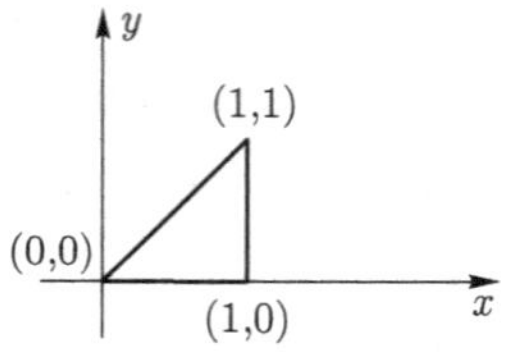

图 6.4-10　单位直角三角形

根据式 (2.3-1a), 对 $f_{\triangle}(x,y)$ 的二维傅里叶变换进行解析计算为

$$\begin{aligned}\mathcal{F}\{f_{\triangle}(x,y)\}=F_{\triangle}(k_x,k_y)&=\iint\limits_{-\infty}^{\infty} f_{\triangle}(x,y)\mathrm{e}^{\mathrm{j}k_x x+\mathrm{j}k_y y}\mathrm{d}x\mathrm{d}y\\ &=\int_0^1\int_0^x \mathrm{e}^{\mathrm{j}k_x x+\mathrm{j}k_y y}\mathrm{d}y\mathrm{d}x\\ &=\begin{cases}\dfrac{1}{2}, & k_x=k_y=0\\ \dfrac{1-\mathrm{e}^{\mathrm{j}k_y}}{k_y^2}+\dfrac{\mathrm{j}}{k_y}, & k_x=0,k_y\neq 0\\ \dfrac{\mathrm{e}^{\mathrm{j}k_x}-1}{k_x^2}-\dfrac{\mathrm{j}\mathrm{e}^{\mathrm{j}k_x}}{k_x}, & k_x\neq 0,k_y=0\\ \dfrac{1-\mathrm{e}^{\mathrm{j}k_y}}{k_y^2}-\dfrac{\mathrm{j}}{k_y}, & k_x=-k_y,k_y\neq 0\\ \dfrac{\mathrm{e}^{\mathrm{j}k_x}-1}{k_xk_y}+\dfrac{1-\mathrm{e}^{\mathrm{j}(k_x+k_y)}}{k_y(k_x+k_y)}, & \text{其他}.\end{cases}\end{aligned} \tag{6.4-31}$$

coordinates introduces *nonlinear mapping* between the two spectra [see Eq. (6.4-20)]. For this reason, interpolation is often used to ensure proper matching between the two sampling grids due to rotation. In addition, interpolation is unique for each polygon due to its geometry with respect to the hologram plane. In short, interpolation slows down the overall calculation speed of the hologram.

Ahrenberg et al. (2008) pioneered a method to analytically compute the spectrum of an arbitrary polygon, that is, $\mathcal{F}\{u_{s,i}\}$, through the use of *affine transformation.* As a result of the work, there is no need to perform FFT for each polygon and the subsequent interpolation needed in the traditional method. In addition, each spectrum of the polygon can be pre-calculated. The central idea of the analytical method is that we can relate the Fourier transform (or the spectrum) of an arbitrary triangle to that of a *unit right triangle* (to be defined next), which is known analytically. The relation of the two spectra is established through the use of 2-D affine transformation. Once the spectrum of $u_{s,i}$ is obtained analytically, rotation of the spectrum to the parallel local coordinate system is through the use of Eq. (6.4-28), and finally diffraction between the two parallel planes is performed to have the polygon field on the hologram.

A) Spectrum of a Unit Right Triangle

Figure 6.4-10 shows a *unit right triangle* Δ with the coordinates of the three vertices as $(0,0)$, $(1,0)$, and $(1,1)$. The *2-D unit triangle function* is then defined as

$$f_{\Delta}(x,y)=\begin{cases}1, & \text{if } (x,y) \text{ lies inside } \Delta\\ 0, & \text{otherwise.}\end{cases} \tag{6.4-30}$$

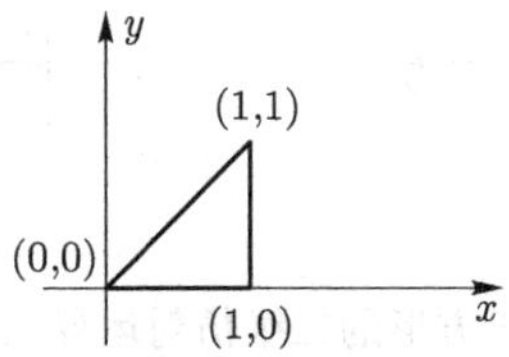

Figure 6.4-10 Unit right triangle

The 2-D Fourier transform of $f_{\Delta}(x,y)$, according to Eq. (2.3-1a), is evaluated analytically and given by

$$\mathcal{F}\{f_{\Delta}(x,y)\}=F_{\Delta}(k_x,k_y)=\iint_{-\infty}^{\infty} f_{\Delta}(x,y)\mathrm{e}^{\mathrm{j}k_x x+\mathrm{j}k_y y}\mathrm{d}x\mathrm{d}y=\int_0^1\int_0^x \mathrm{e}^{\mathrm{j}k_x x+\mathrm{j}k_y y}\mathrm{d}y\mathrm{d}x$$

$$=\begin{cases}\dfrac{1}{2}, & k_x=k_y=0\\ \dfrac{1-\mathrm{e}^{\mathrm{j}k_y}}{k_y^2}+\dfrac{\mathrm{j}}{k_y}, & k_x=0, k_y\neq 0\\ \dfrac{\mathrm{e}^{\mathrm{j}k_x}-1}{k_x^2}-\dfrac{\mathrm{j}\mathrm{e}^{\mathrm{j}k_x}}{k_x}, & k_x\neq 0, k_y=0\\ \dfrac{1-\mathrm{e}^{\mathrm{j}k_y}}{k_y^2}-\dfrac{\mathrm{j}}{k_y}, & k_x=-k_y, k_y\neq 0\\ \dfrac{\mathrm{e}^{\mathrm{j}k_x}-1}{k_x k_y}+\dfrac{1-\mathrm{e}^{\mathrm{j}(k_x+k_y)}}{k_y(k_x+k_y)}, & \text{elsewhere.}\end{cases} \tag{6.4-31}$$

根据上述 $f_{\triangle}(x,y)$ 的频谱解析表达式, 接下来讨论如何利用仿射变换将任意三角形的频谱与 $F_{\triangle}(k_x,k_y)$ 联系起来.

B) 仿射变换

在二维仿射变换 (*affine transformation*) 的基础上, 导出倾斜局部坐标系下任意三角形的解析频谱. 二维仿射变换是一类二维几何变换 (*geometrical transformation*), 它可以将输入坐标 (x_{in},y_{in}) 映射到输出坐标 (x_{out},y_{out}), 根据为

$$\begin{pmatrix} x_{out} \\ y_{out} \end{pmatrix} = \begin{pmatrix} a_{11} & a_{12} \\ a_{21} & a_{22} \end{pmatrix} \begin{pmatrix} x_{in} \\ y_{in} \end{pmatrix} + \begin{pmatrix} a_{13} \\ a_{23} \end{pmatrix}. \tag{6.4-32}$$

举个例子, $\begin{pmatrix} a_{11} & a_{12} \\ a_{21} & a_{22} \end{pmatrix} = \begin{pmatrix} 1 & 0 \\ 0 & 1 \end{pmatrix}$ 且 $\begin{pmatrix} a_{13} \\ a_{23} \end{pmatrix} = \begin{pmatrix} x_0 \\ y_0 \end{pmatrix}$ 表示平移 (*translation*). 该平移为一种几何变换, 它将图像中的每一点在给定方向上移动相同的距离. 另一个例子为, $\begin{pmatrix} a_{11} & a_{12} \\ a_{21} & a_{22} \end{pmatrix} = \begin{pmatrix} \cos\theta & \sin\theta \\ -\sin\theta & \cos\theta \end{pmatrix}$ 和 $\begin{pmatrix} a_{13} \\ a_{23} \end{pmatrix} = \begin{pmatrix} 0 \\ 0 \end{pmatrix}$ 表示将图像围绕原点旋转一个 θ 角. 图 6.4-11 所示为对单位正方形进行的二维仿射变换操作的举例. 其他典型的仿射变换有缩放 (*scaling*)、镜像 (*reflection*) 和错切 (*shear*) (见习题 6.10). 所有这些操作一般用式 (6.4-32) 表示的仿射变换来表示.

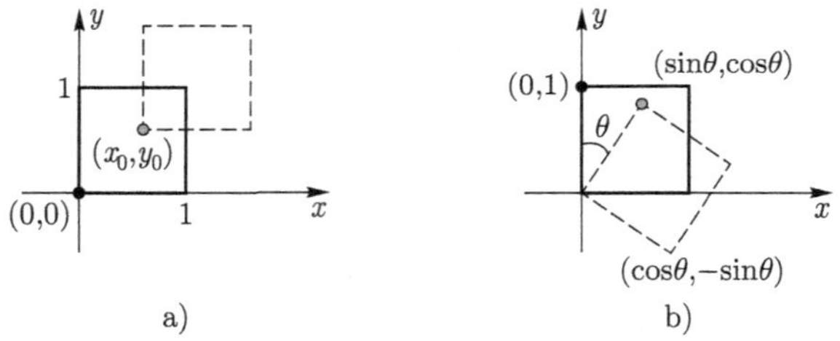

图 6.4-11 单位正方形的二维仿射运算: a) 平移; b) 旋转

图 6.4-12a) 所示为一个在源坐标系中顶点坐标为 (x_1,y_1), (x_2,y_2) 和 (x_3,y_3) 的任意三角形 $f_\Gamma(x_s,y_s)$. 在图 6.4-12b) 中, 在 xy 坐标系下, 给出了一个单位直角三角形. 因此, 联系这两个坐标系的仿射变换可以写为

$$\begin{pmatrix} x_s \\ y_s \end{pmatrix} = \begin{pmatrix} a_{11} & a_{12} \\ a_{21} & a_{22} \end{pmatrix} \begin{pmatrix} x \\ y \end{pmatrix} + \begin{pmatrix} a_{13} \\ a_{23} \end{pmatrix}. \tag{6.4-33}$$

对于此特殊情况, 求其元素 a_{ij}.

建立两个三角形顶点之间的两两对应关系, 有以下三个矩阵方程

$$\begin{pmatrix} x_1 \\ y_1 \end{pmatrix} = \begin{pmatrix} a_{11} & a_{12} \\ a_{21} & a_{22} \end{pmatrix} \begin{pmatrix} 0 \\ 0 \end{pmatrix} + \begin{pmatrix} a_{13} \\ a_{23} \end{pmatrix},$$

$$\begin{pmatrix} x_2 \\ y_2 \end{pmatrix} = \begin{pmatrix} a_{11} & a_{12} \\ a_{21} & a_{22} \end{pmatrix} \begin{pmatrix} 1 \\ 0 \end{pmatrix} + \begin{pmatrix} a_{13} \\ a_{23} \end{pmatrix},$$

With the above analytical expression of the spectrum of $f_\Delta(x, y)$, we next discuss how to relate the spectrum of a general triangle to $F_\Delta(k_x, k_y)$ using an affine transformation.

B) Affine Transformation

On the basis of 2-D *affine transformation*, we derive an analytical frequency spectrum of a general triangle on the tilted local coordinate system. An affine transformation is a class of 2-D *geometrical transformations* that maps input coordinates (x_{in}, y_{in}) into output coordinates (x_{out}, y_{out}) according to

$$\begin{pmatrix} x_{out} \\ y_{out} \end{pmatrix} = \begin{pmatrix} a_{11} & a_{12} \\ a_{21} & a_{22} \end{pmatrix} \begin{pmatrix} x_{in} \\ y_{in} \end{pmatrix} + \begin{pmatrix} a_{13} \\ a_{23} \end{pmatrix}. \tag{6.4-32}$$

As an example, $\begin{pmatrix} a_{11} & a_{12} \\ a_{21} & a_{22} \end{pmatrix} = \begin{pmatrix} 1 & 0 \\ 0 & 1 \end{pmatrix}$ and $\begin{pmatrix} a_{13} \\ a_{23} \end{pmatrix} = \begin{pmatrix} x_0 \\ y_0 \end{pmatrix}$ represent *translation*. A translation is a geometric transformation that moves every point of an image by the same distance in a given direction. As another example, $\begin{pmatrix} a_{11} & a_{12} \\ a_{21} & a_{22} \end{pmatrix} = \begin{pmatrix} \cos\theta & \sin\theta \\ -\sin\theta & \cos\theta \end{pmatrix}$ and $\begin{pmatrix} a_{13} \\ a_{23} \end{pmatrix} = \begin{pmatrix} 0 \\ 0 \end{pmatrix}$ represent rotating the image about the origin by an angle of θ. Figure 6.4-11 illustrates the 2-D affine transformation operations on a unit square for the two examples. Other typical affine transformations are *scaling*, *reflection*, and *shear* (see P.6.10). All these operations are represented in general by the affine transformation represented by Eq. (6.4-32).

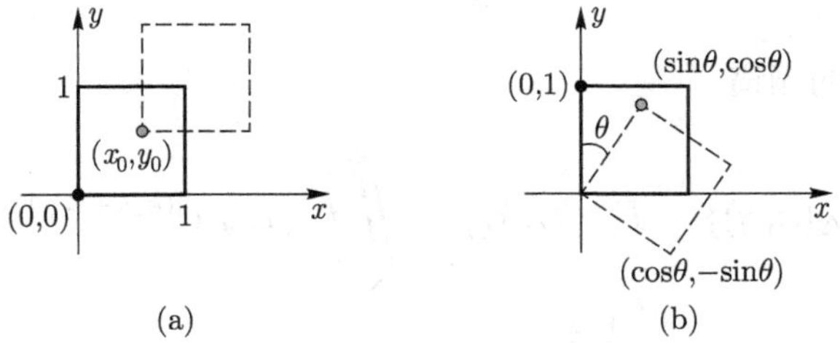

Figure 6.4-11 2-D affine operations on a unit square. (a) Translation and (b) rotation

Figure 6.4-12a shows an arbitrary triangle $f_\Gamma(x_s, y_s)$ on the source coordinate system with the vertex coordinates as (x_1, y_1), (x_2, y_2), and (x_3, y_3). In Figure 6.4-12b, we also show a unit right triangle on the xy-coordinates. Therefore, the affine transform relating the two coordinates can be written as

$$\begin{pmatrix} x_s \\ y_s \end{pmatrix} = \begin{pmatrix} a_{11} & a_{12} \\ a_{21} & a_{22} \end{pmatrix} \begin{pmatrix} x \\ y \end{pmatrix} + \begin{pmatrix} a_{13} \\ a_{23} \end{pmatrix}. \tag{6.4-33}$$

Let us find the elements a_{ij} for this particular situation.

Setting up a pairwise correspondence of the vertices between the two triangles, we have the following three matrix equations:

$$\begin{pmatrix} x_1 \\ y_1 \end{pmatrix} = \begin{pmatrix} a_{11} & a_{12} \\ a_{21} & a_{22} \end{pmatrix} \begin{pmatrix} 0 \\ 0 \end{pmatrix} + \begin{pmatrix} a_{13} \\ a_{23} \end{pmatrix},$$

$$\begin{pmatrix} x_2 \\ y_2 \end{pmatrix} = \begin{pmatrix} a_{11} & a_{12} \\ a_{21} & a_{22} \end{pmatrix} \begin{pmatrix} 1 \\ 0 \end{pmatrix} + \begin{pmatrix} a_{13} \\ a_{23} \end{pmatrix},$$

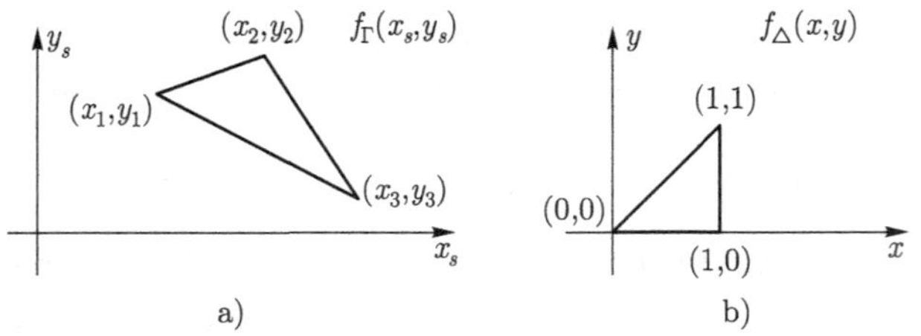

图 6.4-12 a) 源坐标系 (x_s, y_s) 中的任意三角形; b) xy 坐标系中的单位直角三角形

和

$$\begin{pmatrix} x_3 \\ y_3 \end{pmatrix} = \begin{pmatrix} a_{11} & a_{12} \\ a_{21} & a_{22} \end{pmatrix} \begin{pmatrix} 1 \\ 1 \end{pmatrix} + \begin{pmatrix} a_{13} \\ a_{23} \end{pmatrix}. \tag{6.4-34}$$

从上面的三个矩阵方程中, 得到可以用任意三角形中的顶点坐标来表示矩阵中各元素 a_{ij} 的六个方程, 将求出的 a_{ij} 代入式 (6.4-33), 得到所考虑的两个三角形之间的仿射运算关系

$$\begin{pmatrix} x_s \\ y_s \end{pmatrix} = \begin{pmatrix} x_2 - x_1 & x_3 - x_2 \\ y_2 - y_1 & y_3 - y_2 \end{pmatrix} \begin{pmatrix} x \\ y \end{pmatrix} + \begin{pmatrix} x_1 \\ y_1 \end{pmatrix}. \tag{6.4-35}$$

在建立一般三角形与单位直角三角形的仿射运算之后, 将求出表面函数 $u_s(x_s, y_s) = f_\Gamma(x_s, y_s)$ 的频谱. 从式 (6.4-22) 得, $f_\Gamma(x_s, y_s)$ 的频谱为

$$\mathcal{F}\{f_\Gamma(x_s, y_s)\} = F_\Gamma(k_{sx}, k_{sy}) = \iint\limits_{-\infty}^{\infty} f_\Gamma(x_s, y_s) \mathrm{e}^{\mathrm{j}(k_{sx}x_s + k_{sy}y_s)} \mathrm{d}x_s \mathrm{d}y_s. \tag{6.4-36}$$

类似地, 写出 $f_\triangle(x, y)$ 的频谱

$$\begin{aligned} \mathcal{F}\{f_\triangle(x, y)\} = F_\triangle(k_x, k_y) &= \iint\limits_{-\infty}^{\infty} f_\triangle(x, y) \mathrm{e}^{\mathrm{j}(k_x x + k_y y)} \mathrm{d}x \mathrm{d}y \\ &= \int_0^1 \int_0^x \mathrm{e}^{\mathrm{j}(k_x x + k_y y)} \mathrm{d}x \mathrm{d}y. \end{aligned} \tag{6.4-37}$$

对于与单位直角三角形相关的任意三角形, 利用式 (6.4-33), 且有仿射运算 $x_s = a_{11}x + a_{12}y + a_{13}$ 和 $y_s = a_{21}x + a_{22}y + a_{23}$. 当变量变为 (x, y) 时, 式 (6.4-36) 变为

$$\begin{aligned} F_\Gamma(k_{sx}, k_{sy}) &= \iint\limits_{-\infty}^{\infty} f_\Gamma(x_s, y_s) \mathrm{e}^{\mathrm{j}(k_{sx}x_s + k_{sy}y_s)} \mathrm{d}x_s \mathrm{d}y_s \\ &= \int_0^1 \int_0^x f_\triangle(x, y) \mathrm{e}^{\mathrm{j}[k_{sx}(a_{11}x + a_{12}y + a_{13}) + k_{sy}(a_{21}x + a_{22}y + a_{23})]} |J(x, y)| \mathrm{d}x \mathrm{d}y \\ &= \int_0^1 \int_0^x \mathrm{e}^{\mathrm{j}[k_{sx}(a_{11}x + a_{12}y + a_{13}) + k_{sy}(a_{21}x + a_{22}y + a_{23})]} |J(x, y)| \mathrm{d}x \mathrm{d}y, \end{aligned}$$

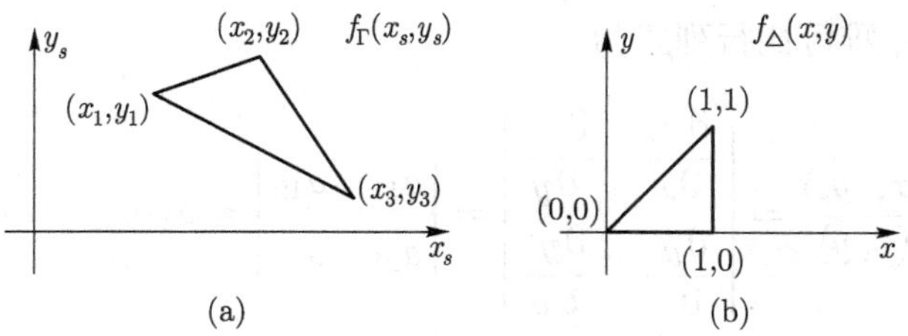

Figure 6.4-12 (a)Arbitrary triangle on the source coordinates (x_s, y_s) and (b) unit right triangle on (x, y) coordinates

and

$$\begin{pmatrix} x_3 \\ y_3 \end{pmatrix} = \begin{pmatrix} a_{11} & a_{12} \\ a_{21} & a_{22} \end{pmatrix} \begin{pmatrix} 1 \\ 1 \end{pmatrix} + \begin{pmatrix} a_{13} \\ a_{23} \end{pmatrix}. \tag{6.4-34}$$

From the three matrix equations, we have the set of six equations allowing us to find the elements a_{ij} in terms of the vertex coordinates of the arbitrary triangle. Upon substituting the found a_{ij} into Eq. (6.4-33), we have the resulting affine operation relating the two triangles under consideration:

$$\begin{pmatrix} x_s \\ y_s \end{pmatrix} = \begin{pmatrix} x_2 - x_1 & x_3 - x_2 \\ y_2 - y_1 & y_3 - y_2 \end{pmatrix} \begin{pmatrix} x \\ y \end{pmatrix} + \begin{pmatrix} x_1 \\ y_1 \end{pmatrix}. \tag{6.4-35}$$

After establishing the affine operation for a general triangle with a unit right triangle, we shall find the spectrum of the surface function $u_s(x_s, y_s) = f_\Gamma(x_s, y_s)$. From Eq. (6.4-22), the spectrum of $f_\Gamma(x_s, y_s)$ is

$$\mathcal{F}\{f_\Gamma(x_s, y_s)\} = F_\Gamma(k_{sx}, k_{sy}) = \iint_{-\infty}^{\infty} f_\Gamma(x_s, y_s) \mathrm{e}^{\mathrm{j}(k_{sx}x_s + k_{sy}y_s)} \mathrm{d}x_s \mathrm{d}y_s. \tag{6.4-36}$$

Similarly, we write the spectrum of $f_\triangle(x, y)$ as

$$\begin{aligned} \mathcal{F}\{f_\triangle(x, y)\} = F_\triangle(k_x, k_y) &= \iint_{-\infty}^{\infty} f_\triangle(x, y) \mathrm{e}^{\mathrm{j}(k_x x + k_y y)} \mathrm{d}x \mathrm{d}y \\ &= \int_0^1 \int_0^x \mathrm{e}^{\mathrm{j}(k_x x + k_y y)} \mathrm{d}x \mathrm{d}y. \end{aligned} \tag{6.4-37}$$

For an arbitrary triangle relating to the unit right triangle, we use Eq. (6.4-33) and have affine operation $x_s = a_{11}x + a_{12}y + a_{13}$ and $y_s = a_{21}x + a_{22}y + a_{23}$. With a change of variables to (x, y), Eq. (6.4-36) becomes

$$\begin{aligned} F_\Gamma(k_{sx}, k_{sy}) &= \iint_{-\infty}^{\infty} f_\Gamma(x_s, y_s) \mathrm{e}^{\mathrm{j}(k_{sx}x_s + k_{sy}y_s)} \mathrm{d}x_s \mathrm{d}y_s \\ &= \int_0^1 \int_0^x f_\triangle(x, y) \mathrm{e}^{\mathrm{j}[k_{sx}(a_{11}x + a_{12}y + a_{13}) + k_{sy}(a_{21}x + a_{22}y + a_{23})]} |J(x, y)| \mathrm{d}x \mathrm{d}y \\ &= \int_0^1 \int_0^x \mathrm{e}^{\mathrm{j}[k_{sx}(a_{11}x + a_{12}y + a_{13}) + k_{sy}(a_{21}x + a_{22}y + a_{23})]} |J(x, y)| \mathrm{d}x \mathrm{d}y, \end{aligned}$$

其中, 根据式 (6.4-33), 雅可比行列式为

$$J(x,y)=\frac{\partial(x_s,y_s)}{\partial(x,y)}=\begin{vmatrix}\dfrac{\partial x_s}{\partial x} & \dfrac{\partial x_s}{\partial y}\\ \dfrac{\partial y_s}{\partial x} & \dfrac{\partial y_s}{\partial y}\end{vmatrix}=\begin{vmatrix}a_{11} & a_{12}\\ a_{21} & a_{22}\end{vmatrix}=a_{11}a_{22}-a_{12}a_{21}. \tag{6.4-38}$$

重新排列上式的 $F_\Gamma(k_{sx},k_{sy})$, 利用式 (6.4-37) 中 $F_\triangle(k_x,k_y)$ 的定义, 有

$$\begin{aligned}F_\Gamma(k_{sx},k_{sy})&=|a_{11}a_{22}-a_{12}a_{21}|\mathrm{e}^{\mathrm{j}(k_{sx}a_{13}+k_{sy}a_{23})}\times\int_0^1\int_0^x \mathrm{e}^{\mathrm{j}[k_{sx}(a_{11}x+a_{12}y)+k_{sy}(a_{21}x+a_{22}y)]}\mathrm{d}x\mathrm{d}y\\&=|a_{11}a_{22}-a_{12}a_{21}|\mathrm{e}^{\mathrm{j}(k_{sx}a_{13}+k_{sy}a_{23})}F_\triangle(a_{11}k_{sx}+a_{21}k_{sy},a_{12}k_{sx}+a_{22}k_{sy}).\end{aligned} \tag{6.4-39}$$

这是 $u_s(x_s,y_s)=f_\Gamma(x_s,y_s)$ 的频谱的解析表达式, 即 $\mathcal{F}\{u_s(x_s,y_s)\}=U_s(k_{sx},k_{sy})=F_\Gamma(k_{sx},k_{sy})$, 因为 $F_\triangle$ 是由式 (6.4-31) 解析给出的. 该频谱是由一个任意三角形得来的. 因此, 根据式 (6.4-29), 即重写如下式, 距离全息图 z_i 处的第 i 个面元的面元场为

$$\begin{aligned}u_i(x,y)=\mathcal{F}^{-1}\Big\{&U_{s,i}(a_{1,i}k_{px}+a_{2,i}k_{py}+a_{3,i}k_{pz},a_{4,i}k_{px}\\&+a_{5,i}k_{py}+a_{6,i}k_{pz};0)|J_i(k_{px},k_{py})|\mathrm{e}^{-\mathrm{j}z_i\sqrt{k_0^2-k_{px}^2-k_{py}^2}}\Big\},\end{aligned} \tag{6.4-40}$$

其中, $U_{s,i}(k_{sx},k_{sy})$ 现在可以解析地给出

$$\begin{aligned}U_{s,i}(k_{sx},k_{sy})&=\mathcal{F}\{u_{s,i}(x_s,y_s;z_s=0)\}=F_{\Gamma_i}(k_{sx},k_{sy})\\&=|a_{11,i}a_{22,i}-a_{12,i}a_{21,i}|\mathrm{e}^{\mathrm{j}(k_{sx}a_{13,i}+k_{sy}a_{23,i})}\\&\quad\times F_\triangle(a_{11,i}k_{sx}+a_{21,i}k_{sy},a_{12,i}k_{sx}+a_{22,i}k_{sy}).\end{aligned}$$

全息图上总的面元场现在可以利用式 (6.4-11) 计算得出. 因此, 在解析法中, 只需进行一次逆 FFT 就可以获得总的面元场, 即全息图上总的物光场.

作为本小节的最后一点, 这里指出, 在讨论这两个基本的面元法, 即传统法与解析法时, 当前还有其他现代的面元法, 这些方法的核心同样基于这两个基本的方法. 比如, 在解析法中, 在旋转之前对源函数进行仿射变换, 使源坐标系平行于全息面, Zhang 等 (2018) 最近提出了一种在二维仿射变换之前进行面元旋转的快速 CGH 方法, Pan 等 (2014) 首次提出一种利用三维仿射变换生成 CGH 的方法, 为了避免传统法中对空间频率网格进行重采样 (即插值的使用) 的必要性, 其他解析法也有探索 [Kim 等 (2008), Jeom 和 Park (2016)].

习题

6.1 如果 f_s 是信号 $g(x)$ 的奈奎斯特采样率, 求出下面每个信号的奈奎斯特采样率:

where, according to Eq. (6.4-33), the Jacobian is

$$J(x,y)=\frac{\partial(x_s,y_s)}{\partial(x,y)}=\begin{vmatrix}\frac{\partial x_s}{\partial x} & \frac{\partial x_s}{\partial y}\\ \frac{\partial y_s}{\partial x} & \frac{\partial y_s}{\partial y}\end{vmatrix}=\begin{vmatrix}a_{11} & a_{12}\\ a_{21} & a_{22}\end{vmatrix}=a_{11}a_{22}-a_{12}a_{21}. \tag{6.4-38}$$

Rearranging the equation $F_\Gamma(k_{sx},k_{sy})$ from above and using the definition of $F_\triangle(k_x,k_y)$ in Eq. (6.4-37), we have

$$\begin{aligned}F_\Gamma(k_{sx},k_{sy})&=|a_{11}a_{22}-a_{12}a_{21}|\mathrm{e}^{\mathrm{j}(k_{sx}a_{13}+k_{sy}a_{23})}\times\int_0^1\int_0^x \mathrm{e}^{\mathrm{j}[k_{sx}(a_{11}x+a_{12}y)+k_{sy}(a_{21}x+a_{22}y)]}\mathrm{d}x\mathrm{d}y\\ &=|a_{11}a_{22}-a_{12}a_{21}|\mathrm{e}^{\mathrm{j}(k_{sx}a_{13}+k_{sy}a_{23})}F_\triangle(a_{11}k_{sx}+a_{21}k_{sy},a_{12}k_{sx}+a_{22}k_{sy}). \end{aligned}\tag{6.4-39}$$

This is an analytical expression of the spectrum of $u_s(x_s,y_s)=f_\Gamma(x_s,y_s)$, that is, $\mathcal{F}\{u_s(x_s,y_s)\}=U_s(k_{sx},k_{sy})=F_\Gamma(k_{sx},k_{sy})$, as $F_\triangle$ is given by Eq. (6.4-31) analytically. The spectrum is due to a single arbitrary triangle. Therefore, the polygon field due to the ith polygon z_i away from the hologram, according to Eq. (6.4-29) which is repeated below, is given by

$$\begin{aligned}u_i(x,y)=\mathcal{F}^{-1}\Big\{&U_{s,i}(a_{1,i}k_{px}+a_{2,i}k_{py}+a_{3,i}k_{pz},a_{4,i}k_{px}\\ &+a_{5,i}k_{py}+a_{6,i}k_{pz};0)|J_i(k_{px},k_{py})|\mathrm{e}^{-\mathrm{j}z_i\sqrt{k_0^2-k_{px}^2-k_{py}^2}}\Big\},\end{aligned}\tag{6.4-40}$$

where $U_{s,i}(k_{sx},k_{sy})$ is now given analytically as

$$\begin{aligned}U_{s,i}(k_{sx},k_{sy})&=\mathcal{F}\{u_{s,i}(x_s,y_s;z_s=0)\}=F_{\Gamma_i}(k_{sx},k_{sy})\\ &=|a_{11,i}a_{22,i}-a_{12,i}a_{21,i}|\mathrm{e}^{\mathrm{j}(k_{sx}a_{13,i}+k_{sy}a_{23,i})}\\ &\quad\times F_\triangle(a_{11,i}k_{sx}+a_{21,i}k_{sy},a_{12,i}k_{sx}+a_{22,i}k_{sy}).\end{aligned}$$

The total polygon field on the hologram can now be computed using Eq. (6.4-11). Hence, in the analytical method, we only need to perform a single inverse FFT to obtain the total polygon field, that is, the total object field, on the hologram.

As a final note to this subsection, we want to point out that while we have discussed two standard polygon-based methods, that is, the traditional method and the analytical method, there are other modern polygon-based methods which build on the idea of the two standard methods. For example, in the analytical method, we take the affine transform of the source function before rotation to align the source coordinate system to be parallel with the hologram plane. Recently, Zhang et al. (2018) have introduced a fast CGH generation method that performs polygon rotation before 2-D affine transformation. Pan et al. (2014) have developed for the first time a method using 3-D affine transformation. Avoiding the necessity for resampling, that is, the use of interpolation, of the spatial frequency grids required in the traditional method, other analytical methods have also been developed [Kim et al. (2008), Yeom and Park (2016)].

Problems

6.1 If f_s is the Nyquist rate for signal $g(x)$, find the Nyquist rate for each of the following signals:

a) $y_a(x)=\dfrac{\mathrm{d}}{\mathrm{d}x}g(x)$.

b) $y_b(x)=g(x)\cos(2\pi f_0 x)$.

c) $y_c(x)=g(ax)$, 其中 a 为一个实的正常数.

d) $y_d(x)=g^2(x)$.

e) $y_e(x)=g^3(x)$.

6.2 假定 B_1 和 B_2 分别为 $g_1(x)$ 和 $g_2(x)$ 的带宽, 求信号 $g(x)=g_1(x)g_2(x)$ 的奈奎斯特采样率.

6.3 对于三步相移全息术, 有以下相移全息图

$$I_0=|\psi_0|^2+|\psi_r|^2+\psi_0\psi_r^*+\psi_0^*\psi_r,$$
$$I_{\pi/2}=|\psi_0+\psi_r\mathrm{e}^{-\mathrm{j}\pi/2}|^2=|\psi_0|^2+|\psi_r|^2+\mathrm{j}\psi_0\psi_r^*-\mathrm{j}\psi_0^*\psi_r,$$

和

$$I_\pi=|\psi_0+\psi_r\mathrm{e}^{-\mathrm{j}\pi}|^2=|\psi_0|^2+|\psi_r|^2-\psi_0\psi_r^*-\psi_0^*\psi_r.$$

证明该物体的复全息图是由下式给出

$$\psi_0=\frac{(1+\mathrm{j})(I_0-I_{\pi/2})+(\mathrm{j}-1)(I_\pi-I_{\pi/2})}{4\psi_r^*}.$$

6.4 对于四步相移全息术, 有以下相移全息图

$$I_0=|\psi_0|^2+|\psi_r|^2+\psi_0\psi_r^*+\psi_0^*\psi_r,$$
$$I_{\pi/2}=|\psi_0|^2+|\psi_r|^2+\mathrm{j}\psi_0\psi_r^*-\mathrm{j}\psi_0^*\psi_r,$$
$$I_\pi=|\psi_0|^2+|\psi_r|^2-\psi_0\psi_r^*-\psi_0^*\psi_r,$$

和

$$I_{3\pi/2}=|\psi_0|^2+|\psi_r|^2-\mathrm{j}\psi_0\psi_r^*+\mathrm{j}\psi_0^*\psi_r.$$

证明该物体的复全息图是由下式给出

$$\psi_0=\frac{(I_0-I_\pi)-\mathrm{j}(I_{\pi/2}-I_{3\pi/2})}{4\psi_r^*}.$$

6.5 图 6.2-3 所示的双光瞳外差扫描系统的相干函数如下所示

$$\Gamma(x'-x'',y'-y'';z'-z'')=\int \mathrm{e}^{\mathrm{j}\frac{k_0}{f}[x_m(x'-x'')+y_m(y'-y'')]}\mathrm{e}^{-\mathrm{j}\frac{k_0(z'-z'')}{2f^2}(x_m^2+y_m^2)}\mathrm{d}x_m\mathrm{d}y_m,$$

证明其值为

$$\Gamma(x'-x'',y'-y'';z-z'')\sim\frac{1}{z'-z''}\mathrm{e}^{\mathrm{j}\frac{k_0[(x'-x'')^2+(y'-y'')^2]}{2(z'-z'')}}.$$

6.6 在编码孔径成像中, 编码孔径被认为具有以下性质

$$C(x)\otimes C(x)=\int_{-\infty}^{\infty}C^*(x')C(x+x')\mathrm{d}x'=\delta(x).$$

证明

$$\int_{-\infty}^{\infty}C^*\left(\frac{x''+x'}{M}\right)C\left(\frac{x+x''}{M}\right)\mathrm{d}x''=\delta\left(\frac{x-x'}{M}\right).$$

(a) $y_a(x) = \frac{\mathrm{d}}{\mathrm{d}x} g(x)$.

(b) $y_b(x) = g(x)\cos(2\pi f_0 x)$.

(c) $y_c(x) = g(ax)$, where a is a real and positive constant.

(d) $y_d(x) = g^2(x)$.

(e) $y_e(x) = g^3(x)$.

6.2 If B_1 and B_2 are the bandwidths of $g_1(x)$ and $g_2(x)$, respectively, find the Nyquist rate for signal $g(x) = g_1(x)g_2(x)$.

6.3 For three-step phase shifting holography, we have the following phase-shifted holograms:

$$I_0 = |\psi_0|^2 + |\psi_r|^2 + \psi_0\psi_r^* + \psi_0^*\psi_r,$$
$$I_{\pi/2} = |\psi_0 + \psi_r \mathrm{e}^{-\mathrm{j}\pi/2}|^2 = |\psi_0|^2 + |\psi_r|^2 + \mathrm{j}\psi_0\psi_r^* - \mathrm{j}\psi_0^*\psi_r,$$

and

$$I_\pi = |\psi_0 + \psi_r \mathrm{e}^{-\mathrm{j}\pi}|^2 = |\psi_0|^2 + |\psi_r|^2 - \psi_0\psi_r^* - \psi_0^*\psi_r.$$

Show that the complex hologram of the object is given by

$$\psi_0 = \frac{(1+\mathrm{j})(I_0 - I_{\pi/2}) + (\mathrm{j}-1)(I_\pi - I_{\pi/2})}{4\psi_r^*}.$$

6.4 For four-step phase shifting holography, we have the following phase-shifted holograms:

$$I_0 = |\psi_0|^2 + |\psi_r|^2 + \psi_0\psi_r^* + \psi_0^*\psi_r,$$
$$I_{\pi/2} = |\psi_0|^2 + |\psi_r|^2 + \mathrm{j}\psi_0\psi_r^* - \mathrm{j}\psi_0^*\psi_r,$$
$$I_\pi = |\psi_0|^2 + |\psi_r|^2 - \psi_0\psi_r^* - \psi_0^*\psi_r,$$

and

$$I_{3\pi/2} = |\psi_0|^2 + |\psi_r|^2 - \mathrm{j}\psi_0\psi_r^* + \mathrm{j}\psi_0^*\psi_r.$$

Show that the complex hologram of the object and it is given by

$$\psi_0 = \frac{(I_0 - I_\pi) - \mathrm{j}(I_{\pi/2} - I_{3\pi/2})}{4\psi_r^*}.$$

6.5 Show that the coherence function of the two-pupil heterodyne scanning system shown in Figure 6.2-3 given below

$$\Gamma(x'-x'', y'-y''; z'-z'') = \int \mathrm{e}^{\mathrm{j}\frac{k_0}{f}[x_m(x'-x'')+y_m(y'-y'')]}\mathrm{e}^{-\mathrm{j}\frac{k_0(z'-z'')}{2f^2}(x_m^2+y_m^2)}\mathrm{d}x_m\mathrm{d}y_m,$$

is evaluated to be

$$\Gamma(x'-x'', y'-y''; z-z'') \sim \frac{1}{z'-z''}\mathrm{e}^{\mathrm{j}\frac{k_0[(x'-x'')^2+(y'-y'')^2]}{2(z'-z'')}}.$$

6.6 In code aperture imaging, the coded aperture is taken to have the following property:

$$C(x) \otimes C(x) = \int_{-\infty}^{\infty} C^*(x')C(x+x')\mathrm{d}x' = \delta(x).$$

Show that

$$\int_{-\infty}^{\infty} C^*\left(\frac{x''+x'}{M}\right) C\left(\frac{x+x''}{M}\right)\mathrm{d}x'' = \delta\left(\frac{x-x'}{M}\right).$$

6.7 在 COACH 中, $G(x,y)$ 为编码相位掩模 (CPM) 在图像传感器平面上的衍射图案, 且具有以下特性

$$G(x,y) \otimes G(x,y) = \delta(x,y).$$

证明

$$G\left(x+\frac{x_i z_h}{f}, y+\frac{y_i z_h}{f}\right) \otimes G(x,y) = \delta\left(x-\frac{x_i z_h}{f}, y-\frac{y_i z_h}{f}\right).$$

6.8 在光学扫描全息术中, 两个扫描光束由下式给出

$$s_1(x,y;z) = \mathcal{F}\{p_1(x,y)\}\big|_{k_x=\frac{k_0 x}{f},k_y=\frac{k_0 y}{f}} * h(x,y;z),$$

和

$$s_2(x,y;z) = \mathcal{F}\{p_2(x,y)\}\big|_{k_x=\frac{k_0 x}{f},k_y=\frac{k_0 y}{f}} * h(x,y;z).$$

定义双光瞳系统的 OTF 为

$$OTF_\Omega(k_x,k_y;z) = \mathcal{F}^*\{s_1(x,y;z)s_2^*(x,y;z)\}.$$

证明 OTF 可以写成两个光瞳 $p_1(x,y)$ 和 $p_2(x,y)$ 的形式

$$OTF_\Omega(k_x,k_y;z) = \mathrm{e}^{\mathrm{j}\frac{z}{2k_0}(k_x^2+k_y^2)}\int p_1^*(x',y')p_2\left(x'+\frac{f}{k_0}k_x, y'+\frac{f}{k_0}k_y\right)\mathrm{e}^{\mathrm{j}\frac{z}{f}(x'k_x+y'k_y)}\mathrm{d}x'\mathrm{d}y'.$$

6.9 证明在面元法进行计算全息的过程中, 源坐标系和平行局部坐标系之间的坐标变换的雅可比行列式为

$$J(k_{px},k_{py}) = \frac{\partial(k_{sx},k_{sy})}{\partial(k_{px},k_{py})} = \frac{(a_2a_6-a_3a_5)k_{px}}{k_{pz}(k_{px},k_{py})} + \frac{(a_3a_4-a_1a_6)k_{py}}{k_{pz}(k_{px},k_{py})} + (a_1a_5-a_2a_4).$$

6.10 通常二维仿射变换可以写成

$$\begin{pmatrix} x_{out} \\ y_{out} \end{pmatrix} = \begin{pmatrix} a_{11} & a_{12} \\ a_{21} & a_{22} \end{pmatrix}\begin{pmatrix} x_{in} \\ y_{in} \end{pmatrix} + \begin{pmatrix} a_{13} \\ a_{23} \end{pmatrix}.$$

对于图题 6.10 中所示的单位正方形图, 当 $\begin{pmatrix} a_{13} \\ a_{23} \end{pmatrix} = \begin{pmatrix} 0 \\ 0 \end{pmatrix}$ 时, 画出以下情况下的仿射变换图.

a) 在 x 方向上的错切 (*shear*):

$$\begin{pmatrix} a_{11} & a_{12} \\ a_{21} & a_{22} \end{pmatrix} = \begin{pmatrix} 1 & \tan\phi \\ 0 & 1 \end{pmatrix}, \text{其中 } \phi < 90°.$$

b) 在 y 方向上的错切:

$$\begin{pmatrix} a_{11} & a_{12} \\ a_{21} & a_{22} \end{pmatrix} = \begin{pmatrix} 1 & 0 \\ \tan\phi & 1 \end{pmatrix}, \text{其中 } \phi < 90°.$$

c) 关于原点 $(0,0)$ 的缩放 (*scaling*):

$$\begin{pmatrix} a_{11} & a_{12} \\ a_{21} & a_{22} \end{pmatrix} = \begin{pmatrix} w & 0 \\ 0 & h \end{pmatrix}, \text{其中 } w \text{ 和 } h \text{ 大于 } 1.$$

6.7 In COACH, $G(x, y)$ is the diffracted pattern of the coded phase mask (CPM) in the plane of the image sensor and has the following property:

$$G(x, y) \otimes G(x, y) = \delta(x, y).$$

Show that

$$G\left(x + \frac{x_i z_h}{f}, y + \frac{y_i z_h}{f}\right) \otimes G(x, y) = \delta\left(x - \frac{x_i z_h}{f}, y - \frac{y_i z_h}{f}\right).$$

6.8 In optical scanning holography, the two scanning beams are given by

$$s_1(x, y; z) = \mathcal{F}\{p_1(x, y)\}|_{k_x = \frac{k_0 x}{f}, k_y = \frac{k_0 y}{f}} * h(x, y; z),$$

and

$$s_2(x, y; z) = \mathcal{F}\{p_2(x, y)\}|_{k_x = \frac{k_0 x}{f}, k_y = \frac{k_0 y}{f}} * h(x, y; z).$$

Defining the OTF of the two-pupil system as

$$OTF_\Omega(k_x, k_y; z) = \mathcal{F}^*\{s_1(x, y; z) s_2^*(x, y; z)\}.$$

show that the OTF can be written in terms of the two pupils $p_1(x, y)$ and $p_2(x, y)$ as

$$OTF_\Omega(k_x, k_y; z) = \mathrm{e}^{\mathrm{j}\frac{z}{2k_0}(k_x^2 + k_y^2)} \int p_1^*(x', y') p_2\left(x' + \frac{f}{k_0} k_x, y' + \frac{f}{k_0} k_y\right) \mathrm{e}^{\mathrm{j}\frac{z}{f}(x' k_x + y' k_y)} \mathrm{d}x' \mathrm{d}y'.$$

6.9 Show that the Jacobian of the coordinate transformation between the source coordinates and the parallel local coordinates is

$$J(k_{px}, k_{py}) = \frac{\partial(k_{sx}, k_{sy})}{\partial(k_{px}, k_{py})} = \frac{(a_2 a_6 - a_3 a_5) k_{px}}{k_{pz}(k_{px}, k_{py})} + \frac{(a_3 a_4 - a_1 a_6) k_{py}}{k_{pz}(k_{px}, k_{py})} + (a_1 a_5 - a_2 a_4).$$

in the polygon-based approach to computer-generated holography.

6.10 The general 2-D affine transformation is written as

$$\begin{pmatrix} x_{out} \\ y_{out} \end{pmatrix} = \begin{pmatrix} a_{11} & a_{12} \\ a_{21} & a_{22} \end{pmatrix} \begin{pmatrix} x_{in} \\ y_{in} \end{pmatrix} + \begin{pmatrix} a_{13} \\ a_{23} \end{pmatrix}.$$

For a unit square figure shown in Figure P.6.10, plot the affine-transformed figure for the following cases when $\begin{pmatrix} a_{13} \\ a_{23} \end{pmatrix} = \begin{pmatrix} 0 \\ 0 \end{pmatrix}$.

(a) Shear in the x-direction:

$$\begin{pmatrix} a_{11} & a_{12} \\ a_{21} & a_{22} \end{pmatrix} = \begin{pmatrix} 1 & \tan\phi \\ 0 & 1 \end{pmatrix}, \text{ where } \phi < 90°.$$

(b) Shear in the y-direction:

$$\begin{pmatrix} a_{11} & a_{12} \\ a_{21} & a_{22} \end{pmatrix} = \begin{pmatrix} 1 & 0 \\ \tan\phi & 1 \end{pmatrix}, \text{ where } \phi < 90°.$$

(c) Scale about the origin (0,0) :

$$\begin{pmatrix} a_{11} & a_{12} \\ a_{21} & a_{22} \end{pmatrix} = \begin{pmatrix} w & 0 \\ 0 & h \end{pmatrix}, \text{ where } w \text{ and } h \text{ are greater than 1.}$$

d) 关于原点 $(0,0)$ 的镜像 (*reflection*):

$$\begin{pmatrix} a_{11} & a_{12} \\ a_{21} & a_{22} \end{pmatrix} = \begin{pmatrix} -1 & 0 \\ 0 & -1 \end{pmatrix}.$$

e) 在 x 轴的镜像:

$$\begin{pmatrix} a_{11} & a_{12} \\ a_{21} & a_{22} \end{pmatrix} = \begin{pmatrix} 1 & 0 \\ 0 & -1 \end{pmatrix}.$$

f) 在 y 轴的镜像:

$$\begin{pmatrix} a_{11} & a_{12} \\ a_{21} & a_{22} \end{pmatrix} = \begin{pmatrix} -1 & 0 \\ 0 & 1 \end{pmatrix}$$

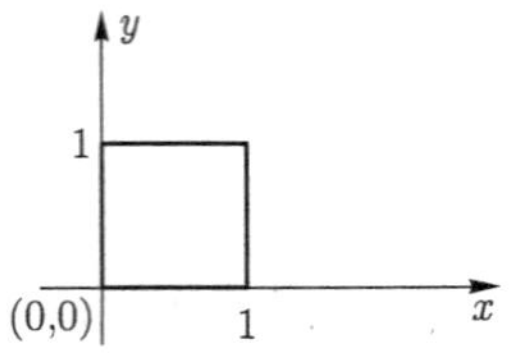

图题 6.10

参考文献

[1] Ahrenberg, L., Benzie, P., Magnor, M. and Watson, J. (2008). "Computer generated holograms from three dimensional meshes using an analytic light transport model," Applied Optics 47, pp. 1567-1574.

[2] Berrang, J. E. (1970). "Television transmission of holograms using a narrow-band video signal," Bell System Technical Journal 49, pp. 879-887.

[3] Burckhardt, C. B. and Enloe, L. H. (1969). "Television transmission of holograms with reduced resolution requirements on the camera tube," Bell System Technical Journal 48, pp. 1529-1535.

[4] Cannon, T. M. and Fenimore, E. E. (1980). "Coded aperture imaging: many holes make light work," Optical Engineering 19, pp. 283-289.

[5] Cochran, G. (1966). "New method of making Fresnel transforms with incoherent light," Journal of the Optical Society of America 56, pp. 1513-1517.

[6] Dicke, R. H. (1968). "Scatter-hole camera for X-rays and gamma rays," The Astrophysical Journal 153, pp. L101-L106.

[7] Enloe, L. H., Murphy, J. A. and Rubinsten, C. B. (1966). "Hologram transmission via television," Bell System Technical Journal 45, pp. 333-335.

[8] Gabor, D. and Goss, P. (1966). "Interference microscope with total wavefront reconstruction," Journal of the Optical Society of America 56, pp. 849-858.

[9] Goodman, J. W. and Lawrence, R. W. (1967). "Digital image formation from electronically detected holograms," Applied Physics Letters 11, pp. 77-79.

[10] Guo, P. and Devaney, A. J. (2004), "Digital microscopy using phase-shifting digital holography with two reference waves," Optics Letters 29, pp. 857-859.

[11] Indebetouw, G., Kim, T., Poon, T.-C. and Schilling, B. W. (1998). "Three-dimensional location of fluorescent inhomogeneities in turbid media by scanning heterodyne holography," Optics Letters 23, pp. 133-137.

[12] Indebetouw, G. and Zhong, W. (2006). "Scanning holographic microscopy of three-dimensional holographic microscopy," Journal of the Optical Society of America A 23, pp. 2657-2661.

(d) Reflection about the origin (0,0):

$$\begin{pmatrix} a_{11} & a_{12} \\ a_{21} & a_{22} \end{pmatrix} = \begin{pmatrix} -1 & 0 \\ 0 & -1 \end{pmatrix}.$$

(e) Reflection about the x-axis:

$$\begin{pmatrix} a_{11} & a_{12} \\ a_{21} & a_{22} \end{pmatrix} = \begin{pmatrix} 1 & 0 \\ 0 & -1 \end{pmatrix}.$$

(f) Reflection about the y-axis:

$$\begin{pmatrix} a_{11} & a_{12} \\ a_{21} & a_{22} \end{pmatrix} = \begin{pmatrix} -1 & 0 \\ 0 & 1 \end{pmatrix}.$$

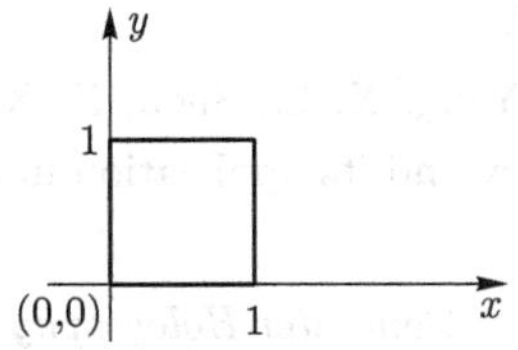

Figure P.6.10

Bibliography

[1] Ahrenberg, L., Benzie, P., Magnor, M. and Watson, J. (2008). "Computer generated holograms from three dimensional meshes using an analytic light transport model," Applied Optics 47, pp. 1567-1574.

[2] Berrang, J. E. (1970). "Television transmission of holograms using a narrow-band video signal," Bell System Technical Journal 49, pp. 879-887.

[3] Burckhardt, C. B. and Enloe, L. H. (1969). "Television transmission of holograms with reduced resolution requirements on the camera tube," Bell System Technical Journal 48, pp. 1529-1535.

[4] Cannon, T. M. and Fenimore, E. E. (1980). "Coded aperture imaging: many holes make light work," Optical Engineering 19, pp. 283-289.

[5] Cochran, G. (1966). "New method of making Fresnel transforms with incoherent light," Journal of the Optical Society of America 56, pp. 1513-1517.

[6] Dicke, R. H. (1968). "Scatter-hole camera for X-rays and gamma rays," The Astrophysical Journal 153, pp. L101-L106.

[7] Enloe, L. H., Murphy, J. A. and Rubinsten, C. B. (1966). "Hologram transmission via television," Bell System Technical Journal 45, pp. 333-335.

[8] Gabor, D. and Goss, P. (1966). "Interference microscope with total wavefront reconstruction," Journal of the Optical Society of America 56, pp. 849-858.

[9] Goodman, J. W. and Lawrence, R. W. (1967). "Digital image formation from electronically detected holograms," Applied Physics Letters 11, pp. 77-79.

[10] Guo, P. and Devaney, A. J. (2004), "Digital microscopy using phase-shifting digital holography with two reference waves," Optics Letters 29, pp. 857-859.

[11] Indebetouw, G., Kim, T., Poon, T.-C. and Schilling, B. W. (1998). "Three-dimensional location of fluorescent inhomogeneities in turbid media by scanning heterodyne holography," Optics Letters 23, pp. 133-137.

[12] Indebetouw, G. and Zhong, W. (2006). "Scanning holographic microscopy of three-dimensional holographic microscopy," Journal of the Optical Society of America A 23, pp. 2657-2661.

[13] Kim, H., Hahn, J. and Lee, B. (2008). "Mathematical modeling of triangle-mesh-modeled three-dimensional surface objects for digital holography," Applied Optics 47, pp. D117-D127.

[14] Kim, T. and Kim, T. (2020). "Coaxial scanning holography," Optics Letters 45, pp. 2046-2049.

[15] Liu, J.-P., Chen, W.-T., Wen, H.-H. and Poon, T.-C. (2020). "Recording of a curved digital hologram for orthoscopic real image reconstruction," Optics Letters 45, pp. 4353-4356.

[16] Liu, J.-P., Tahara, T., Hayasaki, Y. and Poon, T.-C. (2018). "Incoherent digital holography: a review," Applied Sciences 8, pp. 143.

[17] Liu, J.-P. and Poon, T.-C. (2009). "Two-step-only quadrature phase-shifting digital holography," Optics Letters 34, pp. 250-252.

[18] Liu, J.-P., Poon, T.-C., Jhou, G.-S. and Chen, P.-J. (2011). "Comparison of two-, three-, and four-exposure quadrature phase-shifting holography," Applied Optics 50, pp. 2443-2450.

[19] Liu, J.-P., Guo, C.-H., Hsiao, W.-J., Poon, T.-C. and Tsang, P. (2015). "Coherence experiments in single-pixel digital holography," Optics Letters 40, pp. 2366-2369.

[20] Lohmann, A. W. (1965). "Wavefront reconstruction for incoherent objects," Journal of the Optical Society of America 55, pp. 1555-1556.

[21] Meng, X. F., Cai, L. Z., Xu, X. F., Yang, X. L., Shen, X. X., Dong, G. Y. and Wang, Y. R. (2006). "Two-step phase-shifting interferometry and its application in image encryption," Optics Letters 31, pp. 1414-1416.

[22] Matsushima, K. (2020). *Introduction to Computer Holography Creating Computer-Generated Holograms as the Ultimate 3D Image*. Springer, Switzerland.

[23] Mertz, L. (1964). "Metallic beam splitters in interferometry," Journal of the Optical Society of America 54, AD3_7.

[24] Mertz, L. and Young, N. O. (1962). "Fresnel transformations of images," In Habell, K. J., Proceedings of the Conference on Optical Instruments and Techniques, pp. 305-310. Wiley and Sons, New York.

[25] Molesini, G., Bertani, D. and Cetca, M. (1982). "In-line holography with interference filters as Fourier processor," Optica Acta: International Journal of Optics 29, pp. 497-485.

[26] Phan, A.-H., Piao, M.-L., Gil, S.-K. and Kim, N. (2014a). "Generation speed and reconstructed image quality enhancement of a long-depth object using double wavefront recording planes and a GPU," Applied Optics 53, pp. 4817-4824.

[27] Phan, A.-H., Alam, M. A., Jeon, S.-H., Lee, J.-H. and Kim, N. (2014b). "Fast hologram generation of long-depth object using multiple wavefront recording planes," Proc. SPIE Vol. 9006, Practical Holography XXVIII: Materials and Applications, 900612.

[28] Poon, T.-C. (1985). "Scanning holography and two-dimensional image processing by acousto-optic two-pupil synthesis," Journal of the Optical Society of America A 2, pp. 521-527.

[29] Poon, T.-C. (2008). "Scan-free three-dimensional imaging," Nature Photonics 2 , pp. 131-132.

[30] Poon, T. C. and Korpel, A. (1979). "Optical transfer function of an acousto-optic heterodyning image processor," Optics Letters 4, pp. 317-319.

[31] Poon, T.-C. and Banerjee, P. P. (2001). *Contemporary Optical Image Processing with MATLAB®*. Elsevier, Oxford.

[32] Poon, T.-C. and Liu, J.-P. (2014). *Introduction to Modern Digital Holography with MATLAB*. Cambridge University Press, Cambridge.

[33] Poon, T.-C. (2007). *Optical Scanning Holography with MATLAB®*. Springer, New York.

[34] Popescu, G. (2011). *Quantitative Phase Imaging of Cells and Tissues*. Springer, New York.

[35] Rosen, J. and Brooker, G. (2007). "Digital spatially incoherent Fresnel holography," Optics Letters 32, pp. 912-914.

[36] Rosen, J., Anand, V., Rai, M. R., Mukherjee, S. and Bulbul, A. (2019). "Review of 3D imaging by coded aperture correlation holography (COACH)," Applied Sciences 9, pp. 605.

[37] Schilling, B. W. and Poon, T.-C. (1995). "Real-time preprocessing of holographic information," Optical Engineering 34, pp. 3174-3180.

[13] Kim, H., Hahn, J. and Lee, B. (2008). "Mathematical modeling of triangle-mesh-modeled three-dimensional surface objects for digital holography," Applied Optics 47, pp. D117-D127.

[14] Kim, T. and Kim, T. (2020). "Coaxial scanning holography," Optics Letters 45, pp. 2046-2049.

[15] Liu, J.-P., Chen, W.-T., Wen, H.-H. and Poon, T.-C. (2020). "Recording of a curved digital hologram for orthoscopic real image reconstruction," Optics Letters 45, pp. 4353-4356.

[16] Liu, J.-P., Tahara, T., Hayasaki, Y. and Poon, T.-C. (2018). "Incoherent digital holography: a review," Applied Sciences 8, pp. 143.

[17] Liu, J.-P. and Poon, T.-C. (2009). "Two-step-only quadrature phase-shifting digital holography," Optics Letters 34, pp. 250-252.

[18] Liu, J.-P., Poon, T.-C., Jhou, G.-S. and Chen, P.-J. (2011). "Comparison of two-, three-, and four-exposure quadrature phase-shifting holography," Applied Optics 50, pp. 2443-2450.

[19] Liu, J.-P., Guo, C.-H., Hsiao, W.-J., Poon, T.-C. and Tsang, P. (2015). "Coherence experiments in single-pixel digital holography," Optics Letters 40, pp. 2366-2369.

[20] Lohmann, A. W. (1965). "Wavefront reconstruction for incoherent objects," Journal of the Optical Society of America 55, pp. 1555-1556.

[21] Meng, X. F., Cai, L. Z., Xu, X. F., Yang, X. L., Shen, X. X., Dong, G. Y. and Wang, Y. R. (2006). "Two-step phase-shifting interferometry and its application in image encryption," Optics Letters 31, pp. 1414-1416.

[22] Matsushima, K. (2020). *Introduction to Computer Holography Creating Computer-Generated Holograms as the Ultimate 3D Image.* Springer, Switzerland.

[23] Mertz, L. (1964). "Metallic beam splitters in interferometry," Journal of the Optical Society of America 54, AD3_7.

[24] Mertz, L. and Young, N. O. (1962). "Fresnel transformations of images," In Habell, K. J., Proceedings of the Conference on Optical Instruments and Techniques, pp. 305-310. Wiley and Sons, New York.

[25] Molesini, G., Bertani, D. and Cetca, M. (1982). "In-line holography with interference filters as Fourier processor," Optica Acta: International Journal of Optics 29, pp. 497-485.

[26] Phan, A.-H., Piao, M.-L., Gil, S.-K. and Kim, N. (2014a). "Generation speed and reconstructed image quality enhancement of a long-depth object using double wavefront recording planes and a GPU," Applied Optics 53, pp. 4817-4824.

[27] Phan, A.-H., Alam, M. A., Jeon, S.-H., Lee, J.-H. and Kim, N. (2014b). "Fast hologram generation of long-depth object using multiple wavefront recording planes," Proc. SPIE Vol. 9006, Practical Holography XXVIII: Materials and Applications, 900612.

[28] Poon, T.-C. (1985). "Scanning holography and two-dimensional image processing by acousto-optic two-pupil synthesis," Journal of the Optical Society of America A 2, pp. 521-527.

[29] Poon, T.-C. (2008). "Scan-free three-dimensional imaging," Nature Photonics 2 , pp. 131-132.

[30] Poon, T. C. and Korpel, A. (1979). "Optical transfer function of an acousto-optic heterodyning image processor," Optics Letters 4, pp. 317-319.

[31] Poon, T.-C. and Banerjee, P. P. (2001). *Contemporary Optical Image Processing with MATLAB®*. Elsevier, Oxford.

[32] Poon, T.-C. and Liu, J.-P. (2014). *Introduction to Modern Digital Holography with MATLAB.* Cambridge University Press, Cambridge.

[33] Poon, T.-C. (2007). *Optical Scanning Holography with MATLAB®*. Springer, New York.

[34] Popescu, G. (2011). *Quantitative Phase Imaging of Cells and Tissues.* Springer, New York.

[35] Rosen, J. and Brooker, G. (2007). "Digital spatially incoherent Fresnel holography," Optics Letters 32, pp. 912-914.

[36] Rosen, J., Anand, V., Rai, M. R., Mukherjee, S. and Bulbul, A. (2019). "Review of 3D imaging by coded aperture correlation holography (COACH)," Applied Sciences 9, pp. 605.

[37] Schilling, B. W. and Poon, T.-C. (1995). "Real-time preprocessing of holographic information," Optical Engineering 34, pp. 3174-3180.

[38] Schilling, B. W., Poon, T.-C., Indebetouw, G., Storrie, B., Shinoda, K. and Wu, M. (1997). "Three-dimensional holographic fluorescence microscopy," Optics Letters 22, pp. 1506-1508.

[39] Schnars, U. and Jueptner, W. (2005). *Digital Holography: Digital Hologram Recording, Numerical Reconstruction, and Related Techniques.* Springer, Berlin.

[40] Shimobaba, T. and Tto, T. (2019). *Computer Holography Acceleration Algorithms and Hardware Implementations.* CRC Press, Tayler & Francis Group, New York.

[41] Shimobaba, T., Masuda, N. and Ito, T. (2009). "Simple and fast calculation algorithm for computer-generated hologram with wavefront recording plane," Optics Letters 34, pp. 3133-3135.

[42] Shimobaba, T., Nakayama, H., Masuda, N. and Ito, T. (2010). "Rapid calculation algorithm of Fresnel computer-generated-hologram using look-up table and wavefront-recording plane methods for three-dimensional display," Optics Express 18, pp. 19504-19509.

[43] Simpson, R. G., Barrett, H. H., Subach, J. A. and Fisher, H. D. (1975). "Digital processing of annular coded-aperture imagery," Optical Engineering 14, pp. 490-494.

[44] Tsai, C.-M., Sie, H.-Y., Poon, T.-C. and Liu, J.-P. (2021). "Optical scanning holography with a polarization directed flat lens," Applied Optics 60, pp. B113-B118.

[45] Tsang, P. W. M. and Poon, T.-C. (2013). "Review on theory and applications of wavefront recording plane framework in generation and processing of digital holograms," Chinese Optics Letters 11, 010902.

[46] Tsang, P., Cheung, W.-K., Poon, T.-C. and Zhou, C. (2011). "Holographic video at 40 frames per second for 4-million object points," Optics Express 19, pp. 15205-15211.

[47] Tsang, P. W. M., Poon, T.-C. and Wu, Y. M. (2018). "Review of fast methods for point-based computer-generated holography[Invited]," Photonics Research 6, pp. 837-846.

[48] Vijayakumar, A., Kashter, Y., Kelner, R. and Rosen, J. (2016). "Coded aperture correlation holography - a new type of incoherent digital holograms," Optics Express 24, pp. 12430-12441.

[49] Wang, Y., Zhen, Y., Zhang, H. and Zhang, Y. (2004). "Study on digital holography with single phase-shifting operation," Chinese Optics Letters 2, pp. 141-143.

[50] Wu, J., Zhang, H., Zhang, W., Jin, G., Cao, L. and Barbastathis, G. (2020). "Single-shot lensless imaging with Fresnel zone aperture and incoherent illumination," Light: Science & Applications 9, pp. 53.

[51] Yeom, H.-J. and Park, J. H. (2016). "Calculation of reflectance distribution using angular spectrum convolution in mesh-based computer-generated hologram," Optics Express 24, pp. 19801-19813.

[52] Yoneda, N., Satia, Y. and Nomura, T. (2020). "Motionless optical scanning holography," Optics Letters 45, pp. 3184-3187.

[53] Yoneda, N., Satia, Y. and Nomura, T. (2020). "Spatially divided phase-shifting motionless optical scanning holography," OSA Continuum 3, pp. 3523-3555.

[54] Yoshikawa, H., Yamaguchi, T. and Kitayama, R. (2009). "Real-time generation of full color image hologram with compact distance look-up table," In Digital Holography and Three-Dimensional Imaging, OSA Technical Digest (CD) (Optical Society of America, 2009), paper DWC4.

[55] Zhang, Y., Poon, T.-C., Tsang, P. W. M., Wang, R. and Wang, L. (2019). "Review on feature extraction for 3-D incoherent image processing using optical scanning holography," IEEE Transactions on Industrial Informatics 15, pp. 6146-6154.

[56] Zhang, Y., Wang, F., Poon, T.-C., Fan, S. and Xu, W. (2018). "Fast generation of full analytical polygon-based computer-generated holograms," Optics Express 26, pp. 19206-19224.

[57] Zhang, Y., Wang, R., Tsang, P. W. M. and Poon, T.-C. (2020). "Sectioning with edge extraction in optical incoherent imaging processing," OSA Continuum 3, pp. 698-708.

[58] Zhang, Y., Fan, H., Wang, F., Gu, X., Qian, X. and Poon, T.-C. (2022). "Polygon-based computer-generated holography: a review of fundamentals and recent progress [Invited]," Applied Optics 61, pp. B363-B374.

[38] Schilling, B. W., Poon, T.-C., Indebetouw, G., Storrie, B., Shinoda, K. and Wu, M. (1997). "Three-dimensional holographic fluorescence microscopy," Optics Letters 22, pp. 1506-1508.

[39] Schnars, U. and Jueptner, W. (2005). *Digital Holography: Digital Hologram Recording, Numerical Reconstruction, and Related Techniques.* Springer, Berlin.

[40] Shimobaba, T. and Tto, T. (2019). *Computer Holography Acceleration Algorithms and Hardware Implementations.* CRC Press, Tayler & Francis Group, New York.

[41] Shimobaba, T., Masuda, N. and Ito, T. (2009). "Simple and fast calculation algorithm for computer-generated hologram with wavefront recording plane," Optics Letters 34, pp. 3133-3135.

[42] Shimobaba, T., Nakayama, H., Masuda, N. and Ito, T. (2010). "Rapid calculation algorithm of Fresnel computer-generated-hologram using look-up table and wavefront-recording plane methods for three-dimensional display," Optics Express 18, pp. 19504-19509.

[43] Simpson, R. G., Barrett, H. H., Subach, J. A. and Fisher, H. D. (1975). "Digital processing of annular coded-aperture imagery," Optical Engineering 14, pp. 490-494.

[44] Tsai, C.-M., Sie, H.-Y., Poon, T.-C. and Liu, J.-P. (2021). "Optical scanning holography with a polarization directed flat lens," Applied Optics 60, pp. B113-B118.

[45] Tsang, P. W. M. and Poon, T.-C. (2013). "Review on theory and applications of wavefront recording plane framework in generation and processing of digital holograms," Chinese Optics Letters 11, 010902.

[46] Tsang, P., Cheung, W.-K., Poon, T.-C. and Zhou, C. (2011). "Holographic video at 40 frames per second for 4-million object points," Optics Express 19, pp. 15205-15211.

[47] Tsang, P. W. M., Poon, T.-C. and Wu, Y. M. (2018). "Review of fast methods for point-based computer-generated holography [Invited]," Photonics Research 6, pp. 837-846.

[48] Vijayakumar, A., Kashter, Y., Kelner, R. and Rosen, J. (2016). "Coded aperture correlation holography - a new type of incoherent digital holograms," Optics Express 24, pp. 12430-12441.

[49] Wang, Y., Zhen, Y., Zhang, H. and Zhang, Y. (2004). "Study on digital holography with single phase-shifting operation," Chinese Optics Letters 2, pp. 141-143.

[50] Wu, J., Zhang, H., Zhang, W., Jin, G., Cao, L. and Barbastathis, G. (2020). "Single-shot lensless imaging with Fresnel zone aperture and incoherent illumination," Light: Science & Applications 9, pp. 53.

[51] Yeom, H.-J. and Park, J. H. (2016). "Calculation of reflectance distribution using angular spectrum convolution in mesh-based computer-generated hologram," Optics Express 24, pp. 19801-19813.

[52] Yoneda, N., Satia, Y. and Nomura, T. (2020). "Motionless optical scanning holography," Optics Letters 45, pp. 3184-3187.

[53] Yoneda, N., Satia, Y. and Nomura, T. (2020). "Spatially divided phase-shifting motionless optical scanning holography," OSA Continuum 3, pp. 3523-3555.

[54] Yoshikawa, H., Yamaguchi, T. and Kitayama, R. (2009). "Real-time generation of full color image hologram with compact distance look-up table," In Digital Holography and Three-Dimensional Imaging, OSA Technical Digest (CD) (Optical Society of America, 2009), paper DWC4.

[55] Zhang, Y., Poon, T.-C., Tsang, P. W. M., Wang, R. and Wang, L. (2019). "Review on feature extraction for 3-D incoherent image processing using optical scanning holography," IEEE Transactions on Industrial Informatics 15, pp. 6146-6154.

[56] Zhang, Y., Wang, F., Poon, T.-C., Fan, S. and Xu, W. (2018). "Fast generation of full analytical polygon-based computer-generated holograms," Optics Express 26, pp. 19206-19224.

[57] Zhang, Y., Wang, R., Tsang, P. W. M. and Poon, T.-C. (2020). "Sectioning with edge extraction in optical incoherent imaging processing," OSA Continuum 3, pp. 698-708.

[58] Zhang, Y., Fan, H., Wang, F., Gu, X., Qian, X. and Poon, T.-C. (2022). "Polygon-based computer-generated holography: a review of fundamentals and recent progress [Invited]," Applied Optics 61, pp. B363-B374.

第七章　空间光调制器光信息处理

在现代光学处理和显示应用中, 对实时器件的需求越来越大, 这类器件一般被称为空间光调制器 (*spatial light modulator*, SLM)。典型的空间光调制器有声光调制器 (*acousto-optic modulator*, AOM), 电光调制器 (*electro-optic modulator*, EOM) 和液晶显示器 (*liquid crystal display*)。本章将集中讨论这些类型的调制器。

7.1　基于声光调制信息处理

本节中, 首先讨论声光效应. 然后将介绍它在激光束调制中的一些应用, 如频率调制 (*frequency modulation*) 和强度调制 (*intensity modulation*). 最后讨论激光束的偏转 (*laser beam deflection*) 和外差 (*heterodyning*).

7.1.1　声光效应

在声光学中, 主要研究声与光之间的相互作用. 由于声光效应, 光波可以被声音调制, 这为光信息处理提供了有力的手段.

声光调制器 (*acousto-optic modulator*, AOM) 或通常被称为布拉格单元 (*Bragg cell*), 是一种空间光调制器, 它由透明的声光介质 (如致密玻璃) 与压电式换能器 (*piezoelectric transducer*) 构成. 在电源的驱动下, 换能器将声波发射到声介质中. 当声波作用到声光介质中时, 相当于在声光介质中有一个移动的有效空间周期等于声波波长 Λ 的相位光栅, 这使得声光介质的折射率会发生周期性的疏密变化, 从而将入射的平面光波衍射为多级衍射光. 这种情况如图 7.1-1 所示, 其中 L 表示换能器的长度.

相位光栅法 (*phase grating approach*)

一种简单的方法是把声光效应看作光通过衍射光栅的衍射. 衍射光栅 (*diffraction grating*) 是一种具有周期性结构的光学元件, 它将光衍射成沿不同方向传播的多束光. 考虑一个透过率函数为 $t(x)$ 的理想薄光栅 [见式 (3.4-2a)], 对于以角度 ϕ_{inc} 入射的单位振幅平面波, 如图 7.1-1 所示, 有

$$\psi_p(x; z = 0^+) = \psi_p(x; z = 0^-)t(x), \tag{7.1-1}$$

其中, $\psi_p(x; z = 0^-) = \mathrm{e}^{-\mathrm{j}k_{0x}x - k_{0z}z}\big|_{z=0^-} = \mathrm{e}^{-\mathrm{j}k_0 \sin\phi_{inc}x}$. 如图 7.1-1 所示, 按照惯例, 角度的测量从 z 轴开始, 沿逆时针方向为正. 由于光栅在 x 方向上呈周期性结构, 其透过率函数可以用傅里叶级数给出 [见式 (2.2-1)]

$$t(x) = \sum_{m=-\infty}^{\infty} T_m \mathrm{e}^{-\mathrm{j}m\frac{2\pi}{\Lambda}x}, \tag{7.1-2}$$

7 Spatial Light Modulators for Processing Optical Information

In modern optical processing and display applications, there are increased needs of a real-time device and such a device is called a spatial light modulator (SLM). Typical examples of SLMs are *acousto-optic modulators* (AOMs), *electro-optic modulators* (EOMs), and *liquid crystal displays.* In this chapter, we will concentrate on these types of modulators.

7.1 Information Processing with Acousto-Optic Modulators

In this section, we first discuss the acousto-optic effect. We will then cover some of its applications in laser beam modulation such as *frequency modulation* and *intensity modulation.* We conclude the section with discussions on *laser beam deflection* and *heterodyning.*

7.1.1 The Acousto-Optic Effect

In *acousto-optics*, we deal with the interaction between sound and light. As a result of the acousto-optic effect, light waves can be modulated by sound, which provides a powerful means for optical processing information.

An *acousto-optic modulator* (AOM) or commonly known as *Bragg cell* is a spatial light modulator that consists of a transparent acoustic medium (such as dense glass) to which a piezoelectric transducer is bonded. Driven by an electrical source, the transducer launches sound waves into the acoustic medium. Periodic compression and rarefaction of the index of refraction of the acoustic medium due to the applied sound waves acts as a moving phase grating with an effective spatial period equal to the wavelength Λ of the sound in the acoustic medium to diffract the incident plane wave of light into multiple diffracted orders. The situation is illustrated in Figure 7.1-1, where L denotes the length of the transducer.

Phase Grating Approach

A simple approach is to consider the acousto-optic effect as diffraction of light by a diffraction grating. A *diffraction grating* is an optical element with a periodic structure that diffracts light into multiple beams propagating in different directions. Consider an ideal thin grating with transparency function [see Eq. (3.4-2a)] given by $t(x)$. For a unit-amplitude plane wave incident at angle ϕ_{inc}, as shown in Figure 7.1-1, we have

$$\psi_p(x; z = 0^+) = \psi_p(x; z = 0^-)t(x), \tag{7.1-1}$$

where $\psi_p(x; z = 0^-) = \mathrm{e}^{-\mathrm{j}k_{0x}x - k_{0z}z}\Big|_{z=0^-} = \mathrm{e}^{-\mathrm{j}k_0 \sin\phi_{inc}x}$. The convention for angles is counterclockwise positive, measured from the z-axis as shown in Figure 7.1-1. Since the grating is a periodic structure in the x-direction, its transparency function can be described using a Fourier series [see Eq. (2.2-1)] given by

$$t(x) = \sum_{m=-\infty}^{\infty} T_m \mathrm{e}^{-\mathrm{j}m\frac{2\pi}{\Lambda}x}, \tag{7.1-2}$$

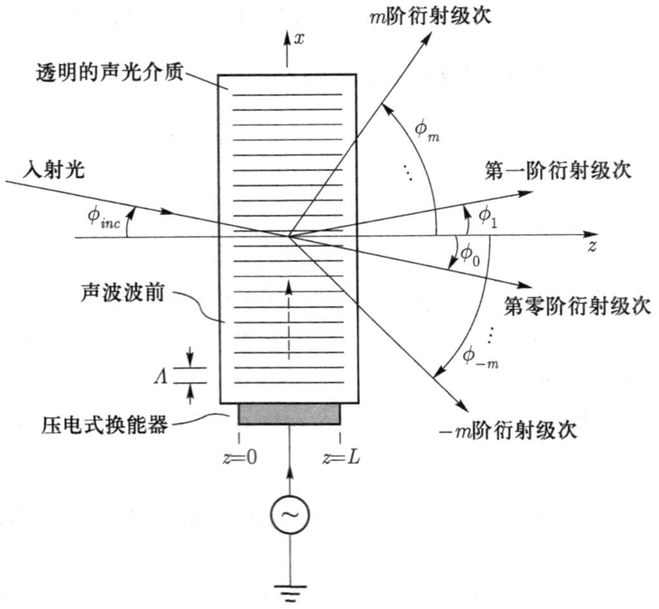

图 7.1-1　声光调制器, 声波作用产生衍射的说明 [Poon 和 Kim (2018)]

其中, T_m 为傅里叶系数, 该傅里叶系数由光栅的类型而定. 可以看出, 指数中使用 "–j" 的规定, 是针对空间函数的 (该规定见 2.3.1 节的讨论). 例如, 对于由大量平行狭缝组成的衍射光栅, 其 T_m 就是 $sinc$ 函数的形式 (见图 2.2-1 中脉冲序列的傅里叶级数和图 2.2-4 中脉冲序列的幅度谱).

紧靠光栅右侧的复振幅变为

$$\psi_p(x;z=0^+) = \mathrm{e}^{-\mathrm{j}k_0\sin\phi_{inc}x}\sum_{m=-\infty}^{\infty}T_m\mathrm{e}^{-\mathrm{j}m\frac{2\pi}{\Lambda}x}$$
$$= \sum_{m=-\infty}^{\infty}T_m\mathrm{e}^{-\mathrm{j}(k_0\sin\phi_{inc}+m\frac{2\pi}{\Lambda})x} = \sum_{m=-\infty}^{\infty}T_m\mathrm{e}^{-\mathrm{j}k_{xm}x},$$

其中

$$k_{xm} = k_0\sin\phi_{inc} + m\frac{2\pi}{\Lambda}. \tag{7.1-3}$$

为了求出距离 z 处的光场分布, 即 $\psi_p(x;z)$, 在 $k_0^2 = k_{xm}^2 + k_{0z}^2$ 的条件下加入相位因子 $\exp(-\mathrm{j}k_{0z}z)$, 进一步注意到, 当波矢量分解为 x 和 z 分量时, 通过这个关系给出相对于 z 轴的传播角度 ϕ_m

$$\sin\phi_m = \frac{k_{xm}}{k_0}, \tag{7.1-4}$$

且 $k_{0z} = k_0\cos\phi_m$. 结合式 (7.1-3) 和式 (7.1-4), 有

$$k_0\sin\phi_m = k_0\sin\phi_{inc} + m\frac{2\pi}{\Lambda},$$

且当 $k_0 = 2\pi/\lambda_0$ 时, 上式变为

$$\sin\phi_m = \sin\phi_{inc} + m\frac{\lambda_0}{\Lambda},\quad m = 0, \pm1, \pm2, \cdots. \tag{7.1-5}$$

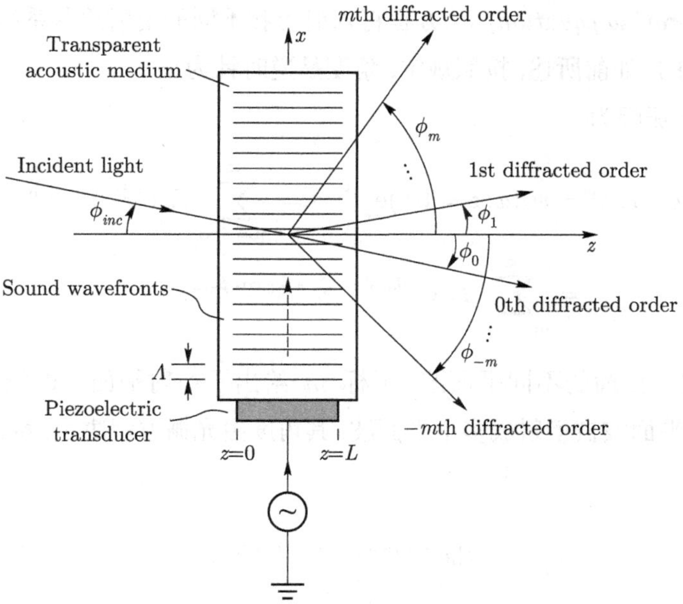

Figure 7.1-1 Acousto-optic modulator, illustrating diffraction by sound. Adapted from Poon and Kim (2018)

where T_m are the Fourier series coefficients depending on the type of grating under consideration. Note that the convention of using "−j" in the exponent is for spatial functions (see discussion in Section 2.3.1 for conventions). For example, for a diffraction grating composed of a large number of parallel slits, its T_m is of the functional form of a *sinc* function (see the Fourier series of a pulse train in Figure 2.2-1 and its spectrum in Figure 2.2-4).

The complex field immediately to the right of the grating now becomes

$$\psi_p(x; z = 0^+) = \mathrm{e}^{-\mathrm{j}k_0 \sin\phi_{inc} x} \sum_{m=-\infty}^{\infty} T_m \mathrm{e}^{-\mathrm{j}m\frac{2\pi}{\Lambda}x}$$

$$= \sum_{m=-\infty}^{\infty} T_m \mathrm{e}^{-\mathrm{j}(k_0 \sin\phi_{inc} + m\frac{2\pi}{\Lambda})x} = \sum_{m=-\infty}^{\infty} T_m \mathrm{e}^{-\mathrm{j}k_{xm}x},$$

where

$$k_{xm} = k_0 \sin\phi_{inc} + m\frac{2\pi}{\Lambda}. \tag{7.1-3}$$

To find the field distribution a distance of z away, that is, $\psi_p(x; z)$, we add phase factor $\exp(-\mathrm{j}k_{0z}z)$ under the condition that $k_0^2 = k_{xm}^2 + k_{0z}^2$. Note further that the wave vector, when resolved into its x-and z-components, gives the angle of propagation ϕ_m with respect to the z-axis through the relationship

$$\sin\phi_m = \frac{k_{xm}}{k_0}, \tag{7.1-4}$$

and $k_{0z} = k_0 \cos\phi_m$. Combining Eqs. (7.1-3) and (7.1-4), we have

$$k_0 \sin\phi_m = k_0 \sin\phi_{inc} + m\frac{2\pi}{\Lambda},$$

and with $k_0 = 2\pi/\lambda_0$, the above equation becomes

$$\sin\phi_m = \sin\phi_{inc} + m\frac{\lambda_0}{\Lambda}, \quad m = 0, \pm1, \pm2, \cdots. \tag{7.1-5}$$

这就是光栅方程 (*grating equation*), 该方程将入射角和不同的衍射角联系起来, 所有的角度都是从横轴开始测量的, 如前所述, 按照规定, 角度是逆时针为正.

因此, z 处的复振幅为

$$\psi_p(x;z)=\psi_p(x;z=0^+)\mathrm{e}^{-\mathrm{j}k_{0z}z}=\sum_{m=-\infty}^{\infty}T_m\mathrm{e}^{-\mathrm{j}k_{xm}x-\mathrm{j}k_{0z}z}$$
$$=\sum_{m=-\infty}^{\infty}T_m\mathrm{e}^{-\mathrm{j}k_0\sin\phi_m x-\mathrm{j}k_0\cos\phi_m z}. \tag{7.1-6}$$

这表示在不同方向上传播的不同平面波. 下标 m 给出了衍射平面波的级次, 物理情况如图 7.1-1 所示, 入射的平面波被衍射成多个平面波, 其角度由光栅方程决定. 特别地, 当 $m=0$ 时, 零级光由下式给出

$$T_0\mathrm{e}^{-\mathrm{j}k_0\sin\phi_0 x-\mathrm{j}k_0\cos\phi_0 z},$$

其中, 根据式 (7.1-5), $\phi_0=\phi_{inc}$, 且该光束与入射的平面波共线.

粒子法 (*particle approach*)

另一种自然且更为准确地考虑平面光波和声波之间相互作用的方法, 即如图 7.1-1 所示, 可以假设 L 足够长, 以产生平的波前传播到声介质. 平面波具有明确定义的动量, 动量为 p 的粒子与其波长之间的关系为 $\lambda=h/p$ (见 1.2 节). 那么, 可以把声波和光波的相互作用看作两个粒子的碰撞, 这样使其相互作用并产生第三个粒子. 光子 (*photon*) 是光的粒子, 而声子 (*phonon*) 是声音的粒子.

在粒子碰撞的过程中, 必须遵守两个守恒定律, 即能量守恒和动量守恒 (*conservation of energy and momentum*). 如果分别用 $\boldsymbol{k}_0$、$\boldsymbol{k}_{+1}$ 和 $\boldsymbol{K}$ 表示入射平面光波、衍射平面光波和声介质中的声平面波的波矢 (*wavevector*) [也被称为传播矢量 (*propagation vector*)], 则动量守恒 (*conservation of momentum*) 的条件可写成

$$\hbar\boldsymbol{k}_{+1}=\hbar\boldsymbol{k}_0+\hbar\boldsymbol{K}, \tag{7.1-7}$$

其中, $\hbar=h/(2\pi)$, 且 h 表示普朗克常量, 将上式除以 $\hbar$, 可得

$$\boldsymbol{k}_{+1}=\boldsymbol{k}_0+\boldsymbol{K}. \tag{7.1-8}$$

相应的能量守恒(*conservation of energy*) 形式为 (除以 $\hbar$ 之后)

$$\omega_{+1}=\omega_0+\varOmega, \tag{7.1-9}$$

其中, ω_0、$\varOmega$ 和 ω_{+1} 分别为入射光、声波和衍射光的角频率. 式 (7.1-8) 和式 (7.1-9) 所描述的相互作用被称为上移相互作用 (*upshifted interaction*). 图 7.1-2a) 为波矢相互作用图. 因为对于所有的实际情况, $|\boldsymbol{K}|\ll|\boldsymbol{k}_0|$, $\boldsymbol{k}_{+1}$ 的大小基本上等于 $\boldsymbol{k}_0$ 的大小, 因此, 所示的波矢动量三

which is known as the *grating equation.* The equation relates the incident angle to the different diffracted angles. All the angles are measured from the horizontal axis and the convention for angles is counterclockwise positive as previously mentioned.

Therefore, the complex field at z is

$$\psi_p(x;z) = \psi_p(x;z=0^+)\mathrm{e}^{-\mathrm{j}k_{0z}z} = \sum_{m=-\infty}^{\infty} T_m \mathrm{e}^{-\mathrm{j}k_{xm}x-\mathrm{j}k_{0z}z}$$
$$= \sum_{m=-\infty}^{\infty} T_m \mathrm{e}^{-\mathrm{j}k_0 \sin\phi_m x-\mathrm{j}k_0\cos\phi_m z}. \tag{7.1-6}$$

This represents different plane waves travelling at different directions. The subscript m gives the order of the diffracted plane wave. The physical situation is shown in Figure 7.1-1, where the incident plane wave is diffracted into many plane waves with their angles governed by the grating equation. In particular, for $m = 0$, that is, the zeroth-order beam is given by

$$T_0 \mathrm{e}^{-\mathrm{j}k_0 \sin\phi_0 x-\mathrm{j}k_0\cos\phi_0 z},$$

where $\phi_0 = \phi_{inc}$ according to Eq. (7.1-5), and the beam is collinear with the incident plane wave.

Particle Approach

Another instinctive and more accurate approach considers the interaction of plane waves of light and sound, that is, we assume that L is sufficiently long in order to produce straight wave fronts propagating into the acoustic medium, as illustrated in Figure 7.1-1. Plane waves have well-defined momenta and particles with momentum p are associated with waves whose wavelength is $\lambda = h/p$ (see Section 1.2). We then can consider the interaction of plane waves of sound and light as a collision of two particles such that they interact and generate a third particle. While *photons* are particles of light, *phonons* are particles of sound.

In the process of particle collision, two conservation laws have to be obeyed, namely the *conservation of energy and momentum.* If we denote the *wavevectors* (also called the *propagation vectors*) of the incident plane wave of light, diffracted plane wave of light, and sound plane wave in the acoustic medium by $\boldsymbol{k}_0$, $\boldsymbol{k}_{+1}$, and $\boldsymbol{K}$, respectively, we can write the condition for *conservation of momentum* as

$$\hbar\boldsymbol{k}_{+1} = \hbar\boldsymbol{k}_0 + \hbar\boldsymbol{K}, \tag{7.1-7}$$

where $\hbar = h/(2\pi)$ and h denotes Planck's constant. Dividing Eq. (7.1-7) by $\hbar$ leads to

$$\boldsymbol{k}_{+1} = \boldsymbol{k}_0 + \boldsymbol{K}. \tag{7.1-8}$$

The corresponding *conservation of energy* takes the form (after division by $\hbar$)

$$\omega_{+1} = \omega_0 + \Omega, \tag{7.1-9}$$

where ω_0, Ω, and ω_{+1} are the radian frequencies of the incident light, sound, and diffracted light, respectively. The interaction described by Eqs. (7.1-8) and (7.1-9) is called the *upshifted interaction.* Figure 7.1-2a shows the wavevector interaction diagram. Because for all practical cases $|\boldsymbol{K}| \ll |\boldsymbol{k}_0|$, the magnitude of $\boldsymbol{k}_{+1}$ is essentially equal to that of $\boldsymbol{k}_0$ and therefore the wavevector momentum triangle shown is nearly isosceles. Figure 7.1-2b describes the

角形几乎是等腰的. 图 7.1-2b) 描述了频率上移的衍射光束. 零级光 (0th-*order beam*) 是沿着与入射光相同的方向传播的光束, 而正一级衍射光 ($+1^{st}$-*order diffracted beam*) 是频率随声频上移了 Ω 的光束.

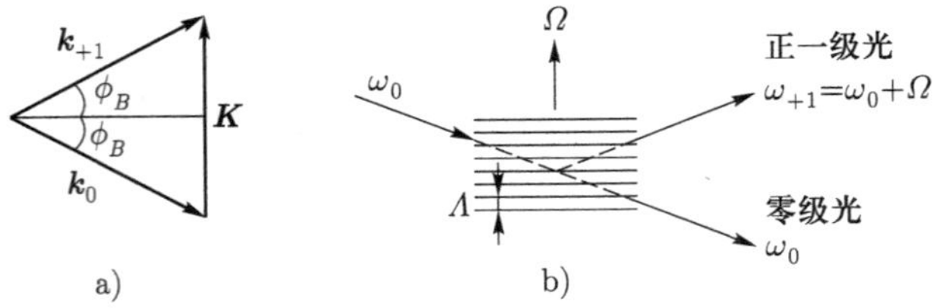

图 7.1-2 频率上移衍射: a) 波矢图; b) 实验结构

现在假设交换入射光和衍射光的方向. 可再次应用守恒定律得到类似式 (7.1-8) 和式 (7.1-9) 的两个方程. 这两个描述下移相互作用 (*downshifted interaction*) 的方程为

$$\boldsymbol{k}_{-1} = \boldsymbol{k}_0 - \boldsymbol{K}, \tag{7.1-10}$$

和

$$\omega_{-1} = \omega_0 - \Omega. \tag{7.1-11}$$

其中, 下标 -1 表示其相互作用为下移. 图 7.1-3a) 和图 7.1-3b) 分别阐述了式 (7.1-10) 和式 (7.1-11), 可以看出, 负一级衍射光 (-1st-*order diffracted beam*) 的频率下降了 Ω.

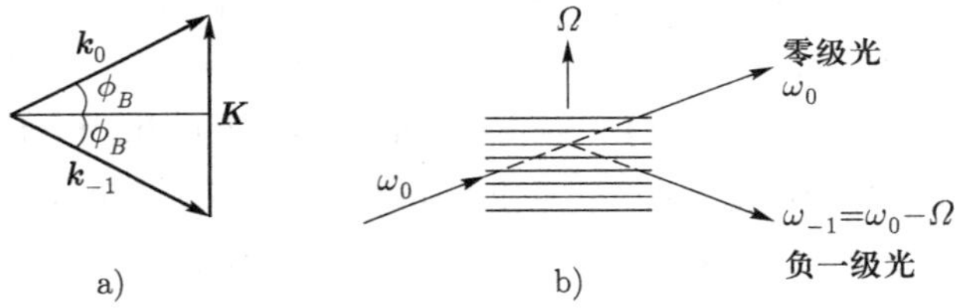

图 7.1-3 频率下移衍射: a) 波矢图; b) 实验结构

可以发现, 图 7.1-2a) 和图 7.1-3a) 中表示的两种情况下相互作用的波矢图都是闭合的, 该闭合示意图表明, 只有一个临界入射角 [布拉格角 (*Bragg angle*)], 才能使平面光波和声波发生相互作用. 从图 7.1-2a) 或图 7.1-3a) 可以发现, 布拉格角 ϕ_B 由下式给出

$$\sin\phi_B = \frac{|\boldsymbol{K}|}{2|\boldsymbol{k}_0|} = \frac{K}{2k_0} = \frac{\lambda_0}{2\Lambda}, \tag{7.1-12}$$

其中, λ_0 为声介质中光的波长, Λ 为声波的波长. 对于商用的声光调制器 (AOM-40 IntraAction 公司), 在其声波的工作频率 $f_s = 40$ MHz, 声波在玻璃中以 $v_s \sim 4000$ m/s 速度传播, 该声波波长 $\Lambda = \dfrac{v_s}{f_s} \sim 0.1$ mm, 如果使用一台氦氖激光器 (空气中其波长为 0.6328 μm), 则在 $n_0 = 1.69$ 的玻璃内部的波长为 $\lambda_0 \sim 0.6328\ \mu\text{m}/n_0 \sim 0.3743\ \mu\text{m}$. 因此, 根据式 (7.1-12), 声介质内部的布拉格角为 $\sim 1.9 \times 10^{-3}$ rad, 或是约为 0.1°; 因此, 在讨论声光效应时, 大多数情况下都做小角度近似. 图 7.1-4 给出了一个工作在 40 MHz 的典型的声光调制器. 在图中,

diffracted beam being upshifted in frequency. The zeroth-order beam is the beam travelling along the same direction of the incident beam, while the +1st-order diffracted beam is the beam with frequency upshifted by the sound frequency Ω.

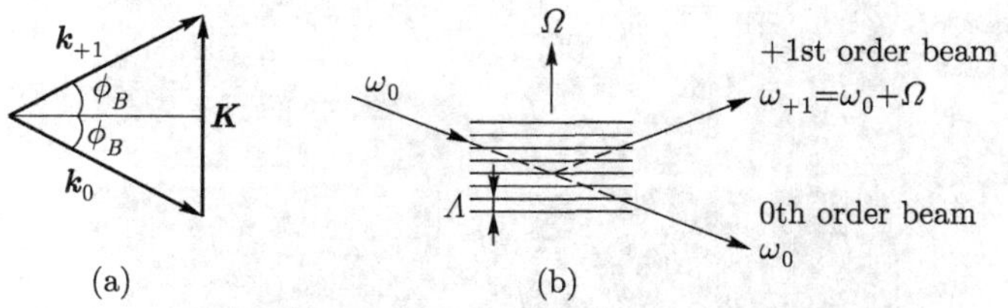

Figure 7.1-2 Upshifted diffraction: (a) wavevector diagram and (b) experimental configuration

Suppose now that we exchange the directions of incident and diffracted light. The conservation laws can be applied again to obtain two equations similar to Eqs. (7.1-8) and (7.1-9). The two equations describing the *downshifted interaction* are

$$\boldsymbol{k}_{-1} = \boldsymbol{k}_0 - \boldsymbol{K}, \tag{7.1-10}$$

and

$$\omega_{-1} = \omega_0 - \Omega, \tag{7.1-11}$$

where the subscript -1s indicate the interaction is downshifted. Figure 7.1-3a and b illustrates Eqs. (7.1-10) and (7.1-11), respectively. Note that the-1st-order diffracted beam is downshifted in frequency by Ω.

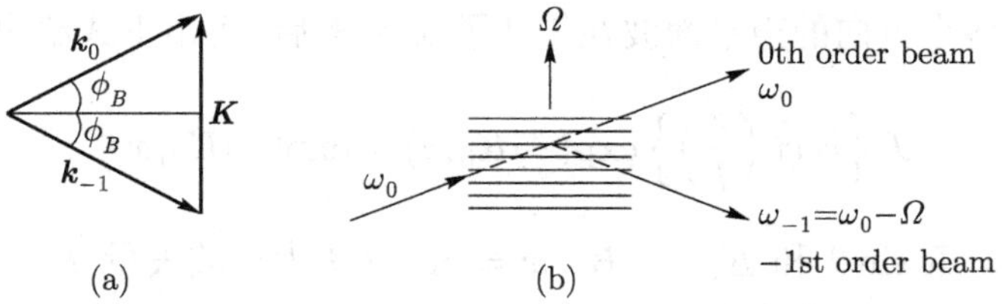

Figure 7.1-3 Downshifted diffraction: (a) wavevector diagram and (b) experimental configuration

We have seen that the wavevector diagrams in Figures 7.1-2a and 7.1-3a are closed for both cases of interaction. The closed diagrams stipulate that there can only one critical incident angle (*Bragg angle*) such that plane waves of light and sound can interact. From Figures 7.1-2a or 7.1-3a, we find that the Bragg angle ϕ_B is given by

$$\sin\phi_B = \frac{|\boldsymbol{K}|}{2|\boldsymbol{k}_0|} = \frac{K}{2k_0} = \frac{\lambda_0}{2\Lambda}, \tag{7.1-12}$$

where λ_0 is the wavelength of light inside the acoustic medium, and Λ is the wavelength of sound. For a commercially available acousto-optic modulator (AOM-40 IntraAction Corporation) operating at sound frequency f_s = 40 MHz and for sound wave traveling in the glass at velocity $v_s \sim 4000$ m/s, the sound wavelength $\Lambda = v_s/f_s \sim 0.1$ mm. If a He-Ne laser is used (its wavelength is about 0.6328 μm in air), the wavelength inside glass is $\lambda_0 \sim 0.6328$ μm$/n_0 \sim 0.3743$ μm with $n_0 = 1.69$. Hence, the Bragg angle, according to Eq. (7.1-12), inside the acoustic medium is $\sim 1.9\times10^{-3}$ radian or about 0.1 degree; therefore, we make small angle approximations in most of the situations in the discussion of the acousto-optic effect. Figure 7.1-4 shows a typical acousto-optic modulator operating at 40 MHz. In the figure, we denote the two diffracted laser spots at a far background. The incident laser

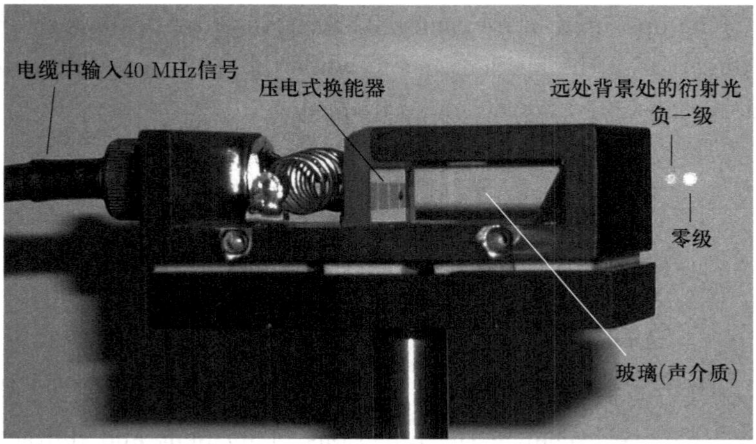

图 7.1-4 IntraAction 公司的典型的声光调制器, AOM-40 型号 [摘自 Poon (2002)]

显示了在远处背景上的两个衍射的激光亮斑. 入射激光束 (不可见, 因为它穿过的是透明的玻璃介质) 沿着换能器长边一侧传播通过玻璃.

由图 7.1-2a) 和图 7.1-3a) 中的闭合图可知, 在声介质中, 平面声波和光波的相互作用存在一定的临界入射角 ($\phi_{inc} = \pm\phi_B$), 且其入射光和散射光的方向在角度上相差 $2\phi_B$. 实际上, 即使入射光的方向不完全是布拉格角, 散射也会发生, 但其最大的散射强度会发生在布拉格角处. 其原因是, 对于有限长度的声换能器, 不再有平面波前, 而实际是声波在传播到介质时发生衍射, 随着换能器长度 L 的减小, 声波场将越来越不像一个平面波, 实际上, 现在更适合用声波的平面波角谱的概念来考虑. 假设一个沿 z 方向长度为 L 的换能器, 如图 7.1-1 所示定义, 根据式 (3.4-9), 各个方向的声平面波在传播了 x 距离后, 可由下式给出

$$\mathcal{F}\left\{rect\left(\frac{z}{L}\right)\right\}\exp(-\mathrm{j}K_{0z}z)\times\exp(-\mathrm{j}K_{0x}x),$$

其中, $K_{0z} = \boldsymbol{K}\cdot\hat{\boldsymbol{z}} = K\sin\theta$ 和 $K_{0x} = \boldsymbol{K}\cdot\hat{\boldsymbol{x}} = K\cos\theta$ 为声波矢量 $\boldsymbol{K}$ 的分量, 且 θ 是从 x 轴开始测量的. 故, 忽略某常数, 上式变为

$$\begin{aligned} & sinc\left(\frac{K_{0z}L}{2\pi}\right)\bigg|_{K_{0z}=K\sin\theta}\exp(-\mathrm{j}K\sin\theta\, z-\mathrm{j}K\cos\theta\, x) \\ & \propto sinc\left(\frac{K\sin\theta L}{2\pi}\right)\exp(-\mathrm{j}K\sin\theta\, z-\mathrm{j}K\cos\theta\, x). \end{aligned} \tag{7.1-13}$$

该 $sinc$ 函数为声源的平面波角谱, 其指数项表示声平面波的传播方向.

图 7.1-5a) 画出了以 θ 为变量的角谱绝对值函数图, 对于小角度, 其第一个零点出现在 $\theta = \pm\arcsin(\Lambda/L) \approx \pm\Lambda/L$ 处. 结果表明, 由于声波场的传播, 声场的 $\boldsymbol{K}$ 矢量可以通过角度 $\pm\Lambda/L$ 来定向. 在图 7.1-5b) 中, 给出了波矢图. 其中, 在 $\theta = 0$ 处由角谱得到的 $\boldsymbol{K}_{+1}$ 是产生 $\boldsymbol{k}_{+1}$ 的主要原因. 在图 7.1-5a) 中, 我们用标记 "+1" 的箭头来表示声平面波的振幅 $\boldsymbol{K}_{+1}$, 它是

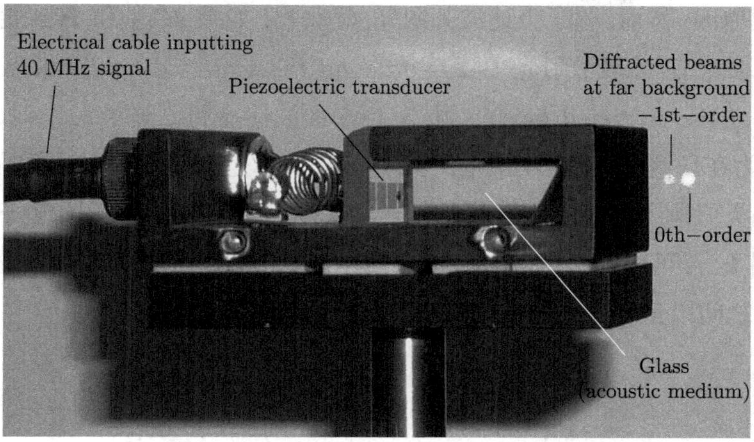

Figure 7.1-4 Typical acousto-optic modulator from IntraAction Corporation, Model AOM-40. Adapted from Poon (2002)

beam (not visible as it traverses across a transparent medium of glass) is travelling along the long dimension of the transducer through the glass.

The closed diagrams in Figures 7.1-2a and 7.1-3a stipulate that there are certain critical angles of incidence ($\phi_{inc} = \pm\phi_B$) in the acoustic medium for plane waves of sound and light to interact, and also that the directions of the incident and scattered light differ in angle by $2\phi_B$. In actuality, scattering occurs even though the direction of incident light is not exactly at the Bragg angle. However, the maximum scattered intensity occurs at the Bragg angle. The reason is that for a finite length of the sound transducer, we no longer have planar wavefronts but rather the sound waves actually diffract as they propagate into the medium. As the length L of the transducer decreases, the sound field will act less and less like a plane wave and, in fact, it is now more appropriate to consider the concept of an angular plane wave spectrum of sound. Assuming a transducer of length L is along the z-direction as defined in Figure 7.1-1, the various plane waves of sound propagating over a distance x, according to Eq. (3.4-9), are given by

$$\mathcal{F}\left\{rect\left(\frac{z}{L}\right)\right\}\exp(-\mathrm{j}K_{0z}z)\times\exp(-\mathrm{j}K_{0x}x),$$

where $K_{0z} = \boldsymbol{K}\cdot\hat{\boldsymbol{z}} = K\sin\theta$ and $K_{0x} = \boldsymbol{K}\cdot\hat{\boldsymbol{x}} = K\cos\theta$ are the components of the sound vector $\boldsymbol{K}$ and θ is measured from the x-axis. Hence, neglecting some constant, the above equation becomes

$$sinc\left(\frac{K_{0z}L}{2\pi}\right)\bigg|_{K_{0z}=K\sin\theta}\exp(-\mathrm{j}K\sin\theta\, z-\mathrm{j}K\cos\theta\, x)$$

$$\propto sinc\left(\frac{K\sin\theta L}{2\pi}\right)\exp(-\mathrm{j}K\sin\theta\, z-\mathrm{j}K\cos\theta\, x). \tag{7.1-13}$$

The *sinc* function is the angular plane wave spectrum of the sound source and the exponential term represents the direction of propagating plane waves of sound.

Figure 7.1-5a plots the absolute value of the angular spectrum as a function of θ and its first zeros are at $\theta = \pm\arcsin(\Lambda/L) \approx \pm\Lambda/L$ for small angles. We see that the $\boldsymbol{K}$-vector of the sound field can be oriented through an angle $\pm\Lambda/L$ due to the spreading of sound. In Figure 7.1-5b, we show the wavevector diagram, where $\boldsymbol{K}_{+1}$ at $\theta = 0$ from the angular spectrum is responsible for the generation of $\boldsymbol{k}_{+1}$. In Figure 7.1-5a, we denote the amplitude of the plane wave of sound $\boldsymbol{K}_{+1}$ by an arrow labelled "+1" that is responsible for the generation

$\boldsymbol{k}_{+1}$ 产生的主要原因. 然而, 除了上述波矢图外, 还得出, 如果 $\boldsymbol{K}_{+2}$ 和 $\boldsymbol{K}_{-1}$ 通过角谱可分别得到, 则可生成 $\boldsymbol{k}_{+2}$ [正二级散射光 (*+2nd-order of the scattered light*)] 和 $\boldsymbol{k}_{-1}$ [负一级散射光 (*−1st-order of the scattered light*)]. 例如, 声平面波传播到 $\theta = -2\phi_B$ 时, 该声平面波振幅 $\boldsymbol{K}_{+2}$ 被表示为在角谱中的箭头处标记 "+2". 如果 $\boldsymbol{K}_{+2}$ 正好在角谱的主瓣内, 就有 $\boldsymbol{K}_{+2}$ 散射 $\boldsymbol{k}_{+1}$ 到 $\boldsymbol{k}_{+2}$. 类似地, 如图 7.1-5a) 所示, $\boldsymbol{K}_{-1}$ 正好落入角谱的主瓣内, 由于下移相互作用, $\boldsymbol{K}_{-1}$ 散射 $\boldsymbol{k}_0$ 到 $\boldsymbol{k}_{-1}$. 可以看出, 如图 7.1-5b) 所示, 散射级次 (*scattered orders*) 按 $2\phi_B$ 分开, 如, $\boldsymbol{k}_{+1}$ 和 $\boldsymbol{k}_{+2}$ 之间以角度 $2\phi_B$ 被隔开.

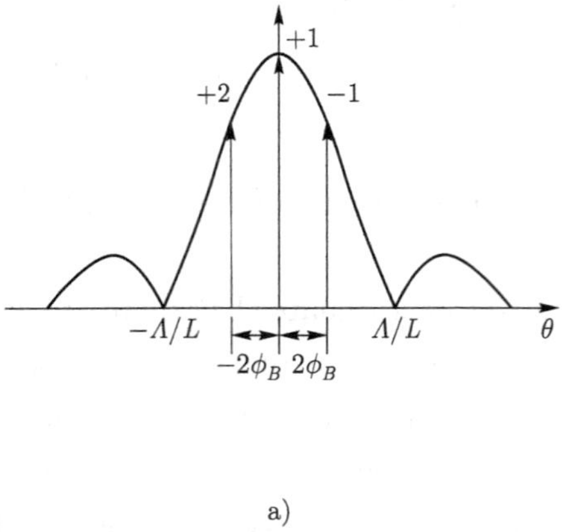

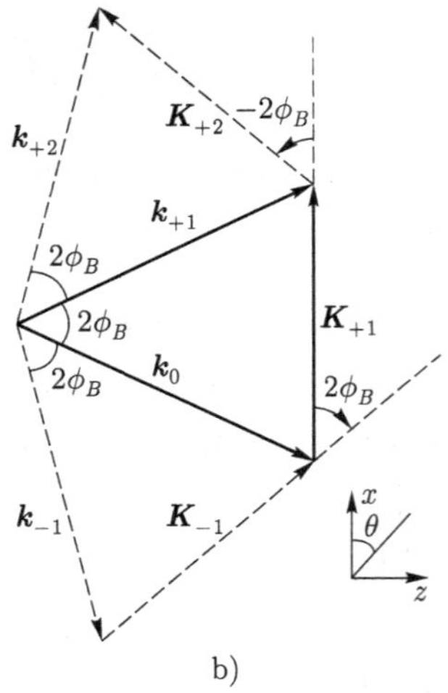

图 7.1-5 a) 声场的角谱; b) 波矢图

为了只产生一个衍射级次的衍射光, 比如 $\boldsymbol{k}_{+1}$, 由图 7.1-5a) 所示的角谱可以清楚地看出, 须加以条件

$$2\phi_B = \frac{\lambda_0}{\Lambda} \gg \frac{\Lambda}{L},$$

或

$$L \gg \frac{\Lambda^2}{\lambda_0}. \tag{7.1-14}$$

如图 7.1-6 所示情况为, 一个沿 $\boldsymbol{K}_{+2}$ 方向的相关的声波矢量要么不存在, 要么存在于其在声波角谱中的值的大小可以忽略不计时 (见图中标注 "+2"的箭头, 其中可以看出, 声平面波的振幅非常小). 如果 L 满足式 (7.1-14), 则声光调制器是可用于布拉格条件 (*Bragg regime*) 的, 这种器件通常被称为布拉格单元 (*Bragg cell*). 然而, 现实中, 完全的能量转换不可能仅存在于两束衍射光束之间, 因为总是存在两个以上的衍射光束.

重新整理式 (7.1-14), 有

$$\frac{\lambda_0 L}{\Lambda^2} \gg 1,$$

或

$$2\pi L\frac{\lambda_0}{\Lambda^2} \gg 2\pi.$$

这里定义声光学中的 Klein-Cook 参数 (*Klein-Cook parameter*) [Klein 和 Cook (1967)] 如下

$$Q = 2\pi L\frac{\lambda_0}{\Lambda^2}. \tag{7.1-15}$$

of $\boldsymbol{k}_{+1}$. However, in addition to that wavevector diagram, we also show that $\boldsymbol{k}_{+2}$ (+2nd-order of the *scattered light*) and $\boldsymbol{k}_{-1}$ (−1st-order of the scattered light) could be generated if $\boldsymbol{K}_{+2}$ and $\boldsymbol{K}_{-1}$ are available from the angular spectrum, respectively. For example, $\boldsymbol{K}_{+2}$ is the sound plane wave travelling at $\theta = -2\phi_B$, and the amplitude of the plane wave $\boldsymbol{K}_{+2}$ is labeled "+2" by an arrow in the angular spectrum. If $\boldsymbol{K}_{+2}$ is well within the main lobe of the angular spectrum, we have $\boldsymbol{K}_{+2}$ scatters $\boldsymbol{k}_{+1}$ into $\boldsymbol{k}_{+2}$. Similarly, $\boldsymbol{K}_{-1}$ scatters $\boldsymbol{k}_0$ into $\boldsymbol{k}_{-1}$ for downshifted interaction if $\boldsymbol{K}_{-1}$ is well within the main lobe of the angular spectrum, as illustrated in Figure 7.1-5a. Note that the *scattered orders* are separated by $2\phi_B$, as shown in Figure 7.1-5b. For example, $\boldsymbol{k}_{+1}$ and $\boldsymbol{k}_{+2}$ are separated by $2\phi_B$.

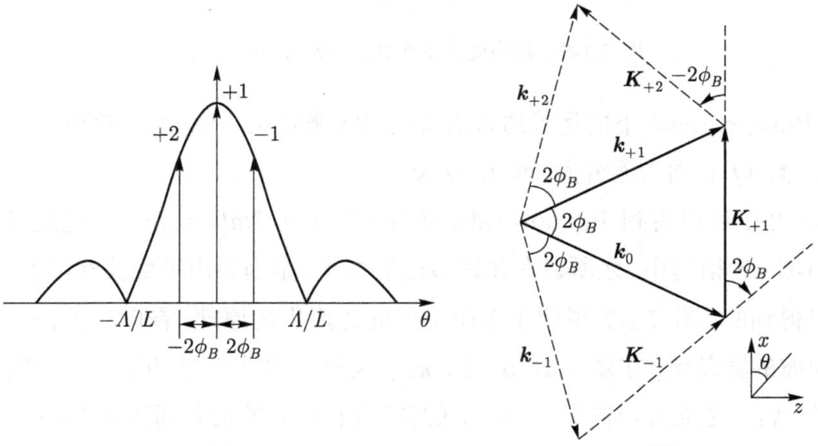

Figure 7.1-5 (a) Angular spectrum of the sound field and (b) wavevector diagrams

In order to have only one diffracted order of light generated, that is, $\boldsymbol{k}_{+1}$, it is clear, from the angular spectrum shown in Figure 7.1-5a, that we have to impose the condition

$$2\phi_B = \frac{\lambda_0}{\Lambda} \gg \frac{\Lambda}{L},$$

or

$$L \gg \frac{\Lambda^2}{\lambda_0}. \tag{7.1-14}$$

The situation is shown in Figure 7.1-6, where a pertinent sound wave vector lies along $\boldsymbol{K}_{+2}$ is either not present or present in negligible amounts in the angular spectrum of the sound (see the arrow labeled "+2" in the figure, where we see that the amplitude of that sound plane wave is very small). If L satisfies Eq. (7.1-14), the acousto-optic modulator is said to operate in the *Bragg regime* and the device is commonly known as the *Bragg cell*. However, physical reality dictates that a complete energy transfer between the two diffracted beams is impossible since there always exists more than two diffracted beams.

If we rearrange Eq. (7.1-14), we have

$$\frac{\lambda_0 L}{\Lambda^2} \gg 1,$$

or

$$2\pi L \frac{\lambda_0}{\Lambda^2} \gg 2\pi.$$

We define the *Klein-Cook parameter* [Klein and Cook (1967)] in acousto-optics as follows:

$$Q = 2\pi L \frac{\lambda_0}{\Lambda^2}. \tag{7.1-15}$$

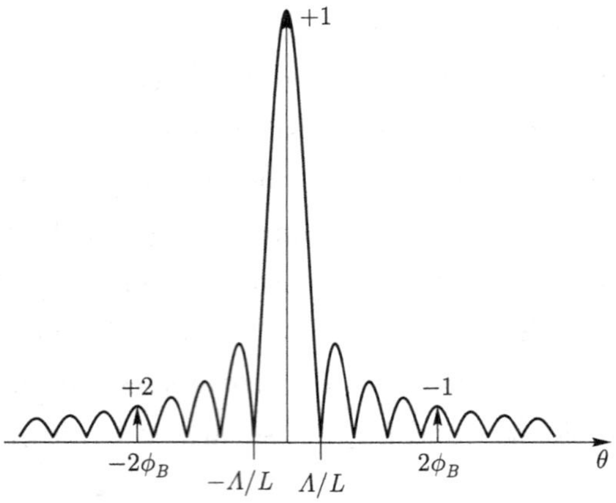

图 7.1-6 长换能器的窄角谱 ($2\phi_B \gg \Lambda/L$).

布拉格条件 (*Bragg regime*) 下的运算通常由 $Q \gg 2\pi$ 来定义, 在理想布拉格条件下, 只有两个衍射级次存在, 且 Q 必须为无穷大, 或 $L \to \infty$.

如果 $Q \ll 2\pi$, 可以得到 Raman-Nath 条件 (*Raman-Nath regime*). 这是 L 足够短的情况. 在 Raman-Nath 衍射中, 之所以存在许多衍射级次, 是因为声平面波在散射所需的不同角度上都是可以得到的. 图 7.1-7 说明了多阶衍射同时产生的原理. 在图 7.1-7a) 中, 给出了多个方向声平面波的宽角谱, 在图 7.1-7b) 中, $\boldsymbol{k}_{+1}$ 是通过声波矢量 $\boldsymbol{K}_{+1}$ 的作用由光波 $\boldsymbol{k}_0$ 的衍射而得到的, $\boldsymbol{k}_{+2}$ 是通过声波矢量 $\boldsymbol{K}_{+2}$ 的作用由 $\boldsymbol{k}_{+1}$ 的衍射而得到的, 以此类推. 其中, $\boldsymbol{K}_{\pm p}(p = 0, \pm 1, \pm 2, \cdots)$ 表示声平面波频谱 (*plane wave spectrum of the sound*) 的各个组成分量.

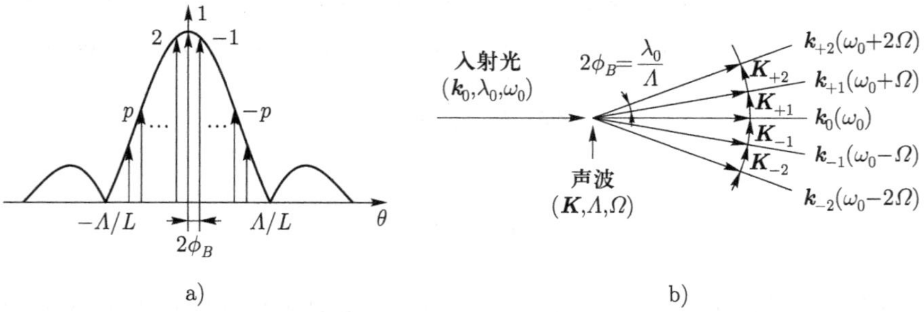

图 7.1-7 拉曼–奈斯衍射 (*Raman-Nath diffraction*): a) 短换能器的宽角谱 ($\Lambda/L \gg 2\phi_B$); b) 多个衍射级次同时产生 [Poon 和 Kim (2018)]

虽然上述讨论的简单粒子法描述了布拉格衍射产生的必要条件, 但它并不能预测声光相互作用过程如何影响不同衍射光束之间的振幅分布. 在下文中, 将描述声光的一般形式.

一般形式

假设, 相互作用发生在一个光学上不均匀 (*optically inhomogeneous*)、非磁性且各向同性 (*nonmagnetic isotropic*) 的无源介质 ($\rho_v = 0, \boldsymbol{J} = 0$) 中, 在该介质中, 具有 μ_0 和时变介电常数 $\tilde{\varepsilon}(\boldsymbol{R}, t)$, 当一束光波入射到该时变介电常数 (*time-varying permittivity*) 的介质上时, 可以写出该时变介电常数为

$$\tilde{\varepsilon}(\boldsymbol{R}, t) = \varepsilon + \varepsilon'(\boldsymbol{R}, t), \tag{7.1-16}$$

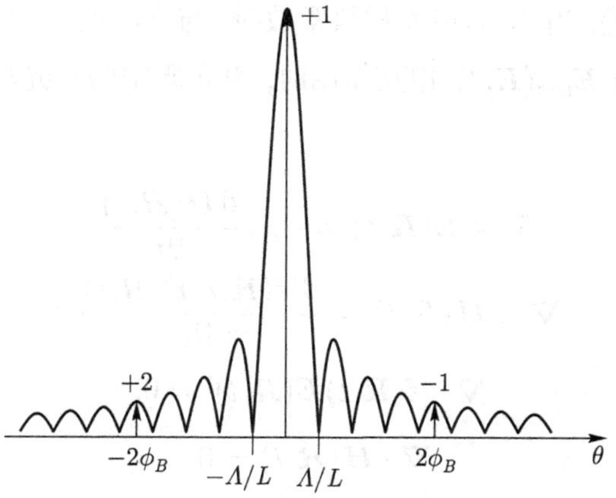

Figure 7.1-6 Narrow angular spectrum for a long transducer ($2\phi_B \gg \Lambda/L$)

Operation in the *Bragg regime* is typically defined by $Q \gg 2\pi$. In the ideal Bragg regime, only two diffracted orders exist, and Q would have to be infinity or $L \to \infty$.

If $Q \ll 2\pi$, we have the *Raman-Nath regime.* That is the case where L is sufficiently short. In Raman-Nath diffraction, many diffracted orders exist because plane waves of sound are available at the various angles required for scattering. The principle of the simultaneous generation of many orders is illustrated in Figure 7.1-7. In Figure 7.1-7a, we show a broad angular spectrum that provides many directions of sound plane waves, and in Figure 7.1-7b we show that $\boldsymbol{k}_{+1}$ is generated through the diffraction of $\boldsymbol{k}_0$ by $\boldsymbol{K}_{+1}$, $\boldsymbol{k}_{+2}$ is generated through the diffraction of $\boldsymbol{k}_{+1}$ by $\boldsymbol{K}_{+2}$, and so on, where $\boldsymbol{K}_{\pm p}$s ($p = 0, \pm 1, \pm 2, \cdots$) denote the various components of the *plane wave spectrum of the sound.*

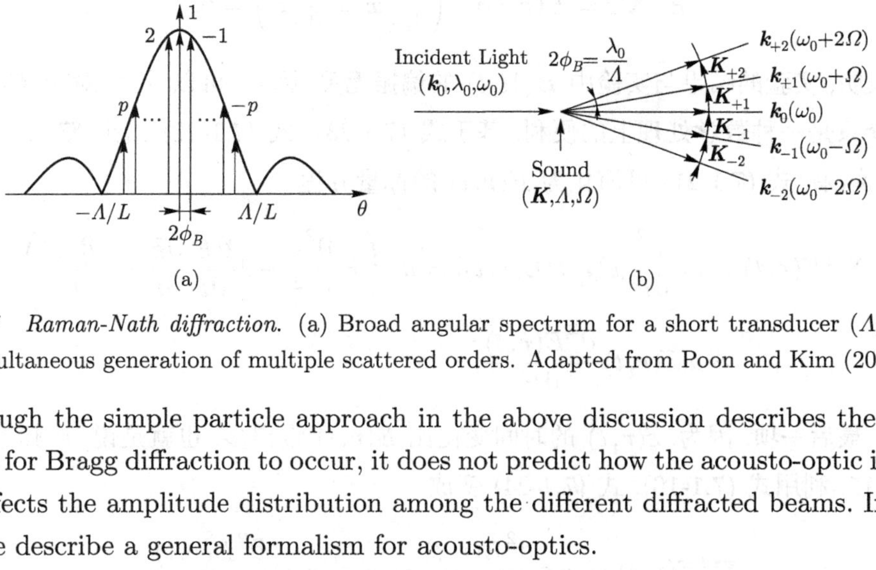

Figure 7.1-7 *Raman-Nath diffraction.* (a) Broad angular spectrum for a short transducer ($\Lambda/L \gg 2\phi_B$) and (b) simultaneous generation of multiple scattered orders. Adapted from Poon and Kim (2018)

Although the simple particle approach in the above discussion describes the necessary conditions for Bragg diffraction to occur, it does not predict how the acousto-optic interaction process affects the amplitude distribution among the different diffracted beams. In the next section, we describe a general formalism for acousto-optics.

General Formalism

We assume that the interaction takes place in an optically inhomogeneous, nonmagnetic isotropic source-free ($\rho_v = 0, \boldsymbol{J_C} = 0$) medium with μ_0 and time-varying permittivity $\widetilde{\varepsilon}(\boldsymbol{R}, t)$ when an optical field is incident on the time-varying permittivity. We can write the time-varying permittivity as

$$\widetilde{\varepsilon}(\boldsymbol{R}, t) = \varepsilon + \varepsilon'(\boldsymbol{R}, t), \tag{7.1-16}$$

其中, $\varepsilon'(\boldsymbol{R},t)=\varepsilon CS(\boldsymbol{R},t)$, 当 C 为声光材料常数时, 与声波振幅 $S(\boldsymbol{R},t)$ 成正比.

当该声波与入射光场 $\boldsymbol{E}_{inc}(\boldsymbol{R},t)$ 相互作用时, 声介质中的总光场 $\boldsymbol{E}(\boldsymbol{R},t)$ 和 $\boldsymbol{H}(\boldsymbol{R},t)$ 必须满足以下麦克斯韦方程

$$\nabla\times\boldsymbol{E}(\boldsymbol{R},t)=-\mu_0\frac{\partial\boldsymbol{H}(\boldsymbol{R},t)}{\partial t}, \tag{7.1-17}$$

$$\nabla\times\boldsymbol{H}(\boldsymbol{R},t)=\frac{\partial[\tilde{\varepsilon}(\boldsymbol{R},t)\boldsymbol{E}(\boldsymbol{R},t)]}{\partial t}, \tag{7.1-18}$$

$$\nabla\cdot[\tilde{\varepsilon}(\boldsymbol{R},t)\boldsymbol{E}(\boldsymbol{R},t)]=0, \tag{7.1-19}$$

$$\nabla\cdot\boldsymbol{H}(\boldsymbol{R},t)=0. \tag{7.1-20}$$

这里来推导 $\boldsymbol{E}(\boldsymbol{R},t)$ 的波动方程. 取式 (7.1-17) 的旋度, 并利用式 (7.1-18), 则 $\boldsymbol{E}(\boldsymbol{R},t)$ 的方程为

$$\nabla\times(\nabla\times\boldsymbol{E})=\nabla(\nabla\cdot\boldsymbol{E})-\nabla^2\boldsymbol{E}=-\mu_0\frac{\partial^2}{\partial t^2}[\tilde{\varepsilon}(\boldsymbol{R},t)\boldsymbol{E}(\boldsymbol{R},t)]. \tag{7.1-21}$$

现在, 从式 (7.1-19), 有

$$\nabla\cdot\tilde{\varepsilon}\boldsymbol{E}=\tilde{\varepsilon}\nabla\cdot\boldsymbol{E}+\boldsymbol{E}\cdot\nabla\tilde{\varepsilon}=0. \tag{7.1-22}$$

由于声光相互作用局限于二维结构, 假设一个二维 (x-z) 声场, 且 $\boldsymbol{E}(\boldsymbol{R},t)$ 沿 y 方向线 (性) 偏振, 即 $\boldsymbol{E}(\boldsymbol{R},t)=E(\boldsymbol{r},t)\hat{\boldsymbol{y}}$, 其中 $\boldsymbol{r}$ 是 x-z 平面上的位置矢量. 根据假设, 有

$$\boldsymbol{E}\cdot\nabla\tilde{\varepsilon}=E(\boldsymbol{r},t)\hat{\boldsymbol{y}}\cdot\left(\frac{\partial\tilde{\varepsilon}}{\partial x}\hat{\boldsymbol{x}}+\frac{\partial\tilde{\varepsilon}}{\partial z}\hat{\boldsymbol{z}}\right)=0. \tag{7.1-23}$$

研究表明, 实验的结果与实验中 $\boldsymbol{E}(\boldsymbol{R},t)$ 的偏振无关. 因此, 可以看出, 对于 $\boldsymbol{E}(\boldsymbol{R},t)$ 的 $\hat{\boldsymbol{y}}$ 偏振的选择只是一种数学处理上的便利. 基于式 (7.1.23), 式 (7.1-22) 给出 $\nabla\cdot\boldsymbol{E}=0$, 因为 $\tilde{\varepsilon}\neq 0$. 由此结果, 式 (7.1-21) 可简化为 $E(\boldsymbol{r},t)$ 的标量方程

$$\begin{aligned}\nabla^2E(\boldsymbol{r},t)&=\mu_0\frac{\partial^2}{\partial t^2}[\tilde{\varepsilon}(\boldsymbol{r},t)E(\boldsymbol{r},t)]=\mu_0\left(E\frac{\partial^2\tilde{\varepsilon}}{\partial t^2}+2\frac{\partial E}{\partial t}\frac{\partial\tilde{\varepsilon}}{\partial t}+\tilde{\varepsilon}\frac{\partial^2E}{\partial t^2}\right)\\&\approx\mu_0\tilde{\varepsilon}(\boldsymbol{r},t)\frac{\partial^2E(\boldsymbol{r},t)}{\partial t^2}.\end{aligned} \tag{7.1-24}$$

这里保留了最后一项, 因为, $\tilde{\varepsilon}(\boldsymbol{r},t)$ 的时间变化比 $E(\boldsymbol{r},t)$ 慢得多, 也就是说, 声频率比光频率低得多. 现在, 利用式 (7.1-16), 式 (7.1-24) 变成

$$\nabla^2E(\boldsymbol{r},t)-\mu_0\varepsilon\frac{\partial^2E(\boldsymbol{r},t)}{\partial t^2}=\mu_0\varepsilon'(\boldsymbol{r},t)\frac{\partial^2E(\boldsymbol{r},t)}{\partial t^2}. \tag{7.1-25}$$

该标量波动方程常用于研究声光中的相互作用.

常规相互作用结构 (*conventional interaction configuration*)

式 (7.1-25) 很难解, 为此, 采用 Korpel-Poon 多平面波理论 (*Korpel-Poon multiple-plane-wave theory*) 来进一步讨论声光效应及其应用 [Poon 和 Kim (2018)]. 考虑如图 7.1-8 所示的

where $\varepsilon'(\boldsymbol{R},t) = \varepsilon CS(\boldsymbol{R},t)$ as it is proportional to the sound field amplitude $S(\boldsymbol{R},t)$ with C being the acousto-optic material constant.

When the sound field interacts with the incident light field $\boldsymbol{E}_{inc}(\boldsymbol{R},t)$, the total light fields $\boldsymbol{E}(\boldsymbol{R},t)$ and $\boldsymbol{H}(\boldsymbol{R},t)$ in the acoustic medium must satisfy Maxwell's equations as follows:

$$\nabla \times \boldsymbol{E}(\boldsymbol{R},t) = -\mu_0 \frac{\partial \boldsymbol{H}(\boldsymbol{R},t)}{\partial t}, \tag{7.1-17}$$

$$\nabla \times \boldsymbol{H}(\boldsymbol{R},t) = \frac{\partial [\tilde{\varepsilon}(\boldsymbol{R},t)\boldsymbol{E}(\boldsymbol{R},t)]}{\partial t}, \tag{7.1-18}$$

$$\nabla \cdot [\tilde{\varepsilon}(\boldsymbol{R},t)\boldsymbol{E}(\boldsymbol{R},t)] = 0, \tag{7.1-19}$$

$$\nabla \cdot \boldsymbol{H}(\boldsymbol{R},t) = 0. \tag{7.1-20}$$

We shall derive the wave equation for $\boldsymbol{E}(\boldsymbol{R},t)$. Taking the curl of Eq. (7.1-17) and using Eq. (7.1-18), the equation for $\boldsymbol{E}(\boldsymbol{R},t)$ becomes

$$\nabla \times (\nabla \times \boldsymbol{E}) = \nabla(\nabla \cdot \boldsymbol{E}) - \nabla^2 \boldsymbol{E} = -\mu_0 \frac{\partial^2}{\partial t^2}[\tilde{\varepsilon}(\boldsymbol{R},t)\boldsymbol{E}(\boldsymbol{R},t)]. \tag{7.1-21}$$

Now, from Eq. (7.1-19), we have

$$\nabla \cdot \tilde{\varepsilon}\boldsymbol{E} = \tilde{\varepsilon}\nabla \cdot \boldsymbol{E} + \boldsymbol{E} \cdot \nabla\tilde{\varepsilon} = 0. \tag{7.1-22}$$

Since the acousto-optic interaction is confined to a two-dimensional (2-D) configuration, we assume a 2-D (x-z) sound field with $\boldsymbol{E}(\boldsymbol{R},t)$ linearly polarized along the y-direction, that is, $\boldsymbol{E}(\boldsymbol{R},t) = E(\boldsymbol{r},t)\hat{\boldsymbol{y}}$, where $\boldsymbol{r}$ is the position vector in the x-z plane. According to the assumptions, we have

$$\boldsymbol{E} \cdot \nabla\tilde{\varepsilon} = E(\boldsymbol{r},t)\hat{\boldsymbol{y}} \cdot \left(\frac{\partial \tilde{\varepsilon}}{\partial x}\hat{\boldsymbol{x}} + \frac{\partial \tilde{\varepsilon}}{\partial z}\hat{\boldsymbol{z}}\right) = 0. \tag{7.1-23}$$

As it turns out experimental results are independent of the polarization used in $\boldsymbol{E}(\boldsymbol{R},t)$ during experiments. We see that the choice of $\hat{\boldsymbol{y}}$ polarization for $\boldsymbol{E}(\boldsymbol{R},t)$ is therefore a mathematical convenience. With Eq. (7.1-23), Eq. (7.1-22) gives us $\nabla \cdot \boldsymbol{E} = 0$ as $\tilde{\varepsilon} \neq 0$. With this result, Eq. (7.1-21) reduces to the following scalar equation in $E(\boldsymbol{r},t)$:

$$\begin{aligned}\nabla^2 E(\boldsymbol{r},t) &= \mu_0 \frac{\partial^2}{\partial t^2}[\tilde{\varepsilon}(\boldsymbol{r},t)E(\boldsymbol{r},t)] = \mu_0\left(E\frac{\partial^2 \tilde{\varepsilon}}{\partial t^2} + 2\frac{\partial E}{\partial t}\frac{\partial \tilde{\varepsilon}}{\partial t} + \tilde{\varepsilon}\frac{\partial^2 E}{\partial t^2}\right) \\ &\approx \mu_0 \tilde{\varepsilon}(\boldsymbol{r},t)\frac{\partial^2 E(\boldsymbol{r},t)}{\partial t^2}.\end{aligned} \tag{7.1-24}$$

where we have retained the last term because the time variation of $\tilde{\varepsilon}(\boldsymbol{r},t)$ is much slower than that of $E(\boldsymbol{r},t)$, that is, the sound frequency is much lower than that of the light. Now, using Eq. (7.1-16), Eq. (7.1-24) becomes

$$\nabla^2 E(\boldsymbol{r},t) - \mu_0\varepsilon\frac{\partial^2 E(\boldsymbol{r},t)}{\partial t^2} = \mu_0\varepsilon'(\boldsymbol{r},t)\frac{\partial^2 E(\boldsymbol{r},t)}{\partial t^2}. \tag{7.1-25}$$

This scalar wave equation is often used to investigate interaction in acousto-optics.

Conventional Interaction Configuration

Equation (7.1-25) is formidable. We shall adopt the *Korpel-Poon multiple-plane-wave theory* for further discussion on the acousto-optic effect and its applications [Poon and Kim (2018)].

常规相互作用结构, 假设入射平面光波为

$$E_{inc}(\boldsymbol{r},t)=\mathrm{Re}[E_{inc}(\boldsymbol{r})\mathrm{e}^{\mathrm{j}\omega_0 t}],$$

且

$$E_{inc}(\boldsymbol{r})=\psi_{inc}\mathrm{e}^{-\mathrm{j}k_0\sin\phi_{inc}x-\mathrm{j}k_0\cos\phi_{inc}z} \tag{7.1-26}$$

是入射角为 ϕ_{inc} 的复振幅, 同样假设逆时针方向为正, 同时假设声波是沿 x 方向传播且其有限宽度为 L 的一个均匀声波, 由于输入到声换能器的电信号形式为 $e(t)=\mathrm{Re}[|e(t)|\mathrm{e}^{\mathrm{j}\theta}\mathrm{e}^{\mathrm{j}\Omega t}]$, 故取时变介电常数形式

$$\varepsilon'(\boldsymbol{r},t)=\varepsilon CS(\boldsymbol{r},t)=\varepsilon C\mathrm{Re}[S(\boldsymbol{r})\mathrm{e}^{\mathrm{j}\Omega t}],$$

其中

$$S(\boldsymbol{r})=A\mathrm{e}^{-\mathrm{j}Kx}, \tag{7.1-27}$$

其中, A 一般可以为复数, $A=|A|\mathrm{e}^{\mathrm{j}\theta}$, 因此, $\varepsilon'(\boldsymbol{r},t)=\varepsilon C|A|\cos(\Omega t-Kx+\theta)$, 这表示沿 x 方向传播的声平面波, 如图 7.1-8 所示. 将 $C|A|$ 与声光调制器所导致的折射率变化 $\Delta n(\boldsymbol{r},t)$ 联系起来.

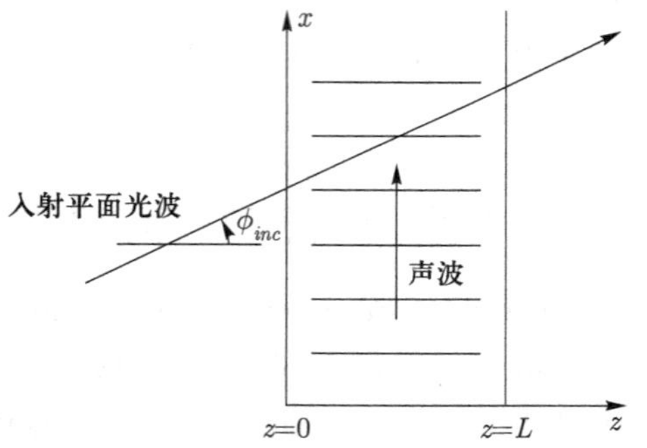

图 7.1-8 平面光波入射下常规尖锐边界 (*sharp boundary*) 相互作用结构

由于

$$\begin{aligned}\widetilde{\varepsilon}(\boldsymbol{r},t)&=\varepsilon_0 n^2(\boldsymbol{r},t)=\varepsilon_0[n_0+\Delta n(\boldsymbol{r},t)]^2\\&\simeq\varepsilon_0 n_0^2\left[1+\frac{2\Delta n(\boldsymbol{r},t)}{n_0}\right]=\varepsilon\left[1+\frac{2\Delta n(\boldsymbol{r},t)}{n_0}\right]=\varepsilon+\varepsilon'(\boldsymbol{r},t),\end{aligned}$$

根据式 (7.1-16), 得

$$\varepsilon'(\boldsymbol{r},t)=\frac{2\varepsilon\Delta n(\boldsymbol{r},t)}{n_0}=\varepsilon C|A|\cos(\Omega t-Kx+\theta).$$

这是振荡频率在 Ω 处的正弦声波. 因此, 可将声介质中折射率的变化明确地表示为

$$\Delta n(\boldsymbol{r},t)=\frac{n_0}{2}C|A|\cos(\Omega t-Kx+\theta)=\Delta n_{\max}\cos(\Omega t-Kx+\theta).$$

这清晰地表明, 当声波沿 x 方向传播时, 声波介质的折射率存在周期性的压缩和伸长现象.

Consider a conventional interaction configuration shown in Figure 7.1-8, where we assume the incident plane wave of light is

$$E_{inc}(\boldsymbol{r},t)=\mathrm{Re}[E_{inc}(\boldsymbol{r})\mathrm{e}^{\mathrm{j}\omega_0 t}],$$

and

$$E_{inc}(\boldsymbol{r})=\psi_{inc}\mathrm{e}^{-\mathrm{j}k_0\sin\phi_{inc}x-\mathrm{j}k_0\cos\phi_{inc}z} \tag{7.1-26}$$

is the complex amplitude with ϕ_{inc} being the incident angle, again assumed counterclockwise positive. We also assume that a uniform sound wave of finite width L with sound propagating along the x-direction. Because of an electrical signal of the form $e(t)=\mathrm{Re}[|e(t)|\mathrm{e}^{\mathrm{j}\theta}\mathrm{e}^{\mathrm{j}\Omega t}]$ fed to the acoustic transducer, we take the time-varying permittivity in the form

$$\varepsilon'(\boldsymbol{r},t)=\varepsilon CS(\boldsymbol{r},t)=\varepsilon C\mathrm{Re}[S(\boldsymbol{r})\mathrm{e}^{\mathrm{j}\Omega t}],$$

where

$$S(\boldsymbol{r})=A\mathrm{e}^{-\mathrm{j}Kx}, \tag{7.1-27}$$

with $A=|A|\mathrm{e}^{\mathrm{j}\theta}$ as A, in general, can be complex. Hence, we find $\varepsilon'(\boldsymbol{r},t)=\varepsilon C|A|\cos(\Omega t-Kx+\theta)$,which represents a plane wave of sound propagating along the x-direction, as shown in Figure 7.1-8. Let us relate $C|A|$ to the refractive index variation $\Delta n(\boldsymbol{r},t)$ induced in the acousto-optic cell.

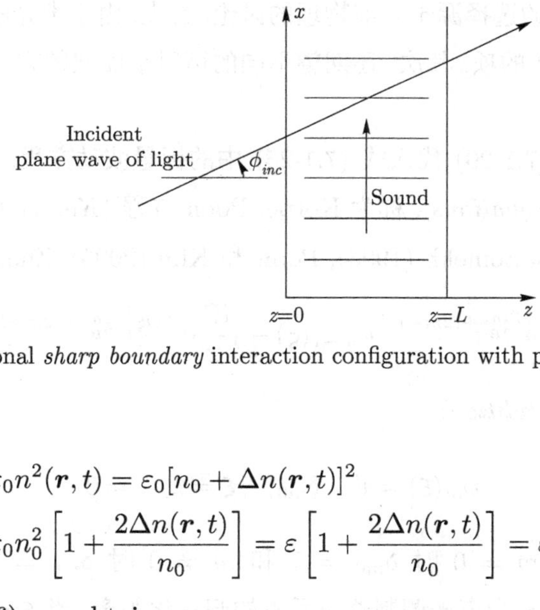

Figure 7.1-8 Conventional *sharp boundary* interaction configuration with plane-wave light incidence

Because

$$\begin{aligned}\widetilde{\varepsilon}(\boldsymbol{r},t)&=\varepsilon_0 n^2(\boldsymbol{r},t)=\varepsilon_0[n_0+\Delta n(\boldsymbol{r},t)]^2\\ &\simeq\varepsilon_0 n_0^2\left[1+\frac{2\Delta n(\boldsymbol{r},t)}{n_0}\right]=\varepsilon\left[1+\frac{2\Delta n(\boldsymbol{r},t)}{n_0}\right]=\varepsilon+\varepsilon'(\boldsymbol{r},t),\end{aligned}$$

according to Eq. (7.1-16), we obtain

$$\varepsilon'(\boldsymbol{r},t)=\frac{2\varepsilon\Delta n(\boldsymbol{r},t)}{n_0}=\varepsilon C|A|\cos(\Omega t-Kx+\theta),$$

for sinusoidal sound wave oscillating at Ω. Therefore, we can express the change of the index of refraction within the acoustic medium explicitly as

$$\Delta n(\boldsymbol{r},t)=\frac{n_0}{2}C|A|\cos(\Omega t-Kx+\theta)=\Delta n_{\max}\cos(\Omega t-Kx+\theta).$$

It clearly shows that there is a periodic compression and rarefaction of the index of refraction of the acoustic medium as the sound wave is travelling along the x-direction.

在多平面波散射理论中, 定义参数

$$\widetilde{\alpha}=\frac{k_0CAL}{2},$$

其强度为

$$|\widetilde{\alpha}|=\alpha=\frac{k_0C|A|L}{2}=k_0(\Delta n_{\max}/n_0)L. \tag{7.1-28}$$

这正比于 $\Delta n_{\max}$. α 表示光通过长度为 L 的声介质的峰值相位延迟 (*peak phase delay*), 在多平面波理论中, 使用峰值相位延迟来表示声音振幅的强度.

对于由式 (7.1-26) 给出的入射平面波, 寻求以下这种形式的解

$$E(\boldsymbol{r},t)=\sum_{m=-\infty}^{\infty}\operatorname{Re}\left[E_m(\boldsymbol{r})\mathrm{e}^{\mathrm{j}(\omega_0+m\Omega)t}\right],$$

其中

$$E_m(\boldsymbol{r})=\psi_m(z,x)\mathrm{e}^{-\mathrm{j}k_0\sin\phi_m x-\mathrm{j}k_0\cos\phi_m z}. \tag{7.1-29}$$

根据式 (7.1-5) 中的光栅方程来选择 ϕ_m. $\psi_m(z,x)$ 为第 m 阶衍射平面波在频率 $\omega_0+m\Omega$ 处的复振幅, 且 $k_0\sin\phi_m$ 和 $k_0\cos\phi_m$ 是其传播矢量的 x 分量和 z 分量. 接下来, 求出第 m 阶衍射平面波 $\psi_m(z,x)$ 的振幅.

可以看出, 解决方案的选择源于之前物理的讨论. 首先, 由于声光的相互作用, 我们期望频率移动, 即包含 $\omega_0+m\Omega$ 的项. 其次, 还期望不同的衍射平面波的方向由光栅方程控制 [见式 (7.1-15)].

将式 (7.1-27) 和式 (7.1-29) 代入式 (7.1-25) 中的标量波动方程, 得到包含 ψ_m 的无限耦合方程 (*infinite coupled equations*), 称为 Korpel-Poon 方程 (*Korpel-Poon equations*) [Korpel 和 Poon (1980), Appel 和 Somekh (1993), Poon 和 Kim (2005), Zhang 等 (2022)]

$$\frac{\mathrm{d}\psi_m(\xi)}{\mathrm{d}\xi}=-\mathrm{j}\frac{\widetilde{\alpha}}{2}\mathrm{e}^{-\mathrm{j}\frac{1}{2}Q\xi\left[\frac{\phi_{inc}}{\phi_B}+(2m-1)\right]}\psi_{m-1}(\xi)-\mathrm{j}\frac{\widetilde{\alpha}^*}{2}\mathrm{e}^{\mathrm{j}\frac{1}{2}Q\xi\left[\frac{\phi_{inc}}{\phi_B}+(2m+1)\right]}\psi_{m+1}(\xi), \tag{7.1-30a}$$

有边界条件 (*boundary condition*)

$$\psi_m(\xi)=\psi_{inc}\delta_{m0},\quad \xi=z/L=0. \tag{7.1-30b}$$

其中, 符号 δ_{m0} 是通过 $m=0$ 时 $\delta_{m0}=1$ 和 $m\neq 0$ 时 $\delta_{m0}=0$ 定义的克罗内克函数 (*Kronecker delta*). 因此, ξ 为声光调制器单元里的归一化距离, 当 $\xi=1$ 时, 其表示该调制器的出射面.

在推导上述耦合方程时, 采用如下几种近似:

1) 因为 $\phi_m\ll 1$, 所以 $\psi_m(z,x)\approx\psi_m(z)$.

2) 根据物理概念, $\omega_0\gg\Omega$.

3) $\psi_m(z)$ 是 z 的缓慢变化的函数, 因为在一个波长的传播距离范围内, ψ_m 的变化很小, 即 $\Delta\psi_m\ll\psi_m$. 用微分形式表示, 即 $\mathrm{d}\psi_m=(\partial\psi_m/\partial z)\Delta z=(\partial\psi_m/\partial z)\lambda_0\ll\psi_m$, 这会变为 $\partial\psi_m/\partial z\ll k_0\psi_m$ 或者 $\partial^2\psi_m(z)/\partial z^2\ll k_0\partial\psi_m(z)/\partial z$.

In the theory of multiple plane-wave scattering, a parameter

$$\widetilde{\alpha} = \frac{k_0 CAL}{2}$$

is defined and its magnitude is

$$|\widetilde{\alpha}| = \alpha = \frac{k_0 C|A|L}{2} = k_0(\Delta n_{\max}/n_0)L, \tag{7.1-28}$$

which is proportional to $\Delta n_{\max}$. α represents the *peak phase delay* of the light through the acoustic medium of length L. We will use the peak phase delay to denote the strength of the sound amplitude in the multiple plane wave theory.

For a plane-wave incidence given by Eq. (7.1-26), we look for a solution of the form

$$E(\boldsymbol{r}, t) = \sum_{m=-\infty}^{\infty} \mathrm{Re}\left[E_m(\boldsymbol{r})\mathrm{e}^{\mathrm{j}(\omega_0 + m\Omega)t}\right],$$

where

$$E_m(\boldsymbol{r}) = \psi_m(z, x)\mathrm{e}^{-\mathrm{j}k_0 \sin\phi_m x - \mathrm{j}k_0 \cos\phi_m z} \tag{7.1-29}$$

with the choice for ϕ_m dictated by the grating equation in Eq. (7.1-5). $\psi_m(z, x)$ is the complex amplitude of the mth diffracted plane wave at frequency $\omega_0 + m\Omega$, and $k_0 \sin\phi_m$ and $k_0 \cos\phi_m$ are the x and z components of its propagation vector. In what follows, we seek to find the amplitude of the mth diffracted plane wave $\psi_m(z, x)$.

Note that the choice of the solution is motivated by previous physical discussions. First, we expect frequency shifting, that is, the term involving $\omega_0 + m\Omega$ due to sound-light interaction. Second, we also expect that the direction of the different diffracted plane waves is governed by the grating equation [see Eq. (7.1-5)].

Upon substituting Eqs. (7.1-27) and (7.1-29) into the scalar wave equation in Eq. (7.1-25), we have the following *infinite coupled equations* involving ψ_m, called the *Korpel-Poon equations* [Poon and Korpel (1980), Appel and Somekh (1993), Poon and Kim (2005), Zhang et al. (2022)]:

$$\frac{\mathrm{d}\psi_m(\xi)}{\mathrm{d}\xi} = -\mathrm{j}\frac{\widetilde{\alpha}}{2}\mathrm{e}^{-\mathrm{j}\frac{1}{2}Q\xi\left[\frac{\phi_{inc}}{\phi_B} + (2m-1)\right]}\psi_{m-1}(\xi) - \mathrm{j}\frac{\widetilde{\alpha}^*}{2}\mathrm{e}^{\mathrm{j}\frac{1}{2}Q\xi\left[\frac{\phi_{inc}}{\phi_B} + (2m+1)\right]}\psi_{m+1}(\xi), \tag{7.1-30a}$$

with

$$\psi_m(\xi) = \psi_{inc}\delta_{m0} \text{ at } \xi = z/L = 0 \tag{7.1-30b}$$

as the boundary conditions. The symbol δ_{m0} is the Kronecker delta defined by $\delta_{m0} = 1$ for $m = 0$ and $\delta_{m0} = 0$ for $m \neq 0$. So ξ is the normalized distance inside the acousto-optic cell, and $\xi = 1$ signifies the exit plane of the cell.

In deriving the above coupled equations, we take on several approximations as follows:

(1) $\phi_m \ll 1$, so that $\psi_m(z, x) \approx \psi_m(z)$.

(2) according to physical intuition, $\omega_0 \gg \Omega$.

(3) $\psi_m(z)$ is a slowly varying function of z in that within a wavelength of the propagation distance,the change in ψ_m is small, that is, $\Delta\psi_m \ll \psi_m$. In differential form, $\Delta\psi_m \ll \psi_m$ becomes $\mathrm{d}\psi_m = (\partial\psi_m/\partial z)\Delta z = (\partial\psi_m/\partial z)\lambda_0 \ll \psi_m$, leading to $\partial\psi_m/\partial z \ll k_0\psi_m$ or equivalently to $\partial^2\psi_m(z)/\partial z^2 \ll k_0\,\partial\psi_m(z)/\partial z$.

作为这些近似的结果, 在式 (7.1-30) 中有一阶耦合微分方程. 详细的推导过程已有过讨论 [Poon 和 Kim (2018)], 这里将这些推导过程作为习题留在本章的最后 (见习题 7.1).

式 (7.1-30) 的物理解释是, 在相互作用中相邻阶之间存在着相互耦合, 即 ψ_m 是 ψ_{m-1} 和 ψ_{m+1} 耦合而成的. 接下来用这些方程来研究 Raman-Nath 衍射和布拉格衍射 (*Bragg diffraction*).

7.1.2 Raman-Nath 衍射和布拉格衍射

本节中, 首先考虑声光效应中两个典型解的情况: 理想 Raman-Nath 衍射和布拉格衍射. 理论上, Raman-Nath 衍射和布拉格衍射是两个极端的例子. 在理想 Raman-Nath 衍射 (*Raman-Nath diffraction*) 中, 声换能器的长度 L 被认为是零, 而在理想布拉格衍射中, 认为 $L\to\infty$. 本节将以对 L 为有限的近布拉格衍射的讨论作为结尾.

理想 Raman-Nath 衍射

在典型情况下, 为垂直入射, 即, $\phi_{inc}=0$ 和 $Q=0$. 当 $Q=0$ 时, 声柱 (*sound column*) 被认为是一个无限薄的相位型光栅. 为简单起见, 取 $\widetilde{\alpha}$ 为实数, 即 $\widetilde{\alpha}=\widetilde{\alpha}^*=\alpha$, 式 (7.1-30a) 变为

$$\frac{\mathrm{d}\psi_m(\xi)}{\mathrm{d}\xi}=-\mathrm{j}\frac{\alpha}{2}[\psi_{m-1}(\xi)+\psi_{m+1}(\xi)], \tag{7.1-31}$$

当 $\psi_m(\xi=0)=\psi_{inc}\delta_{m0}$ 时. 利用贝塞尔函数的递推关系 (*recursion relation for Bessel functions*), 得

$$\frac{\mathrm{d}J_m(x)}{\mathrm{d}x}=\frac{1}{2}[J_{m-1}(x)-J_{m+1}(x)]. \tag{7.1-32}$$

可证明声柱内的解为

$$\psi_m(\xi)=(-\mathrm{j})^m\psi_{inc}J_m(\alpha\xi). \tag{7.1-33}$$

这就是有名的 Raman-Nath 解, J_m 是一个 m 阶贝塞尔函数 (*Bessel function of order m*). 在离开声柱时, 即 $\xi=1$ 时, 各散射级次的振幅可由下式给出

$$\psi_m(\xi=1)=(-\mathrm{j})^m\psi_{inc}J_m(\alpha). \tag{7.1-34}$$

该解满足能量守恒

$$\sum_{m=-\infty}^{\infty}I_m=\sum_{m=-\infty}^{\infty}|\psi_m|^2=|\psi_{inc}|^2\sum_{m=-\infty}^{\infty}J_m^2(\alpha)=|\psi_{inc}|^2.$$

理想布拉格衍射

理想布拉格衍射的特征是只产生两个衍射级次, 即对于 $Q\to\infty$, 当 $\phi_{inc}=\phi_B$ [下移布拉格衍射 (*downshifted Bragg diffraction*)] 或 $\phi_{inc}=-\phi_B$ [上移布拉格衍射 (*upshifted Bragg diffraction*)] 时. 图 7.1-9 给出了上移和下移的布拉格衍射 (*upshifted and downshifted Bragg diffraction*).

对于上移相互作用 (*upshifted interaction*), 即 $\phi_{inc}=-\phi_B$ 时, 有 $m=0$ 和 $m=1$ 衍射级次, 对于小角度, 光栅方程变为

$$\phi_m=\phi_{inc}+m\frac{\lambda_0}{\Lambda}.$$

As a consequence of these approximations, we have first-order coupled differential equations in Eq. (7.1-30). Detailed derivation procedures have been discussed [Poon and Kim (2018)] and we leave these procedures as a problem set at the end of the chapter (see P.7.1).

The physical interpretation of Eq. (7.1-30) is that there is a mutual coupling between neighboring orders in the interaction, that is, ψ_m is being contributed by ψ_{m-1} and ψ_{m+1}. We shall use these equations to investigate Raman-Nath and *Bragg diffraction.*

7.1.2 Raman-Nath and Bragg Diffraction

In this section, we will first consider two classical solutions in acousto-optics: ideal Raman-Nath and Bragg diffraction. In theory, *Raman-Nath diffraction* and *Bragg diffraction* are two extreme cases. In ideal Raman-Nath diffraction, the length of the acoustic transducer is considered to be zero, while $L \to \infty$ is for ideal Bragg diffraction. We then conclude the section with a discussion on near Bragg diffraction, where L is finite.

Ideal Raman-Nath Diffraction

In the classical case, we have normal incidence, that is, $\phi_{inc} = 0$, and $Q = 0$. When $Q = 0$, the sound column is considered to be an infinitely thin phase grating. For simplicity, we take $\tilde{\alpha}$ being real, that is, $\tilde{\alpha} = \tilde{\alpha}^* = \alpha$, Eq. (7.1-30a) becomes

$$\frac{\mathrm{d}\psi_m(\xi)}{\mathrm{d}\xi} = -\mathrm{j}\frac{\alpha}{2}[\psi_{m-1}(\xi) + \psi_{m+1}(\xi)], \tag{7.1-31}$$

with $\psi_m(\xi = 0) = \psi_{inc}\delta_{m0}$. Using the recursion relation for Bessel functions,

$$\frac{\mathrm{d}J_m(x)}{\mathrm{d}x} = \frac{1}{2}[J_{m-1}(x) - J_{m+1}(x)], \tag{7.1-32}$$

we can show that the solutions inside the sound column are

$$\psi_m(\xi) = (-\mathrm{j})^m \psi_{inc} J_m(\alpha\xi), \tag{7.1-33}$$

which is the well-known Raman-Nath solution, and J_m is a *Bessel function of order m.* The amplitudes of the various scattering orders upon leaving the sound column, that is, at $\xi = 1$, are given by

$$\psi_m(\xi = 1) = (-\mathrm{j})^m \psi_{inc} J_m(\alpha), \tag{7.1-34}$$

and the solutions satisfy the conservation of energy

$$\sum_{m=-\infty}^{\infty} I_m = \sum_{m=-\infty}^{\infty} |\psi_m|^2 = |\psi_{inc}|^2 \sum_{m=-\infty}^{\infty} J_m^2(\alpha) = |\psi_{inc}|^2.$$

Ideal Bragg Diffraction

Ideal Bragg diffraction is characterized by the generation of only two scattered orders. This occurs when either $\phi_{inc} = \phi_B$ (*downshifted Bragg diffraction*) or when $\phi_{inc} = -\phi_B$ (upshifted Bragg diffraction) for $Q \to \infty$. In Figure 7.1-9, we show *upshifted and downshifted Bragg diffraction.*

For upshifted interaction, that is, $\phi_{inc} = -\phi_B$, we have diffracted orders $m = 0$ and $m = 1$. For small angles, the grating equation becomes

$$\phi_m = \phi_{inc} + m\frac{\lambda_0}{\Lambda}.$$

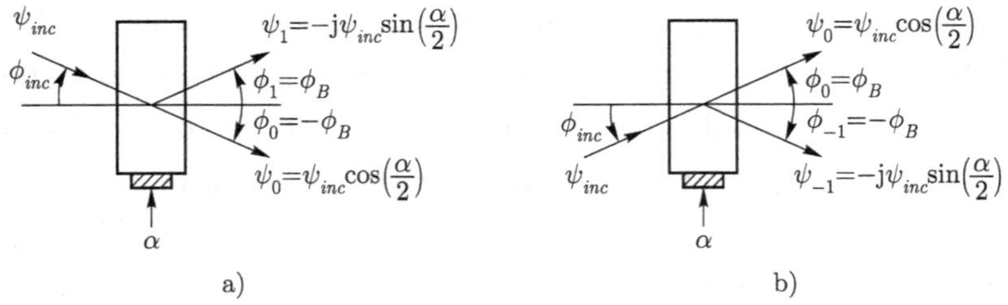

图 7.1-9 布拉格衍射: a) 上移相互作用; b) 下移相互作用

因此, $\phi_0 = \phi_{inc} = -\phi_B$, 及 $\phi_1 = \phi_{inc} + \dfrac{\lambda_0}{\Lambda} = -\phi_B + 2\phi_B = \phi_B$, 如图 7.1-9a) 所示. 注意, 规定角度的逆时针方向为正向, 其物理情况与图 7.1-1 所示相似. 现在, 根据式 (7.1-30), 有如下耦合方程

$$\frac{d\psi_0(\xi)}{d\xi} = -j\frac{\widetilde{\alpha}^*}{2}\psi_1(\xi), \tag{7.1-35a}$$

和

$$\frac{d\psi_1(\xi)}{d\xi} = -j\frac{\widetilde{\alpha}}{2}\psi_0(\xi). \tag{7.1-35b}$$

考虑边界条件式 (7.1-30b), 上述声柱内耦合方程的解为

$$\psi_0(\xi) = \psi_{inc}\cos\left(\frac{\alpha\xi}{2}\right), \tag{7.1-36a}$$

和

$$\psi_1(\xi) = -j\frac{\widetilde{\alpha}}{\alpha}\psi_{inc}\sin\left(\frac{\alpha\xi}{2}\right) = -j\psi_{inc}\sin\left(\frac{\alpha\xi}{2}\right), \tag{7.1-36b}$$

对于 $\widetilde{\alpha} = \widetilde{\alpha}^* = \alpha$, 其中 α 同样是峰值相位延迟, 且与声振幅的强度成正比.

同样, 对于下移相互作用, 即 $\phi_{inc} = \phi_B$ 时, 我们有 $m = 0$ 和 $m = -1$ 衍射级次. 会得到以下的耦合方程

$$\frac{d\psi_0(\xi)}{d\xi} = -j\frac{\widetilde{\alpha}}{2}\psi_{-1}(\xi), \tag{7.1-37a}$$

和

$$\frac{d\psi_{-1}(\xi)}{d\xi} = -j\frac{\widetilde{\alpha}^*}{2}\psi_0(\xi). \tag{7.1-37b}$$

其解为

$$\psi_0(\xi) = \psi_{inc}\cos\left(\frac{\alpha\xi}{2}\right), \tag{7.1-38a}$$

和

$$\psi_{-1}(\xi) = -j\frac{\widetilde{\alpha}^*}{\alpha}\psi_{inc}\sin\left(\frac{\alpha\xi}{2}\right) = -j\psi_{inc}\sin\left(\frac{\alpha\xi}{2}\right), \tag{7.1-38b}$$

对于 $\widetilde{\alpha} = \widetilde{\alpha}^* = \alpha$. 物理情况如图 7.1-9b) 所示. 离开声柱的两个衍射级次的强度, 即在 $\xi = 1$ 的两种情况下, 可由下式给出

$$I_0 = |\psi_0(\xi = 1)|^2 = I_{inc}\cos^2\left(\frac{\alpha}{2}\right), \tag{7.1-39a}$$

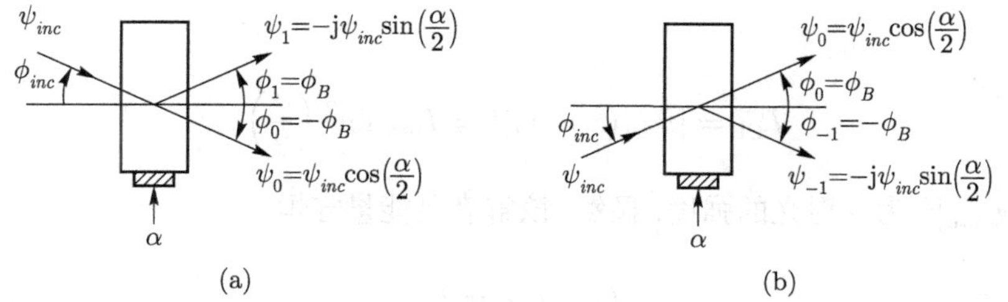

Figure 7.1-9 Bragg diffraction: (a) upshifted interaction and (b) downshifted interaction

Hence, $\phi_0 = \phi_{inc} = -\phi_B$, and $\phi_1 = \phi_{inc} + \dfrac{\lambda_0}{\Lambda} = -\phi_B + 2\phi_B = \phi_B$, as illustrated in Figure 7.1-9a. Remember that the convention for angles is counterclockwise positive. The physical situation is similar to that shown in Figure 7.1-1. Now, according to Eq. (7.1-30), we have the following coupled equations:

$$\frac{d\psi_0(\xi)}{d\xi} = -j\frac{\widetilde{\alpha}^*}{2}\psi_1(\xi), \tag{7.1-35a}$$

and

$$\frac{d\psi_1(\xi)}{d\xi} = -j\frac{\widetilde{\alpha}}{2}\psi_0(\xi). \tag{7.1-35b}$$

The solutions to the above coupled equations inside the sound column, taking the boundary conditions Eq. (7.1-30b) into account, are

$$\psi_0(\xi) = \psi_{inc}\cos\left(\frac{\alpha\xi}{2}\right), \tag{7.1-36a}$$

and

$$\psi_1(\xi) = -j\frac{\widetilde{\alpha}}{\alpha}\psi_{inc}\sin\left(\frac{\alpha\xi}{2}\right) = -j\psi_{inc}\sin\left(\frac{\alpha\xi}{2}\right), \tag{7.1-36b}$$

for $\widetilde{\alpha} = \widetilde{\alpha}^* = \alpha$, where again α is the peak phase delay proportional to the strength of the sound amplitude.

Similarly, for downshifted interaction, that is, $\phi_{inc} = \phi_B$, we have diffracted orders $m = 0$ and $m = -1$. We have the following coupled equations:

$$\frac{d\psi_0(\xi)}{d\xi} = -j\frac{\widetilde{\alpha}}{2}\psi_{-1}(\xi), \tag{7.1-37a}$$

and

$$\frac{d\psi_{-1}(\xi)}{d\xi} = -j\frac{\widetilde{\alpha}^*}{2}\psi_0(\xi). \tag{7.1-37b}$$

The solutions are

$$\psi_0(\xi) = \psi_{inc}\cos\left(\frac{\alpha\xi}{2}\right), \tag{7.1-38a}$$

and

$$\psi_{-1}(\xi) = -j\frac{\widetilde{\alpha}^*}{\alpha}\psi_{inc}\sin\left(\frac{\alpha\xi}{2}\right) = -j\psi_{inc}\sin\left(\frac{\alpha\xi}{2}\right), \tag{7.1-38b}$$

for $\widetilde{\alpha} = \widetilde{\alpha}^* = \alpha$. The physical situation is illustrated in Figure 7.1-9b. The intensities of the two diffracted orders leaving the sound column, that is, at $\xi = 1$, are in both cases given by

$$I_0 = |\psi_0(\xi = 1)|^2 = I_{inc}\cos^2\left(\frac{\alpha}{2}\right), \tag{7.1-39a}$$

和

$$I_{\pm 1} = |\psi_{\pm 1}(\xi = 1)|^2 = I_{inc} \sin^2\left(\frac{\alpha}{2}\right), \tag{7.1-39b}$$

其中, $I_{inc} = |\psi_{inc}|^2$ 为入射光的强度. 显然, 该解满足能量守恒

$$I_0 + I_{\pm 1} = I_{inc}.$$

举例: 能量守恒

直接从耦合方程, 即式 (7.1-35) 或式 (7.1-37), 即可证明能量守恒. 以式 (7.1-35) 为例, 计算

$$\begin{aligned}
&\frac{\mathrm{d}}{\mathrm{d}\xi}\left[|\psi_0(\xi)|^2 + |\psi_1(\xi)|^2\right] = \frac{\mathrm{d}}{\mathrm{d}\xi}\left[\psi_0(\xi)\psi_0^*(\xi) + \psi_1(\xi)\psi_1^*(\xi)\right] \\
&= \frac{\mathrm{d}\psi_0(\xi)}{\mathrm{d}\xi}\psi_0^*(\xi) + \psi_0(\xi)\frac{\mathrm{d}\psi_0^*(\xi)}{\mathrm{d}\xi} + \frac{\mathrm{d}\psi_1(\xi)}{\mathrm{d}\xi}\psi_1^*(\xi) + \psi_1(\xi)\frac{\mathrm{d}\psi_1^*(\xi)}{\mathrm{d}\xi} \\
&= -\mathrm{j}\frac{\widetilde{\alpha}^*}{2}\psi_1(\xi)\psi_0^*(\xi) + \psi_0(\xi)\mathrm{j}\frac{\widetilde{\alpha}}{2}\psi_1^*(\xi) - \mathrm{j}\frac{\widetilde{\alpha}}{2}\psi_0(\xi)\psi_1^*(\xi) + \psi_1(\xi)\mathrm{j}\frac{\widetilde{\alpha}^*}{2}\psi_0^*(\xi) = 0,
\end{aligned}$$

其中, 利用式 (7.1-35) 得出上式的最终结果. 结果表明 $|\psi_0(\xi)|^2 + |\psi_1(\xi)|^2 =$ 常数, 这是能量守恒 (*conservation of energy*) 的另一种表达.

举例: 耦合方程的解

式 (7.1-35a) 和式 (7.1-35b) 为耦合微分方程, 且这些方程可合并为二阶微分方程. 对式 (7.1-35a) 中的 ξ 进行求导, 并结合式 (7.1-35b), 得

$$\frac{\mathrm{d}^2\psi_0(\xi)}{\mathrm{d}\xi^2} = -\mathrm{j}\frac{\widetilde{\alpha}^*}{2}\frac{\mathrm{d}\psi_1(\xi)}{\mathrm{d}\xi} = -\mathrm{j}\frac{\widetilde{\alpha}^*}{2}\left(-\mathrm{j}\frac{\widetilde{\alpha}}{2}\right)\psi_0(\xi) = -\left(\frac{\alpha}{2}\right)^2\psi_0(\xi). \tag{7.1-40}$$

该微分方程的通解 (*general solution*) 是

$$\psi_0(\xi) = A\mathrm{e}^{\mathrm{j}\frac{\alpha}{2}\xi} + B\mathrm{e}^{-\mathrm{j}\frac{\alpha}{2}\xi}. \tag{7.1-41}$$

因为可通过直接代入其二阶微分方程来验证. A 和 B 可通过初始条件 $\psi_0(\xi)|_{\xi=0}$ 和 $\left.\frac{\mathrm{d}\psi_0(\xi)}{\mathrm{d}\xi}\right|_{\xi=0}$ 得到. 根据式 (7.1-30b), 有 $\psi_0(\xi)|_{\xi=0} = \psi_{inc}$, 由式 (7.1-41) 得

$$A + B = \psi_{inc}. \tag{7.1-42a}$$

通过式 (7.1-35a), 可发现, 有第二个初始条件

$$\left.\frac{\mathrm{d}\psi_0(\xi)}{\mathrm{d}\xi}\right|_{\xi=0} = -\mathrm{j}\frac{\widetilde{\alpha}^*}{2}\psi_1(\xi)|_{\xi=0} = 0.$$

此条件再结合式 (7.1-41), 可得出

$$A = B. \tag{7.1-42b}$$

从式 (7.1-42a) 和式 (7.1-42b), 得到 $A = B = \psi_{inc}/2$, 最后得到零级解为

$$\psi_0(\xi) = \left(\frac{\psi_{inc}}{2}\right)\mathrm{e}^{\mathrm{j}\frac{\alpha}{2}\xi} + \left(\frac{\psi_{inc}}{2}\right)\mathrm{e}^{-\mathrm{j}\frac{\alpha}{2}\xi} = \psi_{inc}\cos\left(\frac{\alpha}{2}\xi\right). \tag{7.1-43}$$

and

$$I_{\pm 1} = |\psi_{\pm 1}(\xi = 1)|^2 = I_{inc}\sin^2\left(\frac{\alpha}{2}\right), \tag{7.1-39b}$$

where $I_{inc} = |\psi_{inc}|^2$ is the intensity of the incident light. Clearly, the solutions satisfy the conservation of energy

$$I_0 + I_{\pm 1} = I_{inc}.$$

Example: Conservation of Energy

Directly from the coupled equations, that is, Eqs. (7.1-35) or (7.1-37), we can verity the conservation of energy. Taking Eq. (7.1-35) as an example, we perform

$$\begin{aligned}
&\frac{\mathrm{d}}{\mathrm{d}\xi}\left[|\psi_0(\xi)|^2 + |\psi_1(\xi)|^2\right] = \frac{\mathrm{d}}{\mathrm{d}\xi}\left[\psi_0(\xi)\psi_0^*(\xi) + \psi_1(\xi)\psi_1^*(\xi)\right] \\
&= \frac{\mathrm{d}\psi_0(\xi)}{\mathrm{d}\xi}\psi_0^*(\xi) + \psi_0(\xi)\frac{\mathrm{d}\psi_0^*(\xi)}{\mathrm{d}\xi} + \frac{\mathrm{d}\psi_1(\xi)}{\mathrm{d}\xi}\psi_1^*(\xi) + \psi_1(\xi)\frac{\mathrm{d}\psi_1^*(\xi)}{\mathrm{d}\xi} \\
&= -\mathrm{j}\frac{\widetilde{\alpha}^*}{2}\psi_1(\xi)\psi_0^*(\xi) + \psi_0(\xi)\mathrm{j}\frac{\widetilde{\alpha}}{2}\psi_1^*(\xi) - \mathrm{j}\frac{\widetilde{\alpha}}{2}\psi_0(\xi)\psi_1^*(\xi) + \psi_1(\xi)\mathrm{j}\frac{\widetilde{\alpha}^*}{2}\psi_0^*(\xi) = 0,
\end{aligned}$$

where we have used Eq. (7.1-35) in the above equation to obtain the final result. The result implies that $|\psi_0(\xi)|^2 + |\psi_1(\xi)|^2 =$ constant, an alternative statement for the *conservation of energy*.

Example: Solutions to the Coupled Equations

Equations (7.1-35a) and (7.1-35b) are coupled differential equations and these equations can be combined into a second-order differential equation. Differentiating Eq. (7.1-35a) with respect to ξ and combining with Eq. (7.1-35b), we have

$$\frac{\mathrm{d}^2\psi_0(\xi)}{\mathrm{d}\xi^2} = -\mathrm{j}\frac{\widetilde{\alpha}^*}{2}\frac{\mathrm{d}\psi_1(\xi)}{\mathrm{d}\xi} = -\mathrm{j}\frac{\widetilde{\alpha}^*}{2}\left(-\mathrm{j}\frac{\widetilde{\alpha}}{2}\right)\psi_0(\xi) = -\left(\frac{\alpha}{2}\right)^2\psi_0(\xi). \tag{7.1-40}$$

The *general solution* of the differential equation is

$$\psi_0(\xi) = A\mathrm{e}^{\mathrm{j}\frac{\alpha}{2}\xi} + B\mathrm{e}^{-\mathrm{j}\frac{\alpha}{2}\xi}, \tag{7.1-41}$$

as we can verify by direct substitution into the second-order differential equation. A and B can be found by knowing initial conditions $\psi_0(\xi)|_{\xi=0}$ and $\left.\frac{\mathrm{d}\psi_0(\xi)}{\mathrm{d}\xi}\right|_{\xi=0}$. According to Eq. (7.1-30b), we have $\psi_0(\xi)|_{\xi=0} = \psi_{inc}$, which leads to

$$A + B = \psi_{inc} \tag{7.1-42a}$$

from Eq. (7.1-41). We also find, from Eq. (7.1-35a), the second initial condition:

$$\left.\frac{\mathrm{d}\psi_0(\xi)}{\mathrm{d}\xi}\right|_{\xi=0} = -\mathrm{j}\frac{\widetilde{\alpha}^*}{2}\psi_1(\xi)|_{\xi=0} = 0.$$

This condition together with Eq. (7.1-41) leads to

$$A = B. \tag{7.1-42b}$$

From Eqs. (7.1-42a) and (7.1-42b), we get $A = B = \psi_{inc}/2$, and finally we obtain the solution to the zeroth order as

$$\psi_0(\xi) = \left(\frac{\psi_{inc}}{2}\right)\mathrm{e}^{\mathrm{j}\frac{\alpha}{2}\xi} + \left(\frac{\psi_{inc}}{2}\right)\mathrm{e}^{-\mathrm{j}\frac{\alpha}{2}\xi} = \psi_{inc}\cos\left(\frac{\alpha}{2}\xi\right). \tag{7.1-43}$$

那么, 其一阶解可利用式 (7.1-35a) 得到

$$\psi_1(\xi)=\frac{2}{-\mathrm{j}\widetilde{\alpha}^*}\frac{\mathrm{d}\psi_0(\xi)}{\mathrm{d}\xi}=-\mathrm{j}\frac{\widetilde{\alpha}}{\alpha}\psi_{inc}\sin\left(\frac{\alpha}{2}\xi\right), \tag{7.1-44}$$

即式 (7.1-36b).

近布拉格衍射

近布拉格衍射 (*near Bragg diffraction*) 是指当换能器的长度 L 为有限时, 或相当于 Q 为有限值时. 它的特点是通常产生多个衍射级次. 一般采用式 (7.1-30) 来分析. 这里将讨论两个例子.

情况 1: 入射的平面波偏离布拉格角, 并限制在两个衍射级次

考虑上移的衍射, 让 $\phi_{inc}=-(1+\delta)\phi_B$, 其中 δ 表示为入射平面波偏离布拉格角的量, 参考式 (7.1-30), 为了简便起见, 再次假设 $\widetilde{\alpha}=\widetilde{\alpha}^*=\alpha$, 有

$$\frac{\mathrm{d}\psi_0(\xi)}{\mathrm{d}\xi}=-\mathrm{j}\frac{\alpha}{2}\mathrm{e}^{-\frac{\mathrm{j}\delta Q\xi}{2}}\psi_1(\xi), \tag{7.1-45a}$$

和

$$\frac{\mathrm{d}\psi_1(\xi)}{\mathrm{d}\xi}=-\mathrm{j}\frac{\alpha}{2}\mathrm{e}^{\frac{\mathrm{j}\delta Q\xi}{2}}\psi_0(\xi), \tag{7.1-45b}$$

在初始条件 $\psi_0(\xi=0)=\psi_{inc}$ 和 $\psi_1(\xi=0)=0$ 下. 式 (7.1-45) 的解析解由著名的 Phariseau 公式 (*Phariseau formula*) 给出 [Poon 和 Kim (2018)]

$$\begin{aligned}\psi_0(\xi)=\psi_{inc}\mathrm{e}^{-\frac{\mathrm{j}\delta Q\xi}{4}}\Bigg\{&\cos\left\{\left[(\delta Q/4)^2+(\alpha/2)^2\right]^{\frac{1}{2}}\xi\right\}\\&+\mathrm{j}\frac{\delta Q}{4}\frac{\sin\{[(\delta Q/4)^2+(\alpha/2)^2]^{\frac{1}{2}}\xi\}}{[(\delta Q/4)^2+(\alpha/2)^2]^{\frac{1}{2}}}\Bigg\},\end{aligned} \tag{7.1-46a}$$

和

$$\psi_1(\xi)=\psi_{inc}\mathrm{e}^{\frac{\mathrm{j}\delta Q\xi}{4}}\left\{-\mathrm{j}\frac{\alpha}{2}\frac{\sin\left\{\left[(\delta Q/4)^2+(\alpha/2)^2\right]^{\frac{1}{2}}\xi\right\}}{\left[(\delta Q/4)^2+(\alpha/2)^2\right]^{\frac{1}{2}}}\right\}. \tag{7.1-46b}$$

对于 $\delta=0$, 这些解可简化为理想布拉格衍射 (*ideal Bragg diffraction*) 的解 [见式 (7.1-36)], 且在 $\alpha=n\pi$ $(n=1,2,3,\cdots)$ 处, 两个级次之间存在完全的能量交换, 该情况如图 7.1-10a) 所示, 其中归一化强度 $I_0/I_{inc}=|\psi_0(\xi=1)|^2/|\psi_{inc}|^2$, 且画出了在声柱出口处即 $\xi=1$ 时, $I_1/I_{inc}=|\psi_1(\xi=1)|^2/|\psi_{inc}|^2$ 随 α 变化的曲线, 对于图 7.1-10b) 所示的结果, 使用了 $\delta=0.25$, 即偏离布拉格角为 $0.25\phi_B$, 以及 $Q=14$ (IntraAction 公司的 AOM-40 型号的典型值). δ 即为 m-文件中的 “d”. 可以看出, 当 α 相对比较小, 在 3 的附近时, 从 I_0 到 I_{inc} 的能量转换是准周期的, 且在初始时是转换不完全的, 反之亦然.

The 1st order can then be found by using Eq. (7.1-35a):

$$\psi_1(\xi) = \frac{2}{-\mathrm{j}\tilde{\alpha}^*}\frac{\mathrm{d}\psi_0(\xi)}{\mathrm{d}\xi} = -\mathrm{j}\frac{\tilde{\alpha}}{\alpha}\psi_{inc}\sin\left(\frac{\alpha}{2}\xi\right), \tag{7.1-44}$$

which is Eq. (7.1-36b).

Near Bragg Diffraction

Near Bragg diffraction is when the length of the transducer L is finite, or equivalently Q is finite. It is usually characterized by the generation of multiple diffracted orders. In general, Eq. (7.1-30) is used to analyze the situation. We will discuss two cases as examples.

Case 1: Incident Plane Wave off the Exact Bragg Angle and Limiting to Two Diffracted Orders

We consider upshifted diffraction. We let $\phi_{inc} = -(1+\delta)\phi_B$, where δ represents the deviation of the incident plane wave away from the Bragg angle. With reference to Eq. (7.1-30), we again assume $\tilde{\alpha} = \tilde{\alpha}^* = \alpha$ for simplicity. We have

$$\frac{\mathrm{d}\psi_0(\xi)}{\mathrm{d}\xi} = -\mathrm{j}\frac{\alpha}{2}\mathrm{e}^{-\frac{\mathrm{j}\delta Q\xi}{2}}\psi_1(\xi), \tag{7.1-45a}$$

and

$$\frac{\mathrm{d}\psi_1(\xi)}{\mathrm{d}\xi} = -\mathrm{j}\frac{\alpha}{2}\mathrm{e}^{\frac{\mathrm{j}\delta Q\xi}{2}}\psi_0(\xi), \tag{7.1-45b}$$

with the initial conditions $\psi_0(\xi = 0) = \psi_{inc}$, and $\psi_1(\xi = 0) = 0$. Equation (7.1-45) has analytical solutions given by the well-known *Phariseau formula* [see Poon and Kim (2018)]:

$$\psi_0(\xi) = \psi_{inc}\mathrm{e}^{-\frac{\mathrm{j}\delta Q\xi}{4}}\left\{\cos\left\{\left[(\delta Q/4)^2 + (\alpha/2)^2\right]^{\frac{1}{2}}\xi\right\} + \mathrm{j}\frac{\delta Q}{4}\frac{\sin\{[(\delta Q/4)^2 + (\alpha/2)^2]^{\frac{1}{2}}\xi\}}{[(\delta Q/4)^2 + (\alpha/2)^2]^{\frac{1}{2}}}\right\}, \tag{7.1-46a}$$

and

$$\psi_1(\xi) = \psi_{inc}\mathrm{e}^{\frac{\mathrm{j}\delta Q\xi}{4}}\left\{-\mathrm{j}\frac{\alpha}{2}\frac{\sin\left\{\left[(\delta Q/4)^2 + (\alpha/2)^2\right]^{\frac{1}{2}}\xi\right\}}{[(\delta Q/4)^2 + (\alpha/2)^2]^{\frac{1}{2}}}\right\}. \tag{7.1-46b}$$

For $\delta = 0$, these solutions reduce to those given for *ideal Bragg diffraction* [see Eq. (7.1-36)], and there is a complete power exchange between the two orders at $\alpha = n\pi$ $(n = 1, 2, 3, \cdots)$. The situation is shown in Figure 7.1-10a, where the normalized intensities $I_0/I_{inc} = |\psi_0(\xi = 1)|^2/|\psi_{inc}|^2$ and $I_1/I_{inc} = |\psi_1(\xi = 1)|^2/|\psi_{inc}|^2$ are plotted against α at the exit of the sound column, $\xi = 1$. For the results shown in Figure 7.1-10b, we have used $\delta = 0.25$, that is, a deviation from the Bragg angle of $0.25\phi_B$, and for $Q = 14$ (Typical value for model AOM-40 from IntraAction Corporation). δ is "d" in the m-file. We note that the power transfer from I_0 to I_{inc} and vice versa is quasi-periodic and initial incomplete when α is relatively small around 3.

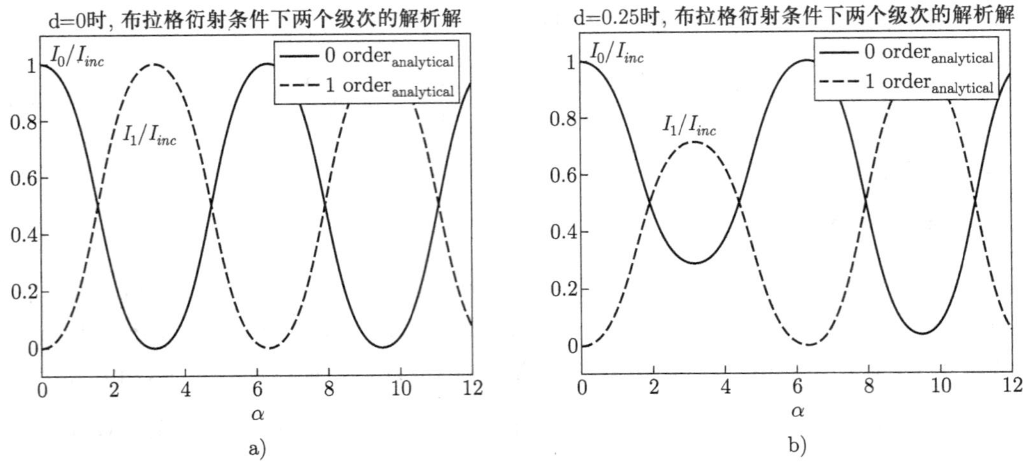

图 7.1-10 a) 完全布拉格衍射 (*pure Bragg diffraction*) 条件下级次之间的功率交换 ($\delta=0$); b) 近布拉格衍射角度条件下的部分功率交换 ($\delta=0.25, Q=14$)

图 7.1-10 是利用下面的 m-文件生成的.

```
==========================================
%AO_Bragg_2order.m, plotting of Eq. (7.1-46)

clear

d=input('delta = ?')
Q=input('Q = ?')
al_e=input('End point of alpha = ?')

n=0;
  for al=0:0.01*pi:al_e
  n=n+1;
  AL(n)=al;

  ps0_a(n)=exp(-j*d*Q/4)*( cos(((d*Q/4)^2+ (AL(n)/2 )^2 )^0.5) +
j*d*Q/4*sin(((d*Q/4)^2+(AL(n)/2) ^2)^0.5)/(((d*Q/4)^2+ (AL(n)/2)^2)^0.5) );
  ps1_a(n)=exp(j*d*Q/4)*-j*AL(n)/2*sin(((d*Q/4)^2+(AL(n)/2)
^2)^0.5)/(((d*Q/4)^2+ (AL(n)/2)^2)^0.5) ;

  end

  plot(AL, ps0_a.*conj(ps0_a), '-.', AL, ps1_a.*conj(ps1_a), '--')
  title('Two-order analytical solutions in Bragg diffraction, d=0.25')
  %xlabel('alpha')
  axis([0 al_e -0.1 1.1])
  legend('0 order_a_n_a_l_y_t_i_c_a_l', '1 order_a_n_a_l_y_t_i_c_a_l')
  grid on
==========================================
```

情况 2: 有限的 Q 并限于四个衍射级次

为了在近布拉格条件 (*near Bragg regime*) 得到更精确的结果, 计算中应包含许多更高级次的衍射项. 这里限制在只有四阶的情况下, 即 $2\geqslant m\geqslant -1$. 对于 $\tilde{\alpha}=\tilde{\alpha}^*=\alpha$, 由式 (7.1-30) 可以得到以下的耦合方程

$$\begin{aligned}\frac{\mathrm{d}\psi_2(\xi)}{\mathrm{d}\xi}&=-\mathrm{j}\frac{\alpha}{2}\mathrm{e}^{-\mathrm{j}\frac{1}{2}Q\xi\left(\frac{\phi_{inc}}{\phi_B}+3\right)}\psi_1(\xi),\\ \frac{\mathrm{d}\psi_1(\xi)}{\mathrm{d}\xi}&=-\mathrm{j}\frac{\alpha}{2}\mathrm{e}^{-\mathrm{j}\frac{1}{2}Q\xi\left(\frac{\phi_{inc}}{\phi_B}+1\right)}\psi_0(\xi)-\mathrm{j}\frac{\alpha}{2}\mathrm{e}^{\mathrm{j}\frac{1}{2}Q\xi\left(\frac{\phi_{inc}}{\phi_B}+3\right)}\psi_2(\xi),\\ \frac{\mathrm{d}\psi_0(\xi)}{\mathrm{d}\xi}&=-\mathrm{j}\frac{\alpha}{2}\mathrm{e}^{-\mathrm{j}\frac{1}{2}Q\xi\left(\frac{\phi_{inc}}{\phi_B}-1\right)}\psi_{-1}(\xi)-\mathrm{j}\frac{\alpha}{2}\mathrm{e}^{\mathrm{j}\frac{1}{2}Q\xi\left(\frac{\phi_{inc}}{\phi_B}+1\right)}\psi_1(\xi),\\ \frac{\mathrm{d}\psi_{-1}(\xi)}{\mathrm{d}\xi}&=-\mathrm{j}\frac{\alpha}{2}\mathrm{e}^{\mathrm{j}\frac{1}{2}Q\xi\left(\frac{\phi_{inc}}{\phi_B}-1\right)}\psi_0(\xi).\end{aligned}\tag{7.1-47}$$

这里给出了一些数值结果, 对于 $\phi_{inc}=-(1+\delta)\phi_B$ 和 $Q=14$. 运行 Bragg_regime_4.m 文件, 在 $\delta=0$ 和 $\delta=0.25$ 时, 其结果分别如图 7.1-11a) 和图 7.1-11b) 所示. 图 7.1-11a) 和图 7.1-11b) 的结果可以分别与图 7.1-10a) 和图 7.1-10b) 的结果直接比较. 具体来说, 要强调的是, 当 α 升高并接近 10 时, 另外两个级次, 即 I_2 和 I_{-1} 开始出现, 如图 7.1-11a) 所示. 而在图 7.1-11b) 中, 对于偏离布拉格角衍射 (*off-Bragg angle diffraction*) 的情况, 当 $\delta=0.25$ 时, I_2 和 I_{-1} 开始出现得更早, 大概在 $\alpha\approx4$ 位置处. 在图 7.1-11c) 中, 可以看出, 所有归一化强度的总和为 1, 即 $\sum\limits_{m=-1}^{m=2}I_m/I_{inc}=1$, 这使得数值计算有了一定的保障.

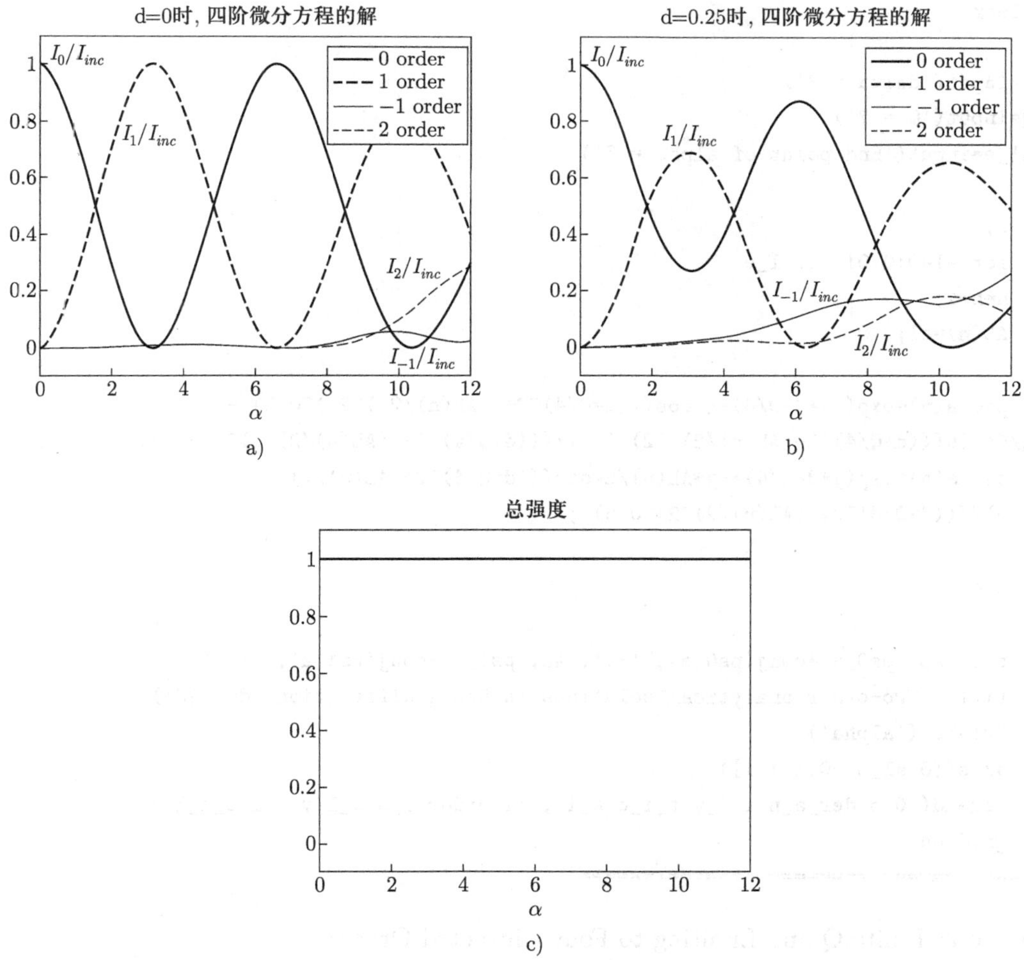

图 7.1-11　a) 在近布拉格衍射中四个级次之间的功率交换 ($\delta=0, Q=14$); b) 在近布拉格衍射中四个级次之间的功率交换 ($\delta=0.25, Q=14$); c) 归一化强度的求和

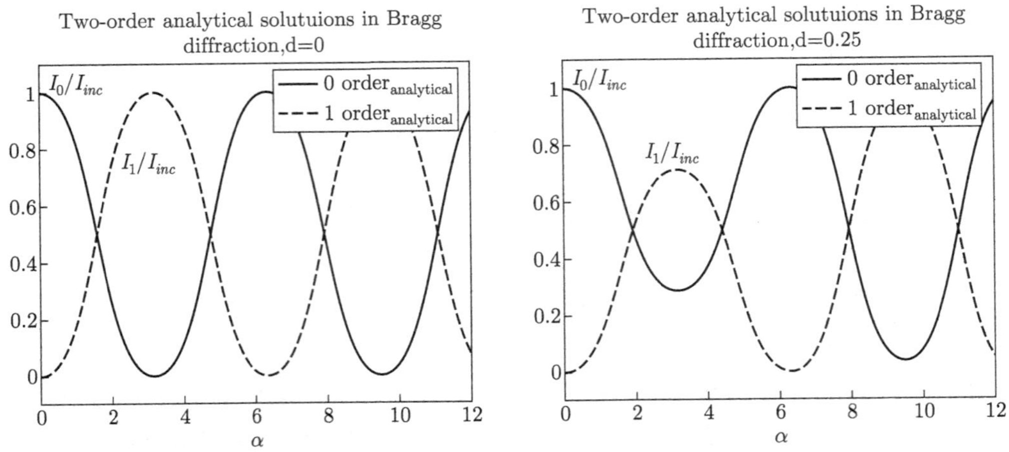

(a) (b)

Figure 7.1-10 (a) Power exchange between orders in *pure Bragg diffraction* ($\delta = 0$) and (b) partial power exchange in near Bragg diffraction ($\delta = 0.25, Q = 14$).

The above plots are generated using the m-file shown below.

```
=========================================
   %AO_Bragg_2order.m, plotting of Eq.\ (7.1-46)

   clear

   d=input('delta = ?')
   Q=input('Q = ?')
   al_e=input('End point of alpha = ?')

   n=0;
     for al=0:0.01*pi:al_e
     n=n+1;
     AL(n)=al;

     ps0_a(n)=exp(-j*d*Q/4)*( cos(((d*Q/4)^2+ (AL(n)/2 )^2 )^0.5) +
j*d*Q/4*sin(((d*Q/4)^2+(AL(n)/2) ^2)^0.5)/(((d*Q/4)^2+ (AL(n)/2)^2)^0.5) );
     ps1_a(n)=exp(j*d*Q/4)*-j*AL(n)/2*sin(((d*Q/4)^2+(AL(n)/2)
^2)^0.5)/(((d*Q/4)^2+ (AL(n)/2)^2)^0.5) ;

     end

     plot(AL, ps0_a.*conj(ps0_a), '-.', AL, ps1_a.*conj(ps1_a), '--')
     title('Two-order analytical solutions in Bragg diffraction, d=0.25')
     %xlabel('alpha')
     axis([0 al_e -0.1 1.1])
     legend('0 order_a_n_a_l_y_t_i_c_a_l', '1 order_a_n_a_l_y_t_i_c_a_l')
     grid on
=========================================
```

Case 2: Finite Q and Limiting to Four Diffracted Orders

For a more accurate result in the *near Bragg regime*, many higher diffracted orders should be included in the calculations. We restrict ourselves to the case of only four orders, that is, $2 \geqslant m \geqslant -1$. We obtain the following coupled equations from Eq. (7.1-30) for $\widetilde{\alpha} = \widetilde{\alpha}^* = \alpha$:

$$\frac{d\psi_2(\xi)}{d\xi} = -j\frac{\alpha}{2}e^{-j\frac{1}{2}Q\xi\left(\frac{\phi_{inc}}{\phi_B}+3\right)}\psi_1(\xi),$$
$$\frac{d\psi_1(\xi)}{d\xi} = -j\frac{\alpha}{2}e^{-j\frac{1}{2}Q\xi\left(\frac{\phi_{inc}}{\phi_B}+1\right)}\psi_0(\xi) - j\frac{\alpha}{2}e^{j\frac{1}{2}Q\xi\left(\frac{\phi_{inc}}{\phi_B}+3\right)}\psi_2(\xi),$$
$$\frac{d\psi_0(\xi)}{d\xi} = -j\frac{\alpha}{2}e^{-j\frac{1}{2}Q\xi\left(\frac{\phi_{inc}}{\phi_B}-1\right)}\psi_{-1}(\xi) - j\frac{\alpha}{2}e^{j\frac{1}{2}Q\xi\left(\frac{\phi_{inc}}{\phi_B}+1\right)}\psi_1(\xi),$$
$$\frac{d\psi_{-1}(\xi)}{d\xi} = -j\frac{\alpha}{2}e^{j\frac{1}{2}Q\xi\left(\frac{\phi_{inc}}{\phi_B}-1\right)}\psi_0(\xi). \tag{7.1-47}$$

We demonstrate some numerical results. For $\phi_{inc} = -(1+\delta)\phi_B$ and $Q = 14$, we run Bragg_regime_4.m for $\delta = 0$ and 0.25 with results shown in Figure 7.1-11a and b, respectively. The results in Figure 7.1-11a and b can be directly compared with those given in Figure 7.1-10a and b, respectively. Specifically, we want to point out that as α gets higher toward around 10, the other two orders, that is, I_2 and I_{-1}, begin to appear in Figure 7.1-11a. I_2 and I_{-1} begin to appear around much earlier at $\alpha \approx 4$ for *off-Bragg angle diffraction* when $\delta = 0.25$, as shown in Figure 7.1-11b. In Figure 7.1-11c, we show the sum of all normalized intensities to be unity, that is, $\sum_{m=-1}^{m=2} I_m/I_{inc} = 1$. This gives us some assurance of the numerical calculations.

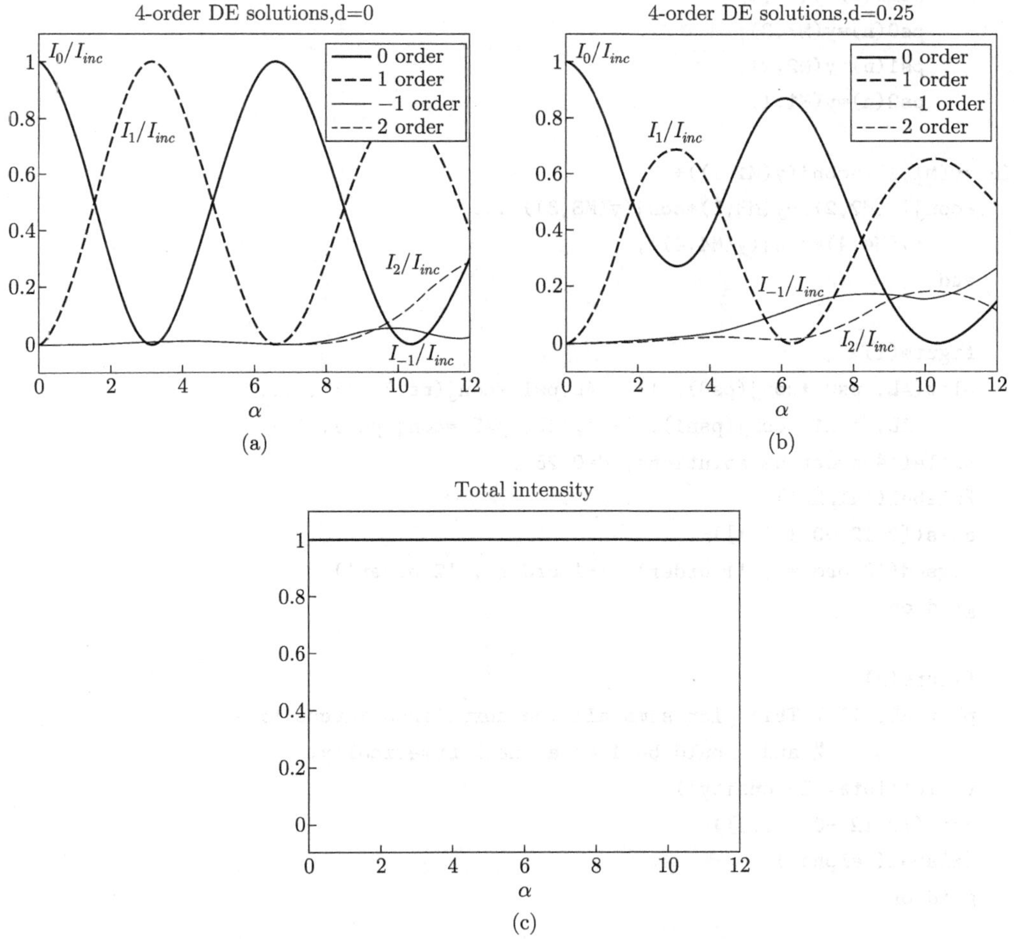

Figure 7.1-11 (a) Power exchange between four orders in near Bragg diffraction ($\delta = 0, Q = 14$), (b) power exchange between four orders in near Bragg diffraction ($\delta = 0.25, Q = 14$), and (c) sum of normalized intensities

```
=========================================
    % Bragg_regime_4.m
    % Adapted from "Engineering Optics with MATLAB, 2nd ed."
    % by T.-C. Poon and T. Kim, World Scientific (2018), Table 4.1.
    % Near Bragg regime involving 4 diffracted orders
    clear

    d=input('delta =')
    Q=input('Q = ')
    n=0;

    for al=0:0.01*pi:14
            n=n+1;
            AL(n)=al;
            [nz,y]=ode45('AO_B4', [0 1], [0 0 1 0], [], d, al, Q) ;

            [M1 N1]=size(y(:,1));
            [M2 N2]=size(y(:,2));
            [M3 N3]=size(y(:,3));
            [M4 N4]=size(y(:,4));

            psn1(n)=y(M4,4);
            ps0(n)=y(M3,3);
            ps1(n)=y(M2,2);
            ps2(n)=y(M1,1);

    I(n)=y(M1,1).*conj(y(M1,1))+
y(M2,2).*conj(y(M2,2))+y(M3,3)*conj(y(M3,3)) ...
            +y(M4,4)*conj(y(M4,4));
        end

        figure(1)
        plot(AL, ps0.*conj(ps0), '-', AL,ps1.*conj(ps1), ':', ...
             AL, psn1.*conj(psn1), '-.', AL, ps2.*conj(ps2), '--')
        title('4-order DE solutions, d=0.25')
        %xlabel('alpha')
        axis([0 12 -0.1 1.1])
        legend('0 order', '1 order', '-1 order', '2 order')
        grid on

        figure(2)
        plot(AL, I) % This plot sums all the normalized intensities
                    % and should be 1 as a check numerically.
        title('Total Intensity')
        axis([0 12 -0.1 1.1])
        %xlabel('alpha')
        grid on

    %AO_B4.m  %creating MATLAB Function for use in Bragg_regime_4.m

    function dy=AO_B4(nz,y,options,d,a,Q)
```

==

```
% Bragg_regime_4.m
% Adapted from "Engineering Optics with MATLAB, 2nd ed."
% by T.-C. Poon and T. Kim, World Scientific (2018), Table 4.1.
% Near Bragg regime involving 4 diffracted orders
clear

d=input('delta =')
Q=input('Q = ')
n=0;

for al=0:0.01*pi:14
        n=n+1;
        AL(n)=al;
        [nz,y]=ode45('AO_B4', [0 1], [0 0 1 0], [], d, al, Q) ;

        [M1 N1]=size(y(:,1));
        [M2 N2]=size(y(:,2));
        [M3 N3]=size(y(:,3));
        [M4 N4]=size(y(:,4));

        psn1(n)=y(M4,4);
        ps0(n)=y(M3,3);
        ps1(n)=y(M2,2);
        ps2(n)=y(M1,1);

    I(n)=y(M1,1).*conj(y(M1,1))+
y(M2,2).*conj(y(M2,2))+y(M3,3)*conj(y(M3,3)) ...
        +y(M4,4)*conj(y(M4,4));
    end

    figure(1)
    plot(AL, ps0.*conj(ps0), '-', AL,ps1.*conj(ps1), ':', ...
        AL, psn1.*conj(psn1), '-.', AL, ps2.*conj(ps2), '--')
    title('4-order DE solutions, d=0.25')
    %xlabel('alpha')
    axis([0 12 -0.1 1.1])
    legend('0 order', '1 order', '-1 order', '2 order')
    grid on

    figure(2)
    plot(AL, I) % This plot sums all the normalized intensities
               % and should be 1 as a check numerically.
    title('Total Intensity')
    axis([0 12 -0.1 1.1])
    %xlabel('alpha')
    grid on

%AO_B4.m  %creating MATLAB Function for use in Bragg_regime_4.m

function dy=AO_B4(nz,y,options,d,a,Q)
```

```
    dy=zeros(4,1);  %a column vector

    % -1<= m <=2
    % d=delta
    % nz=normalized z

    %m=2 -> y(1)
    %m=1 -> y(2)
    %m=0 -> y(3)
    %m=-1 -> y(4)

    dy(1)=-j*a/2*y(2)*exp(-j*Q/2*nz*(-(1+d)+3))  +  0          ;

    dy(2)=-j*a/2*y(3)*exp(-j*Q/2*nz*(-(1+d)+1))  +  -
j*a/2*y(1)*exp(j*Q/2*nz*(-(1+d)+3));

    dy(3)=-j*a/2*y(4)*exp(-j*Q/2*nz*(-(1+d)-1))  +  -
j*a/2*y(2)*exp(j*Q/2*nz*(-(1+d)+1));

    dy(4)=0  +  -j*a/2*y(3)*exp(j*Q/2*nz*(-(1+d)-1));

    return
========================================
```

7.1.3　声光效应的典型应用

激光强度调制 (*laser intensity modulation*)

光调制 (*optical modulation*) 是通过改变光波的振幅、强度、频率、相位或偏振来传递信息信号 (如音频信号) 的过程. 其相反过程, 即从光波中提取信息信号, 则被称为解调 (*demodulation*). 本节中, 利用声光效应讨论激光强度调制. 通过改变声波的振幅, 即通过 α, 可实现对衍射激光束的强度调制 (*intensity modulation*). 通过观察式 (7.1-39) 可以清晰地发现这一事实. 在强度调制中, 激光束的强度随信息信号 $m(t)$ [在通信理论中被称为调制信号 (*modulating signal*)] 成比例地变化. 由此产生的调制强度 $I(t)$ 被称为已调制信号 (*modulated signal*). 对于一个给定的信号 $m(t)$, 我们首先需要通过一个 AM 调制器 (*AM modulator*) 生成所谓的振幅调制 (*amplitude modulation*, AM) 信号. 在振幅调制过程中, 频率和相位固定的一个正弦信号的振幅随 $m(t)$ 成比例变化.

一个基本的 AM 调制器由一个电子乘法器 (*electronic multiplier*) 和一个加法器 (*adder*) 组成, 如图 7.1-12 所示. 相应地, AM 调制器的输出为一个振幅调制的 (*amplitude-modulated*) 信号 $y_{\mathrm{AM}}(t) = [A_c + m(t)]\cos(\Omega t)$, 其中正弦信号 $\cos(\Omega t)$ 被称为 AM 信号的载波 (*carrier*), 且在标准的 AM 中, $A_c > |m(t)|_{\max}$. 然后, 该已调制信号被用来驱动声光调制器的声换能器. 因此, 当 $\alpha \propto |A|$ 时, $\alpha(t) \propto A_c + m(t)$ [见式 (7.1-28)]. 为方便起见, 取 $\alpha = A_c + m(t)$, 并将其画出来, 其中 $m(t)$ 为一正弦曲线, 而 A_c 是正弦平方强度曲线上 P 点处的恒定偏置. 情况如图 7.1-13 所示. 该声光调制器应偏置到线性工作区域时产生 50% 强度传输的点处.

```
    dy=zeros(4,1);  %a column vector

    % -1<= m <=2
    % d=delta
    % nz=normalized z

    %m=2 -> y(1)
    %m=1 -> y(2)
    %m=0 -> y(3)
    %m=-1 -> y(4)

    dy(1)=-j*a/2*y(2)*exp(-j*Q/2*nz*(-(1+d)+3))  +  0          ;

    dy(2)=-j*a/2*y(3)*exp(-j*Q/2*nz*(-(1+d)+1))  +  -
j*a/2*y(1)*exp(j*Q/2*nz*(-(1+d)+3));

    dy(3)=-j*a/2*y(4)*exp(-j*Q/2*nz*(-(1+d)-1))  +  -
j*a/2*y(2)*exp(j*Q/2*nz*(-(1+d)+1));

    dy(4)=0  +  -j*a/2*y(3)*exp(j*Q/2*nz*(-(1+d)-1));

    return
=========================================
```

7.1.3 Typical Applications of the Acousto-Optic Effect

Laser Intensity Modulation

Optical modulation is the process of changing the amplitude, intensity, frequency, phase, or polarization of an optical wave to convey an information signal (such as an audio signal). The reverse process, that is, extracting the information signal from the optical wave, is called *demodulation.* In this section, we discuss laser intensity modulation by use of the acousto-optic effect. By changing the amplitude of the sound, that is, though α, we can achieve *intensity modulation* of the diffracted laser beams. This fact is clearly shown by inspecting Eqs. (7.1-39a) and (7.1-39b). In intensity modulation, the intensity of the laser beam is varied in proportion to the information signal $m(t)$ (known as the *modulating signal* in communication theory). The resulting modulated intensity $I(t)$ is known as the *modulated signal.* For a given $m(t)$, we first need to generate the so-called *amplitude modulation* (AM) signal through an AM modulator. In amplitude modulation, the amplitude of a sinusoidal signal, whose frequency and phase are fixed, is varied in proportion to $m(t)$.

A basic AM modulator consists of an electronic multiplier and an adder, as shown in Figure 7.1-12 schematically. Accordingly, the output of the AM modulator is an amplitude-modulated signal $y_{\mathrm{AM}}(t) = [A_c + m(t)]\cos(\Omega t)$, where the sinusoidal $\cos(\Omega t)$ is known as the *carrier* of the AM signal, and $A_c > |m(t)|_{\max}$ in standard AM. This modulated signal is then used to drive the acoustic transducer of the acousto-optic modulator. Hence, $\alpha(t) \propto [A_c + m(t)]$ as $\alpha \propto |A|$ [see Eq. (7.1-28)]. For convenience, we take $\alpha = [A_c + m(t)]$ and plot it for $m(t)$ as a sinusoidal variation with A_c being a constant bias on point P of the sine squared intensity curve. The situation is shown in Figure 7.1-13. The acousto-optic modulator should be biased to the point which results in a 50% intensity transmission for linear operation.

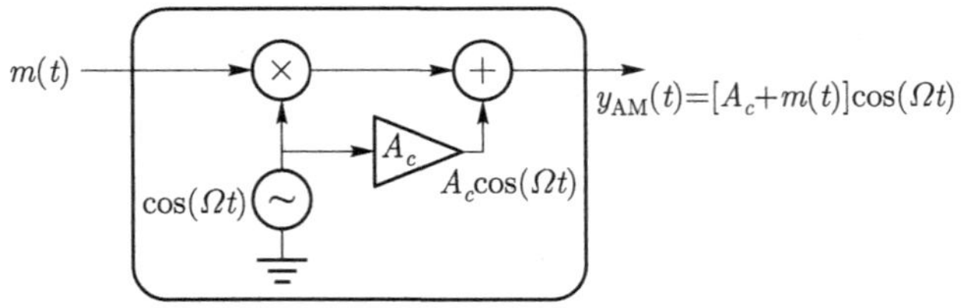

图 7.1-12 AM 调制器: 其三角形表示增益为 A_c 的放大器的电路符号

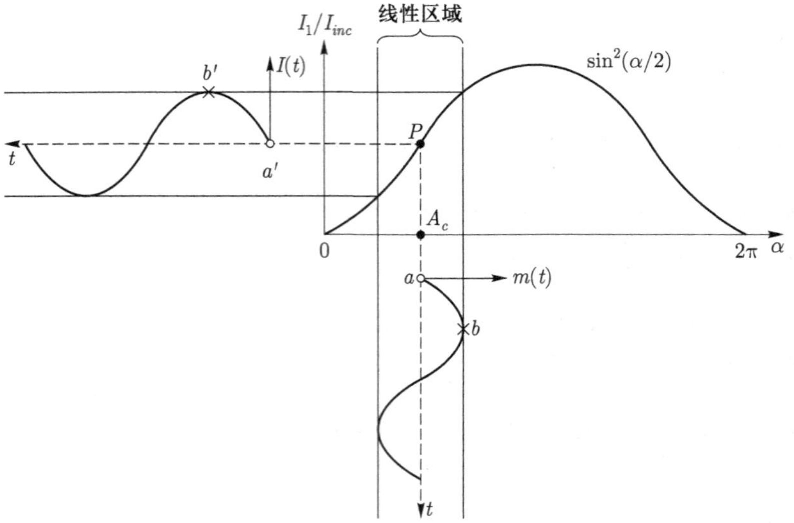

图 7.1-13 声光强度调制原理: 一级光束的调制信号 $m(t)$ 和调制强度 $I(t)$ 之间的关系

α 的变化会对一阶光束的强度 $I(t)$ 产生一个时变强度, 如图所见, 例如, 在曲线 $m(t)$ 上的点 a 和 b 分别在曲线 $I(t)$ 上给出输出强度 a' 和 b'. 为了解调 $I(t)$, 只需将调制的激光束指向一个光电探测器, 根据式 (4.4-3), 光电探测器输出的电流 $i(t)$ 与入射强度成正比. 因此, $i(t) \propto I(t) \propto m(t)$, 图 7.1-14a) 给出了用于演示强度调制的实验装置. $m(t)$ 为音频信号. 图 7.1-14b) 给出了两个光电探测器 PD1 和 PD0 的输出. 可以看出, 两个电信号之间有一个负号. 其原因是, 零阶衍射级次在余弦平方强度, 即 $\cos^2\left(\dfrac{\alpha}{2}\right)$ 曲线范围内工作, 故, 两个衍射级次在线性区域内利用相反的斜率 (*opposite slopes*) 来处理信息.

声光激光光束偏转器 (*acousto-optic laser beam deflector*)

在强度调制中, 主要改变声波信号的振幅. 在激光偏转 (*laser deflection*) 的应用中, 主要改变声波的频率, 这种情况如图 7.1-15 所示.

将一级光束和零级光束之间的夹角定义为偏转角 (*defection angle*) ϕ_d

$$\phi_d = 2\phi_B \approx \frac{\lambda_0}{\Lambda} = \frac{\lambda_0 f_s}{v_s},$$

由于布拉格角一般都很小. 通过改变声波的频率 f_s, 可改变其偏转角度, 从而控制光束的传播

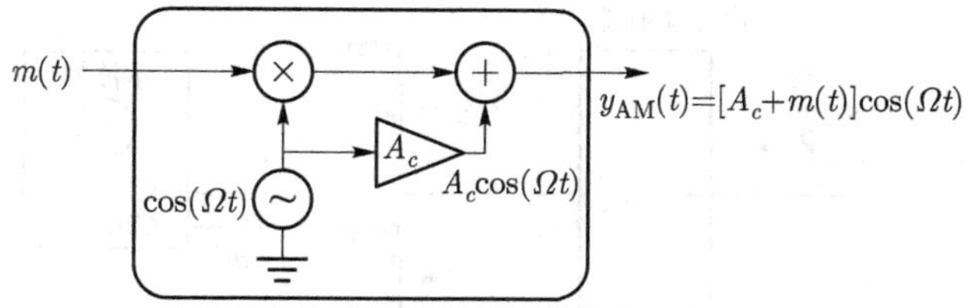

Figure 7.1-12 AM modulator: The triangle represents a circuit symbol of an amplifier of gain A_c

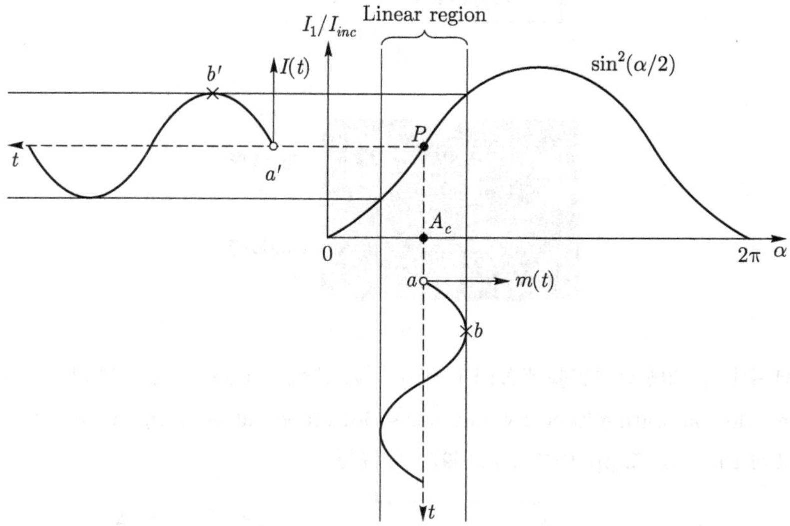

Figure 7.1-13 Acousto-optic intensity modulation principle: relationship between the modulating signal, $m(t)$, and the modulated intensity, $I(t)$, of the first-order beam

The variation of α produces a time-varying intensity on the intensity of the first-order beam $I(t)$, as we see, for example, that points a and b on the $m(t)$ curve give intensities a' and b' as outputs on the $I(t)$ curve, respectively. To demodulate $I(t)$, we simply direct the modulated laser beam toward a photodetector. According to Eq. (4.4-3), the current output $i(t)$ from the photodetector is proportional to the incident intensity; therefore, $i(t) \propto I(t) \propto m(t)$. Figure 7.1-14a shows an experimental setup used to demonstrate intensity modulation. $m(t)$ is an audio signal. Figure 7.1-14b shows the output of the two photodetectors PD1 and PD0. Note that there is a negative sign between the two electrical signals. The reason is that the two diffracted orders process information in the linear regions with opposite slopes as the zeroth-order diffracted beam works with the cosine squared intensity curve, that is, $\cos^2\left(\frac{\alpha}{2}\right)$.

Acousto-Optic Laser Beam Deflector

In intensity modulation, we change the amplitude of the sound signal. For applications in *laser deflection*, we change the frequency of the sound. The situation is shown in Figure 7.1-15.

We define the angle between the first-order beam and the zeroth-order beam as the *defection angle* ϕ_d:

$$\phi_d = 2\phi_B \approx \frac{\lambda_0}{\Lambda} = \frac{\lambda_0 f_s}{v_s}$$

as the Bragg angle is in general small. By changing the frequency of the sound wave f_s, we can change the deflection angle, thereby controlling the direction of the light beam propagation.

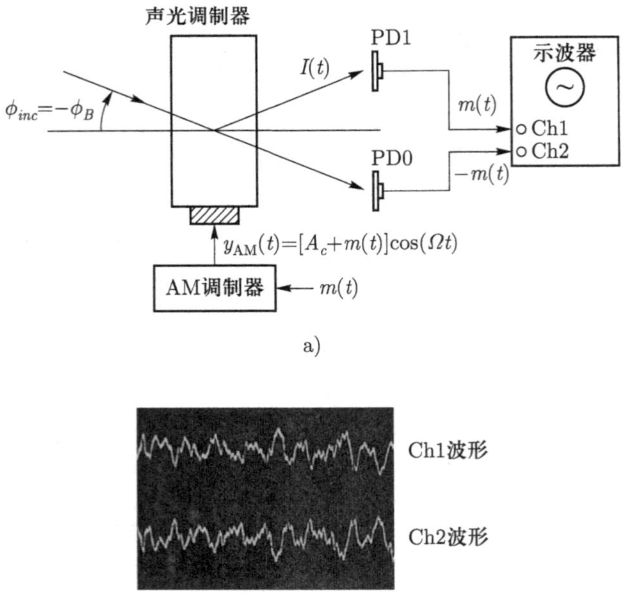

图 7.1-14　a) AM 中调制和解调的实验装置; b) 解调信号. 转载自 Poon et al., Modern optical signal processing experiments demonstrating intensity and pulse-width modulation using an acousto-optic modulator, American Journal of Physics 65, pp. 917-925 (1997), 经许可

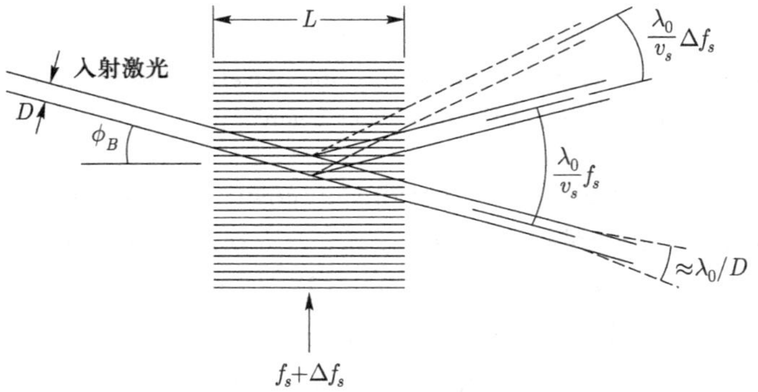

图 7.1-15　声光激光光束偏转器

方向. 声波频率的变化 Δf_s 引起的偏转角的变化 $\Delta\phi_d$ 为

$$\Delta\phi_d = \Delta(2\phi_B) = \frac{\lambda_0}{v_s}\Delta f_s. \tag{7.1-48}$$

对于光偏转器 (*optical deflector*), 考虑其可分辨角的数量 (*number of resolvable angles*) N 是有意义的, 它被定义为偏转角的范围 $\Delta\phi_d$ 与扫描激光束的扩散角 (*angular spread*) $\Delta\phi \approx \lambda_0/D$ 的比值, 即

$$N = \frac{\Delta\phi_d}{\Delta\phi} = \frac{\frac{\lambda_0}{v_s}\Delta f_s}{\frac{\lambda_0}{D}} = \tau\Delta f_s, \tag{7.1-49}$$

其中, $\tau = D/v_s$ 是声波通过激光束的传输时间 (*transit time*), D 为激光束的直径. 当 $D = 1$ cm, 且当声波以 $v_s \sim 4000$ m/s 的速度在玻璃中传播时, 则 $\tau = 2.5$ μs. 如果我们要求 $N = 500$, 则需要 $\Delta f_s = 200$ MHz.

声光偏转器由于没有运动部件, 相比机械扫描器件, 可精确实现高速的角度偏转. 然而, 与基于反射镜的扫描振镜相比, 可分辨的点数较少.

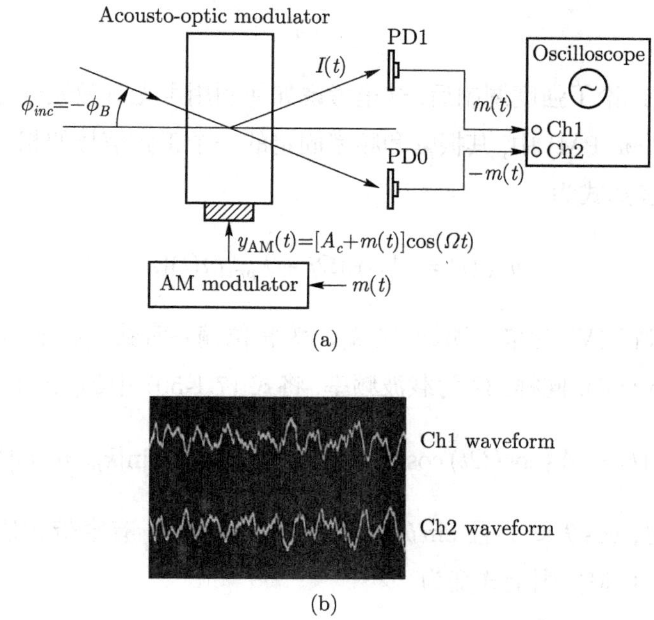

Figure 7.1-14 (a) Experimental setup for AM modulation and demodulation and (b) demodulated signals. Reprinted from Poon et al., Modern optical signal processing experiments demonstrating intensity and pulse-width modulation using an acousto-optic modulator, American Journal of Physics 65, pp. 917-925 (1997), with permission

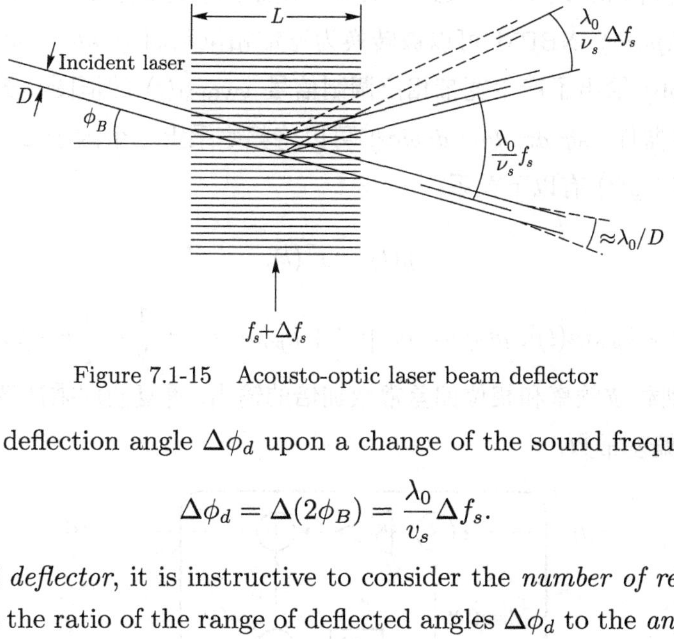

Figure 7.1-15 Acousto-optic laser beam deflector

The change in the deflection angle $\Delta\phi_d$ upon a change of the sound frequency Δf_s is

$$\Delta\phi_d = \Delta(2\phi_B) = \frac{\lambda_0}{v_s}\Delta f_s. \tag{7.1-48}$$

For an *optical deflector*, it is instructive to consider the *number of resolvable angles* N, which is defined as the ratio of the range of deflected angles $\Delta\phi_d$ to the *angular spread* of the scanning laser beam $\Delta\phi \approx \lambda_0/D$, that is,

$$N = \frac{\Delta\phi_d}{\Delta\phi} = \frac{\frac{\lambda_0}{v_s}\Delta f_s}{\frac{\lambda_0}{D}} = \tau\Delta f_s, \tag{7.1-49}$$

where $\tau = D/v_s$ is the *transit time* of the sound through the laser beam as D is the diameter of the laser beam. For $D = 1$ cm and for sound wave traveling in the glass at velocity $v_s \sim 4000$ m/s, then $\tau = 2.5$ μs. If we require that $N = 500$, then $\Delta f_s = 200$ MHz.

Since acousto-optic deflectors do not include moving parts, high angular deflection velocities can be achieved accurately, when compared to mechanical scanners. However, the number of resolvable spots is smaller as compared to its mirror-based scanner counterpart.

相位调制

在利用声光效应讨论了强度调制后，这里考虑如何利用声光对激光进行相位调制. 在相位调制 (*phase modulation*, PM) 中, 其振幅和频率固定的一个正弦信号的相位与 $m(t)$ 成比例变化, 相位调制的传统表达式为

$$y_{\mathrm{PM}}(t) = A\cos[\Omega t + k_{pm}m(t)], \tag{7.1-50}$$

其中, 如果 $m(t)$ 以伏特 (V) 为单位测量, 则 k_{pm} 为相位偏差常数 (*phase-deviation constant*), 单位为弧度/伏特 (rad/V), 同样, Ω 为载波频率. 将式 (7.1-50) 中的余弦函数展开, 得

$$y_{\mathrm{PM}}(t) = A\left\{\cos(\Omega t)\cos[k_{pm}m(t)] - \sin(\Omega t)\sin[k_{pm}m(t)]\right\}.$$

当 $|k_{pm}m(t)| \ll 1$ 时, $\cos\theta \approx 1$ 且 $\sin\theta \approx \theta$. 在这种情况下, 有窄带相位调制 (*narrowband phase modulation*, NBPM), 则上式变为

$$y_{\mathrm{NBPM}}(t) = A[\cos(\Omega t) - k_{pm}m(t)\sin(\Omega t)]. \tag{7.1-51}$$

图 7.1-16a) 所示为一个窄带相位调制器的框图. 图中, 带 90° 的方块表示一个移相器, 该移相器将载波信号移相 90°. 为了生成式 (7.1-50) 所给的表达式, 必须使 k_{pm} 变大. 利用倍频器 (*frequency multiplier*), NBPM 可以被转换为宽带相位调制 (*wideband phase modulation*, WBPM). 图 7.1-16b) 给出了产生宽带相位调制信号 $y_{\mathrm{WBPM}}(t)$ 的框图. 倍频器是一种非线性器件. 比如, 平方律器件 (*square-law device*) 可以将频率乘上一个因子 2. 对于一个平方律器件, 输入 $x(t)$ 和输出 $y(t)$ 有以下关系

$$y(t) = x^2(t).$$

对于 $x(t) = \cos[\Omega t + k_{pm}m(t)]$, $y(t) = \cos^2[\Omega t + k_{pm}m(t)] = \dfrac{1}{2} + \dfrac{1}{2}\cos[2\Omega t + 2k_{pm}m(t)]$, 其直流项被滤除, 得到载波频率和相位偏差常数加倍的输出. 重复使用乘法器 (*multiplier*) 会得到一个较大的相位偏差常数.

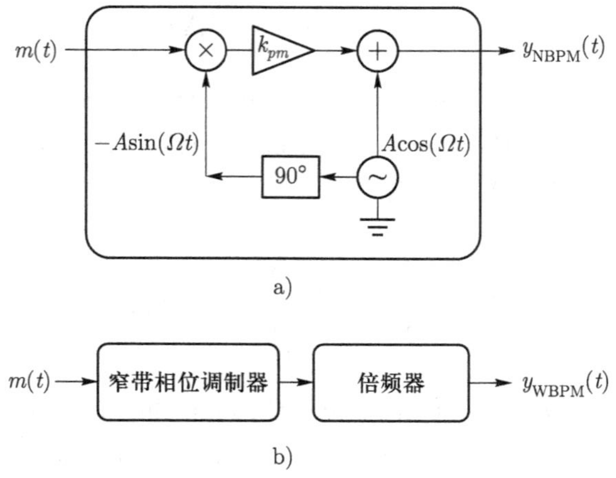

图 7.1-16 a) 窄带相位调制器; b) 宽带相位调制器

利用声光效应对激光进行相位调制 $m(t)$ 时, 我们将式 (7.1-50) 所示形式的相位调制电信

Phase Modulation

While we have discussed intensity modulation using the acousto-optic effect, let us see how to use acousto-optics to phase modulate a laser. In *phase modulation* (PM), the phase of a sinusoidal signal, whose amplitude and frequency are fixed, is varied in proportion to $m(t)$. The conventional expression for phase modulation is

$$y_{\rm PM}(t) = A\cos[\Omega t + k_{pm}m(t)], \qquad (7.1\text{-}50)$$

where k_{pm} is the *phase-deviation constant* in radian/volt if $m(t)$ is measured in volts. Again, Ω is the carrier frequency. We can expand the cosine function in Eq. (7.1-50) to have

$$y_{\rm PM}(t) = A\left\{\cos(\Omega t)\cos[k_{pm}m(t)] - \sin(\Omega t)\sin[k_{pm}m(t)]\right\}.$$

When $|k_{pm}m(t)| \ll 1$, $\cos\theta \approx 1$ and $\sin\theta \approx \theta$. In that case, we have *narrowband phase modulation* (NBPM) and the above equation becomes

$$y_{\rm NBPM}(t) = A[\cos(\Omega t) - k_{pm}m(t)\sin(\Omega t)]. \qquad (7.1\text{-}51)$$

Figure 7.1-16a shows a block diagram of a narrowband phase modulator. In the diagram, the block with 90° denotes a phase shifter that phase shifts the carrier signal by 90°. In order to generate an expression given by Eq. (7.1-50), k_{pm} must be made large. NBPM can be converted to *wideband phase modulation* (WBPM) by using frequency multipliers. Figure 7.1-16b illustrates the block diagram of generating a wideband phase modulation (WBPM) signal, $y_{\rm WBPM}(t)$. A *frequency multiplier* is a nonlinear device. For example, a *square-law device* can multiply the frequency of a factor of 2. For a square-law device, the input $x(t)$ and the output $y(t)$ are related by

$$y(t) = x^2(t).$$

For $x(t) = \cos[\Omega t + k_{pm}m(t)]$, $y(t) = \cos^2[\Omega t + k_{pm}m(t)] = \frac{1}{2} + \frac{1}{2}\cos[2\Omega t + 2k_{pm}m(t)]$. The DC term is filtered out to give the output with carrier frequency and phase-deviation constant doubled. Repeated use of *multipliers* gives us a large phase-deviation constant.

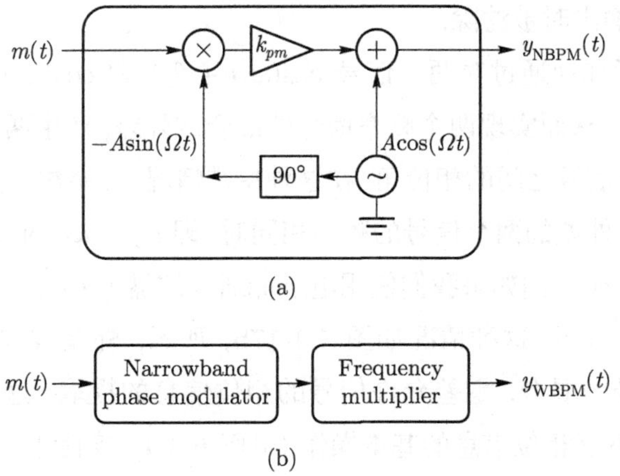

Figure 7.1-16 (a) Narrowband phase modulator and (b) wideband phase modulator

For phase modulation of a laser by $m(t)$ using the acousto-optic effect, we feed a phase modulated electrical signal of the form given by Eq. (7.1-50) to the acoustic transducer. Under

号输入声换能器. 这种情况下, 根据式 (7.1-27), 对于纯相位调制, $\widetilde{\alpha} \propto A = \mathrm{e}^{\mathrm{j}k_{pm}m(t)}$ [见式 (7.1-28)], 该声场模型为 $A = |A|\mathrm{e}^{\mathrm{j}\theta} = \mathrm{e}^{\mathrm{j}k_{pm}m(t)}$.

现在我们来考虑上移的布拉格衍射. 假设使用适度的声压, 即 $\alpha \ll 1$, 由式 (7.1-36a) 和式 (7.1-36b), 声柱出口处 ($\xi = 1$) 的衍射级次可近似为

$$\psi_0(\xi = 1) = \psi_{inc}\cos\left(\frac{\alpha}{2}\right) \approx \psi_{inc}, \tag{7.1-52a}$$

和

$$\psi_1(\xi = 1) = -\mathrm{j}\frac{\widetilde{\alpha}}{\alpha}\psi_{inc}\sin\left(\frac{\alpha}{2}\right) \approx -\mathrm{j}\psi_{inc}\frac{\widetilde{\alpha}}{2} \propto -\mathrm{j}\psi_{inc}\mathrm{e}^{\mathrm{j}k_{pm}m(t)}. \tag{7.1-52b}$$

可以看出, 一级光携带相位调制. 根据式 (7.1-29), 声柱出口处的一级衍射平面波的完整表达式为 (上移相互作用时, $\phi_1 = \phi_B$)

$$\begin{aligned}&\psi_m(z = L, x)\mathrm{e}^{-\mathrm{j}k_0\sin\phi_m x - \mathrm{j}k_0\cos\phi_m L}\big|_{m=1}\mathrm{e}^{\mathrm{j}(\omega_0+\Omega)t}\\ &= \psi_1(z = L, x)\mathrm{e}^{-\mathrm{j}k_0\sin\phi_B x - \mathrm{j}k_0\cos\phi_B L}\mathrm{e}^{\mathrm{j}(\omega_0+\Omega)t}.\end{aligned}$$

由式 (7.1-52b), 可知 $\psi_1(z = L, x) \approx \psi_1(\xi = 1)$, 方程变为

$$\begin{aligned}&\psi_1(\xi = 1)\mathrm{e}^{-\mathrm{j}k_0\sin\phi_B x - \mathrm{j}k_0\cos\phi_B L}\mathrm{e}^{\mathrm{j}(\omega_0+\Omega)t}\\ &\propto -\mathrm{j}\psi_{inc}\mathrm{e}^{\mathrm{j}k_{pm}m(t)}\mathrm{e}^{-\mathrm{j}k_0\sin\phi_B x}\mathrm{e}^{\mathrm{j}(\omega_0+\Omega)t}.\end{aligned} \tag{7.1-53}$$

这是一个调相平面波 (*phase-modulated plane wave*), 即被调制信号 $m(t)$ 所调制.

外差

外差 (*heterodyning*), 也被称为混频 (*frequency mixing*), 是一种频率转换过程. 外差技术起源于无线电工程. 外差原理是在 20 世纪 10 年代末由无线电工程师在试验无线电真空管时发现的. 对于两个不同频率的信号进行外差或混合的简单情况, 其结果信号产生两个新的频率, 即为两个原始频率的和与差. 外差是通过将两个正弦信号应用到一个非线性器件来实现的. 这两个新的频率被称为外差频率 (*heterodyne frequency*). 通常, 只需要其中一个外差频率即可, 而另一个信号在输出时被滤除.

图 7.1-17a) 说明了如何通过对两个信号 $\cos(\omega_1 t + \theta_1)$ 和 $\cos(\omega_2 t)$ 进行简单的电子倍增 (*electronic multiplying*), 从而实现两个频率信号的混合或外差, 产生两个新的频率 $\omega_1 + \omega_2$ 和 $\omega_1 - \omega_2$, 其中 θ_1 是两个信号之间的相位角. 在这种特殊情况下, 乘法器是实现外差作用的非线性器件. 可以看出, 当要外差的两个信号的频率相同时, 即 $\omega_1 = \omega_2$ 时, $\cos(\omega_1 t + \theta_1)$ 的相位信息可以被提取从而获得 $\cos\theta_1$ [如果我们使用电子低通滤波器 (*electronic low-pass filter*, LPF) 来滤除该项 $\cos(2\omega_1 t + \theta_1)$]. 这种情况如图 7.1-17b) 所示, 外差被称为零差 (*homodyning*), 用于两个相同频率信号的混合. 零差允许信号的相位信息的提取, 这是锁相放大器 (*lock-in amplifier*) 从电信号中恢复相位信息的基本操作 (见图 6.2-4). 实际上, 已经在光学扫描全息术中利用了该原理来提取全息信息 (见 6.2.2 节).

the situation, the sound field, according to Eq. (7.1-27), is modelled by $A = |A|e^{j\theta} = e^{jk_{pm}m(t)}$ for pure phase modulation, giving $\widetilde{\alpha} \propto A = e^{jk_{pm}m(t)}$ [see Eq. (7.1-28)].

Let us now consider upshifted Bragg diffraction. Assuming a modest sound pressure is used,that is, $\alpha \ll 1$, the diffracted orders at the exit of the sound column ($\xi = 1$) from Eqs. (7.1-36a) and (7.1-36b) are approximated by

$$\psi_0(\xi = 1) = \psi_{inc} \cos\left(\frac{\alpha}{2}\right) \approx \psi_{inc}, \tag{7.1-52a}$$

and

$$\psi_1(\xi = 1) = -j\frac{\widetilde{\alpha}}{\alpha}\psi_{inc} \sin\left(\frac{\alpha}{2}\right) \approx -j\psi_{inc}\frac{\widetilde{\alpha}}{2} \propto -j\psi_{inc}e^{jk_{pm}m(t)}. \tag{7.1-52b}$$

Note that the first-order light carries the phase modulation. According to Eq. (7.1-29), the full expression of the first-order diffracted plane wave at the exit of the sound column is ($\phi_1 = \phi_B$ for upshifted interaction)

$$\begin{aligned}&\psi_m(z = L, x)e^{-jk_0 \sin\phi_m x - jk_0 \cos\phi_m L}\Big|_{m=1} e^{j(\omega_0+\Omega)t}\\ &= \psi_1(z = L, x)e^{-jk_0 \sin\phi_B x - jk_0 \cos\phi_B L}e^{j(\omega_0+\Omega)t}.\end{aligned}$$

Since we have found $\psi_1(z = L, x) \approx \psi_1(\xi = 1)$ from Eq. (7.1-52b), the equation becomes

$$\begin{aligned}&\psi_1(\xi = 1)e^{-jk_0 \sin\phi_B x - jk_0 \cos\phi_B L}e^{j(\omega_0+\Omega)t}\\ &\propto -j\psi_{inc}e^{jk_{pm}m(t)}e^{-jk_0 \sin\phi_B x}e^{j(\omega_0+\Omega)t}.\end{aligned} \tag{7.1-53}$$

This is a phase-modulated plane wave, which is modulated by modulating signal $m(t)$.

Heterodyning

Heterodyning, also known as *frequency mixing*, is a frequency translation process. Heterodyning has its root in radio engineering. The principle of heterodyning was discovered in the late 1910s by radio engineers experimenting with radio vacuum tubes. For a simple case when signals of two different frequencies are heterodyned or mixed, the resulting signal produces two new frequencies, the sum and difference of the two original frequencies. Heterodyning is performed by applying the two sinusoidal signals to a nonlinear device. The two new frequencies are called *heterodyne frequencies*. Typically, only one of the heterodyne frequencies is required and the other signal is filtered out of the output.

Figure 7.1-17a illustrates how signals of two frequencies are mixed or heterodyned to produce two new frequencies, $\omega_1 + \omega_2$ and $\omega_1 - \omega_2$, by simply *electronic multiplying* the two signals $\cos(\omega_1 t + \theta_1)$ and $\cos(\omega_2 t)$, where θ_1 is the phase angle between the two signals. In this particular case, the multiplier is the nonlinear device to achieve the heterodyne operation. Note that when the frequencies of the two signals to be heterodyned are the same, that is, $\omega_1 = \omega_2$, the phase information of $\cos(\omega_1 t + \theta_1)$ can be extracted to obtain $\cos\theta_1$ if we use an *electronic low-pass filter* (LPF) to filter out the term $\cos(2\omega_1 t + \theta_1)$. This is shown in Figure 7.1-17b. Heterodyning is referred to as *homodyning* for the mixing of two signals of the same frequency. Homodyning allows the extraction of the phase information of the signal and this is the basic operation of a *lock-in amplifier* to retrieve the phase information from an electrical signal (see Figure 6.2-4). Indeed, we have used the principle to extract the holographic information in optical scanning holography (see Section 6.2.2).

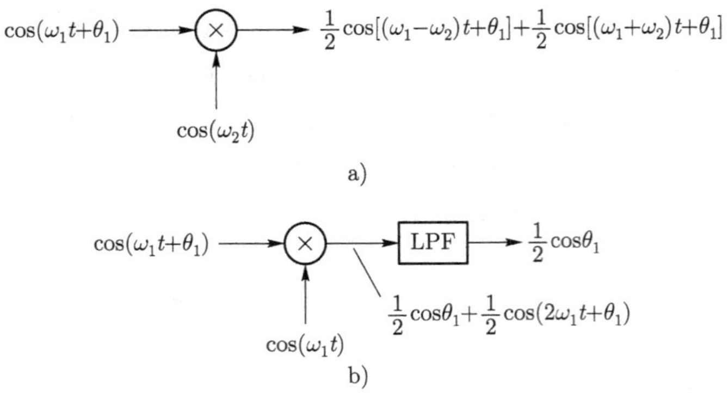

图 7.1-17　a) 外差; b) 零差和相位恢复

现在来讨论外差原理是如何被应用在光学中的. 考虑具有不同时间频率的两个平面波在光电探测器 (*photodetector*) 表面上的外差, 其情况如图 7.1-18 所示, 其中, ϕ 为两个平面波之间的夹角. 根据式 (3.3-9), 一个频率为 $\omega_0+\varOmega$ 的平面波, 一般被表示为

$$\mathrm{e}^{-\mathrm{j}(k_{0x}x+k_{0y}y+k_{0z}z)}\mathrm{e}^{\mathrm{j}(\omega_0+\varOmega)t}.$$

对于以角 ϕ 传播的平面波, 如图 7.1-18 所示, 当光电探测器位于 $z=0$ 处时, 可以得到, $k_{0x}=k_0\hat{\boldsymbol{k}}\cdot\hat{\boldsymbol{x}}=\cos(90^\circ-\phi)=k_0\sin\phi$ 和 $k_{0y}=0$ [也可以见式 (5.1-13)]. 因此, 根据式 (7.1-53), 光电探测器表面上携带信息的平面波 (*information-carrying plane wave*) [也称为信号平面波 (*signal plane wave*)] 一般可以表示为

$$A_s\mathrm{e}^{\mathrm{j}[(\omega_0+\varOmega)t+k_{pm}m(t)]}\mathrm{e}^{-\mathrm{j}k_0\sin\phi\, x}. \tag{7.1-54}$$

其中, $m(t)$ 是嵌入平面波相位中的信息. 另一个频率为 ω_0 的平面波垂直入射到光电探测器上, 且由式 $A_r\mathrm{e}^{\mathrm{j}\omega_0 t}$ 给出, 相对于信号平面波, 它被称为参考平面波 (*reference plane wave*).

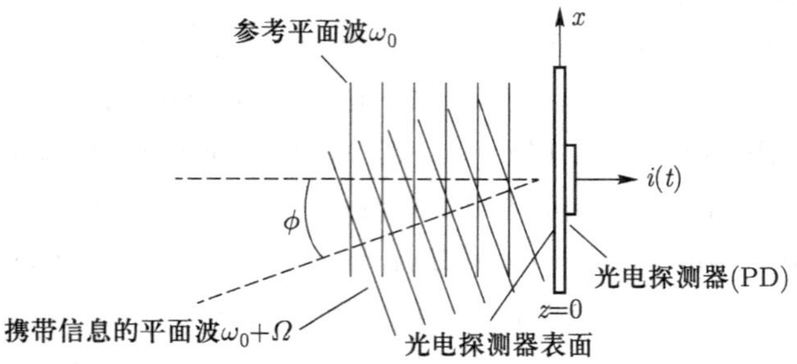

图 7.1-18　光外差

这两个平面波干涉后, 在光电探测器表面上得到总的光场为

$$\psi_t=A_r\mathrm{e}^{\mathrm{j}\omega_0 t}+A_s\mathrm{e}^{\mathrm{j}[(\omega_0+\varOmega)t+k_{pm}m(t)]}\mathrm{e}^{-\mathrm{j}k_0\sin\phi\, x}.$$

由于光电探测器是一种对强度响应 [同样见式 (4.4-3)] 的平方律探测器 (*square-law detector*), 因此, 该光电探测器产生的电流为

$$i(t)\propto\int_D|\psi_t|^2\mathrm{d}x\mathrm{d}y=\int_D|A_r\mathrm{e}^{\mathrm{j}\omega_0 t}+A_s\mathrm{e}^{\mathrm{j}[(\omega_0+\varOmega)t+k_{pm}m(t)]}\mathrm{e}^{-\mathrm{j}k_0\sin\phi\, x}|^2\mathrm{d}x\mathrm{d}y. \tag{7.1-55}$$

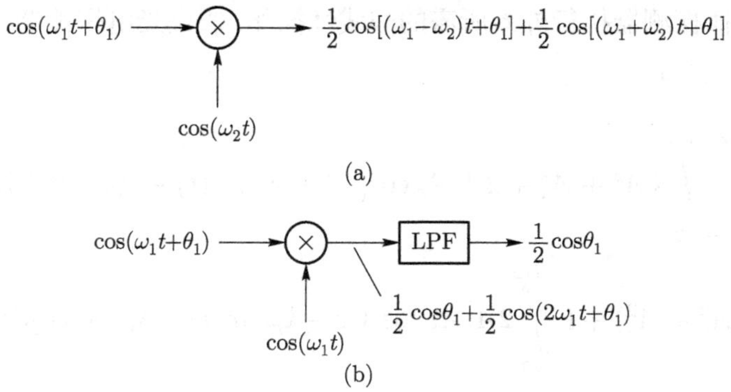

Figure 7.1-17 (a) Heterodyning and (b) homodyning and phase retrieval

Let us now discuss how the principle of heterodyning is used in optics. We consider heterodyning of two plane waves with different temporal frequencies on the surface of a *photodetector*. The situation is shown in Figure 7.1-18, where ϕ is the angle between the two plane waves. According to Eq. (3.3-9), a plane wave of frequency $\omega_0 + \Omega$ in general is expressed as

$$\mathrm{e}^{-\mathrm{j}(k_{0x}x+k_{0y}y+k_{0z}z)}\mathrm{e}^{\mathrm{j}(\omega_0+\Omega)t}.$$

For plane wave propagating at angle ϕ, as shown in Figure 7.1-18, and with the photodetector located at $z = 0$, we have $k_{0x} = k_0\hat{\boldsymbol{k}}\cdot\hat{\boldsymbol{x}} = \cos(90^\circ - \phi) = k_0 \sin\phi$ and $k_{0y} = 0$ [also see Eq. (5.1-13)]. Hence, the *information-carrying plane wave* (also called *signal plane wave*) on the photodetector surface can be expressed in general as, according to Eq. (7.1-53),

$$A_s\mathrm{e}^{\mathrm{j}[(\omega_0+\Omega)t+k_{pm}m(t)]}\mathrm{e}^{-\mathrm{j}k_0\sin\phi\, x}. \tag{7.1-54}$$

where $m(t)$ is the information embedded on the phase of the plane wave. The other plane wave of frequency ω_0 is incident normally to the photodetector and is given by $A_r\mathrm{e}^{\mathrm{j}\omega_0 t}$, which is called a *reference plane wave* in contrast to the signal plane wave.

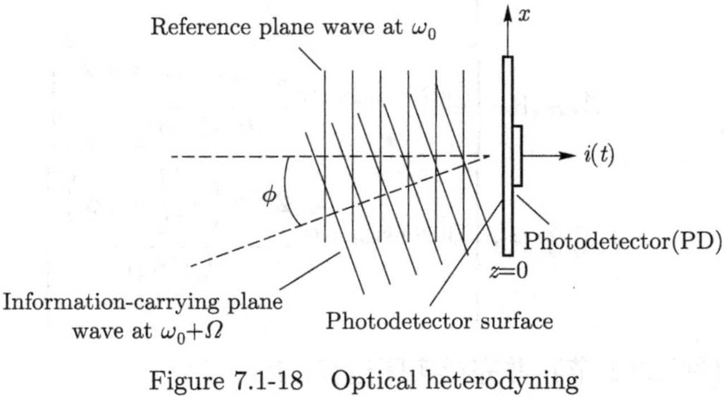

Figure 7.1-18 Optical heterodyning

The two plane waves interfere, giving the total field on the surface of the photodetector as

$$\psi_t = A_r\mathrm{e}^{\mathrm{j}\omega_0 t} + A_s\mathrm{e}^{\mathrm{j}[(\omega_0+\Omega)t+k_{pm}m(t)]}\mathrm{e}^{-\mathrm{j}k_0\sin\phi\, x}.$$

Since the photodetector is a *square-law detector* that responds to the intensity [also see Eq. (4.4-3)], the current from the photodetector is

$$i(t) \propto \int_D |\psi_t|^2\mathrm{d}x\mathrm{d}y = \int_D |A_r\mathrm{e}^{\mathrm{j}\omega_0 t} + A_s\mathrm{e}^{\mathrm{j}[(\omega_0+\Omega)t+k_{pm}m(t)]}\mathrm{e}^{-\mathrm{j}k_0\sin\phi\, x}|^2\mathrm{d}x\mathrm{d}y. \tag{7.1-55}$$

故, 在光学中, 平方律探测器执行外差所需的非线性操作. 假设光电探测器具有 $l \times l$ 的均匀光敏面, 则式 (7.1-55) 变为

$$
\begin{aligned}
i(t) &\propto \int_{-l/2}^{l/2}\int_{-l/2}^{l/2}\{A_r^2 + A_s^2 + 2A_rA_s\cos[\Omega t + k_{pm}m(t) - k_0 x\sin\phi]\}\mathrm{d}x\mathrm{d}y \\
&\propto l^2(A_r^2 + A_s^2) + l\int_{-l/2}^{l/2} 2A_rA_s\cos[\Omega t + k_{pm}m(t) - k_0 x\sin\phi]\mathrm{d}x,
\end{aligned} \tag{7.1-56}
$$

其中, 为了简单起见, 假设两个平面波的振幅 A_r 和 A_s 为实数. 式 (7.1-56) 的第一项为直流电流 (*direct current*, DC), 第二项是频率为 Ω 的交流电流 (*alternating current*, AC), 被称为外差电流 (*heterodyne current*). 可以看出, 原本嵌入在信号平面波相位中的信息内容 $m(t)$ 现在被保留并转移到外差电流的相位中. 事实上, 这种光学上的保相技术 (*phase-preserving technique*) 被称为光学外差法 (*optical heterodyning*), 这让我们想到第五章中讨论的全息记录. 相反, 如果不使用参考平面波, 则信号平面波信号所携带的信息内容将会丢失. 显然, 如果设 $A_r = 0$, 则式 (7.1-56) 就变为 $i(t) \propto A_s^2$, 即变成与信号平面波强度成正比的直流电流.

现在来进一步考虑电流的交流部分, 即式 (7.1-56) 所给出的外差电流. 频率 Ω 的外差电流为

$$
\begin{aligned}
i_\Omega(t) &\propto \int_{-l/2}^{l/2} A_rA_s\cos[\Omega t + k_{pm}m(t) - k_0\sin\phi\, x]\mathrm{d}x \\
&\propto A_rA_s\int_{-l/2}^{l/2}\mathrm{Re}[\mathrm{e}^{\mathrm{j}[\Omega t + k_{pm}m(t) - k_0\sin\phi\, x]}]\mathrm{d}x \\
&= A_rA_s\mathrm{Re}\left[\mathrm{e}^{\mathrm{j}[\Omega t + k_{pm}m(t)]}\int_{-l/2}^{l/2}\mathrm{e}^{-\mathrm{j}k_0\sin\phi\, x}\mathrm{d}x\right] \\
&= A_rA_s\mathrm{Re}\left[\mathrm{e}^{\mathrm{j}[\Omega t + k_{pm}m(t)]}\left.\frac{\mathrm{e}^{-\mathrm{j}k_0\sin\phi\, x}}{-\mathrm{j}k_0\sin\phi}\right|_{x=-l/2}^{x=l/2}\right] \\
&= A_rA_s\mathrm{Re}\left[\mathrm{e}^{\mathrm{j}[\Omega t + k_{pm}m(t)]}\frac{2\sin\left(k_0\dfrac{l}{2}\sin\phi\right)}{k_0\sin\phi}\right].
\end{aligned}
$$

用 *sinc* 函数的定义 (见 2.2.1 节), 并完成实际运算, 上式变为

$$
i_\Omega(t) \propto lA_rA_s sinc\left(\frac{k_0 l}{2\pi}\sin\phi\right)\cos[\Omega t + k_{pm}m(t)]. \tag{7.1-57}
$$

结果表明, 外差电流的振幅由一个 *sinc* 函数控制, 其幅角为对小角度有 $\dfrac{k_0 l}{2\pi}\sin\phi \approx l\phi/\lambda_0$ 的函数, 当 $\phi = 0$ 时, 即两个平面波完全平行时, 该外差电流最大, 并随着 $sinc\left(\dfrac{k_0 l\phi}{2\pi}\right)$ 下降, 当 *sinc* 函数的幅角为 1 时, 没有外差电流. 即

Hence, in optics, the "square-law detector" performs the nonlinear operation needed for heterodyning. Assuming the photodetector has a uniform active area of $l \times l$, Eq. (7.1-55) becomes

$$i(t) \propto \int_{-l/2}^{l/2}\int_{-l/2}^{l/2} \{A_r^2 + A_s^2 + 2A_rA_s\cos[\Omega t + k_{pm}m(t) - k_0 x\sin\phi]\}\mathrm{d}x\mathrm{d}y$$

$$\propto l^2(A_r^2 + A_s^2) + l\int_{-l/2}^{l/2} 2A_rA_s\cos[\Omega t + k_{pm}m(t) - k_0 x\sin\phi]\mathrm{d}x, \qquad (7.1\text{-}56)$$

where we have assumed the amplitudes of the two plane waves A_r and A_s are real for simplicity. The first term of Eq. (7.1-56) is a DC current, and the second team is an AC current at frequency Ω, which is known as the *heterodyne current.* Note that the information content $m(t)$ originally embedded in the phase of the signal plane wave has now been preserved and transferred to the phase of the heterodyne current. Indeed, this *phase-preserving technique* in optics is known as *optical heterodyning*, reminiscent of holographic recording discussed in Chapter 5. In contrast, if the reference plane wave were not used, the information content carried by the signal plane wave would be lost. Clearly, if we set $A_r = 0$, Eq. (7.1-56) becomes $i(t) \propto A_s^2$, a DC current proportional to the intensity of the signal plane wave.

Let us now consider the AC part of the current further, that is, the heterodyne current given by Eq. (7.1-56). The heterodyne current of frequency Ω is given by

$$i_\Omega(t) \propto \int_{-l/2}^{l/2} A_rA_s\cos[\Omega t + k_{pm}m(t) - k_0\sin\phi\, x]\mathrm{d}x$$

$$\propto A_rA_s\int_{-l/2}^{l/2}\mathrm{Re}[\mathrm{e}^{\mathrm{j}[\Omega t + k_{pm}m(t) - k_0\sin\phi\, x]}]\mathrm{d}x$$

$$= A_rA_s\mathrm{Re}\left[\mathrm{e}^{\mathrm{j}[\Omega t + k_{pm}m(t)]}\int_{-l/2}^{l/2}\mathrm{e}^{-\mathrm{j}k_0\sin\phi\, x}\mathrm{d}x\right]$$

$$= A_rA_s\mathrm{Re}\left[\mathrm{e}^{\mathrm{j}[\Omega t + k_{pm}m(t)]}\left.\frac{\mathrm{e}^{-\mathrm{j}k_0\sin\phi\, x}}{-\mathrm{j}k_0\sin\phi}\right|_{x=-l/2}^{x=l/2}\right]$$

$$= A_rA_s\mathrm{Re}\left[\mathrm{e}^{\mathrm{j}[\Omega t + k_{pm}m(t)]}\frac{2\sin\left(k_0\frac{l}{2}\sin\phi\right)}{k_0\sin\phi}\right].$$

Using the definition of the *sinc* function (see Section 2.2.1) and completing the real operation, the above equation becomes

$$i_\Omega(t) \propto lA_rA_s\,sinc\left(\frac{k_0 l}{2\pi}\sin\phi\right)\cos[\Omega t + k_{pm}m(t)]. \qquad (7.1\text{-}57)$$

The result indicates that the amplitude of the heterodyne current is controlled by a *sinc* function whose argument is a function of $\frac{k_0 l}{2\pi}\sin\phi \approx l\phi/\lambda_0$ for small angles. When $\phi = 0$, that is, the two plane waves are completely aligned, the heterodyne current is at maximum, and falls off as $sinc\left(\frac{k_0 l\phi}{2\pi}\right)$. We have no heterodyne current when the argument of the *sinc*

$$\frac{k_0 l\phi}{2\pi}=1,$$

或

$$\phi=\frac{\lambda_0}{l}. \tag{7.1-58}$$

对尺寸为 $l=1\ \text{cm}$ 的光电探测器和 $\lambda_0=0.6\ \mu\text{m}$ 的红光, 计算其 ϕ 值大约为 0.0023°. 因此, 为了获得一个相当大的外差电流, 两个平面波之间的夹角必须远小于 0.0023°. 所以, 我们需要高精度的光学机械支架来进行角旋转, 以最小化两个平面波之间的角分离, 从而实现高效的外差. 在图 7.1-19 中, 给出了一个简单的说明声光外差作用的实验配置, 当我们在式 (7.1-56) 中取 $\phi=0$ 且 $m(t)=0$ 时, 可恢复声频 Ω.

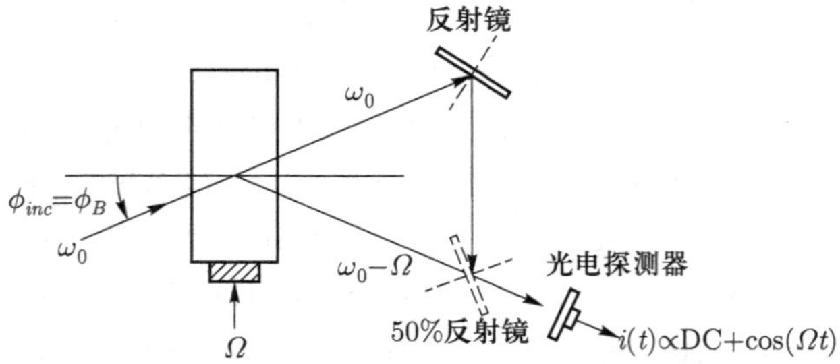

图 7.1-19 声光外差检测

7.2 电光调制器信息处理

因为电光器件通常处理的是入射光波的偏振, 首先介绍偏振的概念.

7.2.1 光的偏振

偏振 (*polarization*) 发生于矢量场中. 在电磁波中, 矢量场有电场和磁场. 实际上, 偏振描述了在空间中某一给定点处, 电场强度矢量 $\boldsymbol{E}$ 的端点, 随时间变化的轨迹. 由于磁场 $\boldsymbol{H}$ 的方向与电场 $\boldsymbol{E}$ 的方向有关, 因此, 没有必要对磁场进行单独的描述. 考虑一束沿 $+z$ 方向传播的平面波, 且其电场的方向在 x-y 平面上. 根据式 (3.3-19a), 有

$$\boldsymbol{E}=E_{0x}\mathrm{e}^{\mathrm{j}(\omega_0 t-k_0 z)}\hat{\boldsymbol{x}}+E_{0y}\mathrm{e}^{\mathrm{j}(\omega_0 t-k_0 z)}\hat{\boldsymbol{y}}, \tag{7.2-1a}$$

其中, E_{0x} 和 E_{0y} 通常为复数, 可以被写为

$$E_{0x}=|E_{0x}|\mathrm{e}^{-\mathrm{j}\phi_x},\quad E_{0y}=|E_{0y}|\mathrm{e}^{-\mathrm{j}\phi_y},$$

或

$$E_{0x}=|E_{0x}|,\quad E_{0y}=|E_{0y}|\mathrm{e}^{-\mathrm{j}\phi_0}, \tag{7.2-1b}$$

其中, $\phi_0=\phi_y-\phi_x$ 为电场 x 分量和 y 分量之间的相对相移. 其电场的物理量 (*physical quantity*) 为

$$\mathrm{Re}[\boldsymbol{E}]=\mathrm{Re}\left[|E_{0x}|\mathrm{e}^{\mathrm{j}(\omega_0 t-k_0 z)}\hat{\boldsymbol{x}}+|E_{0y}|\mathrm{e}^{-\mathrm{j}\phi_0}\mathrm{e}^{\mathrm{j}(\omega_0 t-k_0 z)}\hat{\boldsymbol{y}}\right]. \tag{7.2-2}$$

function is 1, that is,

$$\frac{k_0 l \phi}{2\pi} = 1,$$

or

$$\phi = \frac{\lambda_0}{l}. \tag{7.1-58}$$

For the size of the photodetector $l = 1$ cm and red laser of $\lambda_0 = 0.6$ μm, ϕ is calculated to be around 0.0023°. Therefore, in order to have a sizable heterodyne current, the angle between the two plane waves must be much less than 0.0023°. We, therefore, need to have high-precision opto-mechanical mounts for angular rotation to minimize the angular separation between the two plane waves for efficient heterodyning. In Figure 7.1-19, we show a simple experimental configuration to illustrate acousto-optic heterodyning, which recovers the sound frequency Ω as we take $\phi = 0$ in Eq. (7.1-56) with $m(t) = 0$.

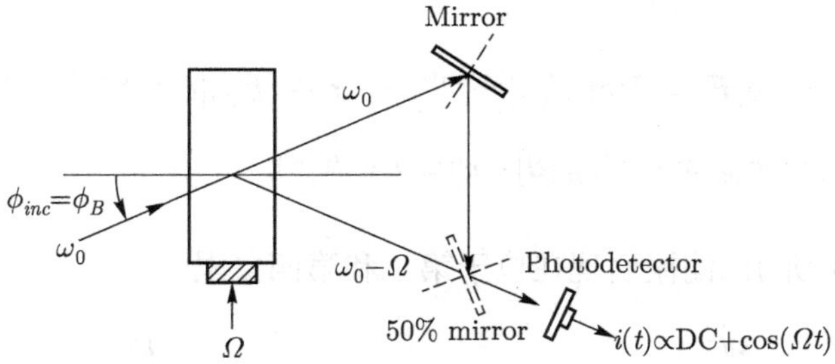

Figure 7.1-19 Acousto-optic heterodyne detection

7.2 Information Processing with Electro-Optic Modulators

We will first introduce the concept of polarization as electro-optic devices typically manipulate polarization of an incoming optical wave.

7.2.1 Polarization of Light

Polarization occurs for vector fields. In electromagnetic waves, the vectors are the electric and magnetic fields. Indeed, polarization describes the locus of the tip of the $\boldsymbol{E}$ vector at a given point in space as time advances. Since the direction of the magnetic field $\boldsymbol{H}$ is related to that of $\boldsymbol{E}$, a separate description of the magnetic field is, therefore, not necessary. Let us consider a plane wave propagating in the $+z$ direction and the electric field is oriented in the x-y plane. According to Eq. (3.3-19a), we have

$$\boldsymbol{E} = E_{0x}\mathrm{e}^{\mathrm{j}(\omega_0 t - k_0 z)}\hat{\boldsymbol{x}} + E_{0y}\mathrm{e}^{\mathrm{j}(\omega_0 t - k_0 z)}\hat{\boldsymbol{y}}, \tag{7.2-1a}$$

E_{0x} and E_{0y} are complex in general and can be written as

$$E_{0x} = |E_{0x}|\mathrm{e}^{-\mathrm{j}\phi_x} \text{ and } E_{0y} = |E_{0y}|\mathrm{e}^{-\mathrm{j}\phi_y},$$

or

$$E_{0x} = |E_{0x}| \text{ and } E_{0y} = |E_{0y}|\mathrm{e}^{-\mathrm{j}\phi_0}, \tag{7.2-1b}$$

where $\phi_0 = \phi_y - \phi_x$ is the relative phase shift between the x- and y-components of the electric field. The physical quantity of the electric field is

$$\mathrm{Re}[\boldsymbol{E}] = \mathrm{Re}\left[|E_{0x}|\mathrm{e}^{\mathrm{j}(\omega_0 t - k_0 z)}\hat{\boldsymbol{x}} + |E_{0y}|\mathrm{e}^{-\mathrm{j}\phi_0}\mathrm{e}^{\mathrm{j}(\omega_0 t - k_0 z)}\hat{\boldsymbol{y}}\right]. \tag{7.2-2}$$

线偏振

考虑 $\phi_0=0$ 或 $\phi_0=\pm\pi$ 的情况. 当 $\phi_0=0$ 时, 根据式 (7.2-2), 有

$$\mathcal{E}(t)=\mathrm{Re}[\boldsymbol{E}]=[|E_{0x}|\hat{\boldsymbol{x}}+|E_{0y}|\hat{\boldsymbol{y}}]\cos(\omega_0 t-k_0 z). \tag{7.2-3}$$

当该波沿着 $+z$ 方向传播时, 其 x 分量和 y 分量是同相的, 且 $\mathcal{E}$ 被固定在 x-y 平面上. 由于余弦函数的值随时间从 1 到 -1 变化, 如图 7.2-1a) 所示, $\mathcal{E}$ 或位于第一象限, 或位于第三象限, 但它总是被固定在一个被称为偏振面 (*plane of polarization*) 的平面上. 该偏振面包含了传播方向和电场矢量, 在这个平面上, 电场随时间变化进行振荡. 由式 (7.2-3) 给出的平面波被称为线偏振 (*linear polarization*). 特别地, 当 $|E_{0y}|=0$ 时, 该平面波变为 $|E_{0x}|\hat{\boldsymbol{x}}\cos(\omega_0 t-k_0 z)$, 它沿 x 方向呈线偏振, 通常被称为 x 偏振 (*x-polarized*). 同样地, 当 $\phi_0=\pm\pi$ 时, 根据式 (7.2-2), 有

$$\begin{aligned}\mathcal{E}(t)&=\mathrm{Re}[\boldsymbol{E}]=\mathrm{Re}[|E_{0x}|\mathrm{e}^{\mathrm{j}(\omega_0 t-k_0 z)}\hat{\boldsymbol{x}}+|E_{0y}|\mathrm{e}^{\mathrm{j}(\omega_0 t-k_0 z)}\mathrm{e}^{\mp\mathrm{j}\pi}\hat{\boldsymbol{y}}]\\&=[|E_{0x}|\hat{\boldsymbol{x}}-|E_{0y}|\hat{\boldsymbol{y}}]\cos(\omega_0 t-k_0 z).\end{aligned} \tag{7.2-4}$$

这种情况如图 7.2-1b) 所示, 偏振面现在位于第二和第四象限.

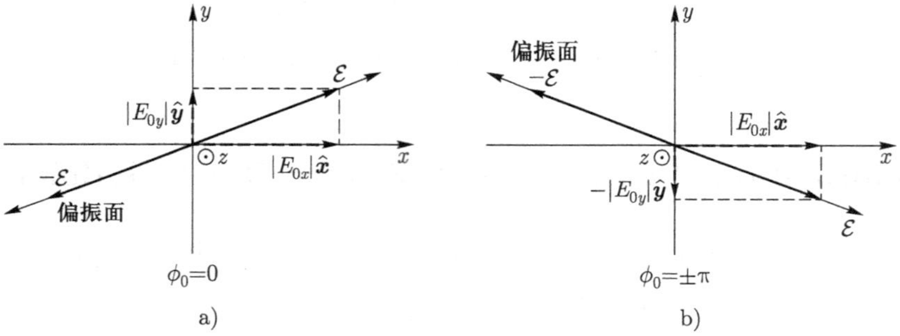

图 7.2-1 线偏振

圆偏振

当 $\phi_0=\pm\pi/2$, 且 $|E_{0x}|=|E_{0y}|=E_0$ 时, 得到圆偏振 (*circular polarization*). 对于 $\phi_0=-\pi/2$, 根据式 (7.2-2), 有

$$\begin{aligned}\mathcal{E}(t)&=\mathrm{Re}[\boldsymbol{E}]=\mathrm{Re}[|E_{0x}|\mathrm{e}^{\mathrm{j}(\omega_0 t-k_0 z)}\hat{\boldsymbol{x}}+|E_{0y}|\mathrm{e}^{\mathrm{j}\pi/2}\mathrm{e}^{\mathrm{j}(\omega_0 t-k_0 z)}\hat{\boldsymbol{y}}]\\&=E_0\hat{\boldsymbol{x}}\cos(\omega_0 t-k_0 z)-E_0\hat{\boldsymbol{y}}\sin(\omega_0 t-k_0 z).\end{aligned} \tag{7.2-5}$$

监测在某一固定位置 $z=z_0=0$ 处, 电场的方向随时间的变化情况, 则上式变为

$$\mathcal{E}(t)=E_0\hat{\boldsymbol{x}}\cos(\omega_0 t)-E_0\hat{\boldsymbol{y}}\sin(\omega_0 t).$$

对于不同的时间, 来求 $\mathcal{E}(t)$, 结果如下

$$\begin{aligned}&t=0,\quad \mathcal{E}(t=0)=E_0\hat{\boldsymbol{x}},\\&t=\frac{\pi}{2\omega_0},\quad \mathcal{E}\left(t=\frac{\pi}{2\omega_0}\right)=-E_0\hat{\boldsymbol{y}},\end{aligned}$$

Linear Polarization

Let us consider the case where $\phi_0 = 0$ or $\phi_0 = \pm\pi$. For $\phi_0 = 0$, according to Eq. (7.2-2), we have

$$\mathcal{E}(t) = \text{Re}[\boldsymbol{E}] = [|E_{0x}|\hat{\boldsymbol{x}} + |E_{0y}|\hat{\boldsymbol{y}}]\cos(\omega_0 t - k_0 z). \tag{7.2-3}$$

The x- and y-components are in phase and $\mathcal{E}$ is fixed in the x-y plane as the wave is propagating along the $+z$ direction. Since the values of the cosine function go from 1 to -1 as time advances, $\mathcal{E}$ is either on the first quadrant or the third quadrant as illustrated in Figure 7.2-1a but it is always fixed on a plane that is referred to as the *plane of polarization.* The plane of polarization contains the direction of propagation and the electric field vector, and on this plane the electric field oscillates as time advances. The plane wave given by Eq. (7.2-3) is said to be *linearly polarized*, in particular, if $|E_{0y}| = 0$, the plane wave becomes $|E_{0x}|\hat{\boldsymbol{x}}\cos(\omega_0 t - k_0 z)$. It is linearly polarized along the x-direction, and often called x-polarized. Similarly, for $\phi_0 = \pm\pi$, according to Eq. (7.2-2), we have

$$\begin{aligned}\mathcal{E}(t) &= \text{Re}[\boldsymbol{E}] = \text{Re}[|E_{0x}|e^{j(\omega_0 t - k_0 z)}\hat{\boldsymbol{x}} + |E_{0y}|e^{j(\omega_0 t - k_0 z)}e^{\mp j\pi}\hat{\boldsymbol{y}}] \\ &= [|E_{0x}|\hat{\boldsymbol{x}} - |E_{0y}|\hat{\boldsymbol{y}}]\cos(\omega_0 t - k_0 z).\end{aligned} \tag{7.2-4}$$

For this case, the situation is shown in Figure 7.2-1b. The plane of polarization now lies on the second and fourth quadrants.

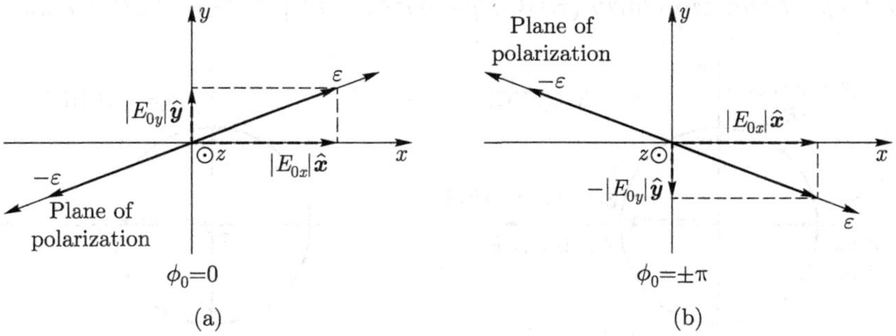

Figure 7.2-1 Linear polarization

Circular Polarization

Circular polarization is obtained when $\phi_0 = \pm\pi/2$, and $|E_{0x}| = |E_{0y}| = E_0$. For $\phi_0 = -\pi/2$, according to Eq. (7.2-2), we have

$$\begin{aligned}\mathcal{E}(t) &= \text{Re}[\boldsymbol{E}] = \text{Re}[|E_{0x}|e^{j(\omega_0 t - k_0 z)}\hat{\boldsymbol{x}} + |E_{0y}|e^{j\pi/2}e^{j(\omega_0 t - k_0 z)}\hat{\boldsymbol{y}}] \\ &= E_0\hat{\boldsymbol{x}}\cos(\omega_0 t - k_0 z) - E_0\hat{\boldsymbol{y}}\sin(\omega_0 t - k_0 z).\end{aligned} \tag{7.2-5}$$

Let us monitor the direction of the electric field at a certain position $z = z_0 = 0$ as time advances. Eq. (7.2-5) then becomes

$$\mathcal{E}(t) = E_0\hat{\boldsymbol{x}}\cos(\omega_0 t) - E_0\hat{\boldsymbol{y}}\sin(\omega_0 t).$$

Let us find $\mathcal{E}(t)$ for different times as follows:

$$\text{At} \quad t = 0, \quad \mathcal{E}(t = 0) = E_0\hat{\boldsymbol{x}},$$

$$t = \frac{\pi}{2\omega_0}, \quad \mathcal{E}\left(t = \frac{\pi}{2\omega_0}\right) = -E_0\hat{\boldsymbol{y}},$$

$$t=\frac{\pi}{\omega_0},\quad \mathcal{E}\left(t=\frac{\pi}{\omega_0}\right)=-E_0\hat{\boldsymbol{x}},$$
$$t=\frac{3\pi}{2\omega_0},\quad \mathcal{E}\left(t=\frac{3\pi}{2\omega_0}\right)=E_0\hat{\boldsymbol{y}},$$

最终

$$t=\frac{2\pi}{\omega_0},\quad \mathcal{E}\left(t=\frac{2\pi}{\omega_0}\right)=E_0\hat{\boldsymbol{x}}.$$

可以看出, 当 $t=0$ 时, 此波为 x 偏振的, 当 $t=\pi/2\omega_0$ 时, 此波为负的 y 偏振的, 以此类推, 直到当 $t=2\pi/\omega_0$ 时, 波重新回到 x 偏振态, 完成一个循环. 这里以红光为例, 当 $\lambda_0=0.6\ \mu\text{m}$ 时, 该激光器的频率为 $f=c/\lambda_0\approx\dfrac{3\times10^8\ \text{m/s}}{0.6\times10^{-6}\ \text{m}}=5\times10^{14}\ \text{Hz}$, 因此, 当波传播时, 它仅需要 $t=2\pi/\omega_0=1/f=2\times10^{-15}\ \text{s}$ 的时间来完成一周. 在图 7.2-2a) 中, 根据不同的时间绘出了 $\mathcal{E}(t)$. 可以看到, 随着时间的变化, 即当波沿 z 方向传播时, $\mathcal{E}(t)$ 的端点轨迹在 x-y 平面上为一个顺时针的圆, 图中用箭头表示, 在 $z=z_0$ 处的正面可观察到. 即得到所谓的顺时针圆偏振波 (*clockwise circularly polarized wave*). 这种波也被称为左旋圆偏振波 [*left-hand circularly* (LHC) *polarized wave*]. 这是因为当左手的拇指指向传播方向, 即 $+z$ 方向时, 其他四个手指指向 $\mathcal{E}(t)$ 的旋转方向. 同样地, 对于 $\phi_0=\pi/2$, 可以证明, $\mathcal{E}(t)$ 在 x-y 平面上的端点轨迹为一个逆时针的圆, 从而得到逆时针圆偏振波 (*counter-clockwise circularly polarized wave*) 或是右旋圆偏振波 [*right-hand circularly* (RHC) *polarized wave*]. 该情况如图 7.2-2b) 所示.

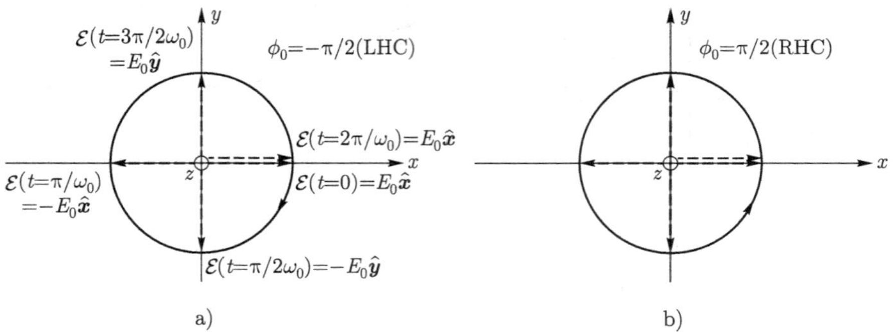

图 7.2-2 a) 左旋圆偏振波; b) 右旋圆偏振波

对于非线偏振或圆偏振的平面波, 得到椭圆偏振 (*elliptical polarization*). 其两个电场分量可能有不同的振幅, 即 $|E_{0x}|\neq|E_{0y}|$, 那么在这两个分量之间也可能含有任意的 ϕ_0.

常见的偏振器件

线偏振片

线偏振片 (*linear polarizer*) 是一种光学元件, 它只允许在其透射轴 (*transmission axis*) 方向上的电场分量通过. 这种选择过程可通过具有选择或各向异性吸收 (*anisotropic absorption*) 来实现, 如使用某些各向异性的材料, 比如众所周知的 Polaroid 公司的 H 偏振片. 这种在吸收时表现出来的各向异性被称为二向色性 (*dichroism*). 该偏振片的制成方法是, 通过在一个方向上拉伸用碘染的聚乙烯醇 (*poly-vinyl alcohol*, PVA) 薄膜, 即可形成高度二向色性的偏振片. 图 7.2-3 所示为透射轴 (箭头表示) 沿 $\hat{\boldsymbol{x}}$ 的线偏振片. 当入射的线偏振波分解成 $\hat{\boldsymbol{x}}$ 和 $\hat{\boldsymbol{y}}$

$$t = \frac{\pi}{\omega_0}, \quad \mathcal{E}\left(t = \frac{\pi}{\omega_0}\right) = -E_0\hat{\boldsymbol{x}},$$

$$t = \frac{3\pi}{2\omega_0}, \quad \mathcal{E}\left(t = \frac{3\pi}{2\omega_0}\right) = E_0\hat{\boldsymbol{y}},$$

and finally

$$t = \frac{2\pi}{\omega_0}, \quad \mathcal{E}\left(t = \frac{2\pi}{\omega_0}\right) = E_0\hat{\boldsymbol{x}}.$$

We see that when $t = 0$, the wave is x-polarized, and when $t = \pi/(2\omega_0)$, the wave is negative y-polarized, and so on until when $t=2\pi/\omega_0$, the wave returns to x-polarization, completing a cycle. Let us take red laser as an example where $\lambda_0=0.6$ μm, and the frequency for the laser is $f=c/\lambda_0 \approx \frac{3\times10^8 \text{ m/s}}{0.6\times10^{-6} \text{ m}}=5\times10^{14}$ Hz. It, therefore, merely takes $t = \frac{2\pi}{\omega_0} = \frac{1}{f} = 2\times10^{-15}$ s to complete a rotation as the wave is propagating. In Figure 7.2-2a, we plot $\mathcal{E}(t)$ according to the different times. We see that as time advances, that is, when the wave is propagating along the z-direction, the tip of $\mathcal{E}(t)$ traces out a clockwise circle in the x-y plane, denoted by an arrow in the figure, as seen head-on at $z = z_0$. We have the so-called *clockwise circularly polarized wave.* Such a wave is also called *left-hand circularly (LHC) polarized.* The reason is that when the thumb of the left hand points along the direction of propagation, that is, the $+z$ direction, the other four fingers point in the direction of rotation of $\mathcal{E}(t)$. By the same token, for $\phi_0 = \pi/2$ we can show that the tip of $\mathcal{E}(t)$ traces out a counter-clockwise circle in the x-y plane. We have the *counter-clockwise circularly polarized wave* or *right-hand circularly (RHC) polarized wave.* The situation is shown in Figure 7.2-2b.

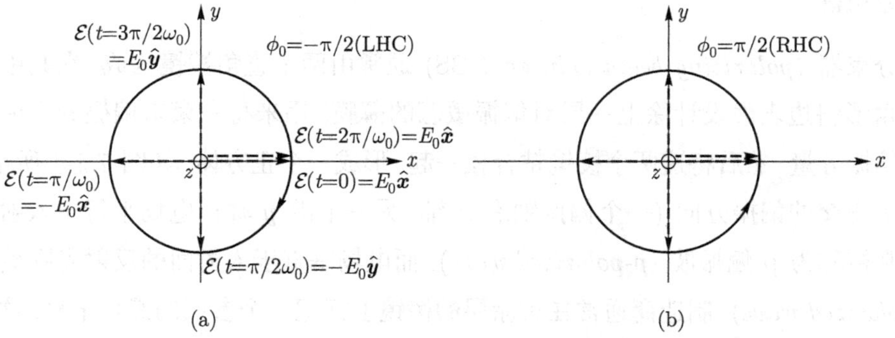

Figure 7.2-2 (a) LHC polarization and (b) RHC polarization

For plane waves that are not linearly or circularly polarized, we have *elliptical polarization.* The two electric field components might have different amplitudes, that is, $|E_{0x}| \neq |E_{0y}|$, and also might contain an arbitrary ϕ_0 between the two components.

Common Polarization Devices

Linear Polarizer

A *linear polarizer* is an optical element that transmits only the component of the electrical field in the direction of its *transmission axis.* This selective process can be done, for example, by selective or *anisotropic absorption* using certain anisotropic materials such as the well-known Polaroid H-sheet. This anisotropy in absorption is called *dichroism.* The sheet is made by stretching in one direction a thin film of *poly-vinyl alcohol* (PVA) stained with iodine and becomes highly dichroic. Figure 7.2-3 shows a linear polarizer with its transmission axis

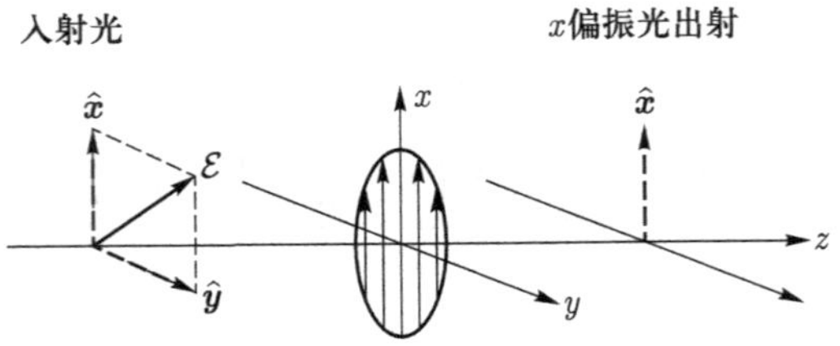

图 7.2-3 透射轴沿 $\hat{x}$ 的线偏振片

两个正交方向时, 从偏振片出射的波变成 x 偏振. 自然光 (*natural light*) (如太阳光和灯泡发出的光) 是由任意偏振的波所组成的, 通过线偏振片的自然光会输出为一束偏振光.

这里用数学方法来描述偏振片的作用. 根据式 (7.2-1), 沿 $+z$ 方向传播的线偏振平面波可以被写为

$$[|E_{0x}|\hat{\boldsymbol{x}}+|E_{0y}|\hat{\boldsymbol{y}}]\mathrm{e}^{\mathrm{j}(\omega_0 t-k_0 z)}.$$

因此, 在线偏振片位置处 ($z=0$), 入射的平面波可以写成

$$\boldsymbol{E}=[|E_{0x}|\hat{\boldsymbol{x}}+|E_{0y}|\hat{\boldsymbol{y}}]\mathrm{e}^{\mathrm{j}\omega_0 t}.$$

由于偏振片的透射轴沿 x 方向, 如图 7.2-3 所示, 在偏振片的出射位置处, 入射光电场的 x 分量变为 (见第三章点乘的使用)

$$(\boldsymbol{E}\cdot\hat{\boldsymbol{x}})\hat{\boldsymbol{x}}=\{[|E_{0x}|\hat{\boldsymbol{x}}+|E_{0y}|\hat{\boldsymbol{y}}]\cdot\hat{\boldsymbol{x}}\}\hat{\boldsymbol{x}}\mathrm{e}^{\mathrm{j}\omega_0 t}=|E_{0x}|\hat{\boldsymbol{x}}\mathrm{e}^{\mathrm{j}\omega_0 t}.$$

偏振分束器

偏振分束器 (*polarizing beam splitter*, PBS) 通常由两个直角棱镜构成. 在其中一个棱镜的直角三角形斜边表面设计涂上一层对偏振敏感的薄膜, 用来将一束未偏振的入射光分成两个正交的偏振分量. 然后将这两个棱镜粘合在一起, 形成一个正方体, 如图 7.2-4 所示. 入射光呈现出两个正交的偏振方向 (一个偏振轴沿 x 轴, 另一个沿 y 轴). 电场平行于入射面的透射偏振光通常被称为 p 偏振波 (*p-polarized wave*), 而电场垂直于入射面的反射光则被称为 s 偏振波 (*s-polarized wave*). 制造商通常在有涂层的棱镜上标记一个参照的黑色记号, 这是用来建议光从一侧透射进有涂层的棱镜中, 从而避免破坏涂层.

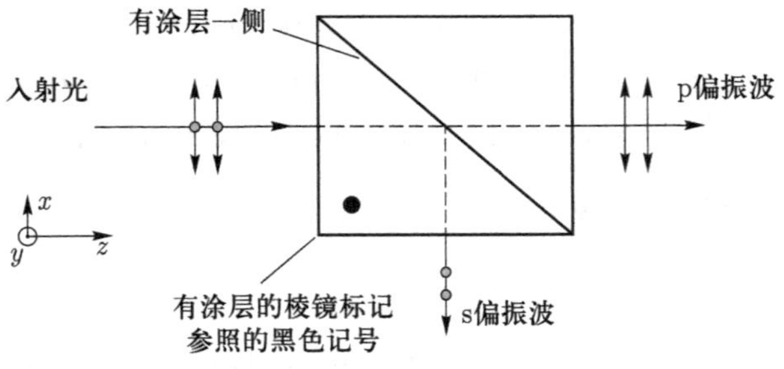

图 7.2-4 偏振光分束器

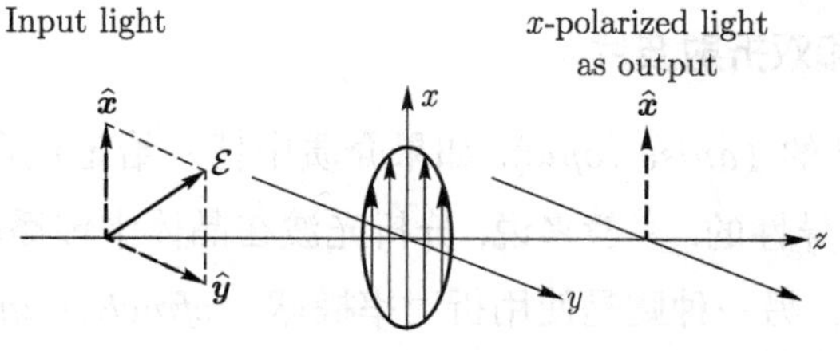

Figure 7.2-3 Linear polarizer with transmission axial along the x-direction

(denoted with arrows) along $\hat{\boldsymbol{x}}$. With the input linear polarized wave decomposed into two orthogonal directions $\hat{\boldsymbol{x}}$ and $\hat{\boldsymbol{y}}$, the output wave from the polarizer has become x-polarized. *Natural light* (such as sunlight and light from bulbs) consists of randomly polarized waves and transmission of natural light through a linear polarizer gives a polarized light as output.

Let us formulate the action of the polarizer mathematically. According to Eq. (7.2-1), a linearly polarized plane wave propagating along the $+z$ direction can be written as

$$[|E_{0x}|\hat{\boldsymbol{x}} + |E_{0y}|\hat{\boldsymbol{y}}]\mathrm{e}^{\mathrm{j}(\omega_0 t - k_0 z)}.$$

Hence, at the location of the linear polarizer ($z = 0$), the incident plane wave can be written as

$$\boldsymbol{E} = [|E_{0x}|\hat{\boldsymbol{x}} + |E_{0y}|\hat{\boldsymbol{y}}]\mathrm{e}^{\mathrm{j}\omega_0 t}.$$

Since the transmission axis of the polarizer is along the x-direction, as shown in Figure 7.2-3, at the exit of the polarizer, the x-component of the incident electric field is (see the use of the dot product in Chapter 3)

$$(\boldsymbol{E} \cdot \hat{\boldsymbol{x}})\hat{\boldsymbol{x}} = \{[|E_{0x}|\hat{\boldsymbol{x}} + |E_{0y}|\hat{\boldsymbol{y}}] \cdot \hat{\boldsymbol{x}}\}\hat{\boldsymbol{x}}\mathrm{e}^{\mathrm{j}\omega_0 t} = |E_{0x}|\hat{\boldsymbol{x}}\mathrm{e}^{\mathrm{j}\omega_0 t}.$$

Polarizing Beam Splitter

Polarizing beam splitters (PBSs) are typically constructed using two right-angle prisms. The hypotenuse surface of one prism is coated with a polarization sensitive thin film designed to separate an incident, unpolarized beam into two orthogonal polarization components. The two prisms are then cemented together to form a cubic shape. Figure 7.2-4 shows the situation. The incident light shows two orthogonal polarization directions (one polarization axis is along the x-axis and the other along the y-axis). The transmitted polarized light with its electric field parallel along the plane of incidence is commonly known as the *p-polarized wave*, while the reflected light whose electrical field is normal to the plane of incidence is called the *s-polarized wave*. Manufactures usually denote a reference black mark to show the coated prism and it is recommended that the light be transmitted into the coated prism in order not to damage the cement.

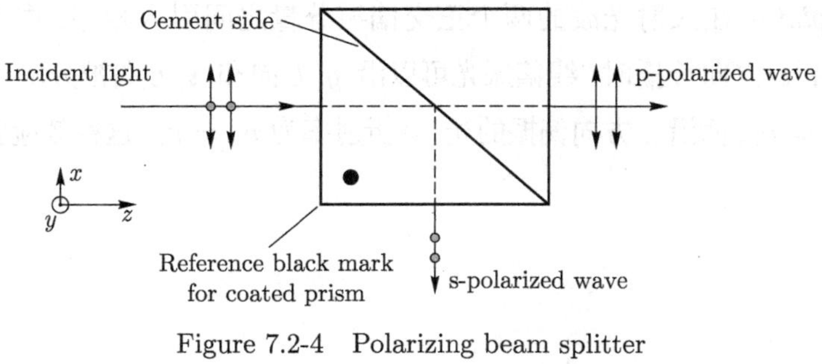

Figure 7.2-4 Polarizing beam splitter

7.2.2 折射率椭球和双折射波片

晶体一般是各向异性的 (*anisotropic*). 如果介质中任一给定点在不同方向上的性质都不同, 那么该介质就是各向异性的. 一般来说, 分析光波在晶体中传播的方法有两种: 一种是采用严格的电磁理论的方法, 另一种则是使用折射率椭球 (*refractive index ellipsoid*) 的方法. 折射率椭球直观、方便, 在实践中经常被使用. 我们可以用折射率椭球来表示晶体的光学性质. 晶体的折射率椭球的方程式可以表示为

$$\frac{x^2}{n_x^2}+\frac{y^2}{n_y^2}+\frac{z^2}{n_z^2}=1, \tag{7.2-6}$$

这里, $n_x^2=\varepsilon_x/\varepsilon_0$, $n_y^2=\varepsilon_y/\varepsilon_0$, $n_z^2=\varepsilon_z/\varepsilon_0$, 其中, $\varepsilon_x,\varepsilon_y,\varepsilon_z$ 表示晶体的主要介电常数. 如图所见, 折射率椭球方便用来描述光波在晶体中的传播. 图 7.2-5 给出了主轴 (*principal axis*) 坐标系统下的折射率椭球, 其中, x,y 和 z 为主轴方向. n_x,n_y 和 n_z 是折射率椭球的折射率 (*refractive indice*). 根据这些折射率, 可以将晶体分为三类.

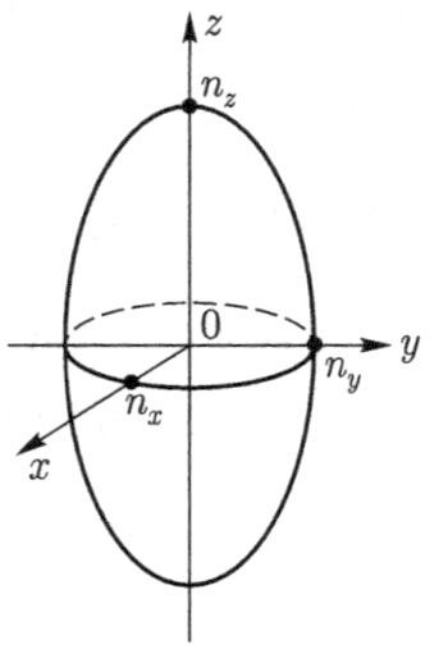

图 7.2-5 折射率椭球

在晶体中, 若其中两个折射率相等, 即 $n_x=n_y\neq n_z$, 这类晶体被称为单轴晶体 (*uniaxial crystal*). 两个相等的折射率通常被称为寻常光折射率 (*ordinary refractive index*) n_o, 即 $n_x=n_y=n_\text{o}$. 与之不同的折射率则被称为非寻常光折射率 (*extraordinary refractive index*) n_e, 即, $n_z=n_\text{e}$. 此外, 如果有 $n_\text{e}>n_\text{o}$, 则此类晶体可称为正单轴晶体 (*positive uniaxial crystal*) [如石英 (*quartz*)], 反之, $n_\text{e}<n_\text{o}$ 时, 则被称为负单轴晶体 (*negative uniaxial crystal*) [如方解石 (*calcite*)]. 由于单轴晶体的其他轴具有相同的折射率, 故单轴晶体的 z 轴被称为光轴 (*optic axis*). 当三个折射率相等时, 即 $n_x=n_y=n_z$, 该晶体为光学各向同性晶体 (*isotropic crystal*) [如钻石 (*diamond*)]. 当三个折射率不同时, 即 $n_x\neq n_y\neq n_z$, 则被称为双轴晶体 (*biaxial*) [如云母 (*mica*)]. 这里将只讨论单轴晶体的应用.

双折射波片 (*birefringent wave plate*)

波片 (*wave plate*) 在入射光波的两个正交偏振分量之间引入相移. 由图 7.2-5 可知, 当光在单轴晶体中沿 x 方向传播时, 线偏振光可以沿 y 方向和 z 方向被分解. 沿 y 方向偏振的光, 其折射率为 $n_y=n_\text{o}$, 而沿 z 方向偏振的光, 其折射率为 $n_z=n_\text{e}$, 这种效应就是在两个偏振波

7.2.2 Index Ellipsoid and Birefringent Wave Plates

Crystals in general are *anisotropic*. The medium is anisotropic if its properties are different in different directions from any given point. In general, there are two methods for analyzing light wave propagating through crystals: the method that employs the rigorous electromagnetic theory and that employs the *refractive index ellipsoid*. The use of the index ellipsoid is intuitive and convenient and is often used in practice. We can express the optical properties of crystals in terms of the refractive index ellipsoid. The equation of the index ellipsoid for a crystal is expressed as

$$\frac{x^2}{n_x^2}+\frac{y^2}{n_y^2}+\frac{z^2}{n_z^2}=1, \tag{7.2-6}$$

where $n_x^2=\varepsilon_x/\varepsilon_0$, $n_y^2=\varepsilon_y/\varepsilon_0$, and $n_z^2=\varepsilon_z/\varepsilon_0$, with ε_x, ε_y, and ε_z denoting the principal dielectric constants of the crystal. As we will see, the index of ellipsoid is used conveniently to describe light wave propagation in the crystal. Figure 7.2-5 shows the index ellipsoid in the coordinate system of the *principal axes*, where x, y, and z represent the directions of the principal axis. n_x, n_y, and n_z are the *refractive indices* of the index ellipsoid. Three crystal classes can be identified in terms of these refractive indices.

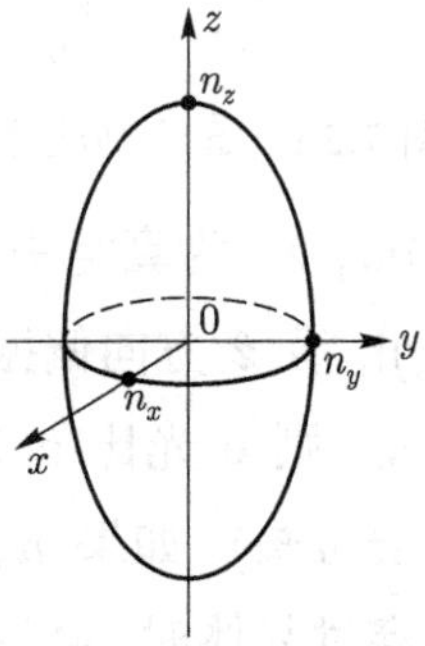

Figure 7.2-5 The index ellipsoid

In crystals where two of the refractive indices are equal, that is, $n_x=n_y\neq n_z$, the crystals are called *uniaxial crystals*. The two identical refractive indices are usually called the *ordinary refractive index* n_o, that is, $n_x=n_y=n_\mathrm{o}$. The distinct refractive index is called the *extraordinary refractive index* n_e, that is, $n_z=n_\mathrm{e}$. Furthermore, the crystal is said to be *positive uniaxial crystal* if $n_\mathrm{e}>n_\mathrm{o}$ (such as *quartz*) and *negative uniaxial crystal* if $n_\mathrm{e}<n_\mathrm{o}$ (such as *calcite*). The z-axis of a uniaxial crystal is called the *optic axis* as the other axes have the same index of refraction. When the three indices are equal, that is, $n_x=n_y=n_z$, the crystal is *optical isotropic* (such as *diamond*). Crystals for which the three indices are different, that is, $n_x\neq n_y\neq n_z$, are called *biaxial* (such as *mica*). We will only discuss the applications of uniaxial crystals.

Birefringent Wave Plates

A *wave plate* introduces a phase shift between two orthogonal polarization components of the incident light wave. It is clear from Figure 7.2-5 that when the light propagates along the x-direction in uniaxial crystals, linearly polarized light can be decomposed along the y- and z-directions. Light polarized along the y-direction sees an index of $n_y=n_\mathrm{o}$, while light polarized along the z-direction sees an index of $n_z=n_\mathrm{e}$. The effect is to introduce a phase shift between the two polarized waves. The crystal is usually cut into a plate, with

之间引入相移. 晶体通常被切割成晶片, 切割方向的选择使得光轴与晶片的表面平行. 此外, 如果晶片表面垂直于 x 主轴, 则该晶片被称为 x 切割 (x-*cut*). 图 7.2-6 所示为厚度是 d 的 x 切割波片 (x-*cut wave plate*). 设线偏振光场在 $x=0$ 处入射到该晶片上, 则

$$\boldsymbol{E}_{inc}=(|E_{0y}|\hat{\boldsymbol{y}}+|E_{0z}|\hat{\boldsymbol{z}})\mathrm{e}^{\mathrm{j}\omega_0 t}. \tag{7.2-7}$$

在图 7.2-6 中, $\theta=\arctan(|E_{0y}|/|E_{0z}|)$, 在波片的出射位置处, 场可以表示为

$$\boldsymbol{E}_{out}=(|E_{0y}|\mathrm{e}^{-\mathrm{j}k_{n_\mathrm{o}}d}\hat{\boldsymbol{y}}+|E_{0z}|\mathrm{e}^{-\mathrm{j}k_{n_\mathrm{e}}d}\hat{\boldsymbol{z}})\mathrm{e}^{\mathrm{j}\omega_0 t}, \tag{7.2-8}$$

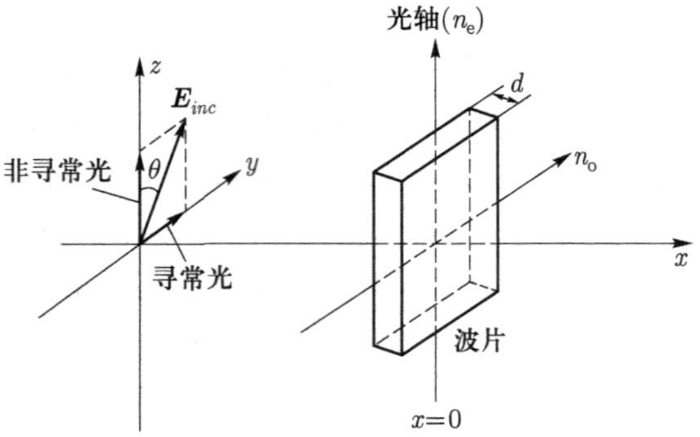

图 7.2-6 x 切割波片

其中, $k_{n_\mathrm{o}}=(2\pi/\lambda_v)n_\mathrm{o}$, $k_{n_\mathrm{e}}=(2\pi/\lambda_v)n_\mathrm{e}$, λ_v 为真空中的波长. 沿 $\hat{\boldsymbol{y}}$ 方向偏振的平面波被称为寻常波 (o 光) [*ordinary wave* (*o-ray*)], 沿 $\hat{\boldsymbol{z}}$ 方向偏振的平面波被称为非寻常波 (e 光) [*extraordinary wave* (*e-ray*)]. 如果 $n_\mathrm{e}>n_\mathrm{o}$, 则 e 光比 o 光传播慢, 因此, z 主轴通常被称为慢轴 (*slow axis*), y 主轴被称为快轴 (*fast axis*). 如果 $n_\mathrm{o}>n_\mathrm{e}$, 则 z 主轴成为快轴, 而 y 轴成为慢轴. 因此, 不同偏振的两种光线穿过晶体时, 会发生不同的相移, 这种现象被称为双折射 (*birefringence*). 可以看出, 单轴晶体的光轴即为光线不会产生双折射的方向. 事实上, 双折射是一类晶体的光学特性, 该类晶体的折射率取决于光的偏振 (e 光或 o 光) 及其光的传播方向. 追迹两束光线之间的相位差, 可将式 (7.2-8) 重写, 有

$$\boldsymbol{E}_{out}\propto(|E_{0y}|\hat{\boldsymbol{y}}+|E_{0z}|\mathrm{e}^{-\mathrm{j}(k_{n_\mathrm{e}}-k_{n_\mathrm{o}})d}\hat{\boldsymbol{z}})\mathrm{e}^{\mathrm{j}\omega_0 t}=(|E_{0y}|\hat{\boldsymbol{y}}+|E_{0z}|\mathrm{e}^{-\mathrm{j}\Delta\phi}\hat{\boldsymbol{z}})\mathrm{e}^{\mathrm{j}\omega_0 t}, \tag{7.2-9}$$

其中, $\Delta\phi=(k_{n_\mathrm{e}}-k_{n_\mathrm{o}})d=\left(\dfrac{2\pi}{\lambda_v}\right)(n_\mathrm{e}-n_\mathrm{o})d$ 表示两束光线之间的相对相移. 将此方程与式 (7.2-2) 进行比较, 可以发现, $\Delta\phi$ 即为 ϕ_0, 那么通过控制波片的厚度 d 即可控制 $\Delta\phi$, 从而可以获得不同类型的波片.

四分之一波片 (QWP)

四分之一波片 (*quarter-wave plate*, QWP) 用来将线偏振光变为圆偏振光 (*circularly polarized light*), 反之亦然. 通过设计正确的厚度, 当 $\Delta\phi=\left(\dfrac{2\pi}{\lambda_v}\right)(n_\mathrm{e}-n_\mathrm{o})d=\pi/2$ 时, 可以得到四分之一波片. 由于 $\Delta\phi=\phi_0=\pi/2$, 且有式 (7.2-9) 中的 $|E_{0y}|=|E_{0z}|$, 根据图 7.2-2b), 经波片后可以得到右旋圆偏振波 [*right-hand circularly* (RHC) *wave*], 具体如图 7.2-7 所示.

the orientation of the cut chosen so that the optic axis is parallel to the surface of the plate. Furthermore, if the plate surface is perpendicular to the principal x-axis, the crystal plate is called *x-cut*. Figure 7.2-6 shows an *x-cut wave plate* of thickness d. Let a linearly polarized optical field incident on the plate at $x = 0$ be

$$\boldsymbol{E}_{inc} = (|E_{0y}|\hat{\boldsymbol{y}} + |E_{0z}|\hat{\boldsymbol{z}})e^{j\omega_0 t}. \tag{7.2-7}$$

In Figure 7.2-6, $\theta = \arctan(|E_{0y}|/|E_{0z}|)$. At the exit of the wave plate, the field is represented by

$$\boldsymbol{E}_{out} = (|E_{0y}|e^{-jk_{n_o}d}\hat{\boldsymbol{y}} + |E_{0z}|e^{-jk_{n_e}d}\hat{\boldsymbol{z}})e^{j\omega_0 t}, \tag{7.2-8}$$

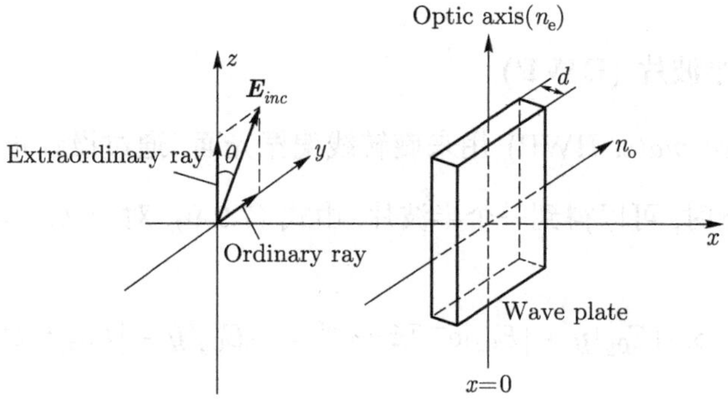

Figure 7.2-6 x-cut wave plate

where $k_{n_o} = (2\pi/\lambda_v)n_o$, and $k_{n_e} = (2\pi/\lambda_v)n_e$ with λ_v being the wavelength in vacuum. The plane wave polarized along $\hat{\boldsymbol{y}}$ is called the *ordinary wave (o-ray)*, and the plane wave polarized along $\hat{\boldsymbol{z}}$ is called the *extraordinary wave (e-ray)*. If $n_e > n_o$, the e-ray travels slower than the o-ray. The z-principal axis is, therefore, often referred to as the *slow axis*, and the y-principal axis is the *fast axis*. If $n_o > n_e$, the z-principal axis becomes the fast axis and the y-axis is the slow axis. Hence, the two rays of different polarizations acquire different phase shifts as they propagate through the crystal, and this phenomenon is called *birefringence*. So, we see that the optic axis of a uniaxial crystal is a direction in which a ray does not suffer birefringence. Indeed, birefringence is the optical property of a crystal having a refractive index that depends on the polarization (e-ray or o-ray) and the direction of propagation of light. Keeping track of the phase difference between the two rays and rewriting Eq. (7.2-8), we have

$$\boldsymbol{E}_{out} \propto (|E_{0y}|\hat{\boldsymbol{y}} + |E_{0z}|e^{-j(k_{n_e}-k_{n_o})d}\hat{\boldsymbol{z}})e^{j\omega_0 t} = (|E_{0y}|\hat{\boldsymbol{y}} + |E_{0z}|e^{-j\Delta\phi}\hat{\boldsymbol{z}})e^{j\omega_0 t}, \tag{7.2-9}$$

where $\Delta\phi = (k_{n_e} - k_{n_o})d = \left(\frac{2\pi}{\lambda_v}\right)(n_e - n_o)d$ is the relative phase shift between the two rays. Comparing this equation with Eq. (7.2-2), we can recognize that $\Delta\phi$ is ϕ_0. By controlling $\Delta\phi$ through the right thickness d of the wave plate, we have different types of wave plate.

Quarter-Wave Plate (QWP)

Quarter-wave plates are used to change linearly polarized light into *circularly polarized light* and vice versa. When $\Delta\phi = \left(\frac{2\pi}{\lambda_v}\right)(n_e - n_o)d = \pi/2$ by designing the correct thickness, we have a quarter-wave plate. Since $\Delta\phi = \phi_0 = \pi/2$, and with $|E_{0y}| = |E_{0z}|$ in Eq. (7.2-9), we have *right-hand circularly* (RHC) *wave* as output from the wave plate, according to Figure 7.2-2b. The situation is shown in Figure 7.2-7.

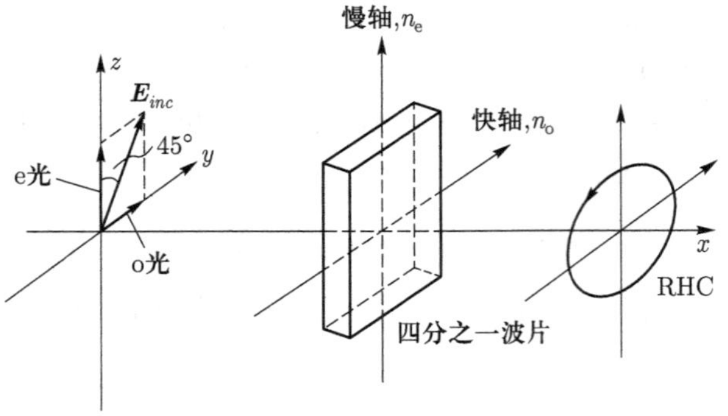

图 7.2-7　四分之一波片

二分之一波片或半波片 (HWP)

半波片 (*half-wave plate*, HWP) 用来旋转线偏振光面. 通过设计正确的厚度, 当 $\Delta\phi = \left(\dfrac{2\pi}{\lambda_v}\right)(n_e - n_o)d = \pi$ 时, 可以得到一个半波片. 由式 (7.2-9), 对于 $\Delta\phi = \pi$, 有

$$\boldsymbol{E}_{out} \propto (|E_{0y}|\hat{\boldsymbol{y}} + |E_{0z}|e^{-j\pi}\hat{\boldsymbol{z}})e^{j\omega_0 t} = (|E_{0y}|\hat{\boldsymbol{y}} - |E_{0z}|\hat{\boldsymbol{z}})e^{j\omega_0 t}.$$

具体情况如图 7.2-8 所示. 可以看出, 半波片可以有效地通过将输入电场经 y 轴旋转角度 2θ, 从而在半波片的出射位置产生 $\boldsymbol{E}_{out}$.

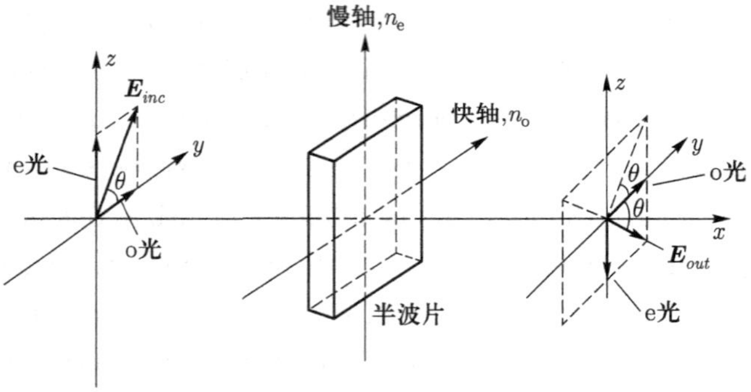

图 7.2-8　半波片

全波片 (FWP)

通过设计正确的厚度, 当 $\Delta\phi = \left(\dfrac{2\pi}{\lambda_v}\right)(n_e - n_o)d = 2\pi$ 时, 可以得到一个全波片 (*full-wave plate*, FWP). 全波片会使通过波片的 o 光和 e 光之间正好附加一个波长的光程差 (或附加一个 2π 的相位差). 由于 $e^{-j\Delta\phi} = e^{-j2\pi} = 1$, 因此, 全波片通常是为特定的波长而设计的, 具有该特定波长的线偏振光在通过相应波片后, 其偏振态不会发生改变. 而其他附近波长的线偏振光, 在经过该波片后会变成椭圆偏振光, 因为对于这些波长, 其 $\Delta\phi$ 不再正好是 2π. 如图 7.2-9 所示, 具有交叉偏振片 (*cross polarizers*) (两个偏振片的透射轴成直角) 的光学系统, 再用一个全波片, 就可以从入射的自然光中滤掉特定颜色的光. 例如, 如果全波片是为绿光所设计的, 那么输出的光会出现一些红色/紫色. 输出时放置的偏振片通常被称为检偏器 (*analyzer*), 因为它被用来检测该光束的偏振态.

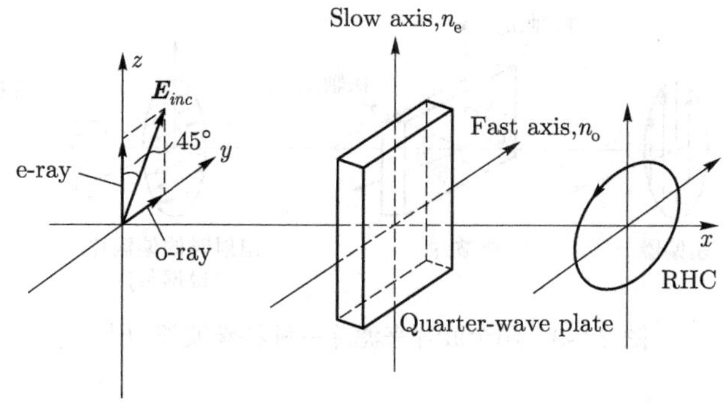

Figure 7.2-7 Quarter-wave plate

Half-Wave Plate (HWP)

Half-wave plates are used to rotate the plane of linearly polarized light. When $\Delta\phi = \left(\frac{2\pi}{\lambda_v}\right)(n_e - n_o)d = \pi$ by designing the correct thickness, we have a half-wave plate. From Eq. (7.2-9), for $\Delta\phi = \pi$, we have

$$\boldsymbol{E}_{out} \propto (|E_{0y}|\hat{\boldsymbol{y}} + |E_{0z}|\mathrm{e}^{-\mathrm{j}\pi}\hat{\boldsymbol{z}})\mathrm{e}^{\mathrm{j}\omega_0 t} = (|E_{0y}|\hat{\boldsymbol{y}} - |E_{0z}|\hat{\boldsymbol{z}})\mathrm{e}^{\mathrm{j}\omega_0 t}.$$

The situation is shown in Figure 7.2-8. We see that effectively the half-wave plate rotates the input electrical field vector through an angle 2θ across the y-axis to give $\boldsymbol{E}_{out}$ at the exit of the half-wave plate.

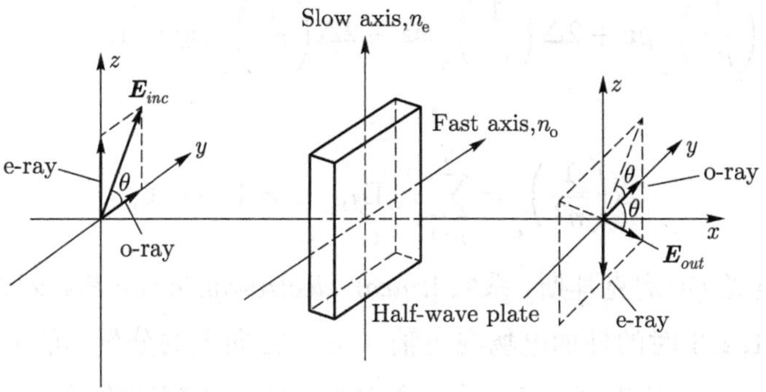

Figure 7.2-8 Half-wave plate

Full-Wave Plates (FWP)

When $\Delta\phi = \left(\frac{2\pi}{\lambda_v}\right)(n_e - n_o)d = 2\pi$ by designing the correct thickness, we have a *full-wave plate*. A full-wave plate introduces exactly an optical path of exactly one wavelength between the o-ray and the e-ray. Since a full-wave plate is typically designed for a particular wavelength as $\mathrm{e}^{-\mathrm{j}\Delta\phi} = \mathrm{e}^{-\mathrm{j}2\pi} = 1$, linearly polarized wave with that wavelength exiting the wave plate remains the same polarization. Linearly polarized light with other wavelengths will, therefore, exit the wave plate elliptically polarized as $\Delta\phi$ is not exactly at 2π for these wavelengths. An optical system with *cross polarizers* (two polarizers with their transmission axes at right angles) along with a full-wave plate shown in Figure 7.2-9 allows the rejection of a particular color from the input natural white light. For example, if the full-wave plate is designed for green light, then the output light would appear somewhat red/violet. The output polarizer is often called an *analyzer* as it is used to determine whether the light has been polarized or not.

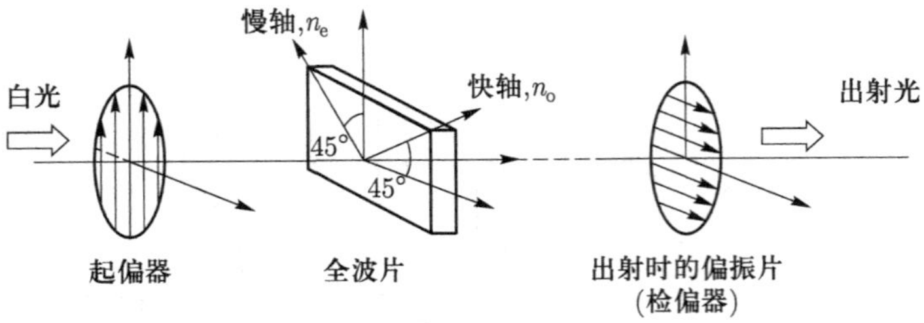

图 7.2-9 用全波片来滤除一种特殊波长的光

7.2.3 电光效应

电光效应 (*electro-optic effect*) 是指在外加电场作用下晶体折射率的变化. 我们将使用折射率椭球法来讨论电光效应, 该方法被广泛用于对通过晶体传播的光波的控制 (如光波的相位调制、强度调制以及偏振).

这里只讨论线性 (或泡克耳斯型) 电光效应 [*linear* (*or Pockels-type*) electro-optic effect], 即该情况下 n_e 和 n_o 的变化与所加的电场成线性比例. 从数学上讲, 该电光效应可以被非常恰当地建模为外加电场引起的折射率椭球的变形. 与其明确规定折射率的变化, 不如用外加电场导致的 $1/n^2$ 的变化, 即用 $\Delta(1/n^2)$, 来描述更为方便. 将分析局限于 Pockels 效应和单轴晶体, 则变形的折射率椭球的表达式变为

$$\left[\frac{1}{n_o^2}+\Delta\left(\frac{1}{n^2}\right)_1\right]x^2+\left[\frac{1}{n_o^2}+\Delta\left(\frac{1}{n^2}\right)_2\right]y^2+\left[\frac{1}{n_e^2}+\Delta\left(\frac{1}{n^2}\right)_3\right]z^2$$
$$+2\Delta\left(\frac{1}{n^2}\right)_4 yz+2\Delta\left(\frac{1}{n^2}\right)_5 xz+2\Delta\left(\frac{1}{n^2}\right)_6 xy=1, \tag{7.2-10}$$

其中

$$\Delta\left(\frac{1}{n^2}\right)_i=\sum_{j=1}^{3} r_{ij}\mathrm{E}_j, \quad i=1,\cdots,6,$$

且 r_{ij} 被称为线性电光 (或泡克耳斯) 系数 [*linear electro-optic* (*or Pockels*) *coefficient*], E_j 这些量分别是当 $j=1,2,3$ 时的外加电场在主轴 x,y,z 方向上的分量. 可以看出, 在式 (7.2-10) 中, 交叉项, 即 yz,xz,xy 的出现, 表示了一个中心在原点的旋转椭球方程. 当外加电场为零, 即 $\mathrm{E}_j=0$ 时, 式 (7.2-10) 简化为式 (7.2-6), 其中 $n_x=n_o=n_y$ 且 $n_z=n_e$. 可以用以下矩阵来表示 $\Delta\left(\frac{1}{n^2}\right)_i$

$$\begin{pmatrix}\Delta\left(\frac{1}{n^2}\right)_1\\ \Delta\left(\frac{1}{n^2}\right)_2\\ \Delta\left(\frac{1}{n^2}\right)_3\\ \Delta\left(\frac{1}{n^2}\right)_4\\ \Delta\left(\frac{1}{n^2}\right)_5\\ \Delta\left(\frac{1}{n^2}\right)_6\end{pmatrix}=\begin{pmatrix}r_{11}&r_{12}&r_{13}\\ r_{21}&r_{22}&r_{23}\\ r_{31}&r_{32}&r_{33}\\ r_{41}&r_{42}&r_{43}\\ r_{51}&r_{52}&r_{53}\\ r_{61}&r_{62}&r_{63}\end{pmatrix}\begin{pmatrix}\mathrm{E}_1\\ \mathrm{E}_2\\ \mathrm{E}_3\end{pmatrix}. \tag{7.2-11}$$

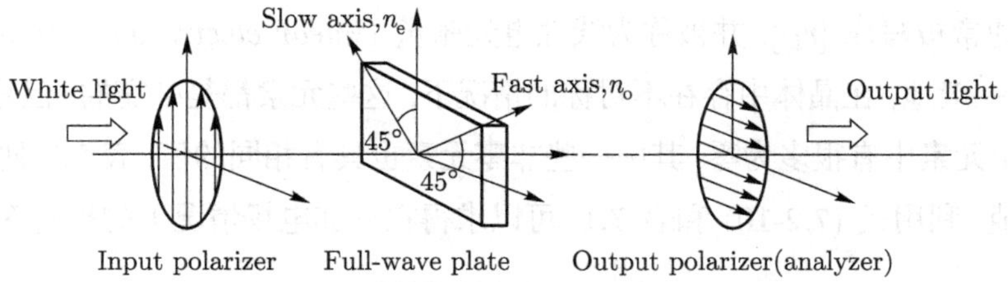

Figure 7.2-9 Application of a full-wave plate to reject a particular color

7.2.3 The Electro-Optic Effect

The *electro-optic effect* is the change of the refractive index of crystals under the presence of external applied electric field. We shall use the index ellipsoid method to discuss the electro-optic effect, which is widely used to realize control over the light wave propagating through the crystals (such as phase modulation, intensity modulation, and polarization of the light wave).

We will only discuss the *linear (or Pockels-type) electro-optic effect*, meaning the case where the change in n_e and n_o is linearly proportional to the applied electric field. Mathematically, the electro-optic effect can be best modelled as a deformation of the index ellipsoid due to an external electric field. Instead of specifying the changes in the refractive indices, it is more convenient to describe the changes in $1/n^2$, that is, $\Delta(1/n^2)$, due to the external electric field. Restricting our analysis to the Pockels effects and uniaxial crystals, the expression for the deformed ellipsoid becomes

$$\left[\frac{1}{n_o^2}+\Delta\left(\frac{1}{n^2}\right)_1\right]x^2+\left[\frac{1}{n_o^2}+\Delta\left(\frac{1}{n^2}\right)_2\right]y^2+\left[\frac{1}{n_e^2}+\Delta\left(\frac{1}{n^2}\right)_3\right]z^2$$
$$+2\Delta\left(\frac{1}{n^2}\right)_4 yz+2\Delta\left(\frac{1}{n^2}\right)_5 xz+2\Delta\left(\frac{1}{n^2}\right)_6 xy=1, \tag{7.2-10}$$

where

$$\Delta\left(\frac{1}{n^2}\right)_i=\sum_{j=1}^{3} r_{ij}\mathrm{E}_j, \quad i=1,\cdots,6,$$

and r_{ij} are called *linear electro-optic (or Pockels) coefficients*. The E_js are the components of the externally applied electric field in the direction of the principal axes x, y, and z for $j=1,2$, and 3, respectively. Note that the appearance of the crossed terms, that is, yz, xz, and xy, in Eq. (7.2-10) signifies the equation of a rotated ellipsoid centered at the origin. When the applied electric field is zero, that is, $\mathrm{E}_j=0$, Eq. (7.2-10) reduces to Eq. (7.2-6), where $n_x=n_o=n_y$ and $n_z=n_e$. We can express $\Delta\left(\frac{1}{n^2}\right)_i$ in matrix form as

$$\begin{pmatrix}\Delta\left(\frac{1}{n^2}\right)_1\\ \Delta\left(\frac{1}{n^2}\right)_2\\ \Delta\left(\frac{1}{n^2}\right)_3\\ \Delta\left(\frac{1}{n^2}\right)_4\\ \Delta\left(\frac{1}{n^2}\right)_5\\ \Delta\left(\frac{1}{n^2}\right)_6\end{pmatrix}=\begin{pmatrix}r_{11} & r_{12} & r_{13}\\ r_{21} & r_{22} & r_{23}\\ r_{31} & r_{32} & r_{33}\\ r_{41} & r_{42} & r_{43}\\ r_{51} & r_{52} & r_{53}\\ r_{61} & r_{62} & r_{63}\end{pmatrix}\begin{pmatrix}\mathrm{E}_1\\ \mathrm{E}_2\\ \mathrm{E}_3\end{pmatrix}. \tag{7.2-11}$$

该 6×3 矩阵通常被写成 $[r_{ij}]$, 并被称为线性电光张量 (*linear electro-optic tensor*). 该张量包含 18 个元素, 一般地, 在晶体中存在不对称的情况下, 这些元素都是必需的. 但是, 取决于晶体的对称性, 18 个元素中有很多为零, 其中一些非零元素也具有相同的值, 表 7.1 列出了一些代表性晶体的非零值. 利用式 (7.2-10) 和表 7.1, 可以求得在外加电场情况下的折射率椭球方程.

表 7.1 泡克耳斯系数

材料	$r_{ij}/(10^{-12}\ \mathrm{m/V})$	$\lambda_v/\mu\mathrm{m}$	折射率
$LiNbO_3$ (铌酸锂, *lithium niobate*)	$r_{13} = r_{23} = 8.6$	0.63	$n_o = 2.2967$
	$r_{33} = 30.8$		
	$r_{22} = -r_{61} = -r_{12} = 3.4$		$n_e = 2.2082$
	$r_{51} = r_{41} = 28$		
KDP (磷酸二氢钾, *potassium dihydrogen phosphate*)	$r_{41} = r_{52} = 8.6$	0.55	$n_o = 1.50737$
	$r_{63} = 10.6$		$n_e = 1.46685$
ADP (磷酸二氢铵, *ammonium dihydrogen phosphate*)	$r_{41} = r_{52} = 2.8$	0.55	$n_o = 1.52$
	$r_{63} = 8.5$		$n_e = 1.48$

举例: 外加电场作用下 KDP 的折射率椭球

在外加电场 $\mathbf{E} = \mathrm{E}_x\hat{\boldsymbol{x}} + \mathrm{E}_y\hat{\boldsymbol{y}} + \mathrm{E}_z\hat{\boldsymbol{z}}$ 的情况下, 我们用常用的 KDP 晶体来做分析, 由表 7.1 可知, 该晶体的电光张量为

$$[r_{ij}] = \begin{pmatrix} 0 & 0 & 0 \\ 0 & 0 & 0 \\ 0 & 0 & 0 \\ r_{41} & 0 & 0 \\ 0 & r_{41} & 0 \\ 0 & 0 & r_{63} \end{pmatrix}. \tag{7.2-12}$$

可以看出, $r_{52} = r_{41}$, 因此, 这些晶体只有两个独立的电光系数, 即 r_{41} 和 r_{63}, 将式 (7.2-12) 代入式 (7.2-11), 有

$$\Delta\left(\frac{1}{n^2}\right)_1 = \Delta\left(\frac{1}{n^2}\right)_2 = \Delta\left(\frac{1}{n^2}\right)_3 = 0,$$

和

$$\Delta\left(\frac{1}{n^2}\right)_4 = r_{41}\mathrm{E}_x, \quad \Delta\left(\frac{1}{n^2}\right)_5 = r_{41}\mathrm{E}_y, \quad \Delta\left(\frac{1}{n^2}\right)_6 = r_{63}\mathrm{E}_z.$$

由这些值, 式 (7.2-10) 变为

$$\frac{x^2}{n_o^2} + \frac{y^2}{n_o^2} + \frac{z^2}{n_e^2} + 2r_{41}\mathrm{E}_x yz + 2r_{41}\mathrm{E}_y xz + 2r_{63}\mathrm{E}_z xy = 1. \tag{7.2-13}$$

举例: 仅在 E_z 下旋转 KDP 的折射率椭球

当只有一个电场沿着 z 主轴作用时, 式 (7.2-13) 变为

$$\frac{x^2}{n_o^2}+\frac{y^2}{n_o^2}+\frac{z^2}{n_e^2}+2r_{63}E_z xy=1. \tag{7.2-14}$$

在方程中, 可以看出, 有一个交叉项 xy 存在, 表明原来的椭球在 x-y 平面上被旋转了. 可找一个新的坐标系, 使式 (7.2-14) 成为坐标系中的“主轴”, 从而求出电场对光传播的影响. 由解析几何可知, 以下旋转矩阵给出了原始坐标 (x, y) 与旋转后的坐标 (x', y') 的关系

$$\begin{pmatrix} x \\ y \end{pmatrix}=\begin{pmatrix} \cos\theta & -\sin\theta \\ \sin\theta & \cos\theta \end{pmatrix}\begin{pmatrix} x' \\ y' \end{pmatrix}, \tag{7.2-15}$$

其中, θ 为旋转角度. 原始坐标系与旋转坐标系的关系如图 7.2-10 所示. 旋转方向遵循右手定则并绕 z 轴旋转.

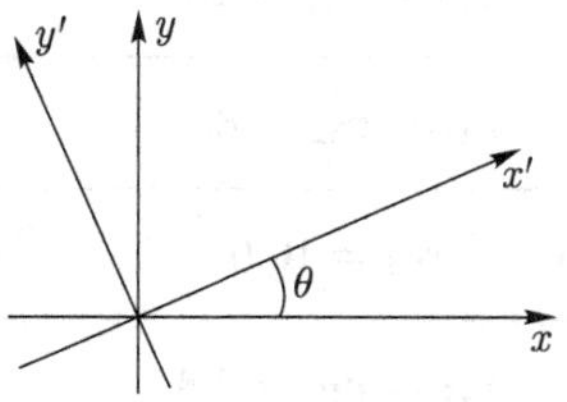

图 7.2-10　xy 坐标系旋转角度 θ 到 $x'y'$ 坐标系

将式 (7.1-15) 代入式 (7.1-14), 得到

$$\left[\frac{1}{n_o^2}+r_{63}E_z\sin(2\theta)\right]x'^2+\left[\frac{1}{n_o^2}-r_{63}E_z\sin(2\theta)\right]y'^2+\frac{1}{n_e^2}z^2+2r_{63}E_z\cos(2\theta)x'y'=1. \tag{7.2-16}$$

为了消去交叉项, 设 $\cos(2\theta)=0$, 即 $\theta=\pm 45°$, 取 $\theta=-45°$ 进行进一步的数学分析, 则式 (7.2-16) 变为

$$\left(\frac{1}{n_o^2}-r_{63}E_z\right)x'^2+\left(\frac{1}{n_o^2}+r_{63}E_z\right)y'^2+\frac{1}{n_e^2}z^2=1. \tag{7.2-17}$$

这是在 KDP 晶体上沿 z 轴施加电场后的折射率椭球, 因此, 该折射率椭球方程与新坐标系中的 x', y' 及 z 主轴相对齐

$$\frac{x'^2}{n_x'^2}+\frac{y'^2}{n_y'^2}+\frac{z^2}{n_e^2}=1,$$

其中

$$\frac{1}{n_x'^2}=\frac{1}{n_o'^2}-r_{63}E_z, \quad \frac{1}{n_y'^2}=\frac{1}{n_o'^2}+r_{63}E_z. \tag{7.2-18}$$

现在估计沿 x' 和 y' 主轴的折射率的值, 从式 (7.2-18), 有

$$n_x'^2=\frac{1}{\dfrac{1}{n_o^2}-r_{63}E_z}=\frac{1}{\dfrac{1}{n_o^2}(1-n_o^2 r_{63}E_z)}.$$

The 6 × 3 matrix is often written as $[r_{ij}]$ and known as the *linear electro-optic tensor*. The tensor contains 18 elements and they are necessary in the most general case, when no symmetry is present in the crystal. But many of the 18 elements are zero and some of the nonzero elements have the same value, depending on the symmetry of the crystal. Table 7.1 lists the nonzero values of some of the representative crystals. Using Eq. (7.2-10) and Table 7.1, we can find the equation of the index ellipsoid in the presence of an external applied field.

Table 7.1 Pockels coefficients

Material	$r_{ij}/(10^{-12}\ \text{m/V})$	$\lambda_v/\mu\text{m}$	Refractive index
$LiNbO_3$ (Lithium Niobate)	$r_{13} = r_{23} = 8.6$	0.63	$n_o = 2.2967$
	$r_{33} = 30.8$		
	$r_{22} = -r_{61} = -r_{12} = 3.4$		$n_e = 2.2082$
	$r_{51} = r_{41} = 28$		
KDP (Potassium dihydrogen phosphate)	$r_{41} = r_{52} = 8.6$	0.55	$n_o = 1.50737$
	$r_{63} = 10.6$		$n_e = 1.46685$
ADP (Ammonium dihydrogen phosphate)	$r_{41} = r_{52} = 2.8$	0.55	$n_o = 1.52$
	$r_{63} = 8.5$		$n_e = 1.48$

Example: Index Ellipsoid of KDP Under the Presence of Applied Electric Field

In the presence of an external electric field $\mathbf{E} = \mathrm{E}_x\hat{\boldsymbol{x}} + \mathrm{E}_y\hat{\boldsymbol{y}} + \mathrm{E}_z\hat{\boldsymbol{z}}$, let us make an analysis with the frequently used KDP crystal. According to Table 7.1, the electrooptic tensor of such a crystal is

$$[r_{ij}] = \begin{pmatrix} 0 & 0 & 0 \\ 0 & 0 & 0 \\ 0 & 0 & 0 \\ r_{41} & 0 & 0 \\ 0 & r_{41} & 0 \\ 0 & 0 & r_{63} \end{pmatrix}. \tag{7.2-12}$$

Note that $r_{52} = r_{41}$. Hence, these crystals only have two independent electro-optic coefficients, that is, r_{41} and r_{63}. Substituting Eq. (7.2-12) into Eq. (7.2-11), we have

$$\Delta\left(\frac{1}{n^2}\right)_1 = \Delta\left(\frac{1}{n^2}\right)_2 = \Delta\left(\frac{1}{n^2}\right)_3 = 0,$$

and

$$\Delta\left(\frac{1}{n^2}\right)_4 = r_{41}\mathrm{E}_x, \quad \Delta\left(\frac{1}{n^2}\right)_5 = r_{41}\mathrm{E}_y, \quad \Delta\left(\frac{1}{n^2}\right)_6 = r_{63}\mathrm{E}_z.$$

With these values, Eq. (7.2-10) becomes

$$\frac{x^2}{n_o^2} + \frac{y^2}{n_o^2} + \frac{z^2}{n_e^2} + 2r_{41}\mathrm{E}_x yz + 2r_{41}\mathrm{E}_y xz + 2r_{63}\mathrm{E}_z xy = 1. \tag{7.2-13}$$

Example: Rotation of the Index Ellipsoid of KDP Under E_z Only

When there is only an electrical field applied along the principal z-axis, Eq. (7.2-13) becomes

$$\frac{x^2}{n_o^2}+\frac{y^2}{n_o^2}+\frac{z^2}{n_e^2}+2r_{63}E_z xy=1. \tag{7.2-14}$$

In Eq. (7.2-14), we see that there is a crossed term xy present, suggesting the original ellipsoid has been rotated in the x-y plane. We can find a new coordinate system so that Eq. (7.2-14) will become the "principal axes" in the coordinate system, allowing us to find the effect of the electrical field on light propagation. From analytic geometry, the following rotation matrix gives the relation between the original coordinates (x, y) to the rotated coordinates (x', y'), where θ is the rotation angle:

$$\begin{pmatrix} x \\ y \end{pmatrix}=\begin{pmatrix} \cos\theta & -\sin\theta \\ \sin\theta & \cos\theta \end{pmatrix}\begin{pmatrix} x' \\ y' \end{pmatrix}. \tag{7.2-15}$$

The relationship between the original and the rotated coordinate systems is shown in Figure 7.2-10. The direction of rotation is according to the right-hand rule with rotation about the z-axis.

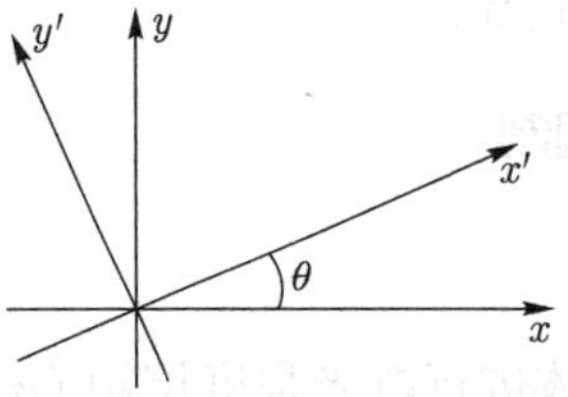

Figure 7.2-10 An xy-coordinate system rotated through an angle θ to an $x'y'$ coordinate system.

Substituting Eq. (7.1-15) into Eq. (7.1-14), we find

$$\left[\frac{1}{n_o^2}+r_{63}E_z\sin(2\theta)\right]x'^2+\left[\frac{1}{n_o^2}-r_{63}E_z\sin(2\theta)\right]y'^2+\frac{1}{n_e^2}z^2+2r_{63}E_z\cos(2\theta)x'y'=1. \tag{7.2-16}$$

To eliminate the crossed term, we let $\cos(2\theta)=0$, which means $\theta=\pm 45°$. We choose to take $\theta=-45°$ for our further mathematical analysis, and then Eq. (7.2-16) becomes

$$\left(\frac{1}{n_o^2}-r_{63}E_z\right)x'^2+\left(\frac{1}{n_o^2}+r_{63}E_z\right)y'^2+\frac{1}{n_e^2}z^2=1. \tag{7.2-17}$$

This is the ellipsoid after an electric field is applied on the KDP crystal along the z-axis. Hence, the index ellipsoid equation is in the coordinate system aligned with the new principal axes, x', y' and z:

$$\frac{x'^2}{n_x'^2}+\frac{y'^2}{n_y'^2}+\frac{z^2}{n_e^2}=1,$$

where

$$\frac{1}{n_x'^2}=\frac{1}{n_o'^2}-r_{63}E_z \text{ and } \frac{1}{n_y'^2}=\frac{1}{n_o'^2}+r_{63}E_z. \tag{7.2-18}$$

Let us now estimate the value of the refractive indices along the x' and y' principal axes. From Eq. (7.2-18),

$$n_x'^2=\frac{1}{\dfrac{1}{n_o^2}-r_{63}E_z}=\frac{1}{\dfrac{1}{n_o^2}(1-n_o^2 r_{63}E_z)}.$$

故, 得

$$n_x' = \frac{n_o}{\sqrt{1-n_o^2 r_{63}\mathrm{E}_z}} \approx \frac{n_o}{1-\dfrac{1}{2}n_o^2 r_{63}\mathrm{E}_z}. \tag{7.2-19}$$

由于 r_{63} 非常小 (大约为 10^{-12} m/V, 见表 7.1), 且因为 ε 很小, 这里用近似 $\sqrt{1-\varepsilon}\approx 1-\varepsilon/2$. 而且, 我们用 $1/(1-\varepsilon)\approx 1+\varepsilon$ 来重写式 (7.2-19), 最终得

$$n_x' \approx n_o + \frac{1}{2}n_o^3 r_{63}\mathrm{E}_z. \tag{7.2-20a}$$

类似地, 对于 n_y', 有

$$n_y' \approx n_o - \frac{1}{2}n_o^3 r_{63}\mathrm{E}_z. \tag{7.2-20b}$$

可以看到, 当一个电场沿着 z 方向作用在 KDP 晶体上时, 该晶体就会从单轴晶体转变为双轴晶体 [见式 (7.2-14) 在 $\mathrm{E}_z=0$ 时的情况], 即, $n_x'\neq n_y'$, 因为 n_x' 和 n_y' 是新坐标系中的主折射率. 该折射率椭球的新主轴被绕 z 轴旋转 45°, 该旋转角度与外加电场的大小无关, 但其折射率的变化与沿 z 轴的电场即 E_z 成正比.

7.2.4　电光强度调制与相位调制

强度调制

这里来研究如何利用 KDP 晶体进行电光强度调制 (*intensity modulation*), 调制系统如图 7.2-11 所示. 它由两个透射轴相互交叉垂直的偏振片中间放置 KDP 晶体所组成. 电场 (E_z) 通过外部电压 $V=\mathrm{E}_z L$ 沿晶体的 z 主轴方向施加, 其中 L 为晶体的厚度, 如图所示. 折射率椭球在 $\hat{\boldsymbol{x}}$-$\hat{\boldsymbol{y}}$ 平面绕新的主轴 $\hat{\boldsymbol{x}}'$ 和 $\hat{\boldsymbol{y}}'$ 旋转 45°, 该旋转与图 7.2-10 中在 $\theta=-45°$ 时所示一致.

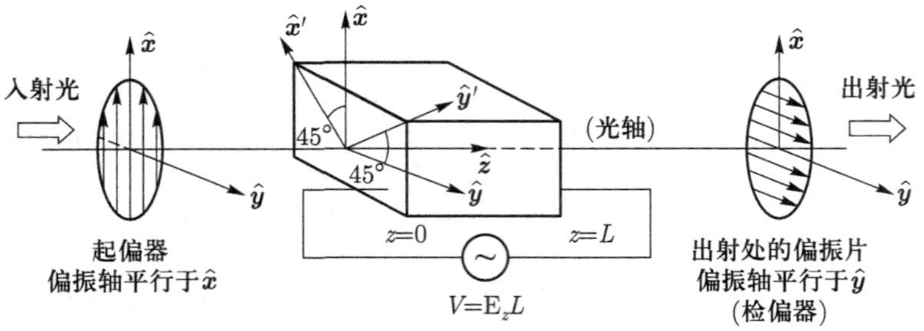

图 7.2-11　电光强度调制系统

输入为 $\hat{\boldsymbol{x}}$ 线偏振光, 即 $\boldsymbol{E}_{inc}=E_0\hat{\boldsymbol{x}}\mathrm{e}^{\mathrm{j}\omega_0 t}$, 它沿着光轴的方向传播, 可以在 $z=0$ 处分解为沿 $\hat{\boldsymbol{x}}'$ 和 $\hat{\boldsymbol{y}}'$ 方向偏振的两个正交分量

$$\boldsymbol{E}_{inc} = \frac{E_0}{\sqrt{2}}(\hat{\boldsymbol{x}}'+\hat{\boldsymbol{y}}')\mathrm{e}^{\mathrm{j}\omega_0 t}.$$

在该晶体的出射位置处, 即 $z=L$ 时, 其场可以表示为

$$\boldsymbol{E}_{exit} = \frac{E_0}{\sqrt{2}}(\mathrm{e}^{-\mathrm{j}k_{n_x'}L}\hat{\boldsymbol{x}}'+\mathrm{e}^{-\mathrm{j}k_{n_y'}L}\hat{\boldsymbol{y}}')\mathrm{e}^{\mathrm{j}\omega_0 t}, \tag{7.2-21}$$

Hence,

$$n_x' = \frac{n_o}{\sqrt{1-n_o^2 r_{63}\mathrm{E}_z}} \approx \frac{n_o}{1-\dfrac{1}{2}n_o^2 r_{63}\mathrm{E}_z}. \tag{7.2-19}$$

as r_{63} is very small (of the order of 10^{-12} m/V, see Table 7.1) and we have used approximation $\sqrt{1-\varepsilon} \approx 1-\varepsilon/2$ for ε small. Furthermore, we use $1/(1-\varepsilon) \approx 1+\varepsilon$ to rewrite Eq. (7.2-19) to finally obtain

$$n_x' \approx n_o + \frac{1}{2}n_o^3 r_{63}\mathrm{E}_z. \tag{7.2-20a}$$

Similarly, for n_y', we have

$$n_y' \approx n_o - \frac{1}{2}n_o^3 r_{63}\mathrm{E}_z. \tag{7.2-20b}$$

We see that when an electric field is applied on the KDP crystal along the z-direction, the crystal is turned from a uniaxial crystal [see. Eq. (7.2-14) for $\mathrm{E}_z=0$] into a biaxial crystal, i.e., $n_x' \neq n_y'$ as n_x' and n_y' are the principal refractive indices in the new coordinate system. The new principal axes of the ellipsoid is rotated around the z-axis an angle of 45°. The rotational angle does not depend on the magnitude of the applied electric field, but the change of the index of refraction is proportional to the electric field along the z-axis, that is, E_z.

7.2.4 Electro-Optic Intensity Modulation and Phase Modulation

Intensity Modulation

Let us investigate how to use the KDP crystal for electro-optic intensity modulation. The modulation system is shown in Figure 7.2-11. It consists of the KDP crystal placed between two crossed polarized whose transmission axes are perpendicular to each other. With the electric field (E_z) applied along the principal z-axis direction of the crystal through an external voltage $V = \mathrm{E}_z L$, where L is the thickness of the crystal, the index ellipsoid has been rotated on the $\hat{\boldsymbol{x}} - \hat{\boldsymbol{y}}$ plane by 45° with the new principal axes $\hat{\boldsymbol{x}}'$ and $\hat{\boldsymbol{y}}'$, as indicated in the figure. This rotation is consistent to that shown in Figure 7.2-10 for $\theta = -45°$.

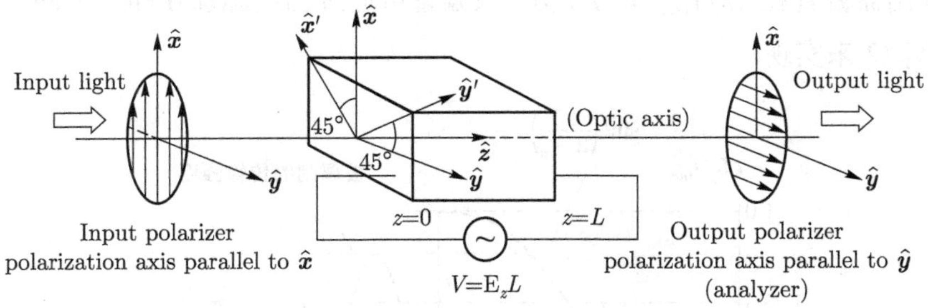

Figure 7.2-11 Electro-optic intensity modulation system

The input $\hat{\boldsymbol{x}}$-linearly polarized light, $\boldsymbol{E}_{inc} = E_0\hat{\boldsymbol{x}}\mathrm{e}^{\mathrm{j}\omega_0 t}$, propagates along the optic axis direction and can be resolved into two orthogonal components polarized along $\hat{\boldsymbol{x}}'$ and $\hat{\boldsymbol{y}}'$ at $z=0$ as

$$\boldsymbol{E}_{inc} = \frac{E_0}{\sqrt{2}}(\hat{\boldsymbol{x}}' + \hat{\boldsymbol{y}}')\mathrm{e}^{\mathrm{j}\omega_0 t}.$$

At the exit of the crystal, that is, $z = L$, the field is represented by

$$\boldsymbol{E}_{exit} = \frac{E_0}{\sqrt{2}}(\mathrm{e}^{-\mathrm{j}k_{n_x'}L}\hat{\boldsymbol{x}}' + \mathrm{e}^{-\mathrm{j}k_{n_y'}L}\hat{\boldsymbol{y}}')\mathrm{e}^{\mathrm{j}\omega_0 t}, \tag{7.2-21}$$

其中, $k_{n'_x}=\dfrac{2\pi}{\lambda_v}n'_x$, 且, $k_{n'_y}=\dfrac{2\pi}{\lambda_v}n'_y$. 这两个分量之间的相位差被称为相位延迟 (*retardation*) $\varPhi$, 可由下式给出

$$\varPhi=(k_{n'_x}-k_{n'_y})L=\frac{2\pi}{\lambda_v}(n'_x-n'_y)L=\frac{2\pi}{\lambda_v}n_o^3r_{63}\mathrm{E}_zL=\frac{2\pi}{\lambda_v}n_o^3r_{63}V. \tag{7.2-22}$$

其中, 这里用了从式 (7.2-20) 中得到的 n'_x 和 n'_y 的值, 在这一点上, 引入一个重要的参数, 称为半波电压 (*half-wave voltage*) V_π. 半波电压是引入一个 π 的相位变化, 即当 $\varPhi=\pi$ 时, 所需要加的电压. 因此, 从式 (7.2-22), 有

$$V_\pi=\frac{\lambda_v}{2n_o^3r_{63}}. \tag{7.2-23}$$

由表 7.1 可知, KDP 晶体在 $\lambda_v=0.55\ \mu\mathrm{m}$ 时, $V_\pi\approx 7.58\ \mathrm{kV}$. 当半波电压较低时, 调制器所需的功率也较低, 降低调制器的半波电压是一项持续进行的研究.

回到图 7.2-11, 来求通过检偏器或输出偏振片的电场, 这是通过求解沿 $\hat{\boldsymbol{y}}$ 方向的 $\boldsymbol{E}_{exit}$ 分量来完成的. 可利用点乘, 检偏器的输出电场为

$$\boldsymbol{E}_{out}=(\boldsymbol{E}_{exit}\cdot\hat{\boldsymbol{y}})\hat{\boldsymbol{y}}=\left[\frac{E_0}{\sqrt{2}}\left(\mathrm{e}^{-\mathrm{j}k_{n'_x}L}\hat{\boldsymbol{x}}'+\mathrm{e}^{-\mathrm{j}k_{n'_y}L}\hat{\boldsymbol{y}}'\right)\mathrm{e}^{\mathrm{j}\omega_0t}\cdot\hat{\boldsymbol{y}}\right]\hat{\boldsymbol{y}}.$$

由于 $\hat{\boldsymbol{x}}'\cdot\hat{\boldsymbol{y}}=\cos(90^\circ+45^\circ)=-1/\sqrt{2}$ 和 $\hat{\boldsymbol{y}}'\cdot\hat{\boldsymbol{y}}=\cos(45^\circ)=1/\sqrt{2}$, 上式变为

$$\boldsymbol{E}_{out}=\frac{E_0}{2}\left(\mathrm{e}^{-\mathrm{j}k_{n'_y}L}-\mathrm{e}^{-\mathrm{j}k_{n'_x}L}\right)\mathrm{e}^{\mathrm{j}\omega_0t}\hat{\boldsymbol{y}}=\frac{E_0}{2}\left(\mathrm{e}^{-\mathrm{j}\frac{2\pi}{\lambda_v}n'_yL}-\mathrm{e}^{-\mathrm{j}\frac{2\pi}{\lambda_v}n'_xL}\right)\mathrm{e}^{\mathrm{j}\omega_0t}\hat{\boldsymbol{y}}.$$

这是一个 $\hat{\boldsymbol{y}}$ 线性极化波, 且其出射光强 $I_0=|\boldsymbol{E}_{out}|^2$ 和入射光强 $I_{inc}=|\boldsymbol{E}_{inc}|^2$ 的比值为

$$\frac{I_0}{I_{inc}}=\sin^2\left[\frac{\pi}{\lambda_v}(n'_x-n'_y)L\right]=\sin^2\left(\frac{\varPhi}{2}\right)=\sin^2\left(\frac{\pi}{2}\frac{V}{V_\pi}\right). \tag{7.2-24}$$

这种正弦平方特性与声光学中使用的强度调制方法类似 (见图 7.1-13). 图 7.2-12 给出了调制信号与被调制的传输强度之间的关系. 同样, 为了获得线性调制, 即已调制强度与调制信号成正比, 需要将调制器偏置在 $I_0/I_{inc}=0.5$ 处. 该偏置可通过已经施加在晶体上的调制信号的外加偏置电压 $V_\pi/2$ 来实现.

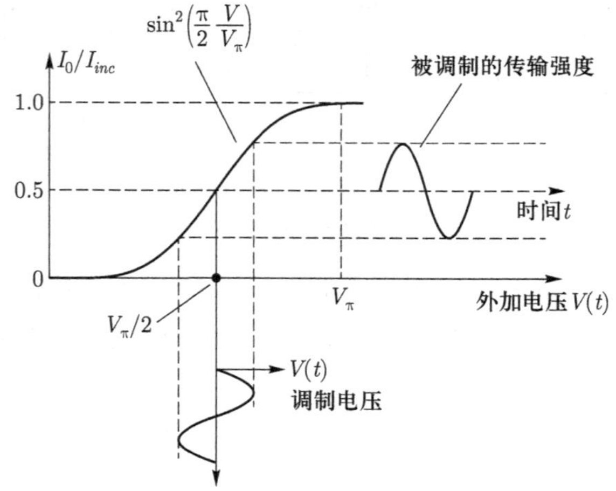

图 7.2-12 调制电压 $V(t)$ 与调制强度的关系

where $k_{n'_x} = 2\pi n'_x/\lambda_v$ and $k_{n'_y} = 2\pi n'_y/\lambda_v$. The phase difference between the two components is called the *retardation* Φ and is given by

$$\Phi = (k_{n'_x} - k_{n'_y})L = \frac{2\pi}{\lambda_v}(n'_x - n'_y)L = \frac{2\pi}{\lambda_v}n_o^3 r_{63}\mathrm{E}_z L = \frac{2\pi}{\lambda_v}n_o^3 r_{63}V, \tag{7.2-22}$$

where we have used the values of n'_x and n'_y from Eq. (7.2-20). At this point, let us introduce an important parameter called the *half-wave voltage* V_π. The half-wave voltage is the voltage required for inducing a phase change of π, that is, when $\Phi = \pi$. Therefore, from Eq. (7.2-22), we have

$$V_\pi = \frac{\lambda_v}{2n_o^3 r_{63}}. \tag{7.2-23}$$

From Table 7.1 and at $\lambda_v = 0.55$ μm, $V_\pi \approx 7.58$ kV for KDP crystals. When the half-wave voltage is low, the power needed for the modulator will be low. Reducing the half-wave voltage of a modulator is an ongoing research.

Returning to Figure 7.2-11, let us find the electrical field passed by the analyzer or the output polarizer. This is done by finding the components of $\boldsymbol{E}_{exit}$ along the $\hat{\boldsymbol{y}}$ direction. We can take advantage of the dot product, and the output electric field from the analyzer is

$$\boldsymbol{E}_{out} = (\boldsymbol{E}_{exit} \cdot \hat{\boldsymbol{y}})\hat{\boldsymbol{y}} = \left[\frac{E_0}{\sqrt{2}}\left(\mathrm{e}^{-\mathrm{j}k_{n'_x}L}\hat{\boldsymbol{x}}' + \mathrm{e}^{-\mathrm{j}k_{n'_y}L}\hat{\boldsymbol{y}}'\right)\mathrm{e}^{\mathrm{j}\omega_0 t} \cdot \hat{\boldsymbol{y}}\right]\hat{\boldsymbol{y}}.$$

Since $\hat{\boldsymbol{x}}' \cdot \hat{\boldsymbol{y}} = \cos(90° + 45°) = -1/\sqrt{2}$ and $\hat{\boldsymbol{y}}' \cdot \hat{\boldsymbol{y}} = \cos(45°) = 1/\sqrt{2}$, the above equation becomes

$$\boldsymbol{E}_{out} = \frac{E_0}{2}\left(\mathrm{e}^{-\mathrm{j}k_{n'_y}L} - \mathrm{e}^{-\mathrm{j}k_{n'_x}L}\right)\mathrm{e}^{\mathrm{j}\omega_0 t}\hat{\boldsymbol{y}} = \frac{E_0}{2}\left(\mathrm{e}^{-\mathrm{j}\frac{2\pi}{\lambda_v}n'_y L} - \mathrm{e}^{-\mathrm{j}\frac{2\pi}{\lambda_v}n'_x L}\right)\mathrm{e}^{\mathrm{j}\omega_0 t}\hat{\boldsymbol{y}}.$$

This is an $\hat{\boldsymbol{y}}$-linearly polarized wave and the ratio of the output intensity, that is, $I_0 = |\boldsymbol{E}_{out}|^2$, and the input intensity, that is, $I_{inc} = |\boldsymbol{E}_{inc}|^2$, is shown to be

$$\frac{I_0}{I_{inc}} = \sin^2\left[\frac{\pi}{\lambda_v}(n'_x - n'_y)L\right] = \sin^2\left(\frac{\Phi}{2}\right) = \sin^2\left(\frac{\pi}{2}\frac{V}{V_\pi}\right). \tag{7.2-24}$$

This sine-squared characteristic is reminiscent of the intensity modulation method used in acousto-optics (see Figure 7.1-13). Figure 7.2-12 illustrates the relationship between the modulating signal and the modulated transmitted intensity. Again, in order to obtain linear modulation, that is, modulated intensity being proportional to the modulating signal, we need to bias the modulator at $I_0/I_{inc} = 0.5$. The bias can be achieved by adding an additional bias voltage $V_\pi/2$ from the modulating signal already applied on the crystal.

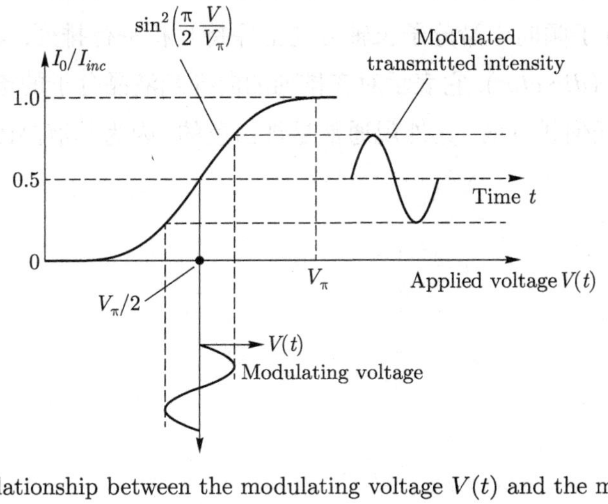

Figure 7.2-12 Relationship between the modulating voltage $V(t)$ and the modulated intensity

相位调制

电光相位调制的原理如图 7.2-13 所示, 系统由一个起偏器和一个 KDP 晶体组成. 偏振片的透射轴可沿着晶体的其中一个 "旋转" 轴, 即 $\hat{\boldsymbol{x}}'$ 或 $\hat{\boldsymbol{y}}'$. 晶体受沿着其光轴上外加电场 E_z 的影响. 图中, 给出了沿 $\hat{\boldsymbol{x}}'$ 轴的起偏器, 沿 z 方向的外加电场不改变其偏振状态, 只改变相位. 在 $z=0$ 处, 设 $\boldsymbol{E}_{inc}=E_0\hat{\boldsymbol{x}}'\mathrm{e}^{\mathrm{j}\omega_0 t}$, 在晶体的出射位置, 有

$$\boldsymbol{E}_{exit}=E_0\mathrm{e}^{-\mathrm{j}k_{n_x'}L}\hat{\boldsymbol{x}}'\mathrm{e}^{\mathrm{j}\omega_0 t}=E_0\mathrm{e}^{-\mathrm{j}\frac{2\pi}{\lambda_v}n_x'L}\hat{\boldsymbol{x}}'\mathrm{e}^{\mathrm{j}\omega_0 t}. \tag{7.2-25}$$

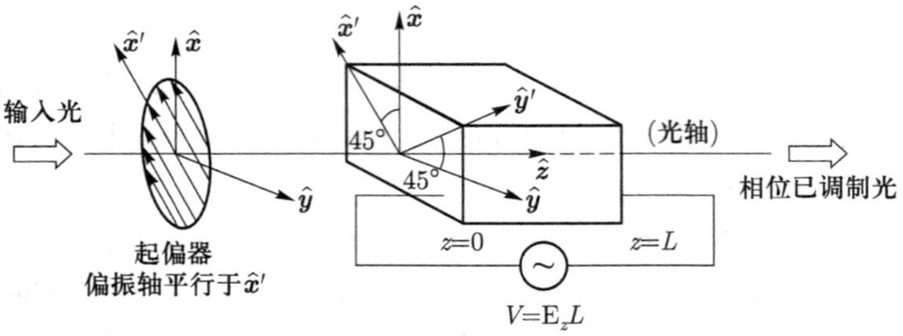

图 7.2-13　电光相位调制原理

根据式 (7.2-20a), 有 $n_x'\approx n_\mathrm{o}+\dfrac{1}{2}n_\mathrm{o}^3r_{63}\mathrm{E}_z$, 因此, 上述信号为相位已调制信号

$$\boldsymbol{E}_{exit}=E_0\mathrm{e}^{-\mathrm{j}\frac{2\pi}{\lambda_v}\left(n_\mathrm{o}+\frac{1}{2}n_\mathrm{o}^3r_{63}\mathrm{E}_z\right)L}\hat{\boldsymbol{x}}'\mathrm{e}^{\mathrm{j}\omega_0 t}=E_0\mathrm{e}^{-\mathrm{j}\frac{2\pi}{\lambda_v}n_\mathrm{o}L}\mathrm{e}^{-\mathrm{j}\frac{\pi}{\lambda_v}n_\mathrm{o}^3r_{63}k_{pm}m(t)}\hat{\boldsymbol{x}}'\mathrm{e}^{\mathrm{j}\omega_0 t}. \tag{7.2-26}$$

由于 $V=\mathrm{E}_zL=k_{pm}m(t)$, 且 $m(t)$ 为调制信号, k_{pm} 为相位偏差常数.

7.3　液晶元件 (盒) 信息处理和显示

7.3.1　液晶

普通的晶体有一个固定的熔点. 在到达熔点时, 晶体会发生从固态到液态的相变. 然而, 有些晶体不会直接从固态变为液态, 而是兼有液体的流动性以及晶体的特性特征. 这种具有过渡相位 (*transitional phase*) 状态的物质称为液晶 (*liquid crystal*, LC). 因此, 液晶具有流体性质, 如流动性和弹性, 但其内部分子的排列依然呈现出结构有序性, 如各向异性. 液晶分子可以被想象为长厚比约为 10, 长度为几纳米的椭圆状或棒状结构. 在所谓的向列相 (*nematic phase*) 晶体中, 随机分布的分子倾向于沿分子长轴方向上保持互相平行排列, 如图 7.3-1 所示. 其中矢量 $\hat{\boldsymbol{n}}$ 被称为指向矢 (*director*), 它表示对齐排列的向列相液晶分子的宏观方向.

当液晶分子没有任何排列时, 其外观通常是乳白色的, 因为其液晶团是无序随机的, 光经

Phase Modulation

Figure 7.2-13 shows a diagram of the principle of electro-optic phase modulation, which consists of an input polarizer and a KDP crystal. The transmission axis of the polarizer can be along one of the "rotated" axes of the crystal, that is, either $\hat{\boldsymbol{x}}'$ or $\hat{\boldsymbol{y}}'$. The crystal is under the influence of the external applied electric field E_z along its optical axis. In the figure, we show that the input polarizer is along the $\hat{\boldsymbol{x}}'$-axis. The applied electric field along the z-direction does not change the state of polarization but only the phase. At $z = 0$, we let $\boldsymbol{E}_{inc} = E_0\hat{\boldsymbol{x}}'\mathrm{e}^{\mathrm{j}\omega_0 t}$. At the exit of the crystal, we have

$$\boldsymbol{E}_{exit} = E_0\mathrm{e}^{-\mathrm{j}k_{n_x'}L}\hat{\boldsymbol{x}}'\mathrm{e}^{\mathrm{j}\omega_0 t} = E_0\mathrm{e}^{-\mathrm{j}\frac{2\pi}{\lambda_v}n_x'L}\hat{\boldsymbol{x}}'\mathrm{e}^{\mathrm{j}\omega_0 t}. \tag{7.2-25}$$

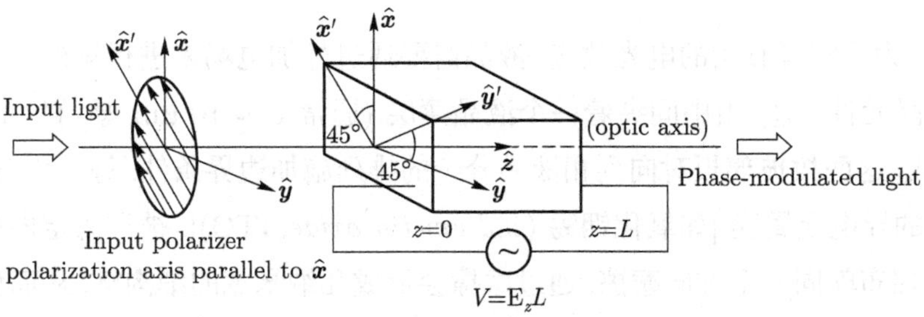

Figure 7.2-13 Principle of electro-optic phase modulation

According to Eq. (7.2-20a), we have $n_x' \approx n_o + \dfrac{1}{2}n_o^3 r_{63}\mathrm{E}_z$. Hence, the above signal is a phase modulated signal:

$$\boldsymbol{E}_{exit} = E_0\mathrm{e}^{-\mathrm{j}\frac{2\pi}{\lambda_v}\left(n_o+\frac{1}{2}n_o^3 r_{63}\mathrm{E}_z\right)L}\hat{\boldsymbol{x}}'\mathrm{e}^{\mathrm{j}\omega_0 t} = E_0\mathrm{e}^{-\mathrm{j}\frac{2\pi}{\lambda_v}n_o L}\mathrm{e}^{-\mathrm{j}\frac{\pi}{\lambda_v}n_o^3 r_{63}k_{pm}m(t)}\hat{\boldsymbol{x}}'\mathrm{e}^{\mathrm{j}\omega_0 t}. \tag{7.2-26}$$

as $V = \mathrm{E}_z L = k_{pm}m(t)$ and $m(t)$ is the modulating signal with k_{pm} being the phase-deviation constant.

7.3 Information Processing and Display with Liquid Crystal Cells

7.3.1 Liquid Crystals

There is a definite melting point for an ordinary crystal. At the melting point, the crystal makes a phase transition from the solid state into the liquid state. Some crystals, however, do not directly change from solid into liquid but process the property of fluidity of liquids as well as the characteristics of a crystal. Materials with this type of *transitional phase* are known as *liquid crystals.* Hence, liquid crystals (LCs) have fluidlike properties such as fluidity and elasticity and yet the arrangement of molecules within them exhibits structural order such as anisotropicity. Liquid crystal molecules can be visualized as ellipsoids or have rodlike shapes and have a length-to-breath ratio of about 10 with a length of several nanometers. In the so-called *nematic phase* of the crystals, the randomly located molecules tend to align themselves with their long axes parallel. The situation is shown in Figure 7.3-1, where the vector $\hat{\boldsymbol{n}}$ is called a *director*, which represents the macroscopic direction of the aligned nematic liquid crystal molecules.

When liquid crystal molecules are contained without any alignment, their appearance is often milky as they scatter light due to the randomness of liquid crystal clusters. In order

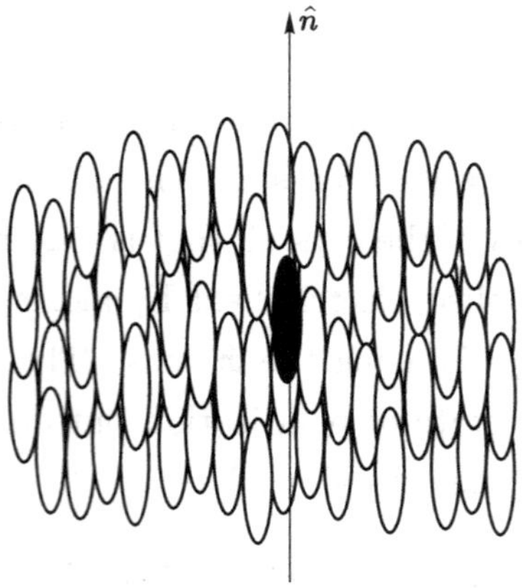

图 7.3-1　向列相液晶的分子序

过时发生散射. 为了实现有用的电光效应, 液晶需要通过外加电场来进行排列.

向列相液晶元件 (盒) 由中间夹着一个液晶薄层 (通常 5 ~ 10 μm 厚) 的两枚玻璃板 (或玻璃基板) 组成. 这两枚玻璃板对向列相液晶分子的排列施加边界条件, 每一片玻璃板上, 涂上一层薄且透明的导电金属膜 [如氧化铟锡 (*indium-tin-oxide*, ITO)], 被称为定向层 (*alignment layer*), 并用细棉布朝同一个方向摩擦, 通过摩擦会形成几纳米宽的细沟槽, 从而使液晶分子实现沿沟槽的方向平行. 这种摩擦方法已被广泛应用于制造大型液晶面板器件. 高质量定向层可通过在玻璃表面用真空沉积蒸镀细密的一氧化硅 (SiO) 薄膜的方式来制备微槽, 从而实现液晶分子的定向排列.

有两种常见的定向排列. 如果每个定向层沿不同的方向进行抛光, 其分子朝向绕垂直于基底方向的轴呈螺旋式旋转, 具体如图 7.3-2a) 所示. 可以看出, LC 单元左侧沿 y 方向被抛光, 而右侧沿 x 轴方向被抛光, 这种情况被称为扭曲定向排列 (*twist alignment*), 因为后玻璃板相对前玻璃板有一个扭曲了的角度. 如果这两块玻璃板之间的定向方向夹角为 90°, 就有所谓的垂直定向排列 (*perpendicular alignment*). 如果定向方向是平行的, 则 LC 分子为平行排列的, 就有平行定向排列 (*parallel alignment*), 如图 7.3-2b) 所示.

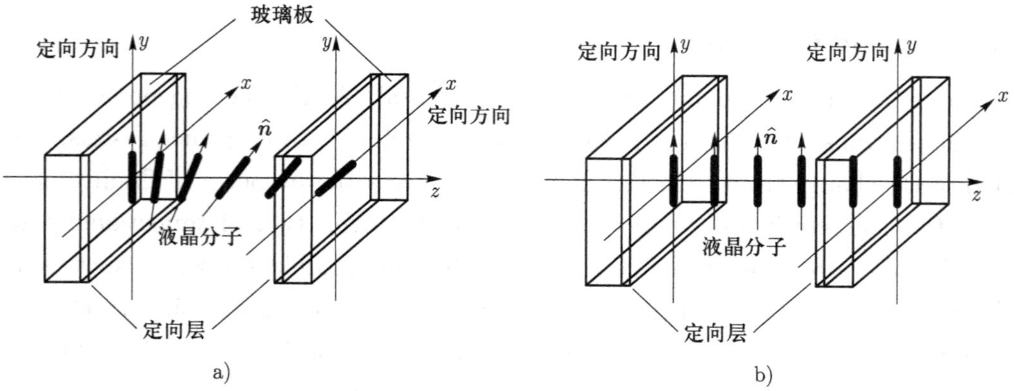

图 7.3-2　a) 扭曲定向排列; b) 平行定向排列

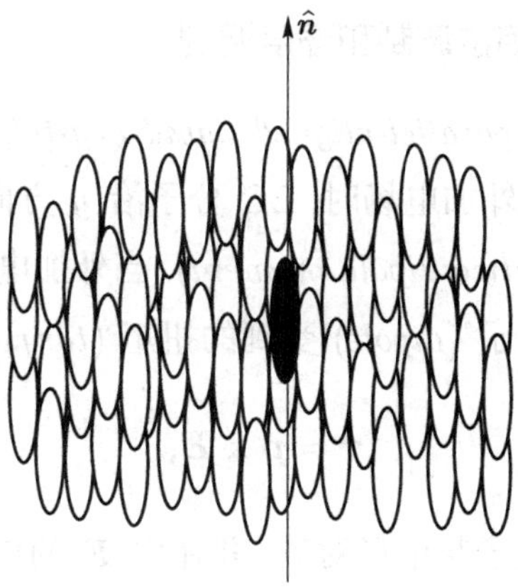

Figure 7.3-1 Molecular ordering of nematic phase of liquid crystal

to realize useful electro-optic effects, liquid crystals need to be aligned though an external applied electrical field.

The nematic liquid-crystal cell is composed of two glass plates (or glass substrates) containing a thin liquid crystal layer (typically 5 ~ 10 μm thick) inside. The two glass plates impose boundary conditions on the alignment of nematic liquid crystal molecules. Each glass plate is coated with a thin electrically conductive but optically transparent metallic film (such as *indium-tin-oxide*, ITO) called the *alignment layer* and then the layer is rubbed with a fine cotton cloth in one direction. Fine grooves about several nanometers wide are formed by rubbing and thus causing the liquid crystal molecules to lie parallel to the direction of the grooves. This rubbing method has been widely used for fabricating large panel liquid crystal devices. High-quality alignment can be made by vacuum deposition of a fine silicon monoxide (SiO) layer to create micro-grooves onto the surface of the glass for aligning LC molecules.

There are two popular alignments. If each alignment layer is polished with different directions, the molecular orientation rotates helically about an axis normal to plates such as the situation shown in Figure 7.3-2a, where we see that the left side of the LC cell is polished along the y-direction and the right side along the x-direction. This situation is known as the *twist alignment* as the back glass plate is twisted with an angle with respect to the front plate. If the alignment directions between the two plates are 90°, we have the so-called *perpendicular alignment.* If the alignment directions are parallel, the LC molecules are parallelly aligned, and we have *parallel alignment.* The parallel alignment is shown in Figure 7.3-2b.

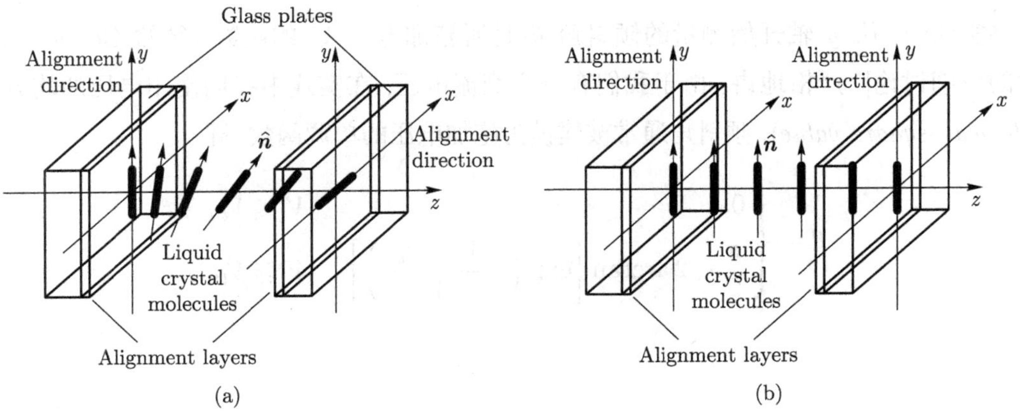

Figure 7.3-2 (a) Twist alignment and (b) parallel alignment

7.3.2　利用液晶元件进行相位调制和强度调制

考虑所谓的平行排列液晶 (*parallel-aligned liquid crystal*, PAL) 元件 (*cell*), 并讨论其调制应用的基本工作原理. 当没有外加电场时, LC 分子沿 y 方向排列. 该 LC 是一种有机化合物, 能产生很强的电偶极矩 (*electric dipole moment*). 当外加电场 $\boldsymbol{E}$ 沿 z 方向作用时, 液晶的偶极矩 $\boldsymbol{p}$ 会产生, 此时诱导偶极子 (*dipole*) 受到的扭矩 (*torque*) $\boldsymbol{\tau}$ 为

$$\boldsymbol{\tau} = \boldsymbol{p} \times \boldsymbol{E}, \tag{7.3-1}$$

且该扭矩沿 x 方向, 试着使偶极子与电场对齐, 并在力 $\boldsymbol{F}$ 的作用下在 y-z 平面按顺时针方向旋转该偶极子, 具体如图 7.3-3a) 所示. 注意, 矢量积的方向由右手定则给出 (参见第三章的矢量积示例). LC 分子偶极子的旋转改变了液晶元件的光学性质, 对此这里将做简短介绍. 在实际应用中, 为了不损坏液晶, LC 元件之间的电压通常由约 1 kHz 的方波函数即几伏特来驱动. 但是, 正如图 7.3-3b) 所示, 改变电极的电压极性, 将不会使偶极子偏向相反的方向, 根据式 (7.3-1), 扭矩依然在 x 方向上.

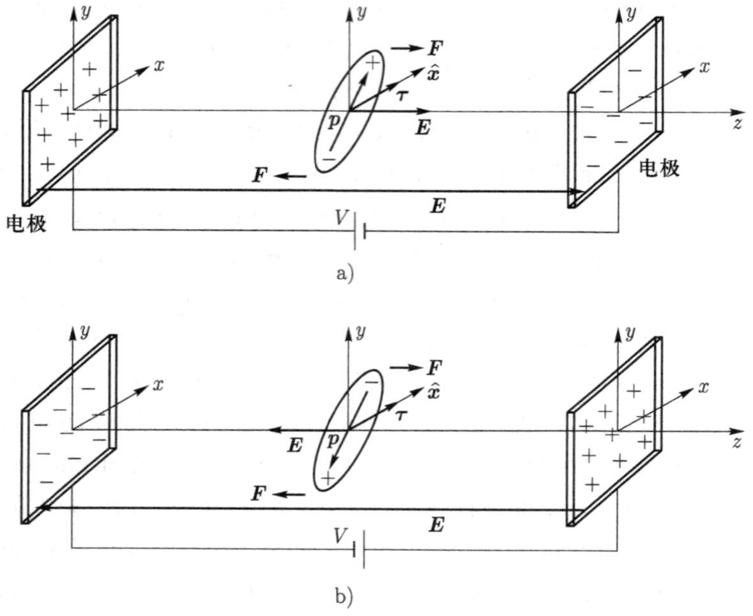

图 7.3-3　外加电场在 LC 分子上产生一个扭矩

事实证明, 从 y 轴开始测量的倾斜角 θ 是所施加电压 V 的函数 [参见 Gennes (1974) 书中的详细讨论]. 严格地讲, 由于我们有一个交流电压, 在实践中我们使用电压的均方根值 (*root-mean-square value*). 倾斜角通常被建模为施加电压的单调函数, 即

$$\theta = \begin{cases} 0, & V \leqslant V_{th} \\ \dfrac{\pi}{2} - 2\arctan\left[\exp\left(-\dfrac{V - V_{th}}{V_0}\right)\right], & V \geqslant V_{th}, \end{cases} \tag{7.3-2}$$

7.3.2 Phase Modulation and Intensity Modulation Using Liquid Crystal Cells

We consider the so-called *parallel-aligned liquid crystal (PAL) cell* and discuss its basic principle of operation for modulation applications. When there is no applied electric field, the LC molecules is aligned along the y-direction. The LC is an organic compound that processes a very strong electric dipole moment. Upon an external applied electric field $\boldsymbol{E}$ along the z-direction, a dipole moment $\boldsymbol{p}$ of the liquid crystal is induced and the dipole experiences a torque $\boldsymbol{\tau}$ given by

$$\boldsymbol{\tau} = \boldsymbol{p} \times \boldsymbol{E}, \tag{7.3-1}$$

and the *torque* is along the x-direction, trying to align the dipole with the electric field and turning clockwise in the y-z plane by the force $\boldsymbol{F}$. The situation is illustrated in Figure 7.3-3a. Remember that the cross product is given with direction given by the right-hand rule (see the cross product example in Chapter 3). The rotation of the dipole of the LC molecules changes the optical properties of the LC cell as we will explain shortly. For practical purposes so as not to damage the liquid crystal, the voltage between the LC cell is usually driven by a few volts of square wave function of about 1 kHz. Changing the voltage polarities of the electrodes, however, would not tilt the dipole in the opposite direction as we can see from Figure 7.3-3b, the torque still remains in the x-direction according to Eq. (7.3-1).

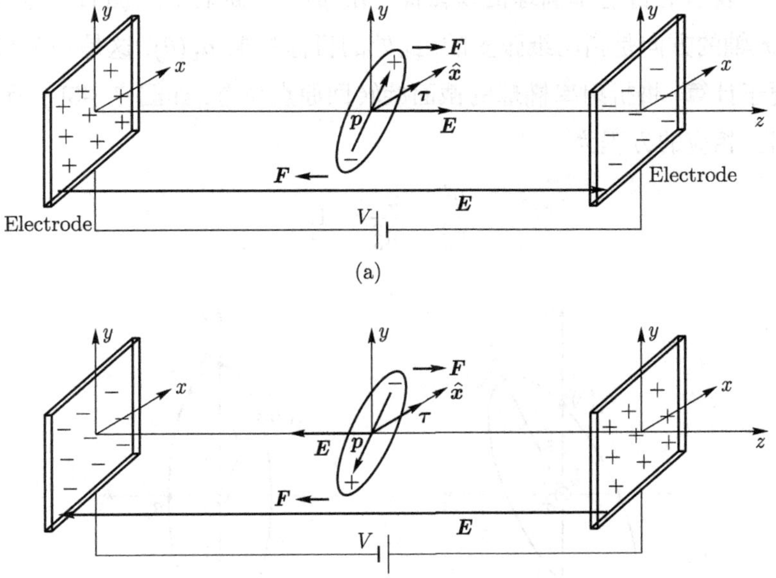

Figure 7.3-3 Applied electric field generating a torque on the LC molecule

As it turns out, the amount of angle tilt θ measured away from the y-axis is a function of the applied voltage V [see detailed discussion from the book by Gennes (1974)]. Strictly speaking, since we have an AC voltage, we use the root-mean-square value of the voltage in practice. The angle tilt is typically modeled as a monotonically function of the applied voltage as

$$\theta = \begin{cases} 0, & V \leqslant V_{th} \\ \dfrac{\pi}{2} - 2\arctan\left[\exp\left(-\dfrac{V - V_{th}}{V_0}\right)\right], & V \geqslant V_{th}, \end{cases} \tag{7.3-2}$$

其中, V_{th} 为 $1\sim2$ V 的阈值电压 (*threshold voltage*), 且低于这个阈值时, 分子不会发生倾斜, V_0 为 $1\sim2$ V 的恒定电压. 对于 $V>V_{th}$, 角度 θ 随着 V 的增大而增大, 直至饱和值 $\pi/2$. 图 7.3-4 绘出了角度 θ 与 $(V-V_{th})/V_0$ 在 $V\geqslant V_{th}$ 时的函数关系, 当电压被切断时, 分子迅速回到加电前的位置.

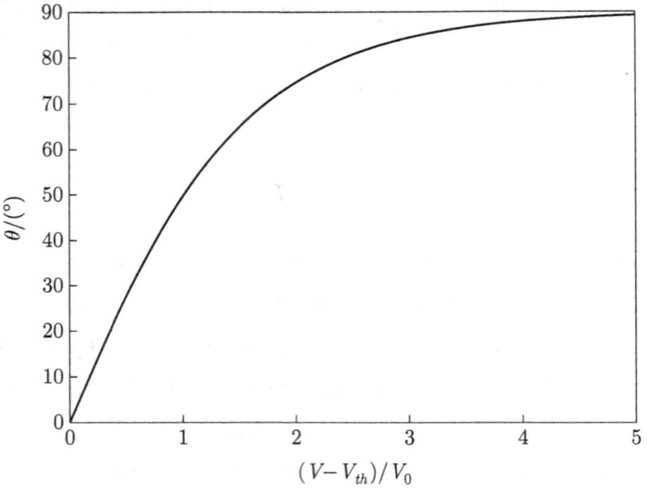

图 7.3-4　倾斜角 θ 与 $(V-V_{th})/V_0$ 的函数关系

这里来求电偶极子在外加电场作用下倾斜后沿 y 轴的折射率. 图 7.3-5a) 给出了倾斜角 (*tilt angle*) θ 表示的折射率椭球的横截面 (在 y-z 平面上). 遵循右手坐标系定则, 圆心处的叉圈表示 x 轴的方向是指向纸张里的. 此处的目标是求 $n_y(\theta)$, 这是一个与倾斜角相关的折射率. 为了便于计算, 将折射率椭球的截面旋转回原点位置, 如图 7.3-5b) 所示. 此时, 该椭圆在 y-z 平面上, 相应的方程为

$$\frac{y^2}{n_{\mathrm{e}}^2}+\frac{z^2}{n_{\mathrm{o}}^2}=1. \tag{7.3-3}$$

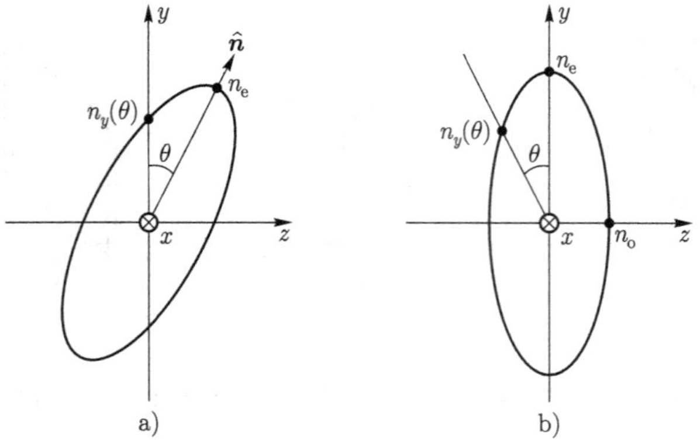

图 7.3-5　折射率椭球截面: a) 由于外加电场产生倾斜角 θ; b) 为了便于计算 $n_y(\theta)$, 旋转回原点位置

通过观察 7.3-5b) 中的几何图形, 发现以下关系

$$n_y^2(\theta)=y^2+z^2, \tag{7.3-4a}$$

和

$$\sin\theta=z/n_y(\theta). \tag{7.3-4b}$$

where V_{th} is a *threshold voltage* of $1 \sim 2$ V and below which no tilting of the molecules occurs and V_0 is a constant voltage of $1 \sim 2$ V. For $V > V_{th}$, the angle θ keeps increasing with V until reaching a saturation value of $\pi/2$. Figure 7.3-4 plots the angle θ in degrees as a function of $(V - V_{th})/V_0$ for $V \geqslant V_{th}$. When the voltage is turned off, the molecules return to their original positions.

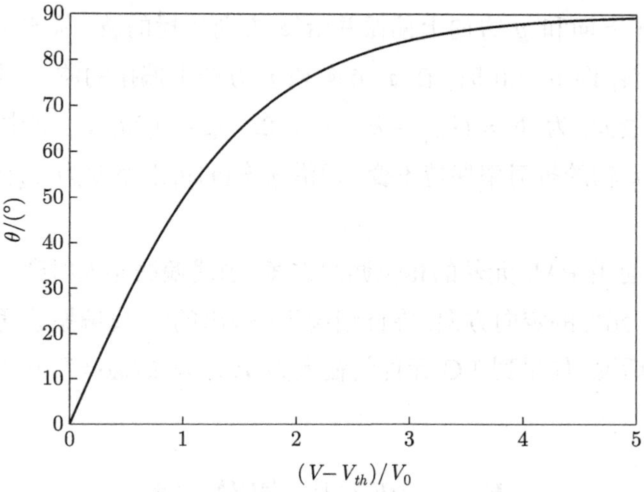

Figure 7.3-4 The tilt angle θ as a function of $(V - V_{th})/V_0$

Let us find the index of refraction along the y-axis after the electric dipole is tilted due to the external applied field. Figure 7.3-5a shows a cross section (in the y-z plane) of the index of ellipsoid with the *tilt angle* θ. The circle with a cross at the center signifies that the direction of the x-axis is into the paper, following the right-handed coordinate system convention. Our objective is to find $n_y(\theta)$, which is a tilt-angle-dependent index of refraction. For ease of calculation, we rotate the cross section of the index of ellipsoid back to the original location, as shown in Figure 7.3-5b. Now, the equation of the ellipse on the y-z plane is

$$\frac{y^2}{n_e^2} + \frac{z^2}{n_o^2} = 1. \tag{7.3-3}$$

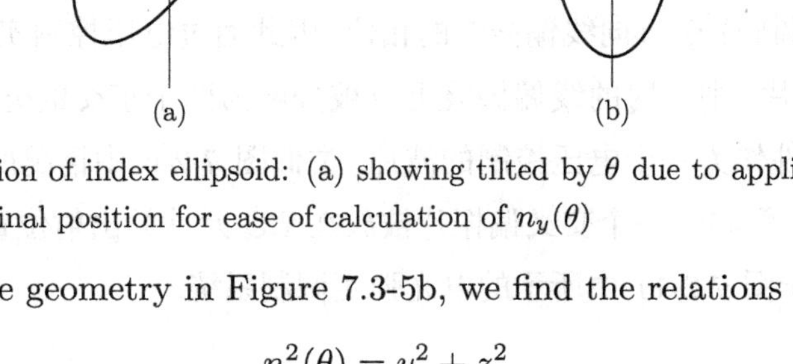

Figure 7.3-5 Cross section of index ellipsoid: (a) showing tilted by θ due to applied electric field and (b) rotating back to the original position for ease of calculation of $n_y(\theta)$

By inspecting the geometry in Figure 7.3-5b, we find the relations

$$n_y^2(\theta) = y^2 + z^2, \tag{7.3-4a}$$

and

$$\sin\theta = z/n_y(\theta). \tag{7.3-4b}$$

结合式 (7.3-3) 和式 (7.3-4), 有

$$\frac{1}{n_y^2(\theta)}=\frac{\cos^2\theta}{n_e^2}+\frac{\sin^2\theta}{n_o^2}. \tag{7.3-5}$$

调制方式

接下来考虑在 x 方向和 y 方向上偏振并沿 z 方向传播的波, 从式 (7.3-5) 和图 7.3-5b) 可以看出, 当没有倾斜, 即 $\theta=0$ 时, 在 x 方向和 y 方向上偏振的波, 其折射率分别为 n_o 和 n_e, 且此时相位延迟最大, 为 $\Phi=(k_{n_y}-k_{n_x})d=2\pi(n_e-n_o)d/\lambda_v$, 其中 d 为 LC 元件的长度. 在倾斜之后, 沿 x 轴的折射率保持不变, 而沿 y 轴的折射率变为 $n_y(\theta)$, 其相位延迟变为 $2\pi[n_y(\theta)-n_o]d/\lambda_v$.

图 7.3-6 所示为使用 PAL 元件的相位调制方案, 当线偏振光入射到一个 PAL 元件上, 其偏振方向平行于 LC 元件的定向方向, 在此情况下, 输出的 y 线偏振波通过 $2\pi n_y(\theta)d/\lambda_v$ 项进行相位调制, 换句话说, 如果到 LC 元件的输入为 $\boldsymbol{E}_{inc}=\hat{\boldsymbol{y}}E_0\mathrm{e}^{\mathrm{j}\omega_0 t}$, 则在该液晶元件的出射位置处, 有

$$\boldsymbol{E}_{exit}=\hat{\boldsymbol{y}}E_0\mathrm{e}^{-\mathrm{j}2\pi n_y(\theta)d/\lambda_v}\mathrm{e}^{\mathrm{j}\omega_0 t}.$$

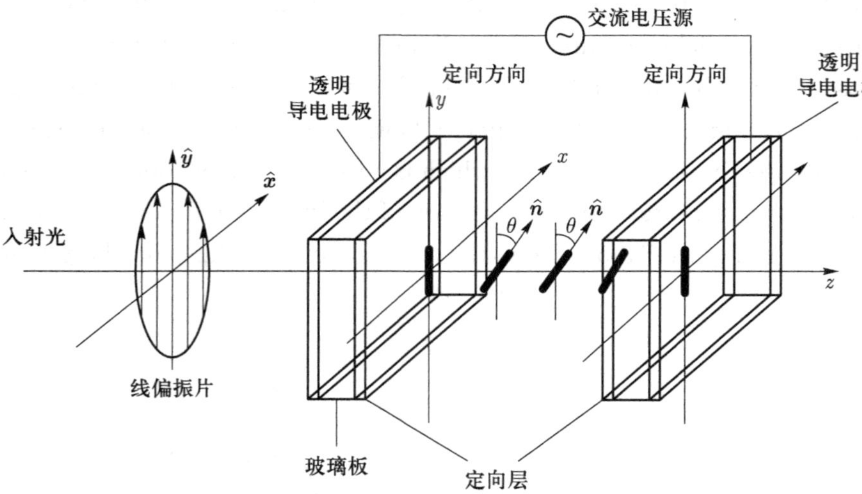

图 7.3-6　使用 PAL 元件的相位调制方案

因此, 出射光只经历了一个单纯的相位变化, 并在 LC 元件的出射位置处保持着相同的偏振态. 在所施加的交流电压下, 通过式 (7.3-2) 控制偶极子的倾斜, 偶极子的倾斜根据式 (7.3-5) 会改变 $n_y(\theta)$, 从而调制沿 $\hat{\boldsymbol{y}}$ 方向线偏振波的相位. 因此通过电压控制实现了相位调制.

如果旋转线偏振片, 则入射的线偏振波可以被分解为两个正交的分量, 即沿 x 方向和 y 方向, 此时该 LC 元件作为一个电压控制的波片, 类似图 7.2-6 中情况所示. 如果该线偏振片从 y 方向开始被旋转了 45°, 一个交叉偏振片被放在 LC 元件的出射位置处, 则可实现电压控制的强度调制, 类似于图 7.2-11 中所示的电光强度调制系统.

Combining Eqs. (7.3-3) and (7.3-4), we find

$$\frac{1}{n_y^2(\theta)} = \frac{\cos^2\theta}{n_e^2} + \frac{\sin^2\theta}{n_o^2}. \tag{7.3-5}$$

Modulation Schemes

Let us consider a wave traveling along the z-direction with polarizations in the x- and y-directions. From Eq. (7.3-5) along with Figure 7.3-5b, we see that when there is no tilt, that is, $\theta = 0$, the waves polarized along in the x- and y-directions have refractive indices n_o and n_e, respectively and the retardation is maximum at $\Phi = (k_{n_y} - k_{n_x})d = 2\pi(n_e - n_o)d/\lambda_v$, where d is length of the LC cell. After the tilt, the index refraction along the y-axis is changed to $n_y(\theta)$ with the index of refraction along the x-axis remaining unchanged. The retardation becomes $2\pi[n_y(\theta) - n_o]d/\lambda_v$.

Figure 7.3-6 shows a phase modulation scheme using a PAL cell. When a linearly polarized light is incident on a PAL cell with the polarization parallel to the direction of the alignment direction of the LC cell and in this case, the output y-linearly polarized wave is phase modulated with the term $2\pi n_y(\theta)d/\lambda_v$. In other words, if the input to the LC cell is $\boldsymbol{E}_{inc} = \hat{\boldsymbol{y}}E_0 e^{j\omega_0 t}$, at the exit of the cell, we have

$$\boldsymbol{E}_{exit} = \hat{\boldsymbol{y}}E_0 e^{-j2\pi n_y(\theta)d/\lambda_v} e^{j\omega_0 t}.$$

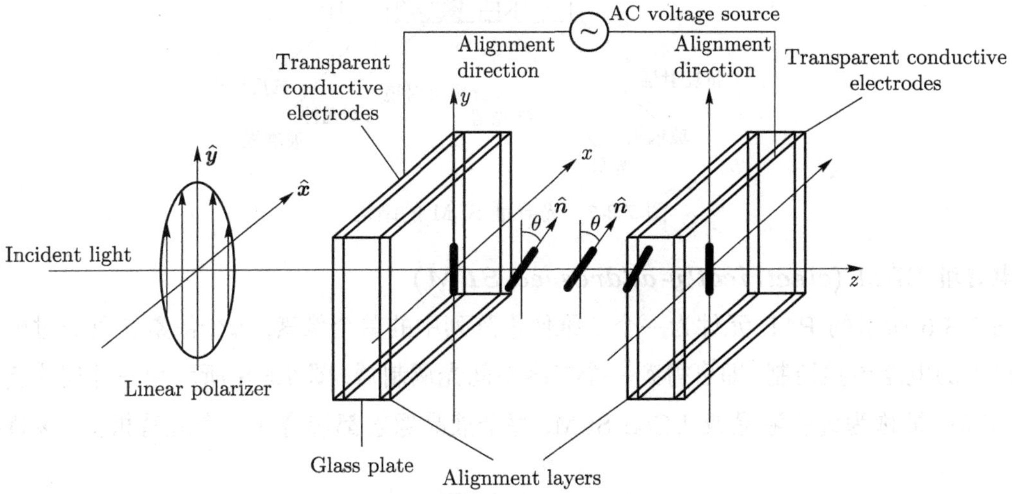

Figure 7.3-6 Phase modulation scheme using a PAL cell

Therefore, the output light experiences a pure phase change and remains the same polarization at the exit of the LC crystal. The applied AC voltage tilts the dipoles governed by Eq. (7.3-2), and the tilt of the dipole changes $n_y(\theta)$ according to Eq. 7.3-5, thereby modulating the phase of the linearly polarized wave along $\hat{\boldsymbol{y}}$. We have voltage-controlled phase modulation.

If the linear polarizer is rotated, the input linearly polarized wave can be decomposed into two orthogonal components, that is, along the x-direction and the y-direction, and the LC cell serve as a voltage-controlled wave plate, similar to the situation shown in Figure 7.2-6. If the linear polarizer is rotated 45° away from the y-direction and a crossed polarizer is placed at the output of the LC cell, we achieve voltage-controlled intensity modulation, reminiscent of the electro-optic intensity modulation system shown in Figure 7.2-11.

基于液晶的空间光调制器 (SLM)

光寻址 SLM (*optically-addressed SLM*)

光寻址 SLM 的结构如图 7.3-7 所示. 电极之间的交流电压为 LC 层提供了一个偏置电场. 在非相干光照射下, 穿过电极的场随着光导层 (*photoconductive layer*) 阻抗 (*impedence*) 的变化而变化. 在该层较暗的区域, 其阻抗很高, 经过 LC 层时会有小的电压下降. 然而, 当该层被照亮时, 其电导率增加, 因此, 阻抗降低, 经过液晶时的电压会增加. 所读出的相干光的相位根据光导层区域上所写入的光或所输入的图像的强度分布而做局部变化. 由于光被反射, 液晶元件的有效长度会加倍, 同时其相位延迟也增加了一倍. 这类器件作为空间非相干–相干光转换器, 为硅基液晶 (LCoS) 技术 [*liquid-crystal-on-silicon* (LCoS) *technology*] 的发明奠定了基础.

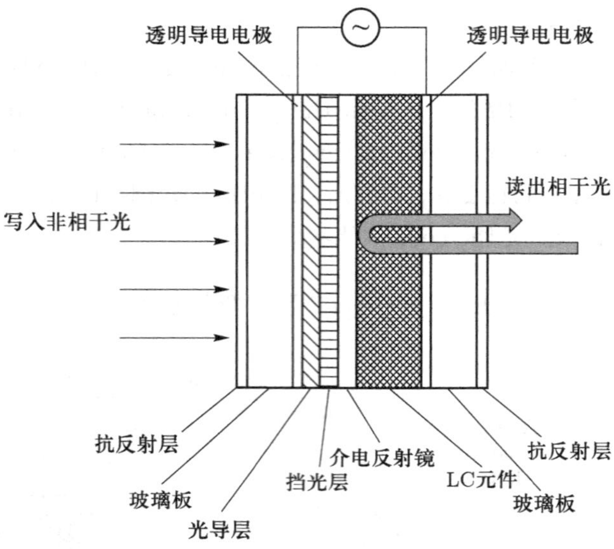

图 7.3-7 光寻址 SLM 的结构

电寻址 SLM (*electrically-addressed SLM*)

图 7.3-6 所示的 PAL 元件为一个二维像素阵列中的单个像素, 每个像素皆可通过电压由计算机中的电路单独控制, 即可得到一个二维空间光调制器. 图 7.3-8 所示为一种现代流行的液晶 SLM, 被称为硅基液晶或 LCoS SLM. 每个液晶像素都接合在一个硅背板上, 该背板是

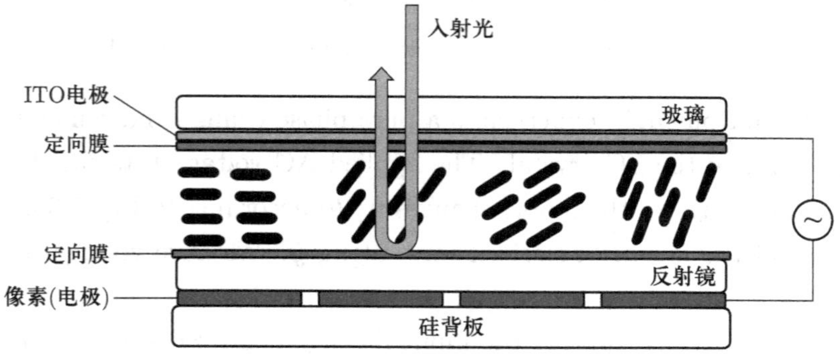

图 7.3-8 LCoS SLM 结构的横截面图

Liquid Crystal-Based Spatial Light Modulators (SLMs)

Optically Addressed SLM

Figure 7.3-7 shows the structure of the SLM. The AC voltage between the electrodes is to provide a bias electric field for the LC layer. The field across the electrodes is then varied according to the change of the *impedance* of the *photoconductive layer* upon illumination by the write incoherent light. In areas where the layer is dark, its impedance is high, and a small voltage is dropped across the LC layer. However, when the layer is illuminated, its conductivity increases and hence, the impedance decreases and the voltage dropped across the liquid increases. The phase of the read-out coherent light is changed locally according to the intensity distribution of the write light or the input image upon the area of the photoconductive layer. Since the light is reflected, the effective cell length is twice and retardation is doubled. This type of device serves as a spatial incoherent-to-coherent light converter and paves the way for the invention of *liquid-crystal-on-silicon* (LCoS) *technology.*

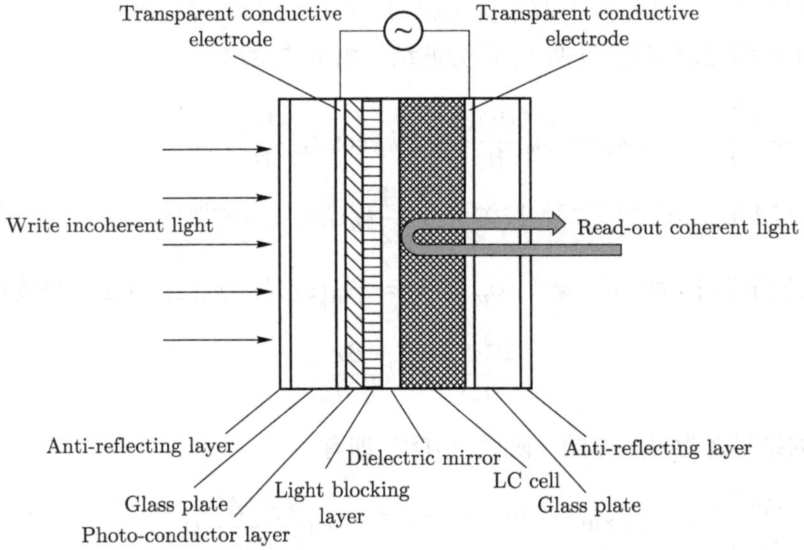

Figure 7.3-7 Structure of the optically addressed SLM

Electrically Addressed SLM

Imagine a PAL cell shown in Figure 7.3-6 is a single pixel in a 2-D array of pixels, and each pixel can be voltage controlled independently by the circuitry in a computer. We have a 2-D spatial light modulator. A popular modern liquid crystal SLM known as liquid crystal on silicon or LCoS SLM is shown in Figure 7.3-8. Each liquid crystal pixel is built on a silicon

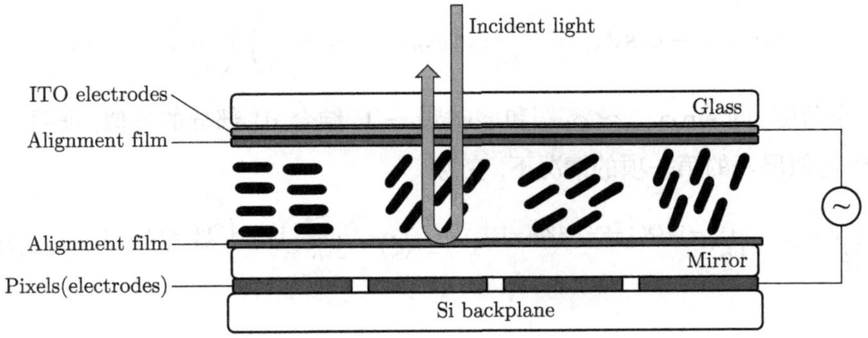

Figure 7.3-8 Cross section of the structure of a LCoS SLM

一个简单的印刷电路板, 它包含与像素化的电极之间的电路连接, 并允许所有连接的计算机板间进行通信.

习题

7.1 在本习题当中, 概括了从式 (7.1-25) 中的标量波动方程获得 Korpel-Poon 方程的步骤.

a) 证明: 通过将式 (7.1-27) 和式 (7.1-29) 代入式 (7.1-25) 的标量方程, 有

$$\nabla^2 E_m(\boldsymbol{r}) + k_0^2 E_m(\boldsymbol{r}) + \frac{1}{2}k_0^2 CS(\boldsymbol{r})E_{m-1}(\boldsymbol{r}) + \frac{1}{2}k_0^2 CS^*(\boldsymbol{r})E_{m+1}(\boldsymbol{r}) = 0,$$

其中, $k_0 = \omega_0\sqrt{\mu_0\varepsilon}$ 为光在介质中的传播常数, $E_m(\boldsymbol{r})$ 为 m 级次的光在频率 $\omega_0 + m\Omega$ 处的矢量振幅, 在推导上式时, 假设 $\omega_0 \gg m\Omega$, 因此, 在推导过程中使用 $\omega_0 + m\Omega \approx \omega_0$.

b) 令 $S(\boldsymbol{r}) = A\mathrm{e}^{-\mathrm{j}Kx}$ [式 (7.1-27)] 且 $E_m(\boldsymbol{r}) = \psi_m(z,x)\mathrm{e}^{-\mathrm{j}k_0\sin\phi_m x - \mathrm{j}k_0\cos\phi_m z}$ [式 (7.1-29)], 结合式 (7.1-5) 中的光栅方程, 证明 a) 部分的结果由下式给出

$$\begin{aligned}&\frac{\partial^2\psi_m}{\partial x^2} + \frac{\partial^2\psi_m}{\partial z^2} - 2\mathrm{j}k_0\sin\phi_m\frac{\partial\psi_m}{\partial x} - 2\mathrm{j}k_0\cos\phi_m\frac{\partial\psi_m}{\partial z}\\&+\frac{1}{2}k_0^2CA^*\psi_{m+1}\mathrm{e}^{-\mathrm{j}k_0z(\cos\phi_{m+1}-\cos\phi_m)} + \frac{1}{2}k_0^2CA\psi_{m-1}\mathrm{e}^{-\mathrm{j}k_0z(\cos\phi_{m-1}-\cos\phi_m)} = 0.\end{aligned}$$

c) 假定入射角很小, 即 $\phi_{inc} \ll 1$, $\psi_m(x,z) \approx \psi_m(z)$, 且 $\psi_m(z)$ 为 z 的慢变函数, 即

$$\frac{\partial^2\psi_m}{\partial z^2} \ll k_0\frac{\partial\psi_m}{\partial z},$$

证明 b) 部分的结果变为 ψ_m 中的一阶微分方程, 即有

$$\frac{\mathrm{d}\psi_m(z)}{\mathrm{d}z} = -\mathrm{j}\frac{k_0CA}{4\cos\phi_m}\psi_{m-1}(z)\mathrm{e}^{-\mathrm{j}k_0z(\cos\phi_{m-1}-\cos\phi_m)} - \mathrm{j}\frac{k_0CA^*}{4\cos\phi_m}\psi_{m+1}(z)\mathrm{e}^{-\mathrm{j}k_0z(\cos\phi_{m+1}-\cos\phi_m)}.$$

该方程为著名的 Raman-Nath 方程 (*Raman-Nath equations*) 的一个变体 [Raman 和 Nath (1935)].

d) 利用光栅方程, 即式 (7.1-5), 证明

$$\cos\phi_{m-1} - \cos\phi_m \approx \frac{K}{k_0}\sin\phi_{inc} + \left(m - \frac{1}{2}\right)\left(\frac{K}{k_0}\right)^2 + \cdots,$$

和

$$\cos\phi_{m+1} - \cos\phi_m \approx -\frac{K}{k_0}\sin\phi_{inc} - \left(m + \frac{1}{2}\right)\left(\frac{K}{k_0}\right)^2 + \cdots.$$

e) 对于小角度, 即 $\sin\phi_{inc} \approx \phi_{inc}$ 和 $\cos\phi_m \approx 1$, 结合 d) 部分的近似, 证明 c) 部分的结果, 在只保留近似展开的第一项的情况下, 变为

$$\frac{\mathrm{d}\psi_m(\xi)}{\mathrm{d}\xi} = -\mathrm{j}\frac{\widetilde{\alpha}}{2}\mathrm{e}^{-\mathrm{j}\frac{1}{2}Q\xi\left[\frac{\phi_{inc}}{\phi_B}+(2m-1)\right]}\psi_{m-1}(\xi) - \mathrm{j}\frac{\widetilde{\alpha}^*}{2}\mathrm{e}^{\mathrm{j}\frac{1}{2}Q\xi\left[\frac{\phi_{inc}}{\phi_B}+(2m+1)\right]}\psi_{m+1}(\xi),$$

backplane. The backplane is simply a printed circuit board containing circuitry connections to pixelated electrodes and allows for communication between all connected computer boards.

Problems

7.1 In this problem, we outline the procedures to obtain the Korpel-Poon equations from the scalar-wave equation in Eq. (7.1-25).

(a) Show that by substituting Eqs. (7.1-27) and (7.1-29) into the scalar equation in Eq. (7.1-25), we have

$$\nabla^2 E_m(\boldsymbol{r}) + k_0^2 E_m(\boldsymbol{r}) + \frac{1}{2}k_0^2 C S(\boldsymbol{r}) E_{m-1}(\boldsymbol{r}) + \frac{1}{2}k_0^2 C S^*(\boldsymbol{r}) E_{m+1}(\boldsymbol{r}) = 0,$$

where $k_0 = \omega_0\sqrt{\mu_0\varepsilon}$ is the propagation constant of the light in the medium. $E_m(\boldsymbol{r})$ is the phasor amplitude of the mth order light at frequency $\omega_0 + m\Omega$. In deriving the above equation, we assume $\omega_0 \gg m\Omega$, and hence we use $\omega_0 + m\Omega \approx \omega_0$ in the derivation process.

(b) Letting $S(\boldsymbol{r}) = A\mathrm{e}^{-\mathrm{j}Kx}$ and $E_m(\boldsymbol{r}) = \psi_m(z,x)\mathrm{e}^{-\mathrm{j}k_0\sin\phi_m x - \mathrm{j}k_0\cos\phi_m z}$ [see Eqs. (7.1-27) and (7.1-29)] along with the grating equation in Eq. (7.1-5), show that the result of part (a) is given by

$$\frac{\partial^2\psi_m}{\partial x^2} + \frac{\partial^2\psi_m}{\partial z^2} - 2\mathrm{j}k_0\sin\phi_m\frac{\partial\psi_m}{\partial x} - 2\mathrm{j}k_0\cos\phi_m\frac{\partial\psi_m}{\partial z}$$
$$+ \frac{1}{2}k_0^2 C A^*\psi_{m+1}\mathrm{e}^{-\mathrm{j}k_0 z(\cos\phi_{m+1} - \cos\phi_m)} + \frac{1}{2}k_0^2 C A\psi_{m-1}\mathrm{e}^{-\mathrm{j}k_0 z(\cos\phi_{m-1} - \cos\phi_m)} = 0.$$

(c) Assuming the incident angle is small, that is, $\phi_{inc} \ll 1$, $\psi_m(x,z) \approx \psi_m(z)$, and $\psi_m(z)$ is a slowly varying function of z, that is,

$$\frac{\partial^2\psi_m}{\partial z^2} \ll k_0\frac{\partial\psi_m}{\partial z},$$

show that the result of part b) becomes a first-order differential equation in ψ_m, and we have

$$\frac{\mathrm{d}\psi_m(z)}{\mathrm{d}z} = -\mathrm{j}\frac{k_0 C A}{4\cos\phi_m}\psi_{m-1}(z)\mathrm{e}^{-\mathrm{j}k_0 z(\cos\phi_{m-1} - \cos\phi_m)} - \mathrm{j}\frac{k_0 C A^*}{4\cos\phi_m}\psi_{m+1}(z)\mathrm{e}^{-\mathrm{j}k_0 z(\cos\phi_{m+1} - \cos\phi_m)}.$$

This equation is one variant of the well-known *Raman-Nath equations* [Raman and Nath (1935)].

(d) Using the grating equation, that is, Eq. (7.1-5), show that

$$\cos\phi_{m-1} - \cos\phi_m \approx \frac{K}{k_0}\sin\phi_{inc} + \left(m - \frac{1}{2}\right)\left(\frac{K}{k_0}\right)^2 + \cdots,$$

and

$$\cos\phi_{m+1} - \cos\phi_m \approx -\frac{K}{k_0}\sin\phi_{inc} - \left(m + \frac{1}{2}\right)\left(\frac{K}{k_0}\right)^2 + \cdots.$$

(e) For small angles, that is, $\sin\phi_{inc} \approx \phi_{inc}$, and $\cos\phi_m \approx 1$, show that the result in part (c) along with the approximations from part (d), where we only keep the first term of the approximation expansion, becomes

$$\frac{\mathrm{d}\psi_m(\xi)}{\mathrm{d}\xi} = -\mathrm{j}\frac{\widetilde{\alpha}}{2}\mathrm{e}^{-\mathrm{j}\frac{1}{2}Q\xi\left[\frac{\phi_{inc}}{\phi_B} + (2m-1)\right]}\psi_{m-1}(\xi) - \mathrm{j}\frac{\widetilde{\alpha}^*}{2}\mathrm{e}^{\mathrm{j}\frac{1}{2}Q\xi\left[\frac{\phi_{inc}}{\phi_B} + (2m+1)\right]}\psi_{m+1}(\xi),$$

其中, $\widetilde{\alpha}=k_0CA/2$, $\xi=z/L$ 且 $Q=L\dfrac{K^2}{k_0}$, 这即为式 (7.1-30).

7.2 一束时间频率为 ω_0 的激光束以布拉格角入射到两个布拉格声光调制器系统中, 如图题 7.2 所示. 频率为 Ω_1 和 Ω_2 的两个正弦信号驱动两个声光调制器. 求激光束 1、激光束 2 和激光束 3 的时间频率.

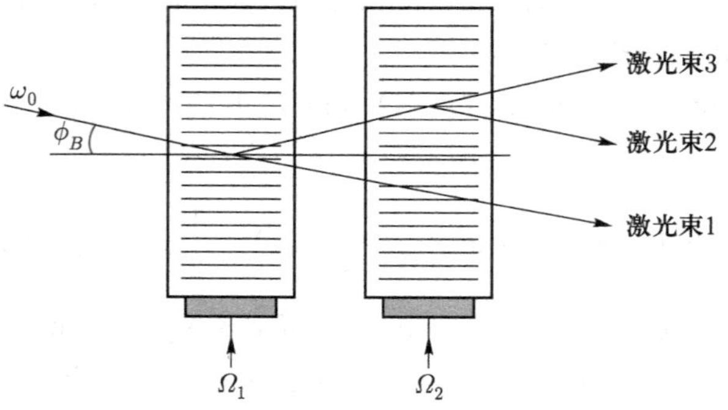

图题 7.2

7.3 一束波长为 0.6326 μm 的红光将会被一个致密玻璃布拉格声光调制器偏转, 其最高的声光频率为 1 GHz, 那么可能对该光束偏转多少 (度)? 假设玻璃中声波的速度为 7.4 km/s, 在计算中, 请勿进行小角度近似.

7.4 证明

$$\psi_m(\xi)=(-\mathrm{j})^m\psi_{inc}J_m(\alpha\xi)$$

为下式的解

$$\frac{\mathrm{d}\psi_m(\xi)}{\mathrm{d}\xi}=-\mathrm{j}\frac{\alpha}{2}[\psi_{m-1}(\xi)+\psi_{m+1}(\xi)].$$

约束条件为 $\psi_m=\psi_{inc}\delta_{m0}$, $\xi=z/L=0$.

7.5 对于近布拉格衍射, Q 是有限大小的, 考虑入射角偏离了布拉格角时, 即 $\phi_{inc}=-(1+\delta)\phi_B$, 其中 δ 表示入射平面波相对布拉格角的偏离, 有以下耦合方程

$$\frac{\mathrm{d}\psi_0(\xi)}{\mathrm{d}\xi}=-\mathrm{j}\frac{\alpha}{2}\mathrm{e}^{-\mathrm{j}\delta Q\xi/2}\psi_1(\xi),$$

和

$$\frac{\mathrm{d}\psi_1(\xi)}{\mathrm{d}\xi}=-\mathrm{j}\frac{\alpha}{2}\mathrm{e}^{\mathrm{j}\delta Q\xi/2}\psi_0(\xi).$$

证明: 当有下式时, 满足能量守恒

$$\frac{\mathrm{d}}{\mathrm{d}\xi}\left[|\psi_1(\xi)|^2+|\psi_2(\xi)|^2\right]=0.$$

7.6 给定一束可能是非偏振或圆偏振的光, 如何确定它的实际偏振态? 可以使用一个波片和一个线偏振片.

7.7 一顺时针圆偏振光入射到厚度为 d 的 x 切割石英上 ($n_{\mathrm{o}}=1.544$, $n_{\mathrm{e}}=1.553$).

a) 假设 x 切割晶体位于 $x=0$ 处, 写该出入射光场的表达式.

where $\widetilde{\alpha} = k_0CA/2$, $\xi = z/L$, and $Q = L\dfrac{K^2}{k_0}$. This is Eq. (7.1-30).

7.2 A laser beam of temporal frequency ω_0 is incident at the Bragg angle to the system of two Bragg acousto-optic modulators, as shown in Figure P.7.2. Two sinusoidal signals at frequencies Ω_1 and Ω_2 drive the two acousto-optic modulators. Find the temporal frequency of laser beam 1, beam 2, and beam 3.

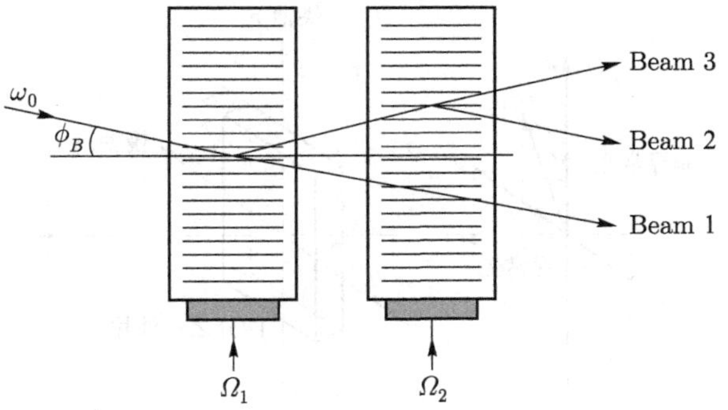

Figure P.7.2

7.3 Suppose that a red laser beam of wavelength 0.6326 μm is to be deflected by a dense-glass Bragg acousto-optic modulator. The highest acoustic frequency is 1 GHz. How much deflection (in degrees) is possible? Assume that the speed of sound in the glass is 7.4 km/s. In your calculation, do not assume the small-angle approximations.

7.4 Show that

$$\psi_m(\xi) = (-\mathrm{j})^m \psi_{inc} J_m(\alpha\xi)$$

are the solutions to

$$\frac{\mathrm{d}\psi_m(\xi)}{\mathrm{d}\xi} = -\mathrm{j}\frac{\alpha}{2}[\psi_{m-1}(\xi) + \psi_{m+1}(\xi)]$$

subject to boundary conditions $\psi_m = \psi_{inc}\delta_{m0}$ at $\xi = z/L = 0$.

7.5 For near Bragg diffraction, Q is finite. Considering the incident angle is off the exact Bragg angle, that is, $\phi_{inc} = -(1+\delta)\phi_B$, where δ represents the deviation of the incident plane wave away from the Bragg angle, we have the following coupled equations:

$$\frac{\mathrm{d}\psi_0(\xi)}{\mathrm{d}\xi} = -\mathrm{j}\frac{\alpha}{2}\mathrm{e}^{-\mathrm{j}\delta Q\xi/2}\psi_1(\xi),$$

and

$$\frac{\mathrm{d}\psi_1(\xi)}{\mathrm{d}\xi} = -\mathrm{j}\frac{\alpha}{2}\mathrm{e}^{\mathrm{j}\delta Q\xi/2}\psi_0(\xi).$$

Show that the conservation of energy is satisfied by showing

$$\frac{\mathrm{d}}{\mathrm{d}\xi}\left[|\psi_1(\xi)|^2 + |\psi_2(\xi)|^2\right] = 0.$$

7.6 Given a beam of light that might be unpolarized or circularly polarized, how might you determine its actual state of polarization? You may use a waveplate and a linear polarizer.

7.7 A clockwise circularly polarized light is incident on an x-cut quartz ($n_\mathrm{o} = 1.544$, $n_\mathrm{e} = 1.553$) of thickness d.

(a) Give an expression of the incident light field, assuming the x-cut crystal is located at $x = 0$.

b) 设计晶体的厚度 d, 使电场的 z 分量和 y 分量之间的相对相位为 $\phi_{zy} = \phi_z - \phi_y = -3\pi$.

c) 给定晶体后光场的表达式, 写出输出光场的偏振态.

7.8 由图 7.2-7, 将四分之一波片旋转 90°, 得到图题 7.8 所示的情况.

a) 描述四分之一波片出射位置处的光场, 用一些数学推导给出证明.

b) 当入射光的电场不是在 45°, 而是在 30° 时, 描述会发生什么.

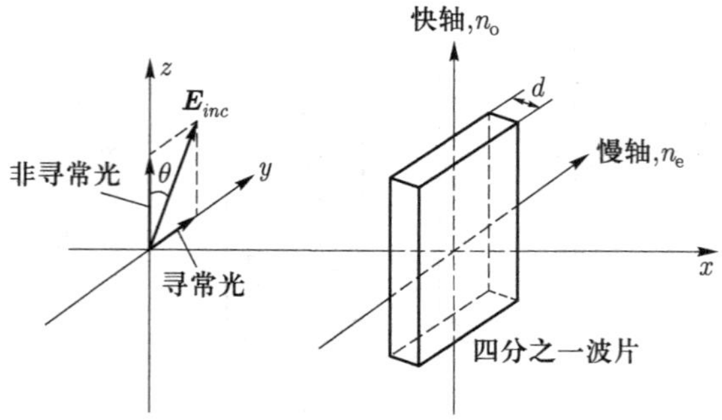

图题 7.8

7.9 a) 求铌酸锂 ($LiNbO_3$) 的电光张量, 并写出类似式 (7.2-12) 的表达式.

b) 求外加电场 $\boldsymbol{E} = E_z\hat{\boldsymbol{z}}$ 的作用下折射率椭球的方程.

c) 假设 $n_o r_{ij} E_z \ll 1$, 且 $n_e r_{ij} E_z \ll 1$, 证明

$$n_x = n_y \approx n_o - \frac{1}{2}n_o^3 r_{13} E_z \quad \text{和} \quad n_z \approx n_o - \frac{1}{2}n_e^3 r_{33} E_z.$$

这表明, 当外加电场沿 z 方向时, 折射率椭球不做任何旋转, 只是其轴的长度发生变化.

参考文献

[1] Appel, R. and Somekh, M. G. (1993). "Series solution for two-frequency Bragg interation using the Korpel-Poon multiple-scattering model," Journal of the Optical Society of America A 10, pp. 466-476.

[2] Banerjee, P. P. and Poon, T.-C. (1991). *Principles of Applied Optics*. Irwin, Illinois.

[3] Brooks P. and Reeve, C. D. (1995). "Limitations in acousto-optic FM demodulators," IEE Proceedings-Optoelectronics 142, pp. 149-156.

[4] de Gennes, P. G. (1974). *The Physics of Liquid Crystals*. Clarendon Press, Oxford.

[5] IntraAction Corp., 3719 Warren Avenue, Bellwood, IL 60104.

[6] Klein, W. R. and Cook, B. D. (1967). "Unified approach to ultrasonic light diffraction," IEEE Transactions on Sonics Ultrasonic SU-14, pp. 123-134.

[7] Korpel, A. (1972). "Acousto-optics," In Applied Solid State Science, Vol. 3. Academic Press, New York.

[8] Korpel, A. and Poon, T.-C. (1980). "Explicit formalism for acousto-optics multiple plane scattering," Journal of the Optical Society of America 70, pp. 817-820.

[9] Phariseau, P. (1956). "On the diffraction of light by progressive supersonic waves," Proceedings of the Indian Academy of Sciences 44, pp. 165-170.

(b) Design the thickness of the crystal d such that the relative phase between the z- and y-components of the electric field $\phi_{zy} = \phi_z - \phi_y = -3\pi$.

(c) Give the expression of the optical field after the crystal. What is the polarization state of the output field.

7.8 From Figure 7.2-7, we rotate the quarter-wave plate by 90° to achieve the situation shown in Figure P.7.8.

(a) Describe the wave field at the exit of the quarter-wave plate. Justify your answer with some mathematical deliberation.

(b) Describe what happens when the incident electric field is not at 45° but at 30°.

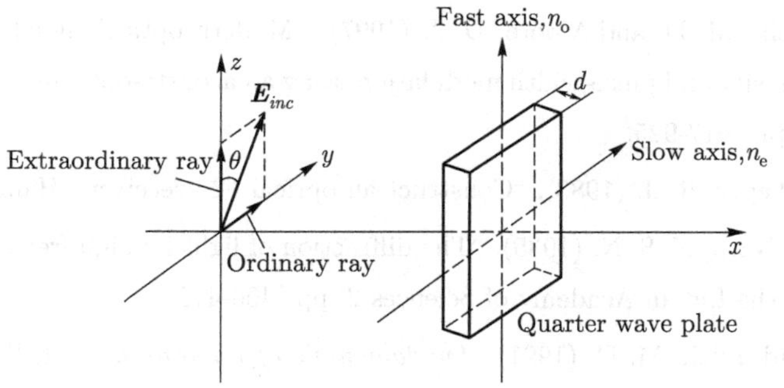

Figure P.7.8

7.9 (a) Find the electro-optic tensor of Lithium Niobate ($LiNbO_3$) and express it similar to Eq. (7.2-12).

(b) Find the equation of the index of ellipsoid in the presence of an external applied electric field $\boldsymbol{E} = E_z\hat{\boldsymbol{z}}$.

(c) Assuming $n_o r_{ij} E_z \ll 1$ and $n_e r_{ij} E_z \ll 1$, show that

$$n_x = n_y \approx n_o - \frac{1}{2} n_o^3 r_{13} E_z \text{ and } n_z \approx n_o - \frac{1}{2} n_e^3 r_{33} E_z.$$

This shows that when the applied electric field is along the z-direction, the ellipsoid does not undergo any rotation but only the lengths of the axes of the ellipsoid change.

Bibliography

[1] Appel, R. and Somekh, M. G. (1993). "Series solution for two-frequency Bragg interation using the Korpel-Poon multiple-scattering model," Journal of the Optical Society of America A 10, pp. 466-476.

[2] Banerjee, P. P. and Poon, T.-C. (1991). *Principles of Applied Optics.* Irwin, Illinois.

[3] Brooks P. and Reeve, C. D. (1995). "Limitations in acousto-optic FM demodulators," IEE Proceedings-Optoelectronics 142, pp. 149-156.

[4] de Gennes, P. G. (1974). *The Physics of Liquid Crystals.* Clarendon Press, Oxford.

[5] IntraAction Corp., 3719 Warren Avenue, Bellwood, IL 60104.

[6] Klein, W. R. and Cook, B. D. (1967). "Unified approach to ultrasonic light diffraction," IEEE Transactions on Sonics Ultrasonic SU-14, pp. 123-134.

[7] Korpel, A. (1972). "Acousto-optics," In Applied Solid State Science, Vol. 3. Academic Press, New York.

[8] Korpel, A. and Poon, T.-C. (1980). "Explicit formalism for acousto-optics multiple plane scattering," Journal of the Optical Society of America 70, pp. 817-820.

[9] Phariseau, P. (1956). "On the diffraction of light by progressive supersonic waves," Proceedings of the Indian Academy of Sciences 44, pp. 165-170.

[10] Pieper, R. J. and Poon, T.-C. (1985). "An acousto-optic FM receiver demonstrating some principles of modern signal processing," IEEE Transactions on Education E-27(3), pp. 11-17.

[11] Poon, T.-C. (2002). "Acousto-optics," In Encyclopedia of Physical Science and Technology, pp. 195-210. Academic Press, New York.

[12] Poon, T.-C. (2005). "Heterodyning," In Encyclopedia of Modern Optics, pp. 201-206. Elsevier Physics.

[13] Poon, T.-C. and Kim, T. (2005). "Acousto-optics with MATLAB®," Proceedings of SPIE 5953, 59530J-1-59530J-12.

[14] Poon, T.-C. and Kim, T. (2018). *Engineering Optics with MATLAB®*, 2nd ed. World Scientific, New Jersey.

[15] Poon, T.-C, McNeill, M. D. and Moore, D. J. (1997). "Modern optical signal processing experiments demonstrating intensity and pulse-width modulation using an acousto-optic modulator," American Journal of Physics 65, pp. 917-925.

[16] Poon, T.-C. and Pieper, R. J. (1983). "Construct an optical FM receiver," Ham Radio, pp. 53-56.

[17] Raman, C. V. and Nath, N. S. N. (1935). "The diffraction of light by high frequency sound waves: Part I.," Proceedings of the Indian Academy of Sciences 2, pp. 406-412.

[18] Saleh, B. E. A. and Teich, M. C. (1991). *Fundamentals of Photonics.* John Wiley & Sons, Inc., New York.

[19] Vander Lugt, A. (1992). *Optical Signal Processing.* John Wiley & Sons, Inc., New York.

[20] Zhang, Y., Fan, H. and Poon, T.-C. (2022). "Optical image processing using acousto-optic modulators as programmable volume holograms: a review [Invited]," Chinese Optics Letters 20, pp. 021101-1.

[10] Pieper, R. J. and Poon, T.-C. (1985). "An acousto-optic FM receiver demonstrating some principles of modern signal processing," IEEE Transactions on Education E-27(3), pp. 11-17.

[11] Poon, T.-C. (2002). "Acousto-optics," In Encyclopedia of Physical Science and Technology, pp. 195-210. Academic Press, New York.

[12] Poon, T.-C. (2005). "Heterodyning," In Encyclopedia of Modern Optics, pp. 201-206. Elsevier Physics.

[13] Poon, T.-C. and Kim, T. (2005). "Acousto-optics with MATLAB®," Proceedings of SPIE 5953, 59530J-1-59530J-12.

[14] Poon, T.-C. and Kim, T. (2018). *Engineering Optics with MATLAB®*, 2nd ed. World Scientific, New Jersey.

[15] Poon, T.-C, McNeill, M. D. and Moore, D. J. (1997). "Modern optical signal processing experiments demonstrating intensity and pulse-width modulation using an acousto-optic modulator," American Journal of Physics 65, pp. 917-925.

[16] Poon, T.-C. and Pieper, R. J. (1983). "Construct an optical FM receiver," Ham Radio, pp. 53-56.

[17] Raman, C. V. and Nath, N. S. N. (1935). "The diffraction of light by high frequency sound waves: Part I.," Proceedings of the Indian Academy of Sciences 2, pp. 406-412.

[18] Saleh, B. E. A. and Teich, M. C. (1991). *Fundamentals of Photonics.* John Wiley & Sons, Inc., New York.

[19] Vander Lugt, A. (1992). *Optical Signal Processing.* John Wiley & Sons, Inc., New York.

[20] Zhang, Y., Fan, H. and Poon, T.-C. (2022). "Optical image processing using acousto-optic modulators as programmable volume holograms: a review [Invited]," Chinese Optics Letters 20, pp. 021101-1.

索引
Index

Y